T0145385

Lecture Notes in Computer Science 13939

Founding Editors

Gerhard Goos
Juris Hartmanis

The series Lecture Notes in Computer Science (LNCS), including its subseries Lecture Notes in Artificial Intelligence (LNAI) and Lecture Notes in Bioinformatics (LNBI), has established itself as a medium for the publication of new developments in computer science and information technology research, teaching, and education.

LNCS enjoys close cooperation with the computer science R & D community, the series counts many renowned academics among its volume editors and paper authors, and collaborates with prestigious societies. Its mission is to serve this international community by providing an invaluable service, mainly focused on the publication of conference and workshop proceedings and postproceedings. LNCS commenced publication in 1973.

Alejandro Frangi · Marleen de Bruijne ·
Demian Wassermann · Nassir Navab
Editors

Information Processing in Medical Imaging

28th International Conference, IPMI 2023
San Carlos de Bariloche, Argentina, June 18–23, 2023
Proceedings

 Springer

Editors
Alejandro Frangi (iD)
University of Leeds
Leeds, UK

Marleen de Bruijne (iD)
University of Copenhagen
Copenhagen, Denmark

Demian Wassermann (iD)
Inria Saclay - Île-de-France Research Centre
Palaiseau, France

Nassir Navab (iD)
Technical University of Munich Garching
Munich, Bayern, Germany

ISSN 0302-9743   ISSN 1611-3349 (electronic)
Lecture Notes in Computer Science
ISBN 978-3-031-34047-5   ISBN 978-3-031-34048-2 (eBook)
https://doi.org/10.1007/978-3-031-34048-2

This Springer imprint is published by the registered company Springer Nature Switzerland AG
The registered company address is: Gewerbestrasse 11, 6330 Cham, Switzerland

Preface

The Information Processing in Medical Imaging (IPMI) Conference series has a long-standing history and tradition. Since 1969, the conference took place on either side of the Atlantic and was organised in Asia for the first time in 2019. IPMI 2023 (https://www.ipmi2023.org), the twenty-eighth conference in the series, was the first in Latin America. The conference was held from June 18th to 23rd at Huinid Pioneros Hotel in San Carlos de Bariloche, Argentina.

Through the years, IPMI became a biennial beacon of foundational computational methods for medical imaging. IPMI's distinctive small-audience format enables extensive discussion time for selected oral presentations. The conference has a retreat-like ambience, which promotes a more intimate exchange and building personal and mentoring relationships across generations from our community.

IPMI 2023 featured traditional study groups, small groups of delegates who scrutinise manuscripts before their oral presentation. This leads to thorough debates enriching the authors and broader audience. Historically, discussions on each paper kick-start with early career delegates in each study group to elicit inclusive participation. Thus, every session becomes a forum for sharing, encouraging, and challenging new ideas.

For this year's edition, we received 240 submissions, of which 169 papers went through the double-blind peer review system, being evaluated by at least three reviewers each. Following the reviewing process, each manuscript having at least one weak accept in its reviews was evaluated by two or more members of the Paper Selection Committee and subsequently discussed at the paper selection meeting. Finally, 63 submissions were accepted for publication and presentation at the conference, resulting in a 26.

Three excellent keynote speakers joined us virtually. Gitta Kutyniok (Ludwig Maximilians Universität München), Petar Veličković (DeepMind), and Jong Chul Ye (Korea Advanced Institute of Science & Technology) enlightened us with various talks on the foundations of modern information processing ranging from the mathematics of machine learning to geometric deep learning and diffusion models in inverse problems in imaging.

The François Erbsmann Prize is awarded every edition to a young scientist of age 35 or below, the first author of a paper, giving their first oral presentation at IPMI. The prize was awarded to one of the 18 eligible oral presenters, and awards were given for the best poster presentations.

The IPMI 2023 Organising Team wants to express special gratitude to the Paper Selection Committee and the 186 reviewers, who did an outstanding job reviewing submissions, consolidating decisions, and designing the programme. We thank Albert Chung, Archana Venkataraman, Diana Mateus, Herve Lombaert, Ipek Oguz, Juan Eugenio Iglesias, Kaleem Siddiqi, Kevin Zhou, and Xavier Pennec, who decided on each paper based on the available reviews and thorough committee discussions. We also thank Andreas Maier, Ipek Oguz, and Archana Venkataraman for their leadership in organising the Study Groups and all the study group leaders. We also express our gratitude to Haoran Dou, Azade Farshad, and Yan Xia for their tireless work as Executive

Assistants to the chairs. We are also indebted to Enzo Ferrante and Ariel Curiale for their engagement from its conception in IPMI 2023 and for their support with all the challenges of organising IPMI across the Atlantic. Finally, we are indebted to the team of Dekon Group, particularly to Mustafa Bay, Meltem Çakmak, and Mehmet Eldegez for their organisational support. This conference would not have been possible without the help of such a great team of colleagues. One of the defining priorities throughout the organisation was to maximise diversity across career stages, gender, and geographical spread in our Organising Committee, Paper Selection Committee, Study Groups, and generally wherever possible.

We have been delighted to organise IPMI 2023 and hope that the scientific and social activities programme made this conference an unforgettable experience. We hope that IPMI 2023, surrounded by the Andes mountain range and unique Patagonian lakes and forests, was an event conducive to scientific exchange and collegial sharing and challenging of ideas, which contributed to international cooperation, scientific progress, and understanding, and to nurturing of the next generation of researchers in information processing in medical imaging.

June 2023

Alejandro Frangi
Marleen de Bruijne
Demian Wassermann
Nassir Navab

Organization

General Chairs

Alejandro F. Frangi   University of Leeds, UK – KU Leuven, Belgium
Marleen de Bruijne   Erasmus MC Rotterdam, The Netherlands
         –University of Copenhagen, Denmark

Program Chairs

Demian Wassermann   Inria Saclay Île-de-France, CEA NeuroSpin,
         Université Paris-Saclay, France
Nassir Navab   Technische Universitat München, Germany

Local Chairs

Enzo Ferrante   CONICET, Universidad Nacional del Litoral,
         Argentina
Ariel Curiale   CONICET, Centro Atómico Bariloche, Instituto
         Balseiro, Argentina

Executive Assistants to Chairs

Haoran Dou   University of Leeds, UK
Azade Farshad   Technische Universitat München, Germany
Yan Xia   University of Leeds, UK

Study Group Chairs

Ipek Oguz   Vanderbilt University, USA
Andreas Maier   University of Erlangen-Nuremberg, Germany
Archana Venkataraman   Boston University, USA

Paper Selection Committee

Archana Venkataraman	Boston University, USA
Diana Mateus	EC Nantes, France
Herve Lombaert	École de Technologie Supérieure, Canada
Ipek Oguz	Vanderbilt University, USA
Juan Eugenio Iglesias	Massachusetts General Hospital & Harvard Medical School, USA
Kaleem Siddiqi	McGill University, Canada
Xavier Pennec	INRIA, France
S. Kevin Zhou	University of Science and Technology of China, China
Albert Chung	University of Exeter, UK

Sponsor's Liaison

Yanwu Xu	Baidu, China

Information Processing In Medical Imaging Board

Christensen, G.	University of Iowa, USA
Chung, A.	University of Exeter, UK
de Bruijne, M.	Erasmus MC Rotterdam, The Netherlands
Duncan, J.S.	Yale University, USA
Frangi, A.F.	University of Leeds, UK – KU Leuven, Belgium
Golland, P.	Massachusetts Institute of Technology, USA
Leahy, R.M.	University of Southern California, USA
Noble, A.J.	Oxford University, UK
Ourselin, S.	King's College London, UK
Pizer, S.M.	University of North Carolina at Chapel Hill, USA
Prince, J.L.	Johns Hopkins University, USA
Sommer, S.	University of Copenhagen, Denmark
Styner, M.	University of North Carolina at Chapel Hill, USA
Szekely, G.	ETH Zürich, Switzerland
Taylor, C.	University of Manchester, UK
Todd-Pokropek, A.	University College London, UK
Wells, W.M. III	Harvard Medical School, USA

Reviewers

Sukesh Adiga Vasudeva
Dawood Al Chanti
Shadi Albarqouni
Daniel Alexander
Michela Antonelli
Alexis Arnaudon
John Ashburner
Aravind Ashok Nair
Suyash Awate
Mohammad Farid Azampour
Ulas Bagci
Monami Banerjee
Mathilde Bateson
Kayhan Batmanghelich
Pierre-Louis Bazin
Benjamin Billot
Sylvain Bouix
Sami Brandt
Katharina Breininger
Anna Calissano
M. Jorge Cardoso
Owen Carmichael
Catie Chang
Hao Chen
Chen Chen
Chao Chen
Xiaoran Chen
Gary Christensen
Moo Chung
Dorin Comaniciu
Tessa Cook
Tim Cootes
Vedrana Dahl
Adrian Dalca
Erik B. Dam
Benoit Dawant
Herve Delingette
Christian Desrosiers
Jose Dolz
James Duncan
Shireen Elhabian
Kjersti Engan
Ertunc Erdil

Shahrooz Faghihroohi
Aasa Feragen
Enzo Ferrante
P. Thomas Fletcher
Vanessa Gonzalez Duque
Mélanie Gaillochet
Mingchen Gao
Sara Garbarino
Harshvardhan Gazula
James Gee
Sarah Gerard
Guido Gerig
Ben Glocker
Polina Golland
Karthik Gopinath
Daniel Grzech
Boris Gutman
Tobias Heimann
Mattias Heinrich
Dewei Hu
Heng Huang
Yuankai Huo
Jana Hutter
Jae Youn Hwang
Amelia Jiménez-Sánchez
Anand Joshi
Sarang Joshi
Bernhard Kainz
Ali Kamen
Anees Kazi
Matthias Keicher
Mahdieh Khanmohammadi
Boklye Kim
Andrew King
Stefan Klein
Ender Konukoglu
Frithjof Kruggel
Holger Kunze
Jan Kybic
Bennett Landman
Hongwei Li
Xiaoxiao Li
Han Li

Chunfeng Lian
Roxane Licandro
Peirong Liu
Mingxia Liu
Pengbo Liu
Xiaofeng Liu
Marco Lorenzi
Andreas Maier
Carsten Marr
Martin Menten
Bjoern Menze
Daniel Moyer
Arrate Munoz Barrutia
Vishwesh Nath
Mads Nielsen
Marc Niethammer
Lipeng Ning
Beatriz Paniagua
Magdalini Paschali
Rasmus Paulsen
Chantal Pellegrini
Tingying Peng
Jens Petersen
Dzung Pham
Louise Piecuch
Steve Pizer
Marco Pizzolato
Kilian Pohl
Jerry Prince
Dou Qi
Chen Qin
Jagath Rajapakse
Laurent Risser
Karl Rohr
Louis Rouillard
Daniel Rueckert
Ario Sadafi
Olivier Salvado
Babak Samari
Mhd Hasan Sarhan
Dustin Scheinost
Julia Schnabel
Thomas Schultz
Ernst Schwartz
Christof Seiler

Raghavendra Selvan
Maxime Sermesant
Wei Shao
Pengcheng Shi
Yonggang Shi
Kuangyu Shi
Ivor Simpson
Ayushi Sinha
Jayanthi Sivaswamy
Lars Smolders
Roger Soberanis-Mukul
Stefan Sommer
Aristeidis Sotiras
Tyler Spears
Jon Sporring
Anuj Srivastava
Martin Styner
Carole Sudre
Tabish Syed
Oriane Thiery
Matthew Toews
Agnieszka Tomczak
Alice Barbara Tumpach
Francois-Xavier Vialard
Clinton Wang
Linwei Wang
Ce Wang
Yalin Wang
Hongzhi Wang
Demian Wassermann
Sandy Wells
Carl-Fredrik Westin
Ross Whitaker
Jonghye Woo
Marcel Worring
Guorong Wu
Yong Xia
Pew-Thian Yap
Chuyang Ye
Yousef Yeganeh
Sean Young
Paul Yushkevich
Ghada Zamzmi
Xiangrui Zeng
Miaomiao Zhang

Gary Zhang
Yubo Zhang
Qingyu Zhao
Guoyan Zheng
Luping Zhou

Xiahai Zhuang
Veronika Zimmer
Maria A. Zuluaga
Reyer Zwiggelaar
Lilla Zöllei

Contents

Multimodal Learning

Optimization

Reconstruction

Self Supervised Learning

Surface Analysis and Segmentation

Biomarkers

Resolving Quantitative MRI Model Degeneracy with Machine Learning via Training Data Distribution Design

Michele Guerreri[1,2]($\boxtimes$), Sean Epstein[1], Hojjat Azadbakht[2], and Hui Zhang[1]

[1] Department of Computer Science and Centre for Medical Image Computing, University College London, London, UK
m.guerreri@ucl.ac.uk
[2] AINOSTICS Ltd., Manchester, UK

Abstract. Quantitative MRI (qMRI) aims to map tissue properties non-invasively via models that relate these unknown quantities to measured MRI signals. Estimating these unknowns, which has traditionally required model fitting - an often iterative procedure, can now be done with one-shot machine learning (ML) approaches. Such parameter estimation may be complicated by intrinsic qMRI signal model degeneracy: different combinations of tissue properties produce the same signal. Despite their many advantages, it remains unclear whether ML approaches can resolve this issue. Growing empirical evidence appears to suggest ML approaches remain susceptible to model degeneracy. Here we demonstrate under the right circumstances ML can address this issue. Inspired by recent works on the impact of training data distributions on ML-based parameter estimation, we propose to resolve model degeneracy by designing training data distributions. We put forward a classification of model degeneracies and identify one particular kind of degeneracies amenable to the proposed attack. The strategy is demonstrated successfully using the Revised NODDI model with standard multi-shell diffusion MRI data as an exemplar. Our results illustrate the importance of training set design which has the potential to allow accurate estimation of tissue properties with ML.

Keywords: quantitative MRI · machine learning · model degeneracy

1 Introduction

Quantitative magnetic resonance imaging (qMRI) generates images in which voxel intensities are quantitatively related to tissue properties, and promises higher reproducibility and interpretability than conventional qualitative MRI. These advantages come at the cost of requiring parameter estimation (fitting signal models to qMRI data), which underpins some of the key challenges associated with qMRI experiments: (1) high computational cost, due to iterative

A. Frangi et al. (Eds.): IPMI 2023, LNCS 13939, pp. 3–14, 2023.
https://doi.org/10.1007/978-3-031-34048-2_1

maximization of a typically non-concave likelihood function, and (2) long acquisition time, due to increased data requirements to support model fitting.

Some qMRI applications, such as microstructure imaging, present a third challenge: multiple parameter combinations generate the same qMRI signal [8]. With conventional fitting approaches this degeneracy induces unstable parameter estimations, preventing the identification of the correct value. In some cases degeneracy can be solved by enriching the acquisition protocol. However, this is not always possible due to hardware limitations or acquisition time constraints.

Supervised machine learning (ML) has been shown to address two of these three challenges. It is orders of magnitude faster than its conventional counterpart [2,5,12], and can cope with significantly shortened acquisition protocols [5]. The third challenge, signal degeneracy, is as-yet unaddressed, and has been identified as a limitation of existing ML approaches [2].

However, while we would expect shortened acquisition protocols to introduce or exacerbate model degeneracy, the fact that ML methods can handle these protocols [5] suggests they may be robust to degeneracy in some circumstances. Indeed, Bishop and Roach [3] identified signal degeneracy as a limitation to their seminal ML parameter estimation method. They proposed a simple fix: identifying the degenerate regions of parameter space associated with the same signal, and removing all but one of these regions in the training data. They explicitly resolved degeneracy by imposing a prior on the training data: any one-to-many mappings (one signal to many parameters) are manually reduced to one-to-one. This process is effective but limited by the arbitrary choice of prior: the truncation of the training data is not data-driven.

In this paper we propose a method to resolve signal degeneracy encountered in ML parameter estimation. We present a variation of Bishop and Roach's approach, where we truncate the training data distribution in a data-driven manner. Our method requires three conditions commonly encountered in qMRI: (i) signal degeneracy is present when using a commonly-used acquisition protocol; (ii) this degeneracy can be resolved by using a super-sampled acquisition protocol not commonly available or accessible; (iii) the super-sampled protocol reveals that all clinically-observed voxels belong to just one of the conventionally-indistinguishable degenerate parameter-space combinations. If these conditions are met, an optimal ML training dataset can be constructed for the commonly-used acquisition protocol by rejecting degenerate parameter combinations which are not observed when using the super-sampled protocol.

This work demonstrates our method with microstructure imaging, but is applicable to any qMRI application for which the above conditions hold. We show that our method resolves degeneracy associated with applying the Revised NODDI model [6,7] to conventional multi-shell diffusion MRI (dMRI) data. This degeneracy can be resolved by acquiring richer tensor-valued diffusion encoding data [15,16]. We demonstrate that an ML parameter estimation method, trained within our proposed degeneracy-resolving framework, resolves signal degeneracy when applied to an inherently-degenerate conventional multi-shell acquisition protocol. We use a simple feed forward deep learning (DL) model to represent ML. We use synthetic data for training and test on unseen *in vivo* data.

2   Theory

This section first describes the three categories degeneracy can be divided into, then their different impact on ML, and finally the proposed method.

2.1   Impact of Model Degeneracy on ML-Based Parameter Estimation

Traditionally, parameter estimation involves the fitting of a forward model, parameterised by some tissue properties of interest, to measurements. Optimal values for the sought properties are obtained by maximising some likelihood function. With this approach, it is possible that multiple combinations of the properties maximise the likelihood function equally well, a problem known as model degeneracy. While the impact of model degeneracy on conventional parameter estimation is well understood, its effect on ML-based parameter estimation has yet to be fully investigated. Here we characterize this impact by deriving the ML output when a degenerate subset of samples is included during training.

ML-based parameter estimation works by learning a function f_θ, parameterised by its learnable parameter vector θ, that can accurately map some measurements $x \in \mathbb{R}^M$ to the corresponding tissue properties $y \in \mathbb{R}^K$. Given a set of N exemplar pairs $[x_i, y_i]$, $i = 1, ..., N$, as the training samples, ML learns the optimal $\hat{\theta}$ that can best approximate the required mapping. Training algorithms generally minimize the mean squared error (MSE) between the predictions and the corresponding labels for the training set:

$$E_\theta = \frac{1}{N} \sum_{i=1}^{N} \sum_{k=1}^{K} [f_\theta^k(x_i) - y_i^k]^2. \tag{1}$$

Importantly, for any given input vector x, f_θ generates only one output vector y; this function can only represent one-to-one or many-to-one mappings. Degenerate, multi-valued correspondences cannot be represented [3].

Without loss of generality, we will assume the degenerate subset of samples is the first n samples of the training set. For this subset, each sample has a distinct output vector but has the same input which we will denote as x_d:

$$\forall i, j \in \{1, 2, ..., n\}, x_i = x_j = x_d \text{ but } y_i \neq y_j. \tag{2}$$

The MSE loss can then be decomposed into one term involving the degenerate subset and one with the rest of the samples, such that

$$E_\theta = \frac{1}{N} \sum_{i=1}^{n} \sum_{k=1}^{K} \left[f_\theta^k(x_d) - y_i^k \right]^2 + \frac{1}{N} \sum_{i=n+1}^{N} \sum_{k=1}^{K} \left[f_\theta^k(x_i) - y_i^k \right]^2. \tag{3}$$

If we assume the parameterization of f_θ is sufficiently flexible such that each of the two MSE terms can be minimized independently, the trained function f_θ will minimize the term involving the degenerate subset E_θ^d when

$$0 = \frac{\partial E_\theta^d}{\partial \theta} = \frac{2}{N} \sum_{k=1}^{K} \frac{\partial f_\theta^k(x_d)}{\partial \theta} \left[n f_\theta^k(x_d) - \sum_{i=1}^{n} y_i^k \right], \qquad (4)$$

which implies that

$$\forall k \in \{1, 2, ..., K\}, f_\theta^k(x_d) = \frac{1}{n} \sum_{i=1}^{n} y_i^k. \qquad (5)$$

This shows in the presence of degenerate samples during training, the learned function will map the degenerate signal x_d to the empirical mean of the tissue properties over the degenerate subset. This highlights the significant role the training data distribution can play in determining the learned mapping: the empirical mean may be tuned, by choosing a suitable training data distribution, to produce the desired mapping.

Crucially, the impact of this result depends on the nature of the degeneracy which can be broadly classified into three categories: (1) degeneracy that genuinely obscures biological differences, i.e. different underlying tissue properties give rise to the same signal; (2) degeneracy that potentially obscures biological differences, but only one non-degenerate case is observed in real data; (3) degeneracy that corresponds to no biological differences.

With Type (1) degeneracy impossible to resolve and Type (3) irrelevant, Type (2) degeneracy is of particular interest as it can be resolved with ML. Namely, by excluding from the training set the combinations of tissue properties which are degenerate, but which are not observed in real data, at inference time the ML framework should be able to correctly estimate the combination which is typically observed.

2.2   Proposed Method to Solve Degeneracy

Here we describe the proposed procedure to resolve Type (2) degeneracy based on the observations above. First, we describe the two conditions necessary for the method to be feasible, given a qMRI model and a standard acquisition protocol: (i) the presence of a degeneracy when fitting the model to the data acquired with the standard protocol; (ii) the availability of a rich protocol that can resolve the degeneracy and includes the standard protocol as a subset. When both conditions are met, the observed degeneracy can be identified as Type (2).

The proposed method comprises four steps as summarized in Fig. 1: First, for a given qMRI model and the standard protocol, establish the presence of degeneracy using the conventional fitting approach on real data acquired with the standard protocol (Panel A). Second, confirm the absence of degeneracy for the data acquired with the rich protocol using the same conventional fitting approach (Panel B). Third, establish the degeneracy is of Type (2) (Panel C). Fourth, construct an optimized training set by excluding all the degenerate parameter combinations which are not observed in Step two. The optimized parameter distribution can then be used to generate the corresponding MRI signals for the chosen qMRI model and the standard protocol to train a ML model.

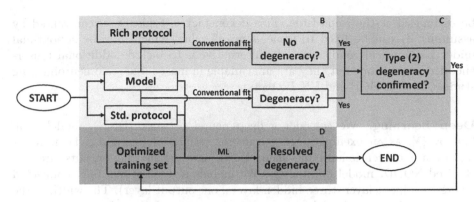

Fig. 1. Schematic of the proposed method to resolve Type (2) degeneracy: A. given a model and a standard protocol, the presence of degeneracy is assessed via conventional fit; B. a rich protocol exists to assess whether degeneracy can be resolved; C. if both A. and B. are verified, degeneracy is confirmed as Type (2); D. if so, an optimized training set can be constructed removing unobserved degeneracies and used for ML training.

3   Methods

This section describes the datasets, implementation and experiments used to demonstrate the proposed method in a real-world application. The qMRI and ML frameworks used to exemplify our approach are detailed. These are followed by a description of the acquisition protocol of the *in vivo* data and the synthesis of simulated data. Finally, we describe the experiments conducted.

3.1   qMRI and ML Frameworks

Revised NODDI. We demonstrate the proposed method with an exemplar microstructure imaging framework: Revised-NODDI [6,7], a recently proposed variation on NODDI [17]. NODDI is a popular neuroimaging technique [4] that aims to quantify neurite morphology using standard dMRI protocols. Revised-NODDI overcomes some of the limitations of the original version [10], and is compatible with the more advanced tensor-encoded dMRI protocols [6].

Similar to NODDI, Revised-NODDI models voxel signals as originating from two compartments: free water and tissue. The tissue signal is further divided into two components: intra-neurite and extra-neurite. For each voxel, Revised-NODDI generates four indices related to microstructural tissue properties. These indices are the neurite density index (NDI), which measures the proportion of the tissue component occupied by neurites; the orientation dispersion index (ODI), which measures the directional variability of neurites; the intra-neurite diffusivity (d_I), which measures the isotropic diffusivity of the neurites; and the free water fraction (FWF), which measures the extent to which the voxel is contaminated by free water.

Revised-NODDI is chosen as the exemplar because, with standard dMRI protocols, the model is known to be degenerate - its four parameters can not be estimated at the same time. This degeneracy is typically circumvented by assuming d_I can be fixed to some *a priori* value, which may be suboptimal under pathological conditions. The alternative is to include additional tensor-encoded acquisitions but they are unavailable in most large-scale neuroimaging studies, such as HCP and UK Biobank.

Deep Learning. We consider a deep neural network (DNN), modeled on Golkov [5], as an exemplar ML parameter estimation method. The network inputs are conventional multi-shell dMRI data, while the outputs are the Revised-NODDI model parameters. We include a total of five fully connected layers (one input layer, three hidden layers, one output layer). The width of the input layer is determined by the acquisition protocol (explained below). The width of the hidden layers is 150. Each hidden layer uses ReLU for activation. A dropout of 0.1 is inserted before the output layer, which has a width of 4 and uses a ReLU for activation. The total number of learnable parameters is $46054 + 150n$ where n is the number of input dMRI data.

3.2   Datasets

in vivo **Data.** We use an *in vivo* data set to determine the presence of model degeneracy and to assess the performance of the proposed method at inference time. We use the publicly available data described in Szczepankicwicz et al. [13]. The dataset comprises scans from one healthy subject acquired with two acquisition methods: a standard multi-shell scheme, known as the linear tensor encoding (LTE) in the terminology of tensor-valued diffusion encoding [15,16] and a non-standard scheme, known as the spherical tensor encoding (STE) which offers orthogonal information to LTE [14]. Each encoding type contains 4 b-values, up to a maximum of 2.0 ms/μm^2. The two schemes are combined to form the rich scheme for our experiments. We refer to this dataset as the "*szcz_DIB*" dataset. To restrict the analysis to white matter (WM), LTE data with b-values up to 1.4 ms/μm^2 was used to estimate diffusion tensors voxel-wise and to subsequently estimate their corresponding fractional anisotropy (FA) values. The resulting FA map was then fed into the FMRIB's Automated Segmentation Tool (FAST) [18] to obtain a WM mask.

Synthetic Data. We use synthetic data to train the DNN model. We generate $N = 10000$ Revised-NODDI WM parameter combinations from a synthetic parameter distribution derived from Guerreri et al [6].

3.3   Implementation

Establishing the Degeneracy for the Standard Protocol. The Revised-NODDI model, that includes d_I as free variable, is known to be degenerate when it is fit to conventional multi-shell dMRI data using conventional fitting [9]. We confirm this degeneracy in the LTE data of

WM from the *in vivo szcs_DIB* dataset, using an adapted version of the MATLAB NODDI toolbox implementation (MATLAB version R2022a). Presence of degeneracy is observed qualitatively by the multiple clusters apparent in the estimated tissue property distribution and spatially-incoherent appearance of the estimated tissue property maps (See Fig. 3). This is verified quantitatively by observing that the initialization of the fitting process from different starting points leads to different tissue property estimation with comparable likelihood for the same voxels, confirming the existence of multiple equivalent local maxima in the likelihood landscape.

Establishing the Lack of Degeneracy for the Rich Protocol. The Revised-NODDI model is known to be non-degenerate when it is fit to data with both LTE and STE [6,7]. We verify this lack of degeneracy associated with the *in vivo szcz_DIB* dataset by fitting WM voxels acquired using LTE and STE together. We use the same fitting approach as above and qualitatively compare the estimated tissue property distribution and the tissue property maps with the ones obtained using LTE data alone (See Fig. 3).

Establishing the Degeneracy as Type (2). This follows directly from above.

Training Datasets Generation. We generate two training datasets to assess whether a non-degenerate training distribution can be used to resolve signal model degeneracy. This process is summarized in Fig. 2.

Both training datasets are generated by initially drawing from the synthetic parameter distribution described above (See Sect. 3.2). The drawn parameters are used to synthesise dMRI signals, using either the *szcz_DIB* standard protocol (LTE alone) or the rich protocol (LTE+STE), and adding Rician noise that matches the real data (SNR = 30). Parameter estimates with the conventional fitting is then obtained for the data simulated with each protocol. By construction, the resulting parameter distribution from the LTE data is degenerate while that from the LTE+STE data is not degenerate. These distributions are subsequently used to generate noisy dMRI signals, using the *szcz_DIB* LTE protocol, and used to train two DNN models. Data are split 80/20 between training and validation.

DNN Training. Network training is performed using an Adam optimizer with learning rate = 0.001. A maximum of 1000 epochs is set with an early stopping of 10 epochs.

3.4   Experiments

We compare the parameters estimated via the two trained DNN models on the *in vivo* data. We expect to observe the estimations from the network trained via the set with degeneracy to be more biased toward the spurious parameter distribution not observed with the rich protocol.

4   Results

Figure 3 shows qualitatively that the degeneracy observed by fitting the Revised-NODDI model to *in vivo* conventional multi-shell LTE data in WM voxels is of Type (2). Left panel shows the estimated tissue properties form two distinct clusters in the tissue-property space. One cluster has higher values of NDI, d_I and ODI than the other. Using different starting points to initialize the conventional fit results in different tissue property estimation with comparable likelihood values, which confirm the presence of degeneracy (results not shown). Right panel of the figure shows the degeneracy can be resolved by fitting the model to a richer data set that includes STE. Crucially, the estimated tissue properties from the rich protocol overlap with only one of the two clusters derived from the standard protocol, suggesting that the values in the other cluster is not observed in real data, thus confirming the observed degeneracy is of Type (2).

Figure 4 shows that training an ML model that excludes the subset of degeneracy parameters not observed in real data can be used to resolve the degeneracy. Left panel shows the tissue properties estimated using a non-degenerate training set, although slightly biased, are close to the parameter distribution derived from the rich protocol with conventional fitting (which we treat as our ground truth). On the other hand, Right panel shows the tissue properties estimated via an ML model that includes the degenerate parameter combinations during training are more biased. The direction of the bias points towards the mean value of the degenerate combinations, as expected from the analysis in the Theory section. The root-mean-squared-errors shown in Table 1 confirm these observations quantitatively.

Table 1. Root mean squared errors of the estimated tissue properties via machine learning. We compare the estimation error from using training strategies which does or does not include degeneracy. C1(2) for Cluster 1(2) (See Fig. 3).

Params.	Conventional fitting			Training set w/ degeneracy			Training set w/o degeneracy		
	C1	C2	All	C1	C2	All	C1	C2	All
NDI	0.08	0.33	0.23	0.10	0.15	0.12	0.06	0.07	0.06
d_I	0.01	0.20	0.17	0.08	0.08	0.08	0.04	0.05	0.04
ODI	0.03	0.02	0.12	0.04	0.06	0.05	0.02	0.02	0.02
FWF	0.07	0.10	0.08	0.05	0.05	0.05	0.04	0.04	0.04

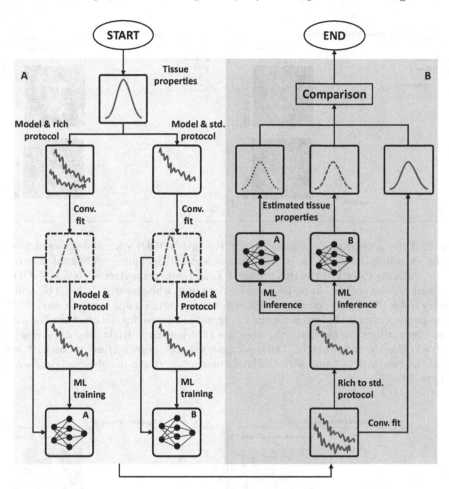

Fig. 2. Schematic of the experimental design used to demonstrate the proposed method. Panel A: training sets generation and machine learning (ML) training. Two sets are generated which include or not include degenerate tissue properties. Panel B: assessing the effect of different training sets on unseen data.

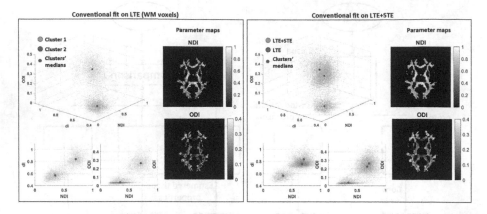

Fig. 3. Type 2 degeneracy assessment for Revised-NODDI with conventional linear tensor encoding (LTE) as the standard protocol and including additional spherical tensor encoding (STE) as the rich protocol. Left: tissue properties of Revised-NODDI obtained via conventional fit on LTE *in vivo* data from white matter (WM). The neurite density index (NDI) and orientation dispersion index (ODI) maps appear 'noisy'. In the tissue-property space the voxels appear to organize into two distinct clusters. Together, these observations demonstrate the presence of degeneracy. Right: the degeneracy is resolved using the rich protocol. The non-degenerate tissue properties overlap with only one of the two clusters found with LTE data alone, confirming the observed degeneracy of Type (2).

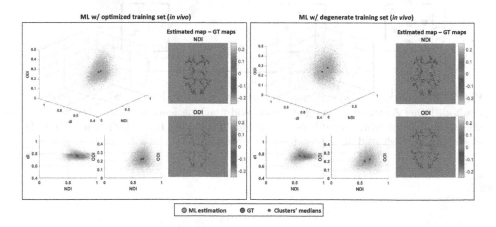

Fig. 4. Comparison of estimated tissue property distributions from different machine learning (ML) training strategies. Left: the estimated properties are obtained using a training set which includes only non-degenerate tissue properties. Right: the estimated properties are obtained using a training which includes degenerate tissue properties. The estimated properties are overlaid on top of the ground-truth (GT) values obtained with the rich protocol (Fig. 3). NDI and ODI difference maps between estimated and GT values are also reported.

5 Discussion and Conclusions

In summary, this paper makes two contributions to the nascent but fast-moving topic of ML-based qMRI parameter estimation. First, we describe a theoretical framework to understand the impact of model degeneracy on this new class of parameter estimation approaches. We show that in general ML-based approaches can not resolve model degeneracy. In contrast to the conventional fitting where model degeneracy leads to unpredictable, thus highly variable, parameter estimates, ML-based approaches instead produce highly consistent but biased estimates. For sufficiently flexible ML models, we show that these biased estimates approach the empirical means of the degenerate training samples, highlighting the key role training data distribution plays in influencing the trained models.

Second, we propose a categorization of model degeneracies that lead to a data-driven scheme of training data distributions. The scheme builds on the idea first proposed in the seminal work of Bishop and Roach [3] which showed how model degeneracy may be mitigated by tuning the training data distribution, in particular, by removing a subset of the training samples so that the remaining samples no longer contain degeneracy. However, the strategy they used to select the subset for removal was subjective. The data-driven selection scheme proposed here is inspired by the notion of image quality transfer [1] where information contained in high-quality images is used to enhance lower-quality images. If a degeneracy is observed in a standard protocol but not in a richer protocol that includes the standard protocol as a subset, this information can be leveraged to select the appropriate subset among the degenerate parameters for training. This observation leads naturally to the proposed categorization of model degeneracies.

This strategy of embedding the information derived from richer, non-standard protocols can also be implemented in a conventional fitting framework as a Bayesian prior distribution, as elegantly demonstrated in [11]. One can consider the present work as an ML-based variant that does not require iterative optimization.

Acknowledgements. Computing resources and support were provided by AINOS-TICS Ltd., enabled through funding by Innovate UK. MG is funded by UKRI under grant MR/W004097/1.

References

1. Alexander, D.C., et al.: Image quality transfer and applications in diffusion MRI. Neuroimage **152**, 283–298 (2017)
2. de Almeida Martins, J.P., et al.: Neural networks for parameter estimation in microstructural MRI: application to a diffusion-relaxation model of white matter. Neuroimage **244**, 118601 (2021)
3. Bishop, C.M., Roach, C.M.: Fast curve fitting using neural networks. J. Rev. Sci. Instrum. **63**(10), 4450–4456 (1992)
4. Cox, S.R., et al.: Ageing and brain white matter structure in 3,513 UK biobank participants. Nat. Commun. **7**(1), 13629 (2016)

5. Golkov, V., et al.: Q-space deep learning: twelve-fold shorter and model-free diffusion MRI scans. IEEE Trans. Med. Imaging **35**(5), 1344–1351 (2016)
6. Guerreri, M., et al.: Revised NODDI model for diffusion MRI data with multiple B-
   tensor encodings. International Society for Magnetic Resonance in Medicine (2018)
7. Guerreri, M., Szczepankiewicz, F., Lampinen, B., Palombo, M., Nilsson, M.,
   Zhang, H.: Tortuosity assumption not the cause of NODDI's incompatibility with
   tensor-valued diffusion encoding. International Society for Magnetic Resonance in
   Medicine (2020)
8. Jelescu, I.O., Veraart, J., Adisetiyo, V., Milla, S.S., Novikov, D.S., Fieremans, E.:
   One diffusion acquisition and different white matter models: how does microstructure change in human early development based on WMTI and NODDI? Neuroimage **107**, 242–256 (2015)
9. Kiselev, V.G., Il'yasov, K.A.: Is the biexponential diffusion biexponential? Magn.
   Reson. Med. **57**(3), 464–469 (2007)
10. Lampinen, B., Szczepankiewicz, F., Martensson, J., van Westen, D., Sundgren,
    P.C., Nilsson, M.: Neurite density imaging versus imaging of microscopic anisotropy
    in diffusion MRI: a model comparison using spherical tensor encoding. Neuroimage
    147, 517–531 (2017)
11. Mozumder, M., Pozo, J.M., Coelho, S., Frangi, A.F.: Population-based Bayesian
    regularization for microstructural diffusion MRI with NODDIDA. Magn. Reson.
    Med. **82**(4), 1553–1565 (2019)
12. Palombo, M., et al.: SANDI: a compartment-based model for non-invasive apparent
    soma and neurite imaging by diffusion MRI. Neuroimage **215**, 116835 (2020)
13. Szczepankiewicz, F., Hoge, S., Westin, C.F.: Linear, planar and spherical tensor-
    valued diffusion MRI data by free waveform encoding in healthy brain, water, oil
    and liquid crystals. Data Brief **25**, 104208 (2019)
14. Szczepankiewicz, F., et al.: The link between diffusion MRI and tumor heterogene-
    ity: mapping cell eccentricity and density by diffusional variance decomposition
    (DIVIDE). Neuroimage **142**, 522–532 (2016)
15. Topgaard, D.: Multidimensional diffusion MRI. J. Magn. Reson. **275**, 98–113
    (2017)
16. Westin, C.F., et al.: Q-space trajectory imaging for multidimensional diffusion MRI
    of the human brain. Neuroimage **135**, 345–362 (2016)
17. Zhang, H., Schneider, T., Wheeler-Kingshott, C.A., Alexander, D.C.: NODDI:
    practical in vivo neurite orientation dispersion and density imaging of the human
    brain. Neuroimage **61**(4), 1000–1016 (2012)
18. Zhang, Y., Brady, M., Smith, S.: Segmentation of brain MR images through a
    hidden Markov random field model and the expectation-maximization algorithm.
    IEEE Trans. Med. Imaging **20**(1), 45–57 (2001)

Subtype and Stage Inference
with Timescales

Alexandra L. Young[1], Leon M. Aksman[2]([✉]), Daniel C. Alexander[3],
Peter A. Wijeratne[3,4],
and for the Alzheimer's Disease Neuroimaging Initiative

[1] Department of Neuroimaging, Institute of Psychiatry, Psychology and
Neuroscience, King's College London, London, UK
`alexandra.young@kcl.ac.uk`
[2] Stevens Neuroimaging and Informatics Institute, Keck School of Medicine,
University of Southern California, Los Angeles, CA, USA
`Leon.Aksman@loni.usc.edu`
[3] Centre for Medical Image Computing, Department of Computer Science, University
College London, London, UK
`d.alexander@ucl.ac.uk`
[4] Department of Informatics, University of Sussex, Brighton, UK
`p.wijeratne@sussex.ac.uk`

Abstract. Neurodegenerative conditions typically have highly heterogeneous trajectories, with variability in both the spatial and temporal
progression of neurological changes. Disentangling the variability in spatiotemporal progression patterns offers major benefits for patient stratification and disease understanding but is a complex methodological
challenge. Here we present Temporal Subtype and Stage Inference (T-
SuStaIn), a technique that uniquely integrates distinct ideas from unsupervised learning: disease progression modelling, clustering, and hidden
Markov modelling. We formulate T-SuStaIn mathematically and devise
an algorithm for inferring the model parameters and uncertainty. We
demonstrate that the combination of disease progression modelling, clustering, and hidden Markov modelling uniquely enables the discovery of
subtypes distinguished not just by ordering of abnormality accumulation, but also timescale. We apply T-SuStaIn to longitudinal volumetric imaging data from the Alzheimer's Disease Neuroimaging Initiative,
deriving spatiotemporal Alzheimer's disease subtypes together with their
timelines of evolution and associated uncertainty. T-SuStaIn has broad
utility across a range of longitudinal clustering problems, both in neurodegenerative conditions and more widely in progressive diseases.

L. M. Aksman—Joint first author.
Data used in preparation of this article were obtained from the Alzheimer's Disease
Neuroimaging Initiative (ADNI) database (adni.loni.usc.edu). As such, the investigators within the ADNI contributed to the design and implementation of ADNI and/or
provided data but did not participate in analysis or writing of this report. A complete
listing of ADNI investigators can be found at: http://adni.loni.usc.edu/wp-content/
uploads/how_to_apply/ADNI_Acknowledgement_List.pdf.

Keywords: Longitudinal clustering · Markov jump process · Disease progression model · Subtyping · Prognosis · Dementia

1   Introduction

Characterising the natural history of a disease is a crucial step towards understanding the underlying disease biology and predicting disease outcomes. In neurodegenerative diseases, such as Alzheimer's disease (AD), this has proven challenging due to the decades-long disease timescale [1], which makes it impractical to chart the disease from start to end in an individual. This problem is exacerbated by the heterogeneity between individuals, with many individuals following an atypical trajectory [2]. Quantitative models of neurodegenerative disease subtypes and their progression timescales provide insights into the heterogeneous patterns of disease changes and enable patient stratification and prediction of future outcomes in clinical trials and healthcare.

Disease progression modelling (e.g., [3–7]) is a form of unsupervised learning that enables the estimation of long-term trajectories of disease change from cross-sectional and short-term longitudinal datasets [8]. The event-based model (EBM) [3] of disease progression describes a disease as a series of events, where each event corresponds to a new biomarker becoming abnormal, enabling the identification of population-level disease progression patterns from cross-sectional datasets. However, the EBM (a) assumes that all individuals follow a common progression pattern; (b) estimates only the sequence, not the timescale, of disease changes; and (c) does not appropriately exploit longitudinal data. The Subtype and Stage Inference (SuStaIn) algorithm [9] places the EBM in a clustering framework, enabling the identification of disease subtypes with distinct disease progression patterns. The temporal event-based model (T-EBM) [10] places the EBM in a hidden Markov modelling framework, appropriately modelling longitudinal data to enable the estimation of disease timescales from short-term longitudinal data.

However, SuStaIn cannot estimate disease timescales and the T-EBM cannot account for disease subtypes. Moreover, SuStaIn inherits the EBM's inability to appropriately exploit longitudinal data, which can hinder model identifiability when degenerate solutions exist, e.g., when trajectories from two subtypes cross over (*crossing subtype trajectories*). The majority of previous techniques for longitudinal clustering of disease subtypes (e.g. [11]) fail to consider heterogeneity in disease stage at baseline (i.e. they do not perform disease progression modelling). Those that do account for disease stage heterogeneity have typically required a large number of observations (approximately 1000 individuals with five time-points each) and have high model complexity [12,13], hindering their utility in medical datasets and in identifying subtypes with low prevalence.

Here we present Temporal Subtype and Stage Inference (T-SuStaIn), a technique that uniquely enables the estimation of disease subtypes with distinct progression patterns and their timescales from short-term longitudinal datasets. We formulate T-SuStaIn mathematically and derive an algorithm for simultaneously inferring disease subtypes, progression patterns, and timescales of progression.

Harnessing the added complexity of combining disease progression modelling, clustering, and hidden Markov modelling in a single framework necessitates the development of a novel constrained transition matrix that restricts the dimensionality of the parameter space by encoding common assumptions of disease progression models. This further enables extension of the inference to use Markov Chain Monte Carlo (MCMC) sampling, providing a joint estimate of the uncertainty in the subtype progression patterns, proportion of individuals belonging to each subtype, and the subtype progression timescales. We show that T-SuStaIn can successfully exploit longitudinal data to recover event sequences that are not identifiable by the cross-sectional SuStaIn algorithm. We demonstrate T-SuStaIn using volumetric structural imaging data from the Alzheimer's Disease Neuroimaging Initiative, identifying AD subtypes with distinct spatiotemporal progression patterns and their timelines.

2 Theory

2.1 Mathematical Model for T-SuStaIn

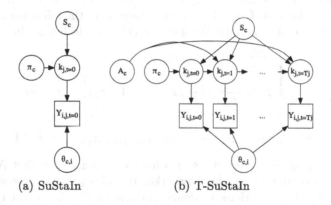

(a) SuStaIn   (b) T-SuStaIn

Fig. 1. Graphical models for SuStaIn and T-SuStaIn. Hidden variables are denoted by circles and observations by squares.

The mathematical model underlying T-SuStaIn is formulated as a mixture of temporal disease progression models, combining ideas from SuStaIn [9] and the T-EBM [10]. In this work we use a piecewise linear z-score model of disease progression [9], which we adapt to have a hidden Markov model formulation that leverages longitudinal datasets and estimates timelines. As such, T-SuStaIn makes the same assumptions as both the T-EBM and SuStaIn, namely: i) monotonic biomarker dynamics at the group level; ii) Markov stage transitions at the individual level; and iii) a mixture of event sequences across the population. Graphical models of SuStaIn and T-SuStain are shown in Fig. 1. We denote the data for each biomarker i and individual j observed at time t as $Y_{i,j,t}$; the initial probability distribution for cluster c as π_c; the transition probability matrix for cluster c with elements $a_{a,b}$ as A_c; the distribution parameters for biomarker i in cluster c as $\theta_{c,i}$; the overall set of model parameters as $\Theta_c = [\pi_c, A_c, \theta_{c,i}]$; the event sequence for cluster c as S_c. Here $S_c = \{s_c(1), ..., s_c(N)\}$ is a permutation of N events that represents the hidden sequence of events defining the state space for a discrete time Markov jump process, where an event is the transition of a biomarker from one z-score to another. We denote the hidden stage for individual j at time t as $k_{j,t}$ and define $0 \leq x \leq 1$ as a dimensionless variable that spans the event space. Following [10], under assumptions (i)–(iii), we can write the equation for the total data likelihood of the T-EBM for cluster c as,

$$P(Y_j|\Theta_c, S_c)$$

$$= \sum_{k_{j,t}=0}^{N} \int_{x=\frac{k}{N+1}}^{x=\frac{k+1}{N+1}} P(k_{j,t=0}|S_c, \pi_c) \prod_{t=1}^{T_j} P(k_{j,t}|k_{j,t-1}, S_c, A_c) \prod_{t=0}^{T_j} \prod_{i=1}^{I} P(Y_{i,j,t}|k_{j,t}, \theta_{c,i}, S_c)dx,$$

(1)

where,

$$P(Y_{i,j,t}|k_{j,t}, \theta_{c,i}, S_c) = \text{NormPDF}(Y_{i,j,t}, \theta_{c,i}).$$

(2)

Following [9], we assume a univariate normal distribution for the data, $Y_i \sim \mathcal{N}(\mu_i, \sigma_i)$, and choose $\theta_{c,i}(x) = [\mu_{c,i}(x), \sigma_{c,i}]$, where $\mu_{c,i}$ and $\sigma_{c,i}$ are the mean and standard deviations of distribution i in cluster c. In the following we drop the c index for notational simplicity. We define $\mu_i(x)$ as a piece-wise linear function,

$$\mu_i(x) = \begin{cases} \frac{z_1}{x_{Ez_1}}x, \ 0 < x \leq x_{Ez_1} \\ z_1 + \frac{z_2-z_1}{x_{Ez_2}-x_{Ez_1}}\left(x - x_{Ez_1}\right), \ x_{Ez_1} < x \leq x_{Ez_2} \\ \vdots \\ z_M + \frac{z_{max}-z_{M_i}}{1-x_{Ez_{Mi}}}\left(x - x_{E_zMi}\right), \ x_{Ez_{Mi}} < x \leq 1. \end{cases}$$

(3)

Here $z_i = z_1, ..., z_{M_i}$ is the set of z-scores for biomarker i such that $N = \sum M_i$, z_{max} is the maximum z-score for biomarker i; and E_{z_i} is the z-score event at $x_{Ez_i} = (k + 1)/(N + 1)$. Here we define the z-scores with respect to the control population, and hence set the standard deviation $\sigma_i = 1$, i.e., the z-scores correspond to the number of standard deviations from the control population.

Following [14], the elements of the transition matrix A_c are defined as,

$$a_{a,b} \equiv P(k_{j,t} = b|k_{j,t-1} = a, S_c, A_c),$$

(4)

the elements of the initial stage probability vector π_a are defined as,

$$\pi_a = P(k_{j,t=0} = a|S_c, \pi_c),$$

(5)

and the expected duration of each stage (sojourn time) δ_k as,

$$\delta_k = \sum_{\delta=1}^{\infty} \delta P_k(\delta) = 1/(1 - a_{kk}),$$

(6)

where a_{kk} are the diagonal elements of A_c.

The mathematical model underlying T-SuStaIn is defined as a mixture of temporal event-based models (Eq. 1),

$$P(Y|\Theta, S) = \prod_{i=1}^{I} \left[\sum_{c=1}^{C} f_c P(Y_j|\Theta_c, S_c) \right], \tag{7}$$

where f_c is the fraction of individuals in subtype c out of a total C clusters.

2.2   Constrained Transition Matrix

In this work we propose a novel constrained form of the transition matrix A_c that aligns with the assumption of sequential transition through disease stages made by disease progression models. In a traditional hidden Markov model, states may be transitioned between in any order. However, in a disease progression model, the states are instead thought of as stages and have a strict order, with individuals transitioning sequentially through each stage. For example, if an individual started at stage 0 at time $t = 0$ and then transitioned to stage 2 at time $t = 1$, a disease progression model would assume that they transition through stage 1 at some point between $t = 0$ and $t = 1$, whereas in a traditional hidden Markov model there is no such assumption (individuals can move instantaneously from any state to any other state). We encode this idea by assuming that, for $i < j < k$, the probability of transitioning from stage i to stage k is equal to the probability of transitioning from stage i to stage j and then from stage j to stage k, i.e. $a_{ik} = a_{ij}a_{jk}$ for $i < j < k$ (assuming a_{ij} and a_{jk} are independent). This generalises to $a_{ik} = \prod_{j=i+1}^{k} a_{j-1,j}$. Disease progression models also assume monotonic progression, which we enforce by using an upper triangular transition matrix to only allow forward transitions between stages. Following on from these two assumptions, we can derive an analytical solution to the transition matrix A_c that depends only on a transition probability vector $\alpha_c = (a_{00}, \ldots, a_{NN})$ encoding the diagonal of the transition matrix.

To do this we derive analytical solutions for the first off-diagonal elements $(a_{0,1}, a_{1,2}, \ldots, a_{N-2,N-1}, a_{N-1,N})$ that depend only on the elements of the transition probability vector $\alpha_c = (a_{00}, \ldots, a_{NN})$. The first off-diagonal elements can then be used to compute the rest of the elements in the upper triangle of the transition matrix using $a_{ik} = \prod_{j=i+1}^{k} a_{j-1,j}$. From this and the monotonicity assumption we have:

$$A_c = \begin{pmatrix} a_{0,0} & a_{0,1} & \prod_{j=1}^{2} a_{j-1,j} & \cdots & \prod_{j=1}^{N-2} a_{j-1,j} & \prod_{j=1}^{N-1} a_{j-1,j} & \prod_{j=1}^{N} a_{j-1,j} \\ 0 & a_{1,1} & a_{1,2} & \cdots & \prod_{j=2}^{N-2} a_{j-1,j} & \prod_{j=2}^{N-1} a_{j-1,j} & \prod_{j=2}^{N} a_{j-1,j} \\ \vdots & \vdots & \vdots & \ddots & \vdots & \vdots & \vdots \\ 0 & 0 & 0 & \cdots & a_{N-2,N-2} & a_{N-2,N-1} & \prod_{j=N-1}^{N} a_{j-1,j} \\ 0 & 0 & 0 & \cdots & 0 & a_{N-1,N-1} & a_{N-1,N} \\ 0 & 0 & 0 & \cdots & 0 & 0 & a_{N,N} \end{pmatrix}.$$

As each row must sum to 1 to give a valid transition matrix, in the last row we have $a_{N,N} = 1$. In the second to last row we have $a_{N-1,N-1} + a_{N-1,N} = 1$, which can be rearranged to give $a_{N-1,N} = 1 - a_{N-1,N-1}$. In the third to last row we have $a_{N-2,N-2} + a_{N-2,N-1} + a_{N-2,N} = 1$. Substituting in $a_{N-2,N} = a_{N-2,N-1}a_{N-1,N}$ and $a_{N-1,N} = 1 - a_{N-1,N-1}$ gives $a_{N-2,N-2} + a_{N-2,N-1} + a_{N-2,N-1}(1 - a_{N-1,N-1}) = 1$, which rearranges to give $a_{N-2,N-1} = \frac{1-a_{N-2,N-2}}{2-a_{N-1,N-1}}$. Following the same logic and substitutions, the fourth to last row rearranges to $a_{N-3,N-2} = \frac{1-a_{N-3,N-3}}{2-a_{N-2,N-2}}$. In general we have $a_{i,i+1} = \frac{1-a_{i,i}}{2-a_{i+1,i+1}}$. So under our assumptions the transition matrix A_c only depends on the diagonal of the transition matrix (the transition probability vector α_c) and we have

$$
A_c = \begin{pmatrix}
a_{0,0} & \frac{1-a_{0,0}}{2-a_{1,1}} & \prod_{j=1}^{2} a_{j-1,j} & \cdots & \prod_{j=1}^{N-2} a_{j-1,j} & \prod_{j=1}^{N-1} a_{j-1,j} & \prod_{j=1}^{N} a_{j-1,j} \\
0 & a_{1,1} & \frac{1-a_{1,1}}{2-a_{2,2}} & \cdots & \prod_{j=2}^{N-2} a_{j-1,j} & \prod_{j=2}^{N-1} a_{j-1,j} & \prod_{j=2}^{N} a_{j-1,j} \\
\vdots & \vdots & \vdots & \ddots & \vdots & \vdots & \vdots \\
0 & 0 & 0 & \cdots & a_{N-2,N-2} & \frac{1-a_{N-2,N-2}}{2-a_{N-1,N-1}} & \prod_{j=N-1}^{N} a_{j-1,j} \\
0 & 0 & 0 & \cdots & 0 & a_{N-1,N-1} & 1 - a_{N-1,N-1} \\
0 & 0 & 0 & \cdots & 0 & 0 & 1
\end{pmatrix}.
$$

2.3 Inference

Similarly to [9], we devise a hierarchical framework that sequentially fits an increasing numbers of clusters from a number of randomly chosen initial progression patterns, choosing the optimal number of clusters using cross-validation. We use MCMC sampling to estimate the model parameters and their uncertainty. There are a number of parameters to infer for each cluster $c = 1, ..., C$: the sequence $\overline{S}_c$, fraction of the population in the cluster $\overline{f}_c$, initial probability vector $\overline{\pi}_c$, and the transition probability matrix $\overline{A}_c$ that maximise the total log likelihood, $\mathcal{L}(S_c, f_c, \pi_c, A_c) = \log P(Y; S_c, f_c, \pi_c, A_c)$. We make two assumptions to simplify inference: we use the constrained transition probability matrix described in Sect. 2.2 (parameterised by the vector α_c) and, following the event-based model [3], we assume a uniform initial probability vector π_c. To speed up convergence of the MCMC, we obtain initial estimates of the set of sequences $\hat{S}_c$ and fractions $\hat{f}_c$ using the SuStaIn algorithm with the modified likelihood function $\mathcal{L}(S_c, f_c, \pi_c, A_c) = \log P(Y; S_c, f_c, \pi_c, A_c)$ and α_c set to $\alpha_c = (0.5, ..., 0.5)$. We initialise the MCMC sampling procedure using $\hat{S}_c$, $\hat{f}_c$, and $\alpha_c = (0.5, ..., 0.5)$, sampling the full distribution of the parameters S_c, f_c, and α_c for each cluster.

2.4 Subtyping and Staging

T-SuStaIn can output subtypes and stages of individuals using either cross-sectional or longitudinal observations. In either case, an individual is first assigned to a subtype, then a stage given that subtype. In the case where an individual only has a single (i.e., cross-sectional) observation, their subtype is assigned according to their maximium likelihood subtype. In the case where an individual has multiple (i.e., longitudinal) observations, their subtype is assigned according to their maximum likelihood subtype across all observations. This ensures that individuals stay in the same subtype longitudinally, which is a benefit over cross-sectional SuStaIn.

3   Experiments

3.1   Synthetic Data

We first verify that T-SuStaIn can recover subtype progression patterns and timelines in synthetic datasets with a similar size and number of visits to the ADNI dataset. We simulate data directly from the mathematical model underlying T-SuStaIn to enable a direct comparison of parameter estimates and thus perform a sanity check that the algorithm can recover trajectories under idealised conditions. We generate 10 synthetic datasets, setting the number of subtypes to two and randomly generating a progression pattern for each subtype, with 75% of individuals belonging to the first subtype, and 25% belonging to the second. We set the number of subjects to 250, with three visits per subject, giving a total number of data points of 750. We set the number of biomarkers to three, the number of z-scores to three (1, 2, and 3), the maximum z-score to 4, and assume the z-scores evolve from a minimum of 0 with a standard deviation of 1. We set the transition probability for each z-score event to $a = 0.2$ for all biomarkers in each subtype, corresponding to an average transition time of 1.25 years.

3.2   Crossing Subtype Trajectories

Crossing subtype trajectories have a stage in the middle of each trajectory where the two subtypes look identical (see example in Fig. 2). In this case cross-sectional SuStaIn cannot disentangle which beginning and end of each trajectory belong to which subtype. However, T-SuStaIn should be able to disentangle the trajectories by observing the trajectories of individuals before and after the cross-over stage. We run a set of simulations that specifically test the performance of T-SuStaIn for inferring crossing subtype trajectories compared to the SuStaIn algorithm in [9], which only handles cross-sectional data. To ensure any improvements are not simply due to an increase in the number of data points, we use the same number of data points in each case. Specifically, we run SuStaIn on five simulated datasets of 2500 subjects with cross-sectional data only, and T-SuStaIn on five simulated datasets of 500 subjects with 5 time-points. We simulate data from two subtypes across three biomarkers with the progression patterns shown in Fig. 2, assuming that the first subtype has a 60% prevalance, and the second a 40% prevalence. We simulate two z-score events per biomarker ($z = 1$ and $z = 2$), a maximum z-score of 3, and assume the z-scores evolve from a minimum of 0 with a standard deviation of 1. The transition probability is set to $a = \frac{1}{3}$ for all biomarkers in each subtype, corresponding to an expected transition time of 1.5 years per stage. Consequently an individual with five time points would be expected to transition three stages on average over the course of their five follow-ups.

3.3   Alzheimer's Disease Dataset

Data used in the preparation of this article were obtained from the Alzheimer's Disease Neuroimaging Initiative (ADNI) database (adni.loni.usc.edu)[1] We selected 308 participants from ADNI [15], including individuals who have complete data for two or more consecutive yearly visits, an MCI or AD diagnosis at baseline, and are APOE4 positive (one or more APOE4 alleles). This gave a set of 808 data points in total (average of 2.62 visits per person) with follow-ups spaced at one year intervals. We applied T-SuStaIn to regional MRI volumes from the hippocampus, temporal lobe and other cortical regions (sum of all cortical regions except those in the temporal lobe), correcting for age, sex, years of education, scanner field strength (1.5T vs. 3T) and intracranial volume (ICV) using a control population of 220 APOE4-negative controls. To do this we built a linear regression model for each region, with regional volume as the dependent variable and the above covariates as the independent variables. We then residualized each region (true value minus predicted value from regression) and z-scored the residuals using the controls' means and standard deviations. We used these z-scored residuals as the biomarker inputs to T-SuStaIn. We ran T-SuStaIn using 4 startpoints and 1E5 MCMC iterations, and following [9] identified the optimal number of clusters using the cross-validation information criterion (CVIC).

3.4   Positional Variance Diagrams

Subtype progression patterns are plotted using positional variance diagrams, which visualise the sequence for each subtype and the uncertainty in that sequence. The colours represent different z-score events, with red corresponding to a z-score of 1, magenta a z-score of 2, and blue a z-score of 3. Each square visualises the probability a particular z-score event occurs at that particular stage, ranging from 0 in white to 1 in red $(z = 1)$, magenta $(z = 2)$ or blue $(z = 3)$.

3.5   Event Timelines

The most probable timelines and their uncertainty are obtained from the MCMC samples of the transition matrix for each subtype by using kernel density estimation to fit a non-parametric distribution to the samples and hence obtain descriptive statistics of the mode and full width at half maximum (FWHM).

4   Results

4.1   T-SuStaIn Can Recover Event Sequences and Timelines in Synthetic Datasets of Similar Size to ADNI

We simulated 10 datasets of a similar size to ADNI (250 subjects with three time points), with two subtypes with a prevalence of 75% and 25% and randomly

---

[1] For further information see: www.adni-info.org.

chosen progression patterns. The kendall tau distance between the ground truth and estimated subtype progression patterns was 0.91 (sd = 0.20) for Subtype 1 and 0.57 (sd = 0.20) for Subtype 2 (kendall tau of 1 indicates maximum similarity, and -1 indicates maximum dissimilarity). T-SuStaIn estimated an average transition probability of 0.25 (sd = 0.06) for Subtype 1 and 0.25 (sd = 0.04) for Subtype 2.

4.2   T-SuStaIn Can Infer Crossing Trajectories in Synthetic Datasets

Figure 2 illustrates an example dataset in which T-SuStaIn infers crossing subtype trajectories when cross-sectional SuStaIn fails. Across five simulated datasets, T-SuStaIn estimated the correct subtype progression patterns in all five simulations. Cross-sectional SuStaIn estimated the correct subtype progression patterns in only two of five simulations, consistent with our expectation that cross-sectional SuStaIn should estimate the correct progression by chance 50% of the time given that there are two possible stage 4-6 progression patterns that could be randomly appended to either of the stage 1-3 progression patterns. T-SuStaIn can leverage longitudinal data from individuals who move from stages 1–3 to stages 4–6 to link the patterns together correctly.

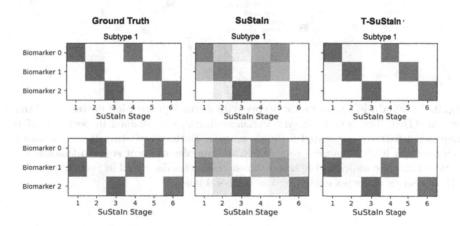

Fig. 2. Example outputs of SuStaIn vs. T-SuStaIn for recovering crossing trajectories using synthetic data. Each progression pattern is visualised using a positional variance diagram (see Sect. 3.4). In the ground truth the subtypes cross at stage 3, at which point all the biomarkers have reached a z-score of 1. However, the two subtypes have distinct progression patterns before and after this cross-over point. These progression patterns can be inferred by T-SuStaIn, but not SuStaIn.

4.3　T-SuStaIn Identifies Two Subtypes with Distinct Progression Patterns in ADNI

Figure 3 shows the two subtypes inferred by T-SuStaIn in ADNI. As with SuStaIn, each subtype has a distinct progression pattern, but T-SuStaIn further estimates an event transition matrix (and therefore a distinct timeline). The first subtype (86% prevalence) has early hippocampal atrophy, followed by temporal lobe atrophy and then widespread cortical atrophy. The second subtype (14% prevalence) has temporal and cortical atrophy at earlier stages. We hypothesise that the first subtype reflects previously described 'typical' AD subtypes and the second 'cortical' AD subtypes [2,9,11].

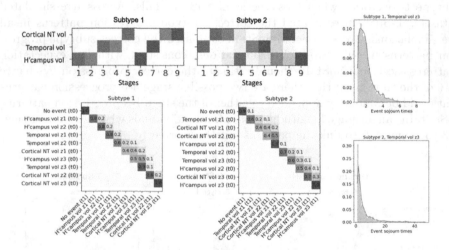

Fig. 3. T-SuStaIn identifies two subtypes with distinct progression patterns and timelines in ADNI. Top row: positional variance diagrams (explained in Sect. 3.4) of the progression patterns estimated for each subtype. Bottom row: transition matrices for each subtype. Right column: example MCMC samples of event sojourn times in each subtype (blue line indicates the kernel density estimates described in Sect. 3.5). Cortical NT: all regions in cortex excluding the temporal lobe.

4.4　Each ADNI Subtype Has a Distinct Timeline

Figure 4 shows the event timelines inferred by T-SuStaIn in ADNI. T-SuStaIn infers that the overall timeline of Subtype 1 is longer than Subtype 2, consistent with previous studies indicating faster progression of cortical AD subtypes [16].

5　Discussion

We introduced T-SuStaIn, a longitudinal discrete clustering technique that disentangles spatial and temporal heterogeneity in progressive diseases. The

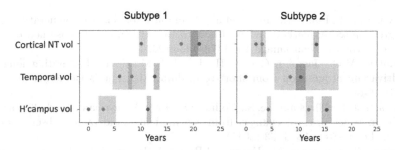

Fig. 4. Visualisation of ADNI subtype timelines. The colours represent different z-score events (red: $z = 1$, magenta: $z = 2$, blue: $z = 3$), with the dots representing when the event occurs (mode described in Sect. 3.5), and the boxes representing the uncertainty in the timing of the event (FWHM described in Sect. 3.5). (Color figure online)

strengths of T-SuStaIn lie in its ability to infer interpretable temporal subtypes from reasonably sized datasets of order 100 individuals, and to provide improved identifiability over cross-sectional SuStaIn, for example in the case of crossing trajectories. Whilst our results support T-SuStaIn's broad potential clinical utility, we acknowledge that its current formulation limits its use to data with fixed-time intervals; future work will allow for variable-time intervals [17]. Although we focused on complete structural MRI data in this work, T-SuStaIn can readily use any type of dynamic biomarker data and accounting for missing data is straightforward, e.g., [10]. As with SuStaIn, various disease progression models can be used with T-SuStaIn to model alternative data types [3,9,18,19]. As such, T-SuStaIn will be able to infer longitudinal subtypes from short-term multi-modal datasets with irregular sampling and missing data, further extending its use in improving disease understanding and patient stratification.

Acknowledgements. ALY is supported by an MRC Skills Development Fellowship (MR/T027800/1). Data collection and sharing for this project was funded by the Alzheimer's Disease Neuroimaging Initiative (ADNI) (National Institutes of Health Grant U01 AG024904) and DOD ADNI (Department of Defense award number W81XWH-12-2-0012) (For a full list of ADNI funders see: https://adni.loni.usc.edu/wp-content/uploads/how_to_apply/ADNI_Data_Use_Agreement.pdf).

References

1. Villemagne, V.L., Burnham, S., Bourgeat, P., et al.: Amyloid deposition, neurodegeneration, and cognitive decline in sporadic Alzheimer's disease: a prospective cohort study. Lancet Neurol. **12**, 357–367 (2013)
2. Vogel, J.W., Young, A.L., Oxtoby, N.P., et al.: Four distinct trajectories of tau deposition identified in Alzheimer's disease. Nat. Med. **27**, 871–881 (2021)
3. Fonteijn, H.M., Modat, M., Clarkson, M.J., et al.: An event-based model for disease progression and its application in familial Alzheimer's disease and Huntington's disease. Neuroimage **60**, 1880–1889 (2012)

4. Jedynak, B.M., Lang, A., Liu, B., et al.: A computational neurodegenerative disease progression score: method and results with the Alzheimer's disease neuroimaging initiative cohort. Neuroimage **15**, 1478–1486 (2012)
5. Donohue, M.C., Jacqmin Gadda, H., Le Goff, M., et al.: Estimating long-term multivariate progression from short-term data. Alzheimer's Dementia **10**, S400–S410 (2014)
6. Schiratti, J.B., Allassonnière, S., Colliot, O., et al.: A Bayesian mixed-effects model to learn trajectories of changes from repeated manifold-valued observations. J. Mach. Learn. Res. **18**, 1–33 (2017)
7. Lorenzi, M., Filippone, M., Frisoni, G.B., et al.: Probabilistic disease progression modeling to characterize diagnostic uncertainty: application to staging and prediction in Alzheimer's disease. Neuroimage **190**, 56–68 (2019)
8. Oxtoby, N.P., Alexander, D.C.: Imaging plus X: multimodal models of neurodegenerative disease. Curr. Opin. Neurol. **30**(4), 371–379 (2019)
9. Young, A.L., Marinescu, R.V., Oxtoby, N.P., et al.: Uncovering the heterogeneity and temporal complexity of neurodegenerative diseases with subtype and stage inference. Nat. Commun. **9** (2018)
10. Wijeratne, P.A., Alexander, D.C.: Learning transition times in event sequences: the temporal event-based model of disease progression. In: Feragen, A., Sommer, S., Schnabel, J., Nielsen, M. (eds.) IPMI 2021. LNCS, vol. 12729, pp. 583–595. Springer, Cham (2021). https://doi.org/10.1007/978-3-030-78191-0_45
11. Poulakis, K., Pereira, J.B., Muehlboeck, J.S., et al.: Multi-cohort and longitudinal Bayesian clustering study of stage and subtype in Alzheimer's disease. Nat. Commun. **13** (2022)
12. Poulet, P.-E., Durrleman, S.: Mixture modeling for identifying subtypes in disease course mapping. In: Feragen, A., Sommer, S., Schnabel, J., Nielsen, M. (eds.) IPMI 2021. LNCS, vol. 12729, pp. 571–582. Springer, Cham (2021). https://doi.org/10.1007/978-3-030-78191-0_44
13. Chen, I.Y., Krishnan, R.G., Sontag, D.: Clustering interval-censored time-series for disease phenotyping. arXiv (2021)
14. Rabiner, L.R.: A tutorial on hidden Markov models and selected applications in speech recognition. Proc. IEEE **77**, 257–286 (1989)
15. Mueller, S.G., Weiner, M.W., Thal, L.J., et al.: The Alzheimer's disease neuroimaging initiative. Neuroimaging Clin. N. Am. **15**, 869–877 (2005)
16. Ferreira, D., Nordberg, A., Westman, E.: Biological subtypes of Alzheimer disease: a systematic review and meta-analysis. Neurology **94** (2020)
17. Metzner, P., Horenko, I., Schütte, C.: Generator estimation of Markov jump processes based on incomplete observations non-equidistant in time. Phys. Rev. E Stat. Nonlinear Soft Matter Phys. **76** (2007)
18. Aksman, L.M., Wijeratne, P.A., Oxtoby, N.P., et al.: pySuStaIn: a python implementation of the subtype and stage inference algorithm. SoftwareX **16**, 100811 (2021)
19. Young, A.L., Vogel, J.W., Aksman, L.M., et al.: Ordinal sustain: subtype and stage inference for clinical scores, visual ratings, and other ordinal data. Front. Artif. Intell. **4**, 613261 (2021)

Brain Connectomics

HoloBrain: A Harmonic Holography for Self-organized Brain Function

Huan Liu[1], Tingting Dan[1], Zhuobin Huang[1], Defu Yang[2], Won Hwa Kim[3], Minjeong Kim[4], Paul Laurienti[5], and Guorong Wu[1,6(✉)]

[1] Department of Psychiatry, University of North Carolina at Chapel Hill, Chapel Hill, NC 27599, USA
grwu@med.unc.edu
[2] Intelligent Information Processing Laboratory and School of Automation, Hangzhou Dianzi University, Hangzhou 310018, China
[3] Computer Science and Engineering/Graduate School of AI, POSTECH, Pohang 37673, South Korea
[4] Department of Computer Science, University of North Carolina at Greensboro, Greensboro, NC 27402, USA
[5] Department of Radiology, Wake Forest School of Medicine, Winston Salem, NC 27157, USA
[6] Department of Computer Science, University of North Carolina at Chapel Hill, Chapel Hill, NC 27599, USA

Abstract. Functional neuroimaging technology offers a new window to snapshot the transient neural activity *in-vivo*. Although tremendous efforts have been made to characterize spontaneous functional fluctuations, little attention has been paid to the functional mechanisms of neural interactions. Inspired by the notion of holography, we propose an explainable machine learning approach to establishing a novel underpinning of self-organized cross-frequency coupling (CFC) through the lens of brain wave interference on top of the network topology. Specifically, we conceptualize that the interaction between ubiquitous neural activities and a collection of reference harmonic wavelets forms a region-adaptive interference pattern that captures cross-frequency coupling of remarkable neuronal oscillations. In this regard, assembling whole-brain CFC patterns under the constraint of brain wiring mechanisms constitutes a *HoloBrain* mapping that records a wide spectrum of spontaneous neural activities. Since each local interference pattern is a symmetric and positive-definite (SPD) matrix, we tailor a deep model of *HoloBrain* (coined *DeepHoloBrain*) to infer the latent feature representations on the Riemannian manifold of SPD matrices for predicting brain states and recognizing disease connectomes. We have applied *DeepHoloBrain* to the Human Connectome Project and several dementia-related datasets. Compared with current state-of-the-art deep models, our *DeepHoloBrain* not only improves the recognition/prediction accuracy but also sheds new light on understanding the neurobiological mechanisms of brain function and cognition.

Keywords: Cross Frequency Coupling · Harmonic Wavelets · SPD Manifold · Deep Neural Network

---

H. Liu and T. Dan—These authors contributed equally to this work.

A. Frangi et al. (Eds.): IPMI 2023, LNCS 13939, pp. 29–40, 2023.
https://doi.org/10.1007/978-3-031-34048-2_3

1 Introduction

Recent developments of *in-vivo* neuroimaging technology, such as functional magnetic resonance imaging (fMRI), allow us to investigate the connectivity between distinct regions in the brain, where we used to study each region separately. Since the research paradigm has been shifted to investigate region-to-region interactions, network neuroscience comes to the stage, which conceptualizes the brain as a connectome–an interactive network map where distinct brain regions synchronize their neuroactivities via myriads of interconnecting nerve fibers [1]. Network neuroscience provides a simple yet elegant system-level approach to understanding how neural circuits support brain function, how they constrain each other, and how they differ across individuals, which sheds new light on gaining insight into cognitive science and uncovering system-level principles of disease mechanisms.

The presumption of functional connectivity (FC) is a statistical dependency between the time series measured neurophysiological signals, which allows us to study whole-brain functional connections as a complex wiring system of the brain network [2]. In this context, graph theory is introduced to describe the global and local topological properties [3]. Recently, the research interest has shifted to dynamic functional connectivities, where the network topology changes over time, even in the resting state [4]. Despite great success in understanding cognition and behavior from a network neuroscience perspective, the neurobiological mechanism of functional co-activation is still largely unknown.

Mounting evidence in neuroscience shows that neuronal oscillations (aka. Brain waves), presenting across a broad frequency spectrum, might emerge remarkable rhythmic changes that support the functional mechanism of the interaction between different frequency bands, a phenomenon termed cross-frequency coupling (CFC) [5]. CFC has been reported in many cortical and subcortical regions in multiple species, yielding distinct signatures in neural dynamics [6]. In computer vision, holography is a stereo-imaging technique that generates a hologram by superimposing a reference beam on the wavefront of interest [7]. The resulting hologram is an interference pattern that can be recorded on a physical medium. Taking together, we present a proof-of-concept approach, called *HoloBrain*, to computationally "record" the cross-frequency couplings of time-evolving interference patterns that are formed by superimposing the harmonic wavelets (with predefined oscillation frequencies) on the subject-specific neural activities.

From the data structure perspective, *HoloBrain* is a graph consisting of nodes (corresponding to brain regions) and edges (weighted by functional co-activations). In contrast to conventional graph embedding vectors, *HoloBrain* uses a symmetric and positive-definite (SPD) matrix to encode all possible CFCs at each brain region (node). In this context, a new machine-learning challenge arises, that is, how to find the most relevant and explainable CFC feature representations throughout the brain network for predicting brain states and recognizing disease brain connectomes. To address this challenge, we formulate the machine-learning problem as a message-exchanging process on a compact subgraph, where we progressively refine the CFC feature representations via the learned random walks on the subgraph. Since the CFC holds the unique data geometry, we further present a manifold-based deep model (coined *DeepHoloBrain*) to infer the explainable CFC features on the Riemannian manifold of SPD matrices.

We have elucidated the neuroscience insight of *HoloBrain* via a set of group comparison studies. Meanwhile, we have evaluated the clinical value of *DeepHoloBrain* in recognizing brain states on Human Connectome Project (HCP) data and diagnosing Alzheimer's disease (AD), obsessive-compulsive disorder (OCD), and schizophrenia (SZ). Compared to the current "black-box deep" models, our *DeepHoloBrain* not only improves the prediction/recognition accuracy using functional neuroimaging data but also offers an explainable solution with great biophysics and neuroscience insight.

2   Methods

Suppose we partition a brain into N distinct regions. At each region, we observe the mean time course of BOLD (blood-oxygen-level-dependent) signal $x_i(t) \in \mathcal{R}$ ($i = 1, \ldots, N$) where t is a continuous variable of time. Thus, we form a discretized time-varying signal matrix $X(t) = [x_t]_{t=1}^T \in \mathcal{R}^{N \times T}$, where $x_t \in \mathcal{R}^N$ is a column vector of whole-brain BOLD signal snapshot at time t. In this context, we can represent the brain network as a graph $\mathcal{G} = (V, W)$ with N nodes (brain regions) $V = \{v_i | i = 1, \ldots, N\}$ and the adjacency matrix $W = [w_{ij}]_{i,j=1}^N \in \mathcal{R}^{N \times N}$ describing FC strength (measured by Pearson's correlation). Let $L = D - W$ as the underlying Laplacian matrix, where D is a diagonal matrix of the total connectivity degree at each node. In the following, we first introduce the neuroscience insight and physics principles of our new model *HoloBrain* for studying brain functions. After that, we present an explainable deep model for *HoloBrain* using graph neural network and Riemannian manifold techniques.

2.1   *HoloBrain*: A Physics-Informed Model for Brain Function Analysis

Our work is inspired by frequency-specific harmonic waves [8], where the oscillation patterns of harmonic waves act as the basis functions to express the neural activities in the graph spectrum domain (constrained by brain network topology). By applying SVD on graph Laplacian matrix L, the top K eigenvectors $\phi = [\phi_k]_{k=1}^K$ (corresponding to the first K smallest eigenvalues) are used as subject-specific harmonic waves.

Harmonic Wavelets. Furthermore, we extend the harmonic technique to the regime of harmonic wavelets where the region-adaptive oscillation patterns allow us to characterize localized neural activities. To do so, we follow the approximation method in [9] to construct the harmonic wavelets for each node v_i in two steps. First, we generate a subgraph mask $u_i \in \mathcal{R}^N$ where all the nodes are connected to the underlying node v_i within h hops. Thus, u_i is an index vector where $u_i(j) = 1$ denotes region v_j is in the subgraph centered at v_i and $u_i(j) = 0$ otherwise. Second, we estimate a collection of harmonic wavelets $\psi_i = [\psi_i^s]_{s=1}^S$ across frequency s by:

$$\min_{\psi_i} tr\left(\psi_i^T L \psi_i\right) + tr\left(\psi_i^T diag(1 - u_i)\psi_i\right) + tr\left(\psi_i^T \phi^T \phi \psi_i\right), s.t., \psi_i^T \psi_i = I. \quad (1)$$

Minimizing the first trace norm with the orthogonal constraint leads deterministic solution of harmonic waves in [8]. Since we sought to localize each harmonic wavelet within a subgraph, the second trace norm is used to encourage $\|\psi_i\|^2$ to be zero out of the subgraph. In order to achieve complementary basis functions, we introduce the third trace norm to make the harmonic wavelets ψ_i orthogonal to the global harmonic wave ϕ. Due to the linearity of the trace norms, we unify three trace terms in Eq. 1 into one trace term with a matrix $\Pi_i = L + diag(1 - u_i) + \phi^T \phi$. Thus, the optimization of Eq. 1 is boiled down to the eigendecomposition of the matrix Π_i.

Examples of harmonic wavelets are shown at the top-left of Fig. 1. It is clear that each harmonic wavelet captures the oscillation patterns (indicated by colored arrows, blue for positive and red for negative values) within a local network topology centered at the underlying brain region, where the waves oscillate slowly in the low-frequency and rapidly in the high-frequency band.

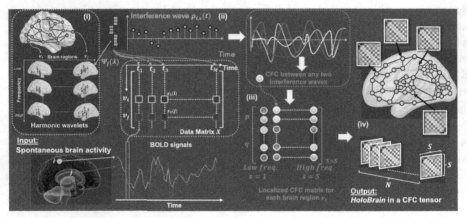

Fig. 1. The workflow of constructing *HoloBrain* for each subject, which consists of four major steps: (i) estimate harmonic wavelets, (ii) calculate interference waves, (iii) construct local CFC matrix, and (iv) assemble localized CFC matrices into a tensor, yielding *HoloBrain*.

Construction of *HoloBrain*. As shown in the middle of Fig. 1, we regard the snapshot of the whole-brain BOLD signal x_t as a time-dependent wave of subject-specific neural activities. Likewise, each harmonic wavelet ψ_i^s is considered as a predefined reference wave corresponding to a neuronal oscillation frequency. In this context, we define an interference wave $\rho_{i,s}(t) = (x_t)^T \cdot \psi_i^s$, which quantifies the dynamic engagement effect of superimposing the reference harmonic wavelets and the subject-specific neural activities.

Following the notion of CFC, we can construct a local CFC matrix for each region v_i by $C_i = \left[c_i^{pq} \right]_{p,q=1}^S$, where each local CFC pattern $c_i^{pq} = \left(\rho_{i,p} \right)^T \cdot \rho_{i,q}$ captures the synchronization between interface waves $\rho_{i,p}$ at p^{th} frequency and $\rho_{i,q}$ at q^{th} frequency band. To that end, *HoloBrain* is defined as an assembly of CFC matrices throughout the brain as $\mathcal{H} = [C_i]_{i=1}^N$, where $\mathcal{H}$ is formulated as a $S \times S \times N$ tensor.

Insight of *HoloBrain*. From a *Neuroscience* perspective, we position each harmonic wavelet as a predefined neuronal oscillation. In this regard, a set of harmonic wavelets at the region v_i constitute a spectrum of brain rhythms with varying frequencies along the brain cortex. Since the neuronal activity associated with stimulus processing might differ depending on its timing relative to the ongoing oscillation [5], the interaction between harmonic wavelets and BOLD signal snapshot at time t indicates the contribution of the underlying harmonic wavelet ψ_i^s to the local cortical excitability. The proposition of considering the time-evolving interaction as the frequency-specific interference pattern sets the stage for understanding the physiological mechanism of CFC in the human brain. Since the human brain is a complex system, our *HoloBrain* "records" the footprint of whole-brain neural activities in the representation of evolving CFC.

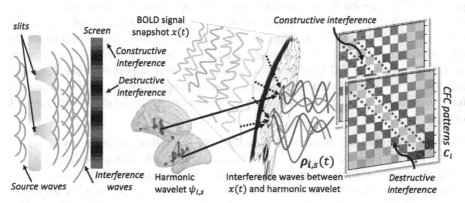

Fig. 2. The physics insight of HoloBrain (right) in analogy to the wave interference principle (left), which both yield constructive and destructive interference patterns on the screen and cross-frequency couplings, respectively.

From a *physics* point of view, the waves from two slits interfere constructively or destructively, as shown in Fig. 2 left. In our *HoloBrain*, each harmonic wavelet acts as the multi-slits where the geometry (e.g., the slit gaps) reflects its oscillation patterns. As shown in the right panel of Fig. 2, we conceptualize the snapshot of BOLD signals $x(t)$ is the source wave of neural activity. In this context, the temporal interaction between a harmonic wavelet and the source wave, i.e., $\rho_{i,s}(t)$, can be regarded as an interference wave. Since each interference wave is associated with a harmonic frequency, we superimpose the interference waves across frequencies and form the $S \times S$ region-specific CFC patterns, which hold the exact nature of interference patterns in the art of 3D hologram [7]. In analogy to the constructive and destructive interferences observed in Young's double slit [10], we also detect a similar phenomenon along the off-diagonal lines in the $S \times S$ CFC matrix (purple dash lines in Fig. 2). Note, the first off-diagonal line (closest the diagonal line) records the CFC patterns having one shift between one pair of frequency bands. The second off-diagonal line captures two-frequency-shift CFC, etc. As a proof-of-concept approach, we sought to examine the neuroscience insight of how these constructive/destructive interferences in our *HoloBrain* underline the cognition status and the separation between healthy and disease connectomes.

From the *machine learning* perspective, the workflow in Fig. 1 is essentially a data representation process where we use the learned harmonic wavelets (as the basis functions) to express the observed BOLD signal at each time point. To that end, the resulting

CFC tensor of *Holobrain* $\mathcal{H}$ can be understood as an expression pattern of location and frequency associated with the evolving brain states.

2.2 *DeepHoloBrain*: An Explainable Deep Model for *HoloBrain*

Challenges. Without loss of generality, we sought to find a non-linear mapping from HoloBrain $\mathcal{H}$ to an outcome label (e.g., disease connectome). There are three major challenges. *First*, the data dimensionality of CFC pattern at each brain region grows quadratic ($\mathcal{O}(S^2)$) with the number of frequency bands. Since the contribution of each brain region varies in different cognitive statuses or disease progression, a compact representation of $\mathcal{H}$ allows us to find the most relevant features and thus enhance the prediction/recognition accuracy. *Second*, each CFC is an instance of SPD matrix. Since the data structure of CFC holds neuroscience insight, it is important to preserve such intrinsic data geometry in feature representation learning. *Third*, since CFC patterns in the *HoloBrain* are topologically wired, the machine-learning model should take network topology into account.

Problem Formulation. The input is the tensor $\mathcal{H}$ of *HoloBrain* associated with a graph topology $\mathcal{G}$. We train a deep model $\mathcal{M}$ to learn the latent low-dimension feature F and a neural circuit $\mathcal{G}_F$ (subgraph of $\mathcal{G}$) that explains the formation of application-specific feature representation F. Due to the consideration of model explainability, we expect the output feature representation F to preserve the geometric structure of $S \times S$ CFC pattern, i.e., $F \in Sym_S^+$ is a data instance of co-activation patterns on the Riemannian manifold of SPD matrices Sym_S^+. Since the conventional graph convolution plus pooling technique [11] has limitations in interpreting the neuroscience insight of learned features, we present a reinforcement learning framework to find the representative CFC pattern F via a set of random walks on the graph $\mathcal{G}$, as decribed next.

Design of our Explainable Deep Model for *HoloBrain*. The overall workflow of our *DeepHoloBrain* is shown in Fig. 3. Following the spirit of the partially observed Markov decision process [12], the agent $\mathcal{A}$ is trained to learn a stochastic policy that, at each step l, maps the history of past interactions with the environment (existing nodes $v^{(1)}$, $v^{(2)}, \ldots, v^{(l-1)}$ in the subgraph $\mathcal{G}_F$) to a probability distribution over actions for the current step l. At each step l, the agent performs two actions: (i) update CFC patterns in the *HoloBrain* through a re-wiring process (steering the feature representation learning), and (ii) add node $v^{(l)}$ to the subgraph $\mathcal{G}_F$ that supports the formation of the representative CFC pattern F (for model explainability). In this context of reinforcement learning, our *DeepHoloBrain* is made of the following learning components (indicated by color in Fig. 3).

(1) **Sensor.** As explained in the "Action" module, the actions from the agent $\mathcal{A}$ include the lth candidate node $v^{(l)}$ and a vector of re-wiring message $m^{(l)} = \left[m_j^{(l)} \right]_{j=1}^{N} \in \mathcal{R}^N$.

Here $v^{(l)}$ is an index function that returns the node index in the lth step. Suppose $v^{(l)} = v_i$. Thus, each element $m_j^{(l)}$ indicates the interaction between C_i and other CFC pattern C_j. Furthermore, we use $\mathcal{G}_F^{(l)} = \{v^{(1)}, v^{(2)}, \ldots, v^{(l)}\}$ denote the set of node indexes selected by the agent in the past l steps. By integrating the re-wiring message $m^{(l)}$ with the connectivity profile (i^{th} row in the adjacency matrix W) at node v_i, we define the external state of the agent $E^{(l)}$ by: $E^{(l)} = \sum_{j=1}^{N} m_j^{(l)} w_{ij} C_j$.

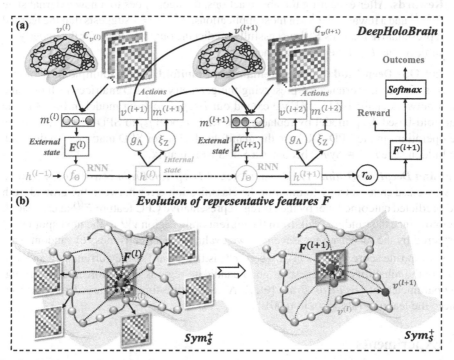

Fig. 3. (a) The overall design of *DeepHoloBrain* falls into a reinforcement learning framework, which consists of a sensor (in purple), an internal state (in brown), a set of actions (in red), and a reward function (in green). (b) The learning process of representative CFC features F forms a trajectory on the manifold Sym_S^+ as we sequentially walk from $v^{(l)}$ to $v^{(l+1)}$ at each step of reinforcement learning.

(2) **Internal state.** The agent maintains an internal state h that summarizes the representative CFC pattern extracted from the history of selected nodes $\mathcal{G}_F^{(l)}$. The agent $\mathcal{A}$ perceives the evolving environment (updated representative feature $F^{(l)}$) by deciding how to act (for exchanging information between CFC patterns) and where to deploy the new sensor (for steering the random walk). In this context, we train a recurrent neural network (RNN) to update the internal state by $h^{(l)} = f_\Theta(h^{(l-1)}, E^{(l)})$, where f_Θ denotes the RNN with trainable parameters Θ.

(3) **Actions.** At each step, the output of RNN is used to predict the node index $v^{(l+1)}$ of the next candidate node in $\mathcal{G}_F^{(l+1)}$ and a re-wiring message $m^{(l+1)}$ that contributes to the change of the environment state. Specifically, the node index $v^{(l+1)}$ is determined stochastically from a distribution parameterized by the random walk network, i.e., $v^{(l+1)} \sim p(v|g_\Lambda(h^{(l)}))$, where g denotes a neural network with trainable parameters Λ. Similarly, the re-wiring vector is drawn from a distribution conditioned on the

output of the attention network $\xi_Z(\boldsymbol{h}^{(l)})$, i.e., $m^{(l+1)} \sim p(m|\xi_Z(\boldsymbol{h}^{(l)}))$, where Z is the attention network parameters. Based on the current hidden state $\boldsymbol{h}^{(l)}$, we further train a prediction network r_ω to learn the representative features $\boldsymbol{F}^{(l)}$, where the output is a softmax classifier for outcome prediction (e.g., recognizing disease connectomes).

(4) **Rewards.** After executing the above actions, the agent goes to a new external state $\boldsymbol{E}^{(l+1)}$ and a reward $\tau^{(l+1)}$. In most applications, $\tau^{(l+1)} = 1$ suggests that the learned representative features $\boldsymbol{F}^{(l)}$ successfully predict the outcome label for the underlying subject after $l + 1$ steps and 0 otherwise.

Extend Our Deep Model to the Riemannian Manifold. Since multiple lines of evidence show the importance of preserving the geometry of SPD matrices in functional brain network analysis [13, 14], we extend our *DeepHoloBrain* model in Fig. 3 to the manifold-based deep model by replacing the RNN model f_Θ to SPD-SRU [15], where the operations in the SPD-SRU use the manifold algebra for SPD matrices. To that end, the hidden state $\boldsymbol{h}^{(l)} \in Sym_S^+$ becomes a $S \times S$ SPD matrix.

Training *DeepHoloBrain*. The driving force of *DeepHoloBrain* consists of two parts. *First*, we use cross entropy to monitor the classification error between the ground truth and predicted outcome label by the current representative CFC feature $\boldsymbol{F}^{(l)}$ at each step l. Second, since the random walk from the current brain region $v^{(l)}$ to the next spot $v^{(l+1)}$ is steered by the re-wiring message $m^{(l)}$, we evaluate the effectiveness of random walk by a composite score of $\tau^{(l)} \cdot m_{v^l}^{(l)}$, where $\tau^{(l)}$ is the reward at the current step and $m_{v^l}^{(l)}$ reflects the reliability of suggesting good candidate regions in reinforcement learning. The parameters of *DeepHoloBrain* $\{\Theta, Z, \Lambda, \omega\}$ are fine-tuned by the Adam optimizer, where the learning rate is set to 0.001.

3 Experiments

In the following experiments, we first investigate the neuroscience insight of *HoloBrain* underlying the relationship between brain function and cognitive status. After that, we evaluate the diagnostic power of *DeepHoloBrain* in recognizing disease connectomes for AD, OCD, and schizophrenia using resting-stage fMRI scans, and recognizing cognitive tasks from task-based fMRI scans in HCP.

Data Description. We evaluate our proposed *HoloBrain* and *DeepHoloBrain* on the task fMRI data from HCP and resting-stage fMRI data from three disease-related datasets. For the task fMRI in HCP, we use Yale's functional atlas [16] which consists of 268 brain parcellations. The task includes 2-back and 0-back task conditions for body, place, face, and tool stimuli, as well as fixation periods. For the resting-state fMRI, the classic AAL atlas [17] is employed. For each fMRI scan, ICA-AROMA [18] is used to remove motion signal artifacts based on temporal and spatial features in the data related to head motion. A band-pass filter (0.009–0.08 Hz) is then applied to each scan. After spatial normalization, we compute the mean BOLD time course for each parcellated brain region. We perform experiments of cognitive normal (CN) vs. disease connectomes classification on ADNI dataset (102 CN vs. 63 AD), OCD study (63 CN vs. 61 OCD), and schizophrenia study (159 CN vs. 182 SZ). Meanwhile, we perform task recognition experiment on 264 subjects with task-fMRI data from HCP.

3.1   Understanding the Neuroscience Insight of *HoloBrain*

Since harmonic wavelets are widely used for signal reconstruction, we determine the number of frequency bands S by evaluating the residual error between the original signals and reconstructed signals with limited bandwidth. By doing so, we fix $S = 10$ in all the following experiments. Thus, each CFC pattern is a 10×10 SPD matrix.

We show the population average of global CFC patterns (i.e., averaging throughout brain regions and across individuals) for each clinic cohort in Fig. 4. There is a clear sign that the population-wise CFC average exhibit remarkable off-diagonal striping patterns, which resembles the constructive/destructive interference phenomena in Young's double slit shown in Fig. 2. In light of this, we hypothesize that the consistency of CFC degree along these striping patterns might offer a new explanation how brain functions emerge diverse cognition and behaviors. Since the striping patterns are visually distinguishable between healthy and disease cohorts, we further speculate that the coherence of maintaining the striping patterns might be an indicator of how brain function is altered as the disease progresses.

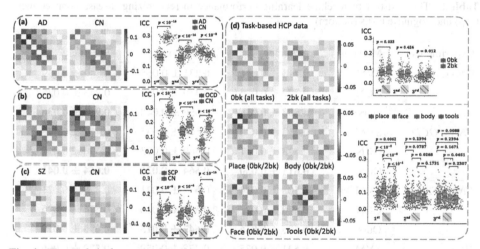

Fig. 4. The statistical power of *HoloBrain* in healthy vs. disease connectome (left) comparison and separating cognitive tasks (right). In each group comparison, we display the average CFC patterns and the quantitative measures that evaluate the discriminative power of *HoloBrain*.

Following this clue, we calculate the interclass correlation coefficient (ICC [19]) for each off-diagonal line, where a higher ICC indicates less variance between CFC degrees under consideration. In Fig. 4, we plot the distribution of ICC for the first three off-diagonal lines for each cohort. It is apparent that the ICC level manifests strong differences between healthy and disease connectomes at the significance level $p < 10^{-6}$. Although the difference between 0-bk and 2-bk tasks is not significant, the CFC patterns show significant differences ($p < 0.005$) in ICC along the first diagonal line among cognitive tasks.

3.2   Evaluating the Clinical Value of *DeepHoloBrain*

In this section, we shift the focus to predicting outcome labels for each functional connectome using the trained *DeepHoloBrain*. The benchmark methods include classic support vector machine (SVM) and graph convolution network (GCN) [11]. For SVM, we vectorize the CFC matrix at each brain region and concatenate them across brain regions. Then we train SVM to predict the outcome from the concatenated vector. For GCN, we consider the vectorized CFC matrix as the graph embedding vector. Then, we train GCN to predict the outcomes based on the topology of the functional connectivities. For all learning-based methods, we use 5-fold cross-validation.

Evaluating the Prediction Accuracy. The prediction results of classifying AD, OCD, and schizophrenia from resting-stage fMRI and recognizing cognitive tasks from task-based fMRI are summarized in Table 1. It is apparent that our *DeepHolo-Brain* outperforms the other counterpart methods in terms of accuracy, sensitivity, and specificity.

Table 1. The statistics of machine learning performance in recognizing disease connectomes (top) and cognitive tasks (bottom).

Experiments	Methods	Accuracy	Sensitivity	Specificity
CN vs. AD	SVM	0.713 ± 0.031	0.231 ± 0.065	0.842 ± 0.039
	GCN	0.772 ± 0.026	0.430 ± 0.056	0.743 ± 0.025
	Ours	$\mathbf{0.808 \pm 0.036}$	$\mathbf{0.650 \pm 0.048}$	$\mathbf{0.941 \pm 0.036}$
CN vs. OCD	SVM	0.513 ± 0.020	0.565 ± 0.035	0.452 ± 0.049
	GCN	0.573 ± 0.011	0.560 ± 0.023	0.530 ± 0.055
	Ours	$\mathbf{0.611 \pm 0.030}$	$\mathbf{0.664 \pm 0.026}$	$\mathbf{0.664 \pm 0.041}$
CN vs. SZ	SVM	0.508 ± 0.014	0.535 ± 0.021	0.478 ± 0.021
	GCN	0.578 ± 0.021	0.563 ± 0.016	0.576 ± 0.025
	Ours	$\mathbf{0.605 \pm 0.016}$	$\mathbf{0.647 \pm 0.015}$	$\mathbf{0.690 \pm 0.017}$
HCP 0-bk vs. 2-bk	SVM	0.521 ± 0.035	0.478 ± 0.059	0.570 ± 0.066
	GCN	0.555 ± 0.016	0.244 ± 0.053	$\mathbf{0.862 \pm 0.041}$
	Ours	$\mathbf{0.595 \pm 0.017}$	$\mathbf{0.567 \pm 0.045}$	0.671 ± 0.047
HCP Tools/Place/ Body/Face	SVM	0.251 ± 0.032	0.155 ± 0.068	0.344 ± 0.078
	GCN	0.272 ± 0.041	0.166 ± 0.056	$\mathbf{0.422 \pm 0.065}$
	Ours	$\mathbf{0.363 \pm 0.034}$	$\mathbf{0.213 \pm 0.044}$	0.396 ± 0.059

Evaluating the Model Explainability. *First*, since our *DeepHoloBrain* model is trained to select the best combination of CFC patterns for connectome classification by reward-guided reinforcement learning, we display the most frequently-visited brain regions and frequently-selected random walks in Fig. 5, for CN vs. AD (top-left), CN vs. OCD (top-right), and CN vs. schizophrenia (bottom-left), respectively. Specifically, we use node size and brightness to reflect the selection frequency, where high frequency indicates the importance in disease classification. Take CN vs. AD classification as an example. Most of the brain regions fall in the default mode network (e.g., precuneus and superior frontal gyrus) and frontoparietal network (e.g., middle frontal gyrus). Meanwhile, we show the population-average subgraph $\mathcal{G}_F$ in each disease classification, which suggests that the CFC patterns along the subgraph reach the best balance between diagnostic power and dimensionality. Furthermore, we display the weighed average of CFC patterns for each clinical cohort in the bottom-right of Fig. 5. Compared to the group comparison results in Fig. 4, it is clear that the constructive and destructive patterns are not only well maintained in the off-diagonal lines in the learned CFD patterns but also greatly enhanced with respect to group-to-group differences.

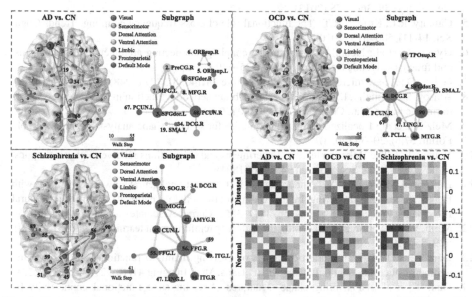

Fig. 5. The most frequently-visited brain regions and subgraph $\mathcal{G}_F$ that are used in CN vs. AD (top-left), CN vs. OCD (top-right), and CN vs. SZ (bottom-left), where the node size and link color indicate the likelihood of being selected in *DeepHoloBrain*. Meanwhile, we display the group average of CFC patterns within the subgraph $\mathcal{G}_F$ for different clinical cohorts in the bottom-right panel.

4   Conclusions

In this work, we propose a principled computational framework to characterize the holography of spontaneous functional connectivities in the human brain. Specifically, our *HoloBrain* technique allows us to elucidate the neuroscience insight of cross-frequency couplings which plays a mechanistic role in neuronal computation, communication, and learning. Furthermore, we tailor a deep model to translate the *HoloBrain* to various neuroscience and clinical applications, such as predicting cognitive states and recognizing disease connectomes.

References

1. Bassett, D.S., Sporns, O.: Network neuroscience. Nat. Neurosci. **20**(3), 353–364 (2017)
2. Airan, R.D., et al.: Factors affecting characterization and localization of interindividual differences in functional connectivity using MRI. Hum. Brain Mapp. **37**(5), 1986–1997 (2016)
3. Rubinov, M., Sporns, O.: Complex network measures of brain connectivity: uses and interpretations. Neuroimage **52**(3), 1059–1069 (2010)
4. Hutchison, R.M., et al.: Dynamic functional connectivity: promise, issues, and interpretations. Neuroimage **80**, 360–378 (2013)
5. Canolty, R.T., Knight, R.T.: The functional role of cross-frequency coupling. Trends Cogn. Sci. **14**(11), 506–515 (2010)
6. Hyafil, A., et al.: Neural cross-frequency coupling: connecting architectures, mechanisms, and functions. Trends Neurosci. **38**(11), 725–740 (2015)
7. Gabor, D.: A new microscopic principle. Nature **161**(4098), 777 (1948)
8. Atasoy, S., Donnelly, I., Pearson, J.: Human brain networks function in connectome-specific harmonic waves. Nat. Commun. **7**(1), 10340 (2016)
9. Melzi, S., et al.: Localized manifold harmonics for spectral shape analysis. Comput. Graph. Forum **37**(6), 20–34 (2018)
10. Young, T.: I. The Bakerian lecture. Experiments and calculations relative to physical optics. Philos. Trans. Roy. Soc. London **94**, 1–16 (1804)
11. Defferrard, M. Bresson, X., Vandergheynst, P.: Convolutional neural networks on graphs with fast localized spectral filtering. In: Proceedings of the 30th International Conference on Neural Information Processing Systems, pp. 3844–3852. Curran Associates Inc., Barcelona (2016)
12. Mnih, V., et al.: Human-level control through deep reinforcement learning. Nature **518**(7540), 529–533 (2015)
13. Dan, T., et al.: Learning brain dynamics of evolving manifold functional MRI data using geometric-attention neural network. IEEE Trans. Med. Imaging **41**(10), 2752–2763 (2022)
14. Dan, T., et al.: Uncovering shape signatures of resting-state functional connectivity by geometric deep learning on Riemannian manifold. Hum. Brain Mapp. **43**(13), 3970–3986 (2022)
15. Chakraborty, R., et al.: A statistical recurrent model on the manifold of symmetric positive definite matrices. In: Neural Information Processing Systems (2018)
16. Shen, X., et al.: Groupwise whole-brain parcellation from resting-state fMRI data for network node identification. Neuroimage **82**, 403–415 (2013)
17. Tzourio-Mazoyer, N., et al.: Automated anatomical labeling of activations in SPM using a macroscopic anatomical parcellation of the MNI MRI single-subject brain. Neuroimage **15**(1), 273–289 (2002)
18. Pruim, R.H.R., et al.: ICA-AROMA: a robust ICA-based strategy for removing motion artifacts from fMRI data. Neuroimage **112**, 267–277 (2015)
19. Koo, T.K., Li, M.Y.: A guideline of selecting and reporting intraclass correlation coefficients for reliability research. J. Chiropr. Med. **15**(2), 155–163 (2016)

Species-Shared and -Specific Brain Functional Connectomes Revealed by Shared-Unique Variational Autoencoder

Li Yang[1], Songyao Zhang[1], Weihan Zhang[1], Jingchao Zhou[2],
Tianyang Zhong[1], Yaonai Wei[1], Xi Jiang[2], Tianming Liu[4], Junwei Han[1],
Yixuan Yuan[3], and Tuo Zhang[1(✉)]

[1] Northwestern Polytechnic University, Xi'an 710071, China
tuozhang@nwpu.edu.cn
[2] University of Electronic Science and Technology, Chengdu 611731, China
[3] City University of Hong Kong, Hong Kong 999077, China
[4] The University of Georgia, Athens, GA 30602, USA

Abstract. A comparative study of large-scale species-shared and -specific functional connectomes, that are respectively inherited or diverged from their common ancestor, is important to the understanding of emergence and evolution of brain functions, cognitions and behaviors. However, recent works largely relied on the scheme that one species is used as the "reference" to which another is aligned and contrasted, whereas a more reasonable "reference" could be related to their common ancestor and unknown. To this end, we proposed a novel method termed shared-unique variational autoencoder (SU-VAE), and applied it to macaque and human MRI datasets to disentangle species-specific variation of functional connectomes from species-shared one. The reconstructed shared and specific connectomes gain supports from reports. The proposed method was further validated by the results that human-specific latent features, in contrast to shared features, better capture the variation of behavior variables unique to human, such as language comprehension. Our studies outperform other methods developed on VAE and linear regression in the aforementioned validation study. Finally, a graph analysis on the identified shared and specific connectomes reveals that, human/macaque -specific connecomtes positively/negatively contribute to the enhancement of network efficiency.

Keywords: Disentangled representation learning · Brain functional connectivity · Species comparison · VAE · Neural networks

1 Introduction

About 28.1 million years ago, macaque monkeys and humans are separated from their common ancestor [1]. One of the greatest differences was found between

A. Frangi et al. (Eds.): IPMI 2023, LNCS 13939, pp. 41–52, 2023.
https://doi.org/10.1007/978-3-031-34048-2_4

their brains, which, rather than scale in size, disproportionally increase the volume, cerebral cortical area as well as the complexity of the underlying wiring diagrams along the phylogenetic tree [2], possibly leading to a divergence or selectivity in brain function [3]. Given that the emergence of brain function was attributed to the orchestration of local and remote cortical areas by means of a densely connected brain network [4], identifying the large-scale species-shared connectomes that could be preserved from their common ancestor and the species-specific ones that contribute to the uniqueness of a particular species, are equally important to understanding the evolution of brain functions and emergence of intelligence, cognition and behavior [5].

Many works have reported the species-preserved organizations of white matter pathways as well as tremendous different pathways between human and macaque brains. However, most of experimental works depend on dissection and micro-scale imaging technique, such that only a small portion of region of interest or a single white matter pathway, such as language pathway, was compared at one time [2], far from a comprehensive understanding of aforementioned large-scale functional connectomes. Neuroimaging, such as functional MRI, in many recent studies demonstrate the possibility to this end, since it records the on-line whole-brain functional dynamics and tremendously increases the dataset size for better quantitative and reproducibility assessment [5]. Nevertheless, these imaging-based comparison studies usually used one species as the "background", such as macaque, with which another is contrasted [6], whereas a more reasonable "background" could be inherited from their common ancestor, and thus unknown to observers. Also, matching one species to another theoretically works on the assumption that we already know which factors we need to match, whereas the huge inter-individual variation induced by a multitude of known and unknown factors [7] could be blended with the desired species-specific "foreground" diverged from the "background", making the detection of both "background" and "foreground" obscure.

Therefore, we focused on comparisons between resting-state functional MRI (rsfMRI) derived connectomes between macaque and human, and developed a novel algorithm based on VAE [8], termed shared-unique variation autoencoder (SU-VAE), that takes as input unpaired subjects from the two species, to disentangle "species-specific" variation of functional connectivity from species-shared one, which were presented as two distant sets of latent features. Figure 1(a) illustrates the basic concept of our proposed algorithm, where sunflower is the shared "background", digit 0s and 1s are two classes of specific "foreground". The reconstructed shared and specific connections were compared with previous reports to investigate their validity. Since we use the human dataset from Human Connectome Project [9], where a spectrum of demography/behavior/cognitive variables are available, a more important validation is investigating whether the cross-subject variation of human-specific features is related to that of variables that tend to be more human-specific, such as language comprehension, whereas the variation of shared features is better related to that of variables found on both species, such as hearing ability. Finally, we investigated the possible roles of the reconstructed shared and specific connections in functional connectomes in respect of graph metrics, such as clustering coefficient and average path length. These analyses might provide more clues to understanding the evolution of brain functional connectomes and the designation of artificial neural networks.

2 Previous Works

2.1 Contrastive and Disentangled Representation Learning

Variational Autoencoders (VAEs) are designed to extract the latent features with variation in a variety of data [8]. Contrastive Variational Autoencoder (CVAE) is interested in modeling latent structure and variation that are enriched in a target dataset compared to background. As a successful application, researchers used CVAE to disentangle Autism-Spectrum-Disorder-specific neuroanatomical variation from variation shared with typical control participants [7]. However, in many applications, there is no appropriate background data to contrast with. Features of background and foreground are both needed to learn, such as genotype/phynotype comparison across species and across diseases. Since finding representations for the task is fundamental in machine learning [10]. Recently, many works focused on learning disentangled representations with VAEs [11], in which each latent feature learns one semantically meaningful factor of variation, while being invariant to other factors. Although there is no widely accepted definition of disentangled representations yet, the main intention is to separate the main factors of variation that are present in provided data distribution. Nevertheless, these methods mainly focus on disentangling the 'factors' of objects, and are unable to cope with the separation of specific and shared content.

2.2 Cross-Species Connectome Comparison

Dirty multi-task regression (DMR) is a data-driven linear method which has been applied to the comparison of large-scale connectomes across multiple species [12], where two weight matrices $\mathbf{S}$ and $\mathbf{U}$ were estimated, respectively, to highlight the connections shared by species or specific to a particular species. In DMR, connections were intrinsically defined to be independent from each other, whereas the interaction among those connections were supposed to be highly nonlinear. Another group of large-scale comparative studies [13] defined vertex-wise connective fingerprint on cortical surfaces and estimated vertex-to-vertex alignment between species. These studies usually need a bridge to quantify similarity of features across species, such as white matter fiber pathways reported to be preserved across species. The abundance of such preserved brain structure in literature largely decides the precision of cross-species alignment. More importantly, all these methods above are in a group-wise manner. The inter-individual variation was not fully taken into account, such that the undesired variation across subjects could bias the comparative analysis [14].

3   Materials and Methods

3.1   Data Sets, Preprocessing and Functional Network Construction

Human Imaging Data. T1-weighted MRI (T1w MRI) and rsfMRI data are from Human Connectome Project (HCP) dataset [9]. Important imaging parameters are: T1w MRI: TR = 2400 ms, TE = 2.14 ms, flip angle = 8°, image matrix = 260 × 311 × 260 and resolution = 0.7 × 0.7 × 0.7 mm^3. RsfMRI: TR = 0.72 s, TE = 33.1 ms, flip angle = 52°, in-plane FOV = 208 × 180 mm^2, matrix = 90 × 104, 220 mm FOV, 72 slices, 1200 time points, 2.0 mm isotropic voxels, BW = 2290 Hz/Px. These data have been minimally pre-processed upon release. Preprocessing of T1w MRI includes skull stripping, tissue-segmentation and white matter surface reconstruction. The surfaces were transferred to the grayordinate system [9], a common surface template, via cross-subject registration to standard volume and surface spaces. It is noted that the aligned surface of each subject was re-sampled such that vertices have cross-subject correspondence. Preprocessing of rsfMRI includes: ICA-denoise, spatial artifact and distortion removal. Within-subject cross-modal registration was applied such that each grayordinate vertex was associated with a preprocessed fMRI signal. Our experiments included 500 samples for training/testing (0.8/0.2). **Macaque Monkey Imaging Data.** T1w MRI and rsfMRI data are from UW-Madison Rhesus MRI dataset[1]. Important imaging parameters are: T1w MRI: TR=11.4/8.65 ms, TE = 5.41/1.89 ms, flip angle = 10°, image matrix = 512 × 248 × 512 and resolution = 0.27 × 0.5 × 0.27 mm^3. The rsfMRI data were preprocessed based on DPARSF[2] including slice timing, realignment, covariant regression, band-pass filtering (0.01–0.1 Hz), and smoothing (FWHM = 4 mm). We fed T1w images to CIVET[3], registering it into the NMT-standardized space using an affine transformation, followed by image resampling and tissue segmentation. The white matter cortical surface was reconstructed using Freesurfer[4]. The surfaces were resampled to 40k vertices to ensure vertex-to-vertex correspondence across subjects. After linear registration between fMRI and T1w MRI via FLIRT[5], we mapped the volume time-series to surface vertices. Our experiments included 60 macaque subjects after quality control.

Functional Network Construction. A functional connectome for each subject of each species was estimated. Brodmann areas (BAs) were used as nodes in this work as a testbed, since 28 BAs were suggested to be common across the two species [12]. As so, the "F99" template macaque surface was warped to individual ones via the surface registration method [15] to transfer the BAs to native spaces. Likewise, the standard MNI152 human template T1w MRI is warped to individual T1 MRI volumes via FSL-fnirt, and BAs was transferred

---

[1] http://fcon_1000.projects.nitrc.org/indi/PRIME/uwmadison.html.
[2] http://rfmri.org/content/dparsf.
[3] https://mcin.ca/technology/civet.
[4] https://www.freesurfer.net.
[5] http://www.fmrib.ox.ac.uk/fsl.

to surfaces in native spaces by means of Workbench[6]. The Pearson correlation coefficient between mean signals from two BAs was computed to construct the functional connectome. Histogram equalization was adopted to macaque group-mean connectome to match the global strength distribution of the human one. Finally, negative connections were removed from individual connectomes.

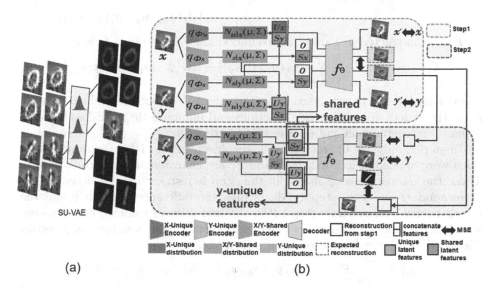

(a)   (b)

Fig. 1. Digits 0s and 1s are used as foreground and sunflower images as background to illustrate the flowchart. (a) With two unpaired images which are from two classes as the inputs, SU-VAE can disentangle the shared and unique latent features between two classes. (b) The architecture of SU-VAE.

3.2 SU-VAE

As shown in Fig. 1(b), the first step of SU-VAE is to derive the shared features between two classes, and the second step use the shared-reconstruction from step 1 to disentangle unique features. In step 1, two unpaired inputs from two classes are denoted by x and y. q_{Φ_u} and q_{Φ_s} represents the encoder of unique and shared features, respectively. $N_{u|x}(\mu, \Sigma)$ represents the distribution of x-unique, $N_{s|x}(\mu, \Sigma)$, $N_{s|y}(\mu, \Sigma)$ represents the distribution of x and y -shared, $N_{u|y}(\mu, \Sigma)$ corresponds to y-unique. U_x and U_y represents the resampled data from unique encoders of x and y, S_x and S_y are for shared ones. We concatenated the U_x and S_y to reconstruct the x' to force the encoder to remove the x-shared features from u_x-encoder. Similarly, U_y and S_x are concatenated to reconstruct y'. We just fed the S_x to the decoder to generate the expected shared-reconstruction,

---

[6] https://www.humanconnectome.org/software/connectome-workbench.

as well as S_y, and force two shared-encoder to learn the distributions close to each other. By far, the model learn the shared features between datasets x and y.

For input y_i , the likelihood lower bound is shown as follows:

$$\mathcal{L}_y\left(\boldsymbol{y}_i\right) \geq \underset{q_{\phi_s}(s_x)q_{\phi_u}(u_y)}{E} \left[f_\theta\left(\boldsymbol{y}_i \mid \boldsymbol{s_x}, \boldsymbol{u_y}\right)\right] - \mathrm{KL}\left(q_{\phi_s}\left(\boldsymbol{s_x} \mid \boldsymbol{x}_i\right) \| p(\boldsymbol{s_x})\right)$$
$$-\mathrm{KL}\left(q_{\phi_u}\left(\boldsymbol{u_y} \mid \boldsymbol{y}_i\right) \| p(\boldsymbol{u_y})\right) \tag{1}$$

For reconstructed y-shared, the likelihood lower bound is as follows:

$$\mathcal{L}_{y-shared}\left(\boldsymbol{y}_i\right) \geq \underset{q_{\phi_s}(s_y)}{E} \left[f_\theta\left(\boldsymbol{y}_i \mid \boldsymbol{0}, \boldsymbol{s_y}\right)\right] - \mathrm{KL}\left(q_{\phi_s}\left(\boldsymbol{s_y} \mid \boldsymbol{y}_i\right) \| p(\boldsymbol{s_y})\right) \tag{2}$$

x and x-shared can be expressed in a similar way. The model is trained to maximize the sum of above objective functions and minimize the MSE-loss between x-shared reconstruction and y-shared reconstruction.

In step 2, only the operations on y are shown in Fig. 1(b), the same operations were applied to x. Firstly, U_y and S_y are concatenated to reconstruct y'. Then, the reconstructed y-shared in this step is restricted to be closed to the corresponding one from step 1. Meanwhile, the difference between y and the reconstructed y-shared from step 1 (the absolute value of y minus shared- reconstruction) is restricted to be as equal the reconstructed y-unique as possible. The likelihood lower bound for input y_i is as follows:

$$\mathcal{L}_y\left(\boldsymbol{y}_i\right) \geq \underset{q_{\phi_s}(s_y)q_{\phi_u}(u_y)}{E} \left[f_\theta\left(\boldsymbol{y}_i \mid \boldsymbol{s_y}, \boldsymbol{u_y}\right)\right] - \mathrm{KL}\left(q_{\phi_s}\left(\boldsymbol{s_y} \mid \boldsymbol{y}_i\right) \| p(\boldsymbol{s_y})\right)$$
$$-\mathrm{KL}\left(q_{\phi_u}\left(\boldsymbol{u_y} \mid \boldsymbol{y}_i\right) \| p(\boldsymbol{u_y})\right) \tag{3}$$

The likelihood lower bound for reconstructed y-unique is as follows:

$$\mathcal{L}_{y-unique}\left(\boldsymbol{y}_i\right) \geq \underset{q_{\phi_u}(u_y)}{\mathbb{E}} \left[f_\theta\left(\boldsymbol{y}_i \mid \boldsymbol{u_y}, \boldsymbol{0}\right)\right] - \mathrm{KL}\left(q_{\phi_u}\left(\boldsymbol{u_y} \mid \boldsymbol{y}_i\right) \| p(\boldsymbol{u_y})\right) \tag{4}$$

By maximizing the sum of objective functions (2), (3), (4), we trained the model to learn the unique and shared latent features between x and y.

3.3   Statistics

Representational similarity analysis (RSA) [7] was adopted to test whether the variation of shared or specific connectome latent features (highlighted by purple texts in Fig. 1(b)) across human subjects correlate with the variation of demographic/behavior/cognitive variables provided in HCP dataset. A pairwise dissimilarity (e.g., Euclidean distance) between subjects, with respect to their shared-/specific- connectome features as well as the variables, was respectively computed to yield dissimilarity matrices. Kendall rank correlation (Kendall coefficient τ) was used to quantify the consistency between the connectome dissimilarity matrix and thes one for each variable. Finally, We adopted graph measures, including average path length, mean clustering coefficient averaged over nodal ones and global efficiency, to characterize the nature of shared and specific connectome structure.

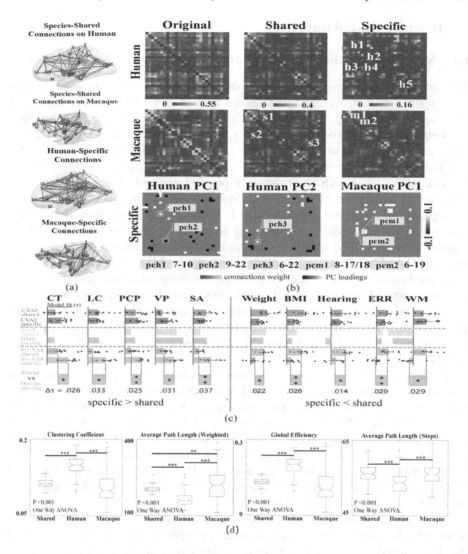

Fig. 2. (a) Illustration of mean species-shared and -specific connectomes on one example human subject and one example macaque subject. (b) Rows 1–2: The mean original functional connectomes, reconstructed shared and specific ones within each species. Red boxes highlight the strong shared- and specific- connections have been reported so far, while purple boxes highlight those have not been reported yet. Row #3: PC loading of specific matrices that account for the variance of all subjects. Boxes index and highlight the connections that are reported to be different between species. (c) Human functional connectome dissimilarity matrices are compared with those of human subjects' demographic/behavior/cognition variables. The functional connectome features were yielded from CVAE, SU-VAE and DMR, respectively, separated by dashed lines. Abbreviations: CT: Cognition Total, LC: Language Comprehension, PCP: Pattern Comparison Processing, VP: Visuospatial Processing, WM: Working Memory, SA: Sustained Attention, BMI: Body Mass Index, ERR: Emotion Recognition Response. (d) Graph metrics of shared and original connectomes. (Color figure online)

4 Results and Analysis

4.1 Validation of the Species -Shared and -Specific Connectomes

The reconstructed shared connectomes and specific ones of human and macaque testing subjects are averaged and shown in Fig. 2(a). The matrix format of them are shown in Fig. 2(b) as well. The shared connectomes are highly similar between the two species, with the pairwise Pearson correlation coefficient across all testing human-macaque shared pairs as 0.93 ± 0.04, whereas the specific matrices are different (0.03 ± 0.07). We compared the reconstructed connections (Fig. 2(b)) with the existing reports. Identified shared- and specific-connections with strong weights, that find supports from reports (red boxes), are reported in the upper panel of Table 1. Some of such connections that do not find supports from reports (purple boxs) are reported in lower panel of Table 1. Most of these identified species-shared and macaque-specific connections can be found in CoCoMac datasets[7], a collection of reported tract-tracing pathways on macaque brains. Note that most reports of the connections are based on white matter structural pathway analysis, such that some of them may not necessarily be strong in the functional connectomes, even though the structure was suggested to be a constraint of functions. Likewise, the identified connections not found in reports could be interpreted in the same manner, or, provide clues for further studies.

Further, the PC (principal component) loadings of all specific connections was computed, since the species-shared components were removed and only the subject variance attributed to species difference was encoded in these PC loadings [7]. PC1 and PC2 maps (that explain 15.1/19.58 percent of variance on human subjects and macaque subjects, respectively) of specific-connection variance are shown in bottom of Fig. 2(b). Those connections reported to be different between species, including connections (BA7-BA10), (BA6/9-BA22) and FEF-Visual cortex (BA8-BA17/18, BA6-BA19) [19,23,25], are well identified by heavy PC loadings (highlighted by indices).

In addition, RSA was used to compute the Kendall τ correlations to investigate whether the variation of latent shared or specific features capture that of the demographic/behavior/cognitive variables across subjects in human dataset (HCP). The results via our method is found in the bottom rows in Fig. 2(c). On the right side, the shared features better capture variations of variables common to both species (higher τ for shared- features), such as hearing, or irrelevant of species comparison, such as body mass. In addition, variations of some of the high-level cognitive properties, such as emotion recognition and working memory, were also found to be better captured by that of shared features, since it has been reported that both monkeys and human adults are capable of recognizing faces from their own species [27] and the ability of visual working memory in delay tasks, such as delayed matching to sample, which require that a memory be held during a brief delay period, has been found on macaque as well [28].

---

[7] http://cocomac.g-node.org.

Table 1. Connections have been reported or not yet. Abbreviation: LF: left frontal operculum, SPF: superior parietal-middle frontal, SV2: somatosensory cortex-V2, FOS: fronto-occipital stream, V2T: V2-Middle Temporal, SM: somatosensory and motor cortices, SS: SMA-preSMA, VC: visual cortices, SMP: SMA-preSMA,TL: temporal lobe, PTP: prefrontal cortex-temporal pole, MTP: premotor-temporal pole, CSP: convergence and segregation of processing streams, VS: visuo-somatomotor, SE: sensory, MO:motor.

Connections supported by reports (red boxs in Fig. 1(b))					
species-shared		human-specific		macaque-specific	
s1.BA1-7	SM [12]	h1.BA6-BA9	LO [18]	m1.BA3-BA1/BA2	SE [24]
s2.BA6-BA8	SS [16]	h2/pch1.BA7-BA10	SPF [19]	m2.BA4-BA6 BA5-BA7	MO [24]
s3.BA17-19 BA20-22	VC [17]	h3.BA2/3/4-BA18/19	SV2 [20] V2M1 [20]	pcm1.BA8-BA17/18	CSP [25]
		h4.BA9-BA18	FO [21]	pcm2.BA6-BA19	VS [26]
		h5.BA18-BA20/21	V2T [22]		
		pch2/3.BA9/6-BA22	MTP [23] PTP [23]		

Connections have not been reported yet (purple boxes in Fig. 1(b))

shared:(BA13-BA14/22, BA5/9/19-BA25/26, BA1/2-BA22)
human-specific:(BA4-BA25/26, BA14-BA18,BA1-BA7)
macaque-specific:(BA17-BA19,BA20-BA21, BA21-BA22,BA9-BA10/BA12,
BA10-BA12,BA25-BA9/BA10/BA12,BA14-BA15/BA16,BA23-BA26)

On the left side, variables, such as language comprehension (a strong measure of crystallized abilities), pattern comparison processing (a measure of speed of processing, which is considered a 'fluid ability'), visuospatial processing (variable short Penn line orientation test) and sustained attention, are more likely to be related to human-specific cognitive abilities. Their variation was better captured by SU-VAE's human-specific features (higher τ for specific- features). We computed the $\Delta\tau$ between each shared- and specific- τ pairs and used paired sample t-tests to quantify the significance of the difference (reported in Fig. 2(c) and Table 2). The $\Delta\tau$ is significant across all demographic/behavior/cognitive variables. Likewise, we report the CVAE and DMR results. Note that, in order to adapt to the CVAE, we used the macaque functional connection matrix as the background data (i.e. shared features) to let the model learn the features of human that are different from macaques (human-specific). As shown in the top row of Fig. 2(c), the features of shared- and specific- features of CVAE showed Kendall correlation contrast $\Delta\tau$ with some of the behavior-unrelated or lower level behavior variables (Table 2), whereas very few $\Delta\tau$ were found to be significant between shared and specific features via t-test (Table 2), suggesting that CVAE features may fail in disentangling the human specific features from shared ones. For DMR, we overlaid the shared weight matrix and specific ones respectively as mask to individual connectomes and zeroed connections with low weights. Unlike CVAE and SU-VAE methods, DMR can only yield a group-wise weight matrix on the cohort. Hence, no statistical comparison between shared τ and specific τ is performed. But it can still be observed that shared features are not consistently correlated better with low level behavior score, such as hearing, or the one of species preserved function, such as working memory. In summary, from the perspective of technique, as shown in Table 2, SU-VAE yields the greatest contrast between shared and specific τs (red font). The contrast is significant (low p-values) across all individual variables via SU-VAE but not for other methods. Also, the contrast between shared and specific τs via SU-VAE is well interpreted as discussed above.

Table 2. Comparison RSA results between three methods on the testing data, Difference of Kendall rank correlation coefficient $\Delta\tau$ between shared and specific features and p-value via t-test are shown. *,**,*** indicates the $p > 0.05$, $p < 0.05$ and $p < 0.01$, respectively. A positive number indicates that the function or behavior is more likely to be human-unique, otherwise it is the species-shared. Red font highlights the largest absolute $\Delta\tau$ of these methods.

Participant properties	CVAE	DMR	Proposed
Cognition Total	0.018(*)	0.027	0.026(***)
Language Comprehention	−0.003(*)	0.030	0.033(***)
Pattern Comparison Processing	0.006(*)	0.011	0.025(**)
Visuospatial Processing	0.014(*)	0.017	0.031(***)
Working Memory	0.006(*)	0.012	−0.029(***)
Sustained Attention	−0.010(*)	0.016	0.037(**)
BMI	−0.028(**)	−0.012	−0.023(**)
Weight	−0.023(**)	−0.007	−0.026(**)
Hearing	−0.008(*)	−	−0.014(**)
Emotion Recognition	−0.003(*)	−0.003	−0.029(**)

4.2 Interpretation from Graph Theory

Figure 2(d) shows the measure of the global graph metrics, including average clustering coefficient, average path length on both original connectomes with weights and binary ones ("steps" for short, 10 percent of top connections reserved) and global efficiency, on the shared connectome, as well as original human and macaque connectomes. In contrast to the shared connectomes, the human-specific ones increase the clustering coefficient and global efficiency and reduce the average path length, whereas the macaque-specific ones reduce them. Given that a small-world network tends to have higher clustering coefficient and shorter average path [29], the human-specific connectomes positively contribute to network efficiency and the build-up of network of small-worldness, whereas macaque-specific ones seem to contribute in an adverse way.

5   Conclusion

A method named SU-VAE was proposed to disentangle species-specific variation of functional connectome from species-shared one, which was validated by comparing the reconstructed species-shared and specific connectomes with reports and a variety of demography/cognitive/behavioural variables. Our studies outperform other state-of-the-arts in these validation studies. Further, a graph analysis on the identified shared and specific connectome reveals that, in contrast to shared connectome, human/macaque -specific connecomtes positively/negatively contribute to the enhancement of network efficiency. The proposed method and experimental results could provide more clues to the understanding of emergence and evolution of brain function, cognition and behavior.

Acknowledgements. This work was supported by the National Natural Science Foundation of China (31971288, U1801265, 61936007, U20B2065, 61976045, 62276050, and 61976045), National Key R&D Program of China (2020AAA0105701), Sichuan Science and Technology Program (2021YJ0247), Innovation and Technology Commission-Innovation and Technology Fund ITS/100/20, Doctor Dissertation of Northwestern Polytechnical University CX2022053.

References

1. Kay, R.F., Ross, C., Williams, B.A.: Anthropoid origins. Science **275**(5301), 797–804 (1997)
2. Mars, R.B., Jbabdi, S., et al.: Diffusion-weighted imaging tractography-based parcellation of the human parietal cortex and comparison with human and macaque resting-state functional connectivity. J. Neurosci. **31**(11), 4087–4100 (2011)
3. Smaers, J.B., Steele, J., Case, C.R., Cowper, A., Amunts, K., Zilles, K.: Primate prefrontal cortex evolution: human brains are the extreme of a lateralized ape trend. Brain Behav. Evol. **77**(2), 67–78 (2011)
4. Axer, M., Amunts, K.: Scale matters: the nested human connectome. Science **378**(6619), 500–504 (2022)
5. de Schotten, M.T., Croxson, P.L., Mars, R.B.: Large-scale comparative neuroimaging: where are we and what do we need?. Cortex, **118**, 188–202 (2019)
6. Raghanti, M.A., Stimpson, C.D., Marcinkiewicz, J.L., Erwin, J.M., Hof, P.R., Sherwood, C.C.: Cortical dopaminergic innervation among humans, chimpanzees, and macaque monkeys: a comparative study. Neuroscience **155**(1), 203–220 (2008)
7. Aglinskas, A., Hartshorne, J.K., Anzellotti, S.: Contrastive machine learning reveals the structure of neuroanatomical variation within autism. Science **376**(6597), 1070–1074 (2022)
8. Kingma, D.P., Welling, M.: Auto-encoding variational bayes. arXiv preprint arXiv:1312.6114 (2013)
9. Van Essen, D.C., et al.: The WU-Minn human connectome project: an overview. Neuroimage **80**, 62–79 (2013)
10. Bengio, Y., Courville, A., Vincent, P.: Representation learning: a review and new perspectives. IEEE Trans. Pattern Anal. Mach. Intell. **35**(8), 1798–1828 (2013)
11. Liu, X., et al.: Learning disentangled representations in the imaging domain. Med. Image Anal. **80**, 102516 (2022)
12. Zhang, T., et al.: Species-shared and -specific structural connections revealed by dirty multi-task regression. In: Martel, A.L., et al. (eds.) MICCAI 2020, Part VII. LNCS, vol. 12267, pp. 94–103. Springer, Cham (2020). https://doi.org/10.1007/978-3-030-59728-3_10

13. Mars, R.B., et al.: Whole brain comparative anatomy using connectivity blueprints. Elife **7**, e35237 (2018)
14. Thiebaut de Schotten, M., Forkel, S.J.: The emergent properties of the connected brain (2022). https://doi.org/10.1126/science.abq2591
15. Yeo, B.T.T., Sabuncu, M.R., Vercauteren, T., Ayache, N., Fischl, B., Golland, P.: Spherical demons: fast diffeomorphic landmark-free surface registration. IEEE Trans. Med. Imaging **29**(3), 650–668 (2010)
16. Rilling, J.K., et al.: The evolution of the arcuate fasciculus revealed with comparative DTI. Nat. Neurosci. **11**(4), 426–428 (2008)
17. de Schotten, M.T., et al.: Monkey to human comparative anatomy of the frontal lobe association tracts. Cortex **48**(1), 82–96 (2012)
18. Hanakawa, T., Honda, M., Sawamoto, N., et al.: The role of rostral Brodmann area 6 in mental-operation tasks: an integrative neuroimaging approach. Cereb. Cortex **12**(11), 1157–1170 (2002)
19. Silk, T., Vance, A., Rinehart, N., et al.: Fronto-parietal activation in attention-deficit hyperactivity disorder, combined type: functional magnetic resonance imaging study. Br. J. Psychiatry **187**(3), 282–283 (2005)
20. Dumontheil, I., Burgess, P.W., et al.: Development of rostral prefrontal cortex and cognitive and behavioural disorders. Dev. Med. Child Neurol. **50**(3), 168–181 (2008)
21. Herbet, G., Moritz-Gasser, S., Duffau, H.: Direct evidence for the contributive role of the right inferior fronto-occipital fasciculus in non-verbal semantic cognition. Brain Struct. Funct. **222**(4), 1597–1610 (2017)
22. Catani, M., Jones, D.K., Donato, R., et al.: Occipito-temporal connections in the human brain. Brain **126**(9), 2093–2107 (2003)
23. Bashwiner, D.M., Wertz, C.J., Flores, R.A., et al.: Musical creativity "revealed" in brain structure: interplay between motor, default mode and limbic networks. Sci. Rep. **6**(1), 1–8 (2016)
24. Carmichael, S.T., Price, J.L.: Sensory and premotor connections of the orbital and medial prefrontal cortex of macaque monkeys. J. Comp. Neurol. **363**(4), 642–664 (1995)
25. Vanduffel, W., et al.: Visual motion processing investigated using contrast agent-enhanced fMRI in awake behaving monkeys. Neuron **32**(4), 565–577 (2001)
26. Shipp, S., Blanton, M., Zeki, S.: A visuo-somatomotor pathway through superior parietal cortex in the macaque monkey: cortical connections of areas V6 and V6A. Eur. J. Neurosci. **10**(10), 3171–3193 (1998)
27. Pascalis, O., Bachevalier, J.: Face recognition in primates: a cross-species study. Behav. Proc. **43**, 87–96 (1998)
28. Miller, E.K., Erickson, C.A., et al.: Neural mechanisms of visual working memory in prefrontal cortex of the macaque. J. Neurosci. **16**, 5154–5167 (1996)
29. Cassar, A.: Coordination and cooperation in local, random and small world networks: experimental evidence. Games Econom. Behav. **58**(2), 209–230 (2007)

mSPD-NN: A Geometrically Aware Neural Framework for Biomarker Discovery from Functional Connectomics Manifolds

Niharika S. D'Souza[1](✉) and Archana Venkataraman[2]

[1] IBM Research, Almaden, San Jose, USA
Niharika.Dsouza@ibm.com
[2] Department of Electrical and Computer Engineering, Johns Hopkins University, Baltimore, USA

Abstract. Connectomics has emerged as a powerful tool in neuroimaging and has spurred recent advancements in statistical and machine learning methods for connectivity data. Despite connectomes inhabiting a matrix manifold, most analytical frameworks ignore the underlying data geometry. This is largely because simple operations, such as mean estimation, do not have easily computable closed-form solutions. We propose a geometrically aware neural framework for connectomes, i.e., the mSPD-NN, designed to estimate the geodesic mean of a collections of symmetric positive definite (SPD) matrices. The mSPD-NN is comprised of bilinear fully connected layers with tied weights and utilizes a novel loss function to optimize the matrix-normal equation arising from Fréchet mean estimation. Via experiments on synthetic data, we demonstrate the efficacy of our mSPD-NN against common alternatives for SPD mean estimation, providing competitive performance in terms of scalability and robustness to noise. We illustrate the real-world flexibility of the mSPD-NN in multiple experiments on rs-fMRI data and demonstrate that it uncovers stable biomarkers associated with subtle network differences among patients with ADHD-ASD comorbidities and healthy controls.

Keywords: Functional Connectomics · SPD Manifolds · Fréchet Mean Estimation · Geometry-Aware Neural Networks

1 Introduction

Resting state functional MRI (rs-fMRI) measures steady state patterns of co-activation [11] (i.e., *connectivity*) as a proxy for communication between brain regions. The 'connectome' is a whole-brain map of these connections, often represented as a correlation or covariance matrix [16] or a network-theoretic object such as adjacency matrix or graph kernel [10]. The rise of connectomics has spurred many analytical frameworks for group-wise diagnostics and biomarker discovery from this data. Early examples include statistical comparisons of connectivity features [16], aggregate network theoretic measures [10], and dimensionality reduction techniques [8,14]. More recently, the field has embraced deep neural networks to learn complex feature representations from both the connectome and the original rs-fMRI time series [2,7,18]. While these approaches have yielded valuable insights, they largely ignore the underlying geometry of the connectivity data. Namely, under a geometric lens, connectomes derived from rs-fMRI data lie on the manifold of symmetric positive definite (SPD) matrices. A major computational bottleneck for developing geometrically-aware generalizations [1,19] is the estimation of the geodesic mean on SPD manifolds. This is a far more challenging problem than statistical estimation in Euclidean data spaces because extensions of elementary operations such as addition, subtraction, and distances on the SPD manifold entail significant computational overhead [17].

The most common approach for estimating the geodesic mean on the SPD manifold is via gradient descent [20]. While this method is computationally efficient, it is highly sensitive to the step size. To mitigate this issue, Riemannian optimization methods [12], the majorization-maximization algorithm [25], and fixed-point iterations [4] can be used. While these extensions have desirable convergence properties, this comes at the cost of increased computational complexity, meaning they do not scale well to higher input dimensionality and larger numbers of samples [3]. In contrast, the work of [3] leverages the approximate joint diagonalization [21] of matrices on the SPD manifold. While this approach provides guaranteed convergence to a fixed point, the accuracy and stability of the optimization is sensitive to the deviation of the data from the assumed common principal component (CPC) generating process. Taken together, existing methods for geodesic mean estimation on the SPD manifold poorly balance accuracy, robustness and computational complexity, which makes them difficult to fold into a larger analytical framework for connectomics data.

We propose a novel end-to-end framework to estimate the geodesic mean of data on the SPD manifold. Our method, the Geometric Neural Network (mSPD-NN), leverages a matrix autoencoder formulation [9] that performs a series of bi-linear transformations on the input SPD matrices. This strategy ensures that the estimated mean remains on the manifold at each iteration. Our loss function for training approximates the first order matrix-normal condition arising from Fréchet mean estimation [17]. Using conventional backpropagation via stochastic optimization, the mSPD-NN automatically learns to estimate the geodesic mean of the input data. We demonstrate the robustness of our framework using simulation studies and show that mSPD-NN can handle input noise and high-dimensional data. Finally, we use the mSPD-NN for various groupwise discrimination tasks (feature selection, classification, clustering) on functional connectivity data and discover consistent biomarkers that distinguish between patients diagnosed with ADHD-Autism comorbidities and healthy controls.

2   Biomarker Discovery from Functional Connectomics Manifolds via the mSPD-NN

Let matrices $\{\Gamma_n\}_{n=1}^N \in \mathcal{M}$ be a collection of N functional connectomes belonging to the manifold $\mathcal{M}$ of Symmetric Positive Definite (SPD) matrices of dimen-

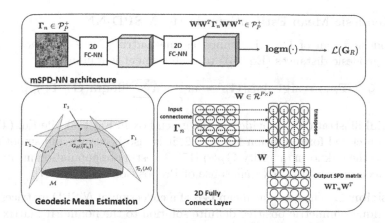

Fig. 1. The mSPD-NN architecture: The input is transformed by a cascade of 2D fully connected layers. The matrix logarithm function is used to obtain the matrix normal form, which serves as the loss function for mSPD-NN during training.

sionality $P \times P$, i.e. $\mathcal{M} \in \mathcal{P}_P^+$ (and a real and smooth Reimannian manifold). We can define an inner product that varies smoothly at each vector $\mathcal{T}_\Gamma(\mathcal{M})$ in the tangent space defined at any point $\Gamma \in \mathcal{M}$. Finally, a *geodesic* denotes the shortest path joining any two points on the manifold along the manifold surface.

Geodesic Mappings: The matrix exponential and the matrix logarithm maps allow us to translate geodesics on the manifold back and forth to the local tangent space at a reference point. The matrix exponential mapping translates a vector $\mathbf{V} \in \mathcal{T}_\Phi(\mathcal{M})$ in the tangent space at $\Phi \in \mathcal{M}$ to a point on the manifold $\Gamma \in \mathcal{M}$ via the geodesic emanating from Φ. Conversely, the matrix logarithm map translates the geodesic between $\Phi \in \mathcal{M}$ to $\Gamma \in \mathcal{M}$ back to the tangent vector $\mathbf{V} \in \mathcal{T}_\Phi(\mathcal{M})$. Mathematically, these operations are parameterized as:

$$\Gamma = \mathbf{Expm}_\Phi(\mathbf{V}) = \Phi^{1/2}\mathbf{expm}(\Phi^{-1/2}\mathbf{V}\Phi^{-1/2})\Phi^{1/2} \tag{1}$$

$$\mathbf{V} = \mathbf{Logm}_\Phi(\Gamma) = \Phi^{1/2}\mathbf{logm}(\Phi^{-1/2}\Gamma\Phi^{-1/2})\Phi^{1/2} \tag{2}$$

Here, $\mathbf{expm}(\cdot)$ and $\mathbf{logm}(\cdot)$ refer to the matrix exponential and logarithm respectively, each requiring an eigenvalue decomposition of the argument matrix, a point-wise transformation of the eigenvalues, and a matrix reconstruction.

Distance Metric: Given two connectomes $\Gamma_1, \Gamma_2 \in \mathcal{M}$, the Fisher Information distance between them is the length of the geodesic connecting the two points:

$$\delta_R(\Gamma_1, \Gamma_2) = ||\mathbf{logm}(\Gamma_1^{-1}\Gamma_2)||_F = ||\mathbf{logm}(\Gamma_2^{-1}\Gamma_1)||_F, \tag{3}$$

where $||\cdot||_F$ denotes the Frobenius norm. The Reimannian norm of Γ is the geodesic distance from the identity matrix $\mathcal{I}$ i.e. $||\Gamma||_R = ||\mathbf{logm}(\Gamma)||_F$

2.1   Geodesic Mean Estimation via the MSPD-NN

The geodesic mean of $\{\boldsymbol{\Gamma}_n\}$ is defined as the matrix $\mathbf{G}_R \in \mathcal{M}$ whose sum of squared geodesic distances (Eq. (3)) to each element is minimal [17].

$$\mathcal{G}_R(\{\boldsymbol{\Gamma}_n\}) = \arg\min_{\mathbf{G}_R} \mathbf{L}(\mathbf{G}_R) = \arg\min_{\mathbf{G}_R} \sum_n \|\mathbf{logm}(\mathbf{G}_R^{-1}\boldsymbol{\Gamma}_n)\|_F^2 \qquad (4)$$

A pictorial illustration is provided in the green box in Fig 1. While Eq. (4) does not have a closed-form solution for $N > 2$, it is also is convex and smooth with respect to the unknown quantity $\mathbf{G}_R(\cdot)$ [17]. To estimate population means from the connectomes, mSPD-NN makes use of Proposition 3.4 from [17].

Proposition 1. The geodesic mean $\mathbf{G}_R$ of a collection of N SPD matrices $\{\boldsymbol{\Gamma}_n\}$ is the unique symmetric positive-definite solution to the nonlinear matrix equation $\sum_n \mathbf{logm}(\mathbf{G}_R^{-1/2}\boldsymbol{\Gamma}_n\mathbf{G}_R^{-1/2}) = \mathbf{0}$. $\mathbf{0}$ is a $P \times P$ matrix of all zeros.

Proof: The proof follows by computing the first order necessary (and here, sufficient) condition for optimality for Eq. (4). First, we express the derivative of a real-valued function of the form $\mathbf{H}(\mathbf{S}(t)) = \frac{1}{2}\|\mathbf{logm}(\mathbf{C}^{-1}\mathbf{S}(t))\|_F^2$ with respect to t. In this expression, the argument $\mathbf{S}(t) = \mathbf{G}_R^{1/2}\mathbf{expm}(t\mathbf{A})\mathbf{G}_R^{1/2}$ is the geodesic arising from $\mathbf{G}_R$ in the direction of $\boldsymbol{\Delta} = \dot{\mathbf{S}}(0) = \mathbf{G}_R^{1/2}\mathbf{A}\mathbf{G}_R^{1/2}$, and the matrix $\mathbf{C} \in \mathcal{P}_P^+$ is a constant SPD matrix of dimension P. By using the cyclic properties of the trace function and the distributive equivalence of $\mathbf{logm}(\mathbf{A}^{-1}[\mathbf{B}]\mathbf{A}) = \mathbf{A}^{-1}[\mathbf{logm}(\mathbf{B})]\mathbf{A}$, we obtain the following condition:

$$\mathbf{H}(\mathbf{S}(t)) = \frac{1}{2}\|\mathbf{logm}(\mathbf{C}^{-1/2}\mathbf{S}(t)\mathbf{C}^{-1/2})\|_F^2$$

By the symmetry of the term $\mathbf{logm}(\mathbf{C}^{-1/2}\mathbf{S}(t)\mathbf{C}^{-1/2})$ we have that:

$$\therefore \frac{d}{dt}\mathbf{H}(\mathbf{S}(t))\Big|_{t=0} = \frac{1}{2}\frac{d}{dt}\mathrm{Tr}\left([\mathbf{logm}(\mathbf{C}^{-1/2}\mathbf{S}(t)\mathbf{C}^{-1/2})]^2\right)\Big|_{t=0}$$

$$\therefore \frac{d}{dt}\mathbf{H}(\mathbf{S}(t))\Big|_{t=0} = \mathrm{Tr}\left([\mathbf{logm}(\mathbf{C}^{-1}\mathbf{G}_R)\mathbf{G}_R^{-1}\boldsymbol{\Delta}]\right) = \mathrm{Tr}[\boldsymbol{\Delta}\mathbf{logm}(\mathbf{C}^{-1}\mathbf{G}_R)\mathbf{G}_R^{-1}]$$

$$\therefore \nabla\mathbf{H} = \mathbf{logm}(\mathbf{C}^{-1}\mathbf{G}_R)\mathbf{G}_R^{-1} = \mathbf{G}_R^{-1}\mathbf{logm}(\mathbf{G}_R\mathbf{C}^{-1})$$

Notice that since $\nabla\mathbf{H}$ is symmetric, it belongs to the tangent space $\mathcal{S}_P$ of $\mathcal{P}_P^+$. Therefore, we express the gradient of $\mathbf{L}(\mathbf{G}_R)$ defined in Eq. (4), as follows:

$$\mathbf{L}(\mathbf{G}_R) = \sum_n \|\mathbf{logm}(\mathbf{G}_R^{-1}\boldsymbol{\Gamma}_n)\|_F^2 \implies \nabla\mathbf{L}(\mathbf{G}_R) = \mathbf{G}_R^{-1}\sum_n \mathbf{logm}(\mathbf{G}_R\boldsymbol{\Gamma}_n^{-1})$$

$$\therefore \arg\min_{\mathbf{G}_R}\mathbf{L}(\mathbf{G}_R) \implies \sum_n \mathbf{logm}(\mathbf{G}_R\boldsymbol{\Gamma}_n^{-1}) = \sum_n \mathbf{logm}(\mathbf{G}_R^{-1/2}\boldsymbol{\Gamma}_n\mathbf{G}_R^{-1/2}) = \mathbf{0}$$

The final step uses the property that $\mathbf{L}(\mathbf{G}_R)$ is a sum of convex functions, with the first order stationary point is the necessary and sufficient condition being the unique minima. *Denoting* $\mathbf{G}_R^{-1/2} = \mathbf{V} \in \mathcal{P}_P^+$, *the matrix multiplications in the argument of the* $\mathbf{logm}(\cdot)$ *term can be efficiently expressed within the feed-forward operations of a neural network with unknown parameters* $\mathbf{V}$. $\qquad\square$

2.2   mSPD-NN Architecture

The mSPD-NN uses the form above to perform geodesic mean estimation. The architecture is illustrated in Fig. 1. The encoder of the mSPD-NN is a 2D fully-connected neural network (FC-NN) [5] layer $\Psi_{\mathrm{enc}}(\cdot) : \mathcal{P}_P^+ \to \mathcal{P}_P^+$ that projects the input matrices Γ_n into a latent representation. This mapping is computed as a cascade of two linear layers with tied weights $\mathbf{W} \in \mathcal{R}^{P \times P}$, i.e., $\Psi_{\mathrm{enc}}(\Gamma_n) = \mathbf{W}\Gamma_n\mathbf{W}^T$ The decoder $\Psi_{dec}(\cdot)$ has the same architecture as the encoder, but with transposed weights $\mathbf{W}^T$. The overall transformation can be written as:

$$\mathrm{mSPD\text{-}NN}(\Gamma_n) = \Psi_{\mathrm{dec}}(\Psi_{\mathrm{enc}}(\Gamma_n)) = \mathbf{W}\mathbf{W}^T(\Gamma_n)\mathbf{W}\mathbf{W}^T = \mathbf{V}(\Gamma_n)\mathbf{V} \quad (5)$$

where $\mathbf{V} \in \mathcal{R}^{P \times P}$ and is symmetric and positive definite by construction. We would like our loss function to minimize Eq. (4) in order to estimate the first order stationary point as $\mathbf{V} = \mathbf{G}_R^{-1/2}$, and therefore devise the following loss:

$$\mathcal{L}(\cdot) = \frac{1}{P^2} \left\| \frac{1}{N} \sum_n \mathrm{logm}\left[\mathbf{W}\mathbf{W}^T(\Gamma_n)\mathbf{W}\mathbf{W}^T\right] \right\|_F^2 \quad (6)$$

Formally, an error of $\mathcal{L}(\cdot) = 0$ implies that the argument satisfies the matrix normal equation exactly under the parameterization $\mathbf{V} = \mathbf{W}\mathbf{W}^T = \mathbf{G}_R^{-1/2}$. Therefore, Eq. (6) allows us to estimate the geodesic mean on the SPD manifold. We utilize standard backpropagation to optimize Eq. (6). From an efficiency standpoint, the mSPD-NN architecture maps onto a relatively shallow neural network. Therefore, this module can be easily integrated into other deep learning inference frameworks for example, for batch normalization on the SPD manifold. This flexibility is the key advantage over classical methods, in which integrating the geodesic mean estimation within a larger framework is not straightforward. Finally, the extension of Eq. (6) to the estimation of a weighted mean (with positive weights $\{w_n\}$) also follows naturally as a multiplier in the summation.

Implementation Details: We train mSPD-NN for a maximum of 100 epochs with an initial learning rate of 0.001 decayed by 0.8 every 50 epochs. The tolerance criteria for the training loss is set at $1e^{-4}$. mSPD-NN implemented in PyTorch (v1.5.1), Python 3.5 and experiments were run on an 4.9 GB Nvidia K80 GPU. We utilize the ADAM optimizer during training and a default PyTorch initialization for the model weights. To ensure that $\mathbf{W}$ is full rank, we add a small bias to the weights, i.e., $\mathbf{W} = \mathbf{W} + \lambda \mathcal{I}_P$ for regularization and stability.

3   Evaluation and Results

3.1   Experiments on Synthetic Data

We evaluate the scalability, robustness, and fidelity of mSPD-NN using simulated data. We compare the mSPD-NN against two popular mean estimation algorithms, the first being the Riemannian gradient descent [20] on the objective in Eq. (4) and the second being the **A**pproximate **J**oint **D**iagonalization **Log**

Euclidean (ALE) mean estimation [3], which directly leverages properties of the common principal components (CPC) data generating process [21].

Our synthetic experiments are built off the CPC model [13]. In this case, each input connectome $\Gamma_n \in \mathcal{R}^{P \times P}$ is derived from a set of components $\mathbf{B} \in \mathcal{R}^{P \times P}$ common to the collection and a set of example specific (and strictly positive) weights across the components $\mathbf{c}_n \in \mathcal{R}^{(+)P \times 1}$. Let the diagonal matrix $\mathbf{C}_n$ be defined as $\mathbf{C}_n = \mathbf{diag}(\mathbf{c}_n) \in \mathcal{R}^{(+)P \times P}$. From here, we have $\Gamma_n = \mathbf{B}\mathbf{C}_n\mathbf{B}^T$.

Evaluating Scalability: In the absence of corrupting noise, the theoretically optimal geodesic mean of the examples $\{\Gamma_n\}_{n=1}^N$ can be computed as: $\mathbf{G}_R^* = \mathbf{B}\ \mathbf{expm}\left[\frac{1}{N}\sum_{n=1}^N \mathbf{logm}(\mathbf{B}^{-1}\Gamma_n\mathbf{B}^{-T})\right]\mathbf{B}^T$ [3]. We evaluate the scalability of each algorithm with respect to the dataset dimensionality P and the number of examples N by comparing its output to this theoretical optimum.

We randomly sample columns of the component matrix $\mathbf{B}$ from a standard normal, i.e., $\mathbf{B}[:,j] \sim \mathcal{N}(\mathbf{0}, \mathcal{I}_P)\ \forall\ j \in \{1, \ldots, P\}$, where $\mathcal{I}_P$ is an identity matrix of dimension P. In parallel, we sample the component weights $\mathbf{c}_{nk}$ according to $\mathbf{c}_{nk}^{1/2} \sim \mathcal{N}(0, 1)\ \forall\ k \in \{1, \ldots, P\}$. To avoid degenerate behavior when the inputs are not full-rank, we clip $\mathbf{c}_{nk}$ to a minimum value of 0.001. We consider two experimental scenarios. In **Experiment 1**, we fix the data dimensionality at $P = 30$ and sweep the dataset size as $N \in \{5, 10, 20, 50, 100, 200\}$. In **Experiment 2**, we fix the dataset size at $N = 20$ and sweep the dimensionality as $P \in \{5, 10, 20, 50, 100, 200\}$. For each parameter setting, we run all three estimation algorithms ten times using different random initializations.

We score performance based on the correctness of the solution and the execution time in seconds. Correctness is measured in two ways. First is the final condition fit $\mathcal{L}(\mathbf{G}_R^{\text{est}})$ from Eq. (6), which quantifies the deviation of the solution from the first order stationary condition (i.e., $\mathcal{L}(\mathbf{G}_R^{\text{est}}) = 0$). Second is the normalized squared Riemannian distance $d_{\text{mean}} = d_R^2(\mathbf{G}_R^{\text{est}}, \mathbf{G}_R^*)/\|\mathbf{G}_R^*\|_R^2$ between the solution and the theoretically optimal mean. Lower values of the condition fit $\mathcal{L}(\mathbf{G}_R)$ and deviation d_{mean} imply a better quality solution.

Figure 2 illustrates the performances of mSPD-NN, gradient descent and ALE mean estimation algorithms. Figures 2(a) and (d) plot the first-order condition fit $\mathcal{L}(\mathbf{G}_R^{\text{est}})$ when varying the dataset size N (Experiment 1) and the matrix dimensionality P (Experiment 2), respectively. Likewise, Figs. 2(b) and (e) plot the recovery performance for each experiment. We observe that the first order condition fit for the mSPD-NN is better than the ALE for all settings, and better than the gradient descent for most settings. We note that the recovery performance of mSPD-NN is better than the baselines in most cases while being a close approximation in the remaining ones. Finally, Figs. 2(c) and (f) illustrate the time to convergence for each algorithm. As seen, the performance of mSPD-NN scales with dataset size but is nearly constant with respect to dimensionality. In all cases, it either beats or is competitive with ALE.

Robustness to Noise: Going one step further, we evaluate the efficacy of the mSPD-NN framework when there is deviation from the ideal CPC generating process. In this case, we add rank-one structured noise to obtain the input data:

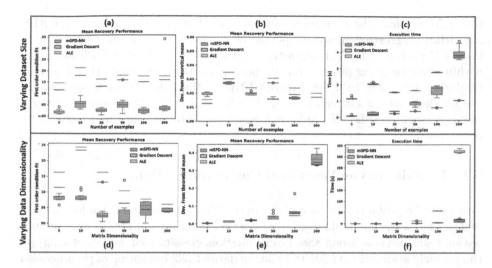

Fig. 2. Evaluating the estimates from mSPD-NN, gradient descent and ALE according to **(a)** *and* **(d)** first-order condition fit (Eq. 6) **(b)** *and* **(e)** deviation from the theoretical solution **(c)** *and* **(f)** execution time for **varying dataset size** N *and* **data dimension** P respectively.

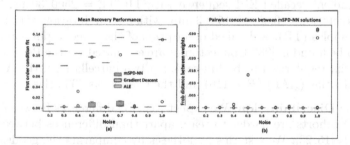

Fig. 3. Performance of the mSPD-NN, gradient descent and ALE estimation under increasing additive noise: **(a)** First order condition fit (Eq. 6) **(b)** Pairwise distance between the recovered mSPD-NN solutions across random initializations.

$\Gamma_n = \mathbf{B}\mathbf{C}_n\mathbf{B}^T + \frac{1}{P}\mathbf{x}_n\mathbf{x}_n^T$. As before, the bases and coefficients are randomly sampled as $\mathbf{B}[:,j] \sim \mathcal{N}(\mathbf{0}, \mathcal{I}_P)$ and $c_{nj}^{1/2} \sim \mathcal{N}(0,1) \ \forall \ j \in \{1, \ldots, P\}$. In a similar vein, the structured noise is generated as $\mathbf{x}_n \sim \mathcal{N}(\mathbf{0}, \sigma^2\mathcal{I}_P) \in \mathcal{R}^{P\times 1}$, with σ^2 controlling the extent of the deviation. For this experiment, we set $P = 30, N = 20$ and vary the noise over the range $[0.2 - 1]$ in increments of 0.1. One caveat in this setup is that the theoretically optimal mean defined previously and cannot be used to evaluate performance. Hence, we report only the first-order condition fit $\mathcal{L}(\mathbf{G}_R)$ We also calculate the pairwise concordance d_{weights} of the final mSPD-NN weights for different initializations.

Figure 3(a) illustrates the first-order condition fit $\mathcal{L}(\mathbf{G}_R^{est})$ across all three methods for increasing noise σ. As seen, $\mathcal{L}(\mathbf{G}_R^{est})$ for the mSPD-NN is consistently lower than the corresponding value for the gradient descent and ALE algorithm, suggesting improved performance despite increasing corruption to the CPC process. The ALE algorithm is designed to utilize the CPC structure within the generating process, but its poor performance suggests that it is particularly susceptible to noise. Figure 3(b) plots the pairwise distances between the geodesic means estimated by mSPD-NN across the 10 random initializations. As seen, mSPD-NN produces a consistent solution, thus underscoring its robustness.

3.2   Experiments on Functional Connectomics Data

Dataset: To probe the efficacy of the mSPD-NN for representation learning on real world matrix manifold data, we experiment on several groupwise discrimination tasks (such as group-wise discrimination, classification and clustering) on the publicly available CNI 2019 Challenge dataset [23] consisting of preprocessed rs-fMRI time series, provided for 158 subjects diagnosed with Attention Deficit Hyperactivity Disorder (ADHD), 92 subjects with Autism Spectrum Disorder (ASD) with an ADHD comorbidity [15], and 257 healthy controls. The scans were acquired on a Phillips 3T Achieva scanner using a single shot, partially parallel, gradient-recalled EPI sequence with TR/TE = 2500/30 ms, flip angle 70, voxel resolution = $3.05 \times 3.15 \times 3$ mm, with a scan duration of either 128 or 156 time samples (TR). A detailed description of the demographics and preprocessing can be found in [23]. Connectomes are estimated via the Pearson's correlation matrix, regularized to be full-rank via two parcellations, the Automated Anatomical Atlas (AAL) ($P = 116$) and the Craddocks 200 atlas ($P = 200$).

Groupwise Discrimination: We expect that FC differences between the ASD and ADHD cohorts are harder to tease apart than differences between ADHD and controls [15,23]. We test this hypothesis by comparing the geodesic means estimated via mSPD-NN for the three cohorts. For robustness, we perform bootstrapped trials for mean estimation by sampling 25 random subjects from a given group (ADHD/ASD/Controls). We then compute the Riemannian distance $d(\mathbf{G}_R(\{\boldsymbol{\Gamma}_{g1}\}), \mathbf{G}_R(\{\boldsymbol{\Gamma}_{g2}\}))$ between the mSPD-NN means associated with groups $g1$ and $g2$. A higher value of $d(\cdot, \cdot)$ implies a better separation between the groups. We also run a Wilcoxon signed rank test on the distribution of $d(\cdot, \cdot)$.

Figure 4 illustrates the pairwise distances between the geodesic means of cohorts $g1 - g2$ across bootstrapped trials (t-SNE representations for the group means are provided in Fig. 5(c)). As a sanity check, we note that the mean estimates across samples within the same cohort (ADHD-ADHD) are closer than those across cohorts (ADHD-controls, ASD-controls, ADHD-ASD). More interestingly, we observe that ADHD-controls separation is consistently larger than that of the ADHD-ASD groups for both parcellations. This result confirms the hypothesis that the overlapping diagnosis for the two classes translates to a reduced separability in the space of FC matrices and indicates that mSPD-NN is able to robustly uncover population level differences in FC.

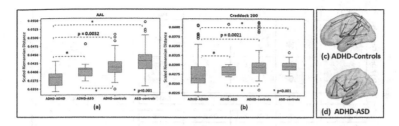

Fig. 4. Groupwise discrimination between the FC matrices estimated via the **(a)** AAL **(b)** Craddock's 200 atlas, for the ADHD/ASD/Controls cohorts according to pairwise distances between the mSPD-NN mean estimates. Results of pairwise connectivity comparisons between group means for **(c)** ADHD-Controls **(d)** ADHD-ASD groups for the AAL parcellation. The red connections are significant differences ($p < 0.001$). (Color figure online)

Classification: Building on the observation that mSPD-NN provides reliable group-separability, we adopt this framework for classification. Using the AAL parcellation, we randomly sample 25 subjects from each class for training, and set aside the rest for evaluation with a 10%/90% validation/test split. We estimate the geodesic mean for each group across the training samples via 10 bootstrapped trials, in which we sub-sample 80% of the training subjects from the respective group. Permutation testing is performed on the mean estimates [24], and functional connections (i.e., entries of $\mathbf{G}_R(\{\boldsymbol{\Gamma}_n\})$) that differ with an FDR-corrected threshold of $p < 0.001$ are retained for classification. Finally, a Random Forest classifier is trained on the selected features to classify ADHD vs Controls. The train-validation-test splits are repeated 10 times to compute confidence intervals.

We use classification accuracy and area under the receiver operating curve (AU-ROC) as metrics for evaluation. The mSPD-NN feature selection plus Random Forest approach provides an accuracy of 0.62 ± 0.031 and an AU-ROC of 0.60 ± 0.04 for ADHD-Control classification on the test samples. We note that this approach outperforms all but one method on the CNI challenge leaderboard [23]. Moreover, one focus of the challenge is to observe how models trained on the ADHD vs Control discrimination task translate to ASD (with ADHD comorbidity) vs Control discrimination in a transfer learning setup. Accordingly, we apply the learned classifiers in each split to ASD vs Control classification and obtain an accuracy of 0.54 ± 0.044 and an AU-ROC of 0.53 ± 0.03. This result is on par with the best performing algorithm in the CNI-TL challenge. The drop in accuracy and AU-ROC for the transfer learning task is consistent with the performance profile of all the challenge submissions. These results suggest that despite the comorbidity, connectivity differences between the cohorts are subtle and hard to reliably capture. Nonetheless, the mSPD-NN+RF framework is a first step to underscoring stable, yet interpretable (see below) connectivity patterns that can discriminate between diseased and healthy populations.

Qualitative Analysis: To better understand the group-level connectivity differences, we plot the most consistently selected features (top 10%) from the previous experiment (ADHD-control feature selection) in Fig. 4(c). We utilize

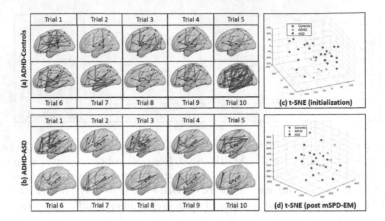

Fig. 5. Pairwise differences between mSPD-NN group means for (a) ADHD-Controls (b) ADHD-ASD groups across bootstrapped trials. Significant differences marked in red ($p < 0.001$). t-SNE plots for group means from experiment on (c) Groupwise Discrimination using mSPD-NN (d) After data-driven clustering via the mSPD-EM. (Color figure online)

the BrainNetViewer Software for visualization. The blue circles are the AAL nodes, while the solid lines denote edges between nodes. We observe that the highlighted connections appear to cluster in the sensorimotor and visual areas of the brain, along with a few temporal lobe contributions. Altered sensorimotor and visual functioning has been previously reported among children and young adults diagnosed with ADHD [6]. Adopting a similar procedure, we additionally highlight differences among the ASD and ADHD cohorts in Fig. 4(d). The selected connections concentrate around the pre-frontal areas of the brain, which is believed to be associated with altered social-emotional regulation in Autism [22]. We additionally provide an extended version of the group connectivity difference results across trials in Fig. 5(a) ADHD vs Controls and (b) ADHD vs ASD. Across train-test-validation splits, we observe that several connectivity differences appear fairly consistently. Overall, the patterns highlighted via statistical comparisons on the mSPD-NN estimates are both robust as well as in line with the physiopathology of ADHD and ASD reported in the literature.

Data-Driven Clustering: Finally, we evaluate the stability of the mapping between the functional connectivity and diagnostic spaces via a geometric clustering experiment. We use the geodesic mean estimates from the groupwise discrimination experiment (generated using the ground truth Controls/ASD/ADHD labels and mSPD-NN) as an initialization and track the shift in the diagnostic assignments upon running an unsupervised Expectation-Maximization (EM) algorithm. At each iteration of the mSPD-EM, the E-Step assigns cluster memberships to a given subject according to the geodesic distance (Eq. (3)) from the cluster centroids, while the M-Step uses the mSPD-NN for recomputing the centroids. Upon convergence, we evaluate the alignment between the inferred clusters and diagnostic labels. To this end, we map each cluster to a diagnostic label according to majority voting, and measure the cluster purity (fraction of cluster members that are correctly assigned). mSPD-EM provides an overall cluster purity of 0.59 ± 0.05 (Controls), 0.52 ± 0.12 (ADHD), ASD 0.51 ± 0.09 (ASD), indicating that there is considerable shift in the assignment of diagnostic labels from ground truth. We also visualise the cluster centroids using t-Stochastic Neighbor Embeddings (t-SNE) at initialization and after convergence of the mSPD-EM in Fig. 5(c) and (d) respectively. We provide 3-D plots to better visualise the cluster separation. Again, we observe that the diagnostic groups overlap considerably and are challenging to separate in the functional connectivity space alone. One possible explanation may be that the distinct neural phenotypes between the disorders are being overwhelemed by other rs-fMRI signatures. Given the migration of diagnostic assignments from the ground truth, the strict diagnostic criteria used to separate the diseased and healthy cohorts group may need to be more critically examined.

4 Conclusion

We have proposed a novel mSPD-NN framework to reliably estimate the geodesic mean of a collection of functional connectivity matrices. Through extensive simulation studies, we demonstrate that the mSPD-NN scales well to high-dimensional data and can handle input noise when compared with current iterative methods. By conducting a series of experiments on group-wise discrimination, feature selection, classification, and clustering, we demonstrate that the mSPD-NN is a reliable framework for discovering consistent group differences between patients diagnosed with ADHD-Autism comorbidities and controls. The mSPD-NN makes minimal assumptions about the data and can potentially be a useful tool to advance data-scientific and clinical research.

Acknowledgements. This work is supported by the National Science Foundation CAREER award 1845430 (PI Venkataraman), the National Institute of Health R01HD108790 (PI Venkataraman) and R01EB029977 (PI Caffo).

References

1. Banerjee, M., Chakraborty, R., Ofori, E., Vaillancourt, D., Vemuri, B.C.: Nonlinear regression on Riemannian manifolds and its applications to neuro-image analysis. In: Navab, N., Hornegger, J., Wells, W.M., Frangi, A.F. (eds.) MICCAI 2015. LNCS, vol. 9349, pp. 719–727. Springer, Cham (2015). https://doi.org/10.1007/978-3-319-24553-9_88
2. Bessadok, A., Mahjoub, M.A., Rekik, I.: Graph neural networks in network neuroscience. IEEE Trans. Pattern Anal. Mach. Intell. **45**, 5833–5848 (2022)
3. Congedo, M., Afsari, B., Barachant, A., Moakher, M.: Approximate joint diagonalization and geometric mean of symmetric positive definite matrices. PLoS ONE **10**(4), e0121423 (2015)

4. Congedo, M., Barachant, A., Koopaei, E.K.: Fixed point algorithms for estimating power means of positive definite matrices. IEEE Trans. Signal Process. **65**(9), 2211–2220 (2017)
5. Dong, Z., et al.: Deep manifold learning of symmetric positive definite matrices with application to face recognition. In: Thirty-First AAAI Conference on Artificial Intelligence (2017)
6. Duerden, E.G., Tannock, R., Dockstader, C.: Altered cortical morphology in sensorimotor processing regions in adolescents and adults with attention-deficit/hyperactivity disorder. Brain Res. **1445**, 82–91 (2012)
7. D'Souza, N.S., Nebel, M.B., Wymbs, N., Mostofsky, S., Venkataraman, A.: Integrating neural networks and dictionary learning for multidimensional clinical characterizations from functional connectomics data. In: Shen, D., et al. (eds.) MICCAI 2019. LNCS, vol. 11766, pp. 709–717. Springer, Cham (2019). https://doi.org/10.1007/978-3-030-32248-9_79
8. D'Souza, N.S., et al.: A joint network optimization framework to predict clinical severity from resting state functional MRI data. Neuroimage **206**, 116314 (2020)
9. D'Souza, N.S., Nebel, M.B., Crocetti, D., Robinson, J., Mostofsky, S., Venkataraman, A.: A matrix autoencoder framework to align the functional and structural connectivity manifolds as guided by behavioral phenotypes. In: de Bruijne, M., et al. (eds.) MICCAI 2021. LNCS, vol. 12907, pp. 625–636. Springer, Cham (2021). https://doi.org/10.1007/978-3-030-87234-2_59
10. Fornito, A., Zalesky, A., Breakspear, M.: Graph analysis of the human connectome: promise, progress, and pitfalls. Neuroimage **80**, 426–444 (2013)
11. Fox, M.D., et al.: Spontaneous fluctuations in brain activity observed with functional magnetic resonance imaging. Nat. Rev. Neuro. **8**(9), 700 (2007)
12. Jeuris, B.: Riemannian optimization for averaging positive definite matrices (2015)
13. Jolliffe, I.T., Cadima, J.: Principal component analysis: a review and recent developments. Philos. Trans. R. Soc. A: Math. Phys. Eng. Sci. **374**(2065), 20150202 (2016)
14. Khosla, M., et al.: Machine learning in resting-state fMRI analysis. Magn. Reson. Imaging **64**, 101–121 (2019)
15. Leitner, Y.: The co-occurrence of autism and attention deficit hyperactivity disorder in children-what do we know? Front. Hum. Neurosci. **8**, 268 (2014)
16. Lindquist, M.A.: The statistical analysis of fMRI data. Stat. Sci. **23**(4), 439–464 (2008)
17. Moakher, M.: A differential geometric approach to the geometric mean of symmetric positive-definite matrices. SIAM J. Matrix Anal. Appl. **26**(3), 735–747 (2005)
18. Nandakumar, N., et al.: A multi-task deep learning framework to localize the eloquent cortex in brain tumor patients using dynamic functional connectivity. In: Kia, S.M., et al. (eds.) MLCN/RNO-AI -2020. LNCS, vol. 12449, pp. 34–44. Springer, Cham (2020). https://doi.org/10.1007/978-3-030-66843-3_4
19. Nguyen, X.S., et al.: A neural network based on SPD manifold learning for skeleton-based hand gesture recognition. In: Proceedings of the IEEE/CVF Conference on Computer Vision and Pattern Recognition, pp. 12036–12045 (2019)
20. Pennec, X., Fillard, P., Ayache, N.: A Riemannian framework for tensor computing. Int. J. Comput. Vision **66**(1), 41–66 (2006)
21. Pham, D.T.: Joint approximate diagonalization of positive definite Hermitian matrices. SIAM J. Matrix Anal. Appl. **22**(4), 1136–1152 (2001)
22. Pouw, L.B., et al.: The link between emotion regulation, social functioning, and depression in boys with ASD. Res. Autism Spectr. Disord. **7**(4), 549–556 (2013)

23. Schirmer, M.D., et al.: Neuropsychiatric disease classification using functional connectomics-results of the connectomics in neuroimaging transfer learning challenge. Med. Image Anal. **70**, 101972 (2021)
24. Zalesky, A., Fornito, A., Bullmore, E.T.: Network-based statistic: identifying differences in brain networks. Neuroimage **53**(4), 1197–1207 (2010)
25. Zhang, T.: A majorization-minimization algorithm for the karcher mean of positive definite matrices. arXiv preprint arXiv:1312.4654 (2013)

Computer-Aided Diagnosis/Surgery

Diffusion Model Based Semi-supervised Learning on Brain Hemorrhage Images for Efficient Midline Shift Quantification

Shizhan Gong[1], Cheng Chen[1], Yuqi Gong[1], Nga Yan Chan[2],
Wenao Ma[1], Calvin Hoi-Kwan Mak[3], Jill Abrigo[2], and Qi Dou[1($\boxtimes$)]

[1] Department of Computer Science and Engineering,
The Chinese University of Hong Kong, Hong Kong, China
`qidou@cuhk.edu.hk`
[2] Department of Imaging and Interventional Radiology,
The Chinese University of Hong Kong, Hong Kong, China
[3] Queen Elizabeth Hospital, Hong Kong, China

Abstract. Brain midline shift (MLS) is one of the most critical factors
to be considered for clinical diagnosis and treatment decision-making for
intracranial hemorrhage. Existing computational methods on MLS quantification not only require intensive labeling in millimeter-level measurement but also suffer from poor performance due to their dependence on
specific landmarks or simplified anatomical assumptions. In this paper,
we propose a novel semi-supervised framework to accurately measure the
scale of MLS from head CT scans. We formulate the MLS measurement
task as a deformation estimation problem and solve it using a few MLS
slices with sparse labels. Meanwhile, with the help of diffusion models, we
are able to use a great number of unlabeled MLS data and 2793 non-MLS
cases for representation learning and regularization. The extracted representation reflects how the image is different from a non-MLS image and
regularization serves an important role in the sparse-to-dense refinement
of the deformation field. Our experiment on a real clinical brain hemorrhage dataset has achieved state-of-the-art performance and can generate
interpretable deformation fields. Our code is available at: https://github.
com/med-air/DiffusionMLS.

Keywords: Computer-aided diagnosis · Semi-supervised learning ·
Diffusion models · Intracranial hemorrhage

1 Introduction

Intracranial hemorrhage (ICH) refers to brain bleeding within the skull, a serious medical emergency that would cause severe disability or even death [1].
A characteristic symptom of severe ICH is brain midline shift (MLS), which
is the lateral displacement of midline cerebral structures (see Fig. 1). MLS is
an important and quantifiable indicator of the severity of mass effects and the

© The Author(s), under exclusive license to Springer Nature Switzerland AG 2023
A. Frangi et al. (Eds.): IPMI 2023, LNCS 13939, pp. 69–81, 2023.
https://doi.org/10.1007/978-3-031-34048-2_6

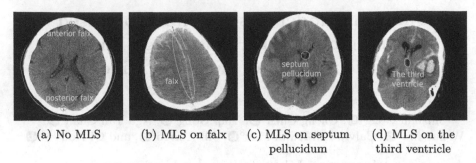

(a) No MLS	(b) MLS on falx	(c) MLS on septum pellucidum	(d) MLS on the third ventricle

Fig. 1. Examples of head CT scans to illustrate how radiologists measure MLS. Dash red line connecting the anterior falx and posterior falx denote a hypothetical normal midline. Blue circles denote the shifted landmarks. Perpendicular red lines from the shifted landmarks to normal midline are measured as MLS scale. (Color figure online)

urgency of intervention [2,3,9]. For instance, the 5 mm (mm) threshold of MLS is frequently used for immediate intervention and close monitoring [4]. MLS quantification demands high accuracy and efficiency, which is difficult to achieve with manual quantification, especially in emergencies, due to the variability in shift regions, unclear landmark boundaries, and non-standard scanning pose. An automated MLS quantification algorithm that can immediately and accurately quantify MLS is highly desirable to identify urgent patients for timely treatment.

To measure MLS, clinicians usually first identify a few CT slices with large shifts and then measure and identify the maximum deviation of landmarks such as the septum pellucidum, third ventricle, or falx from their normal counterpart as the final MLS distance (see examples in Fig. 1). Such a clinical fashion of MLS quantification can be difficult to be translated into a well-defined automation process. Currently, there are only limited studies on automated MLS quantification, using different strategies and varied labeling requirements. Nguyen et al. proposed a landmark-based method that relies on anatomical markers to determine the location of the deformed midline [9]. However, this method can only apply to cases where MLS appears on these specific marker regions. Liao et al. adopted a symmetric-based method to seek a curve connecting all deformed structures [10], which is difficult to generalize due to over-simplified anatomical assumptions and sensitivity to patients' scan poses. A few recent works try to overcome these limitations by using stronger supervision with dense labeling. Some studies formulated MLS quantification as a midline segmentation task [5–7], by delineating the intact midline as labels to supervise the training of segmentation models. Another study designed a hemisphere segmentation task to quantify MLS [8], which requires pixel-wise annotation for each slice. However, obtaining such dense annotations is very costly and time-consuming, while may not be necessary for MLS quantification.

To tackle these limitations, we propose to fit MLS quantification into a deformation prediction problem by using semi-supervised learning (SSL) with only limited annotations. Our framework avoids the strong dependency on specific landmarks or over-simplified assumptions in previous methods while not increasing the labeling efforts. We aim to use only sparse and weak labels as ground truth supervisions, which are just one shifted landmark and its normal counterpart on a limited number of slices provided by radiologists, but we try to fully exploit the unlabeled slices and non-MLS data to impose extra regularization for the sparse-to-dense extension. Existing SSL methods typically use a partially trained model with labeled data to generate pseudo labels for unlabeled data, assuming that labeled and unlabeled data are generally similar. These methods can be sub-optimal in our case as labeled slices of MLS usually present the largest deformation while unlabeled slices contain only minor or no deformation. Instead, we propose our SSL strategy by generating a corresponding non-MLS image for each unlabeled MLS slice with generative models and regularizing that the deformation field should warp the generated non-MLS images into the original MLS ones. However, as we only have volume-wise labels for MLS and non-MLS classification, it can be difficult to train a slice-wise discriminator as required by many generative models such as GANs [12]. Fortunately, the recently proposed diffusion models [15], which prove to have strong power in both distribution learning and image generation without dependency on discriminators, can be a potentially good solution.

In this work, we propose a novel semi-supervised learning framework based on diffusion models to quantify the brain MLS from head CT images with deformation prediction. Our method effectively exploits supervision and regularization from all types of available data including MLS images with sparse ground truth labels, MLS images without labels, and non-MLS images. We validate our method on a real clinical head CT dataset, showing effectiveness of each proposed component. Our contributions include: (1) innovating an effective deformation strategy for brain MLS quantification, (2) incorporating diffusion models as a representation learner to extract features reflecting where and how an MLS image differs from a non-MLS image, and (3) proposing a diffusion model-based semi-supervised framework that can effectively leverage massive unlabelled data to improve the model performance.

2   Methods

Figure 2 illustrates our diffusion model-based semi-supervised learning framework for MLS quantification via deformation prediction. In Sect. 2.1, we introduce our deformation strategy with only sparse supervision. In Sect. 2.2, we propose to incorporate non-MLS data for representation learning. In Sect. 2.3, we describe how to utilize unlabeled MLS images for sparse-to-dense regularization.

2.1   MLS Quantification Through Deformation Estimation

Our proposed deformation strategy for brain MLS quantification aims to find an optimal deformation field ϕ so that an MLS image can be regarded as a hypothetically non-MLS image warped with this deformation field. The deformation

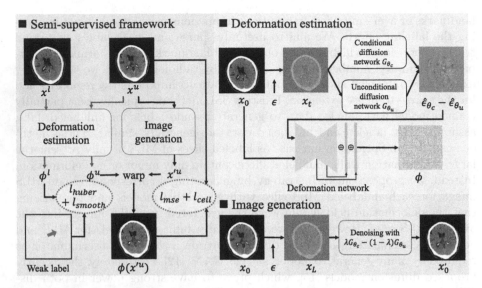

Fig. 2. The pipeline of our proposed semi-supervised deformation strategy for MLS quantification. Sparse labels supervise the labeled image x^l and the unlabeled image x^u is self-supervised with generated negative image x'^u.

field can be parameterized by a function with high complexity so that it does not explicitly rely on a single landmark or over-simplified symmetric assumptions, which naturally overcomes the limitations of existing methods. We apply a learning-based framework to parameterize the deformation field with a U shape neural network. The input to the network is individual 2D slices and the network's output is the stationary velocity field v. The diffeomorphic deformation field ϕ is then calculated through the integration of the velocity field, similarly to VoxelMorph [11] for image registration. The learning process is supervised by sparse deformation ground truth. For each labeled slice, we have the ground truth $\mathbf{y} = (y_1, y_2)$, which is a two-dimensional vector directing from shifted landmark point toward its presumably normal location (the red arrow in Fig. 2). The predicted deformation $\hat{\mathbf{y}}$ is bilinearly interpolated at the shifted landmark point from the deformation field, which is also a two-dimensional vector. To alleviate the influence of a few extremely large deformation points and increase model's robustness, we use Huber loss to measure the similarity between the predicted deformation and the label:

$$l_{\text{huber}}(y_d, \hat{y}_d) = \begin{cases} |y_d - \hat{y}_d|, & |y_d - \hat{y}_d| \geq c, \\ \dfrac{(y_d - \hat{y}_d)^2 + c^2}{2c}, & |y_d - \hat{y}_d| < c. \end{cases} \qquad (1)$$

where $d \in \{1, 2\}$. The hyperparameter c defines the range for absolute error or squared error. We also encourage a smooth deformation field with a diffusion regularizer on the spatial gradients of deformation ϕ to avoid a discontinuous deformation field:

$$l_{\text{smooth}} = \sum_j \sum_k \|\phi_{jk} - \phi_{(j-1)k}\|^2 + \|\phi_{jk} - \phi_{j(k-1)}\|^2, \qquad (2)$$

We apply a coarse-to-fine manner, where velocity fields are generated through upsampling with skip connection to progressively aggregate features of different scales, making the model more adaptive to extremely large deformation.

2.2 Learning Negative Patterns from Non-MLS Images

In order to learn a deformation field to warp a non-MLS image into MLS one, ideally we would need a pair of non-MLS and MLS images for network training, which however does not exist in practice. A naive substitution is to generate a corresponding non-MLS image. However, generated images entail some randomness and often lack important details. Depending too much on such fake inputs can lead to poor robustness. Inspired by the score-matching interpretation of diffusion models [17], we propose to learn the non-MLS distribution from massive amount of negative cases. Given an MLS image, we can evaluate which parts of the image make it different from a non-MLS image. This deviation can serve as latent features that help the deformation network with deformation prediction.

Diffusion models, especially DDPM [14], define a forward diffusion process as the Markov process progressively adding random Gaussian noise to a given image and then trying to approximate the reverse process by a Gaussian distribution. The forward process can be simplified by a one-step sampling: $x_t = \sqrt{\alpha_t}x_0 + \sqrt{1 - \alpha_t}\epsilon$, where $\alpha_t := \prod_{s=0}^{t} 1 - \beta_t$, and β_t are predefined variance schedule. ϵ is sampled from $\mathcal{N}(0, I)$. The mean $\mu_\theta(x_t, t)$ and variance $\Sigma_\theta(x_t, t)$ of the reverse process can be parameterized by neural networks. A popular choice is to re-parameterize $\mu_\theta(x_t, t)$ so that $\hat{\epsilon}_\theta(x_t, t)$ instead of $\mu_\theta(x_t, t)$ is estimated by neural networks to approximate the noise ϵ. Moreover, the output of the diffusion network $\epsilon(x_t, t)$ is actually a scaled score function $\nabla \log p(x_t)$ as it moves the corrupted image towards the opposite direction of the corruption [18].

As a result, through pre-training one unconditional diffusion model trained with all data (denoted as $\mathcal{U}$) and one conditional diffusion model trained with only non-MLS data (denoted as $\mathcal{C}$), the subtraction of two outputs

$$\hat{\epsilon}_{\theta_{\mathcal{U}}}(x_t, t) - \hat{\epsilon}_{\theta_{\mathcal{C}}}(x_t, t) \propto \nabla \log p(x_t|n) - \nabla \log p(x_t) = \nabla \log p(n|x_t), \qquad (3)$$

can be regarded as the gradient of class prediction ($n = 1$ for non-MLS and 0 otherwise) w.r.t to the input image, which reflects how the input images deviate from a non-MLS image. This latent contains information regarding how to transform the MLS positive image into a non-MLS one and therefore is helpful for training the deformation network. Moreover, this feature representation exhibits less fluctuation toward the randomness of the additive noise as both terms are somehow estimations of the stochastic noise, which are then eliminated through subtraction. It is more stable than the predicted noise or generated MSL negative images. For training, we randomly sample t from 0 to the diffusion steps T_{train}, while for inference we fix it to be a certain value. We examine the effects of this value in Sect. 3.4.

2.3   Semi-supervised Deformation Regularization

Deformation estimation is a dense prediction problem, while we only have sparse supervision. This can lead to flickering and poor generalizability if the deformation lacks certain regularization. On the other hand, we have a significant amount of unlabeled data from the MLS volumes that is potentially helpful. Therefore, we propose to include these unlabeled data during training in a semi-supervised manner, so that unlabeled data can provide extra regularization for training or produce additional training examples based on noisy pseudo labels. Many existing semi-supervised methods seek to use the prediction for unlabeled data given by the same or a twin network as pseudo-labels and then supervise the model or impose some regularization with these pseudo-labels. However, these methods hold a strong assumption that labeled and unlabeled data are drawn from the same distribution, which is not true in our case because most labeled data are with large deformation while unlabeled data are with minor or no deformation. Therefore, we want to find another type of pseudo-label to bypass the distribution assumption. As the deformation field is assumed to warp a hypothetically normal image into an MLS one, we generate hypothetically non-MLS images x'_0 using pre-trained diffusion models through a series of denoising steps with classifier-free guidance [16]:

$$\hat{\epsilon}(x_t, t) = \lambda \hat{\epsilon}_{\theta_C}(x_t, t) + (1 - \lambda)\hat{\epsilon}_{\theta_U}(x_t, t), \tag{4}$$

where λ is a hyper-parameter controlling the strength of the guidance. We compare x'_0 warped with the deformation field $\phi(x'_0)$ and calculate its similarity with the original x_0 through MSE loss. As it can be difficult for the generated image to be fully faithful to the original image because the generative process entails a lot of random sampling, this l_{mse} can only serve as noisy supervision. Therefore, instead of generating x'_0 ahead of deformation network training, we generate it in an ad-hoc way (i.e. generating new cases at each iteration) so that the noisy effects can be counteracted.

The final MLS measurement is estimated by calculating the length of the maximum displacement vector from the predicted deformation field, so it is more sensitive to over-estimation. As for unlabelled slices, we still have the prior that its MLS cannot be larger than the MLS of that specific volume δ, we propose to incorporate an additional ceiling loss to punish the over-estimation:

$$l_{\text{ceil}} = \sum_j \sum_k \max(0, ||\phi_{jk}|| - \delta). \tag{5}$$

Overall, the loss term is a combination of supervised loss and unsupervised loss, with a weight term controlling the relative importance of each loss term:

$$l = l_{\text{huber}} + w_1 l_{\text{smooth}} + u(i)(l_{\text{mse}} + w_2 l_{\text{ceil}}), \tag{6}$$

where w_1 and w_2 are two fixed weight terms and $u(i)$ is a time-varying weight term that is expected to gradually increase as the training iteration i progresses so that the training can converge quickly through strong supervision first and then refine and enhance generalizability via unsupervised loss.

3   Experiments and Results

3.1   Data Acquisition and Preprocessing

We retrospectively collected anonymous thick-slice, non-contrast head CT of patients who were admitted with head trauma or stroke symptoms and diagnosed with various subtypes of intracranial hemorrhage, including epidural hemorrhage, subdural hemorrhage, subarachnoid hemorrhage, intraventricular hemorrhage, and intraparenchymal hemorrhage, between July 2019 and December 2019 in the Prince of Wales Hospital, a public hospital under the Hospital Authority of Hong Kong. The ethics approval was obtained from the Joint Chinese University of Hong Kong-New Territories East Cluster Clinical Research ethics committee. The eligible patients comprised 2793 CT volumes, among them 124 are MLS positive cases. The MLS ranges between 2.24 mm and 20.12 mm, with mean value of 8.34 mm and medium value of 8.73 mm. The annotation was performed by two trained physicians and verified by one experienced radiologist (with over 10 years of clinical experience on ICH). The labeling process followed a real clinical measurement pipeline, where the shifted landmark, anterior falx point, and posterior falx point were pointed out, and the length of the vertical line from the landmark to the line connecting the anterior falx point and the posterior falx point was the measured MLS value. For each volume, a few slices with large deformation were separately measured and annotated while the shift of the largest one served as the case-level label. On average, 4 out of 30 slices of each volume were labeled. All slices of non-MLS cases are unlabeled. We discarded the first 8 and the last 5 slices as they are mainly structures irrelevant to MLS. For pre-processing, we adjusted the pixel size of all images to 0.86 mm and then cropped or padded the resulting images to the resolution of 256 × 256 pixels. The HU window was set to 0 and 80. We applied intensity clipping (0.5 and 99.5 percentiles) and min-max normalization (between -1 and 1) to each image. Random rotation between $-15°$ and $15°$ was used for data augmentation.

3.2   Implementation Details

For the diffusion network, we use the network architecture designed in DDPM [15] and set the noise level from 10^{-4} to 2×10^{-2} by linearly scheduling with $T_{\text{train}} = 1000$. For non-MLS image generation, we apply the Denoising Diffusion Implicit Model (DDIM) [13] with 50 steps and set the noise scale to 15 to shorten the generative time. We set the hyper-parameters as $\alpha = 1$, $\beta = 1$, $c = 3$ and $\gamma = 2$. $u(i)$ is set from 1 to 10 with the linear schedule. The diffusion models are trained by the AdamW optimizer with an initial learning rate of 1×10^{-4},

Table 1. Comparison of different methods with 5-fold cross-validation.

Methods	Training data		Volume-wise		Slice-wise	
	Labeled	Unlabeled	MAE↓(mm)	RMSE↓(mm)	MAE↓(mm)	RMSE↓(mm)
Regression	✓		3.91 (2.52)	4.90	3.56 (1.91)	4.16
Deformation	✓		3.80 (2.35)	4.47	2.51 (1.84)	3.17
Mean-Teacher [19]	✓	✓	2.89 (2.26)	3.67	2.43 (1.78)	3.22
CPS [20]	✓	✓	2.72 (2.08)	3.42	2.38 (1.79)	3.15
Ours	✓	✓	**2.43** (2.02)	**3.17**	**2.25** (1.74)	**3.09**

batch size 4, for 2×10^5 iterations. We up-sample the MLS positive data by $10\times$ when training the unconditional diffusion model. The deformation network is trained by the AdamW optimizer with an initial learning rate of 1×10^{-4}, batch size 16, for 100 epochs. All models are implemented with PyTorch 1.12.1 using one Nvidia GeForce RTX 3090 GPU.

3.3   Quantification Accuracy and Deformation Quality

We evaluate the performance of our quantification strategy through mean absolute error (MAE) and root mean square error (RMSE). For volume-wise evaluation, we measure the maximum deformation of each slice of the whole volume and select the largest one as the final result. We also report the slice-wise evaluation based on labeled slices, which reflect how the models perform on slices with large deformation. Since existing MLS estimation methods require different types of labels from ours, it is difficult to directly compare with those methods. We therefore first compare our deformation-based strategy with a regression-based strategy, which uses DenseNet-121 [21] to directly predict the slice-wise MLS. We also compare our proposed semi-supervised learning approach with two popular semi-supervised learning methods: Mean-Teacher [19] and Cross Pseudo Supervision (CPS) [20], which are implemented into our deformation framework. The results are given in Table 1, which are based on 5-fold cross-validations.

From the results, we can see that when only using labeled MLS slices for model learning, our deformation strategy already shows better performance than the regression model. This may attribute to that our deformation model learns the knowledge of both MLS values and locations while a regression model only captures the MLS value information. This difference can be further enlarged if we consider slice-wise performance. Moreover, all three semi-supervised learning methods, i.e., Mean-Teacher, CPS, and ours, consistently improve the performance of deformation prediction, showing the benefits and importance of incorporating unlabeled data into model learning. Our semi-supervised learning method based on diffusion models achieves better quantification results than Mean-Teacher and CPS, significantly reducing the volume-wise MAE from 3.80 mm to 2.43 mm. An interesting observation is that the unlabeled data contribute more to the volume-wise evaluation than the slice-wise evaluation. By inspecting the prediction, we find that the deformation prediction trained

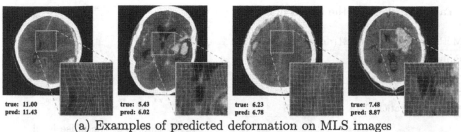

(a) Examples of predicted deformation on MLS images

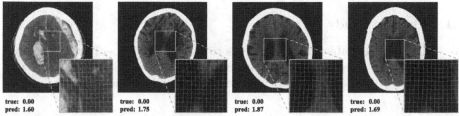

(b) Examples of predicted deformation on non-MLS images

Fig. 3. Predicted deformation on (a) MLS images. (b) non-MLS images. The regions with the largest deformation are highlighted. Slice-wise predicted MSL and ground truth are provided.

with labeled data tends to overestimate the deformation of slices with little or no deformation, which makes the volume-wise prediction error-prone. As most unlabeled data are slices with minor shifts, incorporating these data for semi-supervised learning can impose constraints to avoid large deformation, which greatly improves the model's robustness.

We also visualize the predicted deformation field of several sample cases. From Fig. 3(a), we can see the model can well posit the location where the maximum shift appears and push it to its hypothetically normal counterpart. The largest deformation happens exactly at the site with the maximum shift. To validate the robustness of our model, we also select several patients diagnosed with no MLS and plot the predicted deformation of these samples. As can be seen in Fig. 3(b), our method is able to provide a reasonable prediction for non-MLS images by outputting much smaller values than that for MLS images. Our model's predictions for non-MLS images are not exactly zero are caused on one hand by that even for a completely healthy person, the midline cannot be perfectly aligned due to multiple factors such as scan pose, on the other hand, our models tend to overestimate the shift because we are calculating the maximum deformation as final measurement.

Table 2. Effects of the representation.

Methods	MAE↓(mm)	RMSE↓(mm)
Fully-supervised	3.61	4.47
+ Representation	3.22	3.69
Semi-supervised	2.61	3.24
+ Representation	2.45	3.05

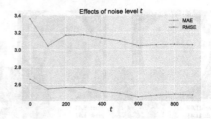

Fig. 4. Effects of the noise level.

3.4 Ablation Study

We conduct several ablation experiments to study the effects of several components in our proposed framework on the model performance. The volume-wise results reported are trained on four folders and tested on one folder.

Effects for Representation Learning. We first conduct ablation studies to verify that the latent feature extracted from the two diffusion models is truly useful for deformation prediction. To this end, we select two deformation models, one trained with only labeled data and the other using semi-supervised learning, and compare their performance with and without the extracted representation as input. The results are given in Table 2. As expected, incorporating the representation can improve the model performance in both cases.

The noise level is an important component of diffusion models. Only with a proper noise level, can the model accurately estimate the deviation of the image toward the negative sample space. Therefore, we do inference with multiple noise levels and compare its effect on model performance. The results are shown in Fig. 4. Our model is very robust towards this hyper-parameter. As long as t is not too small, the model gives very similar performances. The best performance appears in the middle when $t = 600$. This is reasonable as small noise fails to corrupt the original image thus degenerating the performance of score estimation while large noise may obscure too many details of the original image.

Quantity of Unlabeled Images. To verify the usefulness of unlabeled images, we conduct ablation studies on the number of unlabeled images used. For each experiment, we randomly sample 20%, 40%, 60%, and 80% volumes, and we incorporate unlabeled slices of these volumes for semi-supervised training. For the rest volumes, we are only using the labeled slices. We also do one experiment that completely removes the uses of unlabeled images. For each experiment, the pre-trained diffusion models are the same, which uses all the data. In other words, these unlabeled images somehow still contribute to the model training. The results are shown in Fig. 5(a). As can be seen, the model performance and robustness can be enhanced as we incorporate more unlabeled images. This provides strong evidence for our claim that our model truly learns valuable information from unlabeled data.

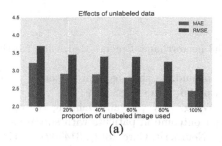
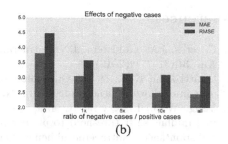

Fig. 5. Results of our ablation experiments in terms of: (a) proportion of unlabeled data used, and (b) proportion of negative data used.

Quantity of Non-MLS Images. To further measure the benefits of including non-MLS cases, we conduct another ablation study on the proportion of non-MLS cases. As currently, the amount of non-MLS cases is much higher than MLS cases, we upsample the MLS cases so that their quantities are approximately the same when training the unconditional diffusion model. For ablation, we first downsample the non-MLS data so that their quantity is $1\times$, $5\times$, and $10\times$ that of the MLS cases, and then upsample the MLS cases to make them balanced. From the results in Fig. 5(b), we find model performance improves with more non-MLS cases incorporated. Increasing non-MLS cases can help train diffusion models and further improve the quality of generated images and extracted feature representations. However, this effect will soon be saturated as the amount of MLS cases is relatively small. This can be a bottleneck for effectively using the non-MLS cases as it is challenging to train unconditional diffusion models with such imbalanced datasets.

4   Conclusions and Future Work

In this paper, we propose a novel framework based on deformation field estimation to automatically measure the brain MLS. The labels we are using are sparse which can greatly alleviate the labeling workload. We also propose a semi-supervised learning strategy based on diffusion models which significantly improves the model performance. Experiments on a clinic dataset show our methods can achieve satisfying performance. We also verify that using unlabeled data and non-MLS cases can truly help improve the model's performance. Our methods have several limitations. First, the model performance highly relies on pre-trained diffusion models. Training diffusion models with extremely imbalanced data requires great effort. Second, the measurement results exhibit randomness due to noise corruption. Finally, the measurement results are prone to overestimation. Our future work will figure out solutions for these limitations.

Acknowledgments. This work was supported by the Hong Kong Innovation and Technology Commission under Project No. ITS/238/21.

References

1. Caceres, J.A., Goldstein, J.N.: Intracranial hemorrhage. Emerg. Med. Clin. North Am. **30**(3), 771 (2012)
2. Quattrocchi, K.B., et al.: Quantification of midline shift as a predictor of poor outcome following head injury. Surg. Neurol. **35**(3), 183–188 (1991)
3. Yang, W.-S., Li, Q., Li, R., Liu, Q.-J., Wang, X.-C., et al.: Defining the optimal midline shift threshold to predict poor outcome in patients with supratentorial spontaneous intracerebral hemorrhage. Neurocrit. Care **28**(3), 314–321 (2017). https://doi.org/10.1007/s12028-017-0483-7
4. Liao, C., Chen, Y., Xiao, F.: Brain midline shift measurement and its automation: a review of techniques and algorithms. Int. J. Biomed. Imaging, 4303161 (2018)
5. Pisov, M., et al.: Incorporating task-specific structural knowledge into CNNs for brain midline shift detection. In: Suzuki, K., et al. (eds.) ML-CDS/IMIMIC -2019. LNCS, vol. 11797, pp. 30–38. Springer, Cham (2019). https://doi.org/10.1007/978-3-030-33850-3_4
6. Wang, S., Liang, K., Li, Y., Yu, Y., Wang, Y.: Context-aware refinement network incorporating structural connectivity prior for brain midline delineation. In: Martel, A.L., et al. (eds.) MICCAI 2020. LNCS, vol. 12267, pp. 208–217. Springer, Cham (2020). https://doi.org/10.1007/978-3-030-59728-3_21
7. Wei, H., et al.: Regression-based line detection network for delineation of largely deformed brain midline. In: Shen, D., et al. (eds.) MICCAI 2019. LNCS, vol. 11766, pp. 839–847. Springer, Cham (2019). https://doi.org/10.1007/978-3-030-32248-9_93
8. Qin, C., et al.: 3D brain midline delineation for hematoma patients. In: de Bruijne, M., et al. (eds.) MICCAI 2021. LNCS, vol. 12905, pp. 510–518. Springer, Cham (2021). https://doi.org/10.1007/978-3-030-87240-3_49
9. Nguyen, N.P., Yoo, Y., Chekkoury, A., Eibenberger, E., et al.: Brain midline shift detection and quantification by a cascaded deep network pipeline on non-contrast computed tomography scans. In: ICCVW (2021)
10. Liao, C., Xiao, F., et al.: Automatic recognition of midline shift on brain CT images. Comput. Biol. Med. **40**, 331–339 (2010)
11. Balakrishnan, G., Zhao, A., Sabuncu, M.R., Guttag, J., Dalca, A.V.: Voxelmorph: a learning framework for deformable medical image registration. IEEE Trans. Med. Imaging **38**(8), 1788–1800 (2019)
12. Goodfellow, I.J., et al.: Generative adversarial networks. In: NeurIPS (2014)
13. Song, J., Meng, C., Ermon, S.: Denoising diffusion implicit models. In: ICLR (2020)
14. Ho, J., et al.: Denoising diffusion probabilistic models. In: NeurIPS (2020)
15. Dhariwal, P., et al.: Diffusion models beat GANs on image synthesis. In: NeurIPS (2021)
16. Ho, J., Salimans, T.: Classifier-free diffusion guidance. In: NeurIPS 2021 Workshop on Deep Generative Models and Downstream Applications (2021)
17. Song, Y., Ermon, S.: Generative modeling by estimating gradients of the data distribution. In: NeurIPS (2019)
18. Luo, C.: Understanding diffusion models: a unified perspective. arXiv preprint arXiv:2208.11970 (2022)

19. Tarvainen, A., Valpala, H.: Mean teachers are better role models: weight-averaged consistency targets improve semi-supervised deep learning results. In: NeurIPS (2017)
20. Chen, X., Yuan, Y., Zeng, G., Wang, J.: Semi-supervised semantic segmentation with cross pseudo supervision. In: CVPR (2021)
21. Huang, G., Liu, Z., van der Maaten, L., Weinberger, K.Q.: Densely connected convolutional networks. In: CVPR (2017)

Don't PANIC: Prototypical Additive Neural Network for Interpretable Classification of Alzheimer's Disease

Tom Nuno Wolf[1,2]([✉]), Sebastian Pölsterl[2], and Christian Wachinger[1,2]

[1] Department of Radiology, Technical University Munich, Munich, Germany
`tom_nuno.wolf@tum.de`
[2] Lab for Artificial Intelligence in Medical Imaging, Ludwig Maximilians University, Munich, Germany

Abstract. Alzheimer's disease (AD) has a complex and multifactorial etiology, which requires integrating information about neuroanatomy, genetics, and cerebrospinal fluid biomarkers for accurate diagnosis. Hence, recent deep learning approaches combined image and tabular information to improve diagnostic performance. However, the black-box nature of such neural networks is still a barrier for clinical applications, in which understanding the decision of a heterogeneous model is integral. We propose PANIC, a prototypical additive neural network for interpretable AD classification that integrates 3D image and tabular data. It is interpretable by design and, thus, avoids the need for post-hoc explanations that try to approximate the decision of a network. Our results demonstrate that PANIC achieves state-of-the-art performance in AD classification, while directly providing local and global explanations. Finally, we show that PANIC extracts biologically meaningful signatures of AD, and satisfies a set of desirable desiderata for trustworthy machine learning. Our implementation is available at https://github.com/ai-med/PANIC.

1 Introduction

It is estimated that the number of people suffering from dementia worldwide will reach 152.8 million by 2050, with Alzheimer's disease (AD) accounting for approximately 60–80% of all cases [20]. Due to large studies, like the Alzheimer's Disease Neuroimaging Initiative (ADNI; [9]), and advances in deep learning, the disease stage of AD can now be predicted relatively accurate [28]. In particular, models utilizing both tabular and image data have shown performance superior to unimodal models [4,5,29]. However, they are considered black-box models, as their decision-making process remains largely opaque. Explaining decisions of Convolutional Neural Networks (CNN) is typically achieved with post-hoc techniques in the form of saliency maps. However, recent studies showed that different post-hoc techniques lead to vastly different explanations of the same

---

T. N. Wolf and S. Pölsterl—These authors contributed equally to this work.

© The Author(s), under exclusive license to Springer Nature Switzerland AG 2023
A. Frangi et al. (Eds.): IPMI 2023, LNCS 13939, pp. 82–94, 2023.
https://doi.org/10.1007/978-3-031-34048-2_7

model [12]. Hence, post-hoc methods do not mimic the true model accurately and have low fidelity [22]. Another drawback of post-hoc techniques is that they provide local interpretability only, i.e., an approximation of the decision of a model for a specific input sample, which cannot explain the overall decision-making of a model. Rudin [22] advocated to overcome these shortcomings with inherently interpretable models, which are interpretable by design. For instance, a logistic regression model is inherently interpretable, because one can infer the decision-making process from the weights of each feature. Moreover, inherently interpretable models do provide both local and global explanations. While there has been progress towards inherently interpretable unimodal deep neural networks (DNNs) [1,11], there is a lack of inherently interpretable heterogeneous DNNs that incorporate both 3D image and tabular data.

In this work, we propose PANIC, a Prototypical Additive Neural Network for Interpretable Classification of AD, that is based on the Generalized Additive Model (GAM). PANIC consists of one neural net for 3D image data, one neural net for each tabular feature, and combines their outputs via summation to yield the final prediction (see Fig. 1). PANIC processes 3D images with an inherently interpretable CNN, as proposed in [11]. The CNN is a similarity-based classifier that reasons by comparing latent features of an input image to a set of class-representative latent features. The latter are representations of specific images from the training data. Thus, its decision-making can be considered similar to the way humans reason. Finally, we show that PANIC is fully transparent, because it is interpretable both locally and globally, and achieves state-of-the-art performance for AD classification.

2 Related Work

Interpretable Models for Tabular Data. Decision trees and linear models, such as logistic regression, are inherently interpretable and have been applied widely [16]. In contrast, multi-layer perceptrons (MLPs) are non-parametric and non-linear, but rely on post-hoc techniques for explanations. A GAM is a non-linear model that is fully interpretable, as its prediction is the sum of the outputs of univariate functions (one for each feature) [15]. Explainable Boosting Machines (EBMs) extend GAMs by allowing pairwise interaction of features. While this may boost performance compared to a standard GAM, the model is harder to interpret, because the number of functions to consider grows quadratically with the number of features. EBMs were used in [24] to predict conversion to AD.

Interpretable Models for Medical Images. ProtoPNet [2] is a case-based interpretable CNN that learns class-specific prototypes and defines the prediction as the weighted sum of the similarities of features, extracted from a given input image, to the learned prototypes. It has been applied in the medical domain for diabetic retinopathy grading [6]. One drawback is that prototypes are restricted by the size of local patches: For example, it cannot learn a *single* prototype to represent hippocampal atrophy, because the hippocampus appears in the left and

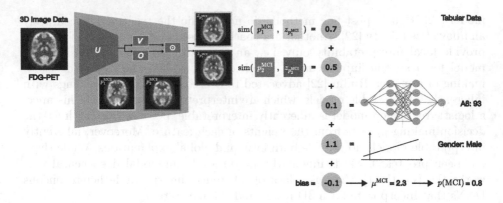

Fig. 1. An exemplary prediction for class MCI with PANIC: 3D FDG-PET images are processed with an interpretable CNN that computes the cosine similarity between latent representations $z_{p_1^{\text{MCI}}}, z_{p_2^{\text{MCI}}}$ and corresponding prototypes $p_1^{\text{MCI}}, p_2^{\text{MCI}}$, as seen on the left. For each categorical feature, such as gender, a linear function is learned. Each continuous feature, such as Aβ is processed with its own MLP. The final prediction is the sum of the outputs of the submodules plus a bias term.

right hemisphere. As a result, the number of prototypes needs to be increased to learn a separate prototype for each hemisphere. The Deformable ProtoPNet [3] allows for multiple fine-grained prototypical parts to extract prototypes, but is bound to a fixed number of prototypical parts that represent a prototype. XProtoNet [11] overcomes this limitation by defining prototypes based on attention masks rather than patches; it has been applied for lung disease classification from radiographic images. Wang et al. [27] used knowledge distillation to guide the training of a ProtoPNet for mammogram classification. However, their final prediction is uninterpretable, because it is the average of the prediction of a ProtoPNet and a black-box CNN. The works in [8,19,21,27,31] proposed interpretable models for medical applications, but in contrast to ProtoPNet, XProtoNet and GAMs, they do not guarantee that explanations are faithful to the model prediction [22].

3   Methods

We propose a prototypical additive neural network for interpretable classification (PANIC) that provides unambiguous local and global explanations for tabular and 3D image data. PANIC leverages the transparency of GAMs by adding functions that measure similarities between an input image and a set of class-specific prototypes, that are latent representations of images from the training data [11].

Let the input consist of N tabular features $x_n \in \mathbb{R}$ ($n \in \{1, \ldots, N\}$), and a 3D grayscale image $\mathcal{I} \in \mathbb{R}^{1 \times H \times D \times W}$. PANIC is a GAM comprising N univariate functions f_n to account for tabular features, and an inherently interpretable

CNN g to account for image data [11]. The latter provides interpretability by learning a set of $K \times C$ class-specific prototypes (C classes, K prototypes per class $c \in \{1, \ldots, C\}$). During inference, the model seeks evidence for the presence of prototypical parts in an image, which can be visualized and interpreted in the image domain. Computing the similarities of prototypes to latent features representing the presence of prototypical parts allows to predict the probability of a sample belonging to class c:

$$p(c \,|\, x_1, \ldots, x_N, \mathcal{I}) = \text{softmax}\,(\mu^c), \quad \mu^c = \beta_0^c + \sum_{n=1}^{N} f_n^c(x_n) + \sum_{k=1}^{K} g_k^c(\mathcal{I}), \quad (1)$$

where $\beta_0^c \in \mathbb{R}$ denotes a bias term, $f_n^c(x_n)$ the class-specific output of a neural additive model for feature n, and $g_k^c(\mathcal{I})$ the similarity between the k-th prototype of class c and the corresponding feature extracted from an input image $\mathcal{I}$. We define the functions f_n^c and g_k^c below.

3.1 Modeling Tabular Data

Tabular data often consists of continuous and discrete-valued features, such as age and genetic alterations. Therefore, we model feature-specific functions f_n^c depending on the type of feature n. If it is continuous, we estimate f_n^c non-parametrically using a multi-layer perceptron (MLP), as proposed in [1]. This assures full interpretability while allowing for non-linear processing of each feature n. If feature n is discrete, we estimate f_n^c parametrically using a linear model, in which case f_n^c is a step function, which is fully interpretable too. Moreover, we explicitly account for missing values by learning a class-conditional missing value indicator s_n^c. To summarize, f_n^c is defined as

$$f_n^c(x_n) = \begin{cases} s_n^c, & \text{if } x_n \text{ is missing,} \\ \beta_n^c x_n, \text{ with } \beta_n^c \in \mathbb{R}, & \text{if } x_n \text{ is categorical} \\ \text{MLP}_n^c(x_n), & \text{otherwise.} \end{cases} \quad (2)$$

Predicting a class with the sum of such univariate functions f_n^c was proposed in [1] as Neural Additive Model (NAM). Following [1], we apply an ℓ_2 penalty on the outputs of $f_n^c(x_n)$:

$$\mathcal{L}_{\text{Tab}}(x_1, \ldots, x_n) = \frac{1}{C} \sum_{c=1}^{C} \sum_{n=1}^{N} [f_n^c(x_n)]^2.$$

We want to emphasize that NAMs retain global interpretability by plotting each univariate function f_n^c over its domain (see Fig. 2). Local interpretability is achieved by evaluating $f_n^c(x_n)$, which equals the contribution of a feature x_n to the prediction of a sample, as defined in Eq. (1) (see Fig. 3 on the left).

3.2 Modeling Image Data

We model 3D image data by defining the function $g_k^c(\mathcal{I})$ in Eq. (1) based on XProtoNet [11], which learns prototypes that can span multiple, disconnected

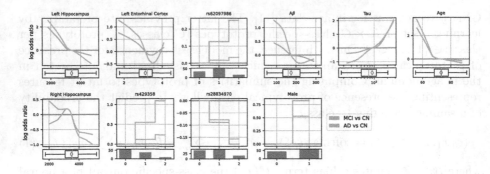

Fig. 2. The plots show the log odds ratio with respect to controls for the top 10 tabular features. Boxplots show the distribution in our data.

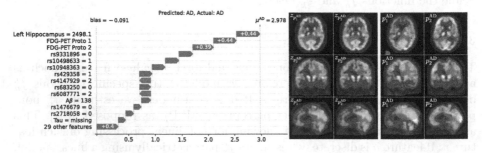

Fig. 3. Explanation for the prediction of a single sample from the test set. Left: Individual contributions to the overall prediction. Right: Input FDG-PET overlayed with the attention maps (green) of the corresponding representations $z_{p_1^{AD}}, z_{p_1^{AD}}$ (columns 1,2), and the attention maps of learned prototypes p_1^{AD}, p_2^{AD} (columns 3,4). (Color figure online)

regions within an image. In XProtoNet, an image is classified based on the cosine similarity between a latent feature vector $z_{p_k^c}$ and learned class-specific prototypes p_k^c, as depicted in the top part of Fig. 1:

$$g_k^c(\mathcal{I}) = \text{sim}(p_k^c, z_{p_k^c}) = \frac{p_k^c \cdot z_{p_k^c}}{\|p_k^c\| \|z_{p_k^c}\|}. \tag{3}$$

A latent feature vector $z_{p_k^c}$ is obtained by passing an image $\mathcal{I}$ into a CNN backbone $\mathcal{U} : \mathbb{R}^{1 \times H \times D \times W} \to \mathbb{R}^{R \times H' \times D' \times W'}$, where R is the number of output channels. The result is passed into two separate modules: (i) the feature extractor $\mathcal{V} : \mathbb{R}^{R \times H' \times D' \times W'} \to \mathbb{R}^{L \times H' \times D' \times W'}$ maps the feature map to the dimensionality of the prototype space L; (ii) the occurrence module $\mathcal{O}^c : \mathbb{R}^{R \times H' \times D' \times W'} \to \mathbb{R}^{K \times H' \times D' \times W'}$ produces K class-specific attention masks. Finally, the latent feature vector $z_{p_k^c}$ is defined as

$$z_{p_k^c} = \text{GAP}[\text{sigmoid}(\mathcal{O}^c(\mathcal{U}(\mathcal{I}))_k) \odot \text{softplus}(\mathcal{V}(\mathcal{U}(\mathcal{I})))], \tag{4}$$

where $\odot$ denotes the Hadamard product, and GAP global average pooling.

Intuitively, $z_{p_k^c}$ represents the GAP-pooled activation maps that a prototype p_k^c would yield if it were present in that image. For visualization, we can upsample the occurrence map $\mathcal{O}^c(\mathcal{U}(\mathcal{I}))_k$ to the input dimensions and overlay it on the input image. The same can be done to visualize prototype p_k^c (see Fig. 3).

Regularization. Training XProtoNet requires regularization with respect to the occurrence module and prototype space [11]: An occurrence and affine loss enforce sparsity and spatial fidelity of the attention masks $\mathcal{O}^c$ with respect to the image domain:

$$\mathcal{L}_{\text{occ}}(\mathcal{I}) = \sum_{c=1}^{C} \|\mathcal{O}^c(\mathcal{U}(\mathcal{I}))\|_1, \quad \mathcal{L}_{\text{affine}}(\mathcal{I}) = \|A(\mathcal{O}^c(\mathcal{U}(\mathcal{I}))) - \mathcal{O}^c(\mathcal{U}(A(\mathcal{I})))\|_1,$$

with A a random affine transformation. Additionally, latent vectors $z_{p_k^c}$ of an image $\mathcal{I}$ with true class label y should be close to prototypes of their respective class, and distant to prototypes of other classes:

$$\mathcal{L}_{\text{clst}}(\mathcal{I}) = -\max_{k,c=y} g_k^c(\mathcal{I}), \quad \mathcal{L}_{\text{sep}}(\mathcal{I}) = \max_{k,c\neq y} g_k^c(\mathcal{I}).$$

3.3   PANIC

As stated in Eq. (1), PANIC is a GAM comprising functions $f_1^c, \ldots, f_N^c$ for tabular data, and functions $g_1^c, \ldots, g_K^c$ for 3D image data (see Eqs. (2) and (3)). Tabular features contribute to the overall prediction in Eq. (1) in terms of the values $f_1^c(x_1), \ldots, f_N^c(x_N)$, while the image contributes in terms of the cosine similarity between the class-specific prototype p_k^c and the latent feature vector $z_{p_k^c}$. By restricting prototypes to contribute to the prediction of a specific class only, we encourage the model to learn discriminative prototypes for each class.

To interpret PANIC locally, we simply consider the outputs of the functions f_n^c, and sum the image-based similarity scores over all prototypes: $\sum_{k=1}^{K} g_k^c(\mathcal{I})$. To interpret the contributions due to the 3D image in detail, we examine the attention map of each prototype, the attention map of the input image, and the similarity score between each prototype and the image (see Fig. 3 on the right). To interpret PANIC globally, we compute the absolute contribution of each function to the per-class logits in Eq. (1), and average it over all samples in the training set, as seen in Fig. 4. In addition, we can directly visualize the function f_n^c learned from the tabular data in terms of the log odds ratio

$$\log \left[\frac{p(c \mid x_1, \ldots, x_n, \ldots, x_N, \mathcal{I})}{p(\text{CN} \mid x_1, \ldots, x_n, \ldots, x_N, \mathcal{I})} \Big/ \frac{p(c \mid, x_1, \ldots, x_n', \ldots, x_N, \mathcal{I})}{p(\text{CN} \mid x_1, \ldots, x_n', \ldots, x_N, \mathcal{I})} \right],$$

where x_n' is the mean value of feature n across all samples for continuous features, and zero for categorical features. As an example, let us consider the AD class. If the log odds ratio for a specific value x_n is positive, it indicates that the odds of being diagnosed as AD, compared to CN, increases. Conversely, if it is negative, the odds of being diagnosed as AD decreases.

Table 1. Statistics for the data used in our experiments.

	CN (N = 379)	Dementia (N = 256)	MCI (N = 610)	Total (N = 1245)
Age				
Mean (SD)	73.5 (5.9)	74.5 (7.9)	72.3 (7.3)	73.1 (7.1)
Range	55.8–90.1	55.1–90.3	55.0–91.4	55.0–91.4
Gender				
Female	193 (50.9%)	104 (40.6%)	253 (41.5%)	550 (44.2%)
Male	186 (49.1%)	152 (59.4%)	357 (58.5%)	695 (55.8%)
Education				
Mean (SD)	16.4 (2.7)	15.4 (2.8)	16.1 (2.7)	16.0 (2.8)
Range	7.0–20.0	4.0–20.0	7.0–20.0	4.0–20.0
MMSE				
Mean (SD)	29.0 (1.2)	23.2 (2.2)	27.8 (1.7)	27.2 (2.7)
Range	24.0–30.0	18.0–29.0	23.0–30.0	18.0–30.0

We train PANIC with the following loss:

$$\mathcal{L}(y, x_1, \ldots, x_n, \mathcal{I}) = \mathcal{L}_{\mathrm{CE}}(y, \hat{y}) + \lambda_1 \mathcal{L}_{\mathrm{Tab}}(x_1, \ldots, x_n) + \lambda_2 \mathcal{L}_{\mathrm{clst}}(\mathcal{I})$$
$$+ \lambda_3 \mathcal{L}_{\mathrm{sep}}(\mathcal{I}) + \lambda_4 \mathcal{L}_{\mathrm{occ}}(\mathcal{I}) + \lambda_5 \mathcal{L}_{\mathrm{affine}}(\mathcal{I}), \tag{5}$$
$$\hat{y} = \arg\max_c \ p(c \,|\, x_1, \ldots, x_n, \mathcal{I}),$$

where $\mathcal{L}_{\mathrm{CE}}$ is the cross-entropy loss, $\lambda_{1,\ldots,5}$ are hyper-parameters, y the true class label, and $\hat{y}$ the prediction of PANIC.

4   Experiments

4.1   Overview

Dataset. Data used in this work was obtained from the ADNI database.[1] We select only baseline visits to avoid data leakage, and use FDG-PET images following the processing pipeline described in [18]. Tabular data consists of the continuous features age, education; the cerobrospinal fluid markers Aβ, Tau, p-Tau; the MRI-derived volumes of the left/right hippocampus and thickness of the left/right entorhinal cortex. The categorical features are gender and 31 AD-related genetic variants identified in [7,13] and having a minor allele frequency of $\geq$5%. Tabular data was standardized using the mean and standard deviation across the training set. Table 1 summarizes our data.

---

[1] https://adni.loni.usc.edu/ [9].

Implementation Details. We train PANIC with the loss in Eq. (5) with AdamW [14] and a cyclic learning rate scheduler [26], with a learning rate of 0.002, and weight decay of 0.0005. We choose a 3D ResNet18 for the CNN backbone $\mathcal{U}$ with $R = 256$ channels in the last ResBlock. The feature extractor $\mathcal{V}$ and the occurrence module $\mathcal{O}$ are CNNs consisting of $1 \times 1 \times 1$ convolutional layers with ReLU activations. We set $K = 2$, $L = 64$, $\lambda_1 = 0.01$ and $\lambda_{2,\ldots,5} = 0.5$, and norm the length of each prototype vector to one. For each continuous tabular feature n, f_n^c is a MLP that shares parameters for the class dependent outputs in (2). Thus, each MLP has 2 layers with 32 neurons, followed by an output layer with C neurons. As opposed to [1], we found it helpful to replace the ExU activations by ReLU activations. We apply spectral normalization [17] to make MLPs Lipschitz continuous. We add dropout with a probability of 0.4 to MLPs, and with a probability of 0.1 to all univariate functions in Eq. (1). The set of affine transformations A comprises all transformations with a scale factor $\in [0.8; 1.2]$ and random rotation $\in [-180°; 180°]$ around the origin. We initialize the weights of $\mathcal{U}$ from a pre-trained model that has been trained on the same training data on the classification task for 100 epochs with early stopping. We cycle between optimizing all parameters of the network and optimizing the parameters of f_n^c only. We only validate the model directly after prototypes p_k^c have been replaced with their closest latent feature vector $z_{p_k^c}$ of samples from the training set of the same class. Otherwise, interpretability on an image level would be lost.

We perform 5-fold cross-validation, based on a data stratification strategy that accounts for sex, age and diagnosis. Each training set is again split such that 64% remain training and 16% are used for hyper-parameter tuning (validation set). We report the mean and standard deviation of the balanced accuracy (BAcc) of the best hyper-parameters found on the validation sets.

4.2 Classification Performance

PANIC achieves $64.0 \pm 4.5\%$ validation BAcc and $60.7 \pm 4.4\%$ test BAcc. We compare PANIC to a black-box model for heterogeneous data, namely DAFT [29]. We carry out a random hyper-parameter search with 100 configurations for learning rate, weight decay and the bottleneck factor of DAFT. DAFT achieves a validation and test BAcc of $60.9 \pm 0.7\%$ and $56.2 \pm 4.5\%$, respectively. This indicates that interpretability does not necessitate a loss in prediction performance.

4.3 Interpretability

PANIC is easy to interpret on a global and local level. Figure 4 summarizes the average feature importance over the training set. It shows that FDG-PET has on average the biggest influence on the prediction, but also that importance can vary across classes. For instance, the SNP rs429358, which is located in the ApoE gene, plays a minor role for the controls, but is highly important for the AD class. This is reassuring, as it is a well known risk factor in AD [25]. The

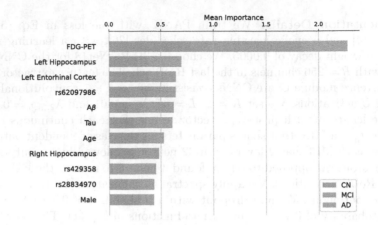

Fig. 4. Ranking of the most import features learned by PANIC. FDG-PET denotes the combined importance of all prototypes related to the FDG-PET image.

overall most important SNP is rs62097986, which is suspected to influence brain volumes early in neurodevelopment [7].

To get a more detailed inside into PANIC, we visualize the log odds ratio with respect to the function f_n^c across the domain of the nine most important tabular features in Fig. 2. We can easily see that PANIC learned that atrophy of the left hippocampus increases the odds of being diagnosed as AD. The volume of the right hippocampus is utilized similarly. For MCI, it appears as PANIC has overfit on outliers with very low right hippocampus volume. Overall, the results for left/right hippocampus agree with the observation that the hippocampus is among the first structures experiencing neurodegeneration in AD [23]. The function for left entorhinal cortex thickness agrees with previous results too [23]. An increase in Aβ measured in CSF is associated with a decreased risk of AD [25], which our model captured correctly. The inverse relationship holds for Tau [25], which PANIC confirmed too. The function learned for age shows a slight decrease in the log odds ratio of AD and MCI, except for patients around 60 years of age, which is due to few data samples for this age range in the training data. We note that the underlying causes that explain the evolution from normal aging to AD remain unknown, but since age is considered a confounder one should control for it [10]. Overall, we observe that PANIC learned a highly non-linear function for the continuous features hippocampus volume, entorhinal thickness, Aβ, Tau, and age, which illustrates that estimating the functions f_n^c via MLPs is effective. In our data, males have a higher incidence of AD (see Table 1), which is reflected in the decision-making of the model too. Our result that rs28834970 decreases the odds for AD does not agree with previous results [13]. However, since PANIC is fully interpretable, we can easily spot this misconception.

Additionally, we can visualize the prototypes by upscaling the attention map specific to each prototype, as produced by the occurrence module, to the input image size and highlighting activations of more than 30% of the maximum value, as proposed in [11] (see Fig. 3 on the right). The axial view of p_1^{AD} shows attention towards the occipital lobe and p_2^{AD} towards one side of the occipital lobe. Atrophy around the ventricles can be seen in FDG-PET [25] and both prototypes p_1^{AD} and p_2^{AD} incorporate this information, as seen in the coronal views. The sagittal views show, that p_2^{AD} focuses on the cerebellum and parts of the occipital lobe. The parietal lobe is clearly activated by the prototype in the sagittal view of p_1^{AD} and was linked to AD previously [25].

We can interpret the decision-making of PANIC for a specific subject by evaluating the contribution of each function with respect to the prediction (see Fig. 3 on the left). The patient was correctly classified as AD, and most evidence supports this (red arrows). The only exceptions are the SNPs rs4147929 and rs6087771, which the models treats as evidence against AD. Hippocampus volume contributed most to the prediction, i.e., was most important. Since, the subject's left hippocampus volume is relatively low (see Fig. 2), this increases our trust in the model's prediction. The subject is heterozygous for rs429358 (ApoE), a well known marker of AD, which the model captured correctly [25]. The four variants rs9331896, rs10498633, rs4147929, and rs4147929 have been associated with nucleus accumbens volume [7], which is involved in episodic memory function [30]. Atrophy of the nucleus accumbens is associated with cognitive impairment [30]. FDG-PET specific image features present in the image show a similarity of 0.44 to the features of prototype p_1^{AD}. It is followed by minor evidence of features similar to prototype p_2^{AD}. During the prediction of the test subject, the network extracted prototypical parts from similar regions. As seen in the axial view, both parts $z_{p_1^{AD}}$ and $z_{p_2^{AD}}$ contain parts of the occipital lobe. Additionally, they cover a large part of the temporal lobe, which has been linked to AD [25].

In summary, the decision-making of PANIC is easy to comprehend and predominantly agrees with current knowledge about AD.

5   Desiderata for Machine Learning Models

We now show that PANIC satisfies four desirable desiderata for machine learning (ML) models, based on the work in [22].

Explanations Must be Faithful to the Underlying Model (Perfect Fidelity). To avoid misconception, an explanation must not mimic the underlying model, but equal the model. The explanations provided by PANIC are the values provided by the functions $f_1^c, \ldots, f_N^c, g_1^c, \ldots, g_K^c$ in Eq. (1). Since PANIC is a GAM, the sum of these values (plus bias) equals the prediction. Hence, explanations of PANIC are faithful to how the model arrived at a specific prediction. We can plot these values to gain local interpretability, as done in Fig. 3.

Explanations Must be Detailed. An explanation must be comprehensive such that it provides all information about what and how information is used by a model (global interpretability). The information learned by PANIC can be described precisely. Since PANIC is based on the sum of univariate functions, we can inspect individual functions to understand what the model learned. Plotting the functions f_n^c over their domain explains the change in odds when the value x_n changes, as seen in Fig. 2. For image data, PANIC uses the similarity between the features extracted from the input image and the K class-specific prototypes. Global interpretability is achieved by visualizing the training image a prototype was mapped to, and its corresponding attention map 3.

A Machine Learning Model Should Help to Improve the Knowledge Discovery Process. For AD, the precise cause of cognitive decline remains elusive. Therefore, ML should help to identify biomarkers and relationships, and inform researchers studying the biological causes of AD. PANIC is a GAM, which means it provides full global interpretability. Therefore, the insights it learned from data are directly accessible, and provide unambiguous feedback to the knowledge discovery process. For instance, we can directly infer what PANIC learned about FDG-PET or a specific genetic mutation, as in Figs. 2 and 4. This establishes a feedback loop connecting ML researchers and researchers studying the biological causes of AD, which will ultimately make diagnosis more accurate.

ML Models Must be Easy to Troubleshoot. If an ML model produces a wrong or unexpected result, we must be able to easily troubleshoot the model. Since PANIC provides local and global interpretability, we can easily do this. We can use a local explanation (see Fig. 3) to precisely determine what the deciding factor for the prediction was and whether it agrees with our current understanding of AD. If we identified the culprit, we can inspect the function f_n^c, in the case of a tabular feature, or the prototypes, in the case of the image data: Suppose, the age of the patient in Fig. 3 was falsely recorded as 30 instead of 71. The contribution of age $f_{\text{age}}^{\text{AD}}$ would increase from -0.138 to 3.28, thus, dominate the prediction by a large margin. Hence, the local explanation would reveal that something is amiss and prompt us to investigate the learned function $f_{\text{age}}^{\text{AD}}$ (see Fig. 2), which is ill-defined for this age.

6   Conclusion

We proposed an inherently interpretable neural network for tabular and 3D image data, and showcased its use for AD classification. We used local and global interpretability properties of PANIC to verify that the decision-making of our model largely agrees with current knowledge about AD, and is easy to troubleshoot. Our model outperformed a state-of-the-art black-box model and satisfies a set of desirable desiderata that establish trustworthiness in PANIC.

Acknowledgements. This research was partially supported by the Bavarian State Ministry of Science and the Arts and coordinated by the bidt, the BMBF (DeepMentia, 031L0200A), the DFG and the LRZ.

References

1. Agarwal, R., Melnick, L., Frosst, N., et al.: Neural additive models: interpretable machine learning with neural nets. In: NeurIPS, vol. 34, pp. 4699–4711 (2021)
2. Chen, C., Li, O., Tao, D., et al.: This looks like that: deep learning for interpretable image recognition. In: NeurIPS, vol. 32 (2019)
3. Donnelly, J., Barnett, A.J., Chen, C.: Deformable ProtoPNet: an interpretable image classifier using deformable prototypes. In: CVPR, pp. 10265–10275 (2022)
4. El-Sappagh, S., Abuhmed, T., Islam, S.M.R., Kwak, K.S.: Multimodal multitask deep learning model for Alzheimer's disease progression detection based on time series data. Neurocomputing **412**, 197–215 (2020)
5. Esmaeilzadeh, S., Belivanis, D.I., Pohl, K.M., Adeli, E.: End-To-end Alzheimer's disease diagnosis and biomarker identification. In: Shi, Y., Suk, H.-I., Liu, M. (eds.) MLMI 2018. LNCS, vol. 11046, pp. 337–345. Springer, Cham (2018). https://doi.org/10.1007/978-3-030-00919-9_39
6. Hesse, L.S., Namburete, A.I.L.: INSightR-Net: interpretable neural network for regression using similarity-based comparisons to prototypical examples. In: Wang, L., Dou, Q., Fletcher, P.T., Speidel, S., Li, S. (eds.) MICCAI 2022. LNCS, vol. 13433, pp. 502–511. Springer, Cham (2022). https://doi.org/10.1007/978-3-031-16437-8_48
7. Hibar, D.P., Stein, J.L., Renteria, M.E., et al.: Common genetic variants influence human subcortical brain structures. Nature **520**(7546), 224–9 (2015)
8. Ilanchezian, I., Kobak, D., Faber, H., Ziemssen, F., Berens, P., Ayhan, M.S.: Interpretable gender classification from retinal fundus images using BagNets. In: de Bruijne, M., et al. (eds.) MICCAI 2021. LNCS, vol. 12903, pp. 477–487. Springer, Cham (2021). https://doi.org/10.1007/978-3-030-87199-4_45
9. Jack, C.R., et al.: The Alzheimer's disease neuroimaging initiative (ADNI): MRI methods. J. Magn. Reson. Imaging **27**(4), 685–691 (2008)
10. Jagust, W.: Imaging the evolution and pathophysiology of Alzheimer disease. Nat. Rev. Neurosci. **19**(11), 687–700 (2018)
11. Kim, E., Kim, S., Seo, M., Yoon, S.: XProtoNet: diagnosis in chest radiography with global and local explanations. In: CVPR, pp. 15719–15728 (2021)
12. Kindermans, P.-J., et al.: The (un)reliability of saliency methods. In: Samek, W., Montavon, G., Vedaldi, A., Hansen, L.K., Müller, K.-R. (eds.) Explainable AI: Interpreting, Explaining and Visualizing Deep Learning. LNCS (LNAI), vol. 11700, pp. 267–280. Springer, Cham (2019). https://doi.org/10.1007/978-3-030-28954-6_14
13. Lambert, J.C., Ibrahim-Verbaas, C.A., Harold, D., et al.: Meta-analysis of 74,046 individuals identifies 11 new susceptibility loci for Alzheimer's disease. Nat. Genet. **45**(12), 1452–1458 (2013)
14. Loshchilov, I., Hutter, F.: Decoupled weight decay regularization. In: ICLR (2019)
15. Lou, Y., Caruana, R., Gehrke, J.: Intelligible models for classification and regression. In: SIGKDD, pp. 150–158 (2012)
16. Martino, F.D., Delmastro, F.: Explainable AI for clinical and remote health applications: a survey on tabular and time series data. Artif. Intell. Rev. **56**, 5261–5315 (2022)
17. Miyato, T., Kataoka, T., Koyama, M., Yoshida, Y.: Spectral normalization for generative adversarial networks. In: ICLR (2018)
18. Narazani, M., Sarasua, I., Pölsterl, S., et al.: Is a PET all you need? a multi-modal study for Alzheimer's disease using 3D CNNs. In: Wang, L., Dou, Q., Fletcher,

P.T., Speidel, S., Li, S. (eds.) MICCAI 2022. LNCS, vol. 13431, pp. 66–76. Springer, Cham (2022). https://doi.org/10.1007/978-3-031-16431-6_7

19. Nguyen, H.D., Clément, M., Mansencal, B., Coupé, P.: Interpretable differential diagnosis for Alzheimer's disease and frontotemporal dementia. In: Wang, L., Dou, Q., Fletcher, P.T., Speidel, S., Li, S. (eds.) MICCAI 2022. LNCS, vol. 13431, pp. 55–65. Springer, Cham (2022). https://doi.org/10.1007/978-3-031-16431-6_6

20. Nichols, E., Steinmetz, J.D., Vollset, S.E., et al.: Estimation of the global prevalence of dementia in 2019 and forecasted prevalence in 2050: an analysis for the global burden of disease study 2019. Lancet Public Health $7(2)$, e105–e125 (2022)

21. Oba, Y., Tezuka, T., Sanuki, M., Wagatsuma, Y.: Interpretable prediction of diabetes from tabular health screening records using an attentional neural network. In: DSAA, pp. 1–11 (2021)

22. Rudin, C.: Stop explaining black box machine learning models for high stakes decisions and use interpretable models instead. Nat. Mach. Intell. $1(5)$, 206–215 (2019)

23. Sabuncu, M.R.: The dynamics of cortical and hippocampal atrophy in Alzheimer disease. Arch. Neurol. $68(8)$, 1040 (2011)

24. Sarica, A., Quattrone, A., Quattrone, A.: Explainable boosting machine for predicting Alzheimer's disease from MRI hippocampal subfields. In: Mahmud, M., Kaiser, M.S., Vassanelli, S., Dai, Q., Zhong, N. (eds.) BI 2021. LNCS (LNAI), vol. 12960, pp. 341–350. Springer, Cham (2021). https://doi.org/10.1007/978-3-030-86993-9_31

25. Scheltens, P., Blennow, K., Breteler, M.M.B., et al.: Alzheimer's disease. Lancet $388(10043)$, 505–517 (2016)

26. Smith, L.N.: Cyclical learning rates for training neural networks. In: WACV, pp. 464–472 (2017)

27. Wang, C., Chen, Y., Liu, Y., et al.: Knowledge distillation to ensemble global and interpretable prototype-based mammogram classification models. In: Wang, L., Dou, Q., Fletcher, P.T., Speidel, S., Li, S. (eds.) Medical Image Computing and Computer Assisted Intervention-MICCAI 2022, vol. 13433, pp. 14–24. Springer, Cham (2022). https://doi.org/10.1007/978-3-031-16437-8_2

28. Wen, J., et al.: Convolutional neural networks for classification of Alzheimer's disease: overview and reproducible evaluation. Med. Image Anal. 63, 101694 (2020)

29. Wolf, T.N., Pölsterl, S., Wachinger, C.: DAFT: a universal module to interweave tabular data and 3D images in CNNs. NeuroImage 260, 119505 (2022)

30. Yi, H.A., Möller, C., Dieleman, N., et al.: Relation between subcortical grey matter atrophy and conversion from mild cognitive impairment to Alzheimer's disease. J. Neurol. Neurosurg. Psychiatry $87(4)$, 425–432 (2015)

31. Yin, C., Liu, S., Shao, R., Yuen, P.C.: Focusing on clinically interpretable features: selective attention regularization for liver biopsy image classification. In: de Bruijne, M., et al. (eds.) MICCAI 2021. LNCS, vol. 12905, pp. 153–162. Springer, Cham (2021). https://doi.org/10.1007/978-3-030-87240-3_15

Filtered Trajectory Recovery: A Continuous Extension to Event-Based Model for Alzheimer's Disease Progression Modeling

Jiangchuan Du[1], Yuan Zhou[2(✉)],
and for the Alzheimer's Disease Neuroimaging Initiative

[1] School of Mathematical Science, Fudan University, Shanghai, China
[2] School of Data Science, Fudan University, Shanghai, China
yuanzhou@fudan.edu.cn

Abstract. Event-based model (EBM) is a flexible way to model the progression of Alzheimer's disease. The core of EBM is to use an ordering of events to build the trajectory of disease progression. The ordering of events is usually inferred from the data using optimization, hence may suffer from local optima and computational complexity. This paper mathematically proves that this ordering can be directedly determined by the cumulative distribution function (CDF) of the data. From this standpoint, we formulate two properties—order preserving property (OPP) and distance preserving property (DPP)—that an estimated trajectory should satisfy. We show that a trajectory that satisfies these two properties is equivalent to a reparametrized version of the true trajectory. Furthermore, we show that one such reparametrized trajectory can be directedly obtained from the CDF of the data by filtering. We call the algorithm filtered trajectory recovery (FTR). Extensive experiments on simulated data and real data from ADNI show that FTR can retrieve a trajectory that provides a better estimation of the event order and stages, disentangle the progression heterogeneity, without assuming a parametric form of the trajectory function. The code is released at https://github.com/Jiangchuan-Du/FTR-master.

Keywords: Alzheimer's Disease · Disease Progression Model · Event-based Model

1 Introduction

Alzheimer's disease is characterized by progressive loss of the neurons in the brain. The disease usually starts years before the symptoms emerge and progresses heterogeneously [1]. One way to quantify this *heterogeneity* is to group the population into subtypes: each individual belongs to a subtype and each

Supplementary Information The online version contains supplementary material available at https://doi.org/10.1007/978-3-031-34048-2_8.

subtype corresponds to a prototypical progression *trajectory*. The progression trajectory can be measured by various biomarkers from neuroimaging scans, clinical scores, cerebrospinal fluid (CSF), etc. *Disease progression models* (DPM) aim to recover the progression trajectory from a collection of short individual time series of these biomarkers. Applying DPMs to discover these trajectories is quite significant in providing biological insight into the underlying disease mechanism. It also enables staging and subtyping the patients, hence benefits patient stratification and facilitates precision medicine in clinical trials.

A general DPM assumes that the data points follow

$$s \sim h(s), \quad \mathbf{x} \sim \mathcal{N}(\mathbf{f}(s), \mathbf{\Sigma}(s))$$

where s is the *stage* (time from the beginning of disease progression), $h : [0,1] \rightarrow \mathbb{R}_+$ is the density function for sampling s, whose domain is the timeline normalized to $[0,1]$. Given sampled s, an observed data point $\mathbf{x} = [x_1, \ldots, x_B]$ is sampled from a Gaussian distribution with mean $\mathbf{f}(s)$ and covariance matrix $\mathbf{\Sigma}(s)$, where B is the number of biomarkers, $\mathbf{f} : [0,1] \rightarrow \mathbb{R}^B : s \mapsto (f_1(s), \ldots, f_B(s))$ is the entire trajectory of disease progression, $\mathbf{\Sigma}(s)$ is the noise covariance matrix at s. Without loss of generality, we assume that the trajectory of each biomarker f_j is a *monotonically increasing* function. Given a set of observed data points, a DPM tries to estimate the underlying trajectory $\mathbf{f}$ and possibly noise $\mathbf{\Sigma}$.

Previous studies that try to solve the problem can be categorized into event-based models (EBM) [2–8], parametric/non-parametric models [9–11], differential equation models [12,13], and deep learning [14,15]. The EBMs model the trajectory as a transition from normal to abnormal using a set of predefined events [2–5,8]. This event-based strategy is extended to model a more flexible piecewise linear function as well as multiple trajectories for quantifying the progression heterogeneity [6,7]. Parametric/non-parametric models characterize the continuous longitudinal progress of biomarkers by a parametric (e.g. sigmoid [9]) or a non-parametric (e.g. Gaussian process [10,11]) function of time and other variates. By assuming that toxic proteins spread along the neural network in the brain, differential equations have been

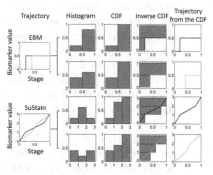

Fig. 1. Motivation of directly estimating the trajectory using the data distribution. The first column shows the trajectory modeled by the original EBM and SuStaIn (see details in Sect. 2). The second, third, and fourth columns show the histograms, CDF, and inverse CDF of the biomarkers, respectively, when h is a uniform distribution.

introduced to model the progression of brain biomarkers [12,13]. The last one, deep learning techniques, in particular recurrent neural networks (RNN), can also predict time series progress in dealing with longitudinal data [14,15].

Despite a plethora of ways to model the trajectory, they more or less made some assumptions on the form of the trajectory function. We observe that a

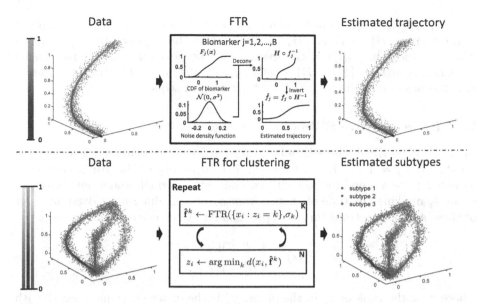

Fig. 2. Overview of the methodology. In FTR (first row), given data (gray dots) generated from a trajectory (time indicated by the colorbar on the left), the CDF of each biomarker $F_j(x)$ is deconvolved by the noise density function $\mathcal{N}(0, \sigma^2)$ to get $H \circ f_j^{-1}$, which is inverted to obtain a reparametrized trajectory $\hat{f}_j = f_j \circ H^{-1}$. Combining all such $\hat{f}_j$'s gives our estimated trajectory. In the second row, FTR is extended to cluster data points from 3 trajectories via a kmeans-like algorithm. (Color figure online)

trajectory is closely related to the *cumulative distribution function* (CDF) of the data in the EBM. For example, Fig. 1 shows two ways in a EBM to model the trajectory of two biomarkers (first column). Under a uniform distribution assumption on h, we see that after calculations of the histogram and CDF, the inverse CDF exactly reflects the true trajectory. This motivates a way to retrieve the trajectory directly from the data distribution.

In this paper, we developed a mathematical theory on using the data distribution for trajectory recovery. First, we show that a closed-form solution to EBM is available by using the CDF of the data (Sect. 2). Then, we extend the trajectory to an arbitrary continuous function by generalizing the idea of EBMs to an *order-preserving property (OPP)* and a *distance-preserving property (DPP)* (Sect. 3). We show that a trajectory satisfying these two properties is equivalent to a reparametrized trajectory (Proposition 2). Finally, we show that a particular reparametrized trajectory can be retrieved by deconvolving the CDF of the observed data (Theorem 3). We call this process *filtered trajectory recovery* (FTR) (see Fig. 2). Extensive experiments on both simulated datasets and a real dataset are conducted to validate the proposed FTR (Sect. 4).

2   Event-Based Model

The EBMs use an order of the events to define the entire trajectory. An *event* is the stage s when a biomarker reaches a specific target e for the first time, i.e.,

$s = \inf\{t : f_j(t) \geq e\}$. An *order* of a sequence of B events $(s_1, \ldots, s_B)$ is a bijection $o : \{1, \ldots, B\} \rightarrow \{1, \ldots, B\}$ that rearranges the sequence to $(s_{o_1}, \ldots, s_{o_B})$ where o_i (denoting $o(i)$) is the event index at position i in the rearranged sequence. A *rank* $r : \{1, \ldots, B\} \rightarrow \{1, \ldots, B\}$ is the inverse function of o. Hence, r_i is the position of the ith event in the rearanged sequence.

2.1 Representative EBMs

In the original EBM, there is only 1 target e for each biomarker, leading to B events $\{s_j : j = 1, \ldots, B\}$ in total [2]. The trajectory of the jth biomarker is assumed to be a step function with two outputs—normal, abnormal—with the event s_j marking the time of change from normal to abnormal. Given an order of these events, the trajectory is approximated by the corresponding ranks:

$$f_j(s) = \begin{cases} \mu_j^0 & \text{if } s < r_j \\ \mu_j^1 & \text{if } s \geq r_j \end{cases}$$

where r_j is the rank of s_j in the order, μ_j^0 is the expected value when the jth biomarker takes a normal value and μ_j^1 represents the expected abnormal value. Note that the timeline is not normalized to 0–1 in this notation. Based on the trajectory definition, a probabilistic model is introduced to estimate the order of biomarker change given a set of data points.

Considering that a step function for approximating the trajectory is far away from satisfactory, an improvement—SuStaIn—is later proposed to approximate the trajectory by piecewise linear functions [6]. In SuStaIn, multiple event targets are defined $e_1 < e_2 < \cdots < e_M$ and they have corresponding events $\{s_{jl} : j = 1, \ldots, B, l = 1, \ldots, M\}$, i.e. $s_{jl} = \inf\{t : f_j(t) \geq e_l\}$. By fixing s_{jM} to be the endpoint on the timeline, SuStaIn uses an order of $S = (s_{11}, \ldots, s_{1,M-1}, s_{21}, \ldots, s_{B,M-1})$ to define the trajectory:

$$f_j(s) = \begin{cases} \frac{e_1 - e_0}{r_{j1} - r_{j0}}(s - r_{j0}) + e_0, & \text{if } r_{j0} \leq s < r_{j1} \\ \quad \vdots \\ \frac{e_M - e_{M-1}}{r_{jM} - r_{j,M-1}}(s - r_{j,M-1}) + e_{M-1}, & \text{if } r_{j,M-1} \leq s \leq r_{jM} \end{cases}$$

where r_{jl} is the rank of s_{jl} in the order for $l \in \{1, \ldots, M-1\}$, $r_{j0} = 0$ and $r_{jM} = B(M-1) + 1$ are the start and end points on the timeline. Similar to the original version, a probabilistic model is introduced to estimate the order of S.

2.2 Closed-Form Solution

Reviewing the above models, we see that to find the order that fits the data, we need to do optimization in a permutation space. When the dimension of the space is large, both the complexity and the robustness of finding the correct order may not be ideal. Following the intuition in Fig. 1, we show in Theorem 1 that a closed-form solution to EBMs is available under mild conditions.

Theorem 1. *Suppose the data point random variable* $\mathbf{x} = [x_1, \ldots, x_B]$ *is generated by*

$$s \sim h(s), \quad x_j \sim \mathcal{N}(f_j(s), \sigma^2),$$

where $h : [0, 1] \to \mathbb{R}_+$ *is a density function and*

$$f_j(s) = \begin{cases} 0, & \text{if } 0 \leq s < a_j, \\ 1, & \text{if } a_j \leq s \leq 1. \end{cases}$$

Then, for any set of change points $\{a_j \in [0, 1]\}$, σ^2 *and* h, *the sequence* $(\hat{a}_1, \ldots, \hat{a}_B)$ *defined by* $\hat{a}_j = F_j(\frac{1}{2})$ *has the same order as* $(a_1, \ldots, a_B)$, *where* F_j *is the CDF of the marginal distribution* $p(x_j)$.

Sketch of Proof. In $F_j(x) = \int_{-\infty}^{x} \int_0^1 p(z; f_j(s), \sigma^2) h(s) ds dz$, the integration over s from 0 to 1 can separated into two ranges: $[0, a_j]$ and $[a_j, 1]$. Using the definition of f_j and this separation, we can show $F_j(x) = H(a_j)\phi(x; 0, \sigma^2) + [1 - H(a_j)]\phi(x; 1, \sigma^2)$, where $\phi(x; 0, \sigma^2)$ is the CDF of $\mathcal{N}(x; 0, \sigma^2)$ and H is the CDF of h. Then we can show $F_j(\frac{1}{2}) - F_i(\frac{1}{2}) > 0$ for any $a_j > a_i$.

Theorem 1 can be extended to monotonically increasing step functions with multiple outputs, e.g. $0, 1, 2, \ldots, M - 1$. A more interesting question is: Can we take the limit to make the output interval arbitrary small such that a continuous function can be estimated? In the next section, we present a way to achieve this.

3 Filtered Trajectory Recovery

Here, we assume that each f_j is strictly monotonically increasing such that the event for a biomarker reaching e can be written as $f_j^{-1}(e)$. Recalling the EBMs, we formulate two properties that an acceptable trajectory should satisfy to achieve our goals: finding the correct order of events and clustering the data points.

3.1 Theory

First, we generalize the concept of retrieving the correct order of the events to an *order-preserving property*. This property ensures that for any events, an estimated trajectory has the same order as the true one that generates the data. In addition, to make the estimated one be able to cluster the data points, we propose a *distance-preserving property*. For an estimated trajectory denoted by $\hat{\mathbf{f}} : [0, 1] \to \mathbb{R}^B : \hat{\mathbf{f}}(s) = (\hat{f}_1(s), \ldots, \hat{f}_B(s))$ and the true one by $\mathbf{f} : [0, 1] \to \mathbb{R}^B : \mathbf{f}(s) = (f_1(s), \ldots, f_B(s))$, these two properties are formalized as:

1. *The order-preserving property (OPP)*: For any j, k, e, e', if $f_j^{-1}(e) < f_k^{-1}(e')$, then $\hat{f}_j^{-1}(e) < \hat{f}_k^{-1}(e')$.
2. *The distance-preserving property (DPP)*: For any $\mathbf{x} \in \mathbb{R}^B$, $d(\mathbf{x}, \mathbf{f}) = d(\mathbf{x}, \hat{\mathbf{f}})$, where $d(\mathbf{x}, \mathbf{f}) = \min_s \|\mathbf{x} - \mathbf{f}(s)\|$.

The order-preserving property implies that for any set of events $\{f_j^{-1}(e_l)\}$, its order is the same as that from $\{\hat{f}_j^{-1}(e_l)\}$ hence $\hat{\mathbf{f}}$ suffices to be used to retrieve the correct order of biomarker change. The distance-preserving property implies that the range of $\hat{\mathbf{f}}$ is the same as the range of $\mathbf{f}$, i.e. the trajectory lies on the same 1-dimensional manifold embedded in the B-dimensional space. This property enables us to use a clustering algorithm to cluster the data points. These two properties together define an admissible trajectory in estimation.

Then we ask: What kind of trajectories would satisfy these two properties? The proposition below suggests that if we reparametrized $\mathbf{f}(s)$ to $\mathbf{f} \circ H^{-1}(t)$ by using a monotonically increasing bijection H, this *reparametrized trajectory* $\mathbf{f} \circ H^{-1}$ would satisfy these two properties.

Proposition 2. *A trajectory* $\hat{\mathbf{f}}$ *satisfies the order-preserving property and the distance-preserving property if and only if* $\hat{\mathbf{f}} = \mathbf{f} \circ H^{-1}$ *where* $H : [0,1] \to [0,1]$ *is a continuous bijection that increases monotonically.*

Sketch of Proof. The sufficiency is straightforward since $f_j^{-1}(e) < f_k^{-1}(e')$ implies $H \circ f_j^{-1}(e) < H \circ f_k^{-1}(e')$. For the necessity, we can define $H_j = \hat{f}_j^{-1} \circ f_j$ for each j and show $H_j = H_k$ by contradiction (assuming $H_j(s_0) > H_k(s_0)$ for some s_0, choose s and s' around s_0 such that $s < s'$, show that $\hat{f}_j^{-1}(e) > \hat{f}_k^{-1}(e')$ where $e = f_j(s)$ and $e' = f_k(s')$).

Now we have identified the space of trajectories that we are interested in. The next natural question is: Can we estimate such a reparametrized trajectory from the data? Note that all H's that are continuous, bijective, and monotonically increasing would suffice for constructing $\mathbf{f} \circ H^{-1}$. For example, let H be the CDF of h, the density function of the stage, i.e. $H(s) = \int_{-\infty}^{s} h(\tau)d\tau$, then $\mathbf{f} \circ H^{-1}$ would satisfy the two properties. An important theoretical result we derive in this paper is that for H being the CDF of h, $\mathbf{f} \circ H^{-1}$ can be directly obtained by deconvolving the marginal CDF of each biomarker:

Theorem 3. *Suppose the data point random variable* $\mathbf{x} = [x_1, \ldots, x_B]$ *is generated by*

$$s \sim h(s), \quad x_j \sim \mathcal{N}(f_j(s), \sigma_j^2)$$

where $f_j : [0,1] \to [f_j(0), f_j(1)]$ *is a continuous bijection that monotonically increases. Then, the marginal CDF of* x_j, F_j, *is a result of convolving* $H \circ \tilde{f}_j^{-1}$ *with the noise density function* $\mathcal{N}(\cdot; 0, \sigma_j^2)$:

$$(H \circ \tilde{f}_j^{-1}) * \mathcal{N}(\cdot; 0, \sigma_j^2) = F_j$$

where H *is the CDF of* h, *and* $\tilde{f}_j^{-1} : \mathbb{R} \to [0,1]$ *is an augmented version of* f_j^{-1} *that extends its domain to* $(-\infty, +\infty)$:

$$\tilde{f}_j^{-1}(y) = \begin{cases} 0, & \text{if } y < f_j(0), \\ f_j^{-1}(y), & \text{if } f_j(0) \le y \le f_j(1), \\ 1, & \text{if } y > f_j(1). \end{cases}$$

In particular, when $\sigma_j = 0$, *we have* $H \circ \tilde{f}_j^{-1} = F_j$.

Sketch of Proof. Let $\phi(x; \mu, \sigma^2)$ be the CDF of $\mathcal{N}(x; \mu, \sigma^2)$, we can show $F_j(x) = \int_0^1 h(s)\phi(x; f_j(s), \sigma_j^2)ds$. Let $y = f_j(s)$, $F_j(x) = \int_{-\infty}^{+\infty} \frac{dH \circ \tilde{f}_j^{-1}(y)}{dy} \phi(x; y, \sigma_j^2)dy$. Using integration by parts, $F_j(x) = \phi(x - y; 0, \sigma_j^2) \cdot H \circ \tilde{f}_j^{-1}(y) \Big|_{-\infty}^{+\infty} - \int_{-\infty}^{+\infty} H \circ \tilde{f}_j^{-1}(y)d\phi(x - y; 0, \sigma_j^2)$ where the first term vanishes and the second is the result.

Theorem 3 verifies our intuition in Fig. 1. If h is a uniform distribution on $[0, 1]$, H would be the identity function, hence f_j is directly related to the CDF F_j via convolution. Moreover, when $\sigma_j = 0$, the CDF is the inverse function of the trajectory f_j. Note that without assuming a uniform distribution on h, the reparametrized trajectory $\{f_j \circ H^{-1}\}$ still satisfies the OPP and DPP, hence suffices for ordering any events and separating the data points. This suggests a new algorithm to estimate the progression trajectory and cluster the data points.

3.2   Algorithm

According to Theorem 3, we only need to deconvolve the marginal CDF F_j using the noise density $\mathcal{N}(\cdot; 0, \sigma_j^2)$ to obtain $H \circ \tilde{f}_j^{-1}$. Then, $H \circ \tilde{f}_j^{-1}$ can be truncated to $H \circ f_j^{-1}$ according to some range $[f_j(0), f_j(1)]$ and the inverses $\{f_j \circ H^{-1} : j = 1, \ldots, B\}$ can be combined to obtain the reparametrized trajectory $\mathbf{f} \circ H^{-1}$ (see Fig. 2). The details are given below.

Let $H \circ \tilde{f}_j^{-1}$, $\mathcal{N}(\cdot; 0, \sigma_j^2)$, and F_j be discretized into $\tilde{\mathbf{s}} = [\mathbf{0}_{l_0}, \mathbf{s}, \mathbf{1}_{l_1}] \in \mathbb{R}^{n+2l}$, $\mathbf{g} \in \mathbb{R}^{2l+1}$, and $\mathbf{F}_j \in \mathbb{R}^n$ respectively, where l_0 and l_1 are the lengths of the leading zeros and trailing ones. The convolution $(H \circ \tilde{f}_j^{-1}) * \mathcal{N}(\cdot; 0, \sigma_j^2) = F_j$ can be discretized into $\mathbf{K}\tilde{\mathbf{s}} = \mathbf{F}_j$, where $\mathbf{K} \in \mathbb{R}^{n \times (n+2l)}$ is a circulant matrix with $\mathbf{K}(i, i : i + 2l) = \mathbf{g}$. Let $\mathbf{K} = [\mathbf{K}_1, \mathbf{K}_2, \mathbf{K}_3]$ where $\mathbf{K}_1$ has l_0 columns, $\mathbf{K}_3$ has l_1 columns. The discretized equation becomes $\mathbf{K}_2\mathbf{s} + \mathbf{K}_3\mathbf{1}_{l_1} = \mathbf{F}_j$, or $\mathbf{As} = \mathbf{b}$ with $\mathbf{A} = \mathbf{K}_2$ and $\mathbf{b} = \mathbf{F}_j - \mathbf{K}_3\mathbf{1}_{l_1}$. Hence, $\mathbf{s}$ can be retrieved by minimizing $\|\mathbf{As} - \mathbf{b}\|$. To keep the trajectory smooth, we add a second derivative Laplacian term $\mathbf{Ls}$ where $\mathbf{L}(i, i : i + 2) = [-1, 2, -1]$. The final objective function is

$$J(\mathbf{s}) = \|\mathbf{As} - \mathbf{b}\|^2 + \lambda\|\mathbf{Ls}\|^2 \ s.t. \ 0 \leq s_j \leq s_{j+1} \leq 1,$$

which is a convex function with inequality constraints and can be solved by standard methods. In this work, we assume $f_j(0)$ is known to be 0 and $f_j(1)$ is unknown. Hence $l_0 = 2l$ while l_1 needs to be searched in a range to minimize $J(\mathbf{s})$. The resulting $\mathbf{s}$ is a discretized version of $H \circ f_j^{-1}$.

This technique can be extended to cluster data points generated from multiple trajectories $\{\mathbf{f}^k : k = 1, 2, \ldots, K\}$. Similar to k-means clustering, we can add a cluster label to each data point and alternately update the reparametrized trajectories, the cluster labels, and the noise variances (see Fig. 2). Note that the estimated trajectories $\hat{\mathbf{f}}^k = \mathbf{f}^k \circ (H^k)^{-1}$ may have different H^ks. To make their timelines be matched, we further reparameterize the trajectories by Euclidean arc length from the origin.

4   Results

4.1   Datasets

We use both simulated datasets and a real dataset for validation. The real dataset is from the Alzheimer's Disease Neuroimaging Initiative (ADNI) [16]. For each dataset, besides running the proposed FTR, we also run its variation without filtering (referred to as *FTR w.o.f.*, i.e. without the deconvolution procedure) for ablation study, and the original EBM (referred to as EBM) [2], the discriminative EBM (referred to as DEBM) [3], and SuStaIn [6] for comparison.

Simulated Datasets. For simulation, we create datasets for two experiments: (i) stage inference with a single trajectory; (ii) subtype and stage inference with multiple trajectories. In both cases, a trajectory is simulated by the following two ways:

1. *Sigmoid function.* The trajectory of the jth biomarker is assumed to follow $f_j(s) = 1/(1 + e^{-\alpha_j(s-\beta_j)})$, where α_j (controlling the slope of increase) is randomly sampled from a Gaussian distribution, $\alpha_j \sim \mathcal{N}(10, 4)$, and β_j (controlling the position of change) is sampled from a uniform distribution, $\beta_j \sim \mathcal{U}([0.1, 0.9])$. We shift each f_j vertically so that it starts at zero.
2. *Event permutation.* Assume event targets $e_l = l$ for $l = 1, 2, 3$ and $e_4 = 5$, and corresponding events $s_{jl} = f_j^{-1}(e_l)$. Randomly order $(s_{11}, \ldots, s_{1,3}, \ldots, s_{B,3})$ with rank r and constraint $r_{jl} < r_{j,l+1}, \forall j, l$. Then, for each j, interpolate $\{(r_{jl}, e_l) : l = 0, 1, \ldots, 4\}$ ($r_{j0} = 0$ and $e_0 = 0$) linearly to obtain f_j.

Given the generated trajectory $\mathbf{f}$, the data points are generated by sampling stage s from a uniform distribution $\mathcal{U}([0, 1])$, and then sampling $\mathcal{N}(\mathbf{f}(s), \sigma^2 \mathbf{I})$ (covariance $(5\sigma)^2 \mathbf{I}$ for event permutation). For multiple trajectories $\{\mathbf{f}^k : k = 1, \ldots, K\}$ ($K = 3$), a data point also has an associated subtype index z, which is sampled from a categorical distribution $z \sim \mathrm{Cat}(1/K, \ldots, 1/K)$ that puts equal probablity to each subtype. A data point is generated by first sampling z, then sampling from trajectory $\mathbf{f}^z$.

The TADPOLE dataset. The Alzheimer's Disease Prediction Of Longitudinal Evolution (TADPOLE) dataset contains preprocessed measurements from brain imaging, CSF, scores from cognitive tests, demographics, and genetics from ADNI 1, ADNI GO, and ADNI 2 [16]. There are 1737 subjects (age: 73.8 ± 7.2, sex: 55% male) in its "D1_D2" dataset, and each subject has multiple visits with an interval of 0.5 or 1 year (also with missing visits). At each visit, a subject is also diagnosed with 3 labels: cognitively normal (CN), mild cognitive impairment (MCI), or AD.

We conducted 2 experiments. The first chooses 7 biomarkers including CSF measurements (ABETA, TAU and p-TAU), brain region volumes (Hippocampus, WholeBrain) and cognitive scores (MMSE, ADAS-Cog-13) for validating a single progression trajectory. In the second one, volume measurements of 84 brain regions from MRI scans are selected to identify subtypes with different spatiotemporal atrophy patterns. We remove the visit records with missing data,

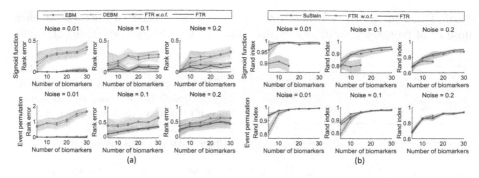

Fig. 3. Comparison of (a) rank error for a single trajectory and (b) Rand index for multiple trajectories in the simulation. Mean with standard deviation (shown in shade) is calculated over 10 randomly generated datasets for each number of biomarkers and each method (5 for SuStaIn). The first and the second rows correspond to using the sigmoid function and event permutation respectively.

leaving 693 subjects with 1812 time points in the first experiment and 1096 subjects with 3929 time points (1262 CN, 1732 MCI and 935 AD) in the second experiment. In both the experiments, we regress out age, sex and intra-cranial volume (ICV) and normalize the biomarker values into z-scores (the first experiment uses the normal group from fitted GMMs in [3] and the second one uses the CN group from the labels).

4.2   Simulation

In the simulation, we randomly generate a dataset containing 3000 data points 10 times for different numbers of biomarkers ($B = 5, 10, \ldots, 30$) and different noise levels ($\sigma = 0.01, 0.1, 0.2$). For the single trajectory experiment, we also ran EBM [2] and DEBM [3] for comparison. For these two EBMs, points with stage $s < 0.1$ were set to CN and the others were set to AD. Since their objective is to estimate the correct order of biomarkers, we used the mean squared error of the ranks of the biomarkers as a metric. The results are shown in Fig. 3(a). We see that FTR has the lowest rank error in both types of trajectories, followed by FTR w.o.f., DEBM and EBM. Note that the inferior performance of EBM and DEBM in the small noise case ($\sigma = 0.01$) may be caused by their small estimated variances from the data.

For the multiple trajectory experiment, we also ran SuStaIn for comparison. For SuStaIn, we set $0.1, 0.5, 0.9, 1$ (resp. $1, 2, 3, 5$) for the event targets in the sigmoid function (resp. event permutation) simulation. For FTR, we ran the clustering algorithm 30 times with random initial label assignments and the best result in terms of reconstruction error was chosen. Since the objective of this task is to cluster the data points, we use Rand index as a metric. The results are shown in Fig. 3(b). We see that again FTR exceeds the baseline method in terms of clustering accuracy in most of the cases. Note that SuStaIn runs much slower than our method hence it was only run on 5 random datasets at each number of biomarkers and also at maximum 15 biomarkers (see Table 1).

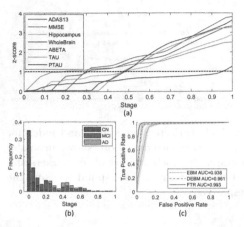

Table 1. Running time (in seconds) of all the methods on a laptop with a Ryzen 7 4800H CPU and 16 GB memory. FTR ran much slower when $K = 3$ because the algorithm was repeated 30 times with random initializations.

Model	# of subtypes	# of biomarkers		
		5	10	15
EBM	$K = 1$	7.59	13.5	13.2
DEBM	$K = 1$	3.17	6.93	20.6
SuStaIn	$K = 3$	1299	6813	31099
FTR w.o.f.	$K = 1$	0.28	0.44	0.60
	$K = 3$	58.1	113	138
FTR	$K = 1$	6.34	12.6	19.7
	$K = 3$	760	1356	2062

Fig. 4. Results of FTR on the 7 biomarkers from TADPOLE. (a) Estimated trajectory. (b) Distribution of estimated stages. (c) Mean ROC curves (standard deviation indicated by shade) for classifying AD from CN using the estimated stages from 5-fold cross validation.

4.3   ADNI

For the 7 biomarker experiment, the normalized z-scores of MMSE, Hippocampus, WholeBrain and ABETA decrease over time hence we add a minus sign to make them increase. The estimated trajectory from FTR is shown in Fig. 4(a). We can see that with a proper event target ($e = 1$), the order of the events—ABETA, MMSE, ADAS13, Hippocampus, TAU, p-TAU, WholeBrain—is exactly the same as that obtained from DEBM [3]. To validate the inferred stages, we split the data into a training set and a test set, trained FTR on the training set, inferred stages on the test set using the estimated trajectory, and classify AD from CN using these stages. The ROC curves from 5-fold cross validation show that FTR outperforms EBM and DEBM in this classification task, suggesting the superiority of our stage inference (see Fig. 4(c)).

The last experiment is on using FTR to identify progressive atrophy patterns in 84 brain regions measured by MRI scans. Following previous studies [1,6], we use 3 subtypes in FTR. The trajectories are visualized by showing stages 0.2, 0.4, 0.6, 0.8, 1 in BrainPainter [17] (see Fig. 5(a)). We see that a typical progression (subtype 2) is identified along with 2 branches (subtype 1 and 3). We also plot the cognitive scores versus time from baseline for each subtype in Fig. 5(b). We can observe that the progression rate of cognitive scores calculated from linear regression is the largest for subtype 3, followed by subtype 2 and 1, which is consistent with the severity of atrophy progression from imaging.

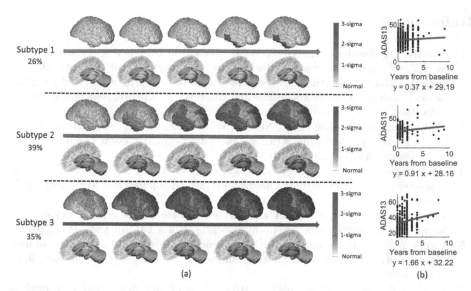

(a)   (b)

Fig. 5. Three subtypes identified by applying FTR on 84 brain regions from TAD-POLE. (a) Visualization of brain atrophy patterns at stage 0.2, 0.4, 0.6, 0.8, 1. (b) Linear regression of clinical symptom progression measured by ADAS-Cog-13.

5   Discussion and Conclusion

The results presented above show that the event order estimated from FTR is the same as that obtained from the EBM while the inferred stages are better to separate AD and CN. This is the first significant finding, which implies the superiority of our continuous trajectory estimation. The second, potentially more interesting, finding is that FTR identifies 3 fine-grained trajectories of spatiotemporal atrophy progression in 84 brain regions. Whether this subtyping scheme has any connection with the previous limbic predominant, typical, hippocampal sparing subtypes [1] will be investigated in the future.

The theory we developed on the relation between the disease progression trajectory and the CDF of the data could also have significant implications. Compared to the previous studies that make a strong assumption on the form of the trajectory function, we only assume that the function is monotonically increasing. This makes our method more flexible to retrieve a wide range of trajectories. Though the current version lacks some features like model selection, uncertainty estimation, we plan to solve these in the future.

In conclusion, this paper mathematically proves that the disease progression trajectory can be directedly recovered from the CDF of the data. Based on this theory, we propose a novel FTR algorithm to estimate the trajectories and find the subtypes. FTR outperforms various state-of-the-art EBMs in our extensive simulation. It identifies a single trajectory using CSF, MRI biomarkers and cognitive scores. It also identifies 3 subtypes that differ in spatiotemporal patterns of progressive atrophy using volumes of 84 brain regions from MRI scans.

References

1. Ferreira, D., Nordberg, A., Westman, E.: Biological subtypes of Alzheimer's disease: a systematic review and meta-analysis. Neurology **94**(10), 436–448 (2020)
2. Fonteijn, H.M., et al.: An event-based model for disease progression and its application in familial Alzheimer's disease and Huntington's disease. NeuroImage **60**(3), 1880–1889 (2012)
3. Venkatraghavan, V., Bron, E.E., Niessen, W.J., Klein, S.: Disease progression timeline estimation for Alzheimer's disease using discriminative event based modeling. NeuroImage **186**, 518–532 (2019)
4. Venkatraghavan, V., et al.: Analyzing the effect of APOE on Alzheimer's disease progression using an event-based model for stratified populations. Neuroimage **227**, 117646 (2021)
5. Van Der Ende, E.L., et al.: A data-driven disease progression model of fluid biomarkers in genetic frontotemporal dementia. Brain **145**(5), 1805–1817 (2022)
6. Young, A.L., et al.: Uncovering the heterogeneity and temporal complexity of neurodegenerative diseases with subtype and stage inference. Nat. Commun. **9**(1), 1–16 (2018)
7. Vogel, J.W., et al.: Four distinct trajectories of tau deposition identified in Alzheimer's disease. Nat. Med. **27**(5), 871–881 (2021)
8. Firth, N.C., et al.: Sequences of cognitive decline in typical Alzheimer's disease and posterior cortical atrophy estimated using a novel event-based model of disease progression. Alzheimer's & Dementia **16**(7), 965–973 (2020)
9. Marinescu, R.V., et al.: DIVE: a spatiotemporal progression model of brain pathology in neurodegenerative disorders. NeuroImage **192**, 166–177 (2019)
10. Lorenzi, M., Filippone, M., Frisoni, G.B., Alexander, D.C., Ourselin, S., et al.: Probabilistic disease progression modeling to characterize diagnostic uncertainty: application to staging and prediction in Alzheimer's disease. NeuroImage **190**, 56–68 (2019)
11. Abi Nader, C., Ayache, N., Robert, P., Lorenzi, M., et al.: Monotonic Gaussian Process for spatio-temporal disease progression modeling in brain imaging data. Neuroimage **205**, 116266 (2020)
12. Garbarino, S., Lorenzi, M., Initiative, A.D.N., et al.: Investigating hypotheses of neurodegeneration by learning dynamical systems of protein propagation in the brain. Neuroimage **235**, 117980 (2021)
13. Raj, A., Kuceyeski, A., Weiner, M.: A network diffusion model of disease progression in dementia. Neuron **73**(6), 1204–1215 (2012)
14. Jung, W., Jun, E., Suk, H.-I., Initiative, A.D.N., et al.: Deep recurrent model for individualized prediction of Alzheimer's disease progression. Neuroimage **237**, 118143 (2021)
15. Nguyen, M., et al.: Predicting Alzheimer's disease progression using deep recurrent neural networks. Neuroimage **222**, 117203 (2020)
16. Marinescu, R.V., et al.: Tadpole challenge: prediction of longitudinal evolution in Alzheimer's disease. arXiv preprint arXiv:1805.03909 (2018)
17. Marinescu, R., Eshaghi, A., Alexander, D.C., Golland, P.: BrainPainter: a software for the visualisation of brain structures, biomarkers and associated pathological processes. arXiv preprint arXiv:1905.08627 (2019)

Live Image-Based Neurosurgical Guidance and Roadmap Generation Using Unsupervised Embedding

Gary Sarwin[1](✉), Alessandro Carretta[2,3], Victor Staartjes[2], Matteo Zoli[3], Diego Mazzatenta[3], Luca Regli[2], Carlo Serra[2], and Ender Konukoglu[1]

[1] Computer Vision Lab, ETH Zurich, Zurich, Switzerland
sarwing@ethz.ch
[2] Department of Neurosurgery, University Hospital of Zurich, Zurich, Switzerland
[3] Department of Biomedical and Neuromotor Sciences (DIBINEM), University of Bologna, Bologna, Italy

Abstract. Advanced minimally invasive neurosurgery navigation relies mainly on Magnetic Resonance Imaging (MRI) guidance. MRI guidance, however, only provides pre-operative information in the majority of the cases. Once the surgery begins, the value of this guidance diminishes to some extent because of the anatomical changes due to surgery. Guidance with live image feedback coming directly from the surgical device, e.g., endoscope, can complement MRI-based navigation or be an alternative if MRI guidance is not feasible. With this motivation, we present a method for live image-only guidance leveraging a large data set of annotated neurosurgical videos. First, we report the performance of a deep learning-based object detection method, YOLO, on detecting anatomical structures in neurosurgical images. Second, we present a method for generating *neurosurgical roadmaps* using unsupervised embedding without assuming exact anatomical matches between patients, presence of an extensive anatomical atlas, or the need for simultaneous localization and mapping. A generated roadmap encodes the *common* anatomical paths taken in surgeries in the training set. At inference, the roadmap can be used to map a surgeon's current location using live image feedback on the path to provide guidance by being able to predict which structures should appear going forward or backward, much like a mapping application. Even though the embedding is not supervised by position information, we show that it is correlated to the location inside the brain and on the surgical path. We trained and evaluated the proposed method with a data set of 166 transsphenoidal adenomectomy procedures.

Keywords: Neuronavigation · Unsupervised Embedding · Endoscopic Surgeries

1 Introduction

Specialists with extensive experience and a specific skill set are required to perform minimally invasive neurosurgeries. During these surgical procedures,

Partially supported by the EANS 2021 Leica Research Grant.

differentiation between anatomical structures, orientation and localization is extremely challenging. On one side, excellent knowledge of the specific anatomy as visualized by the image feedback of the surgical device is required. On the other hand, low contrast, non-rigid deformations, a lack of clear boundaries between anatomical structures, and disruptions such as bleeding, make recognition even for experienced surgeons occasionally very challenging. Various techniques have been developed to help neurosurgeons become oriented and to perform surgery. Computer-assisted neuronavigation has been an important tool and research topic for more than a decade [8,15], but it is still preoperative imaging-based, deeming it unreliable once the arachnoidal cisterns are opened and brain shift occurs [10]. More real-time anatomical guidance can be provided by intraoperative MRI [1,22,23] and ultrasound [2,26]. Orientation has also been greatly enhanced by the application of fluorescent substances such as 5-aminolevulinic acid [7,24]. Awake surgery [9] and electrophysiological neuromonitoring [3,19] can also help navigating around essential brain tissue. These techniques work well and rely on physical traits, other than light reflection. However, they are expensive to implement, require the operating surgeon to become fluent in a new imaging modality, and may require temporarily halting the surgery or retracting surgical instruments to get the intra-operative information [21].

Real-time anatomic recognition based on live image feedback from the surgical device has the potential to address these disadvantages and to act as a reliable tool for intraoperative orientation. This makes the application of machine vision algorithms appealing. The concepts of machine vision can likewise be employed in the neurosurgery operating room to analyze the digital image taken by the micro- or endoscope for automatically identifying the visible anatomic structures and mapping oneself on a planned surgical path [21].

Deep learning applications within the operating room have become more prevalent in recent years. The applications include instrument and robotic tool detection and segmentation [20,28], surgical skill assessment [4], surgical task analysis [13], and procedure automation [25]. Instrument or robotic tool detection and segmentation have been extensively researched for endoscopic procedures owing to the availability of various datasets and challenges [18]. Despite this research on endoscopic videos, the task of anatomic structure detection or segmentation, which could be a foundation for a new approach to neuronavigation, remains relatively unexplored and continues to be a challenge. Note that, anatomy recognition in surgical videos is significantly more challenging than the task of surgical tool detection because of the lack of clear boundaries and differences in color or texture between anatomical structures.

The desire to provide a cheaper real-time solution without relying on additional machines and the improvement of deep learning techniques has driven also the development of vision-based localization methods. Approaches include structure from motion [11] and SLAM [6], such as [14,16], for 3D map reconstruction based on feature correspondence. Many vision-based localization methods rely on landmarks or the challenging task of depth and pose estimation. The main idea behind these methods is to find distinctive landmark positions and follow them across frames for localization, which negatively impacts their performance owing to the low texture, a lack of distinguishable features, non-rigid deformations, and disruptions in endoscopic videos [16]. These methods have mostly been applied to diagnostic procedures, such as colonoscopy, instead of surgical procedures, which pose significant difficulties. Abrupt changes due to the surgical procedure, e.g., bleeding and removal of tissue, make tracking landmarks extremely challenging or even impossible. Therefore, an alternative solution is required to address these challenges.

In this study, a live image-only deep learning approach is proposed to provide guidance during endoscopic neurosurgical procedures. This approach relies on the detection of anatomical structures from RGB images in the form of bounding boxes instead of arbitrary landmarks, as was done in other approaches [5,16], which are difficult to identify and track in the abruptly changing environment during a surgery. The bounding box detections are then used to map a sequence of video frames onto a 1-dimensional trajectory, that represents the surgical path. This allows for localization along the surgical path, and therefore predict anatomical structures in forward or backward directions. The surgical path is learned in an unsupervised manner using an autoencoder architecture from a training set of videos. Therefore, instead of reconstructing a 3D environment and localizing based on landmarks, we rely on a common surgical roadmap and localize ourselves within that map using bounding box detections.

The learned mapping rests on the principle that the visible anatomy and their relative sizes are strongly correlated with the position along the surgical trajectory. Towards this end, bounding box detections capture the presence of structures, their sizes, also relative to each other, and constellations. A simplified representation is shown in Fig. 1. Using bounding box detection of anatomical structures as semantic features mitigates the problem of varying appearance across different patients since bounding box composition is less likely to change across patients than the appearance of the anatomy in RGB images. Furthermore, because the considered anatomical structures only have one instance in every patient, we do not need to rely on tracking of arbitrary structures, e.g., areas with unique appearance compared to their surrounding, which further facilitates dealing with disruptions during surgery, such as bleeding or flushing. We applied the proposed approach on the transsphenoidal adenomectomy procedure, where the surgical path is relatively one-dimensional, as shown in Fig. 2, which makes it well-suited for the proof-of-concept of the suggested method.

2 Methods

2.1 Problem Formulation and Approach

Let $\mathbf{S}_t$ denote an image sequence that consists of endoscopic frames $\mathbf{x}_{t-s:t}$, such as the one shown in Fig. 2, where s represents the sequence length in terms of the number of frames, and $\mathbf{x}_t \in \mathbb{R}^{w \times h \times c}$ is the t-th frame with w, h, and c denoting the width, height, and number of channels, respectively. Our main aim is to

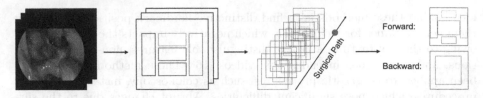

Fig. 1. Simplified representation of the suggested approach. 1. A sequence of input images is processed to detect bounding boxes of anatomical structures. 2. A neural network encodes the sequence of detections into a latent variable that correlates with the position along the surgical path. 3. Given the current position along the surgical path, an estimation of anatomical structures in the forward or backward directions can be obtained, by extrapolating the current value of the latent variable.

embed the sequence $\mathbf{S}_t$ in a 1D latent dimension represented by the variable $\mathbf{z}$. This 1-D latent space represents the surgical path taken from the beginning of the procedure until the final desired anatomy is reached. The approach we take is to determine the anatomical structures visible in the sequence $\mathbf{S}_t$ along the surgical path and map the frame $\mathbf{x}_t$ to the latent space, where effectively the latent space acts as an *implicit* anatomical atlas. We refer to this as an implicit atlas because the position information along the surgical path is not available for construction of the latent space. To achieve this, we perform object detection on all frames $\mathbf{x}_{t-s:t}$ in $\mathbf{S}_t$ and obtain a sequence of detections $\mathbf{c}_{t-s:t}$ that we denote as $\mathbf{C}_t$. A detection $\mathbf{c}_t \in \mathbb{R}^{n\times 5}$ represents the anatomical structures and bounding boxes of the t-th frame, where n denotes the number of different classes in the surgery. More specifically, $\mathbf{c}_t$ consists of a binary variable $\mathbf{y}_t = [y_t^0, \ldots, y_t^n] \in \{0,1\}^n$ denoting the present structures (or classes) in the t-th frame and $\mathbf{b}_t = [\mathbf{b}_t^0, \ldots, \mathbf{b}_t^n]^T \in \mathbb{R}^{n\times 4}$ denoting the respective bounding box coordinates. An autoencoder architecture was used to achieve the embedding, i.e., mapping $\mathbf{C}_t$ to $\mathbf{z}_t$. The encoder maps $\mathbf{C}_t$ to $\mathbf{z}_t$, and the decoder generates $\hat{\mathbf{c}}_t$, which represents the detections of the last frame in a given sequence, given $\mathbf{z}_t$. The model parameters are updated to ensure that $\hat{\mathbf{c}}_t$ fits $\mathbf{c}_t$ on a training set as will be explained in the following.

2.2 Object Detection

Our approach requires being able to detect anatomical structures as bounding boxes in frames from a video. To this end, the object detection part of the pipeline is fulfilled by an iteration of the YOLO network [17]. Specifically, the YOLOv7 network was used [27]. The network was trained on the endoscopic videos in the training set, where frames are sparsely labeled with bounding boxes, which contains 15 different anatomical classes and one surgical instrument class. The trained network was then applied to all the frames of the training videos to create detections of these classes on every frame of the videos. These are then used to train the subsequent autoencoder that models the embedding.

2.3   Embedding

To encode the bounding boxes onto the 1D latent space, the output from the YOLO network was slightly modified to exclude the surgical instrument, because the presence of the instrument in the frame is not necessarily correlated with the frame's position along the surgical path. The autoencoder was designed to reconstruct only the last frame c_t in C_t because z_t is desired to correspond to

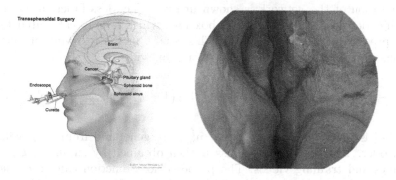

Fig. 2. Left: Transsphenoidal adenomectomy procedure is performed to remove a tumor from the pituitary gland, located at the base of the brain. Through the use of an endoscope and various instruments, the surgeon inserts the instruments into the nostril and crosses the sphenoidal sinus to access the pituitary gland located behind the sella floor. All procedures in the dataset only accessed one nostril to perform the procedure instead of two. Right: A video frame showing only the anatomy. Note that there is lack of clear differences between anatomical structures in such images.

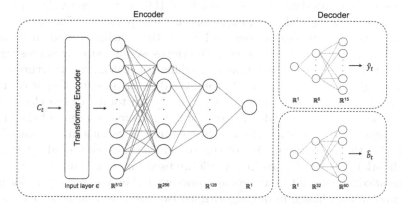

Fig. 3. The model architecture. The model consists of an encoder and two decoders. The encoder consists of a multi-head attention layer, i.e., a transformer encoder, which takes C_t as input, followed by a series of fully connected layers to embed the input in a 1D latent dimension. The two decoders consist of fully connected layers to generate the class probabilities $\hat{y}_t$ and the bounding box coordinates $\hat{b}_t$, respectively.

the current position. However, it takes into account s previous frames to provide more information while determining the latent representation $\mathbf{z}_t$ of an $\mathbf{x}_t$.

The encoder of the autoencoder network consists of multi-head attention layers followed by fully connected layers, which eventually reduce the features to a 1D value. Here a transformer-based encoder is used to encode the temporal information in the sequence of detections. The decoder consists of two fully connected decoders, the first of which generates the class probabilities $\hat{\mathbf{y}}_t$ of $\hat{\mathbf{c}}_t$ and the second generates the corresponding bounding boxes $\hat{\mathbf{b}}_t$. A simplified representation of the network is shown in Fig. 3. The loss function consists of a classification loss and a bounding box loss, which is only calculated for the classes present in the ground truth. This results in the following objective to minimize for the t-th frame in the m-th training video:

$$\mathcal{L}_{m,t} = -\sum_{i=1}^{n} \left(y_{m,t}^i \log\left(\hat{y}_{m,t}^i\right) + \left(1 - y_{m,t}^i\right) \log\left(1 - \hat{y}_{m,t}^i\right) \right) + \sum_{i=1}^{n} y_{m,t}^i \left| \mathbf{b}_{m,t}^i - \hat{\mathbf{b}}_{m,t}^i \right|,$$

where $|\cdot|$ is the l_1 loss and $\hat{y}_{m,t}^i$ and $\hat{\mathbf{n}}_{m,t}^i$ are generated from $\mathbf{z}_{m,t}$ using the autoencoder. The total training loss is then obtained by summing $\mathcal{L}_{m,t}$ over all frames and training videos. The proposed loss function can be considered to correspond to maximizing the joint likelihood of a given $\mathbf{y}$ and $\mathbf{b}$ with a probabilistic model that uses a mixture model for the bounding boxes.

3   Experiments and Results

3.1   Dataset

The object detection dataset used consists of 166 anonymized videos recorded during a transsphenoidal adenomectomy in 166 patients. The videos were recorded using various endoscopes and at multiple facilities, and made available through general research consent. The videos were labeled by neurosurgeons and include 16 different classes, that is, 15 different anatomical structure classes and one surgical instrument class. In total the dataset consists of approximately 19000 labeled frames, and around 3×10^6 frames in total. All the classes have only one instance in every video because of the anatomical nature of the human body, except for the instrument class, because of the various instruments being used during the procedures. Out of the 166 videos, 146 were used for training and validation, and 20 for testing. While we used different centers in our data, we acknowledge that all the centers are concentrated in one geographic location, which may induce biases in our algorithms. However, we also note that we use different endoscopes and they were acquired throughout the last 10 years.

3.2   Implementation Details

The implementation of the YOLO network follows [27] using an input resolution of 1280×1280. The model reached convergence after 125 epochs. To generate the data to train the autoencoder, the object confidence score and intersection-over-union (IoU) threshold were set to 0.25 and 0.45, respectively.

The autoencoder uses a transformer encoder that consists of six transformer encoder layers with five heads and an input size of $s \times 15 \times 5$, where s is set to 64 frames. Subsequently, the dimension of the output of the transformer encoder is reduced by three fully connected layers to 512, 256, and 128 using rectified linear unit (ReLU) activation functions in between. Finally, the last fully connected layer reduces the dimension to 1D and uses a sigmoid activation function to obtain the final latent variable. Furthermore, the two decoders, the class decoder and bounding box decoder, consist of two fully connected layers, increasing the dimension of the latent variable from 1D to 8, 15, and 32, 15×4, respectively. The first layer of both decoders is followed by a ReLU activation function and the final layer by a sigmoid activation function.

For training of the autoencoder, the AdamW optimizer [12] was used in combination with a warm-up scheduler that linearly increases the learning rate from 0 to 1×10^{-4} over 60 epochs. The model was trained for 170 epochs.

3.3   Results

Anatomical Structure Detection: The performance of the YOLO network on the test videos is shown in Table 1, using an IoU threshold for non-maximum suppression of 0.45 and an object confidence threshold of 0.001. The latter is set to 0.001 as this is the common threshold used in other detection works. It is surprising how well YOLO model works on the challenging problem of detecting anatomical structures in endoscopic neurosurgical videos.

Qualitative Assessment of the Embedding: First, to evaluate the learned latent representation, we compute the confidences for every class, i.e., y^i, for different points on the latent space, and plot them in Fig. 4. The confidences are normalized for every class, where the maximum confidence of a class corresponds to the darkest shade of blue, and vice versa. This shows how likely it is to find

Table 1. YOLO detection model results on 20 test videos with an IoU threshold for non-maximum suppression of 0.45 and an object confidence threshold of 0.001.

Class	AP_{50}	$AP_{50:95}$	Class	AP_{50}	$AP_{50:95}$
All (mean)	53.4	26.2	Ostium	43.1	19.4
Septum	78.6	57.9	Instrument	94.4	55.4
SupM	40.6	21.7	Rostrum	17.9	4.70
MidM	63.8	36.7	Sphenoidal Sinus	74.0	40.9
InfM	62.9	33.7	Sella Floor	70.6	27.6
Coana	54.3	22.2	Clival Recess	58.9	23.4
Floor	65.3	31.2	Planum	34.9	15.1
RecSphEthm	41.1	14.9	Osseous Carotis Right	21.6	5.37
Osseous Carotis Left	32.4	9.63			

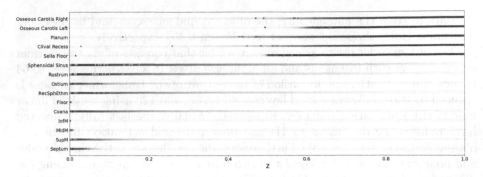

Fig. 4. The normalized generated confidences of each class along the latent space. This visualizes the probability of finding a certain anatomical structure at a specific point in the latent space. Additionally, video frames of twenty test videos responsible for the first appearances of anatomical structures in every video have been encoded and overlaid onto the confidence intervals to demonstrate that their locations correlate with the beginning of these intervals.

an anatomical structure at a certain location in the latent space, resembling a confidence interval for the structure's presence along the surgical path.

Figure 4 shows how the autoencoder encodes and separates anatomical structures along the surgical path. For example, from left ($z = 0$, the start of the surgical path) to right ($z = 1$, the end of the surgical path), it can be seen that the septum is frequently visible at the start of the surgical path, but later it is no longer be visible. Because a sequence encodes to a single point in the latent space, positioning along the surgical path is possible and allows the forecasting of structures in both the forward and backward directions.

Furthermore, twenty test videos were used to validate the spatial embedding of the anatomical structures. For every single one of the videos, the frame of the first appearance of every anatomical structure was noted. To obtain the corresponding z-value for each of the noted frames, a sequence was created from the same frame, using the s previous frames, for all of the noted frames. These sequences were then embedded into the 1D latent dimension to determine whether their locations corresponded to the beginning of the confidence intervals in the latent space where corresponding anatomical structures were expected to start appearing. When examining the encodings of the video data, it is evident that the points are located on the left side of the confidence interval for every class. When considering a path from $z = 0$ to $z = 1$, this demonstrates that the autoencoder is able to accurately map the first appearances of every class in the validation videos to the beginning of the confidence intervals for the classes, showing the network is capable of relative positional embedding.

Figure 5 plots the z-value against time (t) over an entire surgical video, where $t = 1$ denotes the end of the video. The plot on the right shows how the endoscope is frequently extracted going from a certain z-value back to 0, the beginning of the surgical path, instantaneously. Retraction of the endoscope and replacement is common in this surgery and the plot reflects this. Subsequently, swift reinsertion of the endoscope to the region of interest is performed, spending little time at z-values inferior to the ones visited before extraction. Additionally, locations along the latent space are visible where more time is spent than others, such as around $z = 0.2$ and around $z = 0.6$, which correspond to locations where tissue is removed or passageways are created, such as the opening of the sphenoidal sinus. We also note that the z-value also shoots to z=1 at certain times. Z-values from 0.5 to 1.0 actually correspond to a narrow section of the surgical path. However, this narrow section is the crux of the surgery where the surgeon spends more time. Hence the behavior of the model is expected since more time spent in this section leads to a higher number of images, and ultimately, covers a larger section of the latent space.

Lastly, we used the decoder to generate bounding boxes moving along the 1D latent dimension from one end to the other. A GIF showing the bounding boxes of the classes that the model expected to encounter along the surgical path can be found here at the following link: https://gifyu.com/image/SkMIX. Certain classes are expected at different points, and their locations and sizes vary along the latent space. From appearance to disappearance of classes, their bounding boxes grow larger and their centers either move away from the center of the frame to the outer region of the frame or stay centered. This behavior is expected when moving through a tunnel structure, as in an endoscopic procedure. This shows that the latent space can learn a roadmap of the surgery with the expected anatomical structures at any location on the surgical path.

Quantitative Assessment of the Embedding: Beyond qualitative analyses, we performed quantitative analysis to demonstrate that the latent space spatially embeds the surgery path. As there are no ground truth labels on the position of a video frame on the surgical path, a direct quantitative analysis comparing

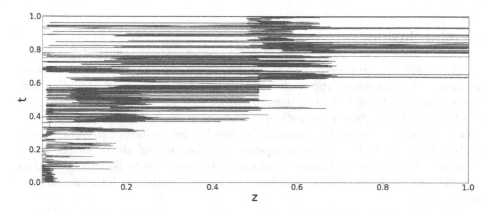

Fig. 5. Z-values over time during a surgical video. Certain z-values are encoded more frequently than others, such as approximately $z = 0.2$ and $z = 0.6$, which is related to the amount of time spent at a certain location during the surgery.

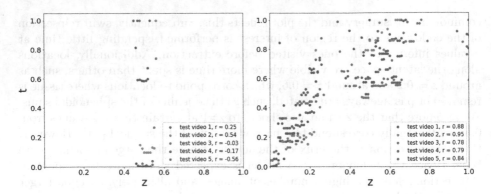

Fig. 6. Latent variable plotted against time of its first encoding for 5 surgical videos. Pearson correlation coefficients for first-time of appearance and z values are given for an untrained (left) and trained (right) model. In these plots, $t = 1$ denotes the end of a video. The untrained model provides a baseline for expected correlation coefficients. High correlation coefficients suggest the embedding captures the relative position on the surgical path.

z-value to ground truth position on the surgical path is not possible. To provide a quantitative evaluation, we make the observation that if the latent space represents the surgical path spatially, frames encoding z-values at the beginning of the path should be encountered in the early stages of the surgery, and vice versa. Therefore, the timestamp t of a sequence responsible for the first encoding of a specific z-value should increase with increasing z-value. This is confirmed by the mean correlation coefficient between t and z for the 20 videos, which is 0.80. Figure 6 shows the relation between t and z for five test videos with their corresponding Pearson correlation coefficients r for an untrained and a trained model.

4   Conclusion

In this study, we propose a novel approach to neuronavigation based on deep learning. The suggested approach is live image-based and uses bounding box detections of anatomical structures to localize itself on a common surgical roadmap that is learned from a dataset containing numerous videos from a specific surgical procedure. The mapping is modeled by the use of an autoencoder architecture and trained without supervision. The method allows for the localization and forecasting of anatomical structures that are to be encountered in forward and backward directions along the surgical path, similar to a mapping application.

The presented work has also some limitations. The main limitation is that we focused on only one surgery in this initial work. Extension to other surgeries is our future research topic. The proposed method can also be combined with SLAM approaches as well as guidance provided by MRI. Both of these directions also form our future work. Another limitation is that the latent dimension only provides relative positional encoding. Going beyond this may require further labels on the real position on the surgical path.

References

1. Berkmann, S., Schlaffer, S., Nimsky, C., Fahlbusch, R., Buchfelder, M.: Intraoperative high-field MRI for transsphenoidal reoperations of nonfunctioning pituitary adenoma. J. Neurosurg. **121**(5), 1166–1175 (2014)
2. Burkhardt, J.K., et al.: High-frequency intra-operative ultrasound-guided surgery of superficial intra-cerebral lesions via a single-burr-hole approach. Ultrasound Med. Biol. **40**(7), 1469–1475 (2014)
3. De Witt Hamer, P.C., Robles, S.G., Zwinderman, A.H., Duffau, H., Berger, M.S.: Impact of intraoperative stimulation brain mapping on glioma surgery outcome: a meta-analysis. J. Clin. Oncol.: Official J. Am. Soc. Clin. Oncol. **30**(20), 2559–2565 (2012)
4. Ismail Fawaz, H., Forestier, G., Weber, J., Idoumghar, L., Muller, P.-A.: Evaluating surgical skills from kinematic data using convolutional neural networks. In: Frangi, A.F., Schnabel, J.A., Davatzikos, C., Alberola-López, C., Fichtinger, G. (eds.) MICCAI 2018. LNCS, vol. 11073, pp. 214–221. Springer, Cham (2018). https://doi.org/10.1007/978-3-030-00937-3_25
5. Grasa, O.G., Bernal, E., Casado, S., Gil, I., Montiel, J.M.: Visual SLAM for hand-held monocular endoscope. IEEE Trans. Med. Imaging **33**(1), 135–146 (2014)
6. Grasa, O.G., Civera, J., Montiel, J.M.: EKF monocular SLAM with relocalization for laparoscopic sequences. In: Proceedings of the IEEE International Conference on Robotics and Automation, pp. 4816–4821 (2011)
7. Hadjipanayis, C.G., Widhalm, G., Stummer, W.: What is the surgical benefit of utilizing 5-aminolevulinic acid for fluorescence-guided surgery of malignant gliomas? Neurosurgery **77**(5), 663–673 (2015)
8. Härtl, R., Lam, K.S., Wang, J., Korge, A., Kandziora, F., Audigé, L.: Worldwide survey on the use of navigation in spine surgery. World Neurosurg. **79**(1), 162–172 (2013)
9. Hervey-Jumper, S.L., et al.: Awake craniotomy to maximize glioma resection: methods and technical nuances over a 27-year period. J. Neurosurg. **123**(2), 325–339 (2015)
10. Iversen, D.H., Wein, W., Lindseth, F., Unsgård, G., Reinertsen, I.: Automatic intraoperative correction of brain shift for accurate neuronavigation. World Neurosurg. **120**, e1071–e1078 (2018)
11. Leonard, S., et al.: Evaluation and stability analysis of video-based navigation system for functional endoscopic sinus surgery on in vivo clinical data. IEEE Trans. Med. Imaging **37**(10), 2185–2195 (2018)
12. Loshchilov, I., Hutter, F.: Decoupled weight decay regularization. In: 7th International Conference on Learning Representations, ICLR 2019 (2017)
13. Luongo, F., Hakim, R., Nguyen, J.H., Anandkumar, A., Hung, A.J.: Deep learning-based computer vision to recognize and classify suturing gestures in robot-assisted surgery. Surgery **169**(5), 1240–1244 (2021)
14. Mahmoud, N., et al.: ORBSLAM-based endoscope tracking and 3D reconstruction. In: Peters, T., et al. (eds.) CARE 2016. LNCS, vol. 10170, pp. 72–83. Springer, Cham (2017). https://doi.org/10.1007/978-3-319-54057-3_7

15. Orringer, D.A., Golby, A., Jolesz, F.: Neuronavigation in the surgical management of brain tumors: current and future trends. Expert Rev. Med. Dev. **9**(5), 491–500 (2012)
16. Ozyoruk, K.B., et al.: EndoSLAM dataset and an unsupervised monocular visual odometry and depth estimation approach for endoscopic videos. Med. Image Anal. **71**, 102058 (2021)
17. Redmon, J., Divvala, S., Girshick, R., Farhadi, A.: You only look once: unified, real-time object detection. In: Proceedings of the IEEE Computer Society Conference on Computer Vision and Pattern Recognition, pp. 779–788 (2015)
18. Rivas-Blanco, I., Perez-Del-Pulgar, C.J., Garcia-Morales, I., Munoz, V.F., Rivas-Blanco, I.: A review on deep learning in minimally invasive surgery. IEEE Access **9**, 48658–48678 (2021)
19. Sanai, N., Mirzadeh, Z., Berger, M.S.: Functional outcome after language mapping for glioma resection. N. Engl. J. Med. **358**(1), 18–27 (2008)
20. Sarikaya, D., Corso, J.J., Guru, K.A.: Detection and localization of robotic tools in robot-assisted surgery videos using deep neural networks for region proposal and detection. IEEE Trans. Med. Imaging **36**(7), 1542–1549 (2017)
21. Staartjes, V.E., et al.: Machine learning in neurosurgery: a global survey. Acta Neurochir. **162**(12), 3081–3091 (2020)
22. Staartjes, V.E., Volokitin, A., Regli, L., Konukoglu, E., Serra, C.: Machine vision for real-time intraoperative anatomic guidance: a proof-of-concept study in endoscopic pituitary surgery. Oper. Neurosurg. (Hagerstown, Md.) **21**(4), 242–247 (2021)
23. Stienen, M.N., Fierstra, J., Pangalu, A., Regli, L., Bozinov, O.: The Zurich checklist for safety in the intraoperative magnetic resonance imaging suite: technical note. Oper. Neurosurg. (Hagerstown, Md.) **16**(6), 756–765 (2019)
24. Stummer, W., Stepp, H., Wiestler, O.D., Pichlmeier, U.: Randomized, prospective double-blinded study comparing 3 different doses of 5-aminolevulinic acid for fluorescence-guided resections of malignant gliomas. Neurosurgery **81**(2), 230–239 (2017)
25. Thananjeyan, B., Garg, A., Krishnan, S., Chen, C., Miller, L., Goldberg, K.: Multilateral surgical pattern cutting in 2D orthotropic gauze with deep reinforcement learning policies for tensioning (2017)
26. Ulrich, N.H., Burkhardt, J.K., Serra, C., Bernays, R.L., Bozinov, O.: Resection of pediatric intracerebral tumors with the aid of intraoperative real-time 3-D ultrasound. Child's Nervous Syst.: ChNS: Official J. Int. Soc. Pediatr. Neurosurg. **28**(1), 101–109 (2012)
27. Wang, C.Y., Bochkovskiy, A., Liao, H.Y.M.: YOLOv7: trainable bag-of-freebies sets new state-of-the-art for real-time object detectors (2022)
28. Wang, S., Raju, A., Huang, J.: Deep learning based multi-label classification for surgical tool presence detection in laparoscopic videos, pp. 620–623 (2017)

Meta-information-Aware Dual-path Transformer for Differential Diagnosis of Multi-type Pancreatic Lesions in Multi-phase CT

Bo Zhou[1,2(✉)], Yingda Xia[1(✉)], Jiawen Yao[1], Le Lu[1], Jingren Zhou[1], Chi Liu[2,3], James S. Duncan[2,3], and Ling Zhang[1]

[1] DAMO Academy, Alibaba Group, Beijing, China
yingda.xia@alibaba-inc.com
[2] Department of Biomedical Engineering, Yale University, New Haven, US
bo.zhou@yale.edu
[3] Department of Radiology and Biomedical Imaging, Yale University, New Haven, US

Abstract. Pancreatic cancer is one of the leading causes of cancer-related death. Accurate detection, segmentation, and differential diagnosis of the full taxonomy of pancreatic lesions, i.e., normal, seven major types of lesions, and "other" lesions, is critical to aid the clinical decision-making of patient management and treatment. However, existing work focus on segmentation and classification for very specific lesion types (PDAC) or groups. Moreover, none of the previous work considers using lesion prevalence-related non-imaging patient information to assist the differential diagnosis. To this end, we develop a meta-information-aware dual-path transformer and exploit the feasibility of classification and segmentation of the full taxonomy of pancreatic lesions. Specifically, the proposed method consists of a CNN-based segmentation path (S-path) and a transformer-based classification path (C-path). The S-path focuses on initial feature extraction by semantic segmentation using a UNet-based network. The C-path utilizes both the extracted features and meta-information for patient-level classification based on stacks of dual-path transformer blocks that enhance the modeling of global contextual information. A large-scale multi-phase CT dataset of 3,096 patients with the pathology-confirmed pancreatic lesion class labels, voxel-wise manual annotations of lesions from radiologists, and patient meta-information, was collected for training and evaluations. Our results show that our method can enable accurate classification and segmentation of the full taxonomy of pancreatic lesions, approaching the accuracy of the radiologist's report and significantly outperforming previous baselines. Results also show that adding the common meta-information, i.e., gender and age, can boost the model's performance, thus demonstrating the importance of meta-information for aiding pancreatic disease diagnosis.

Keywords: Pancreatic Lesion · Dual-path Transformer · Meta-information Aware · Differential Diagnosis

This work was supported by Alibaba Group through Alibaba Research Intern Program.

1   Introduction

Pancreatic cancer is the third leading cause of death among all cancers in the United States, and has the poorest prognosis among all solid malignancies with a 5-year survival rate of about 10% [4]. Early diagnosis and treatment are crucial, which can potentially increase the 5-year survival rate to about 50% [3]. In clinical practice, pancreatic patient management is based on the pancreatic lesion type and the potential of the lesion to become invasive cancer. However, pancreatic lesions are often hard to reach by biopsy needle because of the deep location in the abdomen and the complex structure of surrounding organs and vessels. To this end, accurate imaging-based differential diagnosis of pancreatic lesion type is critical to aid the clinical decision-making of patient management and treatment, e.g., surgery, monitoring, or discharge [11,18]. Multi-phase Computed Tomography (CT) is the first-line imaging tool for pancreatic disease diagnosis. However, accurate differential diagnosis of pancreatic lesions is very challenging because 1) the same type of lesion may have different textures, shapes, and contrast patterns across multi-phase CT, and 2) pancreatic ductal adenocarcinoma (PDAC) accounts for the majority of cases, e.g., >60%, in pathology-confirmed patient population, leading to a long-tail problem.

Most related work in automatic pancreatic CT image analysis focus on segmentation of certain types of pancreatic lesions, e.g., PDAC and pancreatic neuroendocrine tumor (PNET). UNet-based detection-by-segmentation approaches have been extensively studied for the detection of PDAC [16,17,19,20,22] and PNET [21]. Shape-induced information, e.g., tubular structure of dilated duct, is exploited to improve the PDAC detection [9,13]. Graph-based classification network is proposed for pancreatic patient risk stratification and management [18]. There are also recent attempts in the detection and classification of PDAC and nonPDAC using non-contrast CT [14]. However, none of the previous work has yet attempted to address the key clinical need for detection and classification of full taxonomy of pancreatic lesions, i.e., PDAC, PNET, solid pseudopapillary tumor (SPT), intraductal papillary mucinous lesion (IPMN), mucinous cystic lesion (MCN), chronic pancreatitis (CP), serous cystic lesion (SCN) [11], and other rare types that can be further classified into other benign and other malignant. Furthermore, no methods consider adding lesion prevalence-related non-imaging patient information to aid the diagnosis. For example, based on epidemiological data, the incidence of MCN, SCN, and SPT in women is significantly higher than that in men, and MCN, SCN, and SPT has a higher prevalence in young-age, middle-age, and old-age female, respectively [7]. Integrating easily-accessible clinical patient meta-information, e.g., gender and age in the DICOM head, as classification feature inputs could potentially further improve the diagnosis accuracy without needing radiologists' manual input.

To address these challenges and unmet needs, we propose a meta-information-aware dual-path transformer (MDPFormer) for classification and segmentation of the full taxonomy of pancreatic lesions, including normal, seven major types of pancreatic lesions, other malignant, and other benign. Motivated by the recent dual-path design of Mask Transformers [1,12], the proposed MDPFormer consists of a segmentation path (S-path) and a classification path (C-path). The S-path focuses on initial feature extraction by semantic segmentation (normal, PDAC, and nonPDAC) using a CNN-based network. Then, the C-path utilizes both meta-information and the extracted features for individual-level classification (normal, PDAC, PNET, SPT, IPMN, MCN, CP, SCN, other benign, and other malignant) based on stacked dual-path transformer blocks that enhance the modeling of global contextual information. We curated a large-scale multi-phase CT dataset with the pathology-confirmed pancreatic lesion class labels, voxel-wise manual annotations of lesions from radiologists, and patient meta-information. To our knowledge, this model is the most comprehensive to date, and is trained on a labeled dataset (2,372 patients' multi-phase CT scans) larger than that used in previous studies [10,15,18]. We independently test our method on a test set consisting of one whole year of 724 consecutive patients with pancreatic lesions from a high-volume pancreatic cancer center. The experimental results show that our method enables accurate classification and segmentation of the full taxonomy of pancreatic lesions, approaching the accuracy of radiologists' reports (by second-line senior readers via referring to current and previous imaging, patient history, and clinical meta-information). Our method without meta-information input demonstrates superior classification and segmentation performance as compared to previous baselines. Adding the meta-information-aware design further boosts the model's performance, demonstrating the importance of meta-information for improving pancreatic disease diagnosis.

2   Methods

The general pipeline of our method is illustrated in Fig. 1. Our pipeline consists of two stages. In the first stage, we use a localization UNet [2] to segment out the pancreas from the whole CT volume. The sub-volume containing the pancreas is then cropped out based on the segmentation mask. In the second stage, the resized sub-volume is inputted into the meta-information-aware dual-path transformer (MDPFormer) to segment and classify the pancreatic lesions. Details are elaborated in the following sections.

Meta-information-Aware Dual-path Transformer. For classification, we denote $\mathcal{H}_c = \{0, 1, 2, \cdots, 9\}$ for the ten patient/lesion classes, i.e., normal, PDAC, PNET, SPT, IPMN, MCN, CP, SCN, other benign, and other malignant. For segmentation, we group the last eight classes into nonPDAC and denote $\mathcal{H}_s = \{0, 1, 2\}$ for the grouped three patient classes, i.e., normal, PDAC, and nonPDAC. The goal is to enable a more balanced initial class distribution for segmentation, while enabling feature extraction for the full pancreatic lesion taxonomy classification. The training set is thus formulated as $S = \{(X_i, M_i, Y_i, Z_i)|i = 1, 2, \cdots, N\}$, where X_i is the cropped pancreas sub-volume of the i-th patients, M_i is the patient meta information (gender and age), $Y_i \in \mathcal{H}_s$ is the 3-class voxel-wise annotation with the same spatial size as X_i, and $Z_i \in \mathcal{H}_c$ is the 10-class volume-wise label that confirmed by pathology or clinical records.

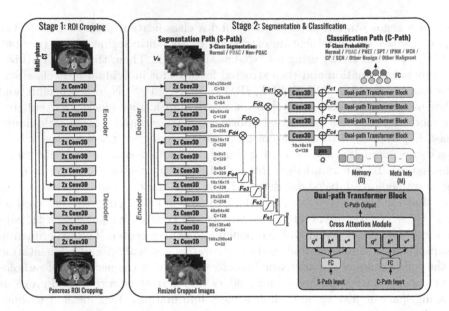

Fig. 1. The overall pipeline and the detailed structure of our MDPFormer. In stage 1, the pancreas sub-volume is cropped based on a coarse pancreas segmentation mask. In stage 2, the resized pancreas sub-volume is inputted into the MDPFormer for segmentation (left path) and classification (right path). The design of dual-path transformer block in the classification path is illustrated on the bottom right (grey box). (Color figure online)

The MDPFormer consists of two paths, including a segmentation path (S-Path) and a classification path (C-Path). The goal of S-path is to extract rich feature representations of the lesion and pancreas at multiple scales by first segmenting the image into three general classes. Given a input X and a segmentation network G_s, we have

$$V_s, F_{d1}, F_{d2}, F_{d3}, F_{d4}, F_{e1}, F_{e2}, F_{e3}, F_{e4} = G_s(X) \qquad (1)$$

where V_s is the segmentation output, $F_{d1}, F_{d2}, F_{d3}, F_{d4}$ are the multi-scale features from the decoder, $F_{e1}, F_{e2}, F_{e3}, F_{e4}$ are the multi-scale features from the encoder. Here, we deploy a 3D UNet [2] as the S-Path backbone network. Instead of directly using the decoder features as C-path input, we combine the multi-scale encoder and decoder features by

$$F_c = f_c(F_d * \sigma(F_e)) + Q \qquad (2)$$

where σ is the sigmoid function for generating attention from the encoder features to guide decoder feature outputs, f_c is a convolution layer that further refines the S-Path feature output, and Q is the learnable position embedding feature that provides position representation to aid the transformer in C-path. F_c is the extracted feature from the S-Path which is used for C-Path input.

The C-Path consists of four consecutive dual-path transformer blocks, where each block takes both the S-Path feature and the global memory feature as inputs. Denote D as the initial 1D memory feature which is randomly initialized learnable parameters [12], we fuse the patient meta-information with the initial memory feature by

$$F_m = [D, M] \tag{3}$$

where D and M are concatenated in the length dimension and M is the meta-information, i.e., patient gender and age, in this work. In each block, we use a cross-attention module to fuse F_m and F_c. First, we compute S-Path queries q^s, keys k^s, and values v^p, by learnable linear projections of the S-Path feature F_s at each feature location. Similarly, queries q^c, keys k^c, and values v^c are computed from C-path global memory feature F_c with another set of projection matrices. The cross-attention output can then be calculated as follows:

$$y^c = softmax(q^c \cdot k^{cs})v^{cs}, \tag{4}$$

$$k^{cs} = \begin{bmatrix} k^c \\ k^s \end{bmatrix}, v^{cs} = \begin{bmatrix} v^c \\ v^s \end{bmatrix}, \tag{5}$$

where $[\cdot]$ is the concatenation operator in the channel dimension to fuse the values and keys from both paths. The output y^c is then inputted into the next block as the F_m memory feature input. Using the C-path feature output from the last dual-path transformer block, we predict the final classification P with two fully connected layers and a softmax. The overall training objective can thus be formulated as:

$$\mathcal{L}_{all} = \mathcal{L}_s(V_s, Y) + \mathcal{L}_c(P, Z) \tag{6}$$

where $\mathcal{L}_s(\cdot)$ is the Dice loss function for segmentation training, and $\mathcal{L}_c(\cdot)$ is the cross-entropy loss for classification training.

3   Experimental Results

Data Preparation. We collected a large-scale multi-phase CT dataset consisting of 3,096 patients from a high-volume pancreatic cancer institution. Each multi-phase CT consists of noncontrast, arterial, and venous phase CT. The data were consecutively collected from 2015–2020. All the 724 patients scanned during 2020 were used as the independent test set, and the rest of the 2,372 patients scanned from 2015–2019 were used as the training set. The training set includes 707 normal, 1,088 PDAC, 110 PNET, 68 SPT, 162 IPMN, 32 MCN, 64 CP, 93 SCN, 48 other benign, and 24 other malignant cases. The test set includes 202 normal, 283 PDAC, 34 PNET, 25 SPT, 73 IPMN, 9 MCN, 29 CP, 38 SCN, 14 other benign, and 17 other malignant cases. All patients with lesions were confirmed by surgical pathology, while normal patients were confirmed by radiology reports and at least 2-year follow-ups. The annotation of lesions was performed collaboratively by an experienced radiologist (with 14 years of specialized experience in pancreatic imaging) and an auto-segmentation model on either arterial or venous phase CT, whichever with better lesion visibility. More specifically, the radiologist first annotates some data to train an auto-segmentation model to segment the remaining data, which is then checked/edited by the radiologist. The CT phases were registered using DEEDS [6]. The gender and age information were extracted from the DICOM head as meta information inputs. The gender is converted to a binary value, i.e., 0 for female and 1 for male. The age is normalized between 0–1 by dividing the value by 100.

Implementation Details. All CT volumes were resampled into $0.68 \times 0.68 \times 3.0$ mm spacing and normalized into zero mean and unit variance. In the training phase of MDPFomer, we cropped the foreground 3D bounding box of the pancreas region, randomly pad small margins on each dimension, and resized the sub-volume into $160 \times 256 \times 40$ ($Y \times X \times Z$) for input. We deployed a 5-fold cross-validation strategy using the 2,372 training set to train and validate five models. During inference, the five models' predictions were ensemble by averaging the prediction results. For each fold, we first pre-trained the S-path network for 1000 epochs, and then trained the whole model in an end-to-end fashion with an SGD optimizer. The initial learning rate was set to 1×10^{-3} with cosine decay, and the batch size was set to 3. The localization UNet in the first stage followed the same training protocol.

Compared Methods and Evaluation Metrics. Our method is compared with two types of baseline approaches. One is the "segmentation for classification (S4C)" method where a segmentation network, i.e., nnUNet [8] or (nn)UNETR [5,8], is first deployed for semantic segmentation of the ten classes on the cropped sub-volume. We then classify the patient based on the class-wise lesion segmentation size. Specifically, if one or multiple lesion classes were presented in the segmentation, we classify the patient to the lesion class with the largest segmentation size; Otherwise, we classify the patient as normal. Note that we implement UNETR [5] in the nnUNet framework [8], called (nn)UNETR, which shows substantially better results than the original UNETR implementation on our data. The other baseline is the CNN-based segmentation-to-classification method. We use the exact same structure of S-path in MDPFormer, and extract all encoder and decoder multi-scale features. Then, we apply global max pooling on each feature map, concatenate them and forward them into two fully connected layers for classification. We also compared our performance with the radiology report, which represents the clinical read performance of second-line senior radiologists (via referring to current and previous imaging, patient history, and clinical information) in the high-volume pancreatic cancer center. The classification performance was evaluated by class-wise accuracy, regular accuracy, and balanced accuracy. The confusion matrices were also reported for detailed evaluation. The segmentation performance was evaluated by the Dice coefficient or score on each class of pancreatic lesion or normal.

3.1 Main Results

The classification results are summarised in Table 1. Comparing DPFormer without meta-information-aware to the previous baseline methods, i.e., UNet-based

Table 1. Evaluation of classification performance on lesion diagnosis (%). Both averaged accuracy (second last row) and balanced accuracy (last row) are reported.

CLASSIFY	nnUNet	(nn)UNETR	SPath+FC	DPFormer	MDPFormer	Report
Normal	96.0	96.2	97.0	**99.0**	99.5	100
PDAC	**94.3**	94.1	**94.3**	**94.3**	96.5	93.3
PNET	38.2	37.5	35.3	**47.1**	47.1	70.6
SPT	64.0	62.8	60.0	**64.0**	72.0	84.0
IPMN	69.9	**68.1**	43.8	60.3	65.8	68.5
MCN	0.0	0.0	11.1	**11.1**	33.3	33.3
CP	6.9	17.2	24.1	**31.0**	44.8	69.0
SCN	44.7	42.1	**50.0**	**50.0**	55.3	42.1
Other-BEN	0.0	0.0	21.4	**28.6**	35.7	35.7
Other-MLG	0.0	0.0	0.0	**11.7**	11.7	17.6
Regular Acc	77.4	77.4	76.2	**79.8**	82.9	84.0
Balance Acc	41.4	41.8	43.6	**49.7**	56.2	61.4

S4C, UNETR-based S4C, and S-Path+FC, we can see that DPFormer can already outperform all the baselines in 9 out of 10 classes and achieve the highest balanced accuracy of 49.71%. In general, it is challenging to use the conventional segmentation approaches to directly segment out the 10 classes and perform classification based on it. For S4C approaches, we can see the classification accuracy of MCN, CP, other benign, and other malignant are all zero or near zero. While S-Path+FC provide slightly better classification result with the additional FC layer for classification, DPFormer with dual path transformer and better feature fusion provides better results. With the meta-information-aware design that incorporates additional gender and age information, our MDPFormer utilizes those easily-accessible tumor-type-related non-imaging information, thus achieving further improved classification balanced accuracy of 56.17%.

The classification results compared to the radiology report are also shown in Fig. 1 and elaborated in Fig. 2. The balanced classification accuracy of the radiology report is 61.41%. Adding meta-information improves our method's balanced classification performance from 49.71% to 56.17%, approaching the performance of the radiology report. Our method also provides better PDAC (96.5% vs. 93.3%) and SCN (55.4% vs. 42.1%) diagnosis accuracy as compared to the reports, which is critical since PDAC is of the highest priority among all pancreatic abnormalities with a 5-year survival rate of approximately 10% and is the most common type (>60% of all pathology-confirmed pancreatic lesions). In general, the radiology reports that perform diagnosis with more meta-information, e.g., patient history, tumor markers, previous report, etc., provide better classification accuracy. Thus, adding additional meta-information may further improve our method's performance. In addition, unlike radiology reports that only give the final diagnosis, our MPDFormer provides both classification probabilities and class-wise lesion segmentation outputs with explainability. Examples of our MDPFormer's classification and segmentation results are shown in Fig. 3.

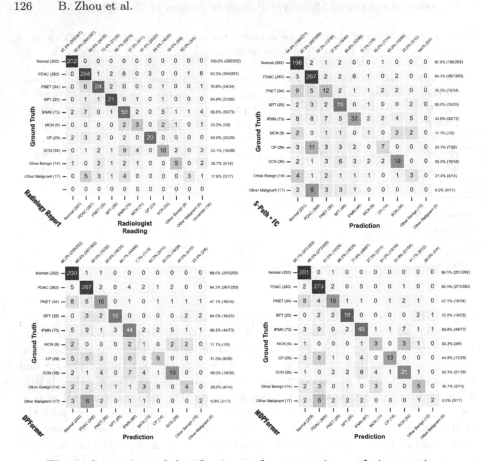

Fig. 2. Comparison of classification performance using confusion matrices.

The accuracy of the "Report" for the normal class is 100% (Table 1 and Fig. 2). This is because our normal cases were selected based on the radiology reports reporting an absence of pancreatic lesions. Actually, the radiologists' specificity for the normal pancreas is 93%–96% in a pancreas CT interpretation setting [10]. Our MDPFormer has a higher specificity (99.5%) than radiologists, making it a reliable detection tool for pancreatic lesions in practice.

Ablative studies for the segmentation performance are summarized in Table 2. For MDPFormer, DPFormer, and SPath+FC, please note that the nonPDAC segmentation class is assigned by the final classification prediction. Similar to the observation from classification evaluations, it is difficult for nnUNet and UNETR to directly perform 10-class segmentation with averaged Dice scores of 0.360 and 0.373 reported, respectively. On the other hand, our MDPFormer can provide significantly better segmentation performances for all 10 normal and lesion classes ($p < 0.001$) and achieve an averaged Dice score of 0.604. Comparing MDPFormer to DPFormer, we can also see that adding the meta-information improves the segmentation performance (averaged Dice of 0.604 versus 0.502).

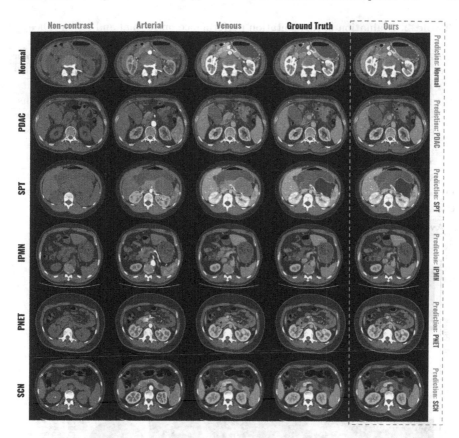

Fig. 3. Examples of classification and segmentation outputs from our MDPFormer. Ground truth lesion classes are annotated on the left and the predicted classes are shown on the right. Segmented pancreas is depicted in Red; lesion in Green or Blue. (Color figure online)

Note that the Dice scores reported in Table 2 are generally higher than that reported in previous studies [15,17,18]. This is mainly because the ground truth annotations are generated semi-automatically. Nevertheless, the above results clearly demonstrate the superiority of our MDPFormer over compared methods.

Next, we provide three patient case studies to show the impact of adding meta-information for classifying the pancreas lesion. The studies are illustrated in Fig. 4, including three patients with MCN, SCN, and SPT, respectively. Using DPFormer without patient meta-information, the MCN, SCN, and SPT were misclassified as other benign, IPMN, and other malignant, respectively. The MDFormer adding the gender and age information to the imaging information provide more accurate tumor probability predictions. For example, for the female 68-year-old patient with SCN, the maximal probability predicted by DPFormer

128   B. Zhou et al.

Table 2. Evaluation of segmentation performance on normal and lesion (Dice).

SEGMENT	nnUNet	(nn)UNETR	SPath+FC	DPFormer	MDPFormer
Normal	0.950±0.118	0.940±0.109	0.951±0.107	**0.953±0.096**	0.958±0.069
PDAC	0.863±0.157	0.860±0.149	**0.866±0.189**	0.865±0.199	0.869±0.196
PNET	0.259±0.302	0.288±0.310	0.352±0.381	**0.355±0.391**	0.456±0.390
SPT	0.513±0.370	0.537±0.352	0.624±0.326	**0.662±0.429**	0.766±0.414
IPMN	0.475±0.304	0.468±0.302	0.515±0.340	**0.518±0.390**	0.598±0.382
MCN	0.071±0.159	0.098±0.189	0.211±0.446	**0.312±0.395**	0.416±0.441
CP	0.051±0.098	0.112±0.253	0.280±0.323	**0.349±0.338**	0.382±0.335
SCN	0.431±0.351	0.428±0.348	0.484±0.303	**0.587±0.441**	0.765±0.438
Other-BEN	0.0±0.0	0.0±0.0	0.227±0.397	**0.293±0.364**	0.459±0.422
Other-MLG	0.0±0.0	0.0±0.0	0.088±0.247	**0.129±0.284**	0.373±0.394
Average	0.361	0.373	0.464	**0.502**	0.604

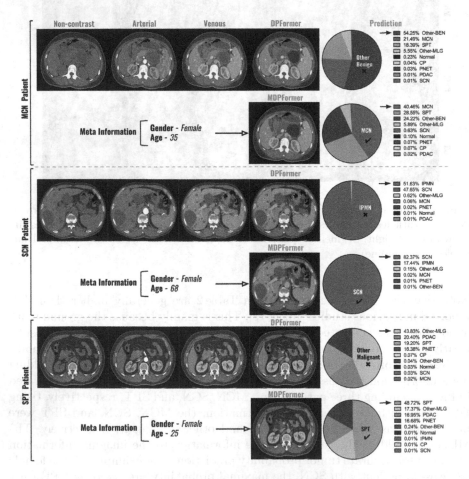

Fig. 4. Case studies of three patients with MCN, SCN, and SPT. The classification probability predictions of DPFormer and MDPFormer models are shown on the right.

is 51.63% for IPMN, while MDPFormer with meta-information provides the maximal probability of 82.37% for SCN.

3.2 Discussion

In this work, we present a meta-information-aware dual-path transformer (MDP-Former) for the classification and segmentation of pancreatic lesions in multi-phase CT. The MDFormer consists of an S-path and C-path, where the S-path focuses on initial feature extraction by group-level segmentation and the C-path utilizes both meta-information and the extracted features for individual-level classification. Compared to previous baselines, our method without meta-information input already shows superior classification and segmentation performance. Adding the meta-information-aware design further boost these performances, demonstrating the importance of meta-information when diagnosing specific pancreatic lesion type. Our MDPFormer is an open framework with several key components adjustable, which could potentially further improve our future performances. First, we used a simple UNet architecture with two consecutive convolution layers at each scale level for feature extraction. Using more advanced segmentation network blocks maybe can provide richer feature representations for better classification and segmentation performances. Second, we only used meta-information of patient gender and age as inputs, which can be automatically extracted from every DICOM data in practice. Adding additional non-imaging information, e.g., family history, symptoms (weight loss, jaundice), and other patient records (CA 19-9 blood test), may further potentially improve MDPFormer to better match the performance of the radiologists who have access to those non-imaging information for diagnosis. Those are important research directions for our future work.

4   Conclusion

This paper presents a new meta-information-aware dual-path transformer for classification and segmentation of the full taxonomy of pancreatic lesions. Our experimental results show that the proposed dual-path transformer can efficiently incorporate the patient meta-information and the extracted image features from the CNN-based segmentation path to make accurate pancreatic lesion classification and segmentation. We demonstrate that our method achieves better performance than previous baselines and approaches the accuracy of radiology reports. Our system could be a useful assistant tool for pancreatic lesion detection, segmentation, and diagnosis in the clinical reading environment.

References

1. Cheng, B., Schwing, A., Kirillov, A.: Per-pixel classification is not all you need for semantic segmentation. Adv. Neural. Inf. Process. Syst. **34**, 17864–17875 (2021)

2. Çiçek, Ö., Abdulkadir, A., Lienkamp, S.S., Brox, T., Ronneberger, O.: 3D U-Net: learning dense volumetric segmentation from sparse annotation. In: Ourselin, S., Joskowicz, L., Sabuncu, M.R., Unal, G., Wells, W. (eds.) MICCAI 2016. LNCS, vol. 9901, pp. 424–432. Springer, Cham (2016). https://doi.org/10.1007/978-3-319-46723-8_49

3. Conroy, T., et al.: FOLFIRINOX or gemcitabine as adjuvant therapy for pancreatic cancer. N. Engl. J. Med. **379**(25), 2395–2406 (2018)

4. Grossberg, A.J., et al.: Multidisciplinary standards of care and recent progress in pancreatic ductal adenocarcinoma. CA: Cancer J. Clin. **70**(5), 375–403 (2020)

5. Hatamizadeh, A., Nath, V., Tang, Y., Yang, D., Roth, H.R., Xu, D.: Swin UNETR: swin transformers for semantic segmentation of brain tumors in MRI images. In: Crimi, A., Bakas, S. (eds.) BrainLes 2021. Lecture Notes in Computer Science, vol. 12962, pp. 272–284. Springer, Cham (2022). https://doi.org/10.1007/978-3-031-08999-2_22

6. Heinrich, M.P., Jenkinson, M., Brady, M., Schnabel, J.A.: MRF-based deformable registration and ventilation estimation of lung CT. IEEE Trans. Med. Imaging **32**(7), 1239–1248 (2013)

7. Hu, F., et al.: Cystic neoplasms of the pancreas: differential diagnosis and radiology correlation. Front. Oncol. **12**, 860740 (2022)

8. Isensee, F., Jaeger, P.F., Kohl, S.A., Petersen, J., Maier-Hein, K.H.: NNU-Net: a self-configuring method for deep learning-based biomedical image segmentation. Nat. Methods **18**(2), 203–211 (2021)

9. Liu, F., Xie, L., Xia, Y., Fishman, E., Yuille, A.: Joint shape representation and classification for detecting PDAC. In: Suk, H.-I., Liu, M., Yan, P., Lian, C. (eds.) MLMI 2019. LNCS, vol. 11861, pp. 212–220. Springer, Cham (2019). https://doi.org/10.1007/978-3-030-32692-0_25

10. Park, H.J., et al.: Deep learning-based detection of solid and cystic pancreatic neoplasms at contrast-enhanced CT. Radiology, 220171 (2022)

11. Springer, S., et al.: A multimodality test to guide the management of patients with a pancreatic cyst. Sci. Transl. Med. **11**(501), eaav4772 (2019)

12. Wang, H., Zhu, Y., Adam, H., Yuille, A., Chen, L.C.: Max-DeepLab: end-to-end panoptic segmentation with mask transformers. In: Proceedings of the IEEE/CVF Conference on Computer Vision and Pattern Recognition, pp. 5463–5474 (2021)

13. Wang, Y., et al.: Deep distance transform for tubular structure segmentation in CT scans. In: Proceedings of the IEEE/CVF Conference on Computer Vision and Pattern Recognition, pp. 3833–3842 (2020)

14. Xia, Y., et al.: Effective pancreatic cancer screening on non-contrast CT scans via anatomy-aware transformers. In: de Bruijne, M., et al. (eds.) MICCAI 2021. LNCS, vol. 12905, pp. 259–269. Springer, Cham (2021). https://doi.org/10.1007/978-3-030-87240-3_25

15. Xia, Y., et al.: The Felix project: deep networks to detect pancreatic neoplasms. medRxiv (2022)

16. Xia, Y., et al.: Detecting pancreatic ductal adenocarcinoma in multi-phase CT scans via alignment ensemble. In: Martel, A.L., et al. (eds.) MICCAI 2020. LNCS, vol. 12263, pp. 285–295. Springer, Cham (2020). https://doi.org/10.1007/978-3-030-59716-0_28

17. Zhang, L., et al.: Robust pancreatic ductal adenocarcinoma segmentation with multi-institutional multi-phase partially-annotated CT scans. In: Martel, A.L., et al. (eds.) MICCAI 2020. LNCS, vol. 12264, pp. 491–500. Springer, Cham (2020). https://doi.org/10.1007/978-3-030-59719-1_48

18. Zhao, T., et al.: 3D graph anatomy geometry-integrated network for pancreatic mass segmentation, diagnosis, and quantitative patient management. In: Proceedings of the IEEE/CVF Conference on Computer Vision and Pattern Recognition, pp. 13743–13752 (2021)
19. Zhou, Y., et al.: Hyper-pairing network for multi-phase pancreatic ductal adenocarcinoma segmentation. In: Shen, D., et al. (eds.) MICCAI 2019. LNCS, vol. 11765, pp. 155–163. Springer, Cham (2019). https://doi.org/10.1007/978-3-030-32245-8_18
20. Zhou, Y., Xie, L., Fishman, E.K., Yuille, A.L.: Deep supervision for pancreatic cyst segmentation in abdominal CT scans. In: Descoteaux, M., Maier-Hein, L., Franz, A., Jannin, P., Collins, D.L., Duchesne, S. (eds.) MICCAI 2017. LNCS, vol. 10435, pp. 222–230. Springer, Cham (2017). https://doi.org/10.1007/978-3-319-66179-7_26
21. Zhu, Z., Lu, Y., Shen, W., Fishman, E.K., Yuille, A.L.: Segmentation for classification of screening pancreatic neuroendocrine tumors. In: Proceedings of the IEEE/CVF International Conference on Computer Vision, pp. 3402–3408 (2021)
22. Zhu, Z., Xia, Y., Xie, L., Fishman, E.K., Yuille, A.L.: Multi-scale coarse-to-fine segmentation for screening pancreatic ductal adenocarcinoma. In: Shen, D. (ed.) MICCAI 2019. LNCS, vol. 11769, pp. 3–12. Springer, Cham (2019). https://doi.org/10.1007/978-3-030-32226-7_1

MetaViT: Metabolism-Aware Vision Transformer for Differential Diagnosis of Parkinsonism with ^{18}F-FDG PET

Lin Zhao[1,2], Hexin Dong[1,3], Ping Wu[4], Jiaying Lu[4], Le Lu[1], Jingren Zhou[1], Tianming Liu[2], Li Zhang[3], Ling Zhang[1], Yuxing Tang[1(✉)], and Chuantao Zuo[4(✉)]

[1] Alibaba Group, Hangzhou, China
yuxing.t@alibaba-inc.com
[2] School of Computing, The University of Georgia, Athens, GA, USA
[3] Peking University, Beijing, China
[4] PET Center, Huashan Hospital, Fudan University, Shanghai, China
zuochuantao@fudan.edu.cn

Abstract. Accurate and early differential diagnosis of parkinsonism (idiopathic Parkinson's disease, multiple system atrophy, and progressive supranuclear palsy) is crucial for informing prognosis and determining treatment strategies. Current automated differential diagnosis methods for ^{18}F-fluorodeoxyglucose (^{18}F-FDG) positron emission tomography (PET) scans, such as convolutional neural networks (CNNs), often focus on local brain regions and do not explicitly model the complex metabolic interactions between distinct brain regions. These interactions, as reflected in FDG PET images, are keys for the differential diagnosis of parkinsonism. Vision transformer (ViT) models are promising in modeling such long-range dependencies, but they may overlook the local metabolic alternations and have not been widely adapted for 3D medical image classification due to data limitations. Therefore, we propose a novel metabolism-aware vision transformer (MetaViT), which uses self-attention and convolution to explicitly characterize both global and local metabolic interactions between interrelated brain regions. A masked image reconstruction task is introduced to guide the MetaViT model to focus on disease-related brain regions, addressing the scarcity of 3D medical imaging data and improving the trustworthiness and interpretability of the resulting model. The proposed framework is evaluated on a 3D FDG PET imaging dataset with 902 subjects, achieving a high accuracy of 97.7% in the differential diagnosis of parkinsonism and outperforming several state-of-the-art CNN and ViT-based approaches.

Keywords: PET · Parkinsonism · Early Differential Diagnosis · Transformer · Masked Image Reconstruction

This work was supported by Alibaba Group through Alibaba Research Intern Program.

1 Introduction

Idiopathic Parkinson's disease (IPD) is among the most common neurodegenerative disorders and has attracted the interest of both clinical and research trials for decades [1,7,9,19]. Accurate and early diagnosis of IPD plays a crucial role in determining potential therapeutic interventions and treatment outcomes [14]. However, it remains challenging due to the large overlap of IPD's symptoms with atypical parkinsonian syndromes like multiple system atrophy (MSA) and progressive supranuclear palsy (PSP) [3,6,19], especially in the early stage. Recently, ^{18}F-fluorodeoxyglucose positron emission tomography (^{18}F-FDG PET) has demonstrated advantages in the differential diagnosis of parkinsonism prior to the development of brain structural damage, by revealing the brain glucose metabolism that indicates the abnormalities of the brain [19,23].

Based on ^{18}F-FDG PET, various computational tools have been developed to exploit the discriminative features from the metabolic patterns of the human brain for early and accurate differential diagnosis. For example, principal component analysis was employed to extract disease-specific features for machine learning algorithms, such as logistic regression [18] and scaled subprofile model [13,16]. Deep learning-based methods have also been widely adopted and applied in the differential diagnosis of parkinsonism [19,23]. For instance, a recent study proposed an IPD Diagnosis Network (PDD-Net) [19] based on a modified 3D deep residual convolutional neural network [5] to provide an end-to-end solution for automatic differentiation and achieved promising results than traditional machine learning-based methods.

Despite the wide adoption and success of the aforementioned techniques, current state-of-the-art differential diagnostic tools remain limited in the sense that they do not effectively model the complex metabolic interactions of interrelated brain regions, which are considered to crucially reflect the differences among parkinsonism in FDG PET images [19]. Most of the previous methods model the interactions in FDG PET images by adapting 3D convolutional neural networks (CNNs). However, CNNs do not explicitly characterize the global metabolic alternations of distinct brain regions, but gradually enlarge the receptive field from local to global through the integration of local information [15]. Recently, vision transformers (ViTs) have become increasingly popular and dominant in image recognition tasks, offering comparable or even superior performance as an alternative to CNNs [2,11]. ViTs divide the entire image into several smaller image patches and model their interactions to aggregate global information, and seem to compensate for the aforementioned shortcoming of CNNs. However, ViTs are inefficient in integrating the local information compared with CNNs such that the local metabolic alternations may not be well characterized and thus degenerate the performance. Meanwhile, due to the lack of inductive biases, optimization of ViTs requires much more training samples than CNNs [2,10], especially in 3D scenarios where the number of parameters of the model far exceeds the number of 3D samples. To overcome this limitation and take advantage of ViTs, a possible way is to rely on fine-grained annotations and to perform the segmentation simultaneously with additional pixel/voxel-level supervision [22]. Nonetheless, PET images measure the metabolic activity of the human brain, whereas segmentation tasks are usually performed on anatomical structures. Therefore, there remains a desperate need for a general and effective framework tailored for the differential diagnosis of parkinsonism using FDG PET images, which combines the advantages of CNNs and ViTs and overcomes their respective drawbacks.

Motivated by this, we propose a novel metabolism-aware vision transformer (MetaViT), as shown in Fig. 1, for accurate differential diagnosis of parkinsonism. Our MetaViT model is specifically designed to explicitly describe both global metabolic interactions of interrelated brain regions and local metabolic alternations within small specific brain regions. To this end, we employ convolution operations to integrate local spatial information, and inter-patch voxel-wise self-attention operations as well as feed-forward layers to mimic the global interactions of metabolism in the brain. Moreover, prior knowledge from the nuclear radiologist is integrated during the model training to guide the MetaViT model to focus more on the brain's regions of interest (ROIs) that are highly correlated with potential metabolic abnormalities through a masked image reconstruction task (Fig. 1(b)). This unique design provides additional supervision for model optimization and compensates for the limited amount of data. In addition, a self-learning strategy is implemented to take advantage of additional data with noisy (clinically possible diagnosis) labels. Experiments on a 3D FDG PET imaging dataset (n = 902) demonstrate the validity and effectiveness of the proposed framework, as well as its superior performance over state-of-the-art CNN and ViT-based methods. We also find that the integration of clinical prior knowledge improves the trustworthiness and interpretability of the resulting model.

The main contributions of our work are summarized as follows:

- We propose a novel metabolism-aware vision transformer (MetaViT) that explicitly characterizes both global and local interactions of glucose metabolism in the brain, demonstrating superior performance than CNN- and ViT-based baselines for 3D FDG PET image classification.
- We propose a masked image reconstruction task to guide MetaViT to focus more on disease-related brain regions that reflect the metabolic changes, which not only compensates for the lacking of sufficient 3D training images but also improves the model's trustworthiness and interpretability.
- Our framework provides a feasible solution to take advantage of ViT for 3D medical image classification, achieving state-of-the-art performance on the differential diagnosis of parkinsonism in FDG PET imaging, suggesting great promise in integrating prior knowledge for 3D medical image classification.

2 Methods

2.1 Metabolism-Aware Vision Transformer

The complex metabolic interactions of interrelated brain regions in the human brain are suggested to be different among parkinsonism in FDG PET images [19]. However, it has not been effectively characterized in previous studies due to

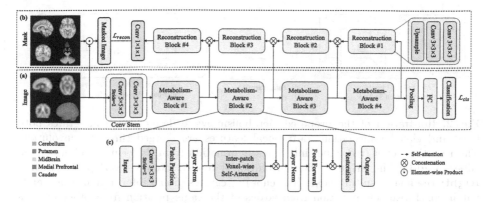

Fig. 1. Illustration of the proposed framework. (a) The architecture of the MetaViT with one convolution stem, four consecutive metabolism-aware blocks, one pooling and one fully-connected layer. (b) Masked image reconstruction branch consisting of four reconstruction blocks. Each reconstruction block is composed of two convolutional layers and one upsampling layer. (c) The constitution of the metabolism-aware block. The convolution integrates the local metabolic alternations while inter-patch voxel-wise self-attention interacts with those global alternations in a data-driven manner.

methodological limitations such as the incompetence of CNNs in modeling long-range and global dependencies explicitly. To address this problem, in this subsection, we propose a novel Metabolism-Aware Vision Transformer (MetaViT tailored for this objective. As illustrated in Fig. 1(a), our design of MetaViT follows a typical hierarchical scheme of CNNs (e.g., ResNet [5]) which consists of a convolutional stem and four consecutive metabolism-aware blocks (Fig. 1(c)) to model the metabolic interactions of interrelated brain regions.

Formally, given an input tensor $\mathbf{X} \in \mathbb{R}^{H \times W \times D \times C}$ where H, W, D, C represent the height, width, depth, and the number of channels, respectively, we firstly model the local dependencies of metabolic alternations by applying a $3 \times 3 \times 3$ convolution with a stride of 2 (Fig. 1c, blue rectangular). We visualize this process in the left panel of Fig. 2. Each voxel (denoted by red cube) in the resulting tensor $\mathbf{X}_{conv} \in \mathbb{R}^{H/2 \times W/2 \times D/2 \times C'}$ integrates the metabolic information of its neighbor voxels (denoted by green cube). Then, the long-range and global dependencies are characterized to enable the interactions across the whole brain. To do so, we introduce an inter-patch voxel-wise self-attention to model long-range dependencies. Specifically, $\mathbf{X}_{conv}$ is divided into N non-overlapping 3D patches $\mathbf{X}_{conv} \in \mathbb{R}^{V \times N \times C'}$ (Fig. 2, middle panel) where $V = h \times w \times d$ is the number of voxels within a patch, h,w,d are the height, width, depth of a patch, respectively, and $N = \frac{H \times W \times D}{8V}$ is the number of patches. For each voxel position $v \in \{1, \cdots, V\}$ within a patch, we then apply multi-head self-attention f_{MSA} and multilayer perceptron f_{MLP} to obtain the $\mathbf{X}_{trans} \in \mathbb{R}^{V \times N \times C'}$:

$$\mathbf{X}'_{trans}(v) = f_{\text{MSA}}(\text{LN}(\mathbf{X}_{conv}(v))) + \mathbf{X}_{conv}(v) \tag{1}$$

$$\mathbf{X}_{trans}(v) = f_{\text{MLP}}(\text{LN}(\mathbf{X}'_{trans}(v))) + \mathbf{X}'_{trans}(v) \tag{2}$$

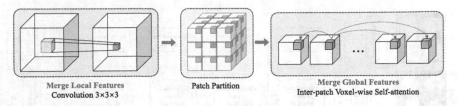

Merge Local Features   Patch Partition   Merge Global Features
Convolution 3×3×3   Inter-patch Voxel-wise Self-attention

Fig. 2. Illustration of the proposed metabolism-aware design which integrates both local and global information. **Left:** local features are merged through convolution. The resulting voxel (red cube) integrates information from its local neighbors (green cube). **Middle:** the output of the convolution is divided into non-overlapping patches. **Right:** the final resulting voxel (red cube) integrates global information using inter-patch voxel-wise self-attention from voxels of the same position from other patches (blue cubes), which already contain their local information. (Color figure online)

where $\mathrm{LN}(\cdot)$ represents the layer normalization. $\mathbf{X}_{trans}$ can be restored as $\mathbf{X}_H \in \mathbb{R}^{H/2 \times W/2 \times D/2 \times C'}$, which is the output for metabolism-aware block. Notably, $\mathbf{X}_{conv}(v)$ integrates the local metabolic information through the 3D convolution operation, and $\mathbf{X}_{trans}(v)$ encodes the global metabolic information across N patches for the vth position in a patch. As illustrated in the right panel of Fig. 2, the voxel in a patch (denoted by the red cube) integrates the information from the same position voxel of other patches (denoted by blue cubes) which already merge their local information. In this way, each voxel in $\mathbf{X}_{trans}$ can interact and infuse the metabolic information from all voxels of FDG PET images in a data-driven manner, which is congruent with our objective. Compared with previous methods, this approach explicitly and efficiently models the metabolic interactions across the brain rather than being limited to local areas. The interrelations of different brain regions are also delineated implicitly through this process.

The MetaViT can be optimized by minimizing the loss function $\mathcal{L}_{cls}$, which is the cross-entropy between the ground truth label y and the predictions $\hat{y}$:

$$\mathcal{L}_{cls} = -\sum_i y_i log(\hat{y}_i) \tag{3}$$

2.2 Masked Image Reconstruction

Optimization of the vision transformer usually requires a large number of training samples. However, in 3D medical imaging scenarios, the amount of data is usually scarce and limited. The intuition is to perform the segmentation task simultaneously, as it would additionally include pixel-level supervision. However, PET images reflect the metabolism rather than the anatomy of the human brain, and hence, the segmentation task is not feasible for our objective.

In this subsection, we introduce a novel masked image reconstruction task to integrate the clinical prior knowledge to guide the model optimization. The main idea of this approach is to utilize the intermediate features from each block of MetaViT to reconstruct the masked original image $X_M = X \odot M$, where M is a binary mask indicating the ROIs of disease-related brain regions and $\odot$ represents the element-wise product. In this way, the model is enforced to learn more features from the brain's ROIs that are highly related to potential metabolic abnormality for a better reconstruction. It actually accelerates the model optimization by utilizing prior knowledge instead of learning from a large amount of data by the model itself.

Specifically, as illustrated in Fig. 1(b), the masked image reconstruction branch consists of several reconstruction blocks. Each block is composed of two $3 \times 3 \times 3$ convolutions and an up-sampling operation with a scale factor of 2. The output features from each metabolic-aware block are firstly concatenated with the output from the previous reconstruction block (except the last one) and then fed into the next reconstruction block. The final output is processed by a $1 \times 1 \times 1$ convolution and considered as the reconstruction of the masked image. The optimization of the masked image reconstruction branch can be performed by minimizing the mean squared error (MSE) between the original masked image X_M and the reconstructed one X'_M:

$$\mathcal{L}_{recon} = \|X_M - X'_M\|_F^2 \tag{4}$$

The final loss function $\mathcal{L}$ for optimizing the whole framework is:

$$\mathcal{L} = \mathcal{L}_{cls} + \lambda \mathcal{L}_{recon} \tag{5}$$

where λ is a scaling factor to balance the two terms.

2.3 Self-learning for Noisy-Labeled Data

In clinical practice, in addition to patient cohorts with definite diagnoses and confirmative diagnoses with follow-up, there are additional Parkinsonian patients with possible diagnoses, where physicians are unable to reach a clinically assured/definite differential diagnosis (considered as "noisy"-labeled data), especially for those in very early stages. To exploit the noisy-labeled data, we employ the self-learning strategy [8,20]. Different from previous self-learning methods of having one teacher, we introduce two additional teachers to benefit from different models' expertise and from the knowledge of the possible diagnoses.

Specifically, we first train the MetaViT model based on masked image reconstruction and a 3D ResNet-18 model [5] on the data with clinically definite diagnosis (ground truth-labeled data). After training, these two teachers are applied to noisy-labeled data to generate predicted labels, and the final pseudo label for these data is obtained by voting among two predicted labels and the noisy label itself. As such, the student model is subsequently forced to learn from the additional knowledge provided by the ResNet-18 and clinically possible diagnoses. Moreover, the student's knowledge is expanded through learning more data that contains more parkinsonism variations, allowing the student to learn beyond his teachers to be capable of classifying more challenging cases.

Lastly, both the noisy-labeled data with the final pseudo labels and the original training data with ground truth labels are used to train the final student model from scratch. Note that such a training strategy is found to be more effective than initializing the student model with the teacher model or first pre-training on noisy-labeled data and then finetuning on original training data [20].

3   Experiments and Results

3.1   Dataset and Pre-Processing

In this study, we adopt the Huashan Parkinsonian PET Imaging (HPPI) dataset (approved by its Institutional Review Board) with 528 subjects (IPD: 277, MSA: 149, PSP: 102; with clinically definite diagnoses) for evaluating the proposed framework and another 374 subjects with clinically possible diagnoses (noisy labels) for self-learning. The brain emission data were acquired 60 min after the injection of approximately 185 MBq ^{18}F-FDG and lasted for 10 min. After the attenuation correction with low-dose CT before the emission scan, scatter, dead time, and random coincidences, the ^{18}F-FDG PET data were reconstructed by the ordered subset expectation maximization method. The PET images were then registered to standard MNI 152 space and smoothed by a 3D Gaussian filter ($\sigma = 5$ mm). Besides, we cropped the PET image with a size of $80 \times 96 \times 80$ and normalized it with a mean of 0 and a standard deviation of 1. We invited an expert nuclear radiologist with >10 years of experience in diagnosing parkinsonism to manually annotate the ROIs on a T1w template in $1 \times 1 \times 1$ mm resolution in MNI 152 space (Fig. 1(b)). The ROIs were then registered to the preprocessed individual PET images and smoothed by a 3D Gaussian filter ($\sigma = 5$ mm) as the final mask for masked image reconstruction.

3.2   Implementation Details and Compared Methods

In our experiments, the proposed model and all compared baselines are trained for 50 epochs with a batch size of 16. We use the AdamW optimizer [12] with $\beta_1 = 0.9$ and $\beta_2 = 0.999$ and a cosine annealing learning rate scheduler with initial learning 10^{-4} and 5 warm-up epochs. The scale factor λ in Eq. (5) is set as 1. The framework is implemented with PyTorch (https://pytorch.org/) deep learning library and the model is trained on a single NVIDIA V100 GPU. We conduct nested five-fold cross-validation. For self-learning, the predicted labels of the noisy-labeled data utilized in each fold are predicted by the MetaViT and 3D ResNet-18 models of the same fold. Note that the pseudo-labeled data is only used in the model training, not in validation.

The compared baselines can be roughly categorized as CNN-based and ViT-based methods. CNN-based category contains PDD-Net [19], two ResNets (ResNet-18 and 50) [5], and visual attention network (VAN) [4]. It is noted that VAN consists of purely convolution operations while the overall architecture follows the design of ViT, which achieves state-of-the-art performance in natural image recognition tasks. The ViT-based class contains a vanilla ViT-Tiny model, Swin Transformer model [11], and two other methods designed especially for small datasets by including the inductive biases from CNN: T2T-ViT [21]and ViT-SD [10]. Considering the limited amount of data and the overfitting problem, we only report the tiny- or small-scale setting with fewer parameters for each model and re-implement them to fit the 3D inputs. The permutation test with 1,000 permutations is used to perform statistical significance ($p < 0.05$ indicating significance) comparisons of classification accuracies between different methods.

3.3   Differential Diagnosis Results

In this subsection, we report the performance of the proposed framework in differential diagnosis to demonstrate its effectiveness and superiority. In Table 1, we report the average accuracy with standard deviation, F1 score, sensitivity, specificity, and the number of trainable parameters for each compared baseline and the proposed method. It is observed that CNN-based methods outperform ViT-based methods, among which vanilla ViT has the worst performance (81.3%). This is probably because inductive biases in CNNs are important for scenarios with limited samples. Because the vanilla ViT model requires much more training samples, in our task, it leaves a large performance margin. The performance of Swin-T is also inferior to CNN-based methods. We assume that the small window partition of the Swin Transformer may be inefficient in modeling the metabolic interactions of the human brain. T2T-ViT and ViT-SD explicitly introduce inductive biases in their architecture design, obtaining great improvement over vanilla ViT (around 10%). Our proposed framework achieves state-of-the-art performance in terms of all evaluation metrics compared to baselines. The accuracy of our framework is significantly ($p<0.001$) higher than all compared methods. Notably, the sensitivity of PSP is improved from around 85% to 96%. In clinical practice, PSP is relatively rare and can be easily misdiagnosed as IPD, while PSP is more malignant than IPD and has totally different treatment protocols. The proposed framework can better support radiologists in making an accurate diagnosis for rare but severe diseases such as PSP. Overall, these results demonstrate the superiority of the proposed framework compared to baselines.

Table 1. The performances of the proposed method and compared baselines for differential diagnosis of parkinsonism. The average accuracy (Acc.) with standard deviation (std), F1 score, sensitivity and specificity over nested five-fold cross-validation are reported. Model sizes in terms of the number of trainable parameters (in M) are shown in the last column. * indicates p-value<0.001 compared to the reference (i.e., **Ours**).

Methods	Acc. $\pm$ std (%)	F1 Score (%)			Sensitivity (%)			Specificity (%)			Params (M)
		IPD	MSA	PSP	IPD	MSA	PSP	IPD	MSA	PSP	
PDD-Net [19]	92.8 ± 2.8*	94.4	95.0	84.3	95.6	94.8	82.7	92.5	98.2	97.2	6.0
ResNet-18 [5]	93.7 ± 1.4*	95.1	95.7	86.1	95.5	95.8	84.1	93.9	98.1	97.4	33.2
ResNet-50 [5]	92.6 ± 1.3*	93.8	95.5	84.4	94.2	94.0	85.8	92.8	99.0	96.0	46.2
VAN-T [4]	92.6 ± 2.1*	94.5	94.8	83.0	96.8	94.0	78.9	91.3	98.1	97.7	5.6
ViT-T [2]	81.3 ± 4.6*	84.7	85.3	63.5	89.5	84.3	56.8	76.7	95.0	95.3	6.2
Swin-T [11]	86.4 ± 4.0*	88.1	88.3	74.7	89.6	88.8	72.5	86.1	97.1	93.8	29.4
T2T-ViT-12 [21]	91.5 ± 3.7*	92.5	94.1	84.8	93.4	93.1	84.6	91.3	98.1	96.5	13.0
ViT-SD-T [10]	92.2 ± 2.3*	93.9	94.8	83.6	93.4	95.3	84.5	94.0	97.6	96.0	17.3
Ours-Backbone	95.5 ± 1.8*	95.8	97.2	90.6	97.4	96.0	88.3	93.6	**99.5**	98.6	8.9
Ours	$\mathbf{97.7\pm1.0}$	**97.9**	**98.3**	**96.1**	**98.1**	**97.9**	**96.1**	**97.7**	**99.5**	**99.1**	18.3

Table 2. Ablation study. The average accuracy with standard deviation (std), p-value, F1 score, sensitivity and specificity over nested five-fold cross-validation with different configurations. Abbreviations: B: backbone; EF: early fusion; IR: image reconstruction; MIR: masked image reconstruction; SL: self-learning; SLE: self-learning with ensemble.

Methods	Acc. $\pm$ std (%)	p	F1 Score (%)			Sensitivity (%)			Specificity (%)		
			IPD	MSA	PSP	IPD	MSA	PSP	IPD	MSA	PSP
a) B	95.5 ± 1.8	Ref	95.8	97.2	90.6	97.4	96.0	88.3	93.6	**99.5**	98.6
b) +EF	95.5 ± 1.1	0.99	95.9	97.5	90.2	97.8	96.5	86.7	93.1	**99.5**	98.8
c) +IR	95.3 ± 2.6	0.97	95.8	96.2	91.3	97.8	94.9	88.7	93.4	99.2	98.8
d) +MIR	96.4 ± 1.6	0.06	96.7	97.7	93.2	97.8	96.8	91.8	95.3	**99.5**	98.8
e) +MIR+SL	97.0 ± 1.5	0.05	97.4	98.0	94.2	97.8	**97.9**	93.6	96.9	99.2	98.8
f) +MIR+SLE	$\mathbf{97.7\pm1.0}$	<0.01	**97.9**	**98.3**	**96.1**	**98.1**	**97.9**	**96.1**	**97.7**	**99.5**	**99.1**

3.4 Ablation Study

In this subsection, we conduct ablation experiments to validate the efficacy of each component in the proposed framework. We report the averaged accuracy, F1 score, sensitivity, and specificity over nested five-fold cross-validation for different configurations in Table 2. Compared with the baseline methods in Table 1, our a) MetaViT backbone outperforms both CNN-based and ViT-based methods, suggesting the advantages of explicitly modeling the metabolic interactions of interrelated brain regions. To evaluate the effectiveness of the masked image reconstruction task, we compare it with two approaches: b) early fusion fuses the mask as an additional channel for the input image, which is the simplest way to integrate prior knowledge with the mask; c) image reconstruction reconstructs the original image rather than the masked image. It is observed that the early fusion strategy is of no help to the diagnosis performance and the original image reconstruction even degenerates the accuracy. In contrast, the d) masked image reconstruction task significantly ($p = 0.06$) improves the accuracy from 95.5% to 96.4%, implying the effectiveness of integrating prior knowledge into the model training. We also observe that self-learning with an additional 374 subjects further contributes to the diagnosis accuracy: f) our final configuration, which generates the pseudo label based on the ensemble of three teachers (MetaViT, ResNet-18, and clinically possible diagnosis), is better than e) the self-learning process only with one teacher (MetaViT). This indicates that self-learning with label ensemble, i.e., learning from multiple teachers, could be a useful strategy to leverage the noisy-labeled data.

3.5   Interpretation of Model Reasoning

We generate the saliency maps of the proposed framework (i.e., MetaViT based on masked image reconstruction and self-learning) using the full-gradient method [17] for interpreting the model reasoning. It assigns a correlation score to each

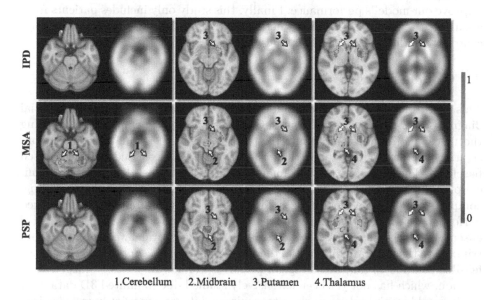

1.Cerebellum   2.Midbrain   3.Putamen   4.Thalamus

Fig. 3. Visualization of average saliency maps of patients with idiopathic Parkinson's disease (IPD), multiple system atrophy (MSA), and progressive supranuclear palsy (PSP) in the testing sets. Each row shows three axial slices of registered MRI and FDG PET averaged using all the subjects with the corresponding disease. We highlight the characteristic regions contributing to the above diseases based on our model. The color corresponds to the correlation score indicating the contribution of a region for each disease (red regions correspond to high scores). The arrows pointed to the most salient brain regions, including 1: cerebellum, 2: midbrain, 3: putamen, 4: thalamus. (Color figure online)

pixel of the original image to show the numerical contribution to the model's decision-making. In Fig. 3, we illustrate average saliency maps (fused with T1-w modality template) and average FDG PET images of patients with IPD, MSA, or PSP in the testing cohort. Our model identifies four brain regions that are associated with the above diseases: the cerebellum, midbrain, putamen, and thalamus. In particular, our model focuses on the putamen for the prediction of all three diseases. Thalamus is linked to both MSA and PSP, while the cerebellum and midbrain are mainly associated with MSA or PSP, respectively. All these brain regions are either correlated with IPD/MSA/PSP pathology or confirmed by previous studies to be related to metabolic changes in IPD/MSA/PSP [18,19].

3.6 Limitations

One limitation of this study could be that it focuses on the major parkinsonian syndromes [18,19]. The ability to detect rare types of disorders (e.g., dementia with Lewy bodies, corticobasal degeneration) might be important in real-world applications. Additionally, although the 3D FDG PET imaging dataset used in this study is one of the largest patient cohorts in the literature, the sample size may still be small for deep learning analysis. More training data may further improve our model's performance. Finally, this study only includes patients from a single center, and it is unclear whether the results can be generalized to other centers with different patient populations.

4   Conclusion

This work presents a novel computational framework for accurate differential diagnosis of parkinsonian syndromes. Our MetaViT design emphasizes modeling the metabolic interactions of interrelated regions in the human brain, demonstrating superior performance than the compared CNN-based and ViT-based baselines. The proposed masked image reconstruction task leverages the clinical prior knowledge of disease-related brain regions to guide the model training, compensating for the lacking of sufficient 3D training samples with improved performances. Exploiting noisy-labeled data based on self-learning with label ensemble further improves the diagnosis accuracy. Our framework not only contributes to the accurate differential diagnosis of parkinsonism but also provides a feasible solution to take advantage of powerful ViT for 3D medical image classification, which has not yet been extensively studied due to limited 3D data. Our study would inspire future work to transpire on integrating prior knowledge into the deep model design and training to improve the reliability and transparency of medical imaging applications.

References

1. Braak, H., Rüb, U., Gai, W., Del Tredici, K.: Idiopathic Parkinson's disease: possible routes by which vulnerable neuronal types may be subject to neuroinvasion by an unknown pathogen. J. Neural Transm. **110**(5), 517–536 (2003)

2. Dosovitskiy, A., et al.: An image is worth 16×16 words: transformers for image recognition at scale. arXiv preprint arXiv:2010.11929 (2020)
3. Eckert, T., et al.: Differentiation of idiopathic Parkinson's disease, multiple system atrophy, progressive supranuclear palsy, and healthy controls using magnetization transfer imaging. Neuroimage **21**(1), 229–235 (2004)
4. Guo, M.H., Lu, C.Z., Liu, Z.N., Cheng, M.M., Hu, S.M.: Visual attention network. arXiv preprint arXiv:2202.09741 (2022)
5. He, K., Zhang, X., Ren, S., Sun, J.: Deep residual learning for image recognition. In: Proceedings of the IEEE Conference on Computer Vision and Pattern Recognition, pp. 770–778 (2016)
6. Hughes, A.J., Daniel, S.E., Ben-Shlomo, Y., Lees, A.J.: The accuracy of diagnosis of parkinsonian syndromes in a specialist movement disorder service. Brain **125**(4), 861–870 (2002)
7. Hughes, A.J., Daniel, S.E., Kilford, L., Lees, A.J.: Accuracy of clinical diagnosis of idiopathic Parkinson's disease: a clinico-pathological study of 100 cases. J. Neurol. Neurosurg. Psychiatry **55**(3), 181–184 (1992)
8. Jumper, J., et al.: Highly accurate protein structure prediction with AlphaFold. Nature **596**(7873), 583–589 (2021)
9. Kish, S.J., Shannak, K., Hornykiewicz, O.: Uneven pattern of dopamine loss in the striatum of patients with idiopathic Parkinson's disease. N. Engl. J. Med. **318**(14), 876–880 (1988)
10. Lee, S.H., Lee, S., Song, B.C.: Vision transformer for small-size datasets. arXiv preprint arXiv:2112.13492 (2021)
11. Liu, Z., et al.: Swin transformer: hierarchical vision transformer using shifted windows. In: Proceedings of the IEEE/CVF International Conference on Computer Vision, pp. 10012–10022 (2021)
12. Loshchilov, I., Hutter, F.: Decoupled weight decay regularization. arXiv preprint arXiv:1711.05101 (2017)
13. Matthews, D.C., et al.: FDG pet Parkinson's disease-related pattern as a biomarker for clinical trials in early stage disease. NeuroImage: Clin. **20**, 572–579 (2018)
14. Pagan, F.L.: Improving outcomes through early diagnosis of Parkinson's disease. Am. J. Manag. Care **18**(7), S176 (2012)
15. Raghu, M., Unterthiner, T., Kornblith, S., Zhang, C., Dosovitskiy, A.: Do vision transformers see like convolutional neural networks? Adv. Neural. Inf. Process. Syst. **34**, 12116–12128 (2021)
16. Spetsieris, P.G., Ma, Y., Dhawan, V., Eidelberg, D.: Differential diagnosis of parkinsonian syndromes using PCA-based functional imaging features. Neuroimage **45**(4), 1241–1252 (2009)
17. Srinivas, S., Fleuret, F.: Full-gradient representation for neural network visualization. In: Advances in Neural Information Processing Systems (NeurIPS) (2019)
18. Tang, C.C., et al.: Differential diagnosis of parkinsonism: a metabolic imaging study using pattern analysis. Lancet Neurol. **9**(2), 149–158 (2010)
19. Wu, P., et al.: Differential diagnosis of parkinsonism based on deep metabolic imaging indices. J. Nucl. Med. **63**, 1741–1747 (2022)
20. Xie, Q., Luong, M.T., Hovy, E., Le, Q.V.: Self-training with noisy student improves ImageNet classification. In: Proceedings of the IEEE/CVF Conference on Computer Vision and Pattern Recognition, pp. 10687–10698 (2020)
21. Yuan, L., et al.: Tokens-to-token ViT: training vision transformers from scratch on imagenet. In: Proceedings of the IEEE/CVF International Conference on Computer Vision, pp. 558–567 (2021)

22. Zhang, L., Wen, Y.: A transformer-based framework for automatic COVID19 diagnosis in chest CTs. In: Proceedings of the IEEE/CVF International Conference on Computer Vision, pp. 513–518 (2021)
23. Zhao, Y., et al.: A 3D deep residual convolutional neural network for differential diagnosis of parkinsonian syndromes on 18 F-FDG pet images. In: 2019 41st Annual International Conference of the IEEE Engineering in Medicine and Biology Society (EMBC), pp. 3531–3534 (2019)

Multi-task Multi-instance Learning for Jointly Diagnosis and Prognosis of Early-Stage Breast Invasive Carcinoma from Whole-Slide Pathological Images

Jianxin Liu, Rongjun Ge, Peng Wan, Qi Zhu, Daoqiang Zhang[✉],
and Wei Shao[✉]

The College of Computer Science and Technology, Nanjing University of Aeronautics
and Astronautics, Nanjing, China
{dqzhang,shaowei20022005}@nuaa.edu.cn

Abstract. With the tremendous progress brought by artificial intelligence, many whole-slide pathological images (WSIs) based machine learning models are designed to estimate the clinical outcome of human cancers. However, most of the existing studies treat the prognosis and diagnosis tasks separately, which overlooks the fact that the diagnosis information indicating the severity of the disease that is highly related to the patients' survival. In addition, it is still challenge to design machine learning models to analyze WSIs since a WSI is of large size but only annotate with coarse labels, it increasingly becomes a research hotspot for the development of automated WSI analysis tools in a scenario without fully annotated data. Based on the above considerations, we propose a multi-task multi-instance (WSI-MTMI) learning method that can simultaneously conduct the prognosis and diagnosis tasks from WSIs. Specifically, inspired by the fact that taking the associations between diagnosis and prognosis tasks can improve the generalization ability for each individual task, we firstly adopt multi-instance learning algorithms to aggregate the patches derived from WSIs by considering both the common and specific task information. Then, we design a novel cross-attention network that can effectively identify useful information shared across different tasks. To evaluate the effectiveness of the proposed method, we test it on the early-stage breast invasive carcinoma (BRCA) derived from the Cancer Genome Atlas project (TCGA), and the experimental results indicate that our method can achieve better performance on both diagnosis and prognosis tasks than the related methods.

Keywords: Multi-task Multi-instance Learning · Pathological images · Cancer diagnosis · Cancer prognosis

1 Introduction

Cancer is the second leading cause of death worldwide. It is reported that the number of affected people has already reached to a high record of 19.3 million

© The Author(s), under exclusive license to Springer Nature Switzerland AG 2023
A. Frangi et al. (Eds.): IPMI 2023, LNCS 13939, pp. 145–157, 2023.
https://doi.org/10.1007/978-3-031-34048-2_12

in 2020 [13]. Thus, the effective and efficient analysis and intervention of human cancer, especially at its early stage, is crucial in treatment selection and planning for each cancer patient.

So far, amounts of bio-markers have been developed and applied to analyze human cancer, including the histopathological images, genetic mutations, epigenetic signatures, and protein markers. Of all these types of biomarkers, the pathological examination remains to be the gold standard for the clinical outcome prediction of human cancers since the whole-slide histopathological images (WSIs) can reveal the degree of tissue cancerization by considering the morphology and distribution of cells within them. Recently, the emerging of digital pathology makes computerized quantitative analysis of histopathology imagery possible, and many computational models have been employed for the prediction of human cancers.

Generally, the existing WSI-based predictions models can be generally divided into two folds i.e., diagnosis and prognosis models. The diagnosis models aim at building classification models to predict the categorical variables (e.g. TNM staging and Nottingham Histologic Score) from WSIs. For example, Zhao et al. [18] designed a self-supervised learning algorithm to extract meaningful features from histopathological images, followed by applying the graph convolutional network (GCN) to drive diagnostic practice. Zhang et al. [17] built a double-tier multi-instance learning framework that can localize the tissue regions to help cancer diagnosis. Moreover, Gao et al. [2] proposed an instance-based Vision Transformer to learn robust representations of histopathological images for the subtype classification task.

On the other hand, besides the diagnosis tasks, many prognosis models were also adopted to perform survival analysis of human cancers on WSIs. For instance, Katzman et al. [4] introduced a Cox proportional hazards deep neural network for modeling interactions between a patient's covariates and treatment effectiveness. Yao et al. [16] proposed Deep Attention Multiple Instance Survival Learning by introducing both Siamese MI-FCN and attention-based MIL pooling to efficiently learn imaging features from the WSI and then aggregate WSI-level information to patient-level. Lv et al. [7] developed a multi-scale histopathological features fusion transformer (MS-Trans) to integrate the relationship of different image patches for survival analysis.

Although much progress has been achieved, most of the existing methods conduct the diagnosis and prognosis tasks separately, without considering their inherent correlation. As a matter of fact, the diagnosis information indicates the extent of the disease severity, which is highly correlated with the patients' clinical outcomes. For instance, the patients in Stage II suffer from more aggressive cancer than those in Stage I, and thus generally with shorter survival time [11]. It is expected that better diagnosis and prognosis performance can be achieved if we take the association between them into consideration. In addition, due to the high resolution of WSI, it is impossible to directly feed the WSIs into the neural network for model training. In recently years, many multi-instance learning (MIL) algorithms focusing on designing proper aggregation strategies to fuse the patch-level prediction results to WSI-level have been proposed to deal with such challenges. However, those MIL methods simply used the pooling strategies or consider the associations among different patches for patch fusion, which overlooks the similarity and variability of different tasks if we directly extend the existing MIL algorithm to multi-task learning scenario.

Based on the above consideration, we propose a Multi-task Multi-instance learning framework for jointly diagnosis and prognosis of early-stage breast invasive carcinoma from whole-slide pathological images. Specifically, basing on the self-attention mechanism, the task similarity in multi-instance learning is captured by sharing a common set of value vectors derived from the sampled patches across all tasks. In addition, our method also extracted the specific key and query vectors from the instance patches, which can capture the variability of the patches across different tasks. Then, we adopt the multi-head attention algorithm to generate the sample-level embedding from the patches. Finally, a novel cross-attention network that can effectively identify useful information shared across different tasks is applied to conduct the diagnosis and prognosis tasks, simultaneously.

To evaluate the effectiveness of the proposed method, we test it on the early-stage breast invasive carcinoma (BRCA) derived from the Cancer Genome Atlas project (TCGA), and the experimental results indicate that our method can achieve better performance on both diagnosis and prognosis tasks than the related methods.

2   Method

We now introduce the proposed WSI based multi-task multi-instance learning architecture (WSI-MTMI) shown in Fig. 1, which can be roughly divided into two steps, 1) extracting patch-level histopathological imaging features and aggregating them into sample-level representations via multi-instance learning module by considering the common and specific task information; and 2) fusing layer features via cross-attention mechanisms which identify useful information shared across different tasks for conducting the diagnosis and prognosis tasks on cancer patients, simultaneously.

Multiple Instance Learning (MIL). Generally, multi-instance learning (MIL) methods take the instances derived from a sample as inputs, $X_i = \left\{x_1^i, \ldots, x_N^i\right\}$, and predicts the sample-level label Y_i. In MIL, instance-level label y_n^i is usually unknow for model training. As for a binary classification task, sample-level label Y_i is positive only if there are at least one positive instance and labelled as negative when none of them is positive, which could be formulated as:

$$Y_i = \begin{cases} 0, & \text{if } \sum y_i = 0, \quad i = 1, \ldots n \\ 1, & \text{otherwise} \end{cases} \tag{1}$$

Furthermore, MIL uses a scoring function $S(X)$ to predict the Y_i:

$$S(X) = g\left(\sum_{x \in X} f(x)\right) \tag{2}$$

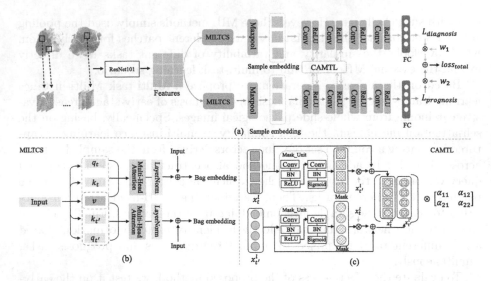

Fig. 1. The overall framework of the WSI-MTMI network (Fig. 1(a)). First, patch-level representation extracted from the pre-trained Resnet-101 network are aggregated by the proposed multi-instance learning module (Fig. 1(b)), which considers the common and specific task information (MILTCS) for multi-task learning; Second, a novel cross-attention network (CAMTL) that can effectively identify useful information shared across different tasks is applied to conduct the diagnosis and prognosis tasks, simultaneously (Fig. 1(c)).

where f is the non-linear transformation operated on the instances level and g is an aggregation operator at the sample level. In this work, we take the full account of the intrinsic relationship among instances for the multi-instance learning.

Multi-instance Learning by Considering the Common and Specific Task Information (MILTCS). In this work, multiple instance learning is casted into a multi-task learning setting, where effective information can be separated into the task-specific and task-shared features across various tasks. In view of this, we develop a self-attention based multi-instance learning module, which captures the common information by a shared value sets while preserves task-specific information by the respective query and key value sets.

As shown in the Fig. 1(b), the input sample of feature $E_m = \{e_{m,1}, \cdots e_{m,n}\} \in R^{d \times n}$ is transformed into task-specific query (key) values q_t (k_t) with the weight matrix W_{qt} (W_{vt}) respectively, and the common value sets v shared across tasks. It is worth noting that the utilization of common value sets encourages the model to exploit the complementary supervision information from different tasks. Meanwhile, this manner also reduces the computation complexity for the sample-level representation learning. For the t_{th} task, self-attention mechanism is implemented as:

$$q_t^h = W_{qt}^h \cdot E_m$$
$$k_t^h = W_{kt}^h \cdot E_m$$
$$SA_t^h = \text{softmax}\left(\frac{q_t^h \cdot k_t^{h^T}}{\sqrt{d/h}}\right) \tag{3}$$

where h is the number of heads in self-attention module. The sample-level embedding B_m^t is calculated as:

$$v_{\text{shared}}^h = W_{v\_\text{shared}}^h \cdot E_m$$
$$\text{head} = SA_t^h \cdot v_{\text{shared}}^h$$
$$MHA(q,k,v)_t = \text{Concat}(\text{ head } t, \ldots, \text{ head } t)W^o \tag{4}$$
$$B_m^t = \text{Maxpool}\left(MHA(q,k,v) + E_m\right)$$

where head_t^h is self-attention of head h, while W^o refers to the multi-head attention weight matrix. We aggregate the patch-level features into bag-level representation via max pooling operation.

Different from previous studies [12], we discard the multi-layer perception layers in the proposed MILTCS module and extend to the multi-task learning setting by specifying the common value sets. Meanwhile, task-specific query and key sets are expected to concentrate on distinct information of different tasks.

Cross-attention Multi-task Learning (CAMTL). As described above, our MILTCS module explicitly models the shared and task-specific information to obtain the sample-embedding. In order to more effectively fuse the features derived from different tasks, as shown in Fig. 1, we propose a novel cross-attention multi-task learning (CAMTL) module that can effectively identify useful information shared across different tasks. We implement such feature fusion operation at multiple layers. Figure 1(b) presents the details of this module consisting of the mask generation module and cross-stitch module [9]. The main motivation behind is that cross-attention mechanism enables to further exploit the collaborative effect of different tasks. The generation of the mask for the opposite task mask_t^l is inspired by [6], We formulate this unit as:

$$\text{mask}_t^l = g_t^l \left(h_t^l \left(x_t^l \right) \right) \tag{5}$$

where x_t^l represents the output of task t at layer l. Like in [6], both h_t^l and g_t^l are composed of 1×1 kernels with batch normalization, followed by a non-linear activation. The mask_t^l is restricted in [0–1]. We could figure that the higher value means that $x_{t'}^l$ has the larger effect on x_t^l at certain dimension. $\text{mask}_t^l \to 1$ equals to $p_t^l = x_t^l + x_{t'}^l$, and $\text{mask}_t^l \to 0$ reduces to an identity map, i.e., no effect imposed on task t from t'. Clearly, this manner provides us a flexible way to extract useful information during feature fusion.

The features of task t are re-calculated via element-wise multiplication between the mask mask_t^l and the features $x_{t'}^l$, and then fused with the raw feature x_t^l by a residual connection. Finally, followed by the cross-stitch module [9], we can derive:

$$\begin{bmatrix} p_t^{l+1} \\ p_{t'}^{l+1} \end{bmatrix} = \begin{bmatrix} \alpha_{11} & \alpha_{12} \\ \alpha_{21} & \alpha_{22} \end{bmatrix} \cdot \begin{bmatrix} \left[\text{mask}_t^l \otimes x_{t'}^l \right] + x_t^l \\ \left[\text{mask}_{t'}^l \otimes x_t^l \right] + x_{t'}^l \end{bmatrix} \tag{6}$$

$$\begin{bmatrix} x_t^{l+1} \\ x_{t'}^{l+1} \end{bmatrix} = \begin{bmatrix} g_t^{l+1} \left(p_t^{l+1} \right) \\ g_t^{l+1} \left(p_{t'}^{l+1} \right) \end{bmatrix} (t \neq t') \tag{7}$$

where p_t^{l+1} is the input of the task t at layer $l+1$, $\otimes$ refers to the element-wise multiplication, α_{ij} is the weight of linear combination, g_t^l is composed of 3×3 kernels, followed by downsampling module to reduce the resolution.

Loss Function. For diagnosis task, we adopt mean square error as loss function as:

$$l_{\text{diagnosis}} = \frac{1}{n} \sum_{i=1}^n (\hat{y}_i - y_i)^2 \tag{8}$$

where the y_i is the ground truth label, $\hat{y}_i$ is the prediction of diagnosis task.

As to prognosis task, the Cox proportional hazard model is applied to predict the patients' clinical outcome [11], and its negative log partial likelihood function can be formulated as:

$$l_{\text{prognosis}} = \sum_{i=1}^N \delta_i \left(\theta^T x_i - \log \sum_{j \in R(t_i)} \exp \left(\theta^T x_j \right) \right) \tag{9}$$

where x_i represents the output of the last layer for the prognosis task and $R(t_i)$ is the risk set at time t_i, which represents the set of patients that are still under risk before time t. In addition, δ_i is an indicator variable. Sample i refers to censored patient if $\delta_i = 0$, otherwise $\delta_i = 1$) patient.

By combining Eq. (8) and Eq. (9), the final objective function for the proposed WSI-MTMI method can be formulated as:

$$l_{\text{total}} = l_{\text{diagnosis}} + \lambda \times l_{\text{prognosis}} \tag{10}$$

where λ is the weight of prognosis task loss.

3   Experiments

To verify the superior performance of our model on the diagnosis and prognosis of cancer patients, we test it on early-stage breast invasive carcinoma (BRCA) dataset public available on The Cancer Genome Atlas (TCGA) database. We show the demographic information of the utilized dataset in Table 1.

Experimental Setting. For each WSI, we randomly sampled 100 non-overlapping adjacent cancer patches with the size of 512*512 from each WSI. Then, we applied the five-fold cross validation strategy to evaluate the performance of our method. Specifically, we iterative selected the patients in one fold as the testing data at each time, while the remaining patients are used for model training. From the training set, we randomly partition 20% samples as validation set to tune model parameters. To fairly compare the prediction results among different methods, we applied the same way to split training and testing dataset. For each patch, we followed the method in [19] that applied the pre-trained ResNet-101 network to extract a 2048-dimensional vectors, followed by applying PCA to reduce its dimension to 1000. The λ is tuned from [1.5, 2.0]. The CAMTL module is consisted of four convolutional layers, and the cross-attention operations are applied on the first two layers. We set the learning rate as 0.07 during the training process of the WSI-MTMI model. We evaluate the performance of the diagnosis task by the measurements of Accuracy, F1-Score, Recall and Precision. For the prognosis task, we take the Concordance Index (according to Eq. (9), $CI = \frac{1}{n} \sum_{i \in \{1..N | \delta_i = 1\}} \sum_{t_j > t_i} I(x_i \theta > x_j \theta))$ and AUC as evaluation metrics. Both the value of CI and AUC range from 0 to 1. The larger CI and AUC value mean the better prediction performance of the model and vice versa.

Evaluate Our WSI-MTMI Method on Diagnosis Classification. In this section, we compare the diagnosis power of the proposed WSI-MTMI with the following five algorithms on BRCA dataset by the measurements of Accuracy, F1-Score, Recall and Precision: (1) DSMIL [5]: A single-task method which designs a dubbed dual-stream multiple instance learning network for WSI classification. (2) Graph Attention MIL(GATMIL) [10]: A single-task method based on Graph Attention MIL for TNM staging prediction. (3)WSI-MID: A variant of the proposed method, which (4) MTAN [6]: A MTL method which using a shared feature network, each task select task-specific feature through attention networks. (5) Cross-stitch [9]: Applying the cross-stitch operation to get shared representations for multi-task learning.(6)NDDR-CNN [1]: A MTL method which apply discriminative dimensionality reduction technique for layer-wise feature fusion across different tasks. The classification results are shown in Table 2.

From Table 2, we can derive the following observations: (1) The multi-task learning algorithms (i.e., MTAN, Cross-stitch and NDDR-CNN) can consistently

Table 1. Demographics and Clinical Characteristics of the Utilized Dataset

Characteristics	Summary	Characteristics	Summary
Patients		Stage	
Censored	571	StageI	125
Non-censored	47	StageII	493
Age(Y)	58.1($\pm$12.9)	Follow-up(M)	27.1($\pm$33.9)

Table 2. The performance of different methods on diagnosis task

Method	Accuracy	F1-Score	Recall	Precision
GATMIL [10]	0.705(±0.039)	0.782(±0.033)	0.791(±0.043)	0.770(±0.037)
WSI-MID	0.709(±0.026)	0.800(±0.027)	0.755(±0.060)	0.826(±0.020)
DSMIL [5]	0.714(±0.025)	0.817(±0.026)	0.808(±0.067)	0.829(±0.031)
MTAN [6]	0.727(±0.044)	0.829(±0.037)	0.841(±0.083)	0.821(±0.025)
Cross-stitch [9]	0.733(±0.037)	0.832(±0.040)	0.815(±0.092)	0.807(±0.090)
NDDR-CNN [1]	0.736(±0.045)	0.830(±0.043)	0.774(±0.090)	0.832(±0.052)
WSI-MTMI	**0.772(±0.038)**	**0.857(±0.035)**	**0.870(±0.091)**	**0.848(±0.031)**

Table 3. The performance of different methods on prognosis prediction task

Method	C-index	AUC
DeepSurv [4]	0.650(±0.022)	0.700(±0.083)
Cross-stitch [9]	0.665(±0.033)	0.762(±0.088)
DeepAttnMIL [16]	0.698(±0.018)	0.724(±0.071)
NDDR-CNN [1]	0.698(±0.016)	0.751(±0.022)
MTAN [6]	0.701(±0.021)	0.764(±0.024)
WSI-MTMI	**0.721(±0.025)**	**0.785(±0.041)**

achieve more accurate discrimination than the single-task algorithms (i.e., GAT-MIL, DSMIL). These results clearly verify the advantage of multi-task learning that can combine information from other tasks in order to obtain better features for cancer diagnosis. (2)The proposed WSI-MTMI can achieve the accuracy of 0.772 and the F1-score of 0.857, which is superior to existing MTL learning methods (i.e., MTAN, Cross-stitch and NDDR-CNN). These results validate the superiorities of capturing the correlation between diagnosis and prognosis tasks can help promote the diagnosis performance.

Evaluate Our WSI-MTMI Method on Prognosis Prediction Task. In this section, we evaluate our proposed method on prognosis prediction task. In addition to the three multi-task methods (i.e., MTAN, Cross-stitch and NDDR-CNN) discussed in the previous section, we also compare the proposed method with two single-task prognostic methods. (1) DeepSurv [4]: a Cox proportional hazards deep learning method. (2) DeepAttnMIL [16]: an attention based MIL method to aggregate WSI-level feature to patient-level for survival learning. We also test the performance of our method on prognosis prediction task. The experimental results are shown in Table 3.

As can be seen from Table 3, WSI-MTMI consistently outperforms the comparing methods on prognosis prediction task. Specifically, our proposed method WSI-MTMI achieves the best average Concordance Index (C-index) of 0.721 and the AUC of our method is also superior to its competitors. These results clearly

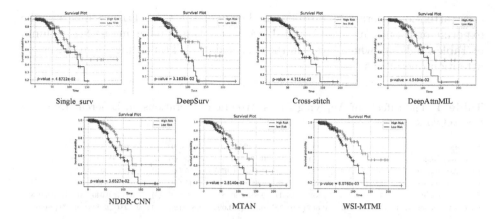

Fig. 2. The survival curves by applying different methods on BRCA datasets

demonstrate that the incorporation of the prior diagnosis information (i.e., TNM stage) can help improve the prognosis performance. In addition, both the experimental results shown in Table 2 and Table 3 indicate that the diagnosis and prognosis tasks are highly correlated with each other, and applying the multi-task learning framework to capture the correlation between different tasks can promote the prediction performance for each individual task.

Evaluate Our WSI-MTMI Method on Patient Stratification Task. In the field of prognosis prediction, another important task is to stratify cancer patients into subgroups with different predicted outcomes, by which we could intervene with personalized treatment plan in the stage of cancer development. In this section, we also compare the prognostic power of different approaches by stratifying cancer patients into 2 subgroups (i.e., the high and low survival risk groups) with different predicted outcomes, with the experimental results shown in Fig. 2. Specifically, the stage method divides all the patients into 2 subgroups according to the TNM stage (i.e., stage I and stage II). For the proposed WSI-MTMI method and other competitors, Gaussian clustering algorithm is adopted to aggregate the patients into 2 subgroups based on the extracted features. Then, we test if these 2 subgroups has significantly different survival outcome using log-rank test [14]. As shown in Fig. 2, the proposed WSI-MTMI method can achieve superior stratification performance (i.e., with p-values of 0.006) than the comparing methods. These results further show the promise of applying task relationship learning framework to identify survival-associated feature shared across different tasks for patient stratification, which opens up the opportunity for building personalized treatment plan in the stage of cancer development.

The Effects of Applying Cross Attention Operations on Different Layers of CAMTL Module. Moreover, we also investigate the effect of applying cross-attention operations on different layers of the CAMTL module, and the experiment results are shown in Table 4, where the number in the first column of Table 4 indicate which layers in the designed CAMTL module adopt the cross-attention operations.

Table 4. The Effects of Applying Cross Attention Operations on Different Layers of the Network

Strategy	Accuracy	F1-Score	Recall	Precision	C-index	AUC
No Cross-Attention	0.729 (±0.078)	0.820 (±0.072)	0.811 (±0.015)	0.850 (±0.044)	0.664 (±0.048)	0.653 (±0.069)
Cross-Attention(3,4)	0.677 (±0.062)	0.780 (±0.041)	0.732 (±0.073)	0.835 (±0.024)	0.673 (±0.026)	0.717 (±0.056)
Cross-Attention(1,2,3,4)	0.736 (±0.060)	0.829 (±0.046)	0.817 (±0.084)	0.844 (±0.022)	0.686 (±0.042)	0.718 (±0.079)
Cross-Attention(1,4)	0.750 (±0.060)	0.838 (±0.051)	0.832 (±0.081)	0.850 (±0.015)	0.689 (±0.046)	0.742 (±0.073)
Cross-Attention(1,2)	**0.772(±0.038)**	**0.857(±0.035)**	**0.870(±0.091)**	**0.848(±0.031)**	**0.721(±0.025)**	**0.771(±0.022)**

As can be seen from Table 4, the diagnosis and prognosis performance will be dramatically decreased if no cross-attention operations are applied, which again validate the fact that take the correlation between diagnosis and prognosis tasks can improve the performance of each individual task. In addition, it is noteworthy that applying the cross-attention operations only on the higher layers (i.e., layer 3 and 4) of CAMTL module will lead to inferior diagnosis and prognosis performance, this is because higher layers contain more task-related information, sharing such information will deteriorate the prediction results for each task. On the other hand, the prediction performance will be largely improved if applying the cross-attention operation on the low layer (i.e., layer 1 and 2) of CAMTL module, one possible reason is that the low-layer cross-attention operations can share common information that is beneficial to both tasks.

Significant Biomarkers Detected by the Proposed WSI-MTMI Method. In this section, we visualize the sample patches that are existed in both high and low survival risk groups in Fig. 3. The patches in high survival risk group (the first line) have large percentage of large nuclei than those in the low survival risk. These results are consistent with the existing discovery that large values of nuclei size usually comes with worse prognosis [15]. In addition, it is worth noting that the nuclei arranged more closer in high survival risk patients, which is in accordance with the fact that the higher tumor cell density will result in worse prognosis and vice versa [15].

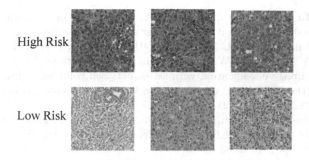

Fig. 3. Sample patches in both high and low survival risk groups.

4   Conclusion

In this paper, we propose a multi-task multi-instance learning for jointly diagnosis and prognosis of early-stage breast invasive carcinoma from Whole-slide pathological images. The main contribution of this study lie in the following two aspects. Firstly, we cast multiple instance learning into a multi-task learning setting and develop a self-attention based multi-instance learning module, which captures the common information by a shared value sets while preserves task-specific information by the respective query and key value sets. Secondly, we propose a cross-attention networks to effectively identify useful information shared across different tasks. Thus, it can avoid negative transfer due to the addition of other information. The experimental results indicate that our method performs better on both diagnosis and prognosis tasks than the existing studies.

Acknowledgement. This Study is Supported by National Natural Science Foundation of China (Nos 6213600, 62272226).

References

1. Gao, Y., Ma, J., Zhao, M., Liu, W., Yuille, A.L.: NDDR-CNN: layerwise feature fusing in multi-task CNNs by neural discriminative dimensionality reduction. In: Proceedings of the IEEE/CVF Conference on Computer Vision and Pattern Recognition, pp. 3205–3214 (2019)
2. Gao, Z., et al.: Instance-based vision transformer for subtyping of papillary renal cell carcinoma in histopathological image. In: de Bruijne, M., et al. (eds.) MICCAI 2021. LNCS, vol. 12908, pp. 299–308. Springer, Cham (2021). https://doi.org/10.1007/978-3-030-87237-3_29
3. Ilse, M., Tomczak, J., Welling, M.: Attention-based deep multiple instance learning. In: International Conference on Machine Learning, pp. 2127–2136. PMLR (2018)

4. Katzman, J.L., Shaham, U., Cloninger, A., Bates, J., Jiang, T., Kluger, Y.: Deep-surv: personalized treatment recommender system using a cox proportional hazards deep neural network. BMC Med. Res. Methodol. **18**(1), 1–12 (2018)
5. Li, B., Li, Y., Eliceiri, K.W.: Dual-stream multiple instance learning network for whole slide image classification with self-supervised contrastive learning. In: Proceedings of the IEEE/CVF Conference on Computer Vision and Pattern Recognition, pp. 14318–14328 (2021)
6. Liu, S., Johns, E., Davison, A.J.: End-to-end multi-task learning with attention. In: Proceedings of the IEEE/CVF Conference on Computer Vision and Pattern Recognition, pp. 1871–1880 (2019)
7. Lv, Z., Lin, Y., Yan, R., Wang, Y., Zhang, F.: Transsurv: transformer-based survival analysis model integrating histopathological images and genomic data for colorectal cancer. In: IEEE/ACM Transactions on Computational Biology and Bioinformatics (2022)
8. Mercan, C., Aksoy, S., Mercan, E., Shapiro, L.G., Weaver, D.L., Elmore, J.G.: Multi-instance multi-label learning for multi-class classification of whole slide breast histopathology images. IEEE Trans. Med. Imaging **37**(1), 316–325 (2017)
9. Misra, I., Shrivastava, A., Gupta, A., Hebert, M.: Cross-stitch networks for multi-task learning. In: Proceedings of the IEEE Conference on Computer Vision and Pattern Recognition, pp. 3994–4003 (2016)
10. Raju, A., Yao, J., Haq, M.M.H., Jonnagaddala, J., Huang, J.: Graph attention multi-instance learning for accurate colorectal cancer staging. In: Martel, A.L., et al. (eds.) MICCAI 2020. LNCS, vol. 12265, pp. 529–539. Springer, Cham (2020). https://doi.org/10.1007/978-3-030-59722-1_51
11. Shao, W., et al.: Ordinal multi-modal feature selection for survival analysis of early-stage renal cancer. In: Frangi, A.F., Schnabel, J.A., Davatzikos, C., Alberola-López, C., Fichtinger, G. (eds.) MICCAI 2018. LNCS, vol. 11071, pp. 648–656. Springer, Cham (2018). https://doi.org/10.1007/978-3-030-00934-2_72
12. Shao, Z., Bian, H., Chen, Y., Wang, Y., Zhang, J., Ji, X., et al.: Transmil: transformer based correlated multiple instance learning for whole slide image classification. Adv. Neural Inf. Process. Syst. **34**, 2136–2147 (2021)
13. Sung, H., et al.: Global cancer statistics 2020: GLOBOCAN estimates of incidence and mortality worldwide for 36 cancers in 185 countries. CA Cancer J. Clin. **71**(3), 209–249 (2021)
14. Wei, L.J., Ying, Z., Lin, D.: Linear regression analysis of censored survival data based on rank tests. Biometrika **77**(4), 845–851 (1990)
15. Wolberg, W.H., Nick Street, W., Mangasarian, O.L.: Importance of nuclear morphology in breast cancer prognosis. Clin. Cancer Res. **5**(11), 3542–3548 (1999)
16. Yao, J., Zhu, X., Jonnagaddala, J., Hawkins, N., Huang, J.: Whole slide images based cancer survival prediction using attention guided deep multiple instance learning networks. Med. Image Anal. **65**, 101789 (2020)
17. Zhang, H., et al.: DTFD-MIL: double-tier feature distillation multiple instance learning for histopathology whole slide image classification. In: Proceedings of the IEEE/CVF Conference on Computer Vision and Pattern Recognition, pp. 18802–18812 (2022)

18. Zhao, Y., et al.: Predicting lymph node metastasis using histopathological images based on multiple instance learning with deep graph convolution. In: Proceedings of the IEEE/CVF Conference on Computer Vision and Pattern Recognition, pp. 4837–4846 (2020)

19. Zuo, Y., et al.: Identify consistent imaging genomic biomarkers for characterizing the survival-associated interactions between tumor-infiltrating lymphocytes and tumors. In: Wang, L., Dou, Q., Fletcher, P.T., Speidel, S., Li, S. (eds.) Medical Image Computing and Computer Assisted Intervention – MICCAI 2022. MICCAI 2022. LNCS, vol. 13432, pp. 222–231. Springer, Cham (2022). https://doi.org/10.1007/978-3-031-16434-7_22

On Fairness of Medical Image Classification with Multiple Sensitive Attributes via Learning Orthogonal Representations

Wenlong Deng[1], Yuan Zhong[2], Qi Dou[2], and Xiaoxiao Li[1]($\boxtimes$)

[1] Department of Electrical and Computer Engineering, The University of British Columbia, Vancouver, BC, Canada
{dwenlong,xiaoxiao.li}@ece.ubc.ca
[2] Department of Computer Science and Engineering, The Chinese University of Hong Kong, Hong Kong, China
{yzhong22,qdou}@cse.cuhk.edu.hk

Abstract. Mitigating the discrimination of machine learning models has gained increasing attention in medical image analysis. However, rare works focus on fair treatments for patients with multiple sensitive demographic attributes, which is a crucial yet challenging problem for real-world clinical applications. In this paper, we propose a novel method for fair representation learning with respect to multi-sensitive attributes. We pursue the independence between target and multi-sensitive representations by achieving orthogonality in the representation space. Concretely, we enforce the column space orthogonality by keeping target information on the complement of a low-rank sensitive space. Furthermore, in the row space, we encourage feature dimensions between target and sensitive representations to be orthogonal. The effectiveness of the proposed method is demonstrated with extensive experiments on the CheXpert dataset. To our best knowledge, this is the first work to mitigate unfairness with respect to multiple sensitive attributes in the field of medical imaging. The code is available at https://github.com/ubc-tea/FCRO-Fair-Classification-Orthogonal-Representation.

1 Introduction

With the increasing application of artificial intelligence systems for medical image diagnosis, it is notably important to ensure fairness of image classification models and investigate concealed model biases that are to-be-encountered in complex real-world situations. Unfortunately, sensitive attributes (e.g., race and gender) accompanied by medical images are prone to be inherently encoded by machine learning models [5], and affect the model's discrimination property [20]. Recently, fair representation learning has shown great potential as it acts as a

---

W. Deng and Y. Zhong—These authors contributed equally to this work.

© The Author(s), under exclusive license to Springer Nature Switzerland AG 2023
A. Frangi et al. (Eds.): IPMI 2023, LNCS 13939, pp. 158–169, 2023.
https://doi.org/10.1007/978-3-031-34048-2_13

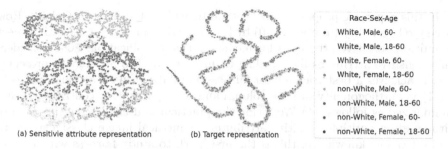

Fig. 1. A t-SNE [10] visualization of (a) sensitive attribute and (b) target representations learned from our proposed methods FCRO on the CheXpert dataset [7]. Sensitive embeddings capture subgroups' variance. We claim FCRO enforces fair classification on the target task by learning orthogonal target representations that are invariant over different demographic attribute combinations.

group parity bottleneck that mitigates discrimination when generalized to downstream tasks. Existing methods [1,4,15,16] have studied the parity between privileged and unprivileged groups upon just a single sensitive attribute, but neglecting the flexibility with respect to multiple sensitive attributes, in which the conjunctions of unprivileged attributes might deteriorate discrimination. This is a crucial yet challenging problem hindering the applicability of machine learning models, especially for medical image classification where patients always have many demographic attributes.

To date, it is still challenging to effectively learn target-related representations which are both fair and flexible to multiple sensitive attributes, regardless of some promising investigations recently. For instance, adversarial methods [1,11] produce robust representations by formulating a min-max game between an encoder that learns class-related representation and an adversary that removes sensitive information from it. Disentanglement-based methods [4] achieve separation by minimizing the mutual information between target and sensitive attribute representations. These methods typically gain efficacy by means of carefully designed objectives. To extend them to the multi-attribute setting, additional loss functions have to be explored, which should handle gradient conflict or interference. Methods using variational autoencoder [3] decompose the latent distributions of target and sensitive and penalize their correlation for disentanglement. However, aligning the distribution of the sensitive attributes is difficult or even intractable given the complex combination of multiple factors. Besides, there are some fairness methods based on causal inference [12] or bi-level optimization [16], which also learn debiased while multi-attributes inflexible representations. Recently, disentanglement is vigorously interpreted as the orthogonality of a decomposed target-sensitive latent representation pair by [15], where they predefine a pair of orthogonal subspaces for target and sensitive attribute representations. However, in the multi-sensitive attributes setting, the dimension of the target space would be continuously compressed and how to solve it is still an open problem.

In this paper, we propose a new method to achieve Fairness via Column-Row space Orthogonality (dubbed FCRO) by learning fair representations with respect to multiple sensitive attributes for medical image classification. FCRO considers multi-sensitive attributes by encoding them into a unified attribute representation. It achieves a best trade-off for fairness and data utility (see illustrations in Fig. 1) via orthogonality in both column and row spaces. Our contributions are summarized as follows: (1) We tackle the practical and challenging problem of fairness given multiple sensitive attributes for medical image classification. To the best of our knowledge, this is the first work to study fairness with respect to multi-sensitive attributes in the field of medical imaging. (2) We relax the independence of target and sensitive attribute representations by orthogonality which can be achieved by our proposed novel column and row losses. (3) We conduct extensive experiments on the CheXpert [7] dataset with over 80,000 chest X-rays. FCRO achieves a superior fairness-utility trade-off over state-of-the-art methods regarding multiple sensitive attributes race, sex, and age.

2   Methodology

2.1   Problem Formulation

Notations. We consider group fairness in this work. While keeping the model utility for the privileged group, group fairness articulates the equality of some statistics like predictive rate between certain groups. Considering a binary classification problem with column vector inputs $x \in \mathcal{X}$, labels $y \in \mathcal{Y} = \{0,1\}$. *Multi-sensitive attributes* $a \in \mathcal{A}$ is vector of m attributes sampled from the conjunction of binary attributes, *i.e.,* Cartesian product of sensitive attributes $\mathcal{A} = \prod_{i \in [m]^\star} A_i$, where $A_i \in \{0,1\}$ is the i-th sensitive attribute. Our training data consist of tuples $\mathcal{D} = \{(x, y, a)\}$. We denote the classification model $f(x) = h_T(\phi_T(x))$ that predicts a class label $\widehat{y}$ given x, where $\phi_T : \mathcal{X} \mapsto \mathbb{R}^d$ is a feature encoder for target embeddings, and $h_T : \mathbb{R}^d \mapsto \mathbb{R}$ is a scoring function. Similarly, we consider a sensitive attribute model $g(x) = \{h_{A_1}(\phi_A(x)), ..., h_{A_m}(\phi_A(x))\}$ that predicts sensitive attributes associated with input x. Given the number of samples n, the input data representation is $X = [x_1, \ldots, x_n]$ and we denote the feature representation $Z_T = \phi_T(X)$, $Z_A = \phi_A(X) \in \mathbb{R}^{d \times n}$.

Fairness with Multiple Sensitive Attributes. A classifier predicts y given an input x by estimating the posterior probability $p(y|x)$. When inputs that are affected by their associated attributes (*i.e.,* $\{A_1, \ldots, A_m\} \to X$) are fed into the network, the posterior probability is written as $p(y|x, a)$. Since biased information from $\mathcal{A}$ is encoded, this can lead to an unfair prediction by the classifier. For example, in the diagnosis of a disease with sensitive attributes age, sex, and race, a biased classifier will result in $p(\widehat{y}|A = \text{male, old, black}) \neq p(\widehat{y}|A = \text{female, young, white})$. In this work, we focus on equalized odds (ED), which is a commonly used and crucial criterion of fair classification in the medical domain

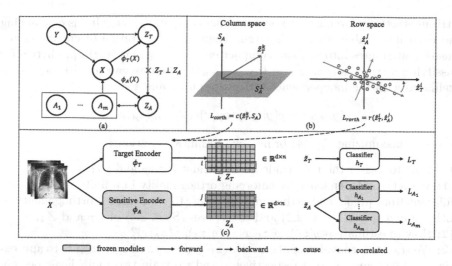

Fig. 2. Overview of our proposed method FCRO. (a) The graphical model of orthogonal representation learning for fair medical image classification with multiple sensitive attributes. (b) The novel column-row space orthogonality. In the column space, we encourage the target model to learn representations in the complement of a low-rank sensitive space. In the row space, we enforce each row vector (feature dimension) of the target and sensitive attribute representations to be orthogonal to each other (c) The overall training pipeline. We use a pre-trained multi-sensitive branch, and propagate orthogonal gradients to target encoder ϕ_T.

[19]. In our case, ED regarding multiple sensitive attributes can be formulated as follows:

$$P(\widehat{Y} = y|A = \pi_1, Y = y) = P(\widehat{Y} = y|A = \pi_2, Y = y), \ \forall \pi_1, \pi_2 \in \mathcal{A}, y \in \mathcal{Y}. \quad (1)$$

A classifier satisfies ED with respect to group A and the target Y if the model prediction $\widehat{Y}$ and A are conditionally independent given Y, *i.e.*, $\widehat{Y} \perp A|Y$.

Fair Representation. To enforce our aforementioned conditions, we follow [15] and introduce target embedding z_T and multi-attribute embedding z_{A_i} that is generated from x. As in the causal structure graph for the classifier depicted in Fig. 2(a), our objective is to find a fair model that maximizes the log-likelihood function of the distribution $p(y, a|x)$, where:

$$p(y, a|x) = \frac{p(y|x, a)p(x|a)p(a)}{p(x)} = p(y|x)p(a|x) \quad (2)$$

$$= p(y|z_T)p(z_T|x) \prod_{i \in [m]} p(a_i|z_{A_i})p(z_{A_i}|x), \quad (3)$$

and we call z_T *fair representation* for the target task (*e.g.*, disease diagnosis). To this end, we aim to maximize Eq. (3) with the conditional independence constraint to train a fair classifier. It is noteworthy that in the multi-sensitive attributes setting, forcing z_T to be independent on all $z_{A_i}, \forall i \in [m]$ is challenging and even intractable when m is large. Therefore, we propose to encode multi-sensitive attributes into a *single* compact encoding z_A that is still predictive for classifying attributes (*i.e.*, $z_A \rightarrow \{a_1, \ldots, a_m\}$). Then we can rewrite Eq. (3) as likelihood with the independence constraint on z_T and z_A:

$$p(y, a|x) = p(y|z_T)p(z_T|x)p(a|z_A)p(z_A|x). \qquad (4)$$

However, maximizing Eq. (4) brings two technical questions:

Q1: How to satisfy the independence constraint for z_T and z_A?
A1: We relax independence by enforcing orthogonality in a finite vector space. Different from predefined orthogonal space in [15], we enforce orthogonality in both column spaces (Sect. 2.2) and row spaces (Sect. 2.3) of Z_T and Z_A.
Q2: How to estimate $p(y|z_T)$, $p(z_T|x)$, $p(a|z_A)$, $p(z_A|x)$?
A2: We train two convolutional encoders $z_T = \phi_T(x)$ and $z_A = \phi_A(x)$ to approximate $p(z_T|x)$ and $p(z_A|x)$ respectively; And we train two multi-layer perception classifiers $y = h_T(z_T)$ and $a = h_A(z_A)$ to approximate $p(y|z_T)$ and $p(a|z_A)$ respectively (Sect. 2.4).

2.2 Column Space Orthogonality

First, we focus on the column space of the target and the sensitive attribute representations. Column space orthogonality aims to learn target representations Z_T that fulfill the following **two aims**: 1) have the least projection onto the sensitive space $\mathcal{S}_A$ and 2) preserve the representation power to predict Y.

Denote the target representation $Z_T = [\tilde{z}_T^1, \tilde{z}_T^2, \ldots, \tilde{z}_T^n]$ and the sensitive attribute representation $Z_A = [\tilde{z}_A^1, \tilde{z}_A^2, \ldots, \tilde{z}_A^n]$, where $\tilde{z}^i \in \mathbb{R}^{d \times 1}$ is a column vector for $i \in [n]$, we represent the column space for Z_T and Z_A as $\mathcal{S}_T = \text{span}(Z_T)$ and $\mathcal{S}_A = \text{span}(Z_A)$ respectively. **Aim 1** can be achieved by forcing $\mathcal{S}_T = \mathcal{S}_A^\perp$. Although both $\tilde{z}_T, \tilde{z}_A \in \mathbb{R}^d$, their coordinates may not be aligned as they are generated from two separate encoders. As a result, if $d \ll \infty$, then there is no straightforward way to achieve $\mathcal{S}_T \perp \mathcal{S}_A$ by directly constraining $\tilde{z}_T^i, \tilde{z}_A^j$ (*e.g.*, forcing $(\tilde{z}_T^i)^\top \tilde{z}_A^j = 0$). **Aim 2** can be achieved by seeking a low-rank representation $\tilde{\mathcal{S}}_A$ for $\mathcal{S}_A$, whose rank is k such that $k \ll d$, because we have $\text{rank}(\mathcal{S}_T) + \text{rank}(\mathcal{S}_A) = d$ if $\mathcal{S}_T = \mathcal{S}_A^\perp$ holds. Then $\mathcal{S}_A^\perp$ would be a high-dimensional space with sufficient representation power for target embeddings. This is especially important when we face multiple sensitive attributes, as the total size of the space is d, and increasing the number of sensitive attributes would limit the capacity of $\mathcal{S}_T$ to learn predictive $\tilde{z}_T$. To this end, we first propose to find the low rank sensitive attribute representation space $\tilde{\mathcal{S}}_A$, and then encourage Z_T to be in $\tilde{\mathcal{S}}_A$'s complement $\tilde{\mathcal{S}}_A^\perp$.

Construct Low-Rank Multi-sensitive Space. We apply Singular Value Decomposition (SVD) on $Z_A = U_A \Sigma_A V_A$ to construct the low-rank space $\tilde{\mathcal{S}}_A$, where $U_A, V_A \in \mathbb{R}^{d \times d}$ are orthogonal matrices with left and right singular vectors $u_A^i \in \mathbb{R}^d$ and $v_A^i \in \mathbb{R}^n$ respectively. And $\Sigma_A \in \mathbb{R}^{d \times n}$ is a diagonal matrix with descending non-negative singular values $\{\delta_A^i\}_{i=1}^{min\{n,d\}}$. Then we extract the most important k left singular vectors to construct $\widetilde{\mathcal{S}}_A = [u_A^1, ..., u_A^k]$, where k controls how much sensitive information to be captured in $\widetilde{\mathcal{S}}_A$. It is notable that $\widetilde{\mathcal{S}}_A$ is agnostic to the number of sensitive attributes because they share the same Z_A. For situations that can not get the whole dataset at once, we follow [8] to select the most important bases from both bases of old iterations and newly constructed ones, thus providing an accumulative low-rank space construction variant to update $\widetilde{\mathcal{S}}_A$ iteratively. As we do not observe significant performance differences between these two variants (Fig. 4(a)), we use and refer to the first one in this paper if there is no special clarification.

Column Orthogonal Loss. With the low-rank space $\widetilde{\mathcal{S}}_A$ for multiple sensitive attributes, we encourage ϕ_T to learn representations in its complement $\widetilde{\mathcal{S}}_A^\perp$. Notice that $\widetilde{\mathcal{S}}_A^\perp$ can also be interpreted as the kernel of the projection onto $\widetilde{\mathcal{S}}_A$, i.e., $\widetilde{\mathcal{S}}_A^\perp = \text{Ker}(\text{proj}_{\widetilde{\mathcal{S}}_A} \widetilde{z}_T)$. Therefore, we achieve column orthogonal loss by minimizing the projection of Z_T to $\widetilde{\mathcal{S}}_A$, which can be defined as:

$$L_{corth} = c(Z_T, \widetilde{\mathcal{S}}_A) = \sum_{i=1}^{n} \frac{\left\| \widetilde{\mathcal{S}}_A^\top \widetilde{z}_T^i \right\|_2^2}{\left\| \widetilde{z}_T^i \right\|_2^2}. \tag{5}$$

As $\widetilde{\mathcal{S}}_A$ is a low-rank space, $\widetilde{\mathcal{S}}_A^\perp$ will have abundant freedom for ϕ_T to extract target information, thus reserving predictive ability.

2.3   Row Space Orthogonality

Then, we study the row space of target and sensitive attribute representations. Row space orthogonality aims to learn target representations Z_T that have the least projection onto the sensitive row space $\widehat{\mathcal{S}}_A$. In other words, we want to ensure orthogonality on each feature dimension between Z_T and Z_A. Denote target representation $Z_T = [\widetilde{z}_T^1; \widetilde{z}_T^2; \ldots; \widetilde{z}_T^d]$ and sensitive attribute representation $Z_A = [\widetilde{z}_A^1; \widetilde{z}_A^2; \ldots; \widetilde{z}_A^d]$, where $\widetilde{z}^i \in \mathbb{R}^{1 \times n}$ is a row vector for $i \in [d]$. We represent row space for target representations and sensitive attribute representations as $\widehat{\mathcal{S}}_T = \text{span}(Z_T^\top)$ and $\widehat{\mathcal{S}}_A = \text{span}(Z_A^\top)$ correspondingly. As the coordinates (i.e., the index of samples) of $\widehat{z}_A$ and $\widehat{z}_T$ are aligned, forcing $\widehat{\mathcal{S}}_T \perp \widehat{\mathcal{S}}_A$ can be directly applied by pushing $\widehat{z}_T^i$ and $\widehat{z}_A^j$ to be orthogonal for arbitrary $i, j \in [d]$.

Unlike column space, the orthogonality here won't affect the utility, since the row vector $\widehat{z}_T$ is not directly correlated to the target y. To be specific, we let pair-wise row vectors $Z_T = [\widehat{z}_T^1, \widehat{z}_T^2, \ldots, \widehat{z}_T^d]$ and $Z_A = [\widehat{z}_A^1, \widehat{z}_A^2, \ldots, \widehat{z}_A^d]$ have a zero inner dot-product. Then for any $i, j \in [d]$, we try to minimize $<\widehat{z}_T^i, \widehat{z}_A^j>$. Here we slightly modify the orthogonality by extra subtracting the mean vector μ_A and μ_T from Z_A and Z_T respectively, where $\mu = \mathbb{E}_{i \in [d]} \widehat{z}^i \in \mathbb{R}^{1 \times n}$. Then orthogonality loss will naturally be integrated into a covariance loss:

$$L_{rorth} = r(Z_T, Z_A) = \frac{1}{d^2} \sum_{i=1}^{d} \sum_{j=1}^{d} \left[(\widehat{z}_T^i - \mu_T)(\widehat{z}_A^j - \mu_A)^\top \right]^2. \tag{6}$$

164   W. Deng et al.

Table 1. CheXpert dataset statistics and group positive rate $p(y = 1|a)$ regarding *pleural effusion* with three sensitive attributes race, sex, and age.

Dataset	#Sample	Group Positive Rate		
		Race	Sex	Age
		(White/Non-white/gap)	(Male/Female/gap)	(>60/≤60/gap)
Original	127130	.410/.393/.017	.405/.408/.003	.440/.359/.081
Augmented	88215	.264/.386/.122	.254/.379/.125	.264/.386/.122

In this way, the resulting loss encourages each feature of Z_T to be independent of features in Z_A thus suppressing the sensitive-encoded covariances that cause the unfairness.

2.4 Overall Training

In this section, we introduce the overall training schema as shown in Fig. 2(c). For the sensitive branch, we noticed that training a shared encoder can pose a risk of sensitive-related features being used for the target classification [4], and we pre-train separate $\{\phi_A, h_{A_1}, ..., h_{A_m}\}$ for multiple sensitive attributes using the sensitive loss as $L_{sens} = \frac{1}{m}\sum_{i\in[m]} L_{A_i}$. Here we use cross-entropy loss as L_{A_i} for the i-th sensitive attribute. Hence $p(z_A|x)$ and $p(a|z_A)$ in Eq. (4) can be obtained. Then, the multi-sensitive space $\mathcal{S}_A$ is constructed as in Sect. 2.2 over the training data. For the target branch, we use cross-entropy loss as our classification objective L_T to supervise the training of ϕ_T and h_T and estimate $p(z_T|x)$ and $p(y|z_T)$ in Eq. (4) respectively. Here we do not make additional constraints to L_T, which means it can be replaced by any other task-specific losses. At last, we apply our column and row orthogonality losses L_{corth} and L_{rorth} to representations as introduced in Sect. 2.2 and Sect. 2.3 along with detached $\mathcal{S}_A$ and Z_A to approximate independence between $p(z_A|x)$ and $p(z_T|x)$. The overall target objective is given as:

$$L_{targ} = L_T + \lambda_c L_{corth} + \lambda_r L_{rorth}, \tag{7}$$

where λ_c and λ_r are hyper-parameters to weigh orthogonality and balance fairness and utility.

3 Experiments

3.1 Setup

Dataset. We adopt CheXpert dataset [7] to predict *Pleural Effusion* in chest X-rays, as it's crucial for chronic obstructive pulmonary disease diagnosis with high incidence. Subgroups are defined based on the following binarized sensitive attributes: *self-reported race* and *ethnicity*, *sex*, and *age*. Note that data bias (positive rate gap) is insignificant in the original dataset (see Table 1, row 'original'). To demonstrate the effectiveness of bias mitigation methods, we amplify

Table 2. Comparasion of predicting *Pleural Effusion* on CheXpert dataset. We report the mean and standard deviation of 5-fold models **trained with multi-sensitive attributes**. AUC is used as the utility metric, and fairness is evaluated using disparities among subgroups defined on multi-sensitive attributes {*Race, Sex, and Age*} *jointly* and each of the attribute *individually*.

Methods	AUC (↑)	Subgroup Disparity (↓)							
		Joint		Race		Sex		Age	
		Δ_{AUC}	Δ_{ED}	Δ_{AUC}	Δ_{ED}	Δ_{AUC}	Δ_{ED}	Δ_{AUC}	Δ_{ED}
ERM [17]	0.863	0.119	0.224	0.018	0.055	0.046	0.142	0.023	0.038
	(.005)	(.017)	(.013)	(.009)	(.017)	(.008)	(.014)	(.004)	(.010)
G-DRO [14]	0.854	0.101	0.187	0.015	0.048	0.034	0.105	0.035	0.051
	(.004)	(.012)	(.034)	(.003)	(.014)	(.010)	(.025)	(.002)	(.010)
JTT [9]	0.834	0.103	0.166	0.019	0.056	0.026	0.079	0.017	0.030
	(.020)	(.017)	(.023)	(.008)	(.016)	(.002)	(.004)	(.006)	(.007)
Adv [18]	0.854	0.089	0.130	0.017	**0.027**	0.022	0.039	0.016	0.023
	(.002)	(.009)	(.018)	(.004)	(**.009**)	(.003)	(.008)	(.004)	(.004)
BR-Net [1]	0.849	0.113	0.200	0.018	0.051	0.037	0.109	0.027	0.039
	(.001)	(.025)	(.023)	(.008)	(.013)	(.012)	(.025)	(.006)	(.006)
PARADE [4]	0.857	0.103	0.193	0.017	0.052	0.042	0.104	0.026	0.031
	(.002)	(.022)	(.032)	(.002)	(.010)	(.006)	(.023)	(.006)	(.011)
Orth [15]	0.856	0.084	0.177	0.012	0.045	0.022	0.083	0.025	0.032
	(.007)	(.022)	(.016)	(.005)	(.012)	(.009)	(.012)	(.006)	(.005)
FCRO (ours)	**0.858**	**0.057**	**0.107**	**0.012**	0.033	**0.015**	**0.024**	**0.013**	**0.019**
	(**.001**)	(.022)	(**.013**)	(**.003**)	(.008)	(**.004**)	(**.008**)	(**.004**)	(**.006**)

the data bias by (1) firstly dividing the data into different groups according to the conjunction of multi-sensitive labels; (2) secondly calculating the positive rate of each subgroup; (3) sampling out patients and increase each subgroup's positive rate gap to 0.12 (see Table 1, row 'augmented'). We resize all images to 224×224 and split the dataset into a 15% test set, and an 85% 5-fold cross-validation set.

Evaluation Metrics. We use the area under the ROC curve (AUC) to evaluate the utility of classifiers. To measure fairness, we follow [13] and compute subgroup disparity with respect to ED (denoted as Δ_{ED}, which is based on true positive rate (TPR) and true negative rate (TNR)) in Eq. (1) as:

$$\Delta_{ED} = \max_{y \in \mathcal{Y}, \pi_1, \pi_2 \in \mathcal{A}} \left| P(\widehat{Y} = y | A = \pi_1, Y = y) - P(\widehat{Y} = y | A = \pi_2, Y = y) \right|. \quad (8)$$

We also follow [20] and compare subgroup disparity regarding AUC (denoted as Δ_{AUC}), which gives a threshold-free fairness metric. Note that we evaluate disparities both *jointly* and *individually*. The *joint* disparities assess multi-sensitive fairness by computing disparity across subgroups defined by the combination of multiple sensitive attributes $\mathcal{A}$. On the other hand, *individual* disparities are calculated based on a specific binary sensitive attribute A_i.

Implementation Details. In our implementation, all methods use the same training protocol. We choose DenseNet-121 [6] as the backbone, but replace

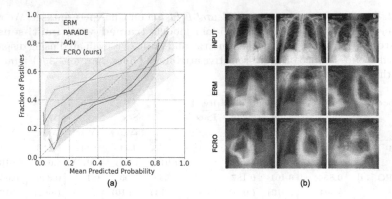

Fig. 3. (a) Subgroup calibration curves. We report the mean (the line) and standard deviation (the shadow around it) of different subgroups defined by the conjunction of race, sex, and age. Larger shadow areas represent more severe unfairness. (b) Class activation map [2] generated from vanilla ERM [17] and FCRO (ours).

the final layer with a linear layer to extract 128-dimensional representations. The optimizer is Adam with learning rate of $1e^{-4}$, and weight decay of $4e^{-4}$. We train for 40 epochs with a batch size of 128. We sweep a range of hyper-parameters for each method and empirically set $\lambda_c = 80$, $\lambda_r = 500$, and $k = 3$ for FCRO. We train models in 5-fold with different random seeds. In each fold, we sort all the validations according to utility and select the best model with the lowest average Δ_{ED} regarding each sensitive attribute among the top 5 utilities.

Baselines. We compare our method with (i) G-DRO [14] and (ii) JTT [9] – methods that seek low worst-group error by minimax optimization on group fairness and target task error, which can be naturally regarded as multi-sensitive fairness methods by defining subgroups with multi-sensitive attributes conjunctions. We also extend recent state-of-the-art fair representation learning methods on single sensitive attributes to multiple ones and compare our method with them, including (iii) Adv [18] and (iv) BR-Net [1] – methods that achieve fair representation via disentanglement using adversarial training, (v) PARADE [4] – a state-of-the-art method that adversarially eliminates mutual information between target and sensitive attribute representations and (vi) Orth [15] – a recent work closest to our method. Orth hard codes the means of both sensitive and target prior distributions to orthogonal means and re-parameterize the encoder output on the orthogonal priors. Besides, we give the result of (vii) ERM [17] – the vanilla classifier trained without any bias mitigation techniques.

3.2   Comparsion with Baselines

Quantitative Results. We summarize quantitative comparisons in Table 2. It can be observed that all the bias mitigation methods can improve fairness compared to ERM [17] at the cost of utility. While ensuring considerable classification accuracy, FCRO achieves significant fairness improvement both *jointly*

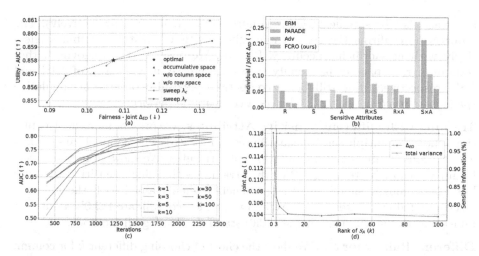

Fig. 4. (a) Fairness-utility trade-off. The perfect point lies in the top left corner. We report ablations and Pareto fronts of the sweep of hyperparameters. (b) Fairness when **trained with various combinations** of three sensitive attributes: race (R), sex (S), and age (A). (c) AUC convergence with different rank k of $\mathcal{S}_A$. (d) Fairness and total variance (the percentage of sensitive information captured by $\mathcal{S}_A$) under different k.

and *individually*, demonstrating the effectiveness of our representation orthogonality motivation. To summarize, compared with the best performance in each metric, FCRO reduced classification disparity on subgroups with *joint* Δ_{AUC} by 2.7% and *joint* Δ_{ED} by 2.3% respectively, and experienced 0.5% Δ_{AUC} and 0.4% Δ_{ED} boosts regarding the average of three sensitive attributes. As medical applications arc sensitive to classification thresholds, we further show calibration curves with the mean and standard deviation of subgroups defined on the conjunction of multiple sensitive attributes in Fig. 3(a). The vanilla ERM [17] suffers from biased calibration among subgroups. Fairness algorithms can help mitigate this, while FCRO shows the most harmonious deviation and the most trustworthy classification. **Qualitative results.** We present the class activation map [2] in Fig. 3(b). We observe that the vanilla ERM [17] model tends to look for sensitive evidence outside the lung regions, *e.g.*, breast, which threatens unfairness. FCRO focuses on the pathology-related part only for fair *Pleural Effusion* classification, which visually confirms the validity of our method.

3.3 Ablation Studies

Loss Modules and Hyperparameters. To evaluate the impact of loss weights and critical components of FCRO, we delve into different choices of hyperparameters and alternative designs. As depicted in Fig. 4(a), we showcase the significance of the key components and the Pareto fronts (*i.e.*, the set of optimal points) curve that results from a range of hyperparameters λ_c and λ_r. Increasing λ_c and λ_r resulted in a small decrease in AUC (0.5% and 0.12%, respectively). However, this was offset by significant gains in fairness (4% and 1.5%, respectively). The observation confirms the goal of managing multiple sensitive attributes effectively without sacrificing accuracy. Besides, We observe that removing either column or row space orthogonality shows a decrease in *joint* Δ_{ED} of 2.4% and 1.8% respectively, but still being competitive. We also observe accumulative space introduced in Sect. 2.2 achieves a comparable performance.

Training with Different Sensitive Attributes. We present an in-depth ablation study on multiple sensitive attributes in Fig. 4(b), where models are trained with various numbers and permutations of attributes. We show all methods perform reasonably better than ERM when trained with a single sensitive attribute but FCRO brought significantly more benefits when trained with the union of discriminated attributes (e.g., Sex × Age), which consolidate FCRO's ability for multi-sensitive attributes fairness. FCRO stand out among all methods.

Different Rank k for $\tilde{\mathcal{S}}_A$. We show the effect of choosing different k for column space orthogonality. As shown in Fig. 4(c), a lower rank k benefits convergence of the model thus improving accuracy, which validates our insights in Sect. 2.2 that lower sensitive space rank will improve the utility of target representations. In Fig. 4(d), we show that $k = 3$ is enough to capture over 95% sensitive information and keep increasing it does not bring too much benefit for fairness, thus we choose $k = 3$ to achieve the best utility-fairness trade off.

4   Conclusion and Future Work

This work studies an essential yet under-explored fairness problem in medical image classification where samples are with sets of sensitive attributes. We formulate this problem mathematically and propose a novel fair representation learning algorithm named FCRO, which pursues orthogonality between sensitive and target representations. Extensive experiments on a large public chest X-rays demonstrate that FCRO significantly boosts the fairness-utility trade-off both *jointly* and *individually*. Moreover, we show that FCRO performs stably under different situations with in-depth ablation studies. For future work, we plan to test the scalability of FCRO on an extremely large number of sensitive attributes.

Acknowledgments. This work is supported in part by the Natural Sciences and Engineering Research Council of Canada (NSERC), Public Safety Canada (NS-5001-22170), in part by NVIDIA Hardware Award, and in part by the Hong Kong Innovation and Technology Commission under Project No. ITS/238/21.

References

1. Adeli, E., et al.: Representation learning with statistical independence to mitigate bias. In: IEEE/CVF Winter Conference on Applications of Computer Vision (2021)
2. Chattopadhay, A., Sarkar, A., Howlader, P., Balasubramanian, V.N.: Grad-CAM++: generalized gradient-based visual explanations for deep convolutional networks. In: IEEE Winter Conference on Applications of Computer Vision (2018)

3. Creager, E., et al.: Flexibly fair representation learning by disentanglement. In: International Conference on Machine Learning, pp. 1436–1445. PMLR (2019)
4. Dullerud, N., Roth, K., Hamidieh, K., Papernot, N., Ghassemi, M.: Is fairness only metric deep? Evaluating and addressing subgroup gaps in deep metric learning. In: The International Conference on Learning Representations (2022)
5. Glocker, B., Jones, C., Bernhardt, M., Winzeck, S.: Algorithmic encoding of protected characteristics in image-based models for disease detection. arXiv preprint arXiv:2110.14755 (2021)
6. Huang, G., Liu, Z., Van Der Maaten, L., Weinberger, K.Q.: Densely connected convolutional networks. In: Proceedings of the IEEE Conference on Computer Vision and Pattern Recognition, pp. 4700–4708 (2017)
7. Irvin, J., et al.: CheXpert: a large chest radiograph dataset with uncertainty labels and expert comparison. In: Proceedings of the AAAI Conference on Artificial Intelligence, vol. 33, pp. 590–597 (2019)
8. Lin, S., Yang, L., Fan, D., Zhang, J.: TRGP: trust region gradient projection for continual learning. In: International Conference on Learning Representations (2022)
9. Liu, E.Z., et al.: Just train twice: improving group robustness without training group information. In: International Conference on Machine Learning (2021)
10. Van der Maaten, L., Hinton, G.: Visualizing data using t-SNE. J. Mach. Learn. Res. 9(11) (2008)
11. Madras, D., Creager, E., Pitassi, T., Zemel, R.: Learning adversarially fair and transferable representations. In: International Conference on Machine Learning, pp. 3384–3393. PMLR (2018)
12. Madras, D., Creager, E., Pitassi, T., Zemel, R.: Fairness through causal awareness: learning causal latent-variable models for biased data. In: Proceedings of the Conference on Fairness, Accountability, and Transparency, pp. 349–358 (2019)
13. Roh, Y., Lee, K., Whang, S.E., Suh, C.: FairBatch: batch selection for model fairness. In: International Conference on Learning Representations (2020)
14. Sagawa, S., Koh, P.W., Hashimoto, T.B., Liang, P.: Distributionally robust neural networks for group shifts: on the importance of regularization for worst-case generalization. In: International Conference on Learning Representations (2019)
15. Sarhan, M.H., Navab, N., Eslami, A., Albarqouni, S.: Fairness by learning orthogonal disentangled representations. In: Vedaldi, A., Bischof, H., Brox, T., Frahm, J.-M. (eds.) ECCV 2020. LNCS, vol. 12374, pp. 746–761. Springer, Cham (2020). https://doi.org/10.1007/978-3-030-58526-6_44
16. Shui, C., et al.: On learning fairness and accuracy on multiple subgroups. In: Advances in Neural Information Processing Systems (2022)
17. Vapnik, V.: Principles of risk minimization for learning theory. In: Advances in Neural Information Processing Systems, vol. 4 (1991)
18. Wadsworth, C., Vera, F., Piech, C.: Achieving fairness through adversarial learning: an application to recidivism prediction. arXiv preprint arXiv:1807.00199 (2018)
19. Xu, Z., Li, J., Yao, Q., Li, H., Shi, X., Zhou, S.K.: A survey of fairness in medical image analysis: concepts, algorithms, evaluations, and challenges. arXiv preprint arXiv:2209.13177 (2022)
20. Zhang, H., Dullerud, N., Roth, K., Oakden-Rayner, L., Pfohl, S., Ghassemi, M.: Improving the fairness of chest X-ray classifiers. In: Conference on Health, Inference, and Learning, pp. 204–233. PMLR (2022)

Pixel-Level Explanation of Multiple Instance Learning Models in Biomedical Single Cell Images

Ario Sadafi[1,2], Oleksandra Adonkina[1,3], Ashkan Khakzar[2], Peter Lienemann[1], Rudolf Matthias Hehr[1], Daniel Rueckert[4], Nassir Navab[2,5], and Carsten Marr[1(✉)]

[1] Institute of AI for Health, Helmholtz Zentrum München - German Research Center for Environmental Health, Neuherberg, Germany
carsten.marr@helmholtz-munich.de
[2] Computer Aided Medical Procedures (CAMP), Technical University of Munich, Munich, Germany
[3] Faculty of Mathematics, Technical University Munich, Munich, Germany
[4] Artificial Intelligence in Healthcare and Medicine, Technical University of Munich, Munich, Germany
[5] Computer Aided Medical Procedures, Johns Hopkins University, Baltimore, USA

Abstract. Explainability is a key requirement for computer-aided diagnosis systems in clinical decision-making. Multiple instance learning with attention pooling provides instance-level explainability, however for many clinical applications a deeper, pixel-level explanation is desirable, but missing so far. In this work, we investigate the use of four attribution methods to explain a multiple instance learning models: GradCAM, Layer-Wise Relevance Propagation (LRP), Information Bottleneck Attribution (IBA), and InputIBA. With this collection of methods, we can derive pixel-level explanations on for the task of diagnosing blood cancer from patients' blood smears. We study two datasets of acute myeloid leukemia with over 100 000 single cell images and observe how each attribution method performs on the multiple instance learning architecture focusing on different properties of the white blood single cells. Additionally, we compare attribution maps with the annotations of a medical expert to see how the model's decision-making differs from the human standard. Our study addresses the challenge of implementing pixel-level explainability in multiple instance learning models and provides insights for clinicians to better understand and trust decisions from computer-aided diagnosis systems.

Keywords: Multiple instance learning · Pixel-level explainability · Blood cancer cytology

A. Sadafi and O. Adonkina—Equal contribution.

A. Frangi et al. (Eds.): IPMI 2023, LNCS 13939, pp. 170–182, 2023.
https://doi.org/10.1007/978-3-031-34048-2_14

1   Introduction

Healthcare systems are challenged by an increasing number of diagnostic requests and a shortage of medical experts. AI can alleviate this problem by providing powerful decision support systems that free medical experts from repetitive, tiring tasks [17]. However, explainability on all levels is required to ensure the proper working of deep learning 'black box' models, and to build trust for the widespread application of health AI.

For decision making that relies on the analysis of hundreds of single instances (e.g. histological patches [4] or single cells [18]), attention-based multiple instance learning (MIL) provides explainability on the instance level [5]. This allows algorithms to highlight suspicious structures in cancer tissue and retrieve prototypical, diagnostic cells in blood or bone marrow smears. In particular in cases where morphological features are unknown, it is of the highest importance to be able to inspect not only high attention instances, but also high attention pixels therein.

A number of different approaches for pixel-level explainability have been proposed and evaluated in the past. Backpropagation based methods such as layerwise relevance propagation (LRP) [15] and guided backpropagation [24] leverage the gradient as attribution. Other methods work with latent features, including GradCAM [21], which utilizes the activations on the final convolution layers, or IBA [20], which measures the predictive information of latent features. These methods are widely used in the medical field to provide some level of explainability: Böhle et at. [3] use LRP to explain the decisions of the neural network on brain MRIs of Alzheimer disease patients; Arnaout et al. [2] propose an ensemble neural network to detect prenatal complex congenital heart disease and use GradCAM to explain the decisions of their expert-level model. Another attribution method, InputIBA [27], has proven to be useful for generating saliency maps for dermatology lesions [11].

Unfortunately, most of these approaches cannot be applied to MIL out of the box. Complex gradient flows and the additional dimension introduced by the bag structure in the MIL model architecture requires adapting explainability algorithms accordingly. Here, we introduce MILPLE, the first multiple instance learning algorithm with pixel-level explainability. We showcase MILPLE (Fig. 1) on two clinical single cell datasets with high relevance for the automatic classification of leukaemia subtypes from patient samples. We adapt GradCAM, LRP, IBA, and InputIBA to a MIL architecture and study the effectiveness of these methods in providing pixel-level explainability for instances. Although the quality of some of the methods seems visually plausible, quantitative analysis shows that there is no silver bullet addressing all challenges. With widespread applications of attention based MIL in different medical tasks, MILPLE helps provide pixel-level explanation using the mentioned algorithms. To foster reproducible research, our code is available on Github https://github.com/marrlab/MILPLE.

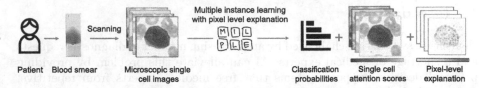

Fig. 1. MILPLE brings pixel-level explainability to multiple instance learning models. We apply MILPLE to two clinical single-cell datasets and showcase its explanatory power for revealing morpho-genetic correlations in blood cancer. In our example, blood smears from over 300 patients suffering from an aggressive leukemic subtype called acute myeloid leukemia (AML) have been digitized and microscopic images of white blood cells have been extracted. AML subtypes are predicted based on the pool of cells, and most important cells are identified based on the MIL attention mechanism, while the most important pixels in each of those are indicated with MILPLE.

2   Methodology

2.1   Multiple Instance Learning

The objective of a multiple instance learning (MIL) model f is analyzing a bag of input instances $B = \{I_1, ..., I_N\}$ and classifying it into one of the classes $c_i \in C$ [12]. In attention-based MIL [7], an attention score $\alpha_k \in A$, $k \in \{1, ..., N\}$ for every instance quantifies the importance of that instance for bag classification:

$$c_i, \alpha_k = f(B). \tag{1}$$

There are two approaches to implement MIL: Instance level and embedding level MIL [25]. We focus on the embedding level MIL, where every input instance is mapped into a low dimensional space via $h_k = f_{\mathrm{emb}}(I_k, \sigma)$ with σ being learned model parameter. By pooling information distributed between the instances, one bag is aggregated into a representative bag feature vector and used for the final classification. Attention pooling [7] provides bag level of explainability and best accuracy in many problems. MIL training can be formulated as

$$\mathcal{L}_{\mathrm{MIL}}(\theta, \sigma) = \mathrm{CE}(c, \hat{c}) \tag{2}$$

with $\hat{c} = f_{\mathrm{MIL}}(H, A; \theta)$, where c is the ground truth label for the whole bag, $H = \{h_1, ..., h_N\}$ are the embedding feature vectors of all instances and CE is the cross entropy loss. θ and σ represent learnable model parameters. Based on the attention scores $\alpha_k \in A$, the bag embedding z is calculated as a weighted average over all of the embedding feature vectors:

$$z = \sum_{k=1}^{N} \alpha_k h_k, \quad \text{where} \quad \alpha_k = \frac{\exp\{w^T \tanh(V h_k^T)\}}{\sum_{j=1}^{N} \exp\{w^T \tanh(V h_j^T)\}}. \tag{3}$$

The parameters V and w are learned in a semi-supervised way during training. With only bag level annotation, instances with the most probable contribution to the classification are given a higher attention score.

2.2   GradCAM

Gradient-weighted Class Activation Mapping (GradCAM) is an explanation technique leveraging the gradient information to localize the most discriminative regions of an input image for a given model prediction. It computes the gradient of the predicted class score with respect to the feature maps of the last convolutional layer and weights each feature map by the corresponding gradient to obtain the class activation map. The class activation map highlights the regions of the input image that are most relevant for the prediction. Blue parts of the map indicate no contribution and red parts indicate high contribution.

2.3   Layer-Wise Relevance Propagation

Layer-wise relevance propagation (LRP) is an explanation technique for deep neural networks which produces pixel-level decomposition of the input by redistributing relevance in the backward pass [14]. Using local redistribution rules a relevance score R_i is assigned to the input variable according to the classifier output $f(x)$:

$$\sum_i R_i^0 = ... = \sum_j R_j^{L-2} = \sum_k R_i^{L-1} = ... = f(x) \tag{4}$$

This backward distribution is lossless, meaning that no relevance is lost in the process while also no additional relevance is introduced at every layer L. A relevance score for every input variable R_i shows the contribution of that variable to the final outcome, which is positive or negative, depending on whether that variable supported the outcome or went against the prediction. The basic rule [14] for LRP is defined as $R_j^{L-1} = \sum_k \frac{a_j w_{jk}}{\sum_j a_j w_{jk}} R_k^L$, where w_{jk} is the weight between the j and k layers and a_j is the activation of neuron j. Eplison rule [14] is an improvement to the basic rule by introducing a positive small ϵ value in the denominator. The ϵ will consume some of the relevance making sparser explanations with less noise. Gamma rule [14] tries to favor positive contributions more by introducing a γ coefficient on positive weights such that the impact on positive weights is controlled with it. As it increases, the effect of positive weights becomes more pronounced. ZBox rule [15] is designed for the input pixel space which is constraint to boxes.

Application to MIL. MIL architectures are a complex combination of different layer types. Fully connected layers are more often used in earlier stages in comparison with normal convolutional neural networks. We tested different combinations of rules. Based on the results and suggestions introduced by Montavon et al. [14], we decided to apply ZBox rule on the first layer for every instance, gamma rule for the feature extractor f_{emb} and epsilon rule on the attention mechanism and final classifier.

2.4   Information Bottleneck Attribution

In contrast to LRP as a back-propagation method, Information bottleneck attribution (IBA) [20] is based on information theory. IBA works by placing a bottleneck on the network to restrict the flow of information by adding noise to the features. A bottleneck on the features F at a given layer can be represented by $Z = \lambda F + (1 - \lambda)\epsilon$ where ϵ is the noise controlled by λ, a mask with the same dimensions as F and elements with values between 0 and 1. The idea is to minimize the mutual information between the input X and Z while maximizing the information between Z and target Y:

$$\max_{\lambda} I(Y, Z) - \beta I(X, Z) \tag{5}$$

Here, β is the Lagrange multiplier controlling the amount of information that passes through the bottleneck. $\mathcal{L}_I$ is an approximation of intractable term $I(X, Z)$:

$$I(X, Z) \approx \mathcal{L}_I = E_F[D_{KL}(P(Z|F) \parallel Q(Z)], \tag{6}$$

where $Q(Z)$ is a normal distribution with estimated mean and variance of F from a batch of samples. Intuitively, $I(Y, T)$ is equivalent to accurate predictions. Thus instead of maximizing it, we can minimize the loss function, cross entropy loss in our case, and therefore information bottleneck can be obtained by using $\mathcal{L} = \beta\mathcal{L}_I + CE$ as the objective.

2.5   Input Information Bottleneck Attribution

The motivation behind InputIBA [27] is to make the information bottleneck optimization in Eq. 5 possible on the input space. IBA as proposed in Eq. 5 and 6 results in an overestimation of mutual information as the bottleneck is applied on earlier layers. The formulation is the most valid when the bottleneck is applied to a deep layer where the Gaussian distribution approximation of activation values is valid [20]. Thus InputIBA proposes a trick where the optimal bottleneck is first computed using Eq. 5. Let us refer to it as Z^*. Then we look for an input bottleneck Z_G that induces the same optimal bottleneck on the deep layer. In order to make the input bottleneck Z_G induce Z^* in deep layers, the following distribution matching is done:

$$\min_{\lambda_G} D[P(f(Z_G))\|P(Z^*)] \tag{7}$$

By optimizing Eq. 7 we find the optimal input bottleneck Z_G^* that induces Z^* in the selected deep layer. InputIBA proceeds to use Z_G^* as a prior for solving the information bottleneck optimization (Eq. 5). The input bottleneck Z_I is conditioned on Z_G as follows: $Z_I = \Lambda Z_G + (1 - \Lambda)\epsilon$, where Λ is the input mask. The final mask Z_I^* is computed by optimizing Eq. 5 on Z_I, and it restricts the flow in the deep layers within limits defined by Z_G^*.

Application to MIL. We had to overcome an obstacle of additional dimension introduced by the bag instances compared to conventional neural networks to apply InputIBA to the MIL structure. In comparison to standard neural networks working with single images, in MIL it is not straightforward to form a batch of bags as convolutions won't handle five dimensions. It is suggested to apply IBA on the deepest layer of the network, however in MIL architectures it seemed that applying IBA on earlier layers yields a better result. After conducting experiments and testing every convolution layers of the resnet backbone, we decided to place the bottleneck at the third convolutional layer where we obtained the best signal compared to other layers. The distance in Eq. 7 is minimized based on an adversarial optimization scheme [27]. The generative adversarial network is trained for each instance in the bag individually. We used $\beta = 40$ to control the amount of information passing through the input bottleneck.

2.6   Quantitative Evaluation of Pixel-Wise Explainability Methods

There is extensive literature studying the quality of the explanations [1,6,8,9, 16,23], but only few quantitative approaches exist. The intuition behind these methods is perturbation of features found to be important and measuring their impact on output to evaluate the quality of the feature attributions.

Insertion/Deletion [19]. Insertion method gradually inserts pixels into the baseline input (zeros) while deletion method removes pixels from input data by replacing them with the baseline value (zero) according to their attribution scores from high to low. While computing the output of the network over different percentage of insertion or deletion a curve is obtained. The area under the curve (AUC) is calculated for every input and averaged over the whole dataset. A higher AUC in insertion means important pixels were inserted first while a lower AUC in deletion means important pixels were removed first.

Remove-and-Retrain [6]. (ROAR) is an empirical measure to approximate the quality of feature attributions by verifying the degradation of the accuracy of a retrained model when the features identified as important are removed from the dataset. The processes is repeated with various percentages of removal. A sharper degradation of the accuracy demonstrates a better identification of important features. Random assignment of importance is defined as a baseline.

3   Experiments

3.1   Dataset

We study the effectiveness of pixel attribution methods on acute meyleod leukimia (AML) subtype recognition tasks using two different datasets: Deep-APL and an in-house AML dataset.

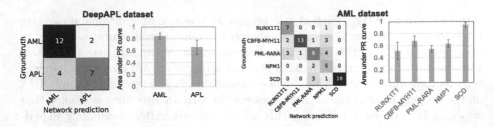

Fig. 2. The confusion matrix and area under the precision recall curve is reported for the two datasets MIL model was trained on. Mean and standard deviation are calculated over 5 independent runs.

DeepAPL [22] is a single cell blood smear dataset consisting of 72 AML and 34 acute promyelocytic leukemia (APL) patients collected at the Johns Hopkins Hospital.

AML dataset is a cohort of 242 patient blood smears from four different prevalent AML genetic subtypes [10]: i) APL with PML::RARA mutation, ii) AML with NPM1 mutation, iii) AML with RUNX1::RUNX1T1 mutation, and iv) AML with CBFB::MYH11 mutation. A fifth group of stem cell donors (SCD) comprises only healthy individuals and is thus used as the control group. Each blood smear contains at least 150 single white blood cell images resulting in a total of 81,214 cells. This dataset is available via TCIA[1].

3.2   Implementation Details

For the backbone of our approach and feature extraction from single cell images, we use the ResNeXt [26] architecture suggested by Matek et al. [13], which is pretrained on the relevant task of single white blood cell classification. Features are extracted from the last convolutional layer of the ResNeXt and passed into the MIL architecture with a second feature extraction step consisting of two convolutional layers with adaptive max-pooling and a fully connected layer. The attention mechanism consists of two fully connected layers and finally, the classifier consists of two fully connected layers. Adam Optimizer with a learning rate of 5×10^4 for DeepAPL and 5×10^5 for the AML dataset with a Nesterov momentum of 0.9 was used. The datasets are split into stratified subsets for train, validation and test in a 60-20-20 percent regime.

3.3   Model Training

The training of the MIL model on the two datasets continues for 40 and 150 epoches, respectively, while the validation loss is monitored. If the validation loss does not decrease for 5 consecutive epochs the training is stopped. We conducted 5 independent runs to train the model. Table 1 shows the mean and standard deviation of accuracy, macro F1 score and area under ROC curve.

---

[1] https://doi.org/10.7937/6ppe-4020.

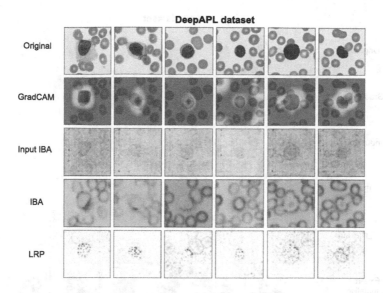

Fig. 3. Pixel-level explanation methods applied to exemplary images from the Deep-APL dataset. For GradCAM blue parts of the map indicate no contribution and red parts indicate high contribution, and similarly, for LRP blue parts indicate negative contribution and red parts indicate positive contribution. In many cases GradCAM and LRP focus on the white blood cells in the center of the image, while IBA focuses also on the red blood cell surrounding it. InputIBA shows a relatively scattered focus. (Color figure online)

3.4 Evaluation of Explanations

Qualitative Evaluation includes inspection of single cell images and comparison with medical expert annotation. Figure 3 and 4 show selected cells from both datasets and pixel-level explanations provided by the four different methods. In Fig. 4, we compare pixel attributions with expert annotations as a medical expert has annotated a small subset of single cells in the AML dataset. Most of the methods detect morphological features defined by the expert as important.

Quantitative Evaluation of the explanations is an essential step for correct understanding of what model focuses on. In order to evaluate the quality of different methods, we performed Insertion/Deletion and ROAR experiments on each of the GradCAM, LRP, IBA, and InputIBA methods as shown in Fig. 5. The performance of the method is highly dependent on the dataset and each time different methods end up to be the most suitable.

3.5 Discussion and Results

Model Performance: We compare our training on DeepAPL with the state-of-the-art method proposed by Sidhom et al. [22] on the dataset. Since the datasets

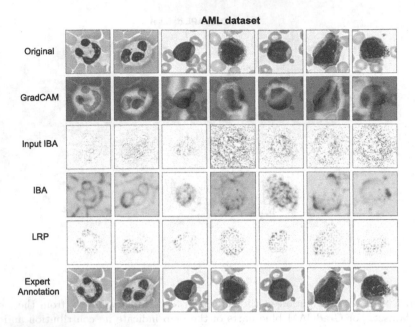

Fig. 4. Pixel-level explanation methods applied on exemplary images from AML dataset. In the first two images, all methods agree on the morphology found relevant by the expert (last row). In the following images, the methods highlight different regions and are only sometimes in concordance with the expert.

are imbalanced, we are reporting the area under precision recall curve for each class as well as the confusion matrix for both datasets to get a better view over the class-wise performance. Figure 2 shows that classification results are robust across the two datasets. On DeepAPL, with no special tailoring of the method to the dataset, we could outperform the state-of-the-art method based on sample analysis cell by cell. MIL takes all cells into consideration and can thus achieve a higher accuracy in the task. On the AML dataset some ambiguity exists between different malignant classes, which is to be expected since AML subtype classification based on cell morphology only is a challenging task even for the medical experts. Model identifies the majority of benign stem cell donors correctly.

Explanations: A close inspection of the pixel explanations from the four different methods reveals fundamental differences (see Fig. 3, 4): For the Deep-APL dataset (Fig. 3) we observe that GradCAM focuses on the white blood cell nucleus in most cases. In some cells however it fails to recognise the cell and instead puts high relevance on background pixels at the image border. Though according to our ROAR results (Fig. 5), removing the white blood cell affects the accuracy significantly, pointing to the fact that the network is using features relevant to them. InputIBA puts most focus on the centre of the image, and thus correctly on the white blood cell. However, pixel attention is spread out

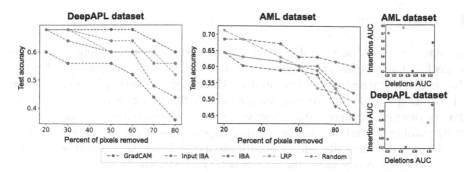

Fig. 5. Remove-and-Retrain (ROAR) experiment (left) and insertion/deletion experiment (right) for both datasets. GradCAM has the best pixel attribution in DeepAPL ROAR experiments, while on the AML dataset, LRP and IBA perform best. Insertion/deletion experiments for DeepAPL support GradCAM. For the AML dataset InputIBA and LRP have the best performance in insertion/deletion experiment.

Table 1. Mean and standard deviation of accuracy, macro F1 score and area under ROC curve is reported for the two blood cancer datasets. Our attention based MIL method outperforms the original DeepAPL method [22].

Data	Method	Accuracy	F1 score	AU ROC
DeepAPL	ours	0.65 ± 0.07	0.63 ± 0.08	0.750 ± 0.078
DeepAPL	Sidhom et al. [22]	–	–	0.739
AML	ours	0.68 ± 0.03	0.65 ± 0.04	0.855 ± 0.037

over the whole image at times (Fig. 3). The ROAR results for InputIBA (Fig. 5) also show that the accuracy drops if corresponding image regions are removed.

On the AML dataset we observe that IBA highlights image regions that correspond to either abnormal cytoplasm (4th, 6th and 7th cell from left, Fig. 4) or to structures in the nucleus (first two cells in Fig. 4). These are particularly interesting since they show that the method is able to retrieve morphological details that escape the human eye (3rd cell: the cell appears to be dark violet in the original images, but IBA is able to focus on morphology therein) and to segment granules, whose structure is relevant for cell type classification (4th cell). The ROAR results from the AML dataset (Fig. 5) show that removing morphological features identified by IBA significantly disrupts accuracy. This signifies that the model relies on these pixel during training. LRP focuses on the white blood cell in the image center and the nucleus therein. We observe that the ROAR results for LRP are not very informative (Fig. 5), and the method performs similarly to random. This might be due to the LRP structure and a problem with the ROAR metric. However, LRP achieves a good score on the Deletion/Insertion metric (Fig. 5). This means that LRP features have an immediate effect on the output of the network.

4   Conclusion

Incorporating pixel-level explainability in multiple instance learning allows us to inspect instances, evaluate the focus of our model, and find morphological details that might be missed by the human eye. All four pixel-level explainability methods we used revealed interesting insights and highlighting morphological details that fit prior expert knowledge. However, more work has to be done on systematically comparing and quantifying clinical expert annotations with explainability predictions, to eventually select appropriate methods for the application at hand, and potentially reveal novel morpho-genetic correlations.

We believe that our study will be instrumental for multiple instance learning applications in health AI. Single-cell data is ideal for method development, since it allows a direct comparison of model prediction and human intuition. However, applied to computational histopathology, where a large amount of digitized data exists, the pixel-level insight into tissue structure at multiple scales might reveal morphological properties previously unrecognized. With novel spatial single-cell RNA sequencing technologies being on the brink of becoming available widely, we expect a high demand for methods like MILPLE.

References

1. Adebayo, J., Gilmer, J., Muelly, M., Goodfellow, I., Hardt, M., Kim, B.: Sanity checks for saliency maps. In: Advances in Neural Information Processing Systems, vol. 31 (2018)
2. Arnaout, R., Curran, L., Zhao, Y., Levine, J.C., Chinn, E., Moon-Grady, A.J.: An ensemble of neural networks provides expert-level prenatal detection of complex congenital heart disease. Nat. Med. **27**(5), 882–891 (2021)
3. Böhle, M., Eitel, F., Weygandt, M., Ritter, K.: Layer-wise relevance propagation for explaining deep neural network decisions in MRI-based Alzheimer's disease classification. Front. Aging Neurosci. **11**, 194 (2019)
4. Campanella, G., et al.: Clinical-grade computational pathology using weakly supervised deep learning on whole slide images. Nat. Med. **25**(8), 1301–1309 (2019)
5. Hehr, M., et al.: Explainable AI identifies diagnostic cells of genetic AML subtypes. PLOS Digit. Health **2**(3), e0000187 (2023)
6. Hooker, S., Erhan, D., Kindermans, P.J., Kim, B.: A benchmark for interpretability methods in deep neural networks. In: Advances in Neural Information Processing Systems, vol. 32 (2019)
7. Ilse, M., Tomczak, J., Welling, M.: Attention-based deep multiple instance learning. In: International Conference on Machine Learning, pp. 2127–2136. PMLR (2018)
8. Khakzar, A., Baselizadeh, S., Khanduja, S., Rupprecht, C., Kim, S.T., Navab, N.: Neural response interpretation through the lens of critical pathways. In: Proceedings of the IEEE/CVF Conference on Computer Vision and Pattern Recognition, pp. 13528–13538 (2021)
9. Khakzar, A., Khorsandi, P., Nobahari, R., Navab, N.: Do explanations explain? Model knows best. In: Proceedings of the IEEE/CVF Conference on Computer Vision and Pattern Recognition, pp. 10244–10253 (2022)

10. Khoury, J.D., et al.: The 5th edition of the world health organization classification of haematolymphoid tumours: myeloid and histiocytic/dendritic neoplasms. Leukemia **36**(7), 1703–1719 (2022)
11. Krammer, S., et al.: Deep learning-based classification of dermatological lesions given a limited amount of labelled data. J. Eur. Acad. Dermatol. Venereol. **36**(12), 2516–2524 (2022)
12. Maron, O., Lozano-Pérez, T.: A framework for multiple-instance learning. In: Advances in Neural Information Processing Systems, vol. 10 (1997)
13. Matek, C., Schwarz, S., Spiekermann, K., Marr, C.: Human-level recognition of blast cells in acute myeloid leukaemia with convolutional neural networks. Nat. Mach. Intell. **1**(11), 538–544 (2019)
14. Montavon, G., Binder, A., Lapuschkin, S., Samek, W., Müller, K.-R.: Layer-wise relevance propagation: an overview. In: Samek, W., Montavon, G., Vedaldi, A., Hansen, L.K., Müller, K.-R. (eds.) Explainable AI: Interpreting, Explaining and Visualizing Deep Learning. LNCS (LNAI), vol. 11700, pp. 193–209. Springer, Cham (2019). https://doi.org/10.1007/978-3-030-28954-6_10
15. Montavon, G., Lapuschkin, S., Binder, A., Samek, W., Müller, K.R.: Explaining nonlinear classification decisions with deep taylor decomposition. Pattern Recogn. **65**, 211–222 (2017)
16. Nie, W., Zhang, Y., Patel, A.: A theoretical explanation for perplexing behaviors of backpropagation-based visualizations. In: International Conference on Machine Learning, pp. 3809–3818. PMLR (2018)
17. Rajpurkar, P., Chen, E., Banerjee, O., Topol, E.J.: AI in health and medicine. Nat. Med. **28**(1), 31–38 (2022)
18. Sadafi, A., et al.: Attention based multiple instance learning for classification of blood cell disorders. In: Martel, A.L., et al. (eds.) MICCAI 2020. LNCS, vol. 12265, pp. 246–256. Springer, Cham (2020). https://doi.org/10.1007/978-3-030-59722-1_24
19. Samek, W., Binder, A., Montavon, G., Lapuschkin, S., Müller, K.R.: Evaluating the visualization of what a deep neural network has learned. IEEE Trans. Neural Netw. Learn. Syst. **28**(11), 2660–2673 (2016)
20. Schulz, K., Sixt, L., Tombari, F., Landgraf, T.: Restricting the flow: information bottlenecks for attribution. arXiv preprint arXiv:2001.00396 (2020)
21. Selvaraju, R.R., Cogswell, M., Das, A., Vedantam, R., Parikh, D., Batra, D.: Gradcam: visual explanations from deep networks via gradient-based localization. In: Proceedings of the IEEE International Conference on Computer Vision, pp. 618–626 (2017)
22. Sidhom, J.W., et al.: Deep learning for diagnosis of acute promyelocytic leukemia via recognition of genomically imprinted morphologic features. NPJ Precis. Oncol. **5**(1), 1–8 (2021)
23. Sixt, L., Granz, M., Landgraf, T.: When explanations lie: why many modified BP attributions fail. In: International Conference on Machine Learning, pp. 9046–9057. PMLR (2020)
24. Springenberg, J.T., Dosovitskiy, A., Brox, T., Riedmiller, M.: Striving for simplicity: the all convolutional net. arXiv preprint arXiv:1412.6806 (2014)
25. Vocaturo, E., Zumpano, E.: Dangerousness of dysplastic nevi: a multiple instance learning solution for early diagnosis. In: 2019 IEEE International Conference on Bioinformatics and Biomedicine (BIBM), pp. 2318–2323. IEEE (2019)

26. Xie, S., Girshick, R., Dollár, P., Tu, Z., He, K.: Aggregated residual transformations for deep neural networks. In: Proceedings of the IEEE Conference on Computer Vision and Pattern Recognition, pp. 1492–1500 (2017)
27. Zhang, Y., et al.: Fine-grained neural network explanation by identifying input features with predictive information. arXiv preprint arXiv: 2110.01471 (2021)

Transient Hemodynamics Prediction Using an Efficient Octree-Based Deep Learning Model

Noah Maul[1,2(✉)], Katharina Zinn[1,2], Fabian Wagner[1], Mareike Thies[1],
Maximilian Rohleder[1,2], Laura Pfaff[1,2], Markus Kowarschik[2],
Annette Birkhold[2], and Andreas Maier[1]

[1] Pattern Recognition Lab, FAU Erlangen-Nürnberg, Erlangen, Germany
noah.maul@fau.de
[2] Siemens Healthcare GmbH, Forchheim, Germany

Abstract. Patient-specific hemodynamics assessment could support diagnosis and treatment of neurovascular diseases. Currently, conventional medical imaging modalities are not able to accurately acquire high-resolution hemodynamic information that would be required to assess complex neurovascular pathologies. Instead, computational fluid dynamics (CFD) simulations can be applied to tomographic reconstructions to obtain clinically relevant information. However, three-dimensional (3D) CFD simulations require enormous computational resources and simulation-related expert knowledge that are usually not available in clinical environments. Recently, deep-learning-based methods have been proposed as CFD surrogates to improve computational efficiency. Nevertheless, the prediction of high-resolution transient CFD simulations for complex vascular geometries poses a challenge to conventional deep learning models. In this work, we present an architecture that is tailored to predict high-resolution (spatial and temporal) velocity fields for complex synthetic vascular geometries. For this, an octree-based spatial discretization is combined with an implicit neural function representation to efficiently handle the prediction of the 3D velocity field for each time step. The presented method is evaluated for the task of cerebral hemodynamics prediction before and during the injection of contrast agent in the internal carotid artery (ICA). Compared to CFD simulations, the velocity field can be estimated with a mean absolute error of 0.024 m s^{-1}, whereas the run time reduces from several hours on a high-performance cluster to a few seconds on a consumer graphical processing unit.

Keywords: Hemodynamics · Octree · Operator learning

1 Introduction

Vascular diseases are globally the leading cause of death [17] and thus optimal medical diagnosis and treatment are desirable. A detailed understanding of vascular abnormalities is essential for treatment planning, which includes

A. Frangi et al. (Eds.): IPMI 2023, LNCS 13939, pp. 183–194, 2023.
https://doi.org/10.1007/978-3-031-34048-2_15

hemodynamic information such as blood velocity and pressure. However, especially for neurovascular pathologies quantitative blood flow information at sufficiently high resolution is difficult to acquire by conventional medical imaging alone. Hence, measurements, e.g., three-dimensional (3D) tomographic reconstructions of the abnormality, are usually coupled with computational methods. These methods include 3D computational fluid dynamics (CFD) simulations, that are based on solving partial differential equations (PDEs) numerically. However, the setup of CFD simulations requires domain-specific knowledge and enormous computational resources, which are usually not available in a clinical environment. Even without the required preprocessing steps, simulations require several hours of runtime on high-performance computing clusters. Recently, deep-learning-based models have been proposed to approximate CFD results with different input data, acting as surrogate models. Once trained, these models allow to reduce runtime during inference [3–5, 21, 23, 30]. However, the presented approaches either regress a desired hemodynamic quantity directly (without estimating the whole 3D velocity or pressure field), work on small volumetric patches (only possible with suitable input data, e.g., magnetic resonance tomography), or are limited to steady-state simulations. To the best of our knowledge, there exists no prior work on high-resolution transient CFD surrogate models for complex vascular geometries. We hypothesize that lacking suitable deep learning architectures and associated data representations for time-resolved volumetric vascular data so far prohibit more accurate methods. Particularly, uniformly spaced voxel volumes are inherently ill-suited for discretizing neurovascular geometries, as the size of the smallest feature defines required resolution and voxel size. Further, the resulting volumes are usually sparse, meaning that a large part of the volume is not covered by (image-able) vascular structures and is, therefore, not contributing to clinically relevant hemodynamics. Convolutional neural networks (CNNs) are in principle well-suited for flow prediction as they exploit local coherence [29]. However, this would require huge computational resources due to the sparse data. Even more challenging is the application of CNNs to transient simulations, where the time dimension must be considered in the architecture, leading to further increase of computational complexity.

In this work, we present a computationally efficient deep learning architecture that is designed to infer high-resolution velocity fields for transient flow simulations in complex 3D vascular geometries. We utilize octree-based CNNs [25, 26], enabling sparse convolutions, to efficiently discretize and process neurovascular geometries with a high spatial resolution. Furthermore, instead of utilizing four-dimensional or auto-regressive architectures, we formulate the regression task as an operator learning problem. This concept has been introduced to solve PDEs in general [13, 14] and has also been applied to medical problems, such as tumor ablation planning [16]. The presented method is evaluated for the task of cerebral hemodynamics prediction based on a 3D digital subtraction angiography (3D DSA) acquisition, which is usually performed for treatment planning of neurovascular procedures.

2   Methods

2.1   Problem Description and Method Overview

In this work, the goal is to train a deep-learning-based CFD surrogate model to approximate the velocity field given the vascular geometry, and the associated boundary and initial conditions, which are the commonly used input data for a hemodynamics CFD simulation. We aim to learn the solution operator for the incompressible Navier-Stokes equations

$$\frac{\partial \mathbf{u}}{\partial t} + \mathbf{u} \cdot \nabla \mathbf{u} = -\nabla p + \nu \nabla^2 \mathbf{u}$$
$$\nabla \cdot \mathbf{u} = 0,$$

(1)

where $\mathbf{u}$ denotes the velocity, p the pressure and ν the kinematic viscosity of a fluid. For this, the following steps are carried out. First, synthetic cerebral vessel trees are generated and preprocessed for simulation. Second, boundary conditions (BCs) are chosen from a representative set and a CFD solver is employed to calculate a set of reference solutions that describe the blood and total flow during injection of a contrast agent (CA). Third, the reference solutions are post-processed, resulting in a dataset that is used to train a neural network. After training, the surrogate model is applied to unseen data.

2.2   Physics Model of 3D DSA Acquisition

Our physics model is designed to describe the physiological blood flow and the effect of the injection of contrast agent (CA) during a neurovascular 3D DSA acquisition. As CA is commonly injected into the ICA, we focus our analysis on this case. Like previous studies [22,24], we assume that the density difference between CA and blood is negligible and model the mixture as a single-phase flow. However, due to resistances downstream, the blood flow rate before injection Q_B and the injection flow rate Q_{CA} do not simply add up but can be described with a mixing factor m [18,22], $Q_T(t) = Q_B(t) + m \cdot Q_{CA}(t)$, where Q_T denotes the total flow rate. As in previous work [22], a mixing factor of 0.3 is chosen. Further, we model the compliance and resistance of the contrast flow through the catheter by an analogous electrical network consisting of a resistor and a capacitor [24]. Therefore, the injection flow rate $Q_{CA}(t)$ can be described by

$$Q_{CA}(t) = \begin{cases} 0 & t < T_S \\ Q_{CA}^{\max} \cdot \left(1 - e^{-(t-T_S)/T_L}\right) & t \geq T_S \end{cases},$$

(2)

where T_S refers to the injection start time, T_L to the time of the lag and $Q_{CA}^{\max}$ to the maximum injection rate that is set to 2.5 mLs^{-1}. Moreover, the physiological blood flow rate should reflect real conditions and is therefore derived from reported values in literature. To generate a set of representative inflow conditions of humans, we follow the approach of Ford et al. [6] and Hoi et al. [8]. They showed how the inflow waveform for the ICA can be modeled by the mean flow rate, cardiac cycle length and age. We select a set of flow rates $\in \{3.4, 4.4, 5.4\}\,\mathrm{mLs}^{-1}$ and cardiac cycle lengths $\in \{785, 885, 985\}\,\mathrm{ms}$ that correspond approximately to the mean $\pm$ one standard deviation. For each combination of the sets, two waveforms (young and elderly) are generated, leading to 18 different inflow waveforms in total. The flow is modeled as laminar and blood as a Newtonian fluid with a kinematic viscosity ν of $3.2 \times 10^{-6}\mathrm{m}^2\mathrm{s}^{-1}$ and a density of $1.06 \times 10^3\mathrm{kg}\,\mathrm{m}^{-3}$. Vessel walls are assumed to be rigid, no-slip and zero-gradient pressure BCs are applied. The outlet BCs are set according to the flow-splitting method [2], which avoids unrealistic zero pressure outlet BCs. The algorithm starts at the inlet (most distal part of the ICA) and determines the flow split ratio between the branches at each bifurcation. When an outlet is reached, the associated flow rate is assigned as the BC.

2.3   Generation of Synthetic Vascular Geometries

The surrogate model is trained and evaluated with a dataset that comprises synthetic neurovascular geometries in combination with physiological BCs. Surface meshes of vessel trees are automatically generated using the 3D modeling software *Blender* (Blender, version 2.92, Blender Foundation) and the *Sapling Tree Gen* addon (Sapling Tree Gen, version 0.3.4, Hale et al.) that is based on a sampling algorithm by Weber et al. [27]. The root branch is modeled as a synthetic ICA, where the radius is chosen uniformly random between 1.62 and 1.98 mm [1]. To cover a large variety of bifurcation types, bifurcation angles are chosen uniformly distributed between 35 and 135° with an additional vertical attraction factor of 2.6 [27]. To ensure a developed flow and avoid backflow at the outlets, flow extensions with an approximate length of five times the respective vessel diameter are added to the inlet and all outlets. Three synthetic trees are depicted in Fig. 1. Centerlines are calculated on the resulting surface mesh and a locally radius-adaptive polyhedral volumetric mesh containing five prismatic boundary layers to capture steep gradients near the vessel wall is generated [9,28].

Fig. 1. Three synthetic vascular trees generated for the dataset.

2.4   CFD Simulation Setup

For each vessel tree, the flow rate, cardiac cycle length and the age is sampled uniformly random and the resulting inflow curve is set as the inflow BC (plug flow velocity profile) for the simulation. The CFD software *OpenFOAM* (Open-FOAM, version 8, The OpenFOAM Foundation) [28] is employed and second order schemes are chosen for space and time discretization. An adaptive implicit time-stepping method with a maximum timestep of 1 ms is employed. Overall, four cardiac cycles are simulated. The first one is used to wash out initial transient effects, while the second one reflects the hemodynamics before contrast injection. The virtual injection of CA starts with the third cardiac cycle and is continued until the end of the simulation. Each simulation is executed utilizing 40 CPUs on a high-performance cluster requiring several hours of runtime.

2.5 Deep Learning Architecture

The proposed deep learning model is designed to infer the velocity field from the geometry and BCs. The model consists of three main building blocks, as illustrated in Fig. 2. In the first block (a), a 3D point cloud representation of the geometry and the BCs (inflow waveform) are processed to compute node features for each point. In the second block (b), an octree [15] is calculated from the point cloud, and passed to an octree-based neural network. In the third block (c), the output of the network is regarded as a continuous neural representation of the velocity field function that can be evaluated at a set of spatiotemporal points in parallel. The individual building blocks are elaborately presented in the following.

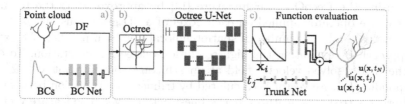

Fig. 2. Overview of the proposed architecture that consists of three buildings blocks: geometry and BC encoding (a), octree construction and octree U-Net inference (b), and neural function evaluation (c).

A) Geometry and Boundary Condition Processing. The input to the model consists of a point cloud representation of the geometry and the BCs. In the dataset, the inflow waveform provides sufficient information to solve the Navier-Stokes equations as the outflow BCs are calculated from the inflow waveform and the geometry (flow-splitting method). Also, the initial conditions are shared across all cases. The distance field (DF) of the point cloud to the tree surface is calculated and the 1D inflow waveform is supplied to a 1D CNN (BC Net). This network contains four blocks, each consisting of a convolution, pooling, and leaky rectified linear unit (LReLU) layer. As a last operation, the output is pooled to a four-dimensional vector by averaging over the feature maps. The feature vector and the DF are concatenated, repeating the same feature vector for all field points, and together form the point cloud features.

B) Octree Construction and U-Net. The octree data structure is built from the spatial information of the point cloud [25] with a maximum depth level of ten, leading to an isotropic node resolution of 0.15 mm on the finest level. At each level until maximum depth, a node is refined if it contains one or more points, such that the adaptive octree resolution can be controlled by the input point cloud. The point features that lie within a node on the finest level are averaged and considered the node features. To learn the neural representation of the velocity field, a U-Net architecture [20] is employed. Wang et al. [25] introduced a convolution layer that acts directly on the octree structure, such that calculations are only performed in spatial locations where necessary (inside vessels). The U-Net consists of three downsampling (strided convolution) and three corresponding upsampling (transposed convolution) steps, starting at the maximum octree depth level down to the seventh level. The four encoder levels consist of two, three, four, and six residual bottleneck blocks [7], respectively. Each decoder level comprises two bottleneck residual blocks. A bottleneck residual block consists of two 3×3 convolutions and one 1×1 convolution for the residual connection. The LReLU is used as an activation function.

C) Neural Function Evaluation. The output of the U-Net is regarded as a continuous neural representation of the velocity field that can be evaluated at arbitrary spatiotemporal points. Meister et al. [16] employ an operator learning approach (mapping between function spaces) to predict a temperature distribution using the DeepONet architecture [14] by approximating the temporal antiderivative operator applied on each voxel of a 3D grid. This allows to efficiently evaluate a large number of spatiotemporal points in parallel. We extend this approach by exploiting the high-resolution octree representation. To evaluate the time-resolved velocity field at an arbitrary spatial point $\mathbf{x} \in \mathbb{R}^3$, the corresponding feature vector is calculated by trilinear interpolation between the neighboring octree node features at the finest octree level. The feature vector is transformed by two fully connected layers and the result is split in three equally sized parts $\mathbf{b_x} = (\mathbf{b_x^1}, \mathbf{b_x^2}, \mathbf{b_x^3})$, $\mathbf{b_x^i} \in \mathbb{R}^d$, that contain the time dynamics information of the three approximated velocity vector components $\hat{\mathbf{u}} = (\hat{u}_1, \hat{u}_2, \hat{u}_3)$ at $\mathbf{x}$. A series of five fully connected layers and LReLU activation functions (trunk net [14]) receives the time point t as an input and result in a latent vector $\mathbf{r_t} \in \mathbb{R}^d$. To evaluate $\hat{\mathbf{u}}(\mathbf{x}, t)$, three dot products $\hat{u}_i(\mathbf{x}, t) = \sum_{k=1}^{d} b_{\mathbf{x},k}^i \, r_{t,k} + c_i$ are calculated, where $i \in \{1, 2, 3\}$ and c_i denotes a learnable bias for each velocity component. The algorithmic steps are also illustrated in Fig. 2. Note that this formulation only requires a single forward pass of the octree U-Net for each full time-resolved velocity field. Further, batches of $\{\mathbf{b_{x_i}}\}$ and $\{\mathbf{r_{t_j}}\}$ can be calculated independently and in parallel.

2.6 Training Setup

To train and test the proposed model for the regression task, a suitable dataset is constructed using the synthetic geometries and simulated hemodynamics data.

Overall, a virtual CA injection is simulated for 45 different virtual vessel trees, as described in Sect. 2.4. The velocity field is saved 30 times per second to be consistent with clinical 3D DSA image data that is usually acquired at 30 frames per second. Due to differing cardiac cycle lengths, this leads to a varying number of samples per simulation. The samples are then split by geometry in training (35), validation (5) and test set (5). For each sample, the cell centers from the volumetric mesh are used as an input point cloud to construct the octree. The same locations are also utilized to evaluate the learned velocity field. As a data augmentation technique during training, the point clouds are randomly rotated and translated, such that a greater variety of octrees can be constructed. To optimize the network parameters, the mean absolute error $\frac{1}{3N}\sum_{j=1}^{N}\sum_{i\in\{1,2,3\}}|\hat{u}_{ij}-u_{ij}|$ between predicted and actual velocity components is calculated for N spatiotemporal points. Ten time points are randomly chosen per simulation and batched to reduce training time. The ADAM optimizer [10] is employed to train the network until convergence and the model with the lowest validation loss is selected for evaluation.

3   Results

The method is evaluated on the test set that contains the simulations of five separate vessel trees. Like in the training phase, the network is queried at the cell centers of the CFD mesh and at 30 time steps per second until the end of the third cardiac cycle. We quantitatively compare the predictions of our method with the CFD simulations.

3.1   Quantitative Evaluation

The overall mean, standard deviation, and median of the absolute error across the whole test set (all spatiotemporal points) is 0.024, 0.043 and 0.010 m s^{-1}, respectively. Further, to avoid time-consuming processing for visualization, the time-averaged velocity field per case is computed for the network prediction and the CFD simulation. In Fig. 3, a regression plot over the time-averaged test set (left), as well as the error distribution for each individual case (right) is depicted. The mean absolute error of the time-averaged velocities is 0.023 m s^{-1} and a coefficient of determination (R^2) of 0.97 is determined for the prediction. For some points, the network tends to underestimate the velocities. The quartiles of the error distribution show that 75% of all velocities can be estimated with an error smaller than 0.040 m s^{-1}.

3.2   Qualitative Evaluation

We qualitatively evaluate the time-averaged velocity distributions for test case five (median of mean absolute error across cases), which is depicted in Fig. 4. Three cross-sectional slices of the velocity magnitude at different locations and streamlines at the first bifurcation are depicted. The first slice is positioned far

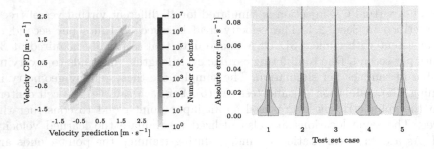

Fig. 3. Statistical evaluation of the predicted velocity field for the test set cases. Due to the large number of points, only the time-averaged velocities are considered. The left figure depicts the joint histogram of predicted and CFD (time-averaged) velocities (all components) over the entire test set (mean absolute error of 0.023 m s^{-1} and R^2 of 0.97). Please note the logarithmic colormap. The right figure displays the error distribution for each individual test case. The y-axis limit is set to the maximum 90-percentile of the cases.

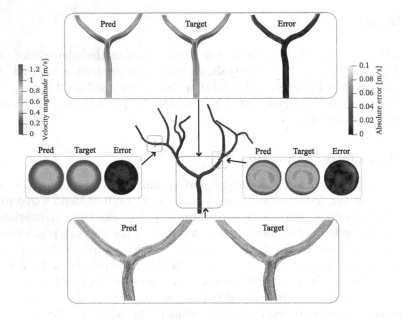

Fig. 4. Evaluation of the (time-averaged) velocity field at three locations for test case five (median of mean absolute error across cases, visualized in Fig. 3). All slices depict the velocity magnitude and the corresponding error between CFD and network prediction for a cross section of the 3D model. Streamline plots provide further information about the velocity field components by visualizing the trajectory of virtual particles.

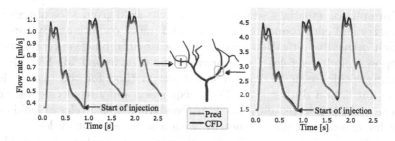

Fig. 5. Comparison of volumetric flow rates at two slices for test case five (median of mean absolute error across cases, visualized in Fig. 3)

downstream after three bifurcations at a position where a less complex flow is expected (small radius and curvature, some distance from the previous bifurcation). To evaluate more complex flow scenarios, an additional slice is placed right before the second bifurcation. The third slice shows the cross section of the first bifurcation. Overall, both slice and streamline plots show reasonable agreement between predicted and actual velocity values as well as distributions. Additionally, the volumetric flow rate over time is compared for the first two slices, which is displayed in Fig. 5. The flow rates agree reasonable, but are slightly underestimated for the three systoles (first and secondary peaks).

3.3 Runtime Evaluation

We evaluate the GPU-time of the model on a NVIDIA Quadro RTX 3000 (6 GB memory) graphical processing unit for one vessel tree from the test set. The overall time can be split in three parts: BC Net and U-Net forward pass t_{net}, spatial function evaluation $t_{spatial}$ and temporal $t_{temporal}$ function evaluation (trunk net forward pass and dot products). The spatial function evaluation comprises the octree interpolation and subsequent network transformation, whereas temporal evaluation refers to the execution of the trunk net and the dot products (visualized in Fig. 2). The spatial evaluation is performed with a batch of 10^6 coordinates. For the temporal evaluation, the trunk net is executed with a batch of 100 timesteps and the dot products are calculated between the output and the 10^6 feature vectors. We measure the runtimes 204.5 ± 2.7 ms, 92.2 ± 4.9 ms and 23.8 ± 1.5 ms for t_{net}, $t_{spatial}$ and $t_{temporal}$ respectively, where mean and standard deviation are calculated across 100 runs for each part. Hence, assuming a velocity prediction for N_s spatial points and N_t temporal points, the overall runtime is approximately $t_{net} + N_s/10^6 \cdot t_{spatial} + N_t/10^2 \cdot t_{temporal}$.

4   Discussion

Computational methods for 3D hemodynamics assessment of neurovascular pathologies require enormous computational resources that are usually not available in clinical environments. We presented a method that is tailored to predict the high-resolution (spatial and temporal) velocity field given a (complex) vascular geometry and corresponding boundary and initial conditions. By combining an explicit octree discretization with an implicit neural function representation, our proposed model enables the approximation of transient 3D hemodynamic simulations within seconds. We evaluated the method for the task of hemodynamics prediction during a 3D DSA acquisition for virtual cerebral vessels trees, where CA is injected into the ICA. Once trained, the velocity field can be inferred for unseen vascular geometries with a mean absolute velocity error of 0.024 ± 0.043 m s^{-1}. Our quantitative and qualitative evaluation showed good agreement between the prediction of our model and the CFD ground truth. Existing approaches for predicting hemodynamics with machine learning surrogate models either regress a derived low-dimensional hemodynamic quantity directly (without outputting 3D velocity or pressure fields) [4], rely on magnetic resonance imaging input [5,21], or predict 3D steady-state simulations [3,12] with fixed BCs. Compared to this, our method allows the prediction of high-resolution unsteady velocity fields for varying BCs. Raissi et al. [19] use physics-informed neural networks (PINNs) to predict high-resolution transient hemodynamics inside an aneurysm assuming that the concentration field of a transported passive scalar can be measured. In contrast to our work, this allows to infer the underlying velocity field without knowledge of the BCs. However, PINNs usually must be retrained in a self-supervised manner for each inference case and are therefore difficult to apply in a medical setup, which is a major advantage of our method.

One consideration for the application of neural networks is the ability to generalize on unseen data. We tested our method on synthetic vessel trees that were not included in the training or validation procedure. The availability of sufficient clinical data is a common problem in medical machine learning, such that synthetically generated data is often used [11]. However, not all flow patterns in anatomical cerebral vessel trees might be covered by synthetic cases. In particular, pathological abnormalities that alter the flow, e.g., stenoses or aneurysms, were not considered in our study. For a clinical application, our method needs to be investigated on real clinical patient data. Our method is not limited to predicting velocity fields for medical applications. It could be applied analogously to predict pressure distributions or other quantities of interest and is suited for any vessel or tube-shaped geometries, even outside the medical field.

5   Conclusion

The computational complexity of CFD simulations restricts patient-specific hemodynamics assessment in the clinical workflow. We presented a deep-learning-based CFD surrogate model tailored to predict the high-resolution spatial and temporal velocity field given a complex vascular geometry and BCs within seconds. We envision that our approach could form the basis for a clinical hemodynamics assessment tool that supports diagnosis of vascular diseases and provides online feedback to clinicians during procedures.

Disclaimer. The concepts and information presented are based on research and are not commercially available.

References

1. Chnafa, C., et al.: Vessel calibre and flow splitting relationships at the internal carotid artery terminal bifurcation. Physiol. Meas. **38**(11), 2044 (2017)
2. Chnafa, C., Brina, O., Pereira, V., Steinman, D.: Better than nothing: a rational approach for minimizing the impact of outflow strategy on cerebrovascular simulations. AJNR Am. J. Neuroradiol. **39**(2), 337–343 (2018)
3. Du, P., Zhu, X., Wang, J.X.: Deep learning-based surrogate model for three-dimensional patient-specific computational fluid dynamics. Phys. Fluids **34**(8), 081906 (2022)
4. Feiger, B., et al.: Accelerating massively parallel hemodynamic models of coarctation of the aorta using neural networks. Sci. Rep. **10**(1), 9508 (2020)
5. Ferdian, E., et al.: 4DFlowNet: super-resolution 4D flow MRI using deep learning and computational fluid dynamics. Front. Phys. **8** (2020)
6. Ford, M.D., Alperin, N., Lee, S.H., Holdsworth, D.W., Steinman, D.A.: Characterization of volumetric flow rate waveforms in the normal internal carotid and vertebral arteries. Physiol. Meas. **26**(4), 477–488 (2005)
7. He, K., Zhang, X., Ren, S., Sun, J.: Deep residual learning for image recognition. In: Proceedings of the CVPR, pp. 770–778 (2016)
8. Hoi, Y., et al.: Characterization of volumetric flow rate waveforms at the carotid bifurcations of older adults. Physiol. Meas. **31**(3), 291–302 (2010)
9. Izzo, R., Steinman, D., Manini, S., Antiga, L.: The vascular modeling toolkit: a python library for the analysis of tubular structures in medical images. J. Open Source Softw. **3**(25), 745 (2018)
10. Kingma, D.P., Ba, J.: Adam: a method for stochastic optimization. In: Proceedings of the ICLR (2015)
11. Li, G., et al.: Prediction of cerebral aneurysm hemodynamics with porous-medium models of flow-diverting stents via deep learning. Front. Physiol. **12**, 733444 (2021)
12. Li, G., et al.: Prediction of 3D cardiovascular hemodynamics before and after coronary artery bypass surgery via deep learning. Commun. Biol. **4**(1), 99 (2021)
13. Li, Z., et al.: Fourier neural operator for parametric partial differential equations. arXiv (2020)
14. Lu, L., Jin, P., Pang, G., Zhang, Z., Karniadakis, G.E.: Learning nonlinear operators via DeepONet based on the universal approximation theorem of operators. Nat. Mach. Intell. **3**(3), 218–229 (2021)
15. Meagher, D.: Geometric modeling using octree encoding. Comput. Graph. Image Process. **19**(2), 129–147 (1982)
16. Meister, F., et al.: Fast automatic liver tumor radiofrequency ablation planning via learned physics model. In: Proceedings of the MICCAI, pp. 167–176 (2022)
17. Mendis, S., Puska, P., Norrving, B., Organization, W.H., Federation, W.H., Organization, W.S.: Global atlas on cardiovascular disease prevention and control (2011)
18. Mulder, G., Bogaerds, A., Rongen, P., van de Vosse, F.: The influence of contrast agent injection on physiological flow in the circle of Willis. Me. Eng. Phys. **33**(2), 195–203 (2011)
19. Raissi, M., Yazdani, A., Karniadakis, G.E.: Hidden fluid mechanics: learning velocity and pressure fields from flow visualizations. Science **367**(6481), 1026–1030 (2020)

20. Ronneberger, O., Fischer, P., Brox, T.: U-Net: convolutional networks for biomed-ical image segmentation. In: Proceedings of the MICCAI, pp. 234–241 (2015)
21. Rutkowski, D.R., Roldán-Alzate, A., Johnson, K.M.: Enhancement of cerebrovas-cular 4d flow MRI velocity fields using machine learning and computational fluid dynamics simulation data. Sci. Rep. **11**(1), 10240 (2021)
22. Sun, Q., Groth, A., Aach, T.: Comprehensive validation of computational fluid dynamics simulations of in-vivo blood flow in patient-specific cerebral aneurysms. Med. Phys. **39**(2), 742–754 (2012)
23. Taebi, A.: Deep learning for computational hemodynamics: a brief review of recent advances. Fluids **7**(6), 197 (2022)
24. Waechter, I., Bredno, J., Hermans, R., Weese, J., Barratt, D.C., Hawkes, D.J.: Model-based blood flow quantification from rotational angiography. Med. Image Anal. **12**(5), 586–602 (2008)
25. Wang, P.S., Liu, Y., Guo, Y.X., Sun, C.Y., Tong, X.: O-CNN: octree-based convo-lutional neural networks for 3D shape analysis. ACM Trans. Graph. **36**(4), 1–11 (2017)
26. Wang, P.S., Liu, Y., Tong, X.: Deep octree-based CNNs with output-guided skip connections for 3D shape and scene completion. In: Proceedings of the CVPRW, pp. 1074–1081 (2020)
27. Weber, J., Penn, J.: Creation and rendering of realistic trees. In: Proceedings of the SIGGRAPH, pp. 119–128. Association for Computing Machinery (1995)
28. Weller, H.G., Tabor, G., Jasak, H., Fureby, C.: A tensorial approach to compu-tational continuum mechanics using object-oriented techniques. Comput. Phys. **12**(6), 620–631 (1998)
29. Xie, Y., Franz, E., Chu, M., Thuerey, N.: tempoGAN: a temporally coherent, volumetric GAN for super-resolution fluid flow. ACM Trans. Graph. **37**(4), 95 (2018)
30. Yuan, X.Y., et al.: Real-time prediction of transarterial drug delivery based on a deep convolutional neural network. Appl. Sci. **12**(20), 10554 (2022)

Weakly Semi-supervised Detection in Lung Ultrasound Videos

Jiahong Ouyang[1], Li Chen[2(✉)], Gary Y. Li[2], Naveen Balaraju[2],
Shubham Patil[2], Courosh Mehanian[3], Sourabh Kulhare[3], Rachel Millin[3],
Kenton W. Gregory[4], Cynthia R. Gregory[4], Meihua Zhu[4], David O. Kessler[5],
Laurie Malia[5], Almaz Dessie[5], Joni Rabiner[5], Di Coneybeare[5], Bo Shopsin[6],
Andrew Hersh[7], Cristian Madar[8], Jeffrey Shupp[9], Laura S. Johnson[9],
Jacob Avila[10], Kristin Dwyer[11], Peter Weimersheimer[12], Balasundar Raju[2],
Jochen Kruecker[2], and Alvin Chen[2]

[1] Stanford University, Stanford, CA, USA
[2] Philips Research North America, Cambridge, MA, USA
li.chen_1@philips.com
[3] Global Health Laboratories, Bellevue, WA, USA
[4] Oregon Health & Science University, Portland, OR, USA
[5] Columbia University Medical Center, New York, NY, USA
[6] New York University, New York, NY, USA
[7] Brooke Army Medical Center, San Antonio, TX, USA
[8] Tripler Army Medical Center, Honolulu, HI, USA
[9] MedStar Washington Hospital Center, Washington, DC, USA
[10] University of Kentucky, Lexington, KY, USA
[11] Warren Alpert Medical School of Brown University, Providence, RI, USA
[12] University of Vermont Larner College of Medicine, Burlington, VT, USA

Abstract. Frame-by-frame annotation of bounding boxes by clinical
experts is often required to train fully supervised object detection models on medical video data. We propose a method for improving object
detection in medical videos through weak supervision from video-level
labels. More concretely, we aggregate individual detection predictions
into video-level predictions and extend a teacher-student training strategy to provide additional supervision via a video-level loss. We also
introduce improvements to the underlying teacher-student framework,
including methods to improve the quality of pseudo-labels based on
weak supervision and adaptive schemes to optimize knowledge transfer between the student and teacher networks. We apply this approach
to the clinically important task of detecting lung consolidations (seen in
respiratory infections such as COVID-19 pneumonia) in medical ultrasound videos. Experiments reveal that our framework improves detection
accuracy and robustness compared to baseline semi-supervised models,
and improves efficiency in data and annotation usage.

Keywords: Weakly Supervised Learning · Semi-Supervised Learning ·
Object Detection · Medical Ultrasound

Work completed during internship at Philips Research North America.

1   Introduction

Despite the remarkable performance of deep learning networks for object detection and other computer vision tasks [2,10,19], most models rely on large-scale annotated training examples, which are often unavailable or burdensome to generate in the medical imaging domain. This is especially true for video-based imaging modalities such as medical ultrasound, where frame-by-frame annotation of bounding boxes or other localization labels is extremely time-consuming and costly, and even more so if annotations must be done by clinical experts.

Reducing the annotation burden of training object detectors on medical images has been a focus of much recent work. Semi-supervised and weakly supervised approaches have been proposed to address the annotation challenge, where unlabeled or inexactly/inaccurately labeled data are used to supplement training, often in combination with a small amount of fully labeled data [1,9,11,13,15,21]. Examples of weak supervision for object detection include point annotations [3,4,14,17,22] and image-level class labels [12,14,17], both of which are applied on individual image frames. However, even these methods of weak supervision may not be practical in the video domain, where hundreds or thousands of image frames requiring interpretation may be collected in a single clinical exam.

In this work, we propose a weakly semi-supervised framework for training object detection models based on video-level supervision, where only a single label is provided for each video. Video-level labels represent a significantly weaker form of supervision than instance- or image-level labels but can be generated much more efficiently. Our approach extends teacher-student models adopted for semi-supervised object detection [11] to the weakly semi-supervised video-based detection task. Our main contributions are as follows:

1. We introduce a simple mechanism during teacher-student training which aggregates individual detections from the teacher (pseudo-labels) into video-level confidence predictions. This allows video-level weak supervision using any standard classification loss.
2. We improve the reliability of pseudo-labels generated during the mutual learning stage by introducing techniques to re-weigh pseudo-labels based on video-level weak supervision.
3. We investigate the learning dynamics between the teacher and student, and propose several improvements to the underlying teacher-student mechanism to increase training stability. These include a method to better initialize models in the "burn-in" stage and a set of adaptive updating schemes to optimize knowledge transfer bidirectionally during mutual learning.

We demonstrate the effectiveness of our approach on the task of detecting lung consolidations in medical ultrasound videos, which is an important step in aiding diagnosis and management of patients with respiratory infections such as bacterial and viral pneumonia (including COVID-19 infection). Computer-aided detection of lung consolidation in medical ultrasound is a uniquely challenging problem where the appearance of pathology varies dramatically across disease types, patient populations, and training levels of the personnel acquiring images. Experimental results on a large, multi-center, clinical ultrasound dataset for lung consolidation demonstrate that the proposed framework leads to improved detection performance, robustness, and training stability compared to existing semi-supervised methods.

2 Related Work

Semi-supervised Object Detection: Semi-supervised object detection aims to utilize large amounts of unlabeled data together with a small set of labeled data. These efforts generally fall into two categories: (1) consistency regularization, which regularizes the prediction of the detector for images undergoing different augmentations [6,7], and (2) pseudo-labeling, where a teacher model is trained on labeled data to generate pseudo-labels for unlabeled data, and a student model is then trained on both the labeled and pseudo-labeled data [11,16,18,20,23]. Unbiased Teacher (UBT) [11] is one of the state-of-the-art methods in this category. Our work is inspired by the framework of UBT, but extends the method to the weakly semi-supervised scenario, where we leverage frame-level pseudo-labels and video-level weak supervision.

Weakly Semi-supervised Object Detection: Weakly semi-supervised object detection is usually based on instance-level weak supervision, e.g., a point on the object [3,4,8,14,17,22], or image-level supervision, e.g., the class of the image [12,14,17]. Video-level labels are a significantly weaker supervisory signal than instance- or frame-level labels, since the only information provided is that the object class exists somewhere in at least one frame of the video, but in which specific frame(s) and at what location(s) is unknown.

3 Method

The basic intuition behind our approach is to simultaneously learn from both frame-level and video-level labels to improve object detection performance. We denote a fully labeled set $\mathcal{D}_f$ and a weakly labeled set $\mathcal{D}_w$ as our training data. The fully labeled set $\mathcal{D}_f = \{x_i^f, y_i^f\}_{i=1}^{N_f}$ comprises a set of N_f frames x_i^f and their paired frame-level annotations y_i^f (i.e., the coordinates of all bounding boxes present in each frame). The weakly labeled set $\mathcal{D}_w = \{x_{j,1:T_j}^w, z_j^w\}_{j=1}^{N_w}$ consists of N_w videos $x_{j,1:T_j}^w$ and their video-level class labels z_j^w (indicating whether a video contains at least one instance of the object class, which is annotated on the whole video of T_j frames). Due to its high accuracy, computational efficiency, and capacity for real-time inference, YOLO-v5 [5] is adopted as the backbone detector. Non-maximum suppression (NMS) is applied on the outputs of YOLO-v5 as the final output of the detector to remove duplicate predictions. Here, we use c to denote the confidence vector (a part of YOLO-v5's output) of all predicted boxes from a frame x.

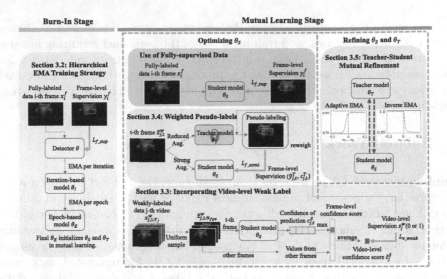

Fig. 1. Overview of the proposed method.

3.1 Teacher-Student Training

We adopt the UBT framework from [11] for the weakly semi-supervised detection task. UBT uses two training stages: a burn-in stage for model initialization and a mutual learning stage for teacher-student training (Fig. 1).

Burn-In Stage: The burn-in stage aims to initialize the detector in a supervised manner using the fully labeled data $\mathcal{D}_f$. The original burn-in method in UBT adopts model weights after a fixed number of early training epochs. However, we observed that detection performance can vary dramatically during training. Here, we improve the training stability during burn-in by applying hierarchical exponential moving average (EMA) updates at each iteration and epoch (yellow block in Fig. 1). Specifically, the iteration-based model θ_I is introduced to transfer knowledge from an initial detection model θ after each iteration (that is, the weights of θ_I are updated per batch). The epoch-based model θ_E is added to transfer information from θ_I after each epoch (weights updated after all batches). This is given by:

$$\begin{cases} \theta_I \leftarrow \alpha_i \theta_I + (1 - \alpha_i)\theta, & \text{for each iteration} \\ \theta_E \leftarrow \alpha_e \theta_E + (1 - \alpha_e)\theta_I, & \text{for each epoch} \end{cases} \tag{1}$$

where α_i and α_e are the iteration- and epoch-based EMA keep rates, respectively. The EMA keep rates define a trade-off between the rate of knowledge being transferred from the preceding model versus the stability of the succeeding model. Major oscillations in model performance are seen while training θ_I alone, even with a carefully selected α_i [5]. In contrast, the addition of θ_E serves to stabilize the training and results in a better-initialized detector after burn-in.

Mutual Learning Stage: The mutual learning stage combines the fully labeled data $\mathcal{D}_f$ and weakly labeled data $\mathcal{D}_w$ for teacher-student training. Both the student and teacher models are initialized from the last checkpoint of θ_E trained in the burn-in stage. During mutual learning, the student θ_S is optimized via back-propagation using a combination of full, weak, and semi-supervised losses (see Sections 3.3 and 3.4), and the teacher θ_T is updated via a gradual EMA transfer of weights from the student. Analogous to the model updates in the burn-in stage, the student is updated with an iteration-based EMA during mutual learning, while the teacher is updated with an epoch-based EMA. An additional mutual refinement scheme is introduced to adaptively adjust the EMA keep rate of the teacher, as well as to conditionally allow transfer of weights back to the student (see Section 3.5). As mutual learning progresses, the accuracy and stability of pseudo-labels produced by the teacher are continuously improved, which in turn improves knowledge distillation to and from the student. At the end of the training, only the teacher θ_T is kept for evaluation and deployment.

3.2   Weakly Semi-supervised Learning

Frame-Level Full Supervision: For both burn-in and mutual learning, fully labeled data $\mathcal{D}_f$ are used in training, with supervision by a detection loss $\mathcal{L}_{f\_sup}$:

$$\mathcal{L}_{f\_sup} = \sum_{i=1}^{N_f} \lambda_{coord} \mathcal{L}_{coord}(\mathcal{T}_s(x_i^f), y_i^f) + \lambda_{conf} \mathcal{L}_{conf}(\mathcal{T}_s(x_i^f), y_i^f), \qquad (2)$$

where $\mathcal{L}_{coord}$ is the bounding box coordinate error, and $\mathcal{L}_{conf}$ is the binary cross-entropy loss between predicted box confidences and corresponding box labels. λ_{coord} and λ_{conf} balance the two losses. $\mathcal{T}_s$ denotes the data augmentation.

Frame-Level Semi-supervision: In the mutual learning stage, the weakly labeled data $\mathcal{D}_w$ are added to allow frame-level semi-supervision based on pseudo-labels from the teacher θ_T. Specifically, we generate sub-clips $\tilde{x}_{j,1:N_{fpv}}^w$ by uniformly sampling N_{fpv} frames from each video $x_{j,1:T_j}^w$ in the weak dataset. These sub-clips are fed through the teacher with reduced augmentation $\mathcal{T}_r$ to obtain box predictions $\hat{y}_{j,1:N_{fpv}}^T$ with confidence scores $c_{j,1:N_{fpv}}^T$ from each frame. Only predicted boxes with confidence above a threshold β are kept as pseudo-labels. We use a second detection loss, similar to Eq. 2, to train the student θ_S against the teacher's pseudo-labels:

$$\mathcal{L}_{f\_semi} = \sum_{j=1}^{N_w} \sum_{t=1}^{N_{fpv}} \left[\lambda_{coord} \mathcal{L}_{coord}(\mathcal{T}_s(x_{j,t}^w), \hat{y}_{j,t}^T) + \lambda_{conf} \mathcal{L}_{conf}(\mathcal{T}_s(x_{j,t}^w), \hat{y}_{j,t}^T) \right] \quad (3)$$

Video-Level Weak Supervision: Finally, we utilize the confidence of boxes predicted by the student $c_{j,1:N_{fpv}}^S$ to obtain a frame-level confidence by computing the maximum confidence among all detected boxes in the frame, i.e.,

$max(c_{j,t}^S)$. A final video-level prediction $\tilde{c}_j^S$ is computed as the averaged frame-level confidence over the sub-clip, i.e., $\tilde{c}_j^S = \frac{1}{N_{fpv}} \sum_{t=1}^{N_{fpv}} max(c_{j,t}^S)$.

Here, we apply a video-level binary cross-entropy classification loss to supervise the student θ_S against the video-level labels z_j^w from the weak data $\mathcal{D}_w$:

$$\mathcal{L}_{v\_weak} = -\sum_{j=1}^{N_w} \left[z_j^w log(\tilde{c}_j^S) + (1 - z_j^w)log(1 - \tilde{c}_j^S) \right] \qquad (4)$$

Combined Loss: The final loss function for training the student model combines the fully supervised detection loss $\mathcal{L}_{f\_sup}$, the frame-level semi-supervised detection loss $\mathcal{L}_{f\_semi}$, and the video-level weakly supervised loss $\mathcal{L}_{v\_weak}$:

$$\mathcal{L} = \lambda_{f\_sup}\mathcal{L}_{f\_sup} + \lambda_{f\_semi}\mathcal{L}_{f\_semi} + \lambda_{v\_weak}\mathcal{L}_{v\_weak} \qquad (5)$$

with λ_{f_sup}, λ_{f_semi}, and λ_{v_weak} to balance the three loss components.

3.3   Weighted Pseudo-labels

A critical factor in the effectiveness of mutual learning is the quantity and quality of pseudo-labels from the teacher θ_T. We propose two pseudo-label re-weighting techniques to increase the number of high-quality pseudo-labels during training.

Weakly Supervised Pseudo-label Filtering: The first approach utilizes the weak video-level label z_j^w to filter false pseudo-labels. For a negative video ($z_j^w = 0$), we can simply remove all pseudo-labels from every frame, which could be considered as re-weighting the pseudo-label as 0. For a positive video ($z_j^w = 1$), if no pseudo-label confidence exceeds β, we keep the pseudo-label with the highest confidence if it exceeds a lower threshold β_l, where $\beta_l < \beta$, which could be considered as re-weighting the pseudo-label from 0 to its confidence.

Soft Pseudo-labels: The second approach assigns a weight to each pseudo-label based on its prediction confidence. That is, we re-weigh the loss component for each pseudo-label $\hat{y}_{j,t}^T$ by the square of its confidence $(c_{j,t}^T)^2$ to create "soft pseudo-labels" $(\hat{y}_{j,t}^T, c_{j,t}^T)$. The semi-supervised detection loss is reformulated as:

$$\mathcal{L}_{coord}(\mathcal{T}_s(x_{j,t}^w), \hat{y}_{j,t}^T) = \sum_{k}^{N_{j,t}} (c_{j,t,k}^T)^2 \mathcal{L}_{coord}(\mathcal{T}_s(x_{j,t}^w), \hat{y}_{j,t}^T)_k$$

$$\mathcal{L}_{conf}(\mathcal{T}_s(x_{j,t}^w), \hat{y}_{j,t}^T) = \sum_{k}^{N_{j,t}} (c_{j,t,k}^T)^2 \mathcal{L}_{conf}(\mathcal{T}_s(x_{j,t}^w), \hat{y}_{j,t}^T)_k \qquad (6)$$

where $N_{j,t}$ is the number of pseudo-labels in frame t of video j. $c_{j,t,k}^T$ denotes the confidence of the k-th pseudo-label in a given frame. $\mathcal{L}_{coord}(\mathcal{T}_s(x_{j,t}^w), \hat{y}_{j,t}^w)_k$ and $\mathcal{L}_{conf}(\mathcal{T}_s(x_{j,t}^w), \hat{y}_{j,t}^w)_k$ are the bounding box coordinate and confidence loss components for the k-th pseudo-label.

3.4   Bidirectional, Adaptive, Teacher-Student Mutual Refinement

The learning dynamics between the teacher and student also play a significant role in determining training stability and model robustness. First, the teacher should be updated at a sufficient rate such that it can catch up to the student before the student overfits, i.e., the EMA keep rate α_e cannot be too large. At the same time, the teacher should be characterized by gradual changes in the training curve as opposed to rapid oscillations. Thus the teacher also cannot be made to update too quickly, i.e., α_e cannot be too small. Finally, the student's training curve should not have sudden drops, for example, due to a bad training batch. Here, we introduce two additional techniques to dynamically balance the rate and direction of knowledge transfer during mutual learning:

Adaptive EMA (Student → Teacher): When using a fixed EMA keep rate α_e, there is a trade-off between training stability and rate of knowledge transfer from the student. Instead, we propose an adaptive EMA keep rate that is conditioned on the relative performance of the teacher and student after each training epoch. We use a sigmoid-shaped function for α_e, given by:

$$\alpha_e = \alpha_{e,min} + (\alpha_{e,max} - \alpha_{e,min}) \cdot \frac{1}{1 + e^{-\tau_0(m_T - m_S) - \tau_1}} \tag{7}$$

where $\alpha_{e,min}$, $\alpha_{e,max}$, τ_0 and τ_1 are hyper-parameters defining the function shape. m_T and m_S denote teacher and student performance on a validation set according to some evaluation metric. The adaptive scheme allows α_e to be dynamically adjusted, that is, α_e is decreased (higher rate of knowledge transfer) as the student outperforms the teacher, and increased (lower rate of knowledge transfer) as the student underperforms compared to the teacher.

Inverse Adaptive EMA (Teacher → Student): To avoid sudden drops in performance by the student during mutual learning, we further introduce a mechanism which allows knowledge transfer in the reversed direction, i.e., from the teacher to the student. We design a similar sigmoidal function for the inverse EMA keep rate α_{inv}, given by:

$$\alpha_{inv} = \begin{cases} 1, & m_T \leq m_S \\ \alpha_{inv,min} + (2 - 2\alpha_{inv,min}) \cdot \frac{1}{1 + e^{-\tau_2(m_S - m_T)}}, & m_T > m_S \end{cases} \tag{8}$$

where $\alpha_{inv,min}$ and τ_2 are hyper-parameters of the function. Here, knowledge transfer to the student is increased (lower α_{inv}) when the teacher outperforms the student, and decreased (higher α_{inv}) when the teacher underperforms.

4   Experiments

4.1   Experimental Settings

Data. An extensive retrospective, multi-center clinical dataset of 7,998 lung ultrasound videos were used in this work. The data were acquired from 420 patients with suspicion of lung consolidation or other related pathology (e.g., pneumonia, pleural effusion) from 8 U.S. clinical sites between 2017 and 2020. The videos were each at least 3 s in length and contained at least 60 frames. 385 (fully labeled training set), 337 (validation set) and 599 (test set) videos were annotated for lung consolidation regions using bounding boxes. All data were partitioned at the subject level. The remaining 6,677 videos were annotated only for the presence or absence of lung consolidation at the video level (weakly labeled set). Annotation was carried out by a multi-center team of expert physicians with medical training in lung ultrasound. Each video was annotated by two experts and adjudicated by a third expert when a disagreement between the first two annotators occurred.

Table 1. Validation and test mAP for fully, semi-, and weakly semi-supervised models. Mean ± standard deviation based on five repeated experiments. All methods were significantly superior to YOLO (first row) and significantly inferior to the proposed method (last row)(paired two-way t-test, p-value < 0.05)

Category	Method	Validation mAP	Test mAP
Fully supervised	YOLO [5]	0.435 ± 0.012	0.412 ± 0.013
	YOLO+HE	0.452 ± 0.015	0.440 ± 0.017
Semi-supervised	YOLO+HE+Unlabeled	0.468 ± 0.005	0.447 ± 0.003
Weakly Semi-supervised	YOLO+HE+Weak	0.505 ± 0.004	0.476 ± 0.002
	YOLO+HE+Weak+Pseudo	0.508 ± 0.004	0.479 ± 0.004
	YOLO+HE+Weak+TSMR	0.515 ± 0.005	0.480 ± 0.004
	YOLO+HE+Weak+Pseudo+TSMR	$\mathbf{0.519 \pm 0.003}$	$\mathbf{0.484 \pm 0.003}$

Implementation Details. We used the PyTorch Ultralytics implementation of the YOLO-v5 object detector [5] with default training settings (Adam optimizer with learning rate of 0.001). The weights for the confidence and coordinate losses were set to $\lambda_{conf} = 1.0$, $\lambda_{coord} = 0.05$. The weights for the frame-level fully supervised, frame-level semi-supervised, and video-level weakly supervised losses were set to $\lambda_{f\_sup} = \lambda_{f\_semi} = 1$, and $\lambda_{v\_weak} = 0.05$. For training from pseudo-labels without re-weighting, the confidence threshold was set to $\beta = 0.5$. Otherwise, hyperparameters were set to $\beta = \beta_l = 0.1$ when using weighted pseudo-labels. To train with a fixed EMA keep rate, we used $\alpha_e = 0.95$. Otherwise, when applying bidirectional adaptive EMA for mutual teacher-student refinement, the hyperparameters were set to $\alpha_{e,min} = 0.75$, $\alpha_{e,max} = 0.99$, $\alpha_{inv,min} = 0.85$, $\tau_0 = 180$, $\tau_1 = 3$, and $\tau_2 = 180$.

Experiments. We first trained a baseline YOLO-v5 detector on the fully supervised data $\mathcal{D}_f$, denoted as **YOLO**. We compared the baseline to a fully supervised training experiment with hierarchical (iteration and epoch-based) EMA training during burn-in, as proposed in Section 3.2; this is denoted by **+HE**. All subsequent experiments involving teacher-student mutual learning were initialized from the same YOLO+HE model checkpoint. We implemented the semi-supervised approach from Unbiased Teacher [11] by adding all videos from the weakly supervised dataset $\mathcal{D}_w$, but without providing video-level labels (i.e., treating these as unlabeled data); this is denoted as **+Unlabeled**. Note, to demonstrate the effectiveness of the proposed method, +Unlabeled a derived version of the Unbiased Teacher [11], using its way of utilizing unlabeled data while keeping the same setting of the model and train strategy as the rest of competing methods. We then introduced our proposed method of weak semi-supervision, described in Section 3.3, by including the video-level labels from $\mathcal{D}_w$, denoted as **+Weak**. Finally, experiments using our methods for pseudo-label re-weighting (Section 3.4) and bidirectional, adaptive, teacher-student mutual refinement (Section 3.5) are denoted as **+Pseudo** and **+TSMR** respectively. We compared mean Average Precision (mAP) on the validation and test sets described above, with each experiment repeated five times to assess repeatability. Experimental results are summarized in Table 1.

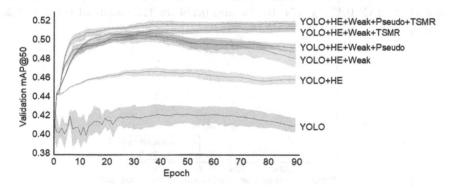

Fig. 2. Learning curves of teacher models θ_T on validation set during teacher-student mutual learning. Solid lines show mean validation mAP across five repeated experiments. Ranges indicate 95% confidence intervals.

4.2   Results and Discussion

Contribution of Hierarchical EMA Training: The baseline, fully-supervised detector (YOLO) achieved validation and test mAP of 0.435 and 0.412, respectively. These improved to 0.452 and 0.440, respectively, with the inclusion of the hierarchical EMA training strategy during burn-in (YOLO+HE).

Contribution of Semi- and Weak (Video-Level) Supervision: The addition of unlabeled data and semi-supervision based on YOLO+HE+Unlabeled improved validation mAP from 0.452 to 0.468 and test mAP from 0.440 to 0.447, which was a statistically significant increase (p-value < 0.05). Furthermore, the standard deviation of mAP values over repeated experiments decreased (0.015 to 0.005 for validation, 0.017 to 0.003 for test), suggesting that the semi-supervised model is more stable and repeatable across runs. This was also reflected in the tighter 95% confidence intervals for validation mAP learning curves across repeated runs (Fig. 2). Model performance again increased with the introduction of weak supervision of video-level labels (YOLO+HE+Weak) (validation mAP 0.505, test mAP 0.476, p-value < 0.05), with corresponding decreases in mAP standard deviation (0.004 in validation, 0.002 in test) and 95% confidence intervals over repeat runs (Fig. 2). To further investigate the contribution of video-level supervision, we trained models with all fully labeled data $\mathcal{D}_f$ but utilized a proportion of video labels z_j^w with the remainder of $\mathcal{D}_w$ treated as unlabeled. mAP improved consistently with increased video-level supervision (Fig. 3).

Contribution of Weighted Pseudo-labels: The use of weighted pseudo-labels (YOLO+HE+Weak+Pseudo) further improved validation mAP from 0.505 to 0.508 and test mAP from 0.476 to 0.479. The re-weighting mechanism eliminated all false positive pseudo-labels in negative videos and increased the number of pseudo-labels to better match the overall number of true labels. In comparison, fixed pseudo-label thresholds resulted in worse detection performance (test mAP 0.477, 0.474, 0.476, and 0.474 for thresholds of 0.1, 0.3, 0.5, and 0.7).

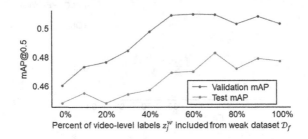

Fig. 3. Contribution of video-level supervision during teacher-student mutual learning. Models were trained with varying proportions of video labels relative to unlabeled data.

Contribution of Bidirectional Teacher-Student Mutual Refinement: Teacher-student mutual refinement (YOLO+HE+Weak+TSMR) further boosted validation mAP from 0.508 to 0.515 and test mAP from 0.479 to 0.480. Model repeatability was improved, as shown in Fig. 2, where variation between experiments was greatly reduced (narrow 95% confidence intervals) and model convergence occurred more quickly. Ablation experiments confirmed that a fixed EMA keep rate α_e was unable to achieve comparable detection performance compared to the proposed bidirectional adaptive EMA updates ($\alpha_e = 0.9$, lower fixed rate: 0.506 and 0.471; $\alpha_e = 0.95$, moderate fixed rate: 0.505 and 0.476; and $\alpha_e = 0.99$, high fixed rate: 0.489 and 0.470, for validation mAP and test mAP respectively).

Final Results for Proposed Method: Finally, best-performing models incorporating all proposed components achieved validation mAP of 0.519 and test mAP of 0.484, which were statistically significant improvements to both fully supervised (0.452 and 0.440) and semi-supervised (0.468 and 0.447) baselines (p-value < 0.05). Furthermore, an extra experiment suggested that comparable test mAP of baseline YOLO detector (0.412) could be achieved by the proposed method using merely one-third of the labeled data (0.395). The reference speed was 2.9 ms per frame on a NVIDIA GeForce RTX 3090 GPU, enabling real-time detection. Examples of lung consolidation detection in ultrasound are seen in Fig. 4, where the proposed method demonstrates successful detection of challenging pathology not identified (or falsely identified) by the baseline YOLO detector.

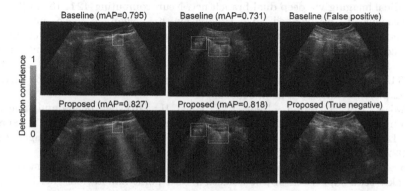

Fig. 4. Lung consolidation detection with baseline YOLO (top) and proposed method (YOLO+HE+Weak+Pseudo+TSMR, bottom). White boxes show expert annotations. Orange boxes show model detections. mAPs calculated for each video are also shown. (Color figure online)

5   Conclusion

This is the first study to introduce a weakly semi-supervised framework for object detection on medical video data. Our method extends a teacher-student training strategy to provide weak supervision via a video-level loss. We also introduce improvements to the underlying teacher-student mutual learning mechanism, including methods to improve the quality of pseudo-labels and optimize knowledge transfer between the student and teacher. Empirical results on a lung ultrasound pathology detection task demonstrate that the framework leads to improved detection accuracy and robustness compared to existing baseline models, while also being more efficient in data and annotation usage. One limitation of the method is the need to empirically select hyperparameters, which could be resolved using adaptive hyperparameter tuning techniques as part of future work. Moreover, we considered the proposed components as orthogonal research directions as other pseudo-label refinement techniques in the state-of-the-art semi-supervised detection methods, which could be further combined in achieving better performance. Lastly, the proposed improvements to the teacher-student mechanism could potentially be adapted for other semi-supervised and weakly supervised learning tasks, including classification and segmentation.

Acknowledgements. We would like to acknowledge the contributions from the following people for their efforts in data curation and annotations: Zohreh Laverriere, Xinliang Zheng (Lia), Annie Cao, Katelyn Hostetler, Yuan Zhang, Amber Halse, James Jones, Jack Lazar, Devjani Das, Tom Kennedy, Lorraine Ng, Penelope Lema, Nick Avitabile.

References

1. Bakalo, R., Goldberger, J., Ben-Ari, R.: Weakly and semi supervised detection in medical imaging via deep dual branch net. Neurocomputing **421**, 15–25 (2021)
2. Bassiouny, R., Mohamed, A., Umapathy, K., Khan, N.: An interpretable object detection-based model for the diagnosis of neonatal lung diseases using ultrasound images. In: 2021 43rd Annual International Conference of the IEEE Engineering in Medicine & Biology Society (EMBC), pp. 3029–3034. IEEE (2021). https://doi.org/10.1109/EMBC46164.2021.9630169, https://ieeexplore.ieee.org/document/9630169/
3. Chai, Z., Lin, H., Luo, L., Heng, P.A., Chen, H.: ORF-net: deep omni-supervised rib fracture detection from chest ct. In: Wang, L., Dou, Q., Fletcher, P.T., Speidel, S., Li, S. (eds.) MICCAI 2022. LNCS, vol. 13433, pp. 238–248. Springer, Cham (2022). https://doi.org/10.1007/978-3-031-16437-8_23
4. Ge, Y., Zhou, Q., Wang, X., Shen, C., Wang, Z., Li, H.: Point-teaching: weakly semi-supervised object detection with point annotations. arXiv preprint arXiv:2206.00274 (2022)
5. Jocher, G., et al.: ultralytics/YOLOv5: v6.0 - YOLOv5n 'Nano' models, Roboflow integration, TensorFlow export, OpenCV DNN support (2021). https://doi.org/10.5281/zenodo.5563715
6. Jeong, J., Lee, S., Kim, J., Kwak, N.: Consistency-based semi-supervised learning for object detection. In: Advances in Neural Information Processing Systems, vol. 32 (2019)
7. Jeong, J., Verma, V., Hyun, M., Kannala, J., Kwak, N.: Interpolation-based semi-supervised learning for object detection. In: Proceedings of the IEEE/CVF Conference on Computer Vision and Pattern Recognition, pp. 11602–11611 (2021)
8. Ji, H., et al.: Point beyond class: a benchmark for weakly semi-supervised abnormality localization in chest X-rays. In: Wang, L., Dou, Q., Fletcher, P.T., Speidel, S., Li, S. (eds.) MICCAI 2022. LNCS, vol. 13433, pp. 249–260. Springer, Cham (2022). https://doi.org/10.1007/978-3-031-16437-8_24
9. Jiao, R., Zhang, Y., Ding, L., Cai, R., Zhang, J.: Learning with limited annotations : a survey on deep semi-supervised learning for medical image segmentation. arXiv, pp. 1–19 (2022)
10. Kulhare, S., et al.: Ultrasound-based detection of lung abnormalities using single shot detection convolutional neural networks. In: Stoyanov, D., et al. (eds.) POCUS/BIVPCS/CuRIOUS/CPM -2018. LNCS, vol. 11042, pp. 65–73. Springer, Cham (2018). https://doi.org/10.1007/978-3-030-01045-4_8
11. Liu, Y.C., et al.: Unbiased teacher for semi-supervised object detection. arXiv preprint arXiv:2102.09480 (2021)
12. Meethal, A., Pedersoli, M., Zhu, Z., Romero, F.P., Granger, E.: Semi-weakly supervised object detection by sampling pseudo ground-truth boxes. arXiv preprint arXiv:2204.00147 (2022)
13. Peng, J., Wang, Y.: Medical image segmentation with limited supervision: a review of deep network models. arXiv, pp. 1–24 (2021)

14. Ren, Z., Yu, Z., Yang, X., Liu, M.-Y., Schwing, A.G., Kautz, J.: UFO^2: a unified framework towards omni-supervised object detection. In: Vedaldi, A., Bischof, H., Brox, T., Frahm, J.-M. (eds.) ECCV 2020. LNCS, vol. 12364, pp. 288–313. Springer, Cham (2020). https://doi.org/10.1007/978-3-030-58529-7_18

15. Shao, F., et al.: Deep learning for weakly-supervised object detection and localization: a survey. Neurocomputing **496**, 192–207 (2022)

16. Tang, Y., Chen, W., Luo, Y., Zhang, Y.: Humble teachers teach better students for semi-supervised object detection. In: Proceedings of the IEEE/CVF Conference on Computer Vision and Pattern Recognition, pp. 3132–3141 (2021)

17. Wang, P., et al.: Omni-DETR: Omni-supervised object detection with transformers. In: Proceedings of the IEEE/CVF Conference on Computer Vision and Pattern Recognition, pp. 9367–9376 (2022)

18. Wang, Z., Li, Y., Guo, Y., Fang, L., Wang, S.: Data-uncertainty guided multi-phase learning for semi-supervised object detection. In: Proceedings of the IEEE/CVF Conference on Computer Vision and Pattern Recognition, pp. 4568–4577 (2021)

19. Xing, W., et al.: Automatic detection of a-line in lung ultrasound images using deep learning and image processing. Med. Phys. **50**, 330–343 (2022)

20. Xu, M., et al.: End-to-end semi-supervised object detection with soft teacher. In: Proceedings of the IEEE/CVF International Conference on Computer Vision, pp. 3060–3069 (2021)

21. Zhang, D., Zeng, W., Guo, G., Fang, C., Cheng, L., Han, J.: Weakly supervised semantic segmentation via alternative self-dual teaching. arXiv preprint arXiv:2112.09459 (2021)

22. Zhang, S., Yu, Z., Liu, L., Wang, X., Zhou, A., Chen, K.: Group R-CNN for weakly semi-supervised object detection with points. In: Proceedings of the IEEE/CVF Conference on Computer Vision and Pattern Recognition, pp. 9417–9426 (2022)

23. Zhou, Q., Yu, C., Wang, Z., Qian, Q., Li, H.: Instant-teaching: an end-to-end semi-supervised object detection framework. In: Proceedings of the IEEE/CVF Conference on Computer Vision and Pattern Recognition, pp. 4081–4090 (2021)

sEBM: Scaling Event Based Models to Predict Disease Progression via Implicit Biomarker Selection and Clustering

Raghav Tandon[1,2] , Anna Kirkpatrick[1,3] , and Cassie S. Mitchell[1,2(✉)]

[1] Laboratory for Pathology Dynamics, Department of Biomedical Engineering, Georgia Institute of Technology and Emory University School of Medicine, Atlanta, GA 30332, USA
{raghav.tandon,akirkpatrick3}@gatech.edu, cassie.mitchell@bme.gatech.edu
[2] Center for Machine Learning, Georgia Institute of Technology, Atlanta, GA 30332, USA
[3] School of Mathematics, Georgia Institute of Technology, Atlanta, GA 30332, USA

Abstract. The Event Based Model (EBM) is a probabilistic generative model to explore biomarker changes occurring as a disease progresses. Disease progression is hypothesized to occur through a sequence of biomarker dysregulation "events". The EBM estimates the biomarker dysregulation event sequence. It computes the data likelihood for a given dysregulation sequence, and subsequently evaluates the posterior distribution on the dysregulation sequence. Since the posterior distribution is intractable, Markov Chain Monte-Carlo is employed to generate samples under the posterior distribution. However, the set of possible sequences increases as $N!$ where N is the number of biomarkers (data dimension) and quickly becomes prohibitively large for effective sampling via MCMC. This work proposes the "scaled EBM" (sEBM) to enable event based modeling on large biomarker sets (e.g. high-dimensional data). First, sEBM implicitly selects a subset of biomarkers useful for modeling disease progression and infers the event sequence only for that subset. Second, sEBM clusters biomarkers with similar positions in the event sequence and only orders the "clusters", with each successive cluster corresponding to the next stage in disease progression. These two modifications used to construct the sEBM method provably reduces the possible space of event sequences by multiple orders of magnitude. The novel modifications are supported by theory and experiments on synthetic and real clinical data provides validation for sEBM to work in higher dimensional settings. Results on synthetic data with known ground truth shows that sEBM outperforms previous EBM variants as data dimensions increase. sEBM was successfully implemented with up to 300 biomarkers, which is a 6-fold increase over previous EBM applications. A real-world clinical application of sEBM is performed using 119 neuroimaging markers from publicly available Alzheimer's Disease Neuroimaging Initiative (ADNI) data to stratify subjects into 6 stages of disease progression. Subjects included cognitively normal (CN), mild cognitive impairment (MCI),

© The Author(s), under exclusive license to Springer Nature Switzerland AG 2023
A. Frangi et al. (Eds.): IPMI 2023, LNCS 13939, pp. 208–221, 2023.
https://doi.org/10.1007/978-3-031-34048-2_17

and Alzheimer's Disease (AD). sEBM stage is differentiated for the 3 groups ($\chi^2 p - value < 4.6e-32$). Increased sEBM stage is a strong predictor of conversion risk to AD ($p - value < 2.3e - 14$) for MCI subjects, as verified with a Cox proportional-hazards model adjusted for age, sex, education and APOE4 status. Like EBM, sEBM does not rely on apriori defined diagnostic labels and only uses cross-sectional data.

Keywords: disease progression modeling · bayesian learning · prognostic biomarker selection · biomarker clustering

1 Introduction

A popular approach to disease progression modeling is the event-based model (EBM) [1,2]. EBM hypothesizes disease progression to occur through a sequence of discrete events, which correspond to biomarker abnormalities without reliance on a priori diagnostic labels or explicit biomarker cut-off or threshold values. The model infers the sequence of events most consistent with data measured from clinical subjects. It has been applied to cross-sectional data from sporadic and familial Alzheimer's disease (AD) [3], Huntington's disease [4], epilepsy [5] and progressive supranuclear palsy [6] to name a few.

The ability to work with cross-sectional data (single measurement per individual) and ease of integrating multiple biomarker types (imaging volumes, fluid, cognitive, etc.) makes EBM a useful model to study disease progression. However, a key challenge to the EBM approach is its scalability to larger biomarker sets. The possible number of event sequence increase as $N!$ (N being the number of biomarkers).

This work presents scaled EBM (sEBM) as an improved event based model that overcomes the challenges of factorial increases in possible event sequences, which inherently occurs in data sets with a large number of biomarkers (high dimensionality data). Throughout this work the classic event-based model in [2] is referred to as EBM, and the modified EBM model based on this work as sEBM (scaled EBM). Permutation complexity refers to the possible number of distinct event-ordering sequences for either of these models.

2 Background

2.1 Event-Based Model (EBM)

EBM for familial cases of Alzheimer's Disease was proposed in [1] and later generalized to sporadic cases in [2], estimates the ordering of biomarker dysregulation events. The model consists of a set of events $E_1, \ldots E_N$ and an ordering $S = (s(1) \ldots s(N))$ which is a permutation of the integers $1, \ldots N$ determining the event ordering $E_{s(1)}, \ldots E_{s(N)}$ representing the sequence of biomarker dysregulation.

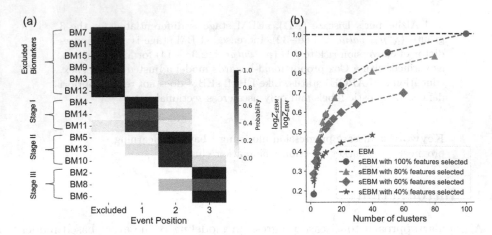

Fig. 1. (a) schematic output from sEBM and (b) reduction in permutation complexity (for the case with equal sized clusters) which can be used to guide the choice of features to be included and number of clusters.

If patient j (whose biomarker measurements are denoted by $X_j = \{x_{1j}, x_{2j} \ldots x_{Nj}\}$ and the full data $X \in \mathbb{R}^{J \times N}$) is at stage k in the progression model, events $E_{s(1)}, \ldots, E_{s(k)}$ have occurred, while events $E_{s(k+1)}, \ldots, E_{s(N)}$ have not. Data likelihood for that patient given an event-ordering S can be written as

$$p(X_j|k, S) = \prod_{i=1}^{k} p(x_{s(i)j}|E_{s(i)}) \times \prod_{i=k+1}^{N} p(x_{s(i)j}|\neg E_{s(i)}) \qquad (1)$$

Since $X_j's$ are independent, the complete data likelihood can be written while marginalizing out the stage k of individual subjects as -

$$p(X|S) = \prod_{j=1}^{J} \sum_{k=0}^{N} p(k) \prod_{i=1}^{k} p(x_{s(i)j}|E_{s(i)}) \times \prod_{i=k+1}^{N} p(x_{s(i)j}|\neg E_{s(i)}) \qquad (2)$$

Bayes' rule can be used to derive a posterior on the event ordering S as

$$p(S|X) = \frac{p(X|S)p(S)}{p(X)} \qquad (3)$$

Model Assumptions. The EBM model in [1,2] makes strong assumptions that are inherited by sEBM. These are - 1) All subjects follow the same disease progression trajectory. Note this assumption has been relaxed in [14,15]. Similar frameworks can be coupled with the proposed sEBM to overcome the rigidity of this assumption and its impact on modeling heterogeneous disease. 2) Independence of likelihood of each biomarker measurement is conditional on event occurrence. 3) Biomarker changes are monotonic and 4) population under study is uniformly sampled across disease stages.

Mixture Models for Computing Data Likelihood Conditioned on Event Occurrence. Computing Eq. (2) requires separate models for $p(x_{s(i)j}|E_{s(i)})$ and $p(x_{s(i)j}|\neg E_{s(i)})$ which is obtained by fitting a two component Gaussian mixture model [2] to all observations of biomarker i in the data ($\{x_{ij}|j = 1 \ldots J\}$). The components of the mixture model correspond to $p(x_{ij}|E_i)$ and $p(x_{ij}|\neg E_i)$.

Sampling Event Ordering S Using MCMC. Since the posterior $p(S|X)$ is intractable, MCMC (Metropolis-Hastings) can be used to generate samples. The MCMC algorithm proceeds as follows - at each iteration, the current ordering S_t swaps two randomly chosen biomarkers to generate a proposed ordering S'. This proposed ordering will be accepted ($S_{t+1} = S'$) with a probability $\min(1, \frac{p(X|S')}{p(X|S_t)})$. The initial state is selected using a greedy-ascent algorithm which accepts a proposed ordering only when $\frac{p(X|S')}{p(X|S_t)} > 1$.

2.2 Challenges Due to Data Dimensionality and Proposed Solutions

The support of the distribution for $P(S|X)$ increases as $N!$ where N is the dimensionality of X. This can lead to severe challenges for effective MCMC sampling of the underlying distribution. Two solutions are proposed within sEBM -

1. Implicit feature selection is performed within sEBM model to subset useful biomarkers and inferring an event ordering S only over them.
2. sEBM associates multiple biomarkers with a single event, instead of a unique event per biomarker. Such clustering of biomarkers into fewer events makes permutations within a cluster no longer count, thereby removing a large number of closely related but distinct sequences.

3 Method

3.1 Implicit Biomarker Selection Within sEBM

Let $C_S \in \{0,1\}^N$ such that $||C_S||^2 = f$ (f out of the N biomarkers are chosen for event ordering). The event ordering S will be a permutation of the integers $1, \ldots, f$. Since both C_S and S are to be inferred, the posterior $p(S|X)$ in Eq. (3) is modified to include C_S. The new posterior becomes $p(C_S, S|X)$ in Eq. (4).

$$p(C_s, S|X) = \frac{p(X|C_s, S)p(C_s, S)}{p(X)} \tag{4}$$

Similar to [1,2], we assume a uniform prior on $P(C_S, S)$. In Eq. (4), $p(X|C_S, S)$ can be written as -

$$\begin{aligned} p(X|C_S, S) = p(X_{\backslash S}, X_S|C_S, S) &= p(X_S|C_S, S)p(X_{\backslash S}|X_S, C_S, S) \\ &= p(X_S|S)p(X_{\backslash S}|X_S) \end{aligned} \tag{5}$$

In Eq. (5), $X_S \in \mathbb{R}^{J \times f}$ refers to the data subset corresponding to the selected f biomarkers whereas $X_{\backslash S} \in \mathbb{R}^{J \times (N-f)}$ corresponds to the data subset of $N - f$ excluded biomarkers (from J individuals). $p(X_S | C_S, S)$ can be simplified to $p(X_S | S)$ (knowing S fixes C_S) and $p(X_{\backslash S} | X_S, C_S, S)$ can be simplified to $p(X_{\backslash S} | X_S)$ since $X_{\backslash S} \perp\!\!\!\perp C_S, S | X_S$. A simple way to look at this independence relation is that no additional information for $X_{\backslash S}$ is derived from the choice of biomarkers (C_S) or their event ordering (S), since they only pertain to X_S.

MCMC samples for $p(C_S, S | X)$ can be generated from Eq. (4) by substituting the expression for $p(X | C_S, S)$ from Eq. (5) and proceeding as follows. At each iteration t, the current state (C_{S_t}, S_t) is perturbed to $(C_{S'}, S')$ by randomly swapping two biomarkers. This swap can either change the event ordering alone or lead to a new set of chosen biomarkers with an event ordering similar to S_t but the included biomarker replacing the excluded biomarker in the proposed event ordering S'. This perturbation is accepted ($S_{t+1} = S'$) with a probability $\min(1, a)$, where $a = \frac{p(X_{S'} | S') p(X_{\backslash S'} | X_{S'})}{p(X_{S_t} | S_t) p(X_{\backslash S_t} | X_{S_t})}$. Each iteration requires two terms to be computed - $p(X_{S'} | S')$ and $p(X_{\backslash S'} | X_{S'})$. These terms can be interpreted as follows -

- $p(X_{S'} | S')$ - The data likelihood over included biomarkers $X_{S'}$ given the chosen event ordering S'. This can be computed as in Eq. (2) (or Eq. (6) which will be introduced in Sect. 3.2)
- $p(X_{\backslash S'} | X_{S'})$ - The data likelihood for the excluded biomarkers given the included biomarkers. It is assumed here that $p(X_{\backslash S_t} | X_{S_t})$ and $p(X_{\backslash S'} | X_{S'})$ are approximately equal since they may differ at most by one out of N data features.

3.2   Implicit Biomarker Clustering Within EBM

The biomarkers used for event ordering can be clustered together into sets with pre-defined sizes using a simple modification on the original EBM. Equation (2) computes the data likelihood by marginalizing all possible disease stages which could be any stage between no biomarker dysregulated ($k = 0$) and all biomarkers dysregulated (final disease stage, $k = N$). Instead of ordering all biomarkers by giving them a unique position along the event cascade, biomarkers could be clustered together. All biomarkers within the same cluster share the same position in the ordering. Positional differences only occur across clusters. This can be introduced in the EBM framework (Eqs. (1) and (2)), by constraining k to only take on specific values which are a function of the individual cluster sizes. Let c_i denote the ith cluster which consists of $|c_i|$ biomarkers. Let the total number of clusters be n. Hence the vector of cluster sizes can be written as $[|c_1| \ldots |c_n|]$ and $\sum_{i=1}^{n} |c_i| = N$. Let $u_z = \sum_{i=1}^{z} |c_i|$ which represents the cumulative sum of cluster sizes until cluster c_z ($1 \le z \le n$) and $U = [0, u_1 \ldots u_n]$. Equation (2) can be constrained to assign events to the clusters c_i instead of finding an event ordering over all biomarkers as shown in Eq. (6) -

$$p(X|S) = \prod_{j=1}^{J} \sum_{k \in U} p(k) \prod_{i=1}^{k} p(x_{s(i)j}|E_{s(i)}) \prod_{i=k+1}^{N} p(x_{s(i)j}|\neg E_{s(i)}) \qquad (6)$$

$k \in U$ represents further discretization of EBM where disease stage advances from no biomarker abnormality ($k = 0$) to all biomarkers abnormal ($k = N$) in steps of $|c_1|, |c_2| \ldots |c_{n-1}|, |c_n|$ biomarkers. This results in fewer possible states to be explored by the MCMC sampler, fewer terms in the marginalization over individual's stage (k), and faster run-times per iteration. The only constraint on c_i is that $1 \leq |c_i| \leq N$, $\sum_{i=1}^{n} |c_i| = N$, and c_i are disjoint.

3.3 Analysis of Reduction in Permutational Complexity for MCMC Sampling via Biomarker Subset Selection and Clustering

Equation (4) requires MCMC sampling over (C_S, S) instead of just S which is the case in Eq. (3). The joint inference over two variables can be easier under certain conditions which are explained here. The possible number of orderings for N biomarkers is $N!$ which can be significantly reduced by clustering biomarkers and only looking at orderings across clusters. This can be easily seen in the case where the N biomarkers are divided equally among c clusters so that each cluster gets m biomarkers ($c \times m = N$ and $N, c, m \in \mathbb{Z}^+$). Equal sized clusters is not a requirement but helps in presenting our case. The possible number of orderings with equal sized clusters is -

$$\underbrace{\binom{N}{m}\binom{N-m}{m}\binom{N-2m}{m} \cdots \binom{m}{m}}_{c} \text{ clusters} = \frac{N!}{m!^c} \qquad (7)$$

Further, if a subset of f biomarkers can be selected from N biomarkers such that only those f are used for event ordering, the number of total possible orderings will be $\binom{N}{f}\frac{f!}{m'!^c}$. Assuming the same number of c clusters as in Eq. (7), the number of biomarkers in this case is given by m' ($c \times m' = f$), and $f, c, m' \in \mathbb{Z}^+$. The feature selection step increases the complexity by a factor of $\binom{N}{f}$ but simultaneously reduces the complexity for ordering the selected biomarkers to $\frac{f!}{m'!^c}$ due to clustering. By choosing f and c appropriately for a given N, it is possible to significantly reduce the overall permutation complexity as shown in Fig. 1b. The y-axis shows the log of permutation complexity under the new model ($Z_{sEBM} = \binom{N}{f}\frac{f!}{m'!^c}$)) divided by the log of permutation complexity from EBM ($Z_{EBM} = N!$), at different f and c (shown for $N = 100$).

3.4 Hyperparameter Selection

There are three important hyperparameters in the sEBM model. 1) Fraction of biomarkers to be included by the model, is recommended to be 0.5 or upwards. The selected biomarkers encode for the excluded biomarkers, thereby allowing their exclusion. Including sufficient number of biomarkers helps in simplifying the acceptance ratio for the MCMC associated with Eqs. (4) and (5). 2) The number of clusters implies the number of disease stages. Popular disease staging procedures can be used as motivation for cluster selection. For example the Braak criterion [11] divides AD into 6 stages. This motivates 5 clusters, as was used in this work in Sect. 4.2. 3) Size of individual clusters. Making individual clusters larger while keeping the number of clusters constant will reduce the underlying permutational complexity. However, the associated risk is lesser information about disease stages with smaller clusters associated to them. Future work will focus on a well formed methodology for hyperparameter selection.

4 Experiments

4.1 Synthetic Data with Known Ground Truth

sEBM is tested on synthetic data generated using the data simulation framework in [7]. The simulation framework generates cross-sectional biomarker data sets from neurodegenerative disease cohorts that reflect the temporal evolution of the disease. The simulation allows for generating data with known ground-truth event ordering which can be compared against an inferred event ordering. The simulation framework assumes that the temporal evolution of the biomarkers is sigmoidal and can be modeled with the equation-

$$z(t, \theta_i) = a_i + \frac{r_i}{1 + \exp\left(-\frac{4}{\tau_i}(t - c_i)\right)} \tag{8}$$

with parameters $\theta_i = (a_i, r_i, \tau_i, c_i)$ for the ith biomarker. a_i is the trajectory minimum for the ith biomarker and $a \sim \mathcal{N}(\mathbf{0}, \Sigma_\alpha)$ with $a_i = a[i]$. Σ_α is a symmetric positive semi-definite matrix to model covariance between a_i (α represents the fraction of non-zero entries in Σ_α and is set to 0.5). $r_i \in [1, 3]$ is the range, $\tau_i \in [3, 6]$ is the gradient, $c_i \in [2, 18]$ is the inflection point. c_i, r_i, τ_i are sampled uniformly from their respective ranges. t represents the time point in the progression trajectory of the individual, and is sampled uniformly from the range $[0, 20]$. A small value for c_i implies that the biomarker changes early on in disease progression and vice-versa. t indicates how far the subject has progressed into the disease and is sampled uniformly across the disease timeline. The simulation dataset are generated using Eq. (8) for 3 different sample sizes (200, 400, 600) and 6 different dimensions (50, 100, 150, 200, 250, 300). At each setting of sample size and dimensions, 6 different dataset are generated using random seeds resulting in 108 ($3 \times 6 \times 6$) different datasets.

Model Baseline Comparisons. sEBM is compared to three other methods on 108 datasets - EBM [2], discriminative EBM [12] and EBM run on data with 50% features removed (largest p-values) using a Mann-Whitney U test. The dEBM model is fit to data using the pyebm package in python, using default parameter choices (the gaussian mixture model algorithm it uses has been shown to be more stable in [12]).

Hyperparameters. In all settings, sEBM is set to implicitly select half of the biomarkers. The number of clusters are set to 5, and cluster sizes are set to be equal in order to facilitate normalization of the Kendall-Tau distance metric (explained below). The underlying MCMC is initialized using the best ordering (highest likelihood) from among 100 greedy initialization run for 800 iterations each. MCMC is then used to generate 2e6 samples of the underlying ordering, of which the first 1.5e6 are discarded as burn-in for mixing of the markov chain.

Comparison to Ground-Truth Ordering. Since ground-truth ordering cannot be known with real cross-sectional data, simulating cross-sectional data with a chosen ground-truth and inferring this ordering using sEBM for comparison is an important exercise. The inferred ordering and the ground-truth ordering are compared by using the Kendall-Tau metric for partial rankings as specified in [8]. It is similar to the general Kendall-Tau distance for full-rankings except that it also accounts for pairs of biomarkers which were assigned same rank in one ordering, but different ranks in the other ordering.

$$d_k^{(p)}(\pi_*, \pi_{gt}) = \sum_{l \prec_{\pi_{gt}} j} \mathbb{1}_{l \succ_{\pi_*} j} + p \sum_{l \cong_{\pi_{gt}} j} \mathbb{1}_{l \not\cong_{\pi_*} j} + p \sum_{l \not\cong_{\pi_{gt}} j} \mathbb{1}_{l \cong_{\pi_*} j} \qquad (9)$$

A few things are to be noted about Eq. (9). It is a distance metric for $p \geq 0.5$ [8], hence in this work we fix $p = 0.5$. $d_k^{(p)}$ can be normalized by finding the maximum distance from the ground-truth sequence, which is obtained by the reverse of the ground-truth sequence ($\pi_{gt}^R(i) = n + 1 - \pi_{gt}(i)$, where $\pi_{gt}^R(i)$ is the position of element i in π_{gt}^R and n is the number of unique positions in the rankings). However, finding π_{gt}^R is not straightforward when there are unequal number of biomarkers at each position. This provides a motivation to use equal sized clusters in our experiments. Last, we convert any full-rankings into partial rankings using specified cluster sizes (for e.g. EBM ordering). This is to facilitate direct comparison with the partial-ranked outputs of sEBM. While Eq. (9) can be easily generalized to full-ranks where it is the same as the adjacent swap distance, that no longer holds true for partial ranks [9]. Hence in all cases, only normalized Kendall-Tau distance for partial-ranks is used. Last, the normalized Kendall-Tau distance is computed using the selected sEBM biomarkers from the maximum likelihood ordering among all MCMC generated samples.

216   R. Tandon et al.

4.2   Real Clinical Data to Assess sEBM

Data previously made available in the TADPOLE benchmarking challenge [10] is
analysed using sEBM. Specifically, file *dfMri_D12.csv* is used. The first recorded
observation of all subjects is divided into a training set (CN and AD, n = 327),
and validation set (MCI, n = 551). The data had 123 features, of which 119
were used by removing 4 which had a constant zero value for all subjects in the
training and validation sets. The longitudinal follow-up data from these subjects
is separately used to study their conversion risk.

Hyperparameters. Number of clusters are set to 5 (as in Sect. 4.1), resulting
in 6 stages which aligns with Braak staging [11]. sEBM is set to implicitly select
approximately half of the data features (60/119) and divide them equally across
5 clusters (12 features per clusters). The MCMC settings are the same as the
one used in experiments with synthetic data (Sect. 4.1).

5   Results

5.1   sEBM Outperforms Baseline EBM Models with Synthetic Data

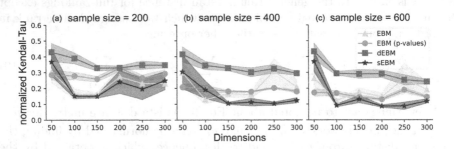

Fig. 2. Performance on simulated data with known ground truth - sEBM is
compared to three different methods- EBM, discriminative EBM (dEBM), and EBM
(p-values) wherein half the features are pre-selected using the Mann Whitney U test.
sEBM shows better scaling ability with data dimensions.

Figure 2 shows sEBM scales better with increasing data dimensions, in comparison to EBM [2], discriminative EBM (dEBM) [12], and EBM (p-values) which
only used half the features to begin with. The shaded regions show the standard
error on the mean from 6 datasets for each sample × dimension setting. sEBM
and all other methods also outperformed [13] (not shown).

5.2   sEBM Shows Distinct Stages for CN, MCI, and AD Subjects

sEBM is used to infer stages on all training subjects (CN and AD subjects, n = 327), and validation subjects (MCI subjects, n = 551) with stages showing difference across the three classes in Fig. 3a (χ^2 p-value $< 4.6e - 32$).

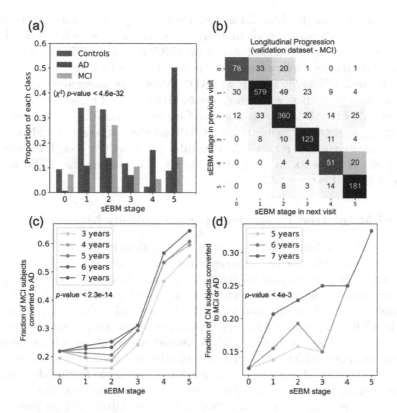

Fig. 3. (a) sEBM assigned stages for CN, MCI and AD subjects (b) show consistency in predicting stages for successive visits (c) For MCI subjects, are strongly associated with risk of converting to AD (d) Same as c, but for CN subjects.

5.3   Longitudinal Validation Shows Consistency of Assigned Staging

The maximum-likelihood event ordering derived from sEBM is used to stage the follow-up visits from all subjects. Subject staging shows longitudinal consistency with most follow-up visits showing same stages or gradual progression to neighboring stages (Fig. 3b). While sEBM works with cross-sectional data, it can be extended to longitudinal data using a temporal version of EBM presented in [16].

5.4   sEBM Accurately Predicts Progression, Conversion to AD

sEBM requires a single clinical baseline measurement from subjects to assign them a stage, which corresponds to disease severity. As expected, CN dominated early stages, AD dominated later stages, and MCI were spread between these two (Fig. 3a). Longitudinal follow-up data shows if the subject's condition worsens at a later visit (e.g. CN to MCI or MCI to AD). Despite not using longitudinal data, sEBM accurately predicts conversion risk to MCI or AD (Fig. 3c). Conversion (3-year) from MCI to AD is modeled using a Cox proportional hazards model using sEBM stage, age, gender, education and APOE4 status (p-value for sEBM stage $< 2.3e-14$ and Hazard ratio $= 1.57$). Similar patterns are seen for conversion (5-year) of CN subjects to MCI or AD in Fig. 3d (p-value for sEBM stage $< 4e-3$ and Hazard ratio $= 1.53$).

5.5   sEBM Inferred Event Ordering is Clinically Meaningful

Stages 1 and 2 indicates early changes to emotional and reward processing regions (nucleus accumbens), cardiovascular risk (right, left vessel), and autonomic and endocrine projections (cingulate), and the classic onset of semantic and non-verbal language processing (pars triangularis, right frontal pole). Stage 3 illustrates changes to cerebrospinal fluid homeostasis (chloroid plexus, third ventricle, CSF) and deeper language processing issues (pars orbitalis). In stage 4, overt memory losses correlate with decreased temporal, entorhinal, and hippocampal volumes, which are commonly associated with AD. Stage 5 indicates stronger ties to reward, emotion, and speech processing via further degeneration of orbitofrontal, temporal lobe structures, amygdala, and the putamen. The excluded features largely contain either highly correlated features functionally represented in the included feature set (e.g. other ventricles) or features that are less associated with AD (white matter, brain stem). Overall, the sEBM predictions align with clinical intuition of AD progression [17]. The exclusion of the cortex volumes was interesting and unexpected given loss of cortex is common in aging and changes of cognition [18] (Fig. 4).

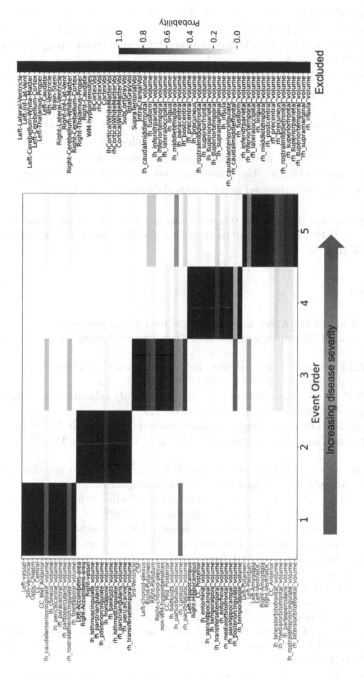

Fig. 4. Maximum likelihood event ordering by sEBM

6   Conclusions and Future Work

sEBM comprised two modifications to EBM to make it scalable with increasing data dimensions - 1) implicit feature selection and; 2) clustering biomarkers into fewer event positions. This drastically reduced the permutation complexity of the underlying MCMC sampling, improved its mixing, and expedited sample generation from the posterior distribution of event orderings. sEBM showed superior performance on synthetic data with known ground-truth in comparison to other similar methods [2,12,13]. On real clinical biomarkers derived from brain MRI, the assigned sEBM stages were well-separated for CN and AD, predictive of conversion risk in MCI subjects, and stable across follow-up clinical visits. While sEBM has some restrictive assumptions such as homogeneity across subjects in disease progression trajectory, these assumptions are inherited from the previous work [1,2] and can be addressed using methods presented in [14,15].

Acknowledgements. Support for this work was provided by National Science Foundation grant 1944247 and National Institutes of Health grants U19AG056169 and 5R01AG070937 to C.M.

References

1. Fonteijn, H.M., et al.: An event-based model for disease progression and its application in familial Alzheimer's disease and Huntington's disease. NeuroImage **60**(3), 1880–1889 (2012)
2. Young, A.L., et al.: A data-driven model of biomarker changes in sporadic Alzheimer's disease. Brain **137**(9), 2564–2577 (2014)
3. Oxtoby, N.P., et al.: Data-driven models of dominantly-inherited Alzheimer's disease progression. Brain **141**(5), 1529–1544 (2018)
4. Byrne, L.M., et al.: Evaluation of mutant huntingtin and neurofilament proteins as potential markers in Huntington's disease. Sci. Transl. Med. **10**(458), eaat7108 (2018)
5. Lopez, S.M., et al.: Event-based modeling in temporal lobe epilepsy demonstrates progressive atrophy from cross-sectional data. Epilepsia (2022)
6. Scotton, W.J., et al.: A data-driven model of brain volume changes in progressive supranuclear palsy. Brain Commun. **4**(3), fcac098 (2022)
7. Young, A.L., Oxtoby, N.P., Ourselin, S., Schott, J.M., Alexander, D.C., Initiative, A.D.N.: A simulation system for biomarker evolution in neurodegenerative disease. Med. Image Anal. **26**(1), 47–56 (2015)
8. Fagin, R., Kumar, R., Mahdian, M., Sivakumar, D., Vee, E.: Comparing partial rankings. SIAM J. Discret. Math. **20**(3), 628–648 (2006)
9. Cicirello, V.A.: Kendall tau sequence distance: extending Kendall tau from ranks to sequences. arXiv preprint arXiv:1905.02752 (2019)
10. Marinescu, R.V., et al.: The Alzheimer's disease prediction of longitudinal evolution (TADPOLE) challenge: results after 1 year follow-up. arXiv preprint arXiv:2002.03419 (2020)
11. Braak, H., Braak, E.: Neuropathological stageing of Alzheimer-related changes. Acta Neuropathol. **82**(4), 239–259 (1991)

12. Venkatraghavan, V., et al.: Disease progression timeline estimation for Alzheimer's disease using discriminative event based modeling. NeuroImage **186**, 518–532 (2019)
13. Firth, N.C., et al.: Sequences of cognitive decline in typical Alzheimer's disease and posterior cortical atrophy estimated using a novel event-based model of disease progression. Alzheimer's Dementia **16**(7), 965–973 (2020)
14. Young, A.L., et al.: Multiple orderings of events in disease progression. In: Ourselin, S., Alexander, D.C., Westin, C.-F., Cardoso, M.J. (eds.) IPMI 2015. LNCS, vol. 9123, pp. 711–722. Springer, Cham (2015). https://doi.org/10.1007/978-3-319-19992-4_56
15. Young, A.L., et al.: Uncovering the heterogeneity and temporal complexity of neurodegenerative diseases with Subtype and Stage Inference. Nat. Commun. **9**(1), 1–16 (2018)
16. Wijeratne, P.A., Alexander, D.C., for the Alzheimer's Disease Neuroimaging Initiative: Learning transition times in event sequences: the temporal event-based model of disease progression. In: Feragen, A., Sommer, S., Schnabel, J., Nielsen, M. (eds.) IPMI 2021. LNCS, vol. 12729, pp. 583–595. Springer, Cham (2021). https://doi.org/10.1007/978-3-030-78191-0_45
17. Salvatore, C., Cerasa, A., Castiglioni, I.: MRI characterizes the progressive course of AD and predicts conversion to Alzheimer's dementia 24 months before probable diagnosis. Front. Aging Neurosci. **10**, 135 (2018)
18. Roe, J.M., et al.: Asymmetric thinning of the cerebral cortex across the adult lifespan is accelerated in Alzheimer's disease. Nat. Commun. **12**(1), 1 (2021)

Domain Adaptation

Source-Free Domain Adaptation for Medical Image Segmentation via Selectively Updated Mean Teacher

Ziqi Wen[ID], Xinru Zhang[ID], and Chuyang Ye[✉][ID]

School of Integrated Circuits and Electronics, Beijing Institute of Technology,
Beijing, China
chuyang.ye@bit.edu.cn

Abstract. Automated medical image segmentation is valuable for disease diagnosis and prognosis, and it has achieved promising performance with deep neural networks. However, a segmentation model trained on a source dataset may not perform well on a different target dataset when the distribution shift or even modality alteration exists between them. To address this problem, domain adaptation techniques can be applied to train the model with the help of the unannotated target dataset. Often when the target data is available, only a segmentation model trained on the source dataset is provided without the source data, and in this case, *source-free domain adaptation* (SFDA) is needed. In this work, we focus on the development of SFDA techniques for medical image segmentation, where the given source model is updated based on the target data. Since no annotations are available for the target dataset, we propose to leverage the consistency of predictions on the target data when different perturbations are made, and adopt the mean teacher framework that can effectively exploit the consistency. Moreover, we assume that the update of the entire model in vanilla mean teacher is suboptimal because when no annotated data is available the knowledge learned for segmentation in the source model can be easily forgotten. Therefore, we propose *selectively updated mean teacher* (SUMT), which seeks to adapt the source model parameters that are sensitive to domain variance and retain the parameters that are invariant to domains. In SUMT, we develop a progressive layer update strategy with channel-wise weight restoration that alleviates forgetting. To evaluate the proposed method, experiments were performed on three datasets, where the source and target data used different modalities for segmentation, or their images were acquired at different sites. The results show that our method improves the segmentation accuracy compared with other SFDA approaches.

Keywords: Source-free domain adaptation · medical image segmentation · selectively updated mean teacher

© The Author(s), under exclusive license to Springer Nature Switzerland AG 2023
A. Frangi et al. (Eds.): IPMI 2023, LNCS 13939, pp. 225–236, 2023.
https://doi.org/10.1007/978-3-031-34048-2_18

1 Introduction

Automated segmentation of medical images can provide a valuable tool for the diagnosis and prognosis of disease and enhance our understanding of disease and treatment planning [13,15]. The use of *deep neural networks* (DNNs) has allowed remarkable improvement of the segmentation accuracy [9]. However, the segmentation model trained on a source dataset may not generalize well to a target dataset that is acquired at a different site on a different scanner due to the domain shift caused by inter-scanner variability [1]. Moreover, the target dataset may even use a different modality for segmenting the same anatomical structure or lesions [2,13], which further increases the difficulty of generalizing the trained model to the target data. In these cases, the segmentation quality for the target dataset can be severely degraded, and it is desirable to develop segmentation approaches that adapt well to different target datasets.

To address the generalization problem, domain adaptation techniques are developed, which exploit both the annotated source training data and unannotated target data [1,2,6,7]. For example, in [6] and [7] adversarial learning is applied to align the features of the source and target domains. In [2], a synergistic fusion of adaptations from both image and feature perspectives is proposed when the source and target domains use different image modalities. AdaEnt [1] uses an additional class ratio predictor for domain adaptation by assuming that the class ratio is invariant between the source and target domains. These methods are shown to allow better adaptation of a DNN-based model to target datasets.

The domain adaptation methods described above assume access to both the annotated source dataset and the unannotated target dataset during model training. However, in real-world scenarios, when the segmentation model is trained with the source data the target data may not be available due to privacy concerns or even not be acquired yet; and it is also not guaranteed that the source dataset can be shared with the target dataset for model retraining. In these cases, *source-free domain adaptation* (SFDA) should be considered, where only the model trained on the source data is provided without the source data, and this given source model is updated based on the unannotated target data. For example, for classification problems, SHOT [12] is developed to align the hypothesis of the source model to the target domain with entropy minimization and diversity regularization, but this method cannot be directly adapted to segmentation; TENT [24] updates the batch statistics and affine parameters in the batch normalization layers of the source model via entropy minimization on the unlabeled target data. More specifically for medical image segmentation, OSUDA is proposed in [13] based on batch normalization statistics under the assumption that scaling and shifting operations in batches are domain shareable. OSUDA explicitly enforces a channel-wise optimization objective, where the domain-specific batch mean and variance are updated incrementally. However, only adapting the batch normalization layers of the source model is generally insufficient for optimal performance. Therefore, the development of SFDA methods for medical image segmentation is still an open problem.

Ideally, SFDA should adapt the domain-specific parameters in the source model according to the target data and retain the domain-invariant parameters. Since no annotations are available for the target dataset, to update the source model, we propose to leverage the consistency of predictions on the target data when different perturbations are made. This idea is common in *semi-supervised learning* (SSL) [5,14,23], where the *mean teacher* (MT) framework [23] has been mostly used for the purpose. However, unlike SSL, in SFDA the source model is updated purely based on the consistency information without any annotated data. This can easily lead to knowledge forgetting, which impairs the domain-invariant knowledge in the source model that is necessary for accurate segmentation.

To avoid this issue, we propose a *selectively updated mean teacher* (SUMT) framework for SFDA-based medical image segmentation. In SUMT, a student model and a teacher model are both initialized by the source model. Since state-of-the-art DNNs for medical image segmentation generally use an encoder-decoder architecture [3,18,21], we also assume that the segmentation model has both an encoder and decoder. First, as earlier layers are more likely to be domain-specific [12], instead of updating all layers in the teacher model, only its encoding layers are updated with *exponential moving average* (EMA) [23] based on the student model. Then, a *channel-wise weight restoration* (CWR) strategy is developed to preserve the domain-invariant knowledge, where the network weights of the encoder of the teacher model that are likely to be domain-invariant are identified, and the identified weights are restored to their initial values. Next, the whole teacher model is updated based on the student model, and CWR is applied to the decoding layers to further alleviate forgetting. Finally, the teacher model is updated again and used for segmentation. To evaluate the proposed method, experiments were performed on three datasets, where the source and target data used different modalities for segmentation or used images acquired at different sites. The results show that our method improves the segmentation accuracy compared with other SFDA approaches.

2 Method

2.1 Problem Formulation and Method Overview

Suppose we are given a DNN-based segmentation model $\mathcal{M}$ with an encoder-decoder architecture trained on a source dataset, but the source training data is not accessible. We seek to perform segmentation on a set $\mathcal{X}$ of N images from a different target dataset, where the i-th target image is denoted by x_i. The target images are acquired differently from the source images, e.g., with different intensity distributions or even different modalities. Due to domain shift, direct application of $\mathcal{M}$ to $\mathcal{X}$ leads to suboptimal performance [12,13,24]. Therefore, the aim of this work is to adapt $\mathcal{M}$ based on $\mathcal{X}$ so that the segmentation performance is improved, which is an SFDA problem.

Since no annotated data is available for the target dataset, we choose to adapt the source model based on the prediction consistency when the target

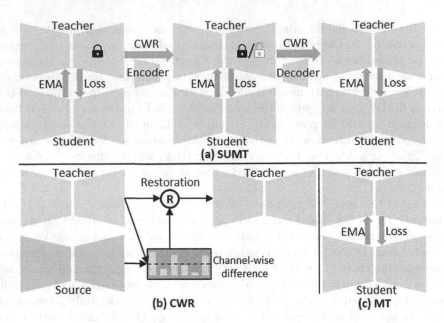

Fig. 1. Method overview: (a) the complete SUMT framework, (b) the CWR strategy in SUMT, and (c) the standard MT for comparison.

data is perturbed, so that the model can accommodate the target domain. The MT framework [23] has been shown to effectively exploit the prediction consistency in the SSL setting. However, in SFDA the knowledge necessary for image segmentation can be easily forgotten if the model is updated solely based on the prediction consistency. For example, the model can simply resort to a degenerate solution that produces the same result for all inputs. Therefore, we propose SUMT that improves upon the MT framework for the SFDA setting. An overview of SUMT is shown in Fig. 1, where the CWR strategy in SUMT is also illustrated and the standard MT is described for comparison. The detailed design of SUMT is presented below.

2.2   Selectively Updated Mean Teacher

To effectively leverage the data from the target domain, SUMT seeks to adapt the source model parameters that are sensitive to domain variance and retain the parameters that are invariant to domains. Like standard MT, in SUMT a teacher model $\mathcal{M}^t$ and a student model $\mathcal{M}^s$ are constructed. $\mathcal{M}^t$ and $\mathcal{M}^s$ share the same network structure, and they are both initialized by the source model $\mathcal{M}$.

In standard MT, $\mathcal{M}^t$ makes predictions on perturbed target images, which are considered pseudo-labels, and $\mathcal{M}^s$ learns from the pseudo-labels based on differently perturbed target images to update the model parameters. Then, all model weights of $\mathcal{M}^t$ are in turn updated based on $\mathcal{M}^s$ with EMA. However, the joint update of all weights can be problematic for SFDA as it may cause forgetting of domain-invariant knowledge for segmentation due to the lack of annotated data. To address this problem, in SUMT we propose to selectively update the model weights of $\mathcal{M}^t$, where the following steps are applied.

Since the earlier layers are more likely to be domain-specific [12], we propose to first update the encoding layers of $\mathcal{M}^t$ while fixing its decoding layers when $\mathcal{M}^t$ is updated based on $\mathcal{M}^s$ with EMA. Formally, we denote the encoders/decoders of $\mathcal{M}$, $\mathcal{M}^t$, and $\mathcal{M}^s$ by E/D, E_t/D_t, and E_s/D_s, respectively. At the t-th iteration of the update, the student model $\mathcal{M}^s$ is updated by minimizing a consistency loss based on the target data. Specifically, suppose the prediction given by $\mathcal{M}^s$ for x_i is c_i, and the corresponding pseudo-label given by $\mathcal{M}^t$ is d_i.[1] The consistency loss $\mathcal{L}_c$ for updating $\mathcal{M}^s$ with fixed $\mathcal{M}^t$ is defined as

$$\mathcal{L}_c = \sum_{i=1}^{N} \left(\mathcal{L}_{ce}(c_i, d_i) + \mathcal{L}_{\text{Dice}}(c_i, d_i) \right), \tag{1}$$

where $\mathcal{L}_{ce}$ and $\mathcal{L}_{\text{Dice}}$ are the cross-entropy loss and Dice loss [20], respectively. Then, the teacher model is updated as

$$E_t \leftarrow E_t \cdot \sigma + E_s \cdot (1 - \sigma) \quad \text{and} \quad D_t \leftarrow D, \tag{2}$$

where σ is the EMA decay rate to be specified. This partial update of $\mathcal{M}^t$ reduces the risk of forgetting high-level semantic knowledge in the decoding layers while adapting the extraction of low-level features in the encoding layers.

In the partial teacher update above, it is still possible that domain-invariant model parameters in the encoder are inappropriately updated. To address this problem, we further propose the CWR strategy which explicitly restores the knowledge from the source model that may need to be retained. Specifically, suppose the model weights of $\mathcal{M}^t$ at the l-th layer associated with the k-th channel are represented as a set $\mathcal{W}_{l,k}^t = \{w_{l,k,p}^t\}_{p=1}^{P_{l,k}}$, where $P_{l,k}$ is the number of these weights and $w_{l,k,p}^t$ is the p-th weight. The amount of the update of $\mathcal{W}_{l,k}^t$ can indicate whether the weights are associated with domain-specific or domain-invariant features. A greater update amount indicates that the channel is likely to focus on the domain-specific feature and contribute to domain adaptation, whereas a smaller amount indicates that the channel tends to extract domain-invariant features and probably should not be forgotten. To measure the amount of the weight update, we compute the difference $d_{l,k}$ between $\mathcal{W}_{l,k}^t$ and the corresponding weights $\mathcal{W}_{l,k} = \{w_{l,k,p}\}_{p=1}^{P_{l,k}}$ in the source model $\mathcal{M}$ as

$$d_{l,k} = \sum_{p=1}^{P_{l,k}} \left| w_{l,k,p}^t - w_{l,k,p} \right|. \tag{3}$$

Based on $d_{l,k}$, we restore the bottom q_l (percentage) of the weights for the l-th layer of $\mathcal{M}^t$ as

$$\mathcal{W}_{l,k}^t \leftarrow \begin{cases} \mathcal{W}_{l,k}, & d_{l,k} \leq H(\mathcal{D}_l, q_l) \\ \mathcal{W}_{l,k}^t, & d_{l,k} > H(\mathcal{D}_l, q_l) \end{cases}, \tag{4}$$

---

[1] Noise perturbation and random flips are applied before the teacher or student prediction as in [16].

where $\mathcal{D}_l$ represents the set of all $d_{l,k}$'s at the l-th layer, and $H(\mathcal{D}_l, q_l)$ sorts these $d_{l,k}$'s in ascending order and returns the value that is ranked q_l. Note that here since D_t is the same as D, no restoration is needed for the decoding layers.

With the restored encoder, we assume that the earlier layers are better adapted to the target data, and the complete teacher model $\mathcal{M}^t$ including its decoder can now be updated with standard MT, where $\mathcal{M}^s$ is reinitialized by $\mathcal{M}^t$. Note that to avoid incorrectly restored weights, a warmup stage that again only updates E_t is inserted before updating the complete teacher model. During the update of the complete teacher model, at the t-th iteration, after $\mathcal{M}^s$ is updated based on $\mathcal{L}_c$, the teacher model is updated with EMA as

$$E_t \leftarrow E_t \cdot \sigma + E_s \cdot (1 - \sigma) \quad \text{and} \quad D_t \leftarrow D_t \cdot \sigma + D_s \cdot (1 - \sigma). \tag{5}$$

Finally, to further avoid knowledge forgetting in the decoder, CWR is applied to the decoding layers of $\mathcal{M}^t$ based on Eq. (4), and the teacher model is then updated again with standard MT using Eq. (5). After convergence, the teacher model is used as the final segmentation model.

2.3   Implementation Details

We focus on 3D segmentation and use the 3D U-Net [3] implemented in SS4L [16] as the backbone segmentation network, which is a popular choice for semi-supervised medical image segmentation [17,18]. Note that variants of U-Net may also be used and integrated with the proposed method, but it is observed that the performance of these variants is usually on par with the original U-Net [8].

The major hyperparameters in the proposed method are the restoration percentages $\{q_l\}_{l=1}^{L}$, where L is the total number of layers and it is equal to ten for the selected 3D U-Net. We set $q_l = 0.1 * l$ because earlier layers tend to be domain-specific and their restoration is less needed. We set the EMA decay rate $\sigma = 0.999$ according to [16]. The other training configurations, such as the optimizer, learning rate, etc., are set to the default specification in [16].

3   Results

3.1   Data Description and Experimental Settings

To evaluate the proposed method, we performed experiments on three datasets. Their details and experimental settings are given below.

BraTS 2018. The first dataset is the publicly available BraTS 2018 dataset [19], and it was used in this work for whole brain tumor segmentation. The BraTS 2018 dataset contains 285 subjects with four modalities of magnetic resonance imaging, including T1w, T2w, T1ce, and FLAIR images. These images are aligned and have the same voxel size of 1 mm isotropic. For each subject, voxel-wise labels for the enhancing tumor, peritumoral edema, and necrotic and non-enhancing tumor core are given, and they were combined to provide the annotation of the whole tumor. We randomly split the dataset into a training set of 200 subjects and a test set of 85 subjects. To investigate the performance of cross-domain segmentation where the training and test sets use different image modalities for segmentation, we considered the following settings: 1) FLAIR for training and T2w for testing, 2) T2w for training and FLAIR for testing, 3) T1w for training and T1ce for testing, and 4) T1ce for training and T1w for testing.

INBT. The second dataset is an in-house dataset for whole brain tumor segmentation, which is referred to as INBT for convenience. The dataset includes 67 annotated FLAIR images acquired on multiple scanners, and they have been skull-stripped with BET [22]. The voxel size of these images ranges from $0.875\,mm \times 0.875\,mm \times 2\,mm$ to $2\,mm \times 2\,mm \times 5\,mm$. We used INBT to investigate the segmentation performance when the same modality was used for segmentation but the training and test images were acquired on different scanners. Specifically, the FLAIR images of the training subjects in the BraTS 2018 dataset were used for model training, and all subjects in INBT were used as the test set.

MSSEG. The last dataset is the MSSEG dataset [4] for segmenting multiple sclerosis lesions. The dataset contains multimodal images of 15 subjects acquired on three scanners (five subjects for each scanner), including Philips Ingenia 3T (PI3T), Siemens Aera 1.5T (SA1.5T), and Siemens Verio 3T (SV3T). These images have been preprocessed with skull-stripping and co-registration [4]. The resolution of the preprocessed images ranges from $0.5\,mm$ to $1.25\,mm$ in each dimension for different subjects. The annotation of multiple sclerosis lesions was performed for each subject by seven independent clinical experts, and their consensus was used as the final annotation. For demonstration, we used the FLAIR modality for segmentation. The MSSEG dataset was used to investigate the segmentation performance when the training and test images were of the same modality but acquired on different scanners. We considered two experimental settings for MSSEG. First, the training images and test images were acquired on scanners of different vendors, where the images acquired on SA1.5T and SV3T were used for training, and the images acquired on PI3T were used for testing. Second, the training images and test images were acquired with different magnetic fields, where the images acquired on PI3T and SV3T were used for training, and the images acquired on SA1.5T were used for testing.

3.2 Evaluation of Segmentation Accuracy

SUMT was applied to the three datasets separately, and it was compared with four other SFDA methods. The first one is *pseudo-labeling* (PL) [10] that generates pseudo-labels on the target data based on the source model and optimizes the segmentation model with the pseudo-labels. The second one is AdaBN [11] that only updates batch normalization statistics based on the target test data during inference. The third one is TENT [24] that updates both batch statistics and affine parameters in the batch normalization layers via entropy minimization on the target data. The fourth one is OSUDA [13] that is designed for medical image segmentation with an adaptive update of batch-wise normalization statistics. Also, the standard MT framework was included for comparison. In addition, direct application of the source model to the target data without SFDA was considered for comparison, and it is referred to as the baseline. For reference, for the BraTS 2018 dataset that has a large number of training subjects, the *upper bound* (UB) performance that was obtained by training the segmentation model with the target modality of the training subjects was also given (e.g., the model was trained with the FLAIR images of the training subjects when segmentation was to be performed on the FLAIR images of the test subjects). The results on the three datasets are presented next individually.

BraTS 2018 for Cross-modality Segmentation. For qualitative evaluation, axial views of representative segmentation results on test scans from the BraTS 2018 dataset are shown in Fig. 2(a) for SUMT and each competing method, together with the image for segmentation and the expert annotation. The results are shown for the different settings of test image modalities. We can see that in these different cases SUMT produced segmentation results that better agree with the annotation than the competing methods.

Next, SUMT was quantitatively evaluated by computing the Dice coefficient between the segmentation results on the test set and expert annotation. The means and standard deviations of the Dice coefficients are summarized in Table 1. In all cases, SUMT outperforms the competing methods with higher Dice coefficients. In addition, with paired Student's t-tests we show that the difference between SUMT and the competing methods is statistically significant, and this is also indicated in Table 1.

Moreover, we investigated the individual benefit of the proposed CWR strategy and progressive layer update. To demonstrate the benefit of CWR, we integrated CWR with the standard MT, where all weights were jointly updated, restored according to Eq. (4), and then updated again. This procedure is referred to as MT-CWR. In addition, to show the necessity of our weight restoration design in Eq. (4), we modified MT-CWR by replacing Eq. (4) with stochastic restoration with the same ratio, and this procedure is referred to as MT-SR. To demonstrate the benefit of the proposed progressive layer update, we integrated it with the standard MT, which is equivalent to the application of the proposed method without CWR, and this procedure is referred to as MT-PLU. The means and standard deviations of the Dice coefficients for these cases are summarized in Table 1 as well. We can see that both MT-CWR and MT-PLU are better than MT but worse than SUMT, which confirms that these two individual contributions and their integration are all beneficial. Besides, MT-CWR is better than MT-SR, which shows the benefit of the proposed restoration design.

INBT for Cross-scanner Segmentation. Qualitative evaluation and quantitative evaluation of the results on INBT for cross-scanner segmentation are given in Fig. 2(b) and Table 2 (the part associated with INBT), respectively.

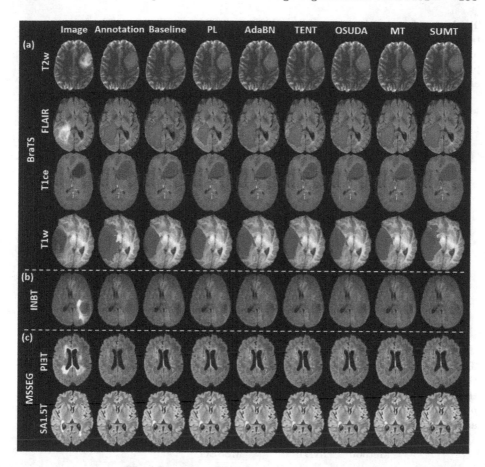

Fig. 2. Axial views of representative segmentation results (red) on test scans for (a) the BraTS 2018 dataset, (b) the INBT dataset, and (c) the MSSEG dataset. The images for segmentation and the expert annotation are also shown for reference. (Color figure online)

From Fig. 2(b), we can see that the segmentation result of SUMT better resembles the expert annotation than the competing methods. Table 2 indicates that SUMT has a higher Dice coefficient than the competing methods and its difference with the competing methods is significant with paired Student's t-tests in four out of six cases. Also, the results of MT-SR, MT-CWR, and MT-PLU are shown in Table 2. The observation for them is consistent with the results of BraTS 2018, where MT-CWR and MT-PLU are better than MT and worse than SUMT, and MT-CWR is better than MT-SR.

MSSEG for Cross-scanner Segmentation. Qualitative evaluation and quantitative evaluation of the segmentation results on the MSSEG dataset are given in Fig. 2(c) and Table 2 (the part associated with MSSEG), respectively,

Table 1. Means and standard deviations of the Dice coefficients (%) of the segmentation results on the test set for the BraTS 2018 dataset. Asterisks indicate that the difference between SUMT and the competing method is statistically significant (*: $p \leq 0.05$, **: $p \leq 0.01$, ***: $p \leq 0.001$) using a paired Student's t-test. The best results are highlighted in bold.

Method	BraTS 2018			
	FLAIR→T2w	T2w→FLAIR	T1w→T1ce	T1ce→T1w
UB	85.0 ± 13.1	81.2 ± 14.8	74.0 ± 19.7	74.6 ± 18.7
Baseline	50.0 ± 30.1***	63.0 ± 26.2***	64.0 ± 20.0***	60.7 ± 28.5***
PL	43.7 ± 33.2***	63.3 ± 25.6***	58.2 ± 25.5***	46.5 ± 28.4***
AdaBN	48.1 ± 30.0***	69.1 ± 24.3***	56.0 ± 25.5***	65.9 ± 24.7**
TENT	48.5 ± 30.4***	72.8 ± 23.4**	56.6 ± 25.0***	64.2 ± 26.0***
OSUDA	47.8 ± 30.5***	72.0 ± 23.7***	55.8 ± 25.9***	64.3 ± 26.1***
MT	53.5 ± 28.6***	74.3 ± 22.7***	65.9 ± 19.8***	67.6 ± 24.8*
MT-SR	54.0 ± 28.1	74.0 ± 24.5	65.4 ± 22.2	66.5 ± 25.4
MT-CWR	54.9 ± 27.7	74.6 ± 23.2	66.4 ± 19.7	68.2 ± 23.1
MT-PLU	55.4 ± 27.0	74.8 ± 21.8	69.0 ± 20.6	70.3 ± 21.3
SUMT	$\mathbf{56.3 \pm 26.7}$	$\mathbf{76.4 \pm 20.4}$	$\mathbf{69.7 \pm 20.0}$	$\mathbf{70.5 \pm 21.5}$

Table 2. Means and standard deviations of the Dice coefficients (%) of the segmentation results on the test set for the INBT and MSSEG datasets. Asterisks indicate that the difference between SUMT and the competing method is statistically significant (*: $p \leq 0.05$, **: $p \leq 0.01$, ***: $p \leq 0.001$) using a paired Student's t-test. The best results are highlighted in bold.

Method	INBT	MSSEG	
		PI3T	SA1.5T
Baseline	63.5 ± 30.5***	60.4 ± 9.6**	38.1 ± 6.5**
PL	70.1 ± 23.2***	57.5 ± 13.2*	46.2 ± 10.0
AdaBN	74.0 ± 22.5*	60.0 ± 9.4*	40.7 ± 9.6**
TENT	74.2 ± 22.2	61.3 ± 9.9**	40.3 ± 8.9**
OSUDA	74.3 ± 22.1	61.6 ± 10.7**	40.4 ± 8.8**
MT	73.3 ± 20.9***	66.3 ± 9.0*	45.6 ± 7.3*
MT-SR	72.7 ± 21.6	65.1 ± 8.7	44.6 ± 7.3
MT-CWR	74.0 ± 19.5	66.3 ± 9.2	45.7 ± 7.6
MT-PLU	75.9 ± 19.6	67.7 ± 8.5	48.0 ± 7.2
SUMT	$\mathbf{76.2 \pm 18.3}$	$\mathbf{68.1 \pm 8.2}$	$\mathbf{48.7 \pm 7.7}$

and the results of MT-SR, MT-CWR, and MT-PLU are shown in Table 2 as well. Like the results of BraTS 2018 and INBT, these results show that SUMT outperforms the competing methods and its difference with the competing methods is significant with paired Student's t-tests in most cases; also, the results of MT-SR, MT-CWR, and MT-PLU in Table 2 confirm the benefit of the integration of CWR and progressive layer update, as well as the weight restoration design.

4 Conclusion

We have proposed SUMT for SFDA-based medical image segmentation. In SUMT, we adapt the mean teacher framework by selectively updating the model parameters to better preserve domain-invariant knowledge. The model update is performed progressively with channel-wise weight restoration. Experimental results on cross-modality and cross-scanner segmentation tasks demonstrate that SUMT outperforms other SFDA methods.

Acknowledgement. This work is supported by the Fundamental Research Funds for the Central Universities.

References

1. Bateson, M., Kervadec, H., Dolz, J., Lombaert, H., Ben Ayed, I.: Source-relaxed domain adaptation for image segmentation. In: Martel, A.L., et al. (eds.) MICCAI 2020. LNCS, vol. 12261, pp. 490–499. Springer, Cham (2020). https://doi.org/10.1007/978-3-030-59710-8_48
2. Chen, C., Dou, Q., Chen, H., Qin, J., Heng, P.A.: Synergistic image and feature adaptation: towards cross-modality domain adaptation for medical image segmentation. In: Proceedings of the AAAI Conference on Artificial Intelligence, vol. 01, pp. 865–872 (2019)
3. Çiçek, Ö., Abdulkadir, A., Lienkamp, S.S., Brox, T., Ronneberger, O.: 3D U-Net: learning dense volumetric segmentation from sparse annotation. In: Ourselin, S., Joskowicz, L., Sabuncu, M.R., Unal, G., Wells, W. (eds.) MICCAI 2016. LNCS, vol. 9901, pp. 424–432. Springer, Cham (2016). https://doi.org/10.1007/978-3-319-46723-8_49
4. Commowick, O., et al.: Multiple sclerosis lesions segmentation from multiple experts: the MICCAI 2016 challenge dataset. Neuroimage **244**, 118589 (2021)
5. Cui, W., et al.: Semi-supervised brain lesion segmentation with an adapted mean teacher model. In: Chung, A.C.S., Gee, J.C., Yushkevich, P.A., Bao, S. (eds.) IPMI 2019. LNCS, vol. 11492, pp. 554–565. Springer, Cham (2019). https://doi.org/10.1007/978-3-030-20351-1_43
6. Ganin, Y., et al.: Domain-adversarial training of neural networks. J. Mach. Learn. Res. **17**(1), 2096-2030 (2016)
7. Ghafoorian, M., et al.: Transfer learning for domain adaptation in MRI: application in brain lesion segmentation. In: Descoteaux, M., Maier-Hein, L., Franz, A., Jannin, P., Collins, D.L., Duchesne, S. (eds.) MICCAI 2017. LNCS, vol. 10435, pp. 516–524. Springer, Cham (2017). https://doi.org/10.1007/978-3-319-66179-7_59
8. Gut, D., Tabor, Z., Szymkowski, M., Rozynek, M., Kucybała, I., Wojciechowski, W.: Benchmarking of deep architectures for segmentation of medical images. IEEE Trans. Med. Imaging **41**(11), 3231–3241 (2022)

9. Isensee, F., Jaeger, P.F., Kohl, S.A., Petersen, J., Maier-Hein, K.H.: nnU-Net: a self-configuring method for deep learning-based biomedical image segmentation. Nat. Methods **18**(2), 203–211 (2021)

10. Lee, D.H.: Pseudo-label: the simple and efficient semi-supervised learning method for deep neural networks. In: ICML Workshop on Challenges in Representation Learning (2013)

11. Li, Y., Wang, N., Shi, J., Hou, X., Liu, J.: Adaptive batch normalization for practical domain adaptation. Pattern Recogn. **80**, 109–117 (2018)

12. Liang, J., Hu, D., Feng, J.: Do we really need to access the source data? Source hypothesis transfer for unsupervised domain adaptation. In: International Conference on Machine Learning, pp. 6028–6039 (2020)

13. Liu, X., Xing, F., Yang, C., El Fakhri, G., Woo, J.: Adapting off-the-shelf source segmenter for target medical image segmentation. In: de Bruijne, M., et al. (eds.) MICCAI 2021. LNCS, vol. 12902, pp. 549–559. Springer, Cham (2021). https://doi.org/10.1007/978-3-030-87196-3_51

14. Liu, Y.C., Ma, C.Y., Kira, Z.: Unbiased teacher v2: semi-supervised object detection for anchor-free and anchor-based detectors. In: Proceedings of the IEEE/CVF Conference on Computer Vision and Pattern Recognition, pp. 9819–9828 (2022)

15. Lu, Q., Ye, C.: Knowledge transfer for few-shot segmentation of novel white matter tracts. In: Feragen, A., Sommer, S., Schnabel, J., Nielsen, M. (eds.) IPMI 2021. LNCS, vol. 12729, pp. 216–227. Springer, Cham (2021). https://doi.org/10.1007/978-3-030-78191-0_17

16. Luo, X.: SSL4MIS (2020). https://github.com/HiLab-git/SSL4MIS

17. Luo, X., Chen, J., Song, T., Wang, G.: Semi-supervised medical image segmentation through dual-task consistency. In: Proceedings of the AAAI Conference on Artificial Intelligence, vol. 35, pp. 8801–8809 (2021)

18. Luo, X., et al.: Efficient semi-supervised gross target volume of nasopharyngeal carcinoma segmentation via uncertainty rectified pyramid consistency. In: de Bruijne, M., et al. (eds.) MICCAI 2021. LNCS, vol. 12902, pp. 318–329. Springer, Cham (2021). https://doi.org/10.1007/978-3-030-87196-3_30

19. Menze, B.H., et al.: The multimodal brain tumor image segmentation benchmark (BRATS). IEEE Trans. Med. Imaging **34**(10), 1993–2024 (2014)

20. Milletari, F., Navab, N., Ahmadi, S.A.: V-Net: fully convolutional neural networks for volumetric medical image segmentation. In: International Conference on 3D Vision, pp. 565–571 (2016)

21. Ronneberger, O., Fischer, P., Brox, T.: U-Net: convolutional networks for biomedical image segmentation. In: Navab, N., Hornegger, J., Wells, W.M., Frangi, A.F. (eds.) MICCAI 2015. LNCS, vol. 9351, pp. 234–241. Springer, Cham (2015). https://doi.org/10.1007/978-3-319-24574-4_28

22. Smith, S.M.: Fast robust automated brain extraction. Hum. Brain Mapp. **17**(3), 143–155 (2002)

23. Tarvainen, A., Valpola, H.: Mean teachers are better role models: weight-averaged consistency targets improve semi-supervised deep learning results. In: Advances in Neural Information Processing Systems, pp. 1195–1204 (2017)

24. Wang, D., Shelhamer, E., Liu, S., Olshausen, B., Darrell, T.: TENT: fully test-time adaptation by entropy minimization. arXiv preprint arXiv:2006.10726 (2020)

UPL-TTA: Uncertainty-Aware Pseudo Label Guided Fully Test Time Adaptation for Fetal Brain Segmentation

Jianghao Wu[1,2] , Ran Gu[1] , Tao Lu[3], Shaoting Zhang[1,2],
and Guotai Wang[1,2(✉)]

[1] School of Mechanical and Electrical Engineering, University of Electronic Science
and Technology of China, Chengdu 611731, China
guotai.wang@uestc.edu.cn
[2] Shanghai Artificial Intelligence Laboratory, Shanghai 200030, China
[3] Department of Radiology, Sichuan Provincial People's Hospital,
University of Electronic Science and Technology of China, Chengdu 610072, China

Abstract. Test Time Adaptation (TTA) is promising to improve a deep learning model's robustness when encountering images from an unseen domain. Existing TTA methods are with low performance due to the insufficient supervision signal from unannotated target domain images, or limited by specific requirements on the pre-training strategy and network structure in the source domain. We aim to separate the pre-training in the source domain and adaptation in the target domain, in order to achieve high-performance and more generalizable TTA without assumptions on the pre-training strategy. To solve this problem, we propose UPL-TTA, an Uncertainty-aware Pseudo Label guided fully Test Time Adaptation method. Specifically, we introduce Test Time Growing (TTG) to duplicate the prediction head of the source model with perturbations at image and feature levels in the target domain. The different predictions obtained in these duplicated prediction heads are used to obtain pseudo labels for the unlabeled target domain images as well as their uncertainty maps, which can identify reliable pseudo labels. Pixels with unreliable pseudo labels are regularized by imposing entropy minimization on the mean prediction of the multiple heads. UPL-TTA was validated bidirectionally on a cross-modality fetal brain segmentation dataset. Compared with no adaptation, it significantly improved the average Dice in the two different target domains by 3.95% and 6.12%, respectively, and outperformed several state-of-the-art TTA methods.

Keywords: Test time adaptation · self-training · fetal brain MRI

1 Introduction

Benefiting from high-precision and large-scale annotations, deep learning with Convolutional Neural Networks (CNNs) has achieved excellent performance in

© The Author(s), under exclusive license to Springer Nature Switzerland AG 2023
A. Frangi et al. (Eds.): IPMI 2023, LNCS 13939, pp. 237–249, 2023.
https://doi.org/10.1007/978-3-031-34048-2_19

medical image segmentation tasks [11]. However, due to the low cross-domain generalizability of existing methods, their performance will decrease largely when applied to images with a new distribution, i.e., an unseen modality [3]. For example, in the practical application, a pre-trained model can hardly maintain robustness when deployed to a new medical center where the data distribution may be different from the training set due to different scanning instruments used or different imaging sequences [5,7,8,20]. Figure 1 shows such an example, where the image intensity and contrast are quite different in two sequences of fetal brain Magnetic Resonance Imaging (MRI): half-Fourier acquisition single-shot turbo spin-echo (HASTE) and true fast imaging with steady state precession (TrueFISP). A model trained with HASTE images has a poor performance on TrueFISP images, and vice versa.

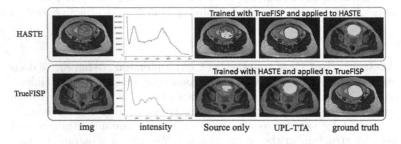

Fig. 1. The domain shift between HASTE and TrueFISP of fetal brain MRI. Our UPL-TTA largely improves the model's robustness on a different sequence at testing time.

Domain Adaptation (DA) is promising to solve the above problem of domain gap between training and testing data [1]. To avoid time-consuming annotations in the target domain, Unsupervised Domain Adaptation (UDA) [14] methods are proposed to align the source and target distributions at image, feature, or output levels [22]. These methods all require simultaneous access to source and target domain data to make the model perform well. However, in practice, source data is often unavailable when the model is deployed to a new center due to the constraints on computation, bandwidth and privacy.

Source-free Domain Adaptation [10,17,19] aims to adapt a pre-trained model to a new target data distribution without access to the source data. In the literature, Test Time Training (TTT) [17] adds an auxiliary branch to predict the rotation by self-supervision, and adapts the shared encoder in the target domain. DTTA [8] and ATTA [5] optimize an auto-encoder during the training of source model to learn shape priors, and align feature distributions for adaptation. However, these methods require the insertion of specific modules, such as an auxiliary prediction branch and auto-encoders, before the training of the source model. They also require that the model should have been pre-trained with a specific strategy in the source domain, which limits their applicability when dealing with a pre-trained model that does not satisfy the training requirements.

In practice, a target domain may be given a pre-trained model that has been trained with an arbitrary strategy. Therefore, it is desirable to achieve fully test time adaptation that does not need the pre-trained model to have a specific structure and training strategy in the source domain [6]. PTBN [12] updates the statistics of Batch Normalization (BN) layers on the target domain data, and TENT [20] tunes BN layers by minimizing the entropy of predictions in the target domain. However, these methods were originally designed for natural images, and they simply assume that the domain shift can be sufficiently alleviated by updating the BN layers, which leads to limited performance in TTA for medical image segmentation [18]. URMA [2] is a method that aids the adaptation process using pseudo labels generated in one branch and uncertainties from multiple branches. However, its pseudo label may contain obvious errors and mislead the model adaptation.

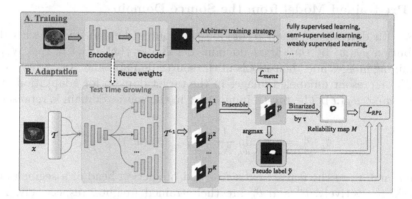

Fig. 2. Overview of our UPL-TTA, where the p^k is the soft prediction of k-th branch, τ is the confidence threshold. It does not require a specific training strategy in the source domain, and uses pseudo labels based on test time growing for adaptation.

In this work, we propose Uncertainty-aware Pseudo Label guided Fully Test Time Adaptation (UPL-TTA) for medical image segmentation, which does not require the pre-trained model to be trained with an extra auxiliary branch or a specific strategy in the source domain before adaptation to a target domain. For a given pre-trained model, we first introduce Test Time Growing (TTG) to duplicate the prediction head (e.g., the decoder in widely used UNet-like CNNs [15]) of the source model several times for the target domain, and add a range of random perturbations (e.g., dropout, spatial transform) to their input image and feature map to obtain several different segmentation predictions. Then pseudo labels for target domain images are obtained by an ensemble of these predictions. To suppress the effect of potentially incorrect pseudo labels, we introduce ensembling-based and MC dropout uncertainty estimation to obtain a reliability map. The pseudo labels of reliable pixels are used to supervise the output of each prediction head, and the predictions of unreliable pixels are regularized by entropy minimization on the average prediction map. Experiments on bidirectional cross-modality adaptation between HASTE and TrueFISP of the fetal brain showed that our UPL-TTA significantly improved the model's performance on the target domain, and outperformed several existing TTA methods.

2   Method

The proposed UPL-TTA framework is depicted in Fig. 2. Without assumptions on the training strategy in the source domain, we duplicate the prediction head of the pre-trained model several times and add perturbations to obtain multiple predictions, which leads to pseudo labels in the unannotated target domain and the corresponding reliability maps to supervise the model for adaptation.

2.1   Pre-trained Model from the Source Domain

Let S with data distribution $\mu_S(x)$ be the source domain and T with data distribution $\mu_T(x)$ be the target domain. Let $\mathbf{X}_S = \{(x_i^s, y_i^s), i = 1, ..., N_s\}$ be the training images and their labels in the source domain, and $\mathbf{X}_T = \{(x_i,), i = 1, ..., N_t\}$ represent unlabeled images in the target domain for adaptation. Note that $\mu_S(x) \neq \mu_T(x)$. The pre-training stage in the source domain is represented as:

$$\theta_g^0, \theta_h^0 = \arg\min_{\theta_g, \theta_h} \frac{1}{N_s} \sum_{i=1}^{N_s} L_s\left(h\big(g(x_i^s)\big), y_i^s\right) \tag{1}$$

where g and h are the feature extractor and prediction head of a segmentation network, respectively. θ_g^0 and θ_h^0 are their trained weights, respectively. L_s is the training loss in the source domain, which might be implemented by fully supervised learning, semi-supervised learning and weakly supervised learning, etc., based on the type of the available labels in the source domain.

2.2   Test-Time Growing for Adaptation

When the source model is deployed to a new center, as access to the source domain data is limited, we consider the problem of adapting the pre-trained model to the target domain based on $\mathbf{X}_T$ and $\{\theta_g^0, \theta_h^0\}$. For the pre-trained feature extractor g and prediction head h, we propose Test-Time Growing (TTG) to duplicate h by K times in the target domain, as shown in Fig. 2. The weights of shared feature extractor g and duplicated prediction heads $\{h^k\}(k = 1, 2, ..., K)$ are initialized as θ_g^0 and θ_h^0, respectively. Note that the weights in different prediction heads will become different due to the inconsistency of the gradients generated by the different predictions under perturbations. An ensemble of these K heads is used to obtain pseudo labels of target domain images that are unannotated. To encourage the different heads to obtain diverse results for better ensemble, we introduce random perturbations on the input image and dropout on features [21].

First, for an input image $x \in \mathcal{R}^{H \times W}$ in the target domain, where H and W are the height and width, respectively, we send it into the network K times, each time with a random spatial transformation and for a different prediction head h^k. The segmentation prediction result for the k-th head is:

$$p^k = \mathcal{T}^{-1} \circ h^k \left(g(\mathcal{T} \circ x) \right) \tag{2}$$

where $\mathcal{T}$ is a random spatial transformation and $\mathcal{T}^{-1}$ is the corresponding inverse transformation. $p^k \in \mathcal{R}^{C \times H \times W}$ is the output segmentation probability map with C channels obtained by Softmax, where C is the class number for segmentation. In this paper, we set $\mathcal{T}$ as random flipping, rotation with $\pi/2$, π and $3\pi/2$.

Second, K different dropout layers are applied in parallel after the feature extractor g, so that the prediction heads take different random subsets of the features as input. We then average across the K different predicted segmentation probability maps for ensemble:

$$\bar{p} = \frac{1}{K} \sum_{k=1}^{K} p^k \tag{3}$$

2.3 Supervision with Reliable Pseudo Labels

Based on the average probability map $\bar{p}$, a pseudo label is obtained by taking the argmax across channels. To reduce noises, it is post-processed by only keeping the largest connected component for each foreground class (e.g., fetal brain segmentation in this work). Then the post-processed pseudo label is converted into a one-hot representation, which is denoted as $\tilde{y} \in \{0,1\}^{C \times H \times W}$. Due to the existence of domain gap, the pseudo labels have a limited accuracy. Directly using the pseudo labels of all pixels for self-training would limit the model's performance.

To deal with these problem, it is important to highlight the reliable pseudo labels and suppress unreliable ones during adaptation. Therefore, we use the uncertainty information of $\bar{p}$ to identify pixels with reliable pseudo labels and only use the reliable region to supervise the model for adaptation. Specifically, a binary reliability map $M \in \{0,1\}^{H \times W}$ is calculated for the pseudo label $\tilde{y}$, and each element in M is defined as:

$$M_n = \begin{cases} 1 & \text{if } \bar{p}_{c^*,n} > \tau \\ 0 & \text{otherwise} \end{cases} \tag{4}$$

where $n = 1, 2, ..., HW$ is the pixel index. $c^* = \arg\max_c(\bar{p}_{c,n})$ is the class with the highest probability for pixel n, and $\bar{p}_{c^*,n}$ represents the confidence for the pseudo label at that pixel. $\tau \in (1/C, 1.0)$ is a confidence threshold.

Then the reliability map is used as a mask to exclude unreliable pixels for supervision, and a Reliable Pseudo Label (RPL) loss is denoted as:

$$\mathcal{L}_{RPL} = \frac{1}{K} \sum_{k=1}^{K} \mathcal{L}_{w-dice}(p^k, \tilde{y}, M) \tag{5}$$

where $\mathcal{L}_{w-dice}$ is the reliability map-weighted Dice loss for the k-th head:

$$\mathcal{L}_{w-dice}(\boldsymbol{p}^k, \tilde{y}, M) = 1 - \frac{1}{Z} \sum_{c=1}^{C} \sum_{n=1}^{HW} \frac{2M_n \boldsymbol{p}_{c,n}^k \tilde{y}_{c,n}}{\boldsymbol{p}_{c,n}^k + \tilde{y}_{c,n} + \epsilon} \qquad (6)$$

where n is the pixel index and $\epsilon = 10^{-5}$ is a small number for numeric stability. $Z = C \sum_n M_n$ is a normalization factor.

2.4 Mean Prediction-Based Entropy Minimization

Entropy minimization is widely used as a regularizer in test time adaptation [9,13,18], which reduces the uncertainty of the system by reducing the entropy of model predictions. However, in our method with multiple prediction heads, applying entropy minimization to each head respectively may lead to sub-optimal results when different heads predict confident while opposite results. For example, in the binary segmentation problem, when branch k predicts a certain pixel being the foreground with a probability of 0.0 and branch $k+1$ predicts it with a foreground probability of 1.0, both branches have the lowest entropy, but their average result has a high entropy. To deal with this problem, we propose to apply entropy minimization to the mean prediction across the K heads:

$$\mathcal{L}_{ment} = -\frac{1}{HW} \sum_{n=1}^{HW} \sum_{c=1}^{C} \bar{\boldsymbol{p}}_{c,n} log(\bar{\boldsymbol{p}}_{c,n}), \qquad (7)$$

where $\bar{\boldsymbol{p}}$ is the mean probability prediction obtained by the K heads of TTG. Compared with minimizing the entropy of each prediction head respectively, minimizing the entropy of their mean prediction $\bar{\boldsymbol{p}}$ can not only reduce the uncertainty of a single prediction head, but also make the predictions of the K heads for the same test sample tend to be consistent, therefore improving the prediction robustness of the model for unseen test samples.

2.5 Adaptation by Self-training

Our adaptation process adopts a self-training paradigm based on the pseudo labels and mean prediction-based entropy minimization. We obtain the average prediction $\bar{\boldsymbol{p}}$, pseudo label $\tilde{y}$ and the reliability map M based TTG for a test sample, and then calculate $\mathcal{L}_{RPL}$ and $\mathcal{L}_{ment}$. The overall loss for TTA is:

$$\mathcal{L} = \mathcal{L}_{RPL} + \lambda \mathcal{L}_{ment}. \qquad (8)$$

where λ is a hyper-parameter to control the weight of $\mathcal{L}_{ment}$.

3   Experiment and Results

3.1   Experimental Details

Dataset. We used a Fetal Brain (FB) segmentation dataset to evaluate our UPL-TTA, and it consisted of fetal brain MRI with two imaging sequences: 1) 68 volumes acquired by HASTE with size of 640×520, in-plane resolution of 0.64 to 0.70 mm and slice-thickness of 6.5–7.15 mm; 2) 44 volumes acquired by TrueFISP with size of 384×312, in-plane resolution of 0.67 to 1.12 mm and thickness of 6.5 mm. The gestational age ranged from 21–33 weeks. As shown in Fig. 1, the intensity distribution and contrast are different in these two sequences, leading to a large domain gap. In addition, the different gestational age leads to varying appearance of the fetal brain, which increases the difficulty for robust segmentation. We performed bidirectional TTA for experiments: 1) HASTE to TrueFISP, where HASTE was used as the source domain and TrueFISP as the target domain; 2) TrueFISP to HASTE. We randomly split the images for each domain into 70%, 10% and 20% for training, validation and testing, respectively, and abandoned the labels of training images in the target domain.

Implementation Details. For preprocessing, we clip the intensities by the 1-st and 99-th percentiles, and linearly normalized them to $[-1,1]$. Each slice was resized to 256×256. Due to the large inter-slice spacing, we used slice-by-slice segmentation with 2D CNNs and stacked the results into a 3D volume. The segmentation network for our method is flexible, and we selected the widely used UNet [15], as most medical image segmentation models are based on UNet-like structures [11,16]. The feature extractor g and prediction head h were implemented by the encoder and decoder of UNet [15], respectively. During pre-training in the source domain, we trained UNet [15] for 400 epochs with Dice loss, Adam optimizer and initial learning rate of 0.01 that was decayed to 90% every 4 epochs. The model weight with the best performance on the validation set in the source domain was used for adaptation. For adaptation in the target domain, we duplicated the decoder of UNet [11] for K times, and updated all the model parameters for 20 epochs with Adam optimizer and a fixed learning rate of 10^{-4}. In the training and adaptation stages, we set all the slices in a single volume as a batch. The hyper-parameter setting was $K = 4$, $\lambda = 1.0$, and $\tau = 0.9$ based on the best performance on the validation set.

During inference, we computed the argmax of the average prediction generated by the K heads, and we did not apply any post-processing to the output. All the experiments were implemented with PyTorch 1.8.1, using an NVIDIA GeForce RTX 2080Ti GPU. For quantitative evaluation of the volumetric segmentation results, we adopted the commonly used Dice score (DSC) and Average Symmetric Surface Distance (ASSD). As the slice thickness is large (6–7.15 mm), we calculated ASSD values with unit of pixel.

3.2   Results

Comparison with Other Methods. Our UPL-TTA was compared with four state-of-the-art test time adaptation methods on the FB dataset: 1) **PTBN** [12]

Table 1. Quantitative comparison of different TTA methods on fetal brain segmentation. † means significant improvement (p-value < 0.05) from "Source only".

Method	HASTE to TrueFISP		TrueFISP to HASTE	
	Dice (%)	ASSD (pixel)	Dice (%)	ASSD (pixel)
Source only	84.09 ± 6.34	1.33 ± 0.49	83.91 ± 7.39	2.31 ± 1.96
Target only	88.85 ± 4.12	0.91 ± 0.30	94.09 ± 3.47	0.50 ± 0.38
PTBN [12]	85.70 ± 4.88	1.85 ± 0.96	85.47 ± 5.65	2.92 ± 2.55
TENT [20]	85.75 ± 3.62	1.60 ± 0.71	$88.21 \pm 5.35†$	1.16 ± 1.17
TTT [17]	85.84 ± 4.52	1.80 ± 0.90	87.20 ± 5.33	2.28 ± 1.99
URMA [2]	84.12 ± 6.82	2.18 ± 1.19	81.05 ± 6.85	5.95 ± 3.9
UPL-TTA (Ours)	$\mathbf{88.04 \pm 4.82†}$	$\mathbf{1.20 \pm 0.74}$	$\mathbf{90.03 \pm 5.28†}$	$\mathbf{0.85 \pm 0.64†}$

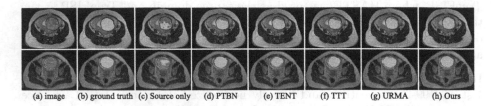

(a) image   (b) ground truth   (c) Source only   (d) PTBN   (e) TENT   (f) TTT   (g) URMA   (h) Ours

Fig. 3. Qualitative comparison of different TTA methods. First row: HASTE to True-FISP. Second row: TrueFISP to HASTE.

that updates batch normalization statistics on the target data during test time; 2) **TENT** [20] that only updates the parameters of batch normalization layers by minimizing the entropy of model predictions on new test data; 3) **TTT** [17] that uses self-supervision for adaptation, where an auxiliary decoder is used in both of the source and target domains to predict the rotation angle of an image; and 4) **URMA** [2] that uses pseudo labels generated in one branch and uncertainties in multiple branches to aid the adaptation process. We also compared our method with two oracle methods: 1) **Source only** where the pre-trained model was directly used for inference on the target domain dataset, and 2) **Target only** where the model was trained with annotated images in the target domain. All the compared methods were implemented with the same backbone (UNet [15]) for a fair comparison.

The quantitative evaluation results of bidirectional TTA are shown in Table 1. It can be observed that **Source only** and **Target only** achieved an average Dice of 84.09% and 88.85%, respectively in HASTE to TrueFISP and 83.90% and 94.09%, respectively in TrueFISP to HASTE, showing the large gap between the two domains. The existing methods only achieved a slight improvement or even a decrease compared with **Source only**, with the Dice values ranging from 84.12% to 85.84% for HASTE to TrueFISP and 81.05% to 88.21% for True-FISP to HASTE, respectively. In contrast, our method largely improved the Dice to 88.04% and 90.03% for the two target domains, respectively. Our method achieved an average ASSD of 1.20 and 0.85 pixels, in the two domains, respectively, which was lower than those of the other TTA methods. The qualitative comparison in Fig. 3 shows that the existing methods tend to achieve under-segmentation of the fetal brain, while our method can successfully segment the entire fetal brain region with high accuracy.

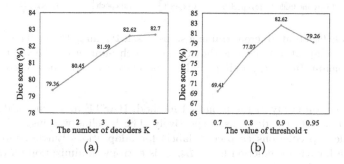

Fig. 4. Performance of our method with different hyper-parameter values on the validation set when HASTE and TrueFISP are the source and target domains, respectively.

Ablation Study. Our UPL-TTA adds two new hyperparameters: the number of duplicated prediction heads K, and the confidence threshold τ to select reliable pseudo labels. We first investigated the effect of K by setting it to 1 to 5 respectively, and the performance on the validation set of TrueFISP is shown in Fig. 4(a). It can be observed that $K = 1$ performed worse than larger K values, showing the superiority of using Test-Time Growing (TTG). As K increased, our method performed progressively better, and $K = 5$ reached a plateau. Therefore, we finally set K to 4 considering the trade-off between performance and memory consumption. Then we investigated the effect of τ. A higher threshold τ will result in a smaller reliable region for each class, which helps avoid the model being misled by inaccurate pseudo label, but a too large τ will make the reliable pseudo label region too small and thus cannot provide sufficient supervision. Quantitative comparison between different τ values in Fig. 4(b) shows that the best performance on the validation set was achieved when $\tau = 0.9$.

We further investigated the effect of each component of our UPL-TTA. The baseline was just using the pre-trained model's predictions as pseudo labels for adaptation, and the introduced components are: 1) single-head entropy minimization (Entropy-min) [4] that is applied to each of the prediction heads respectively; 2) "Reliability map" that uses M to suppress unreliable pseudo labels; 3) Test Time Growing (TTG) that duplicates the prediction head K times with feature dropout; 4) random spatial transformation (T) further introduced to the K heads; and 5) L_{ment} that applies entropy minimization to the mean prediction of the K heads rather than to each head respectively. The quantitative evaluation results are presented in Table 2. We observed that the baseline (74.98%) performed worse than "Source only" (84.09%). Additionally using entropy minimization (83.44%) was still not better than "Source only", which indicated

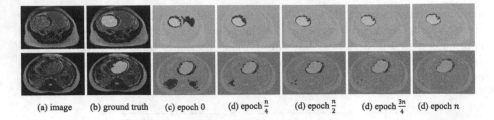

(a) image   (b) ground truth   (c) epoch 0   (d) epoch $\frac{n}{4}$ (d) epoch $\frac{n}{2}$ (d) epoch $\frac{3n}{4}$ (d) epoch n

Fig. 5. Pseudo labels at different training steps in self-training. Epoch 0 means "Source only" (before adaptation) and n is the optimal epoch number on the validation set of the target domain. In (c)–(g), only reliable pseudo labels are encoded by colors. (Color figure online)

Table 2. Ablation study of the proposed method. HASTE and TrueFISP were used as the source and target domains, respectively. The baseline was just using the pre-trained model's predictions as pseudo labels for adaptation. Entropy-min: Entropy minimization for each prediction head. L_{ment} is entropy minimization on the average prediction of the K heads.

Components					Dice (%)	ASSD (pixel)
Entropy-min	Reliability-map	TTG	T	L_{ment}		
					74.98 ± 10.72	4.9 ± 1.37
✓					83.44 ± 7.38	1.39 ± 0.76
✓	✓				85.88 ± 5.49	1.56 ± 1.01
✓	✓	✓			86.77 ± 4.17	0.95 ± 0.24
✓	✓	✓	✓		86.92 ± 5.41	$\mathbf{0.91 \pm 0.36}$
	✓		✓	✓	$\mathbf{88.04 \pm 4.82}$	1.20 ± 0.74

that the pseudo label from a single prediction head contained a lot of misleading information. In contrast, each component of our introduced reliability map, TTG, spatial transformation and L_{ment} led to some improvement, showing the effectiveness of our method.

4   Discussions

In general, a segmentation model contains a feature extractor and a prediction head, and our method duplicates the prediction head via Test-Time Growing (TTG) in the target domain. This paper implemented TTG with an encoder-decoder structure, as most efficient CNNs for medical image segmentation tasks are UNet-like [3,15]. However, our method can be easily applied to other segmentation networks, as it has a minimal assumption on the structure of the pre-trained model and how it was trained in the source domain, which is more general than existing methods like TTT [17] and DTTA [8].

Due to the absence of annotations in the target domain, it is important to obtain effective supervision signal and regularization for the TTA task. Our method uses reliable pseudo labels to deal with the unannotated images, where the TTG improves the quality of pseudo labels, and the introduced reliability map avoids the model being corrupted by inaccurate pseudo labels. URMA [2] also uses pseudo labels to guide the adaptation, but its pseudo labels are obtained from a single decoder, which are less robust than our pseudo labels based on an ensemble of multiple heads. In addition, our mean prediction-based entropy minimization has an implicit consistency regularization on the K prediction heads, which improves the model's robustness against perturbations in the target domain.

Despite UPL-TTA's higher performance than existing TTA methods in the experiment, it is applicable to a moderate domain shift where high-quality pseudo labels can be obtained by TTG. In some other scenarios where the domain gap is extremely large, it may be hard to obtain usable pseudo labels, and our method may not be applicable. In addition, this work only deals with a binary segmentation task, but the pipeline can also be applied for multi-class segmentation and 3D segmentation networks.

5 Conclusion

To summarize, we propose a fully test time adaptation method that adapts the source model to an unannotated target domain without knowing the training strategy of the source model. Without access to source domain images, our proposed uncertainty-aware pseudo label-guided TTA generates multiple prediction outputs for the same sample in the target domain via Test Time Growing (TTG). It generates high-quality pseudo labels and the corresponding reliability maps to provide effective supervision in the unannotated target domain. Pixels with unreliable pseudo labels are further regularized by entropy minimization of the mean prediction across the duplicated heads, which also introduces an implicit consistency regularization. Experiments on bidirectionally cross-modality TTA for fetal brain segmentation showed that our method outperformed several state-of-the-art TTA methods. In the future, it is of interest to implement a 3D version of our method and apply it to other segmentation tasks.

Acknowledgements. This work was supported by National Natural Science Foundation of China (No. 62271115).

References

1. Chen, C., Dou, Q., Chen, H., Qin, J., Heng, P.A.: Synergistic image and feature adaptation: towards cross-modality domain adaptation for medical image segmentation. In: AAAI, pp. 865–872 (2019)
2. Fleuret, F., et al.: Uncertainty reduction for model adaptation in semantic segmentation. In: CVPR, pp. 9613–9623 (2021)
3. Gu, R., Zhang, J., Huang, R., Lei, W., Wang, G., Zhang, S.: Domain composition and attention for unseen-domain generalizable medical image segmentation. In: de Bruijne, M., et al. (eds.) MICCAI 2021. LNCS, vol. 12903, pp. 241–250. Springer, Cham (2021). https://doi.org/10.1007/978-3-030-87199-4_23

4. Hang, W., et al.: Local and global structure-aware entropy regularized mean teacher model for 3D left atrium segmentation. In: Martel, A.L., et al. (eds.) MICCAI 2020. LNCS, vol. 12261, pp. 562–571. Springer, Cham (2020). https://doi.org/10.1007/978-3-030-59710-8_55

5. He, Y., Carass, A., Zuo, L., Dewey, B.E., Prince, J.L.: Autoencoder based self-supervised test-time adaptation for medical image analysis. Med. Image Anal. **72**, 102136 (2021)

6. Hu, M., et al.: Fully test-time adaptation for image segmentation. In: de Bruijne, M., et al. (eds.) MICCAI 2021. LNCS, vol. 12903, pp. 251–260. Springer, Cham (2021). https://doi.org/10.1007/978-3-030-87199-4_24

7. Karani, N., Chaitanya, K., Baumgartner, C., Konukoglu, E.: A lifelong learning approach to brain MR segmentation across scanners and protocols. In: Frangi, A.F., Schnabel, J.A., Davatzikos, C., Alberola-López, C., Fichtinger, G. (eds.) MICCAI 2018. LNCS, vol. 11070, pp. 476–484. Springer, Cham (2018). https://doi.org/10.1007/978-3-030-00928-1_54

8. Karani, N., Erdil, E., Chaitanya, K., Konukoglu, E.: Test-time adaptable neural networks for robust medical image segmentation. Med. Image Anal. **68**, 101907 (2021)

9. Lee, J., Jung, D., Yim, J., Yoon, S.: Confidence score for source-free unsupervised domain adaptation. In: ICML, pp. 12365–12377. PMLR (2022)

10. Li, X., et al.: A free lunch for unsupervised domain adaptive object detection without source data. In: AAAI, vol. 35, pp. 8474–8481 (2021)

11. Litjens, G., et al.: A survey on deep learning in medical image analysis. Med. Image Anal. **42**, 60–88 (2017)

12. Nado, Z., Padhy, S., Sculley, D., D'Amour, A., Lakshminarayanan, B., Snoek, J.: Evaluating prediction-time batch normalization for robustness under covariate shift. arXiv preprint arXiv:2006.10963 (2020)

13. Niu, S., Wu, J., Zhang, Y., Chen, Y., Zheng, S., Zhao, P., Tan, M.: Efficient test-time model adaptation without forgetting. arXiv preprint arXiv:2204.02610 (2022)

14. Pei, C., Wu, F., Huang, L., Zhuang, X.: Disentangle domain features for cross-modality cardiac image segmentation. Med. Image Anal. **71**, 102078 (2021)

15. Ronneberger, O., Fischer, P., Brox, T.: U-net: convolutional networks for biomedical image segmentation. In: Navab, N., Hornegger, J., Wells, W.M., Frangi, A.F. (eds.) MICCAI 2015. LNCS, vol. 9351, pp. 234–241. Springer, Cham (2015). https://doi.org/10.1007/978-3-319-24574-4_28

16. Singh, R., Rani, R.: Semantic segmentation using deep convolutional neural network: a review. In: ICICC (2020)

17. Sun, Y., Wang, X., Liu, Z., Miller, J., Efros, A., Hardt, M.: Test-time training with self-supervision for generalization under distribution shifts. In: International Conference on Machine Learning, pp. 9229–9248. PMLR (2020)

18. Tomar, D., Vray, G., Thiran, J.P., Bozorgtabar, B.: OptTTA: learnable test-time augmentation for source-free medical image segmentation under domain shift. In: Medical Imaging with Deep Learning (2021)

19. Varsavsky, T., Orbes-Arteaga, M., Sudre, C.H., Graham, M.S., Nachev, P., Cardoso, M.J.: Test-time unsupervised domain adaptation. In: Martel, A.L., et al. (eds.) MICCAI 2020. LNCS, vol. 12261, pp. 428–436. Springer, Cham (2020). https://doi.org/10.1007/978-3-030-59710-8_42

20. Wang, D., Shelhamer, E., Liu, S., Olshausen, B., Darrell, T.: Tent: fully test-time adaptation by entropy minimization. In: ICLR (2021)

21. Wang, G., Li, W., Aertsen, M., Deprest, J., Ourselin, S., Vercauteren, T.: Aleatoric uncertainty estimation with test-time augmentation for medical image segmentation with convolutional neural networks. Neurocomputing **338**, 34–45 (2019)
22. Wu, J., Gu, R., Dong, G., Wang, G., Zhang, S.: FPL-UDA: filtered pseudo label-based unsupervised cross-modality adaptation for vestibular schwannoma segmentation. In: ISBI, pp. 1–5. IEEE (2022)

Unsupervised Adaptation of Polyp Segmentation Models via Coarse-to-Fine Self-Supervision

Jiexiang Wang[1] and Chaoqi Chen[2]([✉])

[1] ByteDance, Beijing, China
wangjiexiang@bytedance.com
[2] The University of Hong Kong, Hong Kong, China
cqchen@gmail.com

Abstract. Unsupervised Domain Adaptation (UDA) has attracted a surge of interest over the past decade but is difficult to be used in real-world applications. Considering the privacy-preservation issues and security concerns, in this work, we study a practical problem of Source-Free Domain Adaptation (SFDA), which eliminates the reliance on annotated source data. Current SFDA methods focus on extracting domain knowledge from the source-trained model but neglects the intrinsic structure of the target domain. Moreover, they typically utilize pseudo labels for self-training in the target domain, but suffer from the notorious error accumulation problem. To address these issues, we propose a new SFDA framework, called Region-to-Pixel Adaptation Network (RPANet), which learns the region-level and pixel-level discriminative representations through coarse-to-fine self-supervision. The proposed RPANet consists of two modules, Foreground-aware Contrastive Learning (FCL) and Confidence-Calibrated Pseudo-Labeling (CCPL), which explicitly address the key challenges of "how to distinguish" and "how to refine". To be specific, FCL introduces a supervised contrastive learning paradigm in the region level to contrast different region centroids across different target images, which efficiently involves all pseudo labels while robust to noisy samples. CCPL designs a novel fusion strategy to reduce the overconfidence problem of pseudo labels by fusing two different target predictions without introducing any additional network modules. Extensive experiments on three cross-domain polyp segmentation tasks reveal that RPANet significantly outperforms state-of-the-art SFDA and UDA methods without access to source data, revealing the potential of SFDA in medical applications.

Keywords: Polyp Segmentation · Source-Free Domain Adaptation · Coarse-to-Fine Self-Supervision

1 Introduction

In clinical applications, deep learning models are typically trained on data collected from a small number of hospitals, but with the objective of being deployed

---

J. Wang and C. Chen—Contributed equally to this work.

© The Author(s), under exclusive license to Springer Nature Switzerland AG 2023
A. Frangi et al. (Eds.): IPMI 2023, LNCS 13939, pp. 250–262, 2023.
https://doi.org/10.1007/978-3-031-34048-2_20

across other hospitals to meet a broader range of needs. However, domain shift [30], such as the variations of patient population and imaging conditions, hinders the deployment of well-trained models to a new hospital. This problem has inspired a body of research on UDA [13,29] by explicitly mitigating the distributional shift between a labeled source domain and an unlabeled target domain [3,5,6,8,12,24,27,28,32,33,36,38]. Despite their general efficacy for various tasks, conventional UDA techniques still have two fundamental shortcomings. First, when encountering a new domain, UDA requires joint training of source and target data, which would be cumbersome if the number of source data is large (e.g., high-quality annotated data from well-known medical institutions). Second, given the privacy-preservation issues and security concerns in medical scenarios, medical data from different clinical centers usually need to be kept locally, i.e., the patient data from a local hospital may not be shared to other hospitals. In addition, in terms of memory overhead, source data mostly have larger size than source-trained models [16], which further imposes a great challenge for real-world applications. This motivates us to investigate a highly realistic and challenging setting called *Source-Free Domain Adaptation* (SFDA) [9,21,23], a.k.a. model adaptation, where only a trained source model and an unlabeled target dataset are available during adaptation.

Compared to mainstream UDA methods, which enable knowledge transfer with concurrent access to both source and target samples, SFDA needs to distill the domain knowledge from the fixed source model and adapt it to the target domain. In this case, it is infeasible to leverage the prevailing feature alignment strategies in UDA, such as adversarial learning [12,34], moment matching [27], and relation-based alignment [7], to achieve adaptation. More importantly, we cannot explicitly measure the domain discrepancy between source and target distributions, making the adaptation process brittle to sophisticated adaptation scenarios with significant distributional shifts. To solve this problem, most of the SFDA methods [1,9,19,21,23,25,26] is comprised of two stages, including source pre-training and target adaptation. For example, Bateson et al. [1] develop an entropy minimization term incorporated with a class-ratio prior for preventing trivial solution. Chen et al. [9] propose a denoised pseudo-labeling strategy to improve the performance of self-training by collaboratively using uncertainty estimation and prototype estimation to select reliable pseudo labels. Liu et al. [25] introduce an adaptive batch-wise normalization statistics adaptation framework, which progressively learns the target domain-specific mean and variance and enforces high-order statistics consistency with an adaptive weighting strategy.

In spite of the fruitful progress and promising results, existing SFDA methods suffer from two key challenges. (1) *How to distinguish:* prior efforts typically learn the discriminative power from the source-trained model but ignores the self-supervised ability within the target domain for distinguishing foregrounds and backgrounds, which is crucial for exploring the intra-domain contextual and semantic structures. (1) *How to refine:* the prevailing pseudo-labeling-based approaches [9,23,26] may be confined by false prediction, error accumulation and even trivial solution as the pseudo-labels are initialized by source-trained model and gradually refined under the absence of source supervision. In a nutshell, how to simultaneously utilize domain knowledge from the source domain and mine meaningful self-supervision signals from the target domain are crucial to the success of SFDA, but remains out-of-reach for current methods.

Remedying these issues, we propose a new SFDA framework, called Region-to-Pixel Adaptation Network (RPANet), which unifies region-level and pixel-level representation learning in a *coarse-to-fine* manner. The basic idea is to progressively endow the target segmentation models with the capability of distinguishing foregrounds and backgrounds via self-supervision. Specifically, the proposed RPANet consists of two key modules, Foreground-aware Contrastive Learning (FCL) and Confidence-Calibrated Pseudo-Labeling (CCPL), which respectively address the challenges of "how to distinguish" and "how to refine". FCL develops a supervised contrastive learning paradigm to learn region-level discriminative representations by contrasting different region centroids (computed by pseudo labels) across different target images, which is memory-efficient and robust to noisy pseudo labels. CCPL designs a novel fusion strategy to reduce the overconfidence problem of pseudo labels by fusing two different target predictions without introducing any cumbersome learning modules. Extensive experiments on three cross-domain polyp segmentation tasks demonstrate that our RPANet significantly outperforms state-of-the-art SFDA and UDA methods under the absence of source data.

2   Region-to-Pixel Adaptation Network

In this paper, we investigate the problem of SFDA, where we only have access to a pre-trained source model f_s and an unlabeled target dataset $\mathcal{D}_t = \{x_t | x_t \in \mathbb{R}^{H \times W \times 3}\}$. f_s was trained on labeled source data $\mathcal{D}_s$. The source and target data do not follow the IID assumption. The objective of SFDA is to learn a segmentation model that performs well on the target domain. Figure 1 illustrates an overview of RPANet, which consists of two components, namely, Foreground-aware Contrastive Learning (FCL) and Confidence-Calibrated Pseudo-Labeling (CCPL). In particular, FCL and CCPL are complementary to each other, *i.e.*, FCL enhances the robustness of CCPL by providing region-level discriminative representations as an initialization, while CCPL refines the target pseudo labels to mitigate the bias introduced by FCL.

2.1   Foreground-Aware Contrastive Learning

Contrastive learning [14,18] has achieved compelling results in self-supervised representation learning by making the representations of the positive pair close while keep negative pairs apart. In view of the dense-prediction property of segmentation tasks, some of the prior works [4,35] utilize contrastive learning as a pre-training step. They typically conduct this learning process in the image level (*i.e.*, the augmentations of original image are regarded as positive samples and

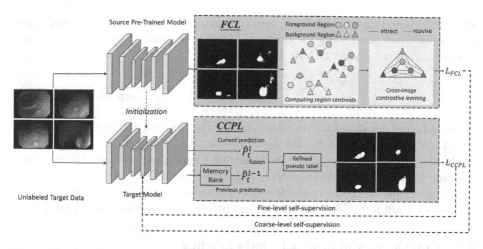

Fig. 1. Overview of our Region-to-Pixel Adaptation Network (RPANet), which includes a pre-trained source model, a target model, and two model adaptation modules (*i.e.,* FCL and CCPL). These two model adaptation modules work in a coarse-to-fine manner to learn discriminative representations in the two levels.

other images from the dataset are regarded as negative samples) while neglecting the holistic context of the entire dataset, *i.e.,* the relations among different real images. The rationale is that these approaches rely on image augmentation techniques to make contrast under the absence of ground-truth annotations. In SFDA, we also have the same dilemma. Instead of following the conventional wisdom, we make use of pseudo labels [20] predicted from the source pre-trained model to perform contrastive learning in a *supervised* way.

Then, an intuitive solution is to utilize pseudo labels to perform pixel-level contrastive learning. However, 1) the pseudo labels may be noisy, and 2) the pixel-level annotations would bring in huge memory overhead for constructing the memory bank. To solve these issues, we propose the FCL module to contrastively learn region-level discriminative representations in a fully supervised way. Here, we use the region centroid to stand for the very region to reduce computational cost. In addition, when computing the centroid, we consider the prediction confidence of different pixels to dynamically assign weights to them.

Formally, the source-trained model f_s takes a target image x_t as input and produces the prediction map $P_t = f_s(x_t)$, and its corresponding feature map is denoted as $F_t \in \mathbb{R}^{H \times W \times K}$. The entropy map I_t is defined as,

$$I_t^{(h,w)} = H(P_t), \tag{1}$$

where $H(\cdot)$ is the entropy function. Then, the region centroid of class k is formulated as,

$$m_k = \frac{\sum_{h,w} F_t \mathbb{1}[P_t^{(h,w)} = k] \cdot (1 - I_t^{(h,w)})}{\sum_{h,w} \mathbb{1}[P_t^{(h,w)} = k]} \tag{2}$$

where $\mathbb{1}(\cdot)$ is an indicator function, and each pixel is re-weighted by its prediction confidence when computing the centroid. After that, the region-level contrastive loss regarding category k can be formulated as,

$$\mathcal{L}_k^{\text{FCL}} = -\frac{1}{|\mathcal{M}_k|} \sum_{m^+ \in M_k} \log \frac{\exp(m \cdot m^+ / \tau)}{\exp(m \cdot m^+ / \tau) + \sum_{m^- \in \mathcal{N}_k} \exp(m \cdot m^- / \tau)} \quad (3)$$

where $\mathcal{M}_k$ and $\mathcal{N}_k$ stand for the collections of the positive and negative samples, for m. m^+ and m^- stand for the certain positive and negative samples, respectively. As its core, FCL leverage the "learning to compare" ability of contrastive learning to make a clear distinction between foreground and background regions as well as model the cross-image foreground/background relations. By doing so, regions that are more likely to be the foregrounds are coarsely highlighted.

2.2 Confidence-Calibrated Pseudo-labeling

Pseudo-labeling [20] has been proved to be a simple yet effective approach for SFDA [9,23,26]. Most of state-of-the-art pseudo-labeling methods prioritize their focus on designing sampling strategies to select pseudo labels with high prediction confidence and fit task-specific properties. However, they ignore the structured dependencies of different pixels inside a single image, which is prone to result in overconfident predictions regarding some local regions. More importantly, in some medical applications, where the foreground and background are highly entangled, these methods may be prohibitively difficult to select trustworthy pseudo labels. For example, in colonoscopy images, polyps and normal tissues are visually similar and have low contrast, greatly impeding the correct assignment of pseudo labels. To address this problem, the proposed CCPL calibrates the outputs soft pseudo labels in the pixel level, which integrates the output of previous and current predictions via a simple yet effective fusion strategy.

Given the output of the target segmentation network $\hat{P}_t = f_t(x_t)$ (pre-softmax logits), the target pseudo-label is updated as follows,

$$\hat{P}_t^l = \alpha \cdot \text{softmax}\left(\frac{\hat{P}_t^l}{\phi(\hat{P}_t^l, \hat{P}_t^{l-1})}\right) + (1 - \alpha) \cdot \text{softmax}\left(\frac{\hat{P}_t^{l-1}}{\phi(\hat{P}_t^l, \hat{P}_t^{l-1})}\right) \quad (4)$$

where $\hat{P}_t^{l-1}$ is the pre-softmax logits of $(l-1)$-th times, l denotes the iteration times, and α (ranged from 0 to 1) is a modulated factor. In practice, we set $\alpha = 0.5$ in all experiments. The function $\phi(a, b) = \sqrt{\sum_i^{|a|} (a_i^2 + b_i^2)}$ is a normalization term, which could smooth the pseudo label probability distribution to avoid overconfident predictions (dominate pixels).

After obtaining reliable pseudo labels, the segmentation network is optimized in a supervised way by minimizing the cross entropy loss,

$$\mathcal{L}_{\text{CCPL}} = -\sum_{h,w} \sum_c \hat{Y}_t^{(h,w,c)} \log(P_t^{(h,w,c)}) \quad (5)$$

where $\hat{Y}_t^{(h,w,c)}$ is the soft pseudo labels.

2.3   Objective Function

First of all, we utilize the weights of source pre-trained model f_s for the initialization of target model. The overall training objective of RPANet, which includes FCL and CCPL, can be formulated as follow,

$$\mathcal{L}_{\text{RPANet}} = \beta \sum_k \mathcal{L}_k^{\text{FCL}} + \gamma \mathcal{L}_{\text{CCPL}} \tag{6}$$

where β and γ are two trade-off parameters.

3   Experiments

3.1   Dataset

We extensively evaluate the proposed method RPANet on the polyp segmentation tasks with three public datasets and an in-house dataset. (1) ClinicDB [2] contains 612 Standard Definition (SD) frames from 31 sequences. (2) ETIS-LARIB [31] contains 196 High Definition (HD) frames from 34 sequences. (3) Kvasir-SEG [17] contains 1000 polyp frames with various resolutions. (4) In-house dataset is collected from a local hospital and contains 5175 frames with polyp. The annotations are sketched by two experienced gastroenterologists. In experiments, considering the number of annotated data in different datasets, we use in-house dataset as the source domain, and the other three public datasets as the target domain respectively.

3.2   Implementation Details and Evaluation Metrics

We adopt DeepLab-v2 [10] as the segmentation model and ResNet101 [15] pre-trained on ImageNet as the backbone network. In the training phase, we use Stochastic Gradient Descent (SGD) with momentum 0.9 as the optimizer. The learning rate is set as 2.5×10^{-4}, which is decayed by a polynomial annealing policy [10]. The batch size is 4 and we train the models for 20 epochs. Following prior works, standard data augmentation techniques are utilized in experiments. For Eq. (3), we set the temperature τ as 0.1. For Eq. (6), we set $\beta = 1$ and $\gamma = 1$ in all experiments. All experiments are conducted on 4 NVIDIA Tesla V100 GPUs with PyTorch deep learning framework. Following [11], we use 6 metrics to quantitatively evaluate the superiority of our method, including mean Dice, mean IoU, weighted Dice metric F_β^w, structure similarity measure S_α, enhanced-alignment metric E_ϕ^{max}, and MAE metric.

3.3   Comparisons with State-of-the-Arts

State-of-the-Arts. We extensively compare the proposed RPANet with state-of-the-art domain adaptive semantic segmentation methods, including Bidirectional Learning (**BDL**) [22], Fourier Domain Adaptation (**FDA**) [37], Historical

Table 1. Quantitative results of different adaptation methods on ClinicDB [2], ETIS-LARIB [31], and Kvasir-SEG [17] datasets.

	Methods	mean Dice	mean IoU	F_β^w	S_α	E_ϕ^{max}	MAE
ClinicDB	w/o adaptation	67.3	55.5	63.9	76.7	82.8	0.047
	BDL [22]	76.8	67.3	73.8	83.3	87.1	0.037
	FDA [37]	78.6	71.1	78.7	85.6	87.6	0.027
	HCL [16]	74.1	64.5	74.5	82.4	85.2	0.034
	DPL [9]	76.3	66.9	75.3	83.7	87.0	0.034
	RPANet w/o FCL	75.7	66.1	74.9	83.2	86.7	0.035
	RPANet w/o CCPL	73.2	62.8	70.6	80.7	86.4	0.041
	RPANet	**80.0**	**71.9**	**80.2**	**86.4**	**89.2**	**0.029**
ETIS-LARIB	w/o adaptation	49.6	41.4	47.2	70.9	70.6	0.026
	BDL [22]	59.3	51.8	57.1	76.3	74.8	0.019
	FDA [37]	61.5	53.5	59.7	77.3	77.0	0.017
	HCL [16]	57.3	49.5	55.4	75.1	73.8	0.024
	DPL [9]	58.8	51.4	56.6	76.1	74.5	0.020
	RPANet w/o FCL	59.1	51.6	56.9	76.3	74.6	0.019
	RPANet w/o CCPL	59.9	52.2	58.1	76.7	75.5	0.017
	RPANet	**63.2**	**55.2**	**61.3**	**78.3**	**77.9**	**0.016**
Kvasir-SEG	w/o adaptation	69.0	57.5	62.4	74.2	79.6	0.109
	BDL [22]	84.2	76.0	82.3	86.7	91.1	0.044
	FDA [37]	84.6	76.5	83.0	87.0	91.4	0.042
	HCL [16]	84.0	75.8	82.4	86.6	91.0	0.043
	DPL [9]	80.4	71.0	79.0	84.1	88.5	0.050
	RPANet w/o FCL	80.5	71.7	76.6	83.5	88.0	0.062
	RPANet w/o CCPL	79.2	69.7	75.3	82.5	87.4	0.063
	RPANet	**85.8**	**78.2**	**84.4**	**87.9**	**92.1**	**0.039**

Contrastive Learning (**HCL**) [16], and Denoised Pseudo-Labeling (**DPL**) [9]. In our polyp segmentation experiments, we use the provided source codes to implement these baseline methods. Note that BDL and FDA are conventional UDA methods, while HCL and DPL are SFDA methods. "W/o adaptation" denotes that the model is trained on the source data and directly applied to the target data without any adaptation.

Table 1 reports the performance comparisons of different adaptation methods on three target domains. The proposed RPANet consistently outperforms all compared methods on all domain adaptation tasks. RPANet significantly promotes the w/o adaptation result from 62.0% to 76.3% on average in terms of mean Dice, which brings about 14.3% improvement. In particular, the performance of HCL and DPL cannot be on par with the conventional UDA methods

Table 2. Further empirical analysis regarding the different design choices of RPANet. Mean Dice is reported on the target domains.

Method	ClinicDB	ETIS-LARIB	Kvasir-SEG
RPANet	80.0	63.2	85.8
Integration with modern SFDA methods			
FCL + HCL [16]	79.3	59.1	85.6
FCL + DPL [9]	78.5	60.1	84.1
CCPL + HCL [16]	78.6	58.8	84.9
CCPL + DPL [9]	77.4	60.3	83.6
Ablation of FCL and CCPL			
FCL w/o entropy regularization	78.9	62.6	85.2
CCPL w/o confidence calibration	78.3	61.8	84.1

(BDL and FDA), revealing the importance and difficulty of adapting polyps segmentation models without access to source data. By contrast, RPANet is almost systematically better than the state-of-the-art UDA methods, highlighting the contributions of our coarse-to-fine self-supervision scheme. To qualitatively illustrate the superiority of our method, we show the examples of polyps segmentation results in Fig. 2 and Fig. 3. We can observe that our RPANet is capable of precisely segmenting polyps, reducing ambiguous predictions, and refining the boundaries between lesions and normal tissues. The justification is that RPANet structurally regularizes the outputs via FCL and CCPL, thereby solving the problem of incomplete and noisy predictions.

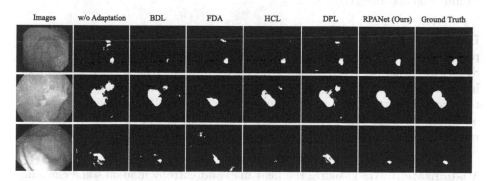

Fig. 2. Qualitative evaluation of different adaptation methods on target domains. From top to bottom: ClinicDB, ETIS-LARIB, Kvasir-SEG.

3.4   Further Empirical Analysis

Ablation Study. To investigate the individual effect of each component (*i.e.*, FCL and CCPL) in our proposed RPANet, we conduct ablation experiments on

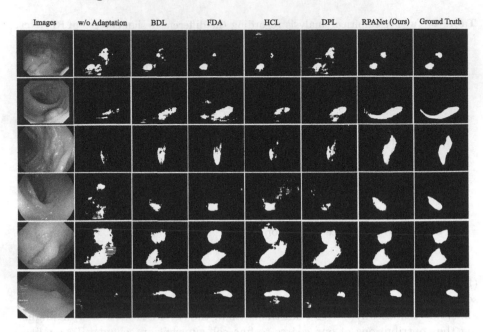

Fig. 3. Qualitative evaluation of different adaptation methods on target domains. From top to bottom: ClinicDB, ETIS-LARIB, Kvasir-SEG.

three domain adaptive polyp segmentation benchmarks. The results are also presented in Table 1. From the table, we can observe that both FCL and CCPL are reasonably designed, and when one of them are removed, the final performance would drop accordingly.

Design Choice. We further explore the alternative design choices for each component to better understand the essence of RPANet. (1) Integration with modern SFDA methods. (2) Ablation of FCL and CCPL. The results are demonstrated in Table 2. We can see that both FCL and CCPL substantially improve state-of-the-art SFDA methods by providing more accurate self-supervision signals. In addition, when the elaborately devised regularization and calibration terms are removed, the performance will be affected to some extent.

Visualization. We visualize the heat map and entropy map on some challenging target images (*i.e.,* the boundaries between the polyp and its surrounding tissues are very ambiguous) as training proceeds in Fig. 4. From the figure, we found that the shapes and boundaries are progressively refined as the training epoch increases, which clearly demonstrates the effectiveness of our coarse-to-fine refinement paradigms. In particular, we can observe that our RPANet achieves competitive results even with very limited training times, such as 8 epochs.

Parameter Sensitivity. We provide the sensitivity analysis regarding hyper-parameters β and γ in Fig. 5. As can be seen, the performance of our RPANet is insensitive to the variations of the value of β and γ on different benchmark datasets, revealing the robustness of the proposed modules. For example, changing β in the range [0.5, 1.5] only incurs small performance variations *i.e.*, 1.6%, 1.1%, and 1.6% in ClinicDB, ETIS-LARIB, and Kvasir-SEG respectively.

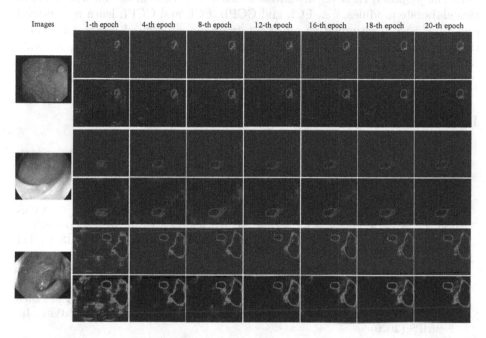

Fig. 4. Visualization results on target images as training proceeds. From top to bottom: ClinicDB, ETIS-LARIB, Kvasir-SEG. In every two rows, **upper:** heat map, **lower:** entropy map. Noting that the brighter the color is, the larger the corresponding value is. (Color figure online)

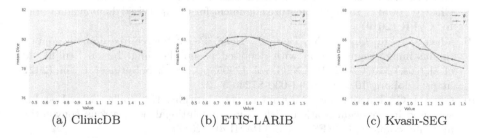

(a) ClinicDB   (b) ETIS-LARIB   (c) Kvasir-SEG

Fig. 5. Sensitivity analysis regarding hyper-parameters β and γ.

4   Conclusion

In this paper, we proposed the RPANet to solve the unsupervised adaptation of polyp segmentation models in the absence of source data. The key idea of our method is to endow the target models with the capability of distinguishing foregrounds and backgrounds via self-supervision in a coarse-to-fine manner. The proposed RPANet instantiates this objective with the incorporation of two elaborate modules, *i.e.*, FCL and CCPL. FCL and CCPL learn region-level and pixel-level discriminative representations via supervised contrastive learning and confidence-calibrated pseudo-label refinement, respectively. Experiments on three cross-domain poly segmentation tasks verified the effectiveness of our method, revealing the possibility of SFDA for real-world medical applications.

References

1. Bateson, M., Kervadec, H., Dolz, J., Lombaert, H., Ben Ayed, I.: Source-relaxed domain adaptation for image segmentation. In: Martel, A.L., et al. (eds.) MICCAI 2020. LNCS, vol. 12261, pp. 490–499. Springer, Cham (2020). https://doi.org/10.1007/978-3-030-59710-8_48
2. Bernal, J., Sánchez, F.J., Fernández-Esparrach, G., Gil, D., Rodríguez, C., Vilariño, F.: WM-DOVA maps for accurate polyp highlighting in colonoscopy: validation vs. saliency maps from physicians. Comput. Med. Imaging Graph. **43**, 99–111 (2015)
3. Bian, C., et al.: Uncertainty-aware domain alignment for anatomical structure segmentation. Med. Image Anal. **64**, 101732 (2020)
4. Chaitanya, K., Erdil, E., Karani, N., Konukoglu, E.: Contrastive learning of global and local features for medical image segmentation with limited annotations. In: NeurIPS (2020)
5. Chen, C., Li, J., Han, X., Liu, X., Yu, Y.: Compound domain generalization via meta-knowledge encoding. In: CVPR (2022)
6. Chen, C., et al.: Relation matters: foreground-aware graph-based relational reasoning for domain adaptive object detection. IEEE TPAMI (2022)
7. Chen, C., Wang, J., Pan, J., Bian, C., Zhang, Z.: GraphSKT: graph-guided structured knowledge transfer for domain adaptive lesion detection. IEEE TMI (2022)
8. Chen, C., et al.: Progressive feature alignment for unsupervised domain adaptation. In: CVPR (2019)
9. Chen, C., Liu, Q., Jin, Y., Dou, Q., Heng, P.-A.: Source-free domain adaptive fundus image segmentation with denoised pseudo-labeling. In: de Bruijne, M., et al. (eds.) MICCAI 2021. LNCS, vol. 12905, pp. 225–235. Springer, Cham (2021). https://doi.org/10.1007/978-3-030-87240-3_22
10. Chen, L.C., Papandreou, G., Kokkinos, I., Murphy, K., Yuille, A.L.: DeepLab: semantic image segmentation with deep convolutional nets, atrous convolution, and fully connected CRFS. IEEE TPAMI **40** (2017)
11. Fan, D.-P., et al.: PraNet: parallel reverse attention network for polyp segmentation. In: Martel, A.L., et al. (eds.) MICCAI 2020. LNCS, vol. 12266, pp. 263–273. Springer, Cham (2020). https://doi.org/10.1007/978-3-030-59725-2_26
12. Ganin, Y., et al.: Domain-adversarial training of neural networks. JMLR **17**(1), 1–35 (2016)

13. Guan, H., Liu, M.: Domain adaptation for medical image analysis: a survey. IEEE. Trans. Biomed. Eng. (2021)
14. He, K., Fan, H., Wu, Y., Xie, S., Girshick, R.: Momentum contrast for unsupervised visual representation learning. In: CVPR (2020)
15. He, K., Zhang, X., Ren, S., Sun, J.: Deep residual learning for image recognition. In: CVPR (2016)
16. Huang, J., Guan, D., Xiao, A., Lu, S.: Model adaptation: historical contrastive learning for unsupervised domain adaptation without source data. In: NeurIPS (2021)
17. Jha, D., et al.: Kvasir-SEG: a segmented polyp dataset. In: Ro, Y.M., et al. (eds.) MMM 2020. LNCS, vol. 11962, pp. 451–462. Springer, Cham (2020). https://doi.org/10.1007/978-3-030-37734-2_37
18. Khosla, P., et al.: Supervised contrastive learning. In: NeurIPS (2020)
19. Kundu, J.N., Venkat, N., Babu, R.V., et al.: Universal source-free domain adaptation. In: CVPR (2020)
20. Lee, D.H., et al.: Pseudo-label: the simple and efficient semi-supervised learning method for deep neural networks. In: Workshop on Challenges in Representation Learning. In: ICML, vol. 3, p. 896 (2013)
21. Li, R., Jiao, Q., Cao, W., Wong, H.S., Wu, S.: Model adaptation: unsupervised domain adaptation without source data. In: CVPR (2020)
22. Li, Y., Yuan, L., Vasconcelos, N.: Bidirectional learning for domain adaptation of semantic segmentation. In: CVPR, pp. 6936–6945 (2019)
23. Liang, J., Hu, D., Feng, J.: Do we really need to access the source data? Source hypothesis transfer for unsupervised domain adaptation. In: ICML (2020)
24. Liu, J., Guo, X., Yuan, Y.: Prototypical interaction graph for unsupervised domain adaptation in surgical instrument segmentation. In: de Bruijne, M., et al. (eds.) MICCAI 2021. LNCS, vol. 12903, pp. 272–281. Springer, Cham (2021). https://doi.org/10.1007/978-3-030-87199-4_26
25. Liu, X., Xing, F., Yang, C., El Fakhri, G., Woo, J.: Adapting off-the-shelf source segmenter for target medical image segmentation. In: de Bruijne, M., et al. (eds.) MICCAI 2021. LNCS, vol. 12902, pp. 549–559. Springer, Cham (2021). https://doi.org/10.1007/978-3-030-87196-3_51
26. Liu, Y., Zhang, W., Wang, J.: Source-free domain adaptation for semantic segmentation. In: CVPR (2021)
27. Long, M., Cao, Y., Wang, J., Jordan, M.: Learning transferable features with deep adaptation networks. In: ICML, pp. 97–105 (2015)
28. Ouyang, C., Kamnitsas, K., Biffi, C., Duan, J., Rueckert, D.: Data efficient unsupervised domain adaptation for cross-modality image segmentation. In: Shen, D., et al. (eds.) MICCAI 2019. LNCS, vol. 11765, pp. 669–677. Springer, Cham (2019). https://doi.org/10.1007/978-3-030-32245-8_74
29. Pan, S.J., Yang, Q.: A survey on transfer learning. IEEE TKDE **22** (2009)
30. Quiñonero-Candela, J., Sugiyama, M., Schwaighofer, A., Lawrence, N.D.: Dataset Shift in Machine Learning. MIT Press, Cambridge (2008)
31. Silva, J., Histace, A., Romain, O., Dray, X., Granado, B.: Toward embedded detection of polyps in WCE images for early diagnosis of colorectal cancer. Int. J. Comput. Assist. Radiol. Surg. **9**(2), 283–293 (2014)
32. Sun, L., Wang, J., Huang, Y., Ding, X., Greenspan, H., Paisley, J.: An adversarial learning approach to medical image synthesis for lesion detection. IEEE JBHI **24** (2020)
33. Tzeng, E., Hoffman, J., Saenko, K., Darrell, T.: Adversarial discriminative domain adaptation. In: CVPR (2017)

34. Wang, J., Huang, H., Chen, C., Ma, W., Huang, Y., Ding, X.: Multi-sequence cardiac MR segmentation with adversarial domain adaptation network. In: Pop, M., et al. (eds.) STACOM 2019. LNCS, vol. 12009, pp. 254–262. Springer, Cham (2020). https://doi.org/10.1007/978-3-030-39074-7_27
35. Wang, X., Zhang, R., Shen, C., Kong, T., Li, L.: Dense contrastive learning for self-supervised visual pre-training. In: CVPR (2021)
36. Xia, Y., et al.: Uncertainty-aware multi-view co-training for semi-supervised medical image segmentation and domain adaptation. Med. Image Anal. **65**, 101766 (2020)
37. Yang, Y., Soatto, S.: FDA: Fourier domain adaptation for semantic segmentation. In: CVPR (2020)
38. Yu, S., et al.: Cross-domain depth estimation network for 3D vessel reconstruction in OCT angiography. In: de Bruijne, M., et al. (eds.) MICCAI 2021. LNCS, vol. 12908, pp. 13–23. Springer, Cham (2021). https://doi.org/10.1007/978-3-030-87237-3_2

Geometric Deep Learning

Edge-Based Graph Neural Networks for Cell-Graph Modeling and Prediction

Tai Hasegawa[1], Helena Arvidsson[2], Nikolce Tudzarovski[2],
Karl Meinke[1(✉)], Rachael V. Sugars[2], and Aravind Ashok Nair[1]

[1] KTH Royal Institute of Technology, Stockholm, Sweden
hasegawa.t.as@m.titech.ac.jp, {karlm,aanair}@kth.se
[2] Karolinska Institutet, Stockholm, Sweden
{helena.arvidsson,nikolce.tudzarovski,rachael.sugars}@ki.se

Abstract. Identification and classification of cell-graph features using graph-neural networks (GNNs) has been shown to be useful in digital pathology. In this work, we consider the role of edge labels in cell-graph modeling, including histological modeling techniques, edge aggregation in GNN architectures, and edge label prediction. We propose EAGNN (Edge Aggregated GNN), a new GNN model that aggregates both node and edge label information to take advantage of topological information about cellular data and facilitate edge label prediction. We introduce new edge label features that improve histological modeling and prediction. We evaluate our EAGNN model for the task of detecting the presence and location of the basement membrane in oral mucosal tissue, as a proof-of-concept application.

Keywords: Digital Pathology · Graph Neural Network · Cell-Graph · Basement Membrane · Oral Mucosa

1 Introduction

To capture intra-cellular and high-level histological relationships in whole-slide images (WSI) of tissue samples, cell-graph models have been considered [13]. In a cell-graph, properties of cells and interactions between them are represented by labelled nodes and edges. Graph neural networks (GNNs) are a specific class of machine learning (ML) algorithm which have been applied to cell-graph models to locate and classify complex histological features [4,16,20]. In this work, we consider the role of edge labels in cell-graph modeling and GNN-based model analysis including: (i) histological modeling using edge labels, (ii) edge aggregation in GNNs, and (iii) edge label classification algorithms.

Our study will mainly focus on new GNN algorithms for aggregating node and edge data, with a view to making edge label predictions in a cell-graph model. We also propose new types of edge features, going beyond the simple geometric distance label found in the literature. These new edge labels are shown to be effective for improved histological analysis by means of ML ablation studies.

We evaluate the new GNN models and edge features on a representative digital pathology task of predicting the presence and location of the basement membrane (BM) in hematoxylin and eosin (H&E) stained oral mucosa samples. Structural properties of the BM play an important role in classifying oral diseases such as chronic graft-versus-host disease (oral cGvHD) [23].

1.1   Contributions of This Work

The main contributions of this work are as follows:

1. We propose EAGNN as a novel message passing model[1] for predicting cell-graph properties. EAGNN aggregates both node and edge label data to yield edge label predictions.
2. We propose two new types of edge classification algorithm, which can be used on the backend of EAGNN to make edge label predictions.
3. We propose three new edge label features for cell-graph analyses: (a) cell density difference, (b) cell entropy difference, and (c) neighborhood overlap similarity.
4. We evaluate different combinations of EAGNN with edge classifiers and edge label features for the prediction of BM location in oral mucosa images. We show that EAGNN can significantly outperform simple node-based aggregation across a wide variety of performance measures.

The organisation of this work is as follows. In Sect. 2 we consider background and related research on GNNs for learning cell-graph models. In Sect. 3 we discuss message passing GNNs and graph data aggregation. In Sect. 4 we define the EAGNN model with edge label aggregation and edge classification algorithms. In Sect. 5 we evaluate EAGNN on the task of predicting BM integrity in healthy and diseased oral tissue images and compare it with simple node-based aggregation. In Sect. 6 we discuss some limitations of this study and address the possible directions to overcome them. Finally, in Sect. 7 we present some conclusions.

2   Background and Related Work

Mapping the structural features of tissues is key to understanding and making a diagnosis of disease severity in digital pathology, for example the BM. An accurate estimation of the BM location and integrity is an important feature in health and disease, but a challenging task. Several studies have considered this problem [6,25,26]. However, most of these methods depend upon pixel level information that fails to capture histological and topological relationships between the BM and various cells locally present in the tissue. In this study, we used cell-graphs and graph-based ML techniques to map oral mucosal tissue and locate the BM in healthy and diseased situations.

---

[1] Official PyTorch implementation of the EAGNN algorithm is publicly available at https://github.com/aravi11/EAGNN.

Generally in the literature on digital pathology, node labels alone have been used for cell-graph modeling and GNN training [1]. To our knowledge, [17] is the only work that has considered the use of cellular interactions in a cell graph for identifying the BM. However the GNN model used in [17] does not utilize the topological information encoded in graph edges during the learning process. Other GNN case studies also utilize edge features such as distance [2,5,21,22] or edge weights [9] for digital pathology. However, in all these approaches the edge features are simple one-dimensional real-valued features. The use of multidimensional edge features for optimal cell-graph representation and prediction is currently heavily under-explored.

3   Graph Neural Networks for Cell-Graph Learning

3.1   Cell-Graphs as Labelled Graph Structures

A cell-graph is a mathematical model of histological tissue features that can represent nuclei and the interactions between nuclei. This model is motivated by the hypothesis that cells in a tissue organize to perform a specific function [27].

Formally, a cell-graph: $G = \left(\mathcal{V}, \ \mathcal{E}, \ l_{\mathcal{V}} : \mathcal{V} \to \mathbb{R}^D, \ l_{\mathcal{E}} : \mathcal{E} \to \mathbb{R}^P \right)$, is a labelled undirected graph where $\mathcal{V} = \{v_1, \ldots, v_n\}$ is a finite set of nodes, and n is the size of the graph. Furthermore, $\mathcal{E} = \{e_1, \ldots, e_m\}$ is a finite set of edges, and each edge $e \in \mathcal{E}$ is an unordered pair of nodes $e = \{v_i, v_j\}$. A node v_j is termed an immediate neighbour of v_i if there exists an edge $\{v_j, v_i\} \in \mathcal{E}$. We let $\mathcal{N}(v)$ denote the set of all immediate neighbours of v. The degree of a node v is the size of its immediate neighbour set, $deg(v) = |\mathcal{N}(v)|$. The functions $l_{\mathcal{V}}$ and $l_{\mathcal{E}}$ are node and edge labelling functions of D and P features respectively.

To apply methods of linear algebra to graph learning problems, an edge set $\mathcal{E}$ can be encoded by an adjacency matrix $\mathbf{A} \in \mathbb{R}^{n \times n}$ of real values, where n is the graph size. The matrix values $\mathbf{A}_{uv}$ and $\mathbf{A}_{vu}$ are both set to 1.0 if there exists an edge between node u and node v in $\mathcal{E}$, otherwise both are set to 0.0. Moreover, the node labelling $l_{\mathcal{V}}$ can be encoded as a node feature matrix $\mathbf{X} \in \mathbb{R}^{n \times D}$ in which case the matrix row $\mathbf{X}_v \in \mathbb{R}^D$ represents the feature vector for node v. Similarly, the edge labelling $l_{\mathcal{E}}$ can be encoded as an edge feature tensor $\mathbf{E} \in \mathbb{R}^{n \times n \times P}$. For each pair of connected nodes $\{u, v\} \in E$ the entry $\mathbf{E}_{uv} \in \mathbb{R}^P$ represents the P-dimensional feature vector of the edge between node u and node v. As a simplifying notation, for any edge feature $1 \le p \le P$ the matrix $\mathbf{E}_p \in \mathbb{R}^{n \times n}$ denotes the projection of $\mathbf{E}$ onto the single edge feature p.

3.2   Graph Neural Networks

A general spatial-based GNN has a layered architecture, where at each layer k, a low-dimensional (d_k-dimensional) representation $\mathbf{h}_u^k \in \mathbb{R}^{d_k}$ of the graph structure around node u is computed. Computation at each layer k normally consists of two stages. Firstly, for each node u, an $AGGREGATE^k$ operation produces an integrated representation $\mathbf{h}^k_{\mathcal{N}(u)}$ of all immediate neighbors $v \in \mathcal{N}(u)$ of u using the representations $\mathbf{h}^{k-1}_v$ from layer $k - 1$ and is represented as,

$$\mathbf{h}^k_{\mathcal{N}(u)} = AGGREGATE^k \left(\{ \mathbf{h}^{k-1}_v, \forall v \in \mathcal{N}(u) \} \right). \tag{1}$$

Secondly, for each node u, a $COMBINE^k$ operation updates the representation $\mathbf{h}^k_u$ of u by combining its previous representation $\mathbf{h}^{k-1}_u$ on layer $k - 1$ with the aggregated representation $\mathbf{h}^k_{\mathcal{N}(u)}$ of all its immediate neighbours $\mathcal{N}(u)$, using a nonlinear function, $\mathbf{h}^k_u = COMBINE^k \left(\mathbf{h}^{k-1}_u, \mathbf{h}^k_{\mathcal{N}(u)} \right)$.

This iterative computation over layers $0 \leq k \leq K$ is initialized by setting $\mathbf{h}^0_u = \mathbf{X}_u$. Spatial variants of GNNs [3,10,18] implement aggregation by matrix multiplication as:

$$\mathbf{H}^k_{agg} = \mathbf{A}\mathbf{H}^{k-1}\mathbf{W}^k_0 \tag{2}$$

where $\mathbf{H}^k_{agg} \in \mathbb{R}^{n \times d_k}$ is the tensor matrix (i.e. stack) of all aggregations $\mathbf{h}^k_{\mathcal{N}(u)}$, $\mathbf{A} \in \mathbb{R}^{n \times n}$ is the adjacency matrix, $\mathbf{H}^{k-1} \in \mathbb{R}^{n \times d_{k-1}}$ is the tensor matrix of representations $\mathbf{h}^{k-1}_v$ on the $k - 1$-th layer, and $\mathbf{W}^k_0 \in \mathbb{R}^{d_{k-1} \times d_k}$ is a matrix of learnable parameters. The combine operation is formulated as:

$$\mathbf{H}^k = \sigma \left(\mathbf{H}^k_{agg} + \mathbf{H}^{k-1}\mathbf{W}^k_1 \right) \tag{3}$$

where $\mathbf{W}^k_1 \in \mathbb{R}^{d_{k-1} \times d_k}$ is a second matrix of learnable parameters, and σ is a nonlinear function applied pointwise, such as ReLU [11]. Finally, after K layers, a low-dimensional node embedding $\mathbf{Z} \in \mathbb{R}^{n \times d_K}$ is obtained as a tensor matrix, $\mathbf{Z} = \mathbf{H}^K$. A widely used spatial GNN using node aggregation alone is GraphSAGE [14].

To solve edge classification problems, we need a low-dimensional edge embedding $\mathbf{Z}_e \in \mathbb{R}^k$ of each edge e. The simplest approach to embed an edge e is to combine the final embeddings of both of its nodes.

4   A GNN Model for Node and Edge Aggregation

Our proposed GNN architecture is depicted in Fig. 1, and can be divided into two stages, node embedding layers and an edge classifier. The node embedding layers derive latent node representation from a cell-graph.

4.1   Node Embedding Layers

Inspired by EGNN(C) [12], we propose an EAGNN layer which incorporates multiple edge features to embed nodes. The major difference between the proposed EAGNN layer and EGNN(C) is the way that edge features are normalised before aggregation. This feature normalization method is explained in detail in Sect. 5.1.

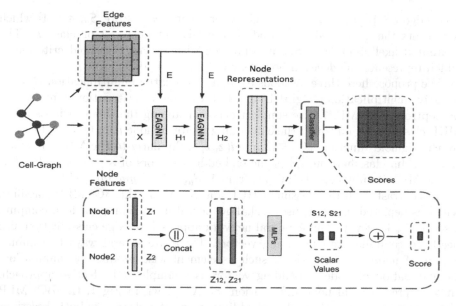

Fig. 1. The overview of the proposed GNN architecture which consists of two EAGNN layers and a classifier. The model extracts node and edge features from a cell-graph, and outputs a score to classify each edge as a BM crossing one or not.

Following the matrix multiplication as in Eq. 2, we formulate the aggregation operation of the proposed model at layer k, named $EAgg^k$, as follows:

$$EAgg^k(\mathbf{E}_p, \mathbf{H}^{k-1}) = \mathbf{E}_p \mathbf{H}^{k-1} \mathbf{W}_0 \tag{4}$$

Then, we combine the previous node representation using the combine operation formulated in Sect. 3.2. We perform these aggregation and combining operations for each edge feature, and concatenate them. Therefore, the formula for the kth EAGNN layer is given by:

$$\mathbf{H}^k = \sigma \left[\|_{p=1}^{P} \left(\mathbf{E}_p \mathbf{H}^{k-1} \mathbf{W}_0^{k-1} + \mathbf{H}^{k-1} \mathbf{W}_1^{k-1} \right) \right] \tag{5}$$

where $\|$ denotes the concatenation operator. As a non-linear function σ we employ the ELU function [7]. Note that this non-linear function is not used in the final layer K of the node embedding layers. As described in Fig. 1, we used two EAGNN layers in our case study of BM identification: our evaluation suggests that two layers are sufficient for this task. After two embedding layers, the node representation $\mathbf{z}_u$ is given by $\mathbf{z}_u = \mathbf{H}_u^2$ for node u.

4.2   Edge Classifiers

Once the node embedding layers have computed the node representations, an edge classifier can partition all edges into two non-overlapping classes A, B. For each edge $e = \{u, v\} \in \mathcal{E}$, the edge classifier computes a score $S_{uv} \in [0, 1]$ which represents the estimated likelihood that e falls into class A or class B. The estimated likelihood score S_{uv} is compared with a class threshold criterion θ, which represents the decision boundary between A and B.

We propose here three methods of edge classification: multiplication (MUL), negative multiplication (NegMUL) and bidirectional concatenation + multilayer perceptron (BC+MLP). The simplest classifier among the three variants is the MUL classifier. It multiplies the embedding vector of node u and node v followed by the sigmoid function: $S_{uv} = sigmoid(\mathbf{z}_u \mathbf{z}_v^T)$. Similarly to the MUL classifier, the NegMul classifier multiplies the embedding vectors of a pair of nodes, but unlike MUL it subtracts the output from 1, $S_{uv} = 1 - sigmoid(\mathbf{z}_u \mathbf{z}_v^T)$.

In contrast with the classifiers MUL and NegMUL, the BC+MLP classifier, which is depicted in Fig. 1, uses a shallow neural network approach to compute a score S_{uv} for each edge. A neural network approach to edge classification also needs to combine the embedding vectors. There are several ways to combine a pair of node representations, such as element-wise product or summation. Concatenation of node embedding vectors is a simple and effective approach, since it preserves all node information [8]. As shown in Fig. 1, the BC+MLP classifier concatenates the node embeddings $\mathbf{z}_u, \mathbf{z}_v$ of nodes u, v in both directions $(\mathbf{z}_{uv} = \mathbf{z}_u \| \mathbf{z}_v, \ \mathbf{z}_{vu} = \mathbf{z}_v \| \mathbf{z}_u)$ to obtain a final score in an orientation-invariant way. The node embedding concatenations $\mathbf{z}_{uv}, \mathbf{z}_{vu}$ are sequentially fed into an MLP to analyse the relationship between the nodes u, v. The scores generated from the MLP layers are combined and passed through the sigmoid activation function to obtain the final score $S_{uv}{}^2$ bounded in $[0, 1]$. A class threshold criterion θ can be chosen as a parameter, and compared with S_{uv} to determine class membership.

5   Evaluation of the EAGNN Model

In this section we compare the performance of EAGNN against the widely used GraphSage GNN which is based on node aggregation alone. The evaluation task is prediction of the BM location and integrity in cell-graph models of oral mucosa samples.

5.1   An Oral Mucosa Cell-Graph Dataset

For supervised training and evaluation of the EAGNN model, a dataset of ground truth cell-graphs was compiled from digitized images of H&E stained oral tissue samples. To compile the dataset, we extracted 42 tiles from WSI of oral mucosal biopsies from nine patients receiving haematopoetic cell transplantation [23]. On each tile, histology experts manually annotated the cell type and x, y centroid co-ordinates for each cell nucleus. Focusing on the cell types $T = \{epithelial, \ fibroblast \ and \ endothelial, \ inflammatory, \ lymphocyte\}$, each nucleus was then modeled by a graph node $v_i \in V$ and labelled by its cell type $N_{type}(v_i) \in T$.

---

[2] Note $S_{uv} = S_{vu}$ and so the proposed method is invariant to node ordering.

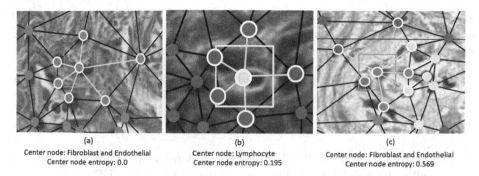

(a) Center node: Fibroblast and Endothelial Center node entropy: 0.0

(b) Center node: Lymphocyte Center node entropy: 0.195

(c) Center node: Fibroblast and Endothelial Center node entropy: 0.569

Fig. 2. Examples of nodes and their entropy values. The center node is highlighted with a rectangular box. The neighboring nodes are highlighted with a white circle and yellow edges. Cell types are represented by four colours: epithelial (red), fibroblast and endothelial (blue), inflammatory (green) and lymphocyte (yellow). (Color figure online)

The location and extent of the BM in each tile was manually annotated using cubic splines. An undirected edge set $\mathcal{E}$ for $\mathcal{V}$ was generated using the Delaunay triangulation method and all edges $e = \{u, v\}$ having a Euclidean length greater than 300 pixels (150 microns) were deleted[3]

For accurate GNN-based prediction of the BM, besides cell type N_{type}, additional node and edge labels were shown to improve the quality of BM prediction.

Additional Node Labels

1. Cell Density: We define the (local) cell density at node $v \in \mathcal{V}$ as the average distance between v and its immediate neighbours $u \in \mathcal{N}(v)$.

$$N_{den}(v) = \frac{1}{|\mathcal{N}(v)|} \sum_{u \in \mathcal{N}(v)} d(v, u). \tag{6}$$

2. Cell Entropy: For defining the (local) cell entropy of a node $v \in \mathcal{V}$, we first calculate the probability $p_v(\tau)$ of finding the cell type $\tau \in T$ in its immediate neighbourhood using Eq. 7. Cell entropy is defined using the Shannon entropy measure in Eq. 8 and vizualised in Fig. 2.

$$p_v(\tau) = \frac{|\{u \in N(v) \ or \ u = v : N_{type}(u) = \tau\}|}{|\mathcal{N}(v)|} \tag{7}$$

$$N_{ent}(v) = - \sum_{\tau \in T} p_v(\tau) \times \ln p_v(\tau) \tag{8}$$

---

[3] We kept the limiting criterion equivalent to 300 pixels to avoid long edges in the cell-graph, as the cell density varies across different parts of a tile.

Additional Edge Labels. For GNN-based prediction of the BM, the use of edge labels significantly increased the accuracy of our model. We make use of five edge labels defined below using edge feature tensor representation (c.f. Sect. 3.1).

1. Node Distance: For each edge $\{u, v\} \in \mathcal{E}(d)$ we use the Euclidean distance between the endpoints as the distance label: $\hat{\mathbf{E}}^{\mathbf{dis}}_{u,v} = d(u, v)$.
2. Cell Density Difference: Recalling the cell density definition of Eq. 6, we can define the difference in cell density along edge $\{u, v\} \in \mathcal{E}(d)$ by: $\hat{\mathbf{E}}^{\mathbf{den}}_{u,v} = N_{den}[v] - N_{den}[u]$
3. Cell Entropy Difference: Similarly to the cell density difference, we can define the cell entropy difference[4] along an edge $\{u, v\} \in \mathcal{E}(d)$ by: $\hat{\mathbf{E}}^{\mathbf{ent}}_{u,v} = |N_{ent}[v] - N_{ent}[u]|$
4. Neighbourhood Overlap Similarity: This measure aims to quantify the overlap between the neighbourhoods of two nodes, which is useful for tasks like edge and community detection [15]. To define the (relativized) overlap between two nodes u and v we use the Sorenson similarity index defined by: $\hat{\mathbf{E}}^{\mathbf{nei}}_{u,v} = \frac{2|\mathcal{N}(u) \cap \mathcal{N}(v)|}{deg(u) + deg(v)}$
5. BM Crossing: For each edge $\{u, v\} \in \mathcal{E}(d)$ we define the binary valued BM crossing measure: $\hat{\mathbf{E}}^{\mathbf{bm}}_{u,v} = 1$ if $\{u, v\}$ crosses the BM, otherwise $\hat{\mathbf{E}}^{\mathbf{bm}}_{u,v} = 0$.

Edge Feature Normalization. To incorporate the edge features we multiply them with the node features during the convolution operations of each EAGNN layer. In order to maintain the feature scale during the matrix multiplication, we normalize the edge feature values. There are several ways to normalize the edge feature values, including the Doubly Stochastic Normalization method proposed in [12]. However the Doubly Stochastic Normalization method assumes all the edge feature values to be non-negative which is not true in our case especially while considering the cell density difference. We normalize the edge feature values by rows as proposed in [24]. The row normalization for feature $1 \le p \le P$ and edge $(u, v) \in \mathcal{E}(d)$ is defined by:

$$\mathbf{E}^{\mathbf{X}}_{uvp} = \frac{\hat{\mathbf{E}}^{\mathbf{X}}_{uvp}}{\sum_{v' \in \mathcal{N}(u)} |\hat{\mathbf{E}}^{\mathbf{X}}_{uv'p}|}. \tag{9}$$

5.2 Training Setup

The cell-graph dataset was divided into two subsets: training (70%) and test data (30%). Some of the training dataset was reserved as a validation set (15%). Table 1 show the population sizes of edge classes in the data set indicating data bias to non-crossing edges. The GNN models were trained by backpropagation to minimize the mean of the binary cross entropy loss function for each mini-batch. The

---

[4] The reason for choosing the absolute difference here was to have non-negative entropy difference value. Ablation studies showed that negative cell entropy differences had an adverse effect on the efficiency of the trained model.

loss function was given by: $\mathbf{loss} = -\frac{1}{N} \sum_{i=1}^{N} [l_i \log(\mathbf{y_{pred}}) + (1 - l_i)\log(1 - \mathbf{y_{pred}})]$, where N is the number of samples in a mini-batch, $l_i \in \{0, 1\}$ and $y_{pred} \in [0, 1]$ are the label and the prediction for sample i respectively. We trained the models for 100 epochs with a batch size of 32. Each batch had 16 BM crossing edges and 16 BM non-crossing edges to avoid the bias derived from data imbalance. To optimize learnable parameters, we used Adam with an initial learning rate set to 0.001 which was scheduled to drop by 0.1 after every 40 epochs. To avoid overfitting on the training data, we used dropout with probability 0.5 for EAGNN layers and BC+MLP classifier, 0.3 for GraphSAGE, and the weight-decay parameter was set to 0.0001.

Table 1. Population sizes of edge classes in the oral mucosa cell-graph dataset.

Dataset	Non-Crossing Edges	Crossing Edges	Total Edges
Training	96195	6221	102416
Validation	20017	1169	21186
Testing	45779	2733	48512
Total	161991	10123	172114

5.3 Discussion

To compensate for data imbalances in the ground truth dataset, we evaluated EAGNN performance on the BM prediction problem using five standard metrics: precision, recall, F1 score, ROC-AUC and accuracy.

Table 2. Comparison of GNN models, edge classifiers, aggregated node features (NF) and edge features (EF) for BM prediction on the oral mucosa cell-graph dataset. (NF specifies aggregated node features: $N_1 = N_{type}$, $N_2 = N_{den}$, and $N_3 = N_{ent}$. EF specifies aggregated edge features: $E_1 = \hat{E}^{dis}$, $E_2 = \hat{E}^{den}$, $E_3 = \hat{E}^{ent}$, and $E_4 = \hat{E}^{nei}$.)

Model	Classifier	NF	EF	Precision	Recall	F1	ROC-AUC	Accuracy
GraphSAGE	MUL	N_1	-	0.3114	0.7494	0.4400	0.9112	0.8925
GraphSAGE	NegMUL	N_1	-	0.7548	0.8547	0.8016	0.9712	0.9762
GraphSAGE	BC+MLP	N_1	-	0.7714	**0.8580**	0.8124	0.9764	0.9777
GraphSAGE	BC+MLP	N_{123}	-	0.7793	0.8426	0.8097	0.9777	0.9781
EAGNN	MUL	N_{123}	$\hat{E}_{1234}$	0.5275	0.5258	0.5267	0.9284	0.9468
EAGNN	NegMUL	N_{123}	$\hat{E}_{1234}$	0.7353	0.8496	0.7883	0.9763	0.9743
EAGNN	BC+MLP	N_{123}	$\hat{E}_{1234}$	**0.8751**	0.7845	0.8273	**0.9822**	0.9816
EAGNN	BC+MLP	N_1	$\hat{E}_{1234}$	0.8548	0.8247	**0.8395**	0.9799	**0.9822**

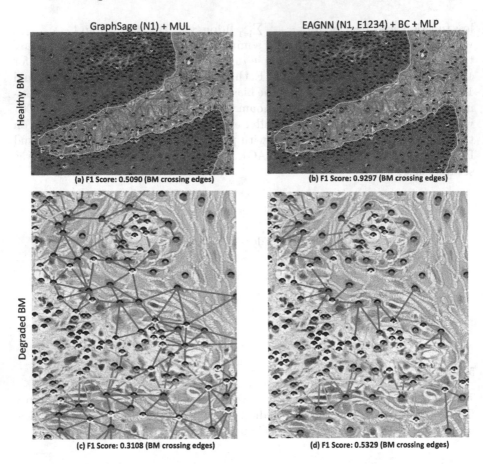

Fig. 3. Comparison of EAGNN and GraphSage predictions for intact and degraded BM tissue samples. The blue line is the BM annotation. The green, yellow and red lines represent true positive, false negative and false positive edge predictions respectively. (Color figure online)

Table 2 summarizes the performance of EAGNN combined with three different types of edge classifiers in comparison with GraphSAGE [14]. The combination GraphSAGE+MUL (row 2) is the baseline architecture which was used in [17] for BM prediction. The combination of EAGNN with the BC+MLPs edge classifier outperforms the other methods in 4 out of 5 metrics. GraphSAGE with BC+MLP resulted in high recall and relatively low precision, while EAGNN with BC+MLP resulted in high precision and relatively low recall. This is caused by the selection of the class threshold criterion θ. Tuning the parameter θ w.r.t F1 score, helped to balance the trade-off between precision and recall. The optimum balance was $\theta = 0.7$.

Figure 3 shows the BM prediction results for the best and worst performing models: EAGNN+BC+MLP ($NF = N_1, EF = E_{1234}$) and GraphSAGE+MUL

($NF = N_1$) respectively[5]. The figure shows BM prediction results for both models on an intact BM (Fig. 3 a, b) and a degraded BM (Fig. 3 c, d) from the test set. Figure 3 c shows that GraphSAGE+Mul erroneously classifies non-crossing edges close to the BM as crossing-edges, as shown by the many red lines. We term this tendency a "halo effect". Notice that EAGNN+BC+MLP does not exhibit this halo effect as the false positives in Fig. 3 d are primarily in the region of broken BM.

6   Limitations

There exist several limitations to this study. One is the limitation of our dataset and prediction task: BM identification. Further cell-graph datasets and prediction tasks are needed to establish the wider value of the edge aggregation, classification and labelling methods proposed here.

Another general limitation is the data imbalance problem inherent in many medical datasets, due to low disease frequency, small sample sizes and the effort of annotation. A specific example of this limitation is the class imbalance problem between BM crossing and non-crossing edges. Since the BM is a thin protein interface localised between the epithelial and connective tissue, non-crossing BM edges constitute a large majority of the edges in each cell graph, as shown in Table 1. Figure 3 shows the difference in F1-scores for BM crossing edges between a healthy tissue sample where the BM is intact and a degraded sample where the BM is broken. This specific imbalance could be addressed by using the focal loss function [19] to reduce class imbalance.

7   Conclusions

In this work, we have proposed a new GNN model EAGNN that aggregates both node and edge label information and is suitable for edge label prediction. We have presented a digital pathology case study of BM prediction showing that aggregation of both node and edge label information can take advantage of the topological information in cell-graphs. In our case study, EAGNN significantly outperformed the widely used GraphSAGE GNN model under several performance measures. Furthermore, we have introduced new edge label features including cell entropy gradient and neighbourhood overlap similarity and shown that these improve the accuracy of BM prediction. Future directions of research include improving EAGNN performance and data augmentation methods to reduce the data imbalance.

Acknowledgements. The authors acknowledge the clinical and technical assistance of Karin Garming-Lergert, Victor Tollemar and Daniella Nordman (Karolinska Institutet) for collection and preparation of biopsies, and annotations. The study was

---

[5] We used Intelligraph tool for cell annotation, graph construction and results evaluation.

financed by grants from Region Stockholm ALF Medicine, Styrgruppen KI/Region Stockholm for Research in Odontology and research funds from Karolinska Institutet and KTH.

Ethics Declarations. Ethical approval for the collection and experimental use of tissue samples from patients attending the Department of Maxillofacial Surgery, Karolinska University Hospital (provided written informed consent) or retrieved retrospectively from Stockholm's Medicine Biobank (Sweden Biobank, access to archived material) has been granted by the Swedish Ethical Review Authority (Etikprövningsmyndigheten). Procedures have been performed according to relevant guidelines, under Registration Numbers 2013/39-31/4, 2014/1184-31/1 and 2019-01295.

References

1. Ahmedt-Aristizabal, D., Armin, M.A., Denman, S., Fookes, C., Petersson, L.: A survey on graph-based deep learning for computational histopathology. Comput. Med. Imaging Graph.: The Official Journal of the Computerized Medical Imaging Society **95**, 102027 (2022)
2. Anand, D., Gadiya, S., Sethi, A.: Histographs: graphs in histopathology. In: Medical Imaging 2020: Digital Pathology, vol. 11320, p. 113200O. International Society for Optics and Photonics (2020)
3. Atwood, J., Towsley, D.: Diffusion-convolutional neural networks. In: Advances in Neural Information Processing Systems, vol. 29 (2016)
4. Aygüneş, B., Aksoy, S., Cinbiş, R.G., Kösemehmetoğlu, K., Önder, S., Üner, A.: Graph convolutional networks for region of interest classification in breast histopathology. In: Medical Imaging 2020: Digital Pathology, vol. 11320, p. 113200K. International Society for Optics and Photonics (2020)
5. Bilgin, C., Demir, C., Nagi, C., Yener, B.: Cell-graph mining for breast tissue modeling and classification. In: 2007 29th Annual International Conference of the IEEE Engineering in Medicine and Biology Society, pp. 5311–5314. IEEE (2007)
6. Cao, L., Lu, Y., Li, C., Yang, W.: Automatic segmentation of pathological glomerular basement membrane in transmission electron microscopy images with random forest stacks. Comput. Math. Methods Med. **2019** (2019)
7. Clevert, D.A., Unterthiner, T., Hochreiter, S.: Fast and accurate deep network learning by exponential linear units (ELUs). arXiv preprint arXiv:1511.07289 (2015)
8. Crichton, G., Guo, Y., Pyysalo, S., Korhonen, A.: Neural networks for link prediction in realistic biomedical graphs: a multi-dimensional evaluation of graph embedding-based approaches. BMC Bioinform. **19**(1), 1–11 (2018)
9. Demir, C., Gultekin, S.H., Yener, B.: Learning the topological properties of brain tumors. IEEE/ACM Trans. Comput. Biol. Bioinf. **2**(3), 262–270 (2005)
10. Gilmer, J., Schoenholz, S.S., Riley, P.F., Vinyals, O., Dahl, G.E.: Neural message passing for quantum chemistry. In: International Conference on Machine Learning, pp. 1263–1272. PMLR (2017)
11. Glorot, X., Bordes, A., Bengio, Y.: Deep sparse rectifier neural networks. In: Proceedings of the Fourteenth International Conference on Artificial Intelligence and Statistics, pp. 315–323. JMLR Workshop and Conference Proceedings (2011)

12. Gong, L., Cheng, Q.: Exploiting edge features for graph neural networks. In: 2019 IEEE/CVF Conference on Computer Vision and Pattern Recognition (CVPR), Los Alamitos, CA, USA, pp. 9203–9211. IEEE Computer Society (2019)
13. Gunduz, C., Yener, B., Gultekin, S.H.: The cell graphs of cancer. Bioinformatics **20**(Suppl. 1), i145–i151 (2004)
14. Hamilton, W., Ying, Z., Leskovec, J.: Inductive representation learning on large graphs. In: Advances in Neural Information Processing Systems, vol. 30 (2017)
15. Hamilton, W.L.: Graph representation learning. Synth. Lect. Artif. Intell. Mach. Learn. **14**(3), 1–159 (2020)
16. Levy, J., Haudenschild, C., Barwick, C., Christensen, B., Vaickus, L.: Topological feature extraction and visualization of whole slide images using graph neural networks. In: BIOCOMPUTING 2021: Proceedings of the Pacific Symposium, pp. 285–296. World Scientific (2020)
17. Nair, A., et al.: A graph neural network framework for mapping histological topology in oral mucosal tissue. BMC Bioinform. **23**(1), 1–21 (2022)
18. Niepert, M., Ahmed, M., Kutzkov, K.: Learning convolutional neural networks for graphs. In: International Conference on Machine Learning, pp. 2014–2023. PMLR (2016)
19. Pasupa, K., Vatathanavaro, S., Tungjitnob, S.: Convolutional neural networks based focal loss for class imbalance problem: a case study of canine red blood cells morphology classification. J. Ambient Intell. Hum. Comput. 1–17 (2020)
20. Pati, P., et al.: HACT-Net: a hierarchical cell-to-tissue graph neural network for histopathological image classification. In: Sudre, C.H., et al. (eds.) UNSURE/GRAIL -2020. LNCS, vol. 12443, pp. 208–219. Springer, Cham (2020). https://doi.org/10.1007/978-3-030-60365-6_20
21. Studer, L., Wallau, J., Dawson, H., Zlobec, I., Fischer, A.: Classification of intestinal gland cell-graphs using graph neural networks. In: 2020 25th International Conference on Pattern Recognition (ICPR), pp. 3636–3643. IEEE (2021)
22. Sureka, M., Patil, A., Anand, D., Sethi, A.: Visualization for histopathology images using graph convolutional neural networks. In: 2020 IEEE 20th International Conference on Bioinformatics and Bioengineering (BIBE), pp. 331–335. IEEE (2020)
23. Tollemar, V., et al.: Histopathological grading of oral mucosal chronic graft-versus-host disease: large cohort analysis. Biol. Blood Marrow Transplant. **26**(10), 1971–1979 (2020)
24. Veličković, P., Cucurull, G., Casanova, A., Romero, A., Liò, P., Bengio, Y.: Graph attention networks. In: International Conference on Learning Representations (2018)
25. Wang, D., Gu, C., Wu, K., Guan, X.: Adversarial neural networks for basal membrane segmentation of microinvasive cervix carcinoma in histopathology images. In: 2017 International Conference on Machine Learning and Cybernetics (ICMLC), vol. 2, pp. 385–389. IEEE (2017)
26. Wu, H.S., Dikman, S., Gil, J.: A semi-automatic algorithm for measurement of basement membrane thickness in kidneys in electron microscopy images. Comput. Methods Programs Biomed. **97**(3), 223–231 (2010)
27. Yener, B.: Cell-graphs: image-driven modeling of structure-function relationship. Commun. ACM **60**(1), 74–84 (2016)

Heterogeneous Graph Convolutional Neural Network via Hodge-Laplacian for Brain Functional Data

Jinghan Huang[1], Moo K. Chung[2], and Anqi Qiu[1,3,4,5,6]($\boxtimes$)

[1] Department of Biomedical Engineering, National University of Singapore, Singapore, Singapore
bieqa@nus.edu.sg
[2] Department of Biostatistics and Medical Informatics, The University of Wisconsin-Madison, Wisconsin, USA
[3] NUS (Suzhou) Research Institute, National University of Singapore, Suzhou, China
[4] Institute of Data Science, National University of Singapore, Singapore, Singapore
[5] The N.1 Institute for Health, National University of Singapore, Singapore, Singapore
[6] Department of Biomedical Engineering, The Johns Hopkins University, Baltimore, USA

Abstract. This study proposes a novel heterogeneous graph convolutional neural network (HGCNN) to handle complex brain fMRI data at regional and across-region levels. We introduce a generic formulation of spectral filters on heterogeneous graphs by introducing the $k-th$ Hodge-Laplacian (HL) operator. In particular, we propose Laguerre polynomial approximations of HL spectral filters and prove that their spatial localization on graphs is related to the polynomial order. Furthermore, based on the bijection property of boundary operators on simplex graphs, we introduce a generic topological graph pooling (TGPool) method that can be used at any dimensional simplices. This study designs HL-node, HL-edge, and HL-HGCNN neural networks to learn signal representation at a graph node, edge levels, and both, respectively. Our experiments employ fMRI from the Adolescent Brain Cognitive Development (ABCD; n = 7693) to predict general intelligence. Our results demonstrate the advantage of the HL-edge network over the HL-node network when functional brain connectivity is considered as features. The HL-HGCNN outperforms the state-of-the-art graph neural networks (GNNs) approaches, such as GAT, BrainGNN, dGCN, BrainNetCNN, and Hypergraph NN. The functional connectivity features learned from the HL-HGCNN are meaningful in interpreting neural circuits related to general intelligence.

1 Introduction

Functional magnetic resonance imaging (fMRI) is one of the non-invasive imaging techniques to measure blood oxygen level dependency (BOLD) signals [8]. The fluctuation of fMRI time series signals can characterize brain activity. The

A. Frangi et al. (Eds.): IPMI 2023, LNCS 13939, pp. 278–290, 2023.
https://doi.org/10.1007/978-3-031-34048-2_22

synchronization of fMRI time series describes the functional connectivity among brain regions for understanding brain functional organization.

There has been a growing interest in using graph neural network (GNN) to learn the features of fMRI time series and functional connectivity that are relevant to cognition or mental disorders [17,23].

GNN often considers a brain functional network as a binary undirected graph, where nodes are brain regions, and edges denote which two brain regions are functionally connected. Functional time series, functional connectivity, or graph metrics (i.e., degree, strength, clustering coefficients, participation, etc.) are defined as a multi-dimensional signal at each node. A substantial body of research implements an convolutional operator over nodes of a graph in the spatial domain, where the convolutional operator computes the fMRI feature of each node via aggregating the features from its neighborhood nodes [17,23]. Various forms of GNN with spatial graph convolution are implemented via 1) introducing an attention mechanism to graph convolution by specifying different weights to different nodes in a neighborhood (GAT, [9]); 2) introducing a clustering-based embedding method over all the nodes and pooling the graph based on the importance of nodes (BrainGNN, [17]); 3) designing an edge-weight-aware message passing mechanism [3]; 4) training dynamic brain functional networks based on updated nodes' features (dGCN, [23]). BrainGNN and dGCN achieve superior performance on Autism Spectrum Disorder (ASD) [17] and attention deficit hyperactivity disorder (ADHD) classification [23]. Graph convolution has also been solved in the spectral domain via the graph Laplacian [2]. For the sake of computational efficiency when graphs are large, the Chebyshev polynomials and other polynomials were introduced to approximate spectral filters for GNN [4,10]. For large graphs, the spectral graph convolution with a polynomial approximation is computationally efficient and spatially localized [10].

Despite the success of the GNN techniques on cognitive prediction and disease classification [17,23], the graph convolution aggregates brain functional features only over nodes and updates features for each node of the graph. Nevertheless, signal transfer from one brain region to another is through their connection, which can, to some extent, be characterized by their functional connectivity. The strength of the connectivity determines which edges signals pass through. Therefore, there is a need for heterogeneous graphs with different types of information attached to nodes, such as functional time series and node efficiency, and edges, such as functional connectivity and path length.

Lately, a few studies have focused on smoothing signals through the topological connection of edges [12,13]. Kawahara et al. [15] proposed BrainNetCNN to aggregate brain functional connectivities among edges. However, brain functional connectivity matrices at each layer are no longer symmetric as the construction nature of the brain functional network. Jo et al. [13] employed a dual graph with the switch of nodes and edges of an original graph so that the GNN approaches described above can be applied (Hypergraph NN). But, the dual graph normally increases the dimensionality of a graph. To overcome this, Jo et al. [13] only considered important edges. Similarly, Jiang et al. [12] introduced convolution with edge-node switching that embeds both nodes and edges to a latent feature space. When graphs are not sparse, the computation of this approach can be intensive. The above-mentioned edge-node switching based model achieved great success on social network and molecular science [12,13], suggesting that GNN approaches on graph edges have advantages when information is defined on graph edges. Thus, it is crucial to consider heterogeneous graphs where multiple types of features are defined on nodes, edges, and etc. This is particularly suitable for brain functional data.

This study develops a novel heterogeneous graph convolutional neural network (HGCNN) simultaneously learning both nodes' and edges' functional features from fMRI data for predicting cognition or mental disorders. The HGCNN is designed to learn 1) nodes' features from their neighborhood nodes' features based on the topological connections of the nodes; 2) edges' features from their neighborhood edges' features based on the topological connections of the edges. To achieve these goals, the HGCNN considers a brain functional network as a simplex graph that allows characterizing node-node, node-edge, edge-edge, and higher-order topology. We develop a generic convolution framework by introducing the Hodge-Laplacian (HL) operator on the simplex graph and designing HL-spectral graph filters to aggregate features among nodes or edges based on their topological connections. In particular, this study takes advantage of spectral graph filters in [4,10] and approximates HL-spectral graph filters using polynomials for spatial locations of these filters. We shall call our HGCNN as HL-HGCNN in the rest of the paper. Unlike the GNNs described above [12,23], this study also introduces a simple graph pooling approach based on its topology such that the HL can be automatically updated for the convolution in successive layers, and the spatial dimension of the graph is reduced. Hence, the HL-HGCNN learns spectral filters along nodes, edges, or higher-dimensional simplex to extract brain functional features.

We illustrate the use of the HL-HGCNN on fMRI time series and functional connectivity to predict general intelligence based on a large-scale adolescent cohort study (Adolescent Brain Cognitive Development (ABCD), n = 7693). We also compare the HL-HGCNN with the state-of-art GNN techniques described above and demonstrate the outstanding performance of the HL-HGCNN. Hence, this study proposes the following novel techniques:

1. a generic graph convolution framework to smooth signals across nodes, edges, or higher-dimensional simplex;
2. spectral filters on nodes, edges, or higher-dimensional simplex via the HL operator;
3. HL-spectral filters with a spatial localization property via polynomial approximations;
4. a spatial pooling operator based on graph topology.

2   Methods

This study designs a heterogeneous graph convolutional neural network via the Hodge-Laplacian operator (HL-HGCNN) that can learn the representation of brain functional features at a node-level and an edge-level based on the graph topology. In the following, we will first introduce a generic graph convolution framework to design spectral filters on nodes and edges to learn node-level and edge-level brain functional representation based on its topology achieved via the HL operator. We will introduce the polynomial approximation of the HL spectral filters to overcome challenges on spatial localization. Finally, we will define an efficient pooling operation based on the graph topology for the graph reduction and update of the HL operator.

2.1 Learning Node-Level and Edge-Level Representation via the Hodge-Laplacian Operator

In this study, the brain functional network is characterized by a heterogeneous graph, $G = \{V, E\}$ with brain regions as nodes, $V = \{v_i\}_{i=1}^{n}$, and their connections as edges, $E = \{e_{ij}\}_{i,j=1,2,\cdots,n}$, as well as functional time series defined on the nodes and functional connectivity defined on the edges. This study aims to design convolutional operations for learning the representation of functional time series at nodes and the representation of functional connectivity at edges based on node-node and edge-edge connections (or the topology of graph G).

Mathematically, nodes and edges are called $0-$ and $1-$dimensional simplex. The topology of G can be characterized by *boundary operator* ∂_k. ∂_1 encodes how two 0-dimensional simplices, or nodes, are connecting to form a 1-dimensional simplex (an edge) [6]. In the graph theory [16], ∂_1 can be represented as a traditional incidence matrix with size $n \times n(n-1)/2$, where nodes are indexed over rows and edges are indexed over columns. Similarly, the second order boundary operator ∂_2 encodes how 1-dimensional simplex, or edges, are connected to form the connections among 3 nodes (2-dimensional simplex or triangle).

The goal of spectral filters is to learn the node-level representation of fMRI features from neighborhood nodes' fMRI features and the edge-level representation of fMRI features from neighborhood edges' fMRI features. The neighborhood information of nodes and edges can be well characterized by the *boundary operators* ∂_k of graph G. It is natural to incorporate the *boundary operators* of graph G in the k-th Hodge-Laplacian (HL) operator defined as

$$\mathcal{L}_k = \partial_{k+1}\partial_{k+1}^{\top} + \partial_k^{\top}\partial_k. \tag{1}$$

When $k = 0$, the 0-th HL operator is

$$\mathcal{L}_0 = \partial_1\partial_1^{\top} \tag{2}$$

over nodes. This special case is equivalent to the standard Graph Laplacian operator, $\mathcal{L}_0 = \Delta$. When $k = 1$, the 1-st HL operator is defined over edges as

$$\mathcal{L}_1 = \partial_2\partial_2^{\top} + \partial_1^{\top}\partial_1. \tag{3}$$

We can obtain orthonormal bases $\psi_k^0, \psi_k^1, \psi_k^2, \cdots$ by solving eigensystem $\mathcal{L}_k\psi_k^j = \lambda_k^j\psi_k^j$. We now consider an HL spectral filter h with spectrum $h(\lambda_k)$ as

$$h(\cdot, \cdot) = \sum_{j=0}^{\infty} h(\lambda_k^j)\psi_k^j(\cdot)\psi_k^j(\cdot). \tag{4}$$

A generic form of spectral filtering of a signal f on the heterogeneous graph G can be defined as

$$f'(\cdot) = h * f(\cdot) = \sum_{j=0}^{\infty} h(\lambda_k^j) c_k^j \psi_k^j(\cdot), \tag{5}$$

where $f(\cdot) = \sum_{j=0}^{\infty} c_k^j \psi_k^j(\cdot)$. When $k = 0$, f is defined on the nodes of graph G. Equation (5) indicates the convolution of a signal f defined on V with a filter h.

Likewise, when $k = 1$, f is defined on the edges of graph G. Equation (5) then indicates the convolution of a signal f defined on E with a filter h. Equation (5) is generic that can be applied to smoothing signals defined on higher-dimensional simplices. Nevertheless, this study considers the heterogeneous graph only with signals defined on nodes and edges (0- and 1-dimensional simplices). In the following, we shall denote these two as "HL-node filtering" and "HL-edge filtering", respectively.

2.2   Laguerre Polynomial Approximation of the HL Spectral Filters

The shape of spectral filters h in Eq. (5) determines how many nodes or edges are aggregated in the filtering process. Our goal of the HL-HGCNN is to design h such as the representation at nodes and edges are learned through their neighborhood. This is challenging in the spectral domain since it requires $h(\lambda)$ with a broad spectrum. In this study, we propose to approximate the filter spectrum $h(\lambda_k)$ in Eq. (5) as the expansion of Laguerre polynomials, T_p, $p = 0, 1, 2, \ldots, P - 1$, such that

$$h(\lambda_k) = \sum_{p=0}^{P-1} \theta_p T_p(\lambda_k), \tag{6}$$

where θ_p is the pth expansion coefficient associated with the pth Laguerre polynomial. T_p can be computed from the recurrence relation of $T_{p+1}(\lambda_k) = \frac{(2p+1-\lambda_k)T_p(\lambda_k)-pT_{p-1}(\lambda_k)}{p+1}$ with $T_0(\lambda_k) = 1$ and $T_1(\lambda_k) = 1 - \lambda_k$.

We can rewrite the convolution in Eq. (5) as

$$f'(\cdot) = h * f(\cdot) = \sum_{p=0}^{P-1} \theta_p T_p(\mathcal{L}_k) f(\cdot). \tag{7}$$

Analog to the spatial localization property of the polynomial approximation of the graph Laplacian (the 0-th HL) spectral filters [4,10,21], the Laguerre polynomial approximation of the 1-st HL spectral filters can also achieve this localization property. Assume two edges, e_{ij} and e_{mn}, on graph G. The shortest distance between e_{ij} and e_{mn} is denoted by $d_G(ij, mn)$ and computed as the minimum number of edges on the path connecting e_{ij} and e_{mn}. Hence, $(\mathcal{L}_1^P)_{e_{ij},e_{mn}} = 0$ if $d_G(ij, mn) > P$, where $\mathcal{L}_1^P$ denotes the P-th power of the 1-st HL. Hence, the spectral filter represented by the P-th order Laguerre polynomials of the 1-st HL is localized within the P-hop edge neighborhood. Therefore, spectral filters in Eq. (6) have the property of spatial localization. This proof can be extended to the k-th HL spectral filters. In Section of Results 3, we will demonstrate this property using simulation data.

2.3   Topological Graph Pooling (TGPool)

The pooling operation has demonstrated its effectiveness on grid-like image data [22]. However, spatial graph pooling is not straightforward, especially for heterogeneous graphs. This study introduces a generic topological graph pooling (TGPool) approach that includes coarsening of the graph, pooling of signals, and an update of the Hodge-Laplacian operator. For this, we take an advantage of the one-to-one correspondence between the *boundary operators* and graph G and define the three operations for pooling based on the *boundary operators*. As the *boundary operators* encode the topology of the graph, our graph pooling is topologically based.

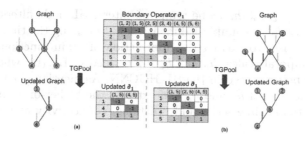

Fig. 1. Topological Graph Pooling (TGPool). Panels (a) and (b) illustrate the topological graph pooling of (a) 0-dimensional (nodes) and (b) 1-dimensional (edges) simplices. The color at each node or edge indicates features and their similarity across nodes or edges. (Color figure online)

For graph coarsening, we generalize the Graclus multilevel clustering algorithm [5] to coarsen the k−dimensional simplices on graph G. We first cluster similar k−dimensional simplices based on their associated features via local normalized cut. At each coarsening level, two neighboring k−dimensional simplices with maximum local normalized cut are matched until all k−dimensional simplices are explored [19]. A balanced binary tree is generated where each k−dimensional simplex has either one (i.e., singleton) or two child k−dimensional simplices. Fake k−dimensional simplices are added to pair with those singletons. The weights of $k + 1$−dimensional simplices involving fake k−dimensional simplices are set as 0. The pooling on this binary tree can be efficiently implemented as a simple 1-dimensional pooling of size 2. Then, two matched k−dimensional simplices are merged as a new k−dimensional simplex by removing the k−dimensional simplex with the lower degree and the $k + 1$−dimensional simplices that are connected to this k−dimensional simplex. To coarsen the graph, we define a new *boundary operator* by deleting the corresponding rows and columns in the boundary operator and computing the HL operators via Eq. 2. Finally, the signal of the new

$k-$dimensional simplex is defined as the average (or max) of the signals at the two $k-$dimensional simplices. Figure 1 illustrates the graph pooling of 0-dimensional and 1-dimensional simplices and the boundary operators of the updated graph after pooling.

2.4 Hodge-Laplacian Heterogeneous Graph Convolutional Neural Network (HL-HGCNN)

We design the HL-HGCNN with the temporal, node, and edge convolutional layers to learn temporal and spatial information of brain functional time series and functional connectivity. Each layer includes the convolution, leaky rectified linear unit (leaky ReLU), and pooling operations. Figure 2 illustrates the overall architecture of the HL-HGCNN model, the temporal, node, and edge convolutional layers.

Filters. Denote h_t, h_v, h_e to be temporal filters, HL-node filters, HL-edge filters, respectively. h_t is a simple 1-dimensional filter along the time domain with different kernel sizes to extract the information of brain functional time series at multiple temporal scales. h_v and h_e are defined in Eq. (6), where θ_p are the parameters to be estimated in the HL-HGCNN. As mentioned earlier, P determines the kernel size of h_v and h_e and extracts the higher-order information of the brain functional time series and functional connectivity at multiple spatial scales.

Leaky ReLU. This study employs leaky rectified linear unit (ReLU) as an activation function, σ, since negative functional time series and functional connectivity are considered biologically meaningful.

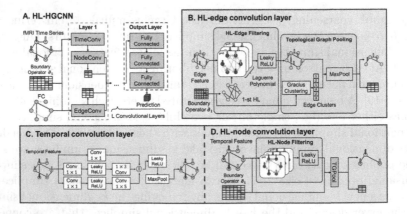

Fig. 2. HL-HGCNN architecture. Panel (A) illustrates the overall architecture of the HL-HGCNN model. Panels (B–D) respectively show the architectures of the HL-edge, temporal, and HL-node convolutional layers.

Pooling. In the temporal convolutional layer, traditional 1-dimensional max pooling operations are applied in the temporal dimension of the functional time series. In the edge and node convolutional layers, TGPool is applied to reduce the dimension of the graph and the dimension of the node and edge signals.

Output Layer. We use one more graph convolutional layer to translate the feature of each node or edge into a scalar. Then, we concatenate the vectorized node and edge representations as the input of the output layer. In this study, the output layers contain fully-connected layers.

2.5   Implementation

$\mathcal{L}_0$ **and** $\mathcal{L}_1$**.** Given a brain functional connectivity matrix, we first build a binary matrix while the element in the connectivity matrix with its absolute value greater than a threshold is assigned as one, otherwise zero. We compute the boundary operator ∂_1 with the size of the number of brain regions and the number of functional connectivities. The i-th row of ∂_1 encodes the functional connection of the i-th vertex and the j-th column of ∂_1 encodes how two vertices are connecting to form an edge [6,7]. Hence, $\mathcal{L}_0 = \partial_1\partial_1^\top$.

According to Eq. (1), the computation of $\mathcal{L}_1$ involves the computation of ∂_2 that characterizes the interaction of edges and triangles. The brain functional connectivity matrix does not form a triangle simplex so the second order boundary operator $\partial_2 = 0$. Hence, $\mathcal{L}_1 = \partial_1^\top\partial_1$.

Optimization. We implement the framework in Python 3.9.13, Pytorch 1.12.1 and PyTorch Geometric 2.1.0 library. The HL-HGCNN is composed of two temporal, node, and edge convolution layers with $\{8,8\}$, $\{16,1\}$, and $\{32,32\}$ filters, respectively. The order of Laguerre polynomials for the 0-th and 1-st HL approximation is set to 3 and 4, respectively. The output layer contains three fully connected layers with 256, 128 and 1 hidden nodes, respectively. Dropout with 0.5 rate is applied to every layer and Leaky ReLU with a leak rate of 0.33 are used in all layers. These model-relevant parameters are determined using greedy search. The HL-HGCNN model is trained using an NVIDIA Tesla V100SXM2 GPU with 32 GB RAM by the ADAM optimizer with a mini-batch size of 32. The initial learning rate is set as 0.005 and decays by 0.95 after every epoch. The weight decay parameter was 0.005.

2.6   ABCD Dataset

This study uses resting-state fMRI (rs-fMRI) images from the ABCD study that is an open-sourced and ongoing study on youth between 9–11 years old (https://abcdstudy.org/). This study uses the same dataset of 7693 subjects and fMRI preprocessing pipeline stated in Huang et al. [11]. A node represents one of 268 brain regions of interest (ROIs) [18] with its averaged time series as node features. Each edge represents the functional connection between any two ROIs with the functional connectivity computed via Pearson's correlation of their averaged time series as edge features. General intelligence is defined as the average of 5 NIH Toolbox cognition scores, including Dimensional Change Card Sort, Flanker, Picture Sequence Memory, List Sorting Working Memory, and Pattern Comparison Processing Speed [1]. General intelligence ranges from 64 to 123 with mean and standard deviation of 95.3 ± 7.3 among 7693 subjects.

3   Results

This section first demonstrates the spatial localization property of HL-edge filters in relation to the order of Laguerre polynomials via simulated data. We then demonstrate the use of HL-edge filtering and its use in GNN for predicting fluid intelligence using the ABCD dataset.

3.1   Spatial Localization of the HL-Edge Filtering via Laguerre Polynomial Approximations

We illustrate the spatial location property of the HL-edge filtering by designing a pulse signal at one edge (Fig. 3(a)) and smoothing it via the HL-edge filter. When applying the HL-edge filter approximated via the 1st-, 2nd-, 3rd-, 4th-order Laguerre polynomials, the filtered signals shown in Fig. 3(b–e) suggest that the spatial localization of the HL-edge filters is determined by the order of Laguerre polynomials. This phenomenon can also be achieved using multi-layer HL-edge filters where each layer contains HL-edge filters approximated using the 1st-order Laguerre polynomial (see Fig. 3(f)).

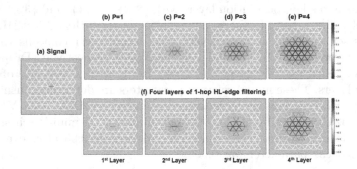

Fig. 3. Spatial localization of the HL-edge filtering. Panel (a) shows the simulated signal only occurring at one edge. Panels (b–e) show the signals filtered using the HL-edge filters with the 1st-, 2nd-, 3rd-, 4th-order Laguerre polynomial approximation, respectively. Panel (f) illustrates the signals generated from the HL-edge convolution networks with 4 layers. Each layer consists of the HL-filter approximated using the 1st-order Laguerre polynomial.

3.2   HL-Node vs. HL-Edge Filters

We aim to examine the advantage of the HL-edge filters over the HL-node filters when fMRI data by nature characterize edge information, such as the functional connectivity. When functional connectivities are defined at a node, they form a vector of the functional connectivities related to this node. In contrast, by nature, the functional connectivity represents the functional connection strength of two brain regions (i.e., edge). Hence, it is a scalar defined at an edge. We design the HL-node network with the two HL-node convolutional layers (see in Fig. 2D) and the output layer with three fully connected layers. Likewise, the HL-edge network with the two HL-edge convolutional layers (see Fig. 2B) and the output layer with three fully connected layers. We employ five-fold cross-validation six times to evaluate the prediction accuracy between predicted and actual general intelligence based on root mean square error (RMSE). Table 1 shows that the HL-edge network has smaller RMSE and performs better than the HL-node network ($p = 1.51 \times 10^{-5}$). This suggests the advantage of the HL-edge filters when features by nature characterize the weights of edges.

3.3   Comparisons with Existing GNN Methods

We now compare our models with the existing state-of-art methods stated above in terms of the prediction accuracy of general intelligence using the ABCD dataset. The first experiment is designed to compare the performance of the HL-node network with that GAT [9], BrainGNN [17], and dGCN [23]. We adopt the architecture of BrainGNN and dGCN from Li et al. [17,23] as both methods were used for fMRI data. The GAT is designed with two graph convolution layers, each consisting of 32 filters and 2-head attention, which is determined via greedy search as implemented in our model. The functional connectivity vector of each region is used as input features. Table 1 suggested that the HL-node network performs better than the GAT ($p = 0.0468$) and BrainGNN (p = 0.0195), and performs equivalently with dGCN ($p = 0.0618$).

Table 1. General intelligence prediction accuracy based on root mean square error (RMSE). p-value is obtained from two-sample t-tests examining the performance of each method in reference to the proposed HL-HGCNN.

	GNN model	RMSE	p-value
GNN with node filtering	**HL-Node network (ours)**	7.134 ± 0.011	4.01×10^{-6}
	GAT [9]	7.165 ± 0.020	1.91×10^{-5}
	BrainGNN [17]	7.144 ± 0.013	1.51×10^{-6}
	dGCN [17]	7.151 ± 0.012	9.83×10^{-6}
GNN with edge filtering	**HL-Edge network (ours)**	7.009 ± 0.012	2.48×10^{-2}
	BrainNetCNN [15]	7.118 ± 0.016	5.34×10^{-6}
	Hypergraph NN [13]	7.051 ± 0.022	3.74×10^{-5}
GNN with node and edge filtering	**HL-HGCNN (ours)**	$\mathbf{6.972 \pm 0.015}$	–

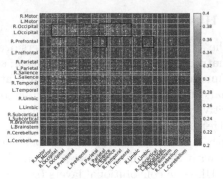

Fig. 4. The saliency map of the brain functional connectivity. Red boxes highlight brain networks with higher weights, indicating greater contributions to the prediction of general intelligence. (Color figure online)

The second experiment compares the HL-edge network with BrainNetCNN [15] and Hypergraph NN [13]. The Hypergraph NN comprises two graph convolution layers with 32 filters and one hypercluster layer after the first graph convolution layer. The BrainNetCNN architecture follows the design in [15]. Table 1 shows that the HL-edge network has smaller RMSE and performs better than the BrainNetCNN ($p = 4.49 \times 10^{-5}$) and Hypergraph NN ($p = 0.0269$).

Finally, our HL-HGCNN integrates heterogeneous types of fMRI data at nodes and edges. Table 1 shows that the HL-HGCNN performs the best compared to all the above methods (all $p < 0.03$).

3.4 Interpretation

We use the graph representation of the final edge convolution layer of the HL-HGCNN to compute the saliency map at the connectivity level. The group-level saliency map is computed by averaging the saliency maps across all the subjects in the dataset. The red boxes in Fig. 4 highlight the functional connectivities of the occipital regions with the prefrontal, parietal, salience, and temporal regions that most contribute to general intelligence. Moreover, our salience map also highlights the functional connectivities of the right prefrontal regions with bilateral parietal regions, which is largely consistent with existing findings on neural activities in the frontal and parietal regions [14,20].

4   Conclusion

This study proposes a novel HL-HGCNN on fMRI time series and functional connectivity for predicting cognitive ability. Our experiments demonstrate the spatial localization property of HL spectral filters approximated via Laguerre polynomials. Moreover, our HL-node, HL-edge, and HL-HGCNN perform better than the existing state-of-art methods for predicting general intelligence, indicating the potential of our method for future prediction and diagnosis based on fMRI. Nevertheless, more experiments on different datasets are needed to further validate the robustness of the proposed model. Our method provides a generic framework that allows learning heterogeneous graph representation on any dimensional simplices, which can be extended to complex graph data. The HL-HGCNN model offers an opportunity to build high-order functional interaction among multiple brain regions, which is our future research direction.

Acknowledgements. This research/project is supported by the Singapore Ministry of Education (Academic research fund Tier 1) and A*STAR (H22P0M0007). Additional funding is provided by the National Science Foundation MDS-2010778, National Institute of Health R01 EB022856, EB02875. This research was also supported by the A*STAR Computational Resource Centre through the use of its high-performance computing facilities.

References

1. Akshoomoff, N., et al.: VIII. NIH Toolbox Cognition Battery (CB): composite scores of crystallized, fluid, and overall cognition. Monogr. Soc. Res. Child Dev. **78**(4), 119–132 (2013)
2. Bruna, J., Zaremba, W., Szlam, A., LeCun, Y.: Spectral networks and locally connected networks on graphs. arXiv preprint arXiv:1312.6203 (2013)
3. Cui, H., Dai, W., Zhu, Y., Li, X., He, L., Yang, C.: Interpretable graph neural networks for connectome-based brain disorder analysis. In: Wang, L., Dou, Q., Fletcher, P.T., Speidel, S., Li, S. (eds.) Medical Image Computing and Computer Assisted Intervention (MICCAI 2022). LNCS, vol. 13438, pp. 375–385. Springer, Cham (2022). https://doi.org/10.1007/978-3-031-16452-1_36
4. Defferrard, M., Bresson, X., Vandergheynst, P.: Convolutional neural networks on graphs with fast localized spectral filtering. In: Advances in Neural Information Processing Systems, pp. 3844–3852 (2016)
5. Dhillon, I.S., Guan, Y., Kulis, B.: Weighted graph cuts without eigenvectors a multilevel approach. IEEE Trans. Pattern Anal. Mach. Intell. **29**(11), 1944–1957 (2007)
6. Edelsbrunner, H., Letscher, D., Zomorodian, A.: Topological persistence and simplification. In: Proceedings 41st Annual Symposium on Foundations of Computer Science, pp. 454–463. IEEE (2000)
7. Edelsbrunner, H., Letscher, D., Zomorodian, A.: Topological persistence and simplification. Discrete Comput. Geom. **28**, 511–533 (2002)
8. Glover, G.H.: Overview of functional magnetic resonance imaging. Neurosurg. Clin. **22**(2), 133–139 (2011)
9. Hu, J., Cao, L., Li, T., Dong, S., Li, P.: GAT-LI: a graph attention network based learning and interpreting method for functional brain network classification. BMC Bioinform. **22**(1), 1–20 (2021)
10. Huang, S.G., Chung, M.K., Qiu, A.: Revisiting convolutional neural network on graphs with polynomial approximations of Laplace–Beltrami spectral filtering. Neural Comput. Appl. **33**(20), 13693–13704 (2021)
11. Huang, S.G., Xia, J., Xu, L., Qiu, A.: Spatio-temporal directed acyclic graph learning with attention mechanisms on brain functional time series and connectivity. Med. Image Anal. **77**, 102370 (2022)

12. Jiang, X., Ji, P., Li, S.: CensNet: convolution with edge-node switching in graph neural networks. In: IJCAI, pp. 2656–2662 (2019)
13. Jo, J., Baek, J., Lee, S., Kim, D., Kang, M., Hwang, S.J.: Edge representation learning with hypergraphs. Adv. Neural. Inf. Process. Syst. **34**, 7534–7546 (2021)
14. Jung, R.E., Haier, R.J.: The parieto-frontal integration theory (P-FIT) of intelligence: converging neuroimaging evidence. Behav. Brain Sci. **30**, 135–154 (2007)
15. Kawahara, J., et al.: BrainNetCNN: convolutional neural networks for brain networks; towards predicting neurodevelopment. Neuroimage **146**, 1038–1049 (2017)
16. Lee, H., Chung, M.K., Kang, H., Lee, D.S.: Hole detection in metabolic connectivity of Alzheimer's disease using k-laplacian. In: Golland, P., Hata, N., Barillot, C., Hornegger, J., Howe, R. (eds.) MICCAI 2014. LNCS, vol. 8675, pp. 297–304. Springer, Cham (2014). https://doi.org/10.1007/978-3-319-10443-0_38
17. Li, X., Duncan, J.: BrainGNN: interpretable brain graph neural network for fMRI analysis. bioRxiv (2020)
18. Shen, X., et al.: Using connectome-based predictive modeling to predict individual behavior from brain connectivity. Nat. Protoc. **12**(3), 506–518 (2017)
19. Shi, J., Malik, J.: Normalized cuts and image segmentation. IEEE Trans. Pattern Anal. Mach. Intell. **22**(8), 888–905 (2000)
20. Song, M., et al.: Brain spontaneous functional connectivity and intelligence. Neuroimage **41**, 1168–1176 (2008)
21. Wee, C.Y., et al.: Cortical graph neural network for AD and MCI diagnosis and transfer learning across populations. NeuroImage Clin. **23**, 101929 (2019)
22. Yu, F., Koltun, V.: Multi-scale context aggregation by dilated convolutions. arXiv preprint arXiv:1511.07122 (2015)
23. Zhao, K., Duka, B., Xie, H., Oathes, D.J., Calhoun, V., Zhang, Y.: A dynamic graph convolutional neural network framework reveals new insights into connectome dysfunctions in ADHD. Neuroimage **246**, 118774 (2022)

Modeling the Shape of the Brain Connectome via Deep Neural Networks

Haocheng Dai[1(✉)], Martin Bauer[2,3], P. Thomas Fletcher[4], and Sarang Joshi[1]

[1] University of Utah, Salt Lake City, USA
haocheng.dai@utah.edu
[2] Florida State University, Tallahassee, USA
[3] University of Vienna, Vienna, Austria
[4] University of Virginia, Charlottesville, USA

Abstract. The goal of diffusion-weighted magnetic resonance imaging (DWI) is to infer the structural connectivity of an individual subject's brain in vivo. To statistically study the variability and differences between normal and abnormal brain connectomes, a mathematical model of the neural connections is required. In this paper, we represent the brain connectome as a Riemannian manifold, which allows us to model neural connections as geodesics. This leads to the challenging problem of estimating a Riemannian metric that is compatible with the DWI data, i.e., a metric such that the geodesic curves represent individual fiber tracts of the connectomics. We reduce this problem to that of solving a highly nonlinear set of partial differential equations (PDEs) and study the applicability of convolutional encoder-decoder neural networks (CEDNNs) for solving this geometrically motivated PDE. Our method achieves excellent performance in the alignment of geodesics with white matter pathways and tackles a long-standing issue in previous geodesic tractography methods: the inability to recover crossing fibers with high fidelity. Code is available at https://github.com/aarentai/Metric-Cnn-3D-IPMI.

1 Introduction

Diffusion-weighted magnetic resonance imaging (DWI) enables the non-invasive study of neural connections within the living human brain. DWI measures the local diffusion of water within axonal bundles, allowing for local directional estimation of neural connections. Long distance structural connectivity of the brain is then inferred by the process of tractography, which estimates white matter tracts via various streamlining algorithms. Deterministic tractography [1] computes the integral curves of the vector field associating the most likely direction of fiber tracts with each voxel. However, the simplest deterministic streamline

H. Dai and S. Joshi were supported by NSF grant DMS-1912030. P. T. Fletcher was supported by NSF grant IIS-2205417. M. Bauer was supported by NSF grants DMS-1912037, DMS-1953244 and by FWF grant FWF-P 35813-N.

A. Frangi et al. (Eds.): IPMI 2023, LNCS 13939, pp. 291–302, 2023.
https://doi.org/10.1007/978-3-031-34048-2_23

tractography is sensitive to imaging noise and also easily confounded in crossing-fiber regions. Various approaches, such as Kalman filtering [6], probabilistic tractography [2], and front propagation [10], have been proposed. The collection of tracts in an individual brain estimated by one or the other methods is referred to as the connectome.

Mathematical Models for the Shape of the Connectome: To study the variability in normal populations and to find differences between neural connections in normal and abnormal brains, we need a precise mathematical model of the connectome. Traditionally, individual fiber tracts have been modeled as smooth curves without any intimate link to the underlying geometry of the brain. In geodesic tractography, as proposed by [3,4,10–12,15,17], the brain is modelled as a compact 3D Riemannian manifold, where length-minimizing curves, or *geodesics*, represent individual fiber tracts. Recall that a Riemannian manifold is a real, differentiable manifold M, equipped with a positive-definite inner product on the tangent space at each point. The shape of the Riemannian manifold (and thus the shape of the geodesics) is determined by the local metric. Smooth Riemannian manifolds with the same topology can have very different shapes because of the differing local metric structure.

Related Work on Geodesic DWI Tractography: DWI is the foundation to model an individual brain as a Riemannian manifold. With the Riemannian-metric-equipped manifold, we can infer the white matter pathways and also the shape of an individual's connectome. O'Donnell et al. [17] first proposed the geodesic tractography algorithm that uses the inverse of the diffusion tensor as the Riemannian metric and treats geodesic curves under the metric as white matter pathways. However, there is a tendency in the inverted-tensor metric for geodesics to easily deviate from the principal eigenvector directions in high-curvature areas. To address this issue, Fletcher et al. [10] enhanced the metric by "sharpening" the inverted-tensor metric, i.e., taking the eigenvalues of the metric tensor to some power so as to increase the anisotropy. But this strategy does not take into account the spatially varying curvature of the vector field, and it increases the sensitivity to noise. Fuster et al. [11] demonstrated that using the adjugate of the diffusion tensor field as the Riemannian metric gave improved geodesic tractography over the inverted sharpened metric while being more robust to imaging noise. In order to strengthen the adherence of geodesics to the white matter pathways, Hao et al. [12] developed an adaptive Riemannian metric by applying a conformal scalar field to the inverse of the diffusion tensor, which necessitates solving a Poisson equation on the Riemannian manifold. Campbell et al. [5] further advanced the Riemannian formulation of structural connectomes by introducing methods for diffeomorphic image registration and atlas building using the Ebin metric on the space of Riemannian metrics. The significant advantage of the Riemannian geometric framework is that it enables the formulation of atlas building as a statistical Fréchet mean estimation problem. Furthermore, the entire toolbox of geometrical statistics can now be applied to the statistical analysis of populations of connectomes, which addresses a current challenge in neuroscience—how to statistically quantify the variability of human brain connectivity and differences in the connectome across populations.

Contributions: The principal aim of this work is to innovate the main building block of the Riemannian formulation for structural connectome atlas building: the estimation of a Riemannian metric, such that the corresponding geodesic tractography provides a faithful description of the tractogram, i.e., that the geodesics of the Riemannian metric follow the integral curves of a set of given vector fields (representing fiber orientations). The existing Riemannian metric estimation techniques for DWI data exhibit two major limitations: the accuracy of the Riemannian metric estimation, i.e., alignment of the geodesics with the fiber tracts, and the fact that they are all based on a single DTI model and are thus not able to consider multiple vector fields simultaneously, which limit their ability to model crossing fibers appropriately. More modern modeling techniques, such as HARDI [24], Q-Ball [23] and DSI [25], are able to infer multiple fiber directions at each point in the brain.

In this paper, we show for the first time how one can leverage deep neural networks (DNNs) to estimate a metric structure of the brain that can accommodate fiber crossings, i.e., multiple fiber directions, which is a natural modeling tool to infer the shape of the brain from DWI. We reduce the problem of estimating a Riemannian metric given tractography estimated from DWI, to that of solving a highly nonlinear set of partial differential equations (PDEs) of the form $\mathcal{L}g(x) = f(x), x \in \Omega \subset \mathbb{R}^n$, where $f(x)$ is the given data, and $\mathcal{L}$ is a non-linear differential operator. This allows us to leverage deep learning frameworks that have been proposed for solving such PDEs precisely, where we use convolution encoder-decoder neural networks (CEDNNs) [29] for representing the solution space. CEDNNs use the universal approximation property of fully-connected convolutional networks to approximate the solution function $g(x)$. The weights of the network are estimated to minimize a cost function which incorporates the PDE in a self-supervised manner and is usually of the form $\|\mathcal{L}g(x) - f(x)\|$. CEDNNs are particularly suited when the input data domain is a structured grid such as in the DWI application.

Using this network architecture and spatially discretized vector fields from any of the plethora of models for local fiber directions, our method achieves excellent performance in terms of geodesic-white-matter-pathway alignment. In particular, we show that the proposed method outperforms any of the previously proposed methods in Riemannian metric estimation for geodesic tractography. In addition to simple deployment and boundary insensitivity, our approach also tackles the long-standing issue in previous methods: the inability to recover crossing fibers with high fidelity. Towards this aim, we exhibit that the metric estimation is able to faithfully represent multiple vector fields as geodesic vector fields of the estimated metric. We inherit the validity of the tracts from the choice of the preferred local directional estimation algorithm, which is explicitly not the focus of this work. The algorithms presented herein are a part of the overall program to study the human brain and its variability in populations.

2   Estimating Riemannian Metrics from Geodesics

In this section, we will introduce a new inverse problem that will be at the center of our approach: the estimation of a Riemannian metric based on the observation of (possibly) multiple fiber directions as the tangents to geodesic curves. We will first recall some definitions and concepts from Riemannian geometry. For further details, we refer to classic textbooks such as [8]. In all of this work, our modeling space is a finite-dimensional manifold M (possibly with boundary). In our application, the topology of the manifold M will be rather trivial and thus we assume in the following that M is a bounded subset of $\mathbb{R}^n$ with $n \in \{2, 3\}$.

Next we introduce the concept of an integral curve: given a vector field $\mathbf{v} \in \mathfrak{X}(M)$, i.e., a map from M to TM, we call a curve $\gamma : \mathbb{R} \to M$ an integral curve of $\mathbf{v}$ if $\partial_t \gamma(t) = \mathbf{v}(\gamma(t))$, i.e., the curve follows the flow lines of the vector field. A Riemannian metric g on M is a family of inner products on each tangent space $T_x M$ that depend smoothly on the base point, $x \in M$. Note that in local coordinates we can identify the Riemannian metric with a field of positive-definite, symmetric matrices $g(x)$, and the inner product between two tangent vectors $v, w \in T_x M$ is simply given by $\langle v, w \rangle_g = v^T g(x) w$. We call a curve between p and q a (minimizing) geodesic if it minimizes the length functional $L(\gamma) = \frac{1}{2} \int_0^1 \sqrt{\langle \partial_t \gamma, \partial_t \gamma \rangle_{g_\gamma}} dt$, where g_γ denotes the Riemannian metric at $\gamma(t)$. For every Riemannian metric there exists a unique connection ∇^g, called the Levi-Civita covariant derivative, which encodes this notion of geodesic curves, i.e., a curve γ is a geodesic if and only if it satisfies the equation $\nabla^g_{\mathbf{v}} \mathbf{v} = \sigma \mathbf{v}$, where $\mathbf{v} = \partial_t \gamma$, and $\sigma = \langle \mathbf{v}, \nabla^g_{\mathbf{v}} \mathbf{v} \rangle_g / \|\mathbf{v}\|^2_g$. We call a vector field a unit geodesic vector field if all its integral curves are geodesics with constant speed, i.e., $\nabla^g_{\mathbf{v}} \mathbf{v} = 0$. Integral curves of both geodesic as well as unit geodesic vector fields are length minimizing—they only differ in their parameterization along the curve. Given a vector field $\mathbf{v}$ we aim to find a Riemannian metric g such that $\mathbf{v}$ is a geodesic vector field. The question under what conditions such a Riemannian metric exists has been intensively investigated, see e.g. [19]. Note that if $\mathbf{v}$ is a non-vanishing geodesic vector field of a metric g, then $\mathbf{v}/\|\mathbf{v}\|_g$ is a unit geodesic vector field. We found, however, that our model is significantly harder to optimize, when insisting on unit geodesic vector fields.

We are now able to formulate the inverse problem studied in this paper as:

Regularized, inexact metric estimation: Given vector fields $\mathbf{v}_i \in \mathfrak{X}(M)$, $i \in \{1, \dots, m\}$, find the Riemannian metric g on M that minimizes the energy functional

$$\mathcal{E}(g) = \sum_{i=1}^m \|\nabla^g_{\mathbf{v}_i} \mathbf{v}_i - \sigma_i \mathbf{v}_i\|_2 + \alpha \operatorname{Reg}(g), \qquad (1)$$

where $\alpha > 0$ is a weight parameter.

Here the first term enforces the condition that the vector fields $\mathbf{v}_i$ are (close to being) geodesic vector fields, while the second term is a regularization parameter that is responsible for the solution selection. We have investigated several regularization terms, such as the Frobenius norm of the difference to the Euclidean metric. In our experience, adding this explicit regularization terms did not improve the performance of the algorithm, which suggests that the implicitly regularization properties via the solution parameterization as a neural network are sufficient for the proposed application.

To get a better understanding of the above loss function, we can write a coordinate expression of ∇^g for a vector field $\mathbf{v}$ as:

$$\nabla^g_{\mathbf{v}} \mathbf{v} = \sum_k \left(\sum_i v^i \frac{\partial v^k}{\partial x^i} + \sum_{i,j} \Gamma^k_{ij} v^i v^j \right) \mathbf{e}_k, \tag{2}$$

where $\mathbf{v} = v^i \mathbf{e}_i$ with $\mathbf{e}_i = \frac{\partial}{\partial x^i}$ being the i-th basis vector. Furthermore, Γ^k_{ij} are the Christoffel symbols, which are defined as $\Gamma^k_{ij} = \frac{1}{2} \sum_{l=1}^n g^{kl} \left(\frac{\partial g_{jl}}{\partial x^i} + \frac{\partial g_{il}}{\partial x^j} - \frac{\partial g_{ij}}{\partial x^l} \right)$, where g_{ij} denotes the entries of the Riemannian metric g, and g^{ij} represents the entries of the inverse of metric g^{-1}.

3   Algorithms and Implementation

Metric Estimation via Neural Networks: We will now present a novel deep learning framework [29] for solving the inverse problem formulated above, which employs a convolutional encoder-decoder neural network (CEDNN) approach to construct the multi-scale features from high-dimensional input. By wrapping the input vector fields and the output Riemannian metric into the loss function, the network is trained to capture the heterogeneous mapping between the given vector fields and the resulting solution (metric). CEDNN takes the whole spatially discretized vector fields as the input and outputs the entire metric field which minimizes Eq. (1).

The CEDNN architecture adopts the dense block [14] paradigm, which furthers the ideas of residual learning in ResNet [13] and bypassing paths in highway networks [21] by concatenating every previous layer's output as the input of the current layer in a feed-forward fashion. The dense connectivity in the dense block improves the information flow in the network, without introducing any optimization difficulty. The encoder contracts the higher-level context and feature of the input, while the decoder commits to recovering the location information to the same scale as the original input fields.

Our CEDNN implementation takes an $m \times n$-channel input, where n is the dimension of the vector fields and m is the number of distinct geodesic vector fields $\mathbf{v}_i$ used for the metric estimation. By sending the concatenated vector fields into the network, CEDNN yields an output of $((n+1) \times n/2 + 1)$-channel tensor. We form the final estimated metric, the $n \times n$ symmetric positive-definite (SPD) matrix g, through eigencomposition: $g = \mathbf{R} \Lambda \mathbf{R}^T$, where the rotation matrix $\mathbf{R}$ follows Rodrigues' rotation formula: $\mathbf{R} = \mathbf{I} + (\sin \theta)\mathbf{K} + (1 - \cos \theta)\mathbf{K}^2$, Λ is a diagonal matrix with positive diagonal entries, $\mathbf{I}$ is an identity matrix, and $\mathbf{K}$ is a skew-symmetric matrix. $\mathbf{R}, \Lambda, \mathbf{K}, \mathbf{I} \in \mathbb{R}^{n \times n}, \theta \in \mathbb{R}$ and we enforce the diagonal entries in Λ to be positive through exponential function. $\Lambda, \mathbf{K}, \theta$ are respectively parameterized by n, $n \times (n - 1)/2$ and 1 real numbers, the sum of which is equivalent to the channel number of the output tensor. This formulation expedites the training speed by 5 folds, compared to forming a SPD matrix via built-in `torch.matrix_exp`. To compute spatial gradients, we adopt a central finite-difference scheme to approximate the derivatives in the loss function as given in Eq. (1). For more details on the architecture and workflow, see Fig. 1.

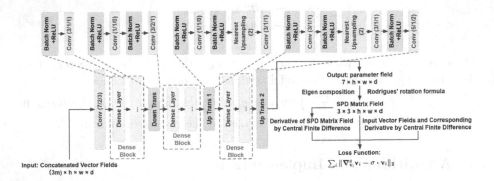

Fig. 1. The architecture of the proposed convolutional encoder-decoder neural networks for 3D solution. Here, h, w, d denote the shape of the input vector fields, and m represents the number of total vector fields. The numbers in a *Conv* box stand for the kernel size, stride, and padding of the convolution, respectively. The number in a *Nearest Upsampling* box indicates the scaling factor.

Geodesic Tractography: Once we have estimated a neural network representation of the metric g, we are able to calculate the Christoffel symbols and then integrate the geodesic equation (Eq. (2)), which directly leads to the corresponding geodesic tractography. Recall that in three-dimension situation, the geodesic Eq. (2) can be written as the following system of second-order ODEs: $\partial_t^2 \gamma^{(k)}(t) + \sum_{i,j=1}^{3} \partial_t \gamma^{(i)}(t) \cdot \Gamma_{i,j}^k(\gamma(t)) \cdot \partial_t \gamma^{(j)}(t) = \sigma \mathbf{v}^{(k)}(\gamma(t))$, where $\partial_t^2 \gamma^{(k)}, \partial_t \gamma^{(k)}$ are the k-th components of the acceleration and velocity vectors and where $\Gamma_{ij}^k(\gamma(t))$ are the Christoffel symbols evaluated at position $\gamma(t)$. Geodesic shooting solves this second-order ODE with a set of initial conditions: the starting position of the geodesic $\gamma(0)$ and its starting velocity $\partial_t \gamma(0)$.

4   Experiments

In all of the experiments presented below, the Adadelta optimizer was employed to minimize the loss function. We also experimented with various other optimizers, including AdaGrad, Adadelta, and Adam. In our experience, the Adadelta optimizer was consistently superior. All computations were carried out on an Nvidia Titan RTX GPU. Under this setting, a 2D example took less than 3 min (3,000 iterations) to achieve convergence, and a 3D example took less than 30 min (10,000 iterations) for convergence.

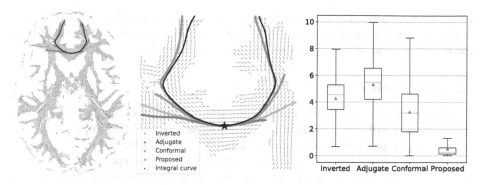

Fig. 2. Left: axial view of a projected vector field derived from HCP 100610 DWI. Central: detailed view of left panel, the geodesics and integral curve start from the star. Right: Box plot of mean min errors between the integral curve and geodesics generated by different methods over 38 HCP brain subjects. Green triangles stand for the mean of the mean min errors, and orange bars represent the median of the mean min errors. (Color figure online)

4.1 Validation and Comparison: 2D Brain Slices

In our first experiment, we compared the geodesic-white-matter-pathway alignment of the proposed method to state-of-the-art geodesic tractography methods: the **inverted** diffusion tensor metric [17], the **adjugate** of the diffusion tensor [11], and the **conformal** metric [12]. We want to emphasize that all these baseline methods are approximating the metric based on a single diffusion tensor image, which yields only one corresponding vector field. Consequently, the crossing fiber estimating ability will not be tested in this section. For easy visualization and interpretability, we performed the comparison on projected 2D brain slices from the Human Connectome Project (HCP) [26], cf. Figure 2. The left and central panels in Fig. 2 demonstrate an example of geodesics shooting from a seed point in the genu of the corpus callosum. The geodesics derived by the other three methods deviate from the integral curve eventually, while ours provides a significantly better alignment to the ground truth.

To quantitatively measure the geodesic-white-matter-pathway alignment over these methods, we tested these algorithms on brain slices from 38 HCP brain subjects. On each brain slice, we uniformly cast 400 seed points in the genu of corpus callosum region, where all the seed points are chosen to be non-grid points with the corresponding vectors being obtained by bi-linear interpolation, i.e., these vectors have never been seen by the network. We then integrated the geodesic and integral curve from each seed point and calculate the error of the geodesic to the corresponding integral curve. To calculate the error of curve Q to curve P, we view P, Q as finite point sets and consider the mean min error between these two sets as $\text{Error}(P, Q) = \frac{1}{|P|} \sum_{p \in P} \min_{q \in Q} \|p - q\|_2$, where P denotes an integral curve and Q a geodesic tractography curve starting at the same seed point. The boxplot in Fig. 2 visualizes the distribution of 15,200 integral-curve-geodesic error sample points across 38 HCP subjects. The boxplot demonstrates our method outperforms the other three methods in terms of both mean and median of mean min errors by a large margin.

4.2 2D Synthetic "Braid"

In this section, we aim to compare the metric estimation ability of CEDNNs with a baseline physics-informed neural networks (PINNs) [18] implementation, where we pay particular attention to the ability of representing crossing fibers. The baseline PINNs use multilayer perceptrons (MLPs) to represent the solution space. The loss function in the baseline PINNs is formulated the same way as in the CEDNN implementation, where the derivative of metric w.r.t. the spatial coordinates was computed by taking advantage of the automatic differentiation engine autograd in PyTorch.

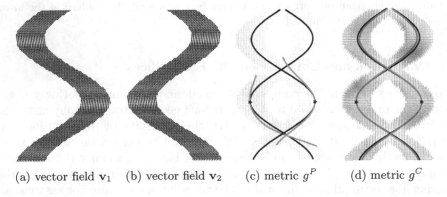

(a) vector field $\mathbf{v}_1$ (b) vector field $\mathbf{v}_2$ (c) metric g^P (d) metric g^C

Fig. 3. (a) input vector field $\mathbf{v}_1$; (b) input vector field $\mathbf{v}_2$; (c) integral curves (black) running on input vector fields and geodesics (indigo) running on PINN-estimated Riemannian metric field g^P at iteration 1000 (background ellipses represent metric tensors), shooting from seed points (star) with the same initial velocity vector as the corresponding integral curve; (d) integral curves (black) and geodesics (indigo) on CEDNN-estimated Riemannian metric field g^C at iteration 1000 (background ellipses represent metric tensors), shooting from seed points (star) with the same initial velocity vector as the corresponding integral curve.

For the experimental data, we synthesized two vector fields in a "braid" pattern of two intertwining pathways (see (a) and (b) in Fig. 3). The central integral curves of the vector bundle are two trigonometric functions: $x_2 = 20\cos(\frac{1}{4\pi}(x_1 - 60)) + 50$ and $x_2 = 20\sin(\frac{1}{4\pi}x_1) + 50$, where x_1, x_2 are spatial coordinates. We then constructed the curve bundle by translating the central integral curve across nine pixels horizontally. The vector field is generated by calculating the tangent vector of the curves at each point, making the curves integral to the vector field. The aim is to estimate a Riemannian metric field such that these curves are geodesic curves.

The CEDNN in our experiment is configured with the following hyper-parameters: the number of dense layers in the three dense blocks are 6, 8, 6, with a growth rate of 16, thereby leading to a total of 747,147 parameters. We used an initial learning rate of 1×10^{-4} for the optimization. Figure 4 presents the loss $\sum_{i=1}^{m} \|\nabla_{\mathbf{v}_i}^g \mathbf{v}_i - \sigma_i \mathbf{v}_i\|_2$ on a log-scale at each iteration. The baseline PINN uses an approximately equivalent amount of parameters as the CEDNN. PINNs with different sizes were extensively explored, however, we did not observe any salient difference in final loss brought by these configurations. In addition, we configured the PINN with Fourier embedding [22] and Siren activation functions [20], which boosted the performance of the PINN in a considerable magnitude, yet still underperforming CEDNN by

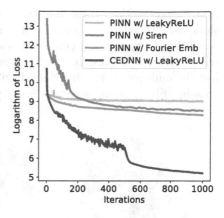

Fig. 4. Log-scaled (base 10) loss convergence comparison between PINN with LeakyReLU activation, with Siren activation, with Fourier embedding and CEDNN with leakyReLU on 2D synthetic "Braid".

orders of magnitudes: in Fig. 4, it is noticeable that the CEDNN converges much faster to a significantly lower residual loss despite the fact that PINN enjoys about the same amount of parameters as the CEDNN. This enforces the conclusions of [16] that the limitation exhibited by PINNs is due to optimization difficulties—irrespective of expressibility of the solution. See also [7], which reports several pitfalls in using PINNs for fluid dynamic simulations.

Figure 3 shows the metric fields g^P (Fig. 3(c)) and g^C (Fig. 3(d)) estimated by PINN and CEDNN respectively and the alignment of ground truth integral curves (black) and geodesics (indigo) associated with the estimated metric: the geodesics on the metric g^C align notably closer to the ground truth integral curves than the one on g^P. In addition to the excellent integral-curve-geodesic alignment, the CEDNN-estimated metric behaves as expected even at the crossing region—the geodesics are not confounded at the crossing.

4.3 3D Brains and Crossing Fibers

In this experiment, we validated our method's ability to estimate 3D crossing-fiber regions in brain DWI from several HCP subjects. We first reconstructed the vector fields through the GQI method [27] in DSI Studio with a diffusion sampling length ratio of 1.25. A whole-brain Riemannian metric was estimated by a CEDNN featuring 40, 30, and 40 dense layers in each dense block. The model was trained with an initial learning rate of 3×10^{-4} for a total of 1×10^4 iterations. The top row in Fig. 5 shows the resulting whole-brain connectome visualized using 3D Slicer via the SlicerDMRI plug-in [28]. There were 138,732 seed points cast in the white matter region for the generation of the geodesics. The color indicates the orientation of the fiber tracts: red (left/right), green (anterior/posterior), blue (superior/inferior). In the bottom row of Fig. 5, we showcase the ability of geodesic tractography with our estimated metric to successfully distinguish two crossing fibers: the forceps minor and frontal projection tracts. We stress that the previous approaches to geodesic tractography do not handle multiple fiber directions in a voxel, and thus cannot correctly handle crossing-fiber regions such as these.

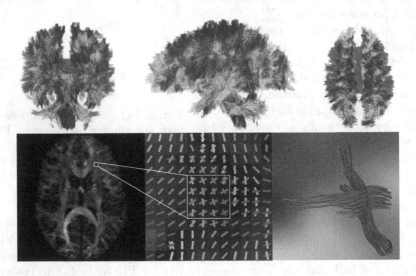

Fig. 5. Top row, left to right: coronal, sagittal, and transversal view of the whole-brain connectome generated by the proposed method. Bottom left: axial view of the same subject. Bottom central: zoom-in of a $4 \times 4 \times 4$ crossing-fiber region. Bottom right: geodesic tractography by proposed method within the same window, the orientation of which matches the vector field in the bottom central panel.

5   Conclusion, Limitations and Future Work

In this paper, we have shown for the first time how to leverage the flexibility of deep learning to model the shape of the human connectome by estimating a Riemannian metric of the brain manifold that faithfully represents the white matter connectivity. We show that our proposed method outperforms any of the previously proposed methods in geodesic tractography by a large margin. In addition our approach solves the long-standing issue of these previous methods: the inability to recover crossing fibers with high fidelity. One limitation of the proposed method is that the generalization ability of the trained model to the unseen data is relatively weak. Nevertheless, we would point out that currently the largest human connectome datasets are in the order of only 1000 s of subjects and thus we do not believe that the adaptivity of the models limits the applicability of our method, as the total training time for a single 3D brain is less than 30 min in our setup. With the ability to robustly and efficiently model the white matter of the brain as a Riemannian manifold, one can directly apply geometrical statistical techniques such as statistical atlas construction [5], principal geodesic analysis [9], and longitudinal regression to precisely study the variability and differences in white matter architecture.

References

1. Basser, P.J., Pajevic, S., Pierpaoli, C., Duda, J., Aldroubi, A.: In vivo fiber tractography using DT-MRI data. Magn. Reson. Med. **44**(4), 625–632 (2000)
2. Behrens, T.E., et al.: Characterization and propagation of uncertainty in diffusion-weighted MR imaging. Magn. Resonance Med. Official J. Int. Soc. Magn. Resonance Med. **50**(5), 1077–1088 (2003)
3. Bihonegn, T., Kaushik, S., Bansal, A., Vojtíšek, L., Slovák, J.: Geodesic fiber tracking in white matter using activation function. Comput. Methods Programs Biomed. **208**, 106283 (2021)
4. Bihonegn, T.T., Bansal, A., Slovák, J., Kaushik, S.: 4th order tensors for multi-fiber resolution and segmentation in white matter. In: 2020 7th International Conference on Biomedical and Bioinformatics Engineering, pp. 36–42 (2020)
5. Campbell, K.M., Dai, H., Su, Z., Bauer, M., Fletcher, P.T., Joshi, S.C.: Structural connectome atlas construction in the space of Riemannian metrics. In: Feragen, A., Sommer, S., Schnabel, J., Nielsen, M. (eds.) IPMI 2021. LNCS, vol. 12729, pp. 291–303. Springer, Cham (2021). https://doi.org/10.1007/978-3-030-78191-0_23
6. Cheng, G., Salehian, H., Forder, J.R., Vemuri, B.C.: Tractography from HARDI using an intrinsic unscented kalman filter. IEEE Trans. Med. Imaging **34**(1), 298–305 (2014)
7. Chuang, P.Y., Barba, L.A.: Experience report of physics-informed neural networks in fluid simulations: pitfalls and frustration. arXiv preprint arXiv:2205.14249 (2022)
8. Do Carmo, M.P., Flaherty Francis, J.: Riemannian Geometry, vol. 6. Springer, New York (1992). https://doi.org/10.1007/978-0-387-29403-2
9. Fletcher, P.T., Lu, C., Pizer, S.M., Joshi, S.: Principal geodesic analysis for the study of nonlinear statistics of shape. IEEE Trans. Med. Imaging **23**(8), 995–1005 (2004)
10. Fletcher, P.T., Tao, R., Jeong, W.-K., Whitaker, R.T.: A volumetric approach to quantifying region-to-region white matter connectivity in diffusion tensor MRI. In: Karssemeijer, N., Lelieveldt, B. (eds.) IPMI 2007. LNCS, vol. 4584, pp. 346–358. Springer, Heidelberg (2007). https://doi.org/10.1007/978-3-540-73273-0_29
11. Fuster, A., Haije, T.D., Tristán-Vega, A., Plantinga, B., Westin, C.F., Florack, L.: Adjugate diffusion tensors for geodesic tractography in white matter. J. Math. Imaging Vis. **54**(1), 1–14 (2016)
12. Hao, X., Whitaker, R.T., Fletcher, P.T.: Adaptive Riemannian metrics for improved geodesic tracking of white matter. In: Székely, G., Hahn, H.K. (eds.) IPMI 2011. LNCS, vol. 6801, pp. 13–24. Springer, Heidelberg (2011). https://doi.org/10.1007/978-3-642-22092-0_2
13. He, K., Zhang, X., Ren, S., Sun, J.: Deep residual learning for image recognition. In: Proceedings of the IEEE Conference on Computer Vision and Pattern Recognition, pp. 770–778 (2016)
14. Huang, G., Liu, Z., Van Der Maaten, L., Weinberger, K.Q.: Densely connected convolutional networks. In: Proceedings of the IEEE Conference on Computer Vision and Pattern Recognition, pp. 4700–4708 (2017)

15. Kaushik, S., Kybic, J., Bansal, A., Bihonegn, T., Slovak, J.: Potential biomarkers from positive definite 4th order tensors in HARDI. In: 2021 IEEE 18th International Symposium on Biomedical Imaging (ISBI), pp. 1003–1006. IEEE (2021)
16. Krishnapriyan, A., Gholami, A., Zhe, S., Kirby, R., Mahoney, M.W.: Characterizing possible failure modes in physics-informed neural networks. In: Ranzato, M., Beygelzimer, A., Dauphin, Y., Liang, P.S., Wortman Vaughan, J. (eds.) Advances in Neural Information Processing Systems, vol. 34, pp. 26548–26560. Curran Associates, Inc. (2021). https://proceedings.neurips.cc/paper_files/paper/2021/file/df438e5206f31600e6ae4af72f2725f1-Paper.pdf
17. O'Donnell, L., Haker, S., Westin, C.-F.: New approaches to estimation of white matter connectivity in diffusion tensor MRI: elliptic PDEs and geodesics in a tensor-warped space. In: Dohi, T., Kikinis, R. (eds.) MICCAI 2002. LNCS, vol. 2488, pp. 459–466. Springer, Heidelberg (2002). https://doi.org/10.1007/3-540-45786-0_57
18. Raissi, M., Perdikaris, P., Karniadakis, G.E.: Physics-informed neural networks: a deep learning framework for solving forward and inverse problems involving nonlinear partial differential equations. J. Comput. Phys. **378**, 686–707 (2019)
19. Rechtman, A.: Existence of periodic orbits for geodesible vector fields on closed 3-manifolds. Ergod. Theory Dynam. Syst. **30**(6), 1817–1841 (2010)
20. Sitzmann, V., Martel, J.N., Bergman, A.W., Lindell, D.B., Wetzstein, G.: Implicit neural representations with periodic activation functions. In: arXiv (2020)
21. Srivastava, R.K., Greff, K., Schmidhuber, J.: Training very deep networks. Adv. Neural Inf. Process. Syst. **28** (2015)
22. Tancik, M., et al.: Fourier features let networks learn high frequency functions in low dimensional domains. Adv. Neural Inf. Process. Syst. **33**, 7537–7547 (2020)
23. Tuch, D.S.: Q-ball imaging. Magn. Resonance Med. Official J. Int. Soc. Magn. Resonance Med. **52**(6), 1358–1372 (2004)
24. Tuch, D.S., Reese, T.G., Wiegell, M.R., Makris, N., Belliveau, J.W., Wedeen, V.J.: High angular resolution diffusion imaging reveals intravoxel white matter fiber heterogeneity. Magn. Resonance Med. Official J. Int. Soc. Magn. Resonance Med. **48**(4), 577–582 (2002)
25. Tuch, D.S., et al.: Diffusion MRI of complex tissue structure. Ph.D. thesis, Massachusetts Institute of Technology (2002)
26. Van Essen, D.C., et al.: The human connectome project: a data acquisition perspective. Neuroimage **62**(4), 2222–2231 (2012)
27. Yeh, F.C., Wedeen, V.J., Tseng, W.Y.I.: Generalized q-sampling imaging. IEEE Trans. Med. Imaging **29**(9), 1626–1635 (2010)
28. Zhang, F., et al.: SlicerDMRI: diffusion MRI and tractography research software for brain cancer surgery planning and visualization. JCO Clin. Cancer Inform. (4), 299–309 (2020)
29. Zhu, Y., Zabaras, N., Koutsourelakis, P.S., Perdikaris, P.: Physics-constrained deep learning for high-dimensional surrogate modeling and uncertainty quantification without labeled data. J. Comput. Phys. **394**, 56–81 (2019)

TetCNN: Convolutional Neural Networks on Tetrahedral Meshes

Mohammad Farazi[1]([✉]), Zhangsihao Yang[1], Wenhui Zhu[1], Peijie Qiu[2], and Yalin Wang[1]

[1] School of Computing and Augmented Intelligence, Arizona State University, Tempe, AZ 85281, USA
mfarazi@asu.edu
[2] McKeley School of Engineering, Washington University in St. Louis, St. Louis, MO 63130, USA

Abstract. Convolutional neural networks (CNN) have been broadly studied on images, videos, graphs, and triangular meshes. However, it has seldom been studied on tetrahedral meshes. Given the merits of using volumetric meshes in applications like brain image analysis, we introduce a novel interpretable graph CNN framework for the tetrahedral mesh structure. Inspired by ChebyNet, our model exploits the volumetric Laplace-Beltrami Operator (LBO) to define filters over commonly used graph Laplacian which lacks the Riemannian metric information of 3D manifolds. For pooling adaptation, we introduce new objective functions for localized minimum cuts in the Graclus algorithm based on the LBO. We employ a piece-wise constant approximation scheme that uses the clustering assignment matrix to estimate the LBO on sampled meshes after each pooling. Finally, adapting the Gradient-weighted Class Activation Mapping algorithm for tetrahedral meshes, we use the obtained heatmaps to visualize discovered regions-of-interest as biomarkers. We demonstrate the effectiveness of our model on cortical tetrahedral meshes from patients with Alzheimer's disease, as there is scientific evidence showing the correlation of cortical thickness to neurodegenerative disease progression. Our results show the superiority of our LBO-based convolution layer and adapted pooling over the conventionally used unitary cortical thickness, graph Laplacian, and point cloud representation.

Keywords: Magnetic resonance imaging · Convolutional neural networks · Tetrahedral meshes · Laplace-Beltrami operator

1 Introduction

Since the emergence of geometric deep learning research, many researchers have sought to develop learning methods on non-Euclidean domains like point clouds, surface meshes, and graphs [3]. In brain magnetic resonance imaging (MRI) analysis, geometric deep learning has been widely employed for applications in brain network analysis, parcellation of brain regions, and brain cortical surface

© The Author(s), under exclusive license to Springer Nature Switzerland AG 2023
A. Frangi et al. (Eds.): IPMI 2023, LNCS 13939, pp. 303–315, 2023.
https://doi.org/10.1007/978-3-031-34048-2_24

analysis [2,5,9,13]. In a benchmark study [9], authors addressed the common limitations of widely used graph neural networks (GNNs) on cortical surface meshes. While the majority of these studies focus on using voxel representation and surface mesh, limitations like limited grid resolution cannot characterize complex geometrical curved surfaces precisely [27]. Cortical thickness is a remarkable AD imaging biomarker; therefore, building learning-based methods will be advantageous by exploiting volumetric meshes over surface meshes since the thickness is inherently embedded in volume [4]. Using volumetric mesh representation also potentially helps with the over-squashing nature of Message Passing Neural Networks (MPNN) by interacting with long-range nodes through interior nodes. Thus, developing efficient volumetric deep learning methods to analyze grey matter morphometry may provide a means to analyze the totality of available brain shape information and open up new opportunities to study brain development and intervention outcomes.

While a manifold can be represented as a graph to employ graph convolutional networks, the majority of GNN models are not suitable for applying to volumetric mesh data. First, the Riemannian metric is absent in a uniform graph representation. Second and foremost, these methods mostly, are not scalable to very large sample sizes like tetrahedral meshes with millions of edges and vertices. Although there are a few methods tailored specifically to work on 3D surface meshes like MeshCNN (Convolutional Neural Networks) [15], these methods are particularly designed for triangular meshes and are not scalable to meshes with a high number of vertices. Consequently, to opt for a method both scalable to tetrahedral meshes and computationally inexpensive, a framework like ChebyNet [6], modified with volumetric LBO over graph Laplacian, is an appropriate candidate to adapt for the tetrahedral meshes.

Although deep learning on mesh has been studied in recent years, few studies have employed explainable methods for qualitative assessment. Generally, to better explain geometric deep learning models, some methods have been proposed in recent years. Gradient-weighted Class Activation Map (Grad-CAM) on graphs [23] is one of the first methods, using the gradient of the target task flowing back to the final convolution layer to create a localization map to visualize important nodes in the prediction task. This technique is commonplace in CNN and GNN models, however, the generalization of the concept is rarely used on surface mesh data [1]. Specifically, Grad-CAM has never been investigated on volumetric meshes, and it is worth generalizing such an explainable technique for the medical image analysis community for a better interpretation of the deep learning model.

Motivated by the prior work [6,16], here we propose to develop Tetrahedral Mesh CNN (TetCNN) to address the issues mentioned above. Using the tetrahedral Laplace-Beltrami operator (LBO) over graph Laplacian, we use the Riemannian metric in tetrahedral meshes to capture intrinsic geometric features. Figure 1 demonstrates that the LBO successfully characterizes the difference between two mesh structures while the graph Laplacian fails. Additionally, we propose novel designs on the pooling layers and adopt the polynomial approximation [14] for computational efficiency. The main contributions of this paper, thus, are summarized as follows: (1) TetCNN is the first of its kind and an exclusive

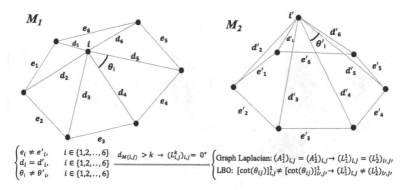

Fig. 1. Illustration of comparison between LBO and graph Laplacian-based spectral filters represented by k^{th} order polynomials of a mesh in 1-ring neighbor of a given vertex (i in M_1, i' in M_2). Based on [14], the Laplacian of kth order polynomials are exactly k localized, therefore, we use $(L_m^i)_{i,j}$ to show the 1-localized Laplacian and $(A_m^i)_{i,j}$ 1-localized adjacency matrix around vertices i and i' in this example. As depicted, two meshes have similar corresponding edge-length within the 1-ring of vertex i and i'. Thus, the **1-localized** graph Laplacian of both meshes is similar while their **1-localized** LBO are different due to differences in cotangent matrix weights. The surface mesh is used for simplified intuition. * is based on Lemma 5.2 in [14].

geometric deep-learning model on tetrahedral meshes. (**2**) We use volumetric LBO to replace graph Laplacian adopted in ChebyNet [6]. (**3**) We re-define the Graclus algorithm [7] used in [1,6] by adapting a localized minimum-cut objective function using the cotangent and mass matrix. (**4**) We approximate the LBO on down-sampled mesh with the piece-wise linear approximation function. This avoids the re-computation of Laplacian in deeper layers. (**5**) We demonstrate the generalization of Grad-CAM to the tetrahedral mesh may be used for biomarker identification. Our extensive experiments demonstrate the effectiveness of our proposed TCNN framework for AD research.

2   Methods

In our LBO-based TetCNN framework, first, we pre-compute the volumetric LBO for each tetrahedral mesh. Secondly, together with the LBO, we feed into the network a set of input features for each vertex, like the 3D coordinates of each vertex. Having built a new graph convolution layer based on the LBO, we need to down-sample the mesh with an efficient down-sampling and pooling layer to learn hierarchical feature representation for the large-sized input data. In Fig. 2, we illustrate the pipeline for the binary classification task by defining specific components of our deep learning model.

2.1   Tetrahedral Laplace Beltrami Operator (LBO)

Let T represent the tetrahedral mesh with a set of vertices $\{v_i\}_{i=1}^n$ where n denotes the total number of vertices, and Δ_{tet} be the volumetric LBO on T,

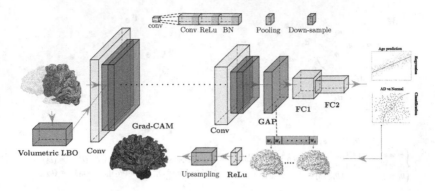

Fig. 2. TetCNN architecture for the classification task. Pre-computed LBO and xyz features are fed to the network with 5 layers. Each layer includes a down-sampling of size 1/4 and a pooling layer afterward except for *"conv5"*, which consists of a global average pooling (GAP). Fully connected (FC) layers and a Sigmoid activation function are used for the binary classification at the end. Grad-CAM is adopted to visualize important biomarkers.

which is a linear differential operator. For a Riemannian manifold, given $f \in C^2$, a real-valued function, the eigen-system of Laplacian is $\Delta_{tet} f = -\lambda f$. The solution to this eigen-system problem can be approximated by a piece-wise linear function f over the tetrahedral mesh T [27]. As proposed in [27], the lumped discrete LBO on T is defined as follows:

$$\Delta f(v_i) = \frac{1}{d_i} \sum_{j \in N(i)} k_{i,j}(f(v_i) - f(v_j)) \tag{1}$$

where $N(i)$ includes the adjacent vertices of vertex v_i, d_i is total tetrahedral volume of all adjacent tetrahedra to vertex v_i, and $k_{i,j}$ is the string constant. Now, we define the stiffness matrix as $A = W - K$ in which $W = diag(w_1, w_2, ..., w_n)$ is the diagonal matrix comprised of weights $w_i = \sum_{j \in N(i)} k_{i,j}$. The detailed definition of $k_{i,j}$ is described in [27]. For A_{ij} we have:

$$A_{i,j} = \begin{cases} k_{i,j} = \frac{1}{12} \sum_{m=1}^{k} l_m^{(i,j)} cot(\theta_m^{(i,j)}), & \text{if } (i,j) \in E. \\ 0, & \text{if } (i,j) \notin E. \\ -\sum_{q \subseteq N(i)} k_{i,q} = -\sum_{q \subseteq N(i)} \frac{1}{12} \sum_{m=1}^{k} l_m^{(i,q)} cot(\theta_m^{(i,q)}), & \text{if } i = j, \end{cases} \tag{2}$$

where $l_m^{(i,j)}$ is the length of the opposite edge to (v_i, v_j) in tetrahedron m sharing (v_i, v_j), $N(i)$ is the set of adjacent vertices to (v_i), E is the set of all edges in T, and finally $\theta_m^{i,j}$ is the diheadral angle of (v_i, v_j) in tetrahedron m. Now, we define the lumped discrete tetrahedral LBO L_{tet} given A and the volume mass matrix D [27]:

$$L_{tet} = D^{-1}A, \tag{3}$$

in which $D = diag(d_1, d_2, ..., d_n)$.

2.2 Spectral Filtering of Mesh Signals with Chebyshev Polynomial Approximation

We define the input signal on the mesh as $x_{in} \in R^N$ and the output of the convolved signal with filter g as $x_{out} \in R^M$. We denote the convolution operator on tetrahedral mesh T with $*_T$. Following the duality property of convolution in the time domain, and having the eigenvalue and eigen-functions of tetrahedral LBO at hand, we define the convolution as:

$$x_{out} = g *_T x_{in} = \Phi((\Phi^T g) \odot (\Phi^T x_{in})) = \Phi f(\Lambda) \Phi^T x_{in}, \qquad (4)$$

in which $\odot$ is the element-wise product, $f(\Lambda)$ is general function based on the eigen-value matrix Λ, and Φ is the eigen-vector matrix. In [6], authors approximated the function f with the linear combination of k-order power of Λ matrix as polynomial filters:

$$f(\Lambda) = \sum_{m=0}^{K-1} \theta_m \Lambda_{tet}^m, \qquad (5)$$

This formulation is localized in space and computationally less expensive than an arbitrary non-parametric filter $f(\Lambda)$. Per [6], the convolution of kernel $f(.)$ centered at vertex i with delta function δ_i given by $(f(L)\delta_i)_j = \sum_k \theta_k (L^k)_{i,j}$ gives the value at vertex j. Interestingly, since the $(L^k)_{i,j}$ is K−localized, i.e., $(L^k)_{i,j} = 0$ if $d(i,j) > K$, the locality is guaranteed with spectral filters approximated with $k - th$ polynomials of LBO [6].

Now, by plugging Eq. 5 in Eq. 4, the convolution can be expressed in terms of the Laplacian itself without any further need to calculate the eigen-functions. Chebyshev polynomials provide a boost in computational efficiency with a closed recursive formulation:

$$x_{out} = \sum_{m=0}^{K} \theta_m T_m(L_{tet}) x_{in}, \qquad (6)$$

where θ_m are a set of learnable model parameters denoting the coefficients of the polynomials, and $T_m \in R^{n*n}$ is the Chebyshev polynomial of order k.

Recursive Formulation of Chebyshev Polynomials. The main idea of using polynomial approximation is to avoid the costly eigendecomposition and multiplication with Φ. Therefore, we parameterize $f(\Lambda_{tet})$ with LBO, i.e., $f(L_{tet})$, using the recursive formulation of Chebyshev polynomials. The cost immediately reduces to $\mathcal{O}(K|\varepsilon|) \ll \mathcal{O}(n^2)$ and is desirable in graph convolution of big graphs and 3D meshes. In Eq. 5, the Chebyshev polynomial T_m can be computed recursively using the form $T_m(x) = 2xT_{m-1}(x) - T_{m-2}$ with $T_0 = 1$ and $T_1 = x$ [6]. Here, all $T_m(k)$ create an orthonormal basis for $L^2([-1,1], \mu)$ with measure μ being $\frac{dy}{\sqrt{1-y^2}}$ in the Hilbert space of square integrable functions. Now given this recurrence, the Eq. 6, $T_m(L_{tet})$ is evaluated at $\tilde{L}_{tet} = \frac{2L_{tet}}{\lambda_{max}} - I$ with the initialization of the recurrence being $\tilde{x}_0 = x_{in}$, $\tilde{x}_1 = \tilde{L}_{tet} x_{in}$ with $\tilde{x}$ representing $T_m(L_{tet}) x_{in}$ in Eq. 6.

2.3   Mesh Coarsening and Pooling Operation

Although graph coarsening and mesh coarsening methods differ, using tetrahedral mesh down-sampling based on methods like Qslim [12] or learning-based methods like [30] are both expensive and infeasible as they are template-based with registered shapes. Here we do not try to register tetrahedral meshes, and the number of vertices varies from mesh to mesh. Therefore, we propose to build a sub-sampling approach similar in [6] but using spectral-aware configuration. The method is similar to Graclus clustering by exploiting the Laplacian and matrix D defined in the previous section.

Defining the Normalized Min-Cut Based on Tetrahedral LBO. Here, the objective function is based on normalized cut acting on vertices in a tetrahedral mesh. We need an affinity value between (v_i, v_j) and $vol(.)$ to capture the volume of each node. For the volume in the normalized cut problem of a simple graph, we use the degree of the node; however, in surface and volumetric meshes, this notion refers to the area and volume of the adjacent surface and tetrahedrons of the vertex, respectively. The proposed affinity or edge distance must be correlated with the A and D in Eq. 3. Thus, the proposed affinity distance as a new objective function for the local normalized cut is:

$$d(v_i, v_j) = -A_{i,j}(\frac{1}{D_{ii}} + \frac{1}{D_{jj}}), \tag{7}$$

Using this clustering objective function, at each step, we decimate the mesh by order of two. Consequently, $D_c(i, i)$ with c denoting the coarsen graph, are updated by the sum of their weights for the new matched vertices. The algorithm repeats until all the vertices are matched. Typically, at each convolution layer, we use two or three consecutive pooling since the size of the tetrahedral mesh is very large.

After coarsening, the challenging part is to match the new set of vertices with that of the previous ones. As proposed in [6], we use the same approach of exploiting a balanced binary tree and rearrangement of vertices by creating necessary fake nodes in the binary tree structure. For an exhaustive description of this approach please see Sect. 2.3 in [6].

Approximation of LBO on Down-Sampled Mesh. After each pooling, we have a coarsened mesh that needs updated LBO to pass it to the new convolution layer. We adopt the piece-wise constant approximation approach [21] where the clustering assignment matrix G is used. This choice of G is the most simple yet efficient one as the matrix is already computed for Graclus clustering. In some literature, they refer to G as the prolongation operator. Now the updated Laplacian $\hat{L}$ can be derived using the following equation for $\hat{L}$:

$$\hat{L} = G^T L G, \tag{8}$$

Grad-CAM for Tetrahedral Mesh. To utilize Grad-CAM for our framework, we need to adopt the Grad-CAM in [23] to our tetrahedral mesh. We use the

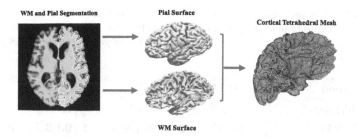

Fig. 3. Procedure of creating a cortical tetrahedral mesh of a closed surface from white and pial surface pre-processed and segmented by FreeSurfer [11].

k-th feature after the GAP layer denoted as f_k which is calculated based on the last layer feature map $X_{k,n}^L$. Here, n and L refer to the n-th node and the last layer of the network, respectively. Now, weights of Grad-CAM for class c of feature k in a tetrahedral mesh are calculated using

$$\alpha_k^{l,c} = \frac{1}{N} \sum_{n=1}^{N} \frac{\partial y^c}{\partial X_{k,n}^L} \tag{9}$$

To calculate the final heat map, we need to apply an activation function like ReLU and an upsampling method to project the weights to our original input mesh. As for upsampling, we use the *KNN* interpolation. The final heat-map of the last layer H is as follows

$$H_c^{L,n} = \text{ReLU}(\sum_k \alpha_k^{l,c} X_{k,n}^L) \tag{10}$$

3   Experimental Results

Data Processing. In our experiment, we study the diagnosis task for Alzheimer's disease. Our dataset contained 116 Alzheimer's disease (AD) patients, and 137 normal controls (NC) from the Alzheimer's Disease Neuroimaging Initiative phase 2 (ADNI-2) baseline initial-visit dataset [18]. All the subjects underwent the whole-brain MRI scan using a 3-Tesla MRI scanner. More details regarding the scans can be found at http://adni.loni.usc.edu/wp-content/uploads/2010/05/ADNI2_GE_3T_22.0_T2.pdf.

Cortical Tetrahedral Mesh Generation. We followed the procedure in [8] to create cortical tetrahedral meshes. First, pial and white surfaces were processed and created by FreeSurfer [11]. To remove self-intersections while combining pial and white surfaces, we repeatedly moved erroneous nodes and their small neighborhood along the inward normal direction by a small step size. This process continued to be done until the intersection was removed. Consequently, we used local smoothing on the modified nodes. Finally, we used TetGen [26] to create tetrahedral meshes of the closed surfaces. Figure 3 illustrates the cortical tetrahedral

Table 1. Classification results between AD vs. NC under different settings and parameters (GL = graph Laplacian, (.) defines the polynomial order k for LBO and GL, E_pool = Euclidean-based pooling). *Cortical thickness generated by FreeSurfer. [†] We use LBO-based pooling.

Method	ACC	SEN	SPE
Thickness*	76.2%	77.0%	78.6%
LBO(1) [†]	**91.7% ± 2.1**	89.1% ± 5.1	**93.3% ± 3.5**
LBO(2)[†]	90.8% ± 2.0	87.5% ± 4.8	92.1% ± 3.1
GL(1)	87.1% ± 1.8	90.4% ± 4.7	89.5% ± 3.1
GL(2)	85.7% ± 2.1	**90.0% ± 4.2**	87.5% ± 2.9
LBO(1)+E_pool	84.1% ± 2.4	83.5% ± 4.9	87.1% ± 3.5
LBO(1)+LBO_pool	91.7% ± 2.1	89.1% ± 5.1	92.1% ± 3.5

Table 2. Classification results between AD vs. NC comparison to the baseline using different data representation. The number of different subjects is also used for fair comparison.

Study	ACC	SEN	SPE	Subject Split
GF-Net [29]	94.1% ± 2.8	93.2% ± 2.4	90.6% ± 2.6	(188, 229)
Qiu et al. [25]	83.4%	76.7%	88.9	(188, 229)
ViT3D [29]	85.5% ± 2.9	87.9% ± 3.6	86.8% ± 3.7	(188, 229)
Huang et al. [16]	90.9% ± 0.6	91.3% ± 0.1	90.7% ± 0.5	(261, 400)
H-FCN [20]	90.5%	90.5%	91.3%	(389, 400)
ResNet3D [29]	87.7% ± 3.5	90.2% ± 2.8	89.7% ± 3.0	(188, 229)
DA-Net [31]	92.4%	91.0%	93.8%	(389, 400)
Ours	91.7% ± 2.1	89.1% ± 5.1	92.1% ± 3.5	(116, 137)

mesh generation process. The number of vertices in all tetrahedral meshes was around $150k$. To validate the robustness of our model, we used the simple xyz coordinate as input features and normalized them using min-max normalization. We avoided using informative features as they may contribute to the final performance rather than the TetCNN itself. We pre-computed the lumped LBO for all meshes and embedded them in our customized data-loader (Table 2).

Classification Model Setup. For comparison between different manifold spectral models, we tested our model based on both tetrahedral LBO and graph Laplacian. For the sake of equal comparison among each setting, we used the same network architecture and hyper-parameters. We used 5 TetCNN layers followed by ReLu activation function [22] and batch-normalization [17]. Before the

Table 3. Comparison of TetCNN with DGCNN [28] and PointNet [24].

	DGCNN [28]	PointNet [24]	TetCNN
ACC	73.45%	77.35%	**91.7%**

two fully connected layers, we applied a GAP to ensure the same size feature space among all mini-batches. We used 10-fold cross-validation and picked 15% of the training set for validation. We set the hyper-parameter k to two different values as it is shown in Table 1. The batch size for all TetCNN experiments was 8, and the loss function used for the model was Cross-Entropy. ADAM optimizer [19] with Learning 10^{-3}, weight decay of 10^{-4}, and number of epochs to 150 were used for training the model. For the AD vs. NC classification performance evaluation, we used three measures accuracy (ACC), sensitivity (SEN), and specificity (SPE). As a benchmark, we also used the FreeSurfer thickness features to train an AdaBoost classifier.

Point Clouds Model Setup for Classification. Point clouds have been widely used in deep learning literature to study manifold data. In our work, we further implemented DGCNN [28] and PointNet [24] as our baseline models to analyze volume data. For both PointNet and DGCNN, we trained the network with batch size 1 to feed the whole data without losing points for a fair comparison. All experiments were implemented in Python 3.7 with Pytorch Geometric 1.8 library [10] using NVIDIA GeForce Titan X GPU.

Age Prediction Setup. In order to further compare the TetCNN using volumetric LBO and its Graph Laplacian counterpart, we used a regression model to see the difference in age prediction. We used the same processed data from the ADNI dataset, but we only trained the model on normal subjects. Further, we tested the trained model on both normal subjects and independent AD subjects to see the accuracy and effect of AD on age prediction. In order for an unbiased age prediction on the AD cohort, we made a test set that matched the age distribution of normal subjects. We used 5-k fold cross-validation. As for AD subjects, we randomly chose 25 subjects to test on the trained model and repeated the experiment 5 times. All the parameters of the network are the same as the classification model except for the last fully connected layer the output dimension is one as we predict a number instead of discrete class labels.

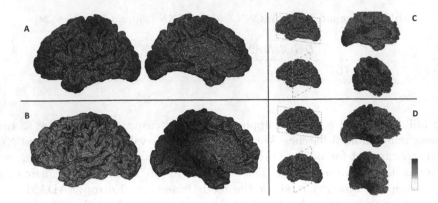

Fig. 4. Grad-CAM results for AD class showing the important regions. Comparison between LBO-based (top) and graph Laplacian (bottom) on the left hemisphere of the brain. A-B From left: Lateral-Medial view. C-D From top: Sagittal-Coronal view. Darker colors show more importance, hence greater weight.

Classification Results. As we see in Table 1, TetCNN with $k = 1$ outperformed any other setting, including graph Laplacian with the same parameter.

We expected the increase in k would result in boosted performance, however, the results are marginally worse. We assume this behavior demonstrates the fact that 1-ring neighbor provides sufficient information that making the receptive field larger does not contribute to more discriminative features, necessarily. Overall, TetCNN with an LBO-based setting outperformed its graph

Table 4. Age prediction result.

Method	RMSE (NC)	RMSE (AD)
LBO(1)	**6.3 ± 0.5** yr	**7.2 ± 0.7** yr
LBO(2)	6.5 ± 0.6 yr	7.4 ± 0.4 yr
GL(1)	7.2 ± 0.4 yr	7.9 ± 0.4 yr
GL(2)	7.1 ± 0.5 yr	8.1 ± 0.5 yr

Laplacian counter-part, presumably, owing to both rich geometric features learned using LBO, as exhaustively depicted in Fig. 1, and efficient spectral-based mesh down-sampling using the proposed objective function in Eq. 7. We also tested our new Graclus based on LBO and compared it to the default localized min-cut based on the Euclidean distance between two vertices. The LBO-based objective function clearly outperformed the one used in [6], which is not suitable for mesh structure as the degree of a node is almost similar in a mesh.

Regarding the comparison with point cloud learning frameworks, our observation in Table 3 shows that DGCNN and PointNet could not provide comparable results with our method due to the lack of deformation sensitivity in point cloud representation. These methods produce state-of-the-art results for the classification of completely distinct objects but fail to compete with mesh structure for learning subtle deformations in volume data.

Lastly, we compared our TetCNN with other methods in the literature that are based on either brain network, surface mesh or voxel-based representations.

Our results, though have a smaller dataset size for training, have comparable performance to state-of-the-art models.

Grad-CAM Results. In Fig. 4, we illustrated the Grad-CAM results on the left grey matter tetrahedral mesh of AD subjects, trained on both LBO (A)and Graph Laplacian-based scheme (B). As illustrated, the important regions for the AD class are different in the two approaches. The identified ROIs from the LBO are more centered at the medial temporal lobe, frontal lobe, and posterior cingulate, areas that are affected by AD. But the ROIs from the graph Laplacian are more scattered, without concise ROIs. Although more validations are desired, the current results demonstrate our interpretable model may identify important AD biomarkers.

Age Prediction Results. We tested our model on a regression task to compare Graph Laplacian and LBO. Furthermore, we aimed to see if the age prediction in AD patients has a larger margin of error with respect to normal subjects. Results in Table 4 show the consistent outperformance of LBO-based TetCNN over its graph Laplacian counterpart. Also, it shows an erroneous prediction of AD patients with a margin of around one year which is predictable due to changes in the cortical thickness of AD patients being more severe.

Complexity. Finally, in terms of computational complexity, the parameterized filter introduced in Eq. 5 addresses the non-locality in space and high learning complexity of $\mathcal{O}(n)$ problem of a non-parametric filter by employing the polynomial approximation of the tetrahedral LBO. Our novel approach reduced the time complexity to the dimension of k, hence $\mathcal{O}(k)$.

4 Conclusion and Future Work

In this study, we proposed a graph neural network based on volumetric LBO with modified pooling and down-sampling for tetrahedral meshes with different sizes. Results show the outperformance of the model to ChebyNet using graph Laplacian. Also, the adapted Grad-CAM for tetrahedral meshes showed regions affected within the surface and volume of the brain cortex in AD patients consistent with the findings in the literature. Our proposed learning framework is general and can be applied to other LBO definitions. In the future, we will also study brain parcellation and segmentation tasks with our LBO-based TetCNN.

Acknowledgement. This work was partially supported by the grants from NIH (R21AG065942, R01EY032125, and R01DE030286).

References

1. Azcona, E.A., et al.: Interpretation of brain morphology in association to Alzheimer's disease dementia classification using graph convolutional networks on triangulated meshes. In: Reuter, M., Wachinger, C., Lombaert, H., Paniagua, B., Goksel, O., Rekik, I. (eds.) ShapeMI 2020. LNCS, vol. 12474, pp. 95–107. Springer, Cham (2020). https://doi.org/10.1007/978-3-030-61056-2_8

2. Bessadok, A., Mahjoub, M.A., Rekik, I.: Brain multigraph prediction using topology-aware adversarial graph neural network. Med. Image Anal. **72**, 102090 (2021)
3. Bronstein, M.M., Bruna, J., Cohen, T., Veličković, P.: Geometric deep learning: grids, groups, graphs, geodesics, and gauges. arXiv preprint arXiv:2104.13478 (2021)
4. Chandran, V., Maquer, G., Gerig, T., Zysset, P., Reyes, M.: Supervised learning for bone shape and cortical thickness estimation from ct images for finite element analysis. Med. Image Anal. **52**, 42–55 (2019)
5. Cucurull, G., et al.: Convolutional neural networks for mesh-based parcellation of the cerebral cortex (2018)
6. Defferrard, M., Bresson, X., Vandergheynst, P.: Convolutional neural networks on graphs with fast localized spectral filtering. Adv. Neural Inf. Process. Syst. **29** (2016)
7. Dhillon, I.S., Guan, Y., Kulis, B.: Weighted graph cuts without eigenvectors a multilevel approach. IEEE Trans. Pattern Anal. **29**(11), 1944–1957 (2007)
8. Fan, Y., Wang, G., Lepore, N., Wang, Y.: A tetrahedron-based heat flux signature for cortical thickness morphometry analysis. In: Frangi, A.F., Schnabel, J.A., Davatzikos, C., Alberola-López, C., Fichtinger, G. (eds.) MICCAI 2018. LNCS, vol. 11072, pp. 420–428. Springer, Cham (2018). https://doi.org/10.1007/978-3-030-00931-1_48
9. Fawaz, A., et al.: Benchmarking geometric deep learning for cortical segmentation and neurodevelopmental phenotype prediction. bioRxiv (2021)
10. Fey, M., Lenssen, J.E.: Fast graph representation learning with pytorch geometric. arXiv preprint arXiv:1903.02428 (2019)
11. Fischl, B.: Freesurfe. Neuroimage **62**(2), 774–781 (2012)
12. Garland, M., Heckbert, P.S.: Surface simplification using quadric error metrics. In: Proceedings of the 24th Annual Conference on Computer Graphics and Interactive Techniques, pp. 209–216 (1997)
13. Gopinath, K., Desrosiers, C., Lombaert, H.: Learnable pooling in graph convolutional networks for brain surface analysis. IEEE Trans. Pattern Anal. Mach. Intell. **44**(2), 864–876 (2020)
14. Hammond, D.K., Vandergheynst, P., Gribonval, R.: Wavelets on graphs via spectral graph theory. Appl. Comput. Harmon. Anal. **30**(2), 129–150 (2011)
15. Hanocka, R., Hertz, A., Fish, N., Giryes, R., Fleishman, S., Cohen-Or, D.: MeshCNN: a network with an edge. ACM Trans. Graph. **38**(4), 1–12 (2019)
16. Huang, S.G., Chung, M.K., Qiu, A.: Revisiting convolutional neural network on graphs with polynomial approximations of Laplace-Beltrami spectral filtering. Neural Comput. Appl. **33**(20), 13693–13704 (2021)
17. Ioffe, S., Szegedy, C.: Batch normalization: accelerating deep network training by reducing internal covariate shift. In: Proceedings of the International Conference on Machine Learning, pp. 448–456. PMLR (2015)
18. Jack Jr., C.R., et al.: The Alzheimer's disease neuroimaging initiative (ADNI): MRI methods. J. Magn. Reson. Imaging **27**(4), 685–691 (2008)
19. Kingma, D.P., Ba, J.: Adam: a method for stochastic optimization. arXiv preprint arXiv:1412.6980 (2014)
20. Lian, C., Liu, M., Zhang, J., Shen, D.: Hierarchical fully convolutional network for joint atrophy localization and Alzheimer's disease diagnosis using structural MRI. IEEE Trans. Pattern Anal. Mach. Intell. **42**(4), 880–893 (2020)
21. Liu, H.T.D., Jacobson, A., Ovsjanikov, M.: Spectral coarsening of geometric operators. arXiv preprint arXiv:1905.05161 (2019)

22. Nair, V., Hinton, G.E.: Rectified linear units improve restricted Boltzmann machines. In: ICML (2010)
23. Pope, P.E., Kolouri, S., Rostami, M., Martin, C.E., Hoffmann, H.: Explainability methods for graph convolutional neural networks. In: Proceedings of the IEEE/CVF Conference on Computer Vision and Pattern Recognition, pp. 10772–10781 (2019)
24. Qi, C.R., Su, H., Mo, K., Guibas, L.J.: PointNet: deep learning on point sets for 3D classification and segmentation. In: Proceedings of the IEEE Computer Society Conference on Computer Vision and Pattern Recognition, pp. 652–660 (2017)
25. Qiu, S., et al.: Development and validation of an interpretable deep learning framework for Alzheimer's disease classification. Brain **143**(6), 1920–1933 (2020)
26. Si, H.: TetGen, a delaunay-based quality tetrahedral mesh generator. ACM Trans. Math. Softw. (TOMS) **41**(2), 1–36 (2015)
27. Wang, G., Wang, Y.: Towards a holistic cortical thickness descriptor: heat kernel-based grey matter morphology signatures. Neuroimage **147**, 360–380 (2017)
28. Wang, Y., Sun, Y., Liu, Z., Sarma, S.E., Bronstein, M.M., Solomon, J.M.: Dynamic graph CNN for learning on point clouds (2019)
29. Zhang, S., et al.: 3D global Fourier network for Alzheimer's disease diagnosis using structural MRI. In: Wang, L., Dou, Q., Fletcher, P.T., Speidel, S., Li, S. (eds.) MICCAI 2022. LNCS, vol. 13431, pp. 34–43. Springer, Cham (2022). https://doi.org/10.1007/978-3-031-16431-6_4
30. Zhou, Y., Wu, C., Li, Z., Cao, C., Ye, Y., Saragih, J., Li, H., Sheikh, Y.: Fully convolutional mesh autoencoder using efficient spatially varying kernels. Adv. Neural. Inf. Process. Syst. **33**, 9251–9262 (2020)
31. Zhu, W., Sun, L., Huang, J., Han, L., Zhang, D.: Dual attention multi-instance deep learning for Alzheimer's disease diagnosis with structural MRI. IEEE Trans. Med. Imaging **40**(9), 2354–2366 (2021)

Groupwise Atlasing

BInGo: Bayesian Intrinsic Groupwise Registration via Explicit Hierarchical Disentanglement

Xin Wang[1,2], Xinzhe Luo[1], and Xiahai Zhuang[1(✉)]

[1] School of Data Science, Fudan University, Shanghai, China
zxh@fudan.edu.cn
[2] Department of Electrical and Computer Engineering, University of Washington, Seattle, USA

Abstract. Multimodal groupwise registration aligns internal structures in a group of medical images. Current approaches to this problem involve developing similarity measures over the joint intensity profile of all images, which may be computationally prohibitive for large image groups and unstable under various conditions. To tackle these issues, we propose BInGo, a general *unsupervised* hierarchical Bayesian framework based on deep learning, to learn *intrinsic* structural representations to measure the similarity of multimodal images. Particularly, a variational auto-encoder with a novel posterior is proposed, which facilitates the *disentanglement* learning of structural representations and spatial transformations, and characterizes the imaging process from the common structure with shape transition and appearance variation. Notably, BInGo is scalable to learn from small groups, whereas being tested for large-scale groupwise registration, thus significantly reducing computational costs. We compared BInGo with five iterative or deep-learning methods on three public intra-subject and intersubject datasets, i.e. BraTS, MS-CMR of the heart, and Learn2Reg abdomen MR-CT, and demonstrated its superior accuracy and computational efficiency, even for very large group sizes (e.g., over 1300 2D images from MS-CMR in each group).

1 Introduction

Multimodal groupwise registration aims to align multimodal images into a common structural space. Unlike conventional pairwise registration which aligns moving images separately to a fixed image, groupwise registration can ameliorate the bias from designating a reference image, by estimating a common space to which all the images are co-registered. Therefore, it has become an essential task in multivariate image analysis, including longitudinal research, atlas construction, motion estimation, and population studies [8,12,16,20].

X. Wang and X. Luo—Equal Contribution.

This work was funded by the National Natural Science Foundation of China (grant No. 61971142 and 62111530195) and Fujian Provincial Natural Science Foundation project (2021J02019).

Conventional iterative methods usually estimate the desired spatial transformations by optimizing intensity-based similarity measures [15,18,22,24,27]. Recently, unsupervised deep-learning methods attempt to realize groupwise registration in an end-to-end fashion [6,9]. For instance in [6], the authors devised a network to optimize the conditional template entropy (CTE) introduced in [24]. These learning-based models estimate conventional similarity measures developed for iterative registration using stochastic gradient methods, and hence may inherit the same suboptimal performance due to insufficient modeling of the image similarity. Besides, these measures rely on correctly characterizing the joint intensity profile over the entire image group, which may be computationally prohibitive and applicability-limited due to large group sizes.

In this work, however, we shift attention from devising similarity metrics to learning the underlying generative imaging process. Specifically, we establish a probabilistic generative model for the observed images, in which transformations and the common structure are disentangled as latent variables. In this way, groupwise registration can be achieved by unsupervised variational auto-encoding: the encoder extracts structural representations of images and then infers the spatial deformations; the decoder imitates the imaging process by reconstructing original images from estimated common structure and the inverse deformations. The contributions of this work can be summarized as follows:

(1) We propose BInGo, a theoretically-grounded unsupervised Bayesian framework for groupwise registration, which learns intrinsic structural similarity of multimodal images by a principled variational distribution, and is capable of disentangling structural representations from image appearance.
(2) BInGo is scalable to be trained with small image groups while being applied to large-scale and variable-size test groups, significantly improving applicability and computational efficiency.
(3) We demonstrate the superiority of BInGo over similarity-based methods on three multimodal intrasubject and intersubject datasets.

To the best of our knowledge, this is the first work that realizes scalable multimodal groupwise registration by disentangled representation learning.

2   Methodology

Let the original image group be $\boldsymbol{X} = (X_m)_{m=1}^{M}$, with M the number of modalities and $X_m : \mathbb{R}^d \supset \Omega_m \to \mathbb{R}$. Groupwise registration aims to find spatial transformations $\boldsymbol{\phi} = (\phi_m : \mathbb{R}^d \supset \Omega \to \Omega_m)_{m=1}^{M}$, such that the aligned images $(X_m \circ \phi_m)_{m=1}^{M}$ share a *common* structural representation $\boldsymbol{Z}$.

We assume X_m is generated from $\boldsymbol{Z}$ and ϕ_m^{-1} by a *transformation-equivariant* imaging process f_m that contains modality-specific appearance information, i.e.

$$X_m = f_m(\boldsymbol{Z} \circ \phi_m^{-1}) = f_m(\boldsymbol{Z}) \circ \phi_m^{-1}. \tag{1}$$

Therefore, a 3-step unsupervised auto-encoding scheme can be formed: I) By modeling f_m^{-1} we could encode *individual* structural representation for X_m as

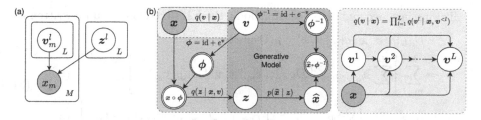

Fig. 1. The proposed Bayesian framework. (a) Graphical model of the imaging process. (b) Inference model for the velocity fields v (orange), the common structural representation z (green), and the reconstruction (blue). Random variables are in circles, deterministic variables are in double circles, and observed variables are shaded. (Color figure online)

$Z \circ \phi_m^{-1} = f_m^{-1}(X_m)$, from which ϕ could be estimated more easily. II) Z could be then inferred from $Z = f_m^{-1}(X_m \circ \phi_m)$ following the equivariance assumption. III) By modeling f_m we could reconstruct X_m from the estimated Z and ϕ_m^{-1}.

The following establish BInGo to realize this scheme by variational and disentangled auto-encoding, thus learning ϕ in a unified and end-to-end fashion.

2.1 Hierarchical Bayesian Inference

We first formalize the estimation of spatial deformations $\phi = (\phi_m)_{m=1}^M$ through Bayesian inference, as illustrated in Fig. 1. Specifically, let $x = (x_m)_{m=1}^M$ be a sample of X. We decompose the latent variables generating x into two *independent* subgroups: 1) the common structural representation z, and 2) the stationary velocity fields $v = (v_m)_{m=1}^M$ that parameterize ϕ [1]. Thus, following the variational Bayes framework, the objective function to maximize is the evidence lower bound (ELBO) of the log-likelihood, which takes the form

$$\mathcal{L}(x) \triangleq \mathbb{E}_{q(z,v \mid x)}\big[\log p(x \mid z, v)\big] \quad D_{\mathrm{KL}}\big[q(z, v \mid x) \,\|\, p(z)\, p(v)\big] \quad (2)$$

where $q(z, v \mid x) = q(v \mid x)\, q(z \mid v, x)$ defines the variational posterior distribution, and D_{KL} is the Kullback-Leibler (KL) divergence. After optimizing the ELBO, the desired velocity fields $\widehat{v}$ can be inferred as the posterior mean.

Furthermore, we express the latent variables with L hierarchical levels [26], i.e. $z = (z^l)_{l=1}^L$ and $v_m = (v_m^l)_{l=1}^L$, and a higher level indicates a finer-scale (larger) resolution. Thus, the total deformation is $\phi_m = \phi_m^1 \circ \cdots \circ \phi_m^L$, where $\phi_m^l = \mathrm{id} + \exp(v_m^l)$. The hierarchical strategy allows to model a complex deformation by several simpler and easier-to-learn ones.

To simplify the KL, we introduce additional independence assumptions: 1) the prior and posterior of velocities factorize as $p(v^l \mid v^{<l}) = p(v^l) = \prod_{m=1}^M p(v_m^l)$ and $q(v^l \mid x, v^{<l}) = \prod_{m=1}^M q(v_m^l \mid x, v^{<l})$, where $< l$ denotes the group of latent variables in levels lower than l, and 2) the common structure z^l can be inferred directly from x and v, i.e. $q(z^l \mid x, v, z^{<l}) = q(z^l \mid x, v)$, since $\{x, v\}$ can determine registered images, thus containing all information about z^l for any l.

Taking into account these assumptions, the KL divergence can be written as

$$
D_{\mathrm{KL}}\big[q(\boldsymbol{v}\,|\,\boldsymbol{x})\,q(\boldsymbol{z}\,|\,\boldsymbol{v},\boldsymbol{x})\,\|\,p(\boldsymbol{v})\,p(\boldsymbol{z})\big]
$$

$$
= \sum_{m=1}^{M}\left[\sum_{l=1}^{L}\mathbb{E}_{q(v_m^{\leq l}\,|\,x)}\Big\{D_{\mathrm{KL}}\big[q(v_m^l\,|\,\boldsymbol{x},\boldsymbol{v}^{<l})\,\|\,p(v_m^l)\big]\Big\}\right] \qquad \text{(i)}
$$

$$
+ \mathbb{E}_{q(\boldsymbol{v}\,|\,\boldsymbol{x})}\left[\sum_{l=1}^{L}\mathbb{E}_{q(z^{<l}\,|\,x,v)}\Big\{D_{\mathrm{KL}}\big[q(\boldsymbol{z}^l\,|\,\boldsymbol{x},\boldsymbol{v})\,\|\,p(\boldsymbol{z}^l\,|\,\boldsymbol{z}^{<l})\big]\Big\}\right] \quad \text{(ii)},
$$

(3)

where we prescribe $p(\boldsymbol{z}^1\,|\,\boldsymbol{z}^{<1}) \triangleq p(\boldsymbol{z}^1)$ and $q(v_m^{\leq 1}\,|\,\boldsymbol{x}) = q(\boldsymbol{z}^{<1}\,|\,\boldsymbol{x},\boldsymbol{v}) \triangleq 1$ for simplicity. The key idea here is: the overall KL divergence is decomposed *w.r.t.* (i) the velocity fields $\boldsymbol{v}$, and (ii) the common structure $\boldsymbol{z}$. The former serves as regularization to ensure diffeomorphism, fulfilled by the constraint introduced in [7]; the latter is intended to estimate the structural similarity among the images, as is detailed in the next subsection.

2.2 Intrinsic Similarity over Structural Representations

For registration, it is crucial to efficiently measure the structural similarity, which is achieved in this work by learning instead of a pre-defined metric. To this end, we propose to learn multilevel "expert" distributions $q_m(\boldsymbol{z}^l\,|\,x_m,\boldsymbol{v}_m)$, which serve as f_m^{-1} (i.e., the inverse of imaging processes) to extract the structural representations of warped images. We further assume $q(\boldsymbol{z}^l\,|\,\boldsymbol{x},\boldsymbol{v})$ and $p(\boldsymbol{z}^l\,|\,\boldsymbol{z}^{<l})$ to be the *geometric* mean [17] and *arithmetic* mean [25] of the experts, respectively:

$$
q(\boldsymbol{z}^l|\,\boldsymbol{x},\boldsymbol{v}) \propto \left[\prod_{m=1}^{M} q_m(\boldsymbol{z}^l|\,x_m,\boldsymbol{v}_m)\right]^{\frac{1}{M}}, \quad p(\boldsymbol{z}^l|\boldsymbol{z}^{<l}) \triangleq \frac{1}{M}\sum_{m=1}^{M} q_m(\boldsymbol{z}^l|\,x_m,\boldsymbol{v}_m). \quad (4)
$$

Therefore, the KL in Eq. (3)(ii) essentially measures the *intrinsic* (dis)similarity, i.e., the dissimilarity of the experts (intrinsic structural representations), whose minimization, as a part of the maximization of the ELBO, encourages the experts to be identical, thus forcing the multilevel posteriors $q(\boldsymbol{z}^l\,|\,\boldsymbol{x},\boldsymbol{v})$ to represent the common structure. Meanwhile, the velocity could be learned in tandem.

We model the experts $q_m(\boldsymbol{z}^l\,|\,x_m,\boldsymbol{v}_m)$ as Gaussians $\mathcal{N}(\mu_m^l,\Sigma_m^l)$, and thus so is the joint posterior, i.e. $q(\boldsymbol{z}^l\,|\,\boldsymbol{x},\boldsymbol{v}) = \mathcal{N}(\mu^l,\Sigma^l)$ with

$$
\Sigma^l = M\cdot\left[\sum_{m=1}^{M}(\Sigma_m^l)^{-1}\right]^{-1}, \quad \mu^l = \frac{\Sigma^l}{M}\cdot\sum_{m=1}^{M}\mu_m^l\,(\Sigma_m^l)^{-1}. \quad (5)
$$

In light of the computational intractability of the KL divergence involving Gaussian mixture distributions, we further exploit its convexity to obtain

$$
D_{\mathrm{KL}}\big[q(\boldsymbol{z}^l\,|\,\boldsymbol{x},\boldsymbol{v})\,\|\,p(\boldsymbol{z}^l\,|\,\boldsymbol{z}^{<l})\big] \leqslant \frac{1}{M}\sum_{m=1}^{M} D_{\mathrm{KL}}\big[q(\boldsymbol{z}^l\,|\,\boldsymbol{x},\boldsymbol{v})\,\|\,q_m(\boldsymbol{z}^l|\,x_m,\boldsymbol{v}_m)\big]. \quad (6)
$$

Hence, the minimization of the KL in Eq. (3)(ii) is actually approximated by minimizing the right-hand side of Eq. (6), which has a closed-form expression.

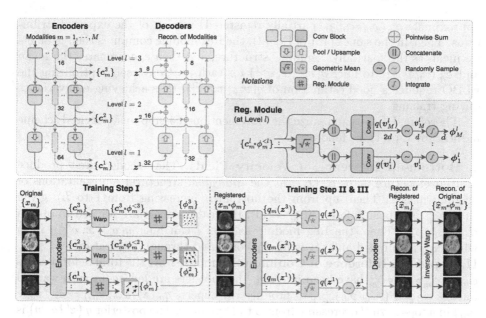

Fig. 2. Architecture ($L = 3$) of BInGo for explicit disentanglement. The channel numbers are indicated around particular feature maps (arrows).

2.3 Explicit Disentanglement with Neural Networks

The ELBO has been decomposed into three terms: reconstruction of the original images (the first term in Eq. (2)), the KL divergence for velocity fields to ensure diffeomorphism (Eq. (3)(i)), and the KL divergence for the intrinsic (dis)similarity to optimize the registration (Eq. (3)(ii)), with weights (hyperparameters) $(\lambda_i)_{i=1}^{3}$. To estimate the ELBO, we propose a dedicated hierarchical variational auto-encoder (VAE) as the inference model in Fig. 1(b), whose architecture with the number of hierarchy $L = 3$ is depicted in Fig. 2. The network learns parameters of $p(\boldsymbol{x} \mid \boldsymbol{z}, \boldsymbol{v}), q\left(\boldsymbol{v}_m^l \mid \boldsymbol{x}, \boldsymbol{v}^{<l}\right)$ and $q_m\left(\boldsymbol{z}^l \mid x_m, \boldsymbol{v}_m\right)$, and the expectations in the ELBO are estimated by Monte Carlo sampling.

The motif of BInGo is to explicitly disentangle the common structure, spatial transformations and appearance information from multimodal images, thus realizing Eq. (1). To this end, the network comprises three types of cooperative modules: 1) M encoders that act as the inverse imaging functions f_m^{-1} to extract multi-level structural representations c_m^l (assumed to be equivariant to any spatial transformation), 2) multi-level registration (Reg) modules that infer spatial transformations from individual structural representations, corresponding to estimating ϕ from $\{\boldsymbol{z} \circ \phi_m^{-1}\}_{m=1}^{M}$, and 3) M decoders, with modality-specific appearance information embedded as parameters, which act as the imaging functions $f_m(\boldsymbol{z})$ to reconstruct registered images from learned common structure.

Particularly, assuming $\phi_m^{<l} \triangleq \phi_m^1 \circ \cdots \circ \phi_m^{l-1}$ have been resolved, the Reg module at level l takes $c_m^l(x_m) \circ \phi_m^{<l}$ as inputs. If equivariance is achieved, they can

324   X. Wang et al.

approximate $c_m^l(x_m \circ \phi_m^{<l})$, "partially registered" version of the expert distributions. Then, in the same spirit of Eq. (4), the Reg module computes their geometric mean as the "partially common" structure, and then concatenates the mean and $c_m^l(x_m) \circ \phi_m^{<l}$ to infer $q(v_m^l \mid x, v^{<l})$ for each m. Conversely, optimizing the ELBO requires a good estimation of $q(v_m^l \mid x, v^{<l})$, thus achieving equivariance during training.

Therefore, BInGo can realize the aforementioned 3-step unsupervised scheme induced by Eq. (1), as follows:

I. Bottom-up Inference of Velocity Fields. The encoders first produce multilevel feature maps $c_m^l(x_m)$ as the *individual* structural representations of original images. Then, up from the bottom (l increasing from 1 to L), given that $\phi_m^{<l}$ have been inferred ($\phi_m^{<1} \triangleq$ id), the feature maps are warped to obtain $c_m^l(x_m) \circ \phi_m^{<l}$. Then, the Reg module at level l can infer $q(v_m^l \mid x, v^{<l})$, from which v_m^l is sampled and ϕ_m^l is finally computed through integration.

II. Top-down Inference of Common Structures. As the total deformations $\phi_m = \phi_m^1 \circ \cdots \circ \phi_m^L$ have been inferred, the encoders are fed *again* with warped images $x \circ \phi \triangleq (x_m \circ \phi_m)_{m=1}^M$ to produce the experts $q_m(z^l \mid x_m, v_m) = c_m^l(x_m \circ \phi_m)$ in a top-down (l decreasing from L to 1) manner. The posterior $q(z^l \mid x, v)$ is then computed by Eq. (4), from which the modality-invariant common structural representation $z = (z^l)_{l=1}^L$ is sampled. Note that we can avoid using encoders twice by warping $c_m^l(x_m)$ to compute the experts as $c_m^l(x_m) \circ \phi_m$ (thanks to equivariance), but this requires more computational cost.

III. Disentangled Auto-Encoding. Based on the common structure z inferred by encoders, the decoders reconstruct the registered images $\hat{x} = (\hat{x}_m)_{m=1}^M$. Then reconstructed original images $\hat{x} \circ \phi^{-1} \triangleq (\hat{x}_m \circ \phi_m^{-1})_{m=1}^M$ are obtained by inverse deformations. We model $p(x \mid z, v)$ as Laplacian centered at $\hat{x} \circ \phi^{-1}$ to estimate the first (reconstruction) term in Eq. (2). In addition, to better disentangle structure and appearance, all the convolutional layers of the VAE are *shared* across modalities, while *domain-specific* batch normalizations (BNs) encode appearance information for each modality [5].

2.4 Towards Large-Scale and Variable-Size Groupwise Registration

Most conventional methods optimize similarity measures defined over the entire image group, making the computation for large groups formidable. Besides, learning-based models that rely on these measures require groups to have the same size, as input and output channel numbers are fixed for images and transformations [9]. These drawbacks significantly reduce their applicability.

However, BInGo is scalable to handle multimodal image groups with various sizes for training and test. As shown in Fig. 3, it can be trained with either bimodal image pairs or image groups of complete modalities, referred to as *partial* or *complete* learning, respectively. Trained in either way, BInGo can be flexibly applied to larger unseen test groups with arbitrary numbers of images, such that computational efficiency of training is substantially boosted.

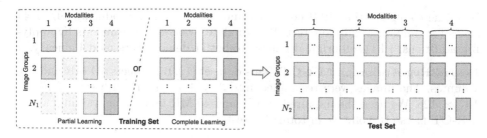

Fig. 3. Partial and complete learning schemes for large-scale and variable-size groupwise registration. Each color denotes a modality, and muted colors indicate missing modalities in certain training groups. (Color figure online)

To this end, each image in the input group (with any size $M' \geqslant 2$) is processed by the encoder of the corresponding modality to produce individual structural representations, based on which the geometric/arithmetic means can still be calculated. Thus, the posteriors, priors, deformations and ELBO are computed as usual during training and inference.

The scalability of our model is conducive in two scenarios:

Intrasubject Images with Missing Modalities. The goal is to co-register multimodal scans for each patient. In practice, there could be many patients with different missing modalities. BInGo can be trained with these heterogeneous datasets via partial learning, while performing groupwise registration for test subjects with complete modalities.

Intersubject Populations. Intersubject groupwise registration, which is crucial in atlas construction and population analysis, could involve plenty of images for every modality. Conventional methods to this problem rely on iterative optimization and suffer from computational complexity scaling with image group size, while BInGo can greatly relieve training burden and be subsequently applied to test groups of complete populations in one shot.

3 Experiments and Results

3.1 Datasets and Preprocessing

BraTS-2021. The dataset provides 3D pre-operative T1, T1Gd, T2 and T2-FLAIR MR scans of patients with glioblastoma [3,19]. We randomly selected 300/50/150 patient sets for training/validation/test. The volumes were downsampled into $2 \times 2 \times 2\,\mathrm{mm}^3$ with ROI of size $80 \times 96 \times 80$. As the images of each patient are pre-registered, we use synthetic free-form deformations (FFDs) with different control point spacings to simulate misalignments.

MS-CMR. The MS-CMRSeg challenge [28] provides cardiac MR sequences LGE, bSSFP, and T2 from 45 patients, which exhibit complementary information of the cardiac structure and pathology. The images were preprocessed by affine co-registration, ROI cropping, and slice selection, producing $39/15/44$ slices for training/validation/test. We simulated worse misalignments by applying additional FFDs on original images to better demonstrate model efficacy.

Learn2Reg Abdomen MR-CT. This dataset [10,13] collects 3D MR and CT volumes. The images were resampled into $3 \times 3 \times 3\,\mathrm{mm}^3$ with size of $128 \times 107 \times 128$. We used the unpaired 50 CT and 40 MR images for training, and 8 MR-CT pairs (16 volumes in total) for test.

3.2   Experimental Setups

Compared Methods. Three types of unsupervised methods for multimodal groupwise registration were compared on the *intrasubject* BraTS and MS-CMR datasets: 1) similarity-based iterative methods using information-theoretic metrics CTE, APE or $\mathcal{X}$-CoReg [18,24,27], 2) similarity-based deep-learning models that optimize CTE or APE using an attention residual U-Net (AttResUNet) [11,21] as the backbone, 3) BInGo, the proposed model learning intrinsic similarity through hierarchical disentanglement. For *intersubject* population groupwise registration on Learn2Reg, the models in 2) are not applicable since they can only be trained with *complete* image groups, whereas BInGo can be trained *partially*. In addition, for BraTS and MS-CMR, all baseline methods used single-level velocity fields as the transformation model; for Learn2Reg, iterative baselines performed rigid, affine and FFD registration successively for optimal accuracy, due to the severe initial misalignments.

Groupwise Unbiasedness. In the experiments, registration was performed in an *unbiased* fashion to avoid transformation degeneracy, i.e., we subtracted the pixel-wise average of the predicted velocity/displacement fields from themselves to make the following necessarily satisfied: $\forall \boldsymbol{\omega} \in \Omega, \sum_m \boldsymbol{v}_m(\boldsymbol{\omega}) = \mathbf{0}$ [2] for any level l in our model; $\frac{1}{M}\sum_m \phi_m(\boldsymbol{\omega}) = \boldsymbol{\omega}$ [4] for the baseline methods.

Implementation Details. In preprocessing, each image was min-max normalized. For network, we set $L = 5$, and each Conv block C_k comprises k Conv-BN-LeakyReLU sequences, where $k = 2\lceil l/2 \rceil + 2$ or 2 for the Reg module at level l or other modules, respectively. Each upsampling involved a linear scaling plus a Conv with kernel size 1. To balance registration and reconstruction, Conv kernel sizes were $1/3$ for decoders/other modules. Momentum of BNs was also set to 0.01 to boost generalizability. The hyperparameters $(\lambda_i)_{i=1}^3$ for Learn2Reg and complete learning on BraTS/MS-CMR were $(80, 8, 150)/(250, 25, 50)/(90, 10, 50)$. Experiments were conducted using Pytorch [23] (training optimizer: Adam [14]; learning rate: 10^{-3}; batch size: 20/1 for MS-CMR/other datasets) on an NVIDIA RTX^{TM} 3090 GPU.

Evaluation Metrics. We reported the Dice similarity coefficient (DSC) averaged over all pairwise combinations of the segmentation masks for registered images. For the BraTS dataset, as the ground-truth misalignments were available, we also reported the groupwise warping index (gWI) [18], which would reduce to zero if the misaligned images were perfectly co-registered.

Table 1. Results on intrasubject image groups from BraTS and MS-CMR. The mean values and standard deviations are presented for the gWI (in voxels) and DSC, with top 2 bolded. The number of parameters (in millions) of each model for each dataset are also reported. Statistically significant improvement ($p < 0.05$ for one-sided paired t-tests) of BInGo with complete† or partial* learning was marked with daggers or asterisks, respectively.

Method	BraTS			MS-CMR	
	DSC ↑	gWI ↓	#Params.	DSC ↑	#Params.
None	$0.610 \pm 0.150^{\dagger*}$	$1.430 \pm 0.644^{\dagger*}$	—	$0.722 \pm 0.101^{\dagger*}$	—
APE [27]	$\mathbf{0.726 \pm 0.078}$	$\mathbf{0.596 \pm 0.149}$	7.373	$0.811 \pm 0.072^{\dagger*}$	0.154
CTE [24]	$0.561 \pm 0.148^{\dagger*}$	$1.087 \pm 0.411^{\dagger*}$	7.373	$0.816 \pm 0.077^{\dagger*}$	0.154
$\mathcal{X}$-CoReg [18]	$0.707 \pm 0.089^{\dagger}$	$0.697 \pm 0.212^{\dagger}$	7.373	$0.840 \pm 0.077^{\dagger*}$	0.154
APE+AttResUNet	$0.693 \pm 0.078^{\dagger}$	$0.757 \pm 0.153^{\dagger*}$	22.955	$0.846 \pm 0.048^{\dagger*}$	8.036
CTE+AttResUNet	$0.659 \pm 0.096^{\dagger*}$	$0.916 \pm 0.210^{\dagger*}$	22.955	$0.874 \pm 0.043^{\dagger*}$	8.036
BInGo (complete†)	$\mathbf{0.717 \pm 0.068}$	$\mathbf{0.596 \pm 0.132}$	13.429	$\mathbf{0.887 \pm 0.033}$	4.516
BInGo (partial*)	0.693 ± 0.075	0.709 ± 0.172	13.429	$\mathbf{0.877 \pm 0.042}$	4.516

Table 2. Results on intersubject population groups from Learn2Reg. The means and standard deviations of DSCs versus the numbers of images in each group are reported.

Method	#Images			
	2	4	8	16
None	0.396 ± 0.168	0.386 ± 0.070	0.319 ± 0.007	0.306 ± 0.000
APE [27]	0.586 ± 0.376	0.574 ± 0.261	N/A	N/A
CTE [24]	0.609 ± 0.306	0.088 ± 0.036	N/A	N/A
$\mathcal{X}$-CoReg [18]	0.675 ± 0.329	0.567 ± 0.224	N/A	N/A
BInGo (partial*)	$\mathbf{0.781 \pm 0.108}$	$\mathbf{0.715 \pm 0.122}$	$\mathbf{0.677 \pm 0.059}$	$\mathbf{0.645 \pm 0.000}$

3.3 Results

Intrasubject Image Groups. Table 1 presents registration accuracy on the BraTS and MS-CMR datasets. BInGo (with complete† or partial* learning) worked consistently better than similarity-based iterative or learning approaches, except for the APE-based iterative method on BraTS. Notably, BInGo could achieve superior performance to deep-learning baselines with only half of training parameters, even though their backbone is more advanced than the vanilla UNet-like structure used in BInGo. Furthermore, the proposed partial learning strategy could perform better than most baselines with significantly lower computational demands, potentiating large-scale end-to-end groupwise registration.

Intersubject Population Image Groups. For Learn2Reg, we merged every K test MR-CT pairs to form larger groups of size $M' = 2K$ as test sets of intersubject populations. Table 2 presents the results versus different M'. When the groups became large, all iterative approaches failed due to excessive GPU cost, while our partial learning strategy worked consistently better and maintained

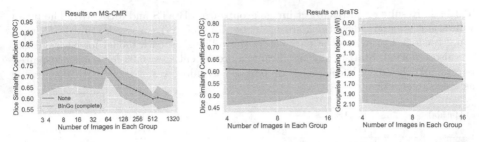

Fig. 4. Results (mean values with one standard deviation bands) before (blue) and after (orange) registration by trained BInGo for images groups with different sizes. (Color figure online)

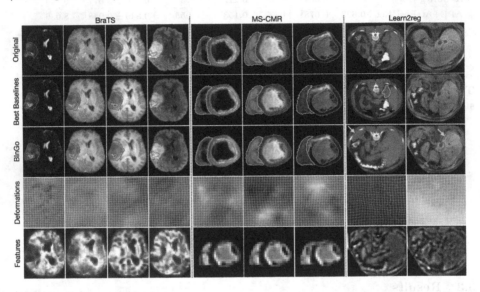

Fig. 5. Example image groups (with segmentation contours overlaid) before or after registration using the best baselines or BInGo. Deformations and example features of registered images from BInGo are also presented. Green and orange arrows indicate some areas of significant improvement (compared to the best baselines) and relatively poor performance, respectively. (Color figure online)

a good performance. Note that the deep-learning baselines are not capable of handling test groups with varying sizes, and thus not presented.

Additional Scalability Tests. For MS-CMR and BraTS, we merged test groups in a similar way to evaluate BInGo (trained via complete learning) on image groups with different sizes. As shown in Fig. 4, with M' increasing, co-registration becomes significantly more difficult (indicated by the worse initial DSC/gWI), whereas BInGo maintained decent performance for ultra-large groups (e.g., over 1300 images from MS-CMR), and achieved even better accuracy for larger 3D image groups from BraTS, showing remarkable robustness on large-scale group-wise registration.

Qualitative Results. We visualized the results from the best baselines (APE/CTE+AttResUNet/$\mathcal{X}$-CoReg for BraTS/MS-CMR/Learn2Reg) and BInGo in Fig. 5. BInGo could achieve better alignment for both large-scale anatomy and local fine structures in most cases, and the predicted deformations reached great smoothness and diffeomorphism. The feature maps from BInGo were nearly modality-invariant and shared similar structures, illustrating successful disentanglement of the common structure, spatial transformations and appearance information. Besides, the relatively poor performance in certain local areas may be due to the too severe initial misalignment, which made different tissues with similar intensities appear in the same location. Still, BInGo performed no worse than the baselines in such regions, and achieved satisfactory accuracy for the entire foreground.

4   Conclusion and Discussion

In this work we have presented BInGo, a generative Bayesian framework for unsupervised multimodal groupwise registration. This new formulation of image registration has achieved comparable performance with similarity-based iterative and unsupervised methods. In particular, we demonstrated that equipped with unique scalability, BInGo could reduce the computational burden of groupwise registration without compromising accuracy. This opens up the possibility to realize learning-based multimodal groupwise registration on a large scale and with various group sizes. A potential limitation of our work is that there may exist performance drop on specific datasets, e.g. the highly challenging abdominal images. Future work includes investigation into the negative factors that may inhibit BInGo from generalizing to large image groups.

References

1. Ashburner, J.: A fast diffeomorphic image registration algorithm. Neuroimage **38**(1), 95–113 (2007)
2. Avants, B., Gee, J.C.: Geodesic estimation for large deformation anatomical shape averaging and interpolation. Neuroimage **23**, S139–S150 (2004)
3. Baid, U., et al.: The rsna-asnr-miccai brats 2021 benchmark on brain tumor segmentation and radiogenomic classification. arXiv preprint arXiv:2107.02314 (2021)
4. Bhatia, K.K., Hajnal, J.V., Puri, B.K., Edwards, A.D., Rueckert, D.: Consistent groupwise non-rigid registration for atlas construction. In: 2nd IEEE International Symposium on Biomedical Imaging: Nano to Macro, pp. 908–911. IEEE (2004)
5. Cang, W.G., You, T., Seo, S., Kwak, S., Han, B.: Domain-specific batch normalization for unsupervised domain adaptation. In: Proceedings of the IEEE/CVF Conference on Computer Vision and Pattern Recognition, pp. 7354–7362 (2019)
6. Che, T., et al.: DGR-Net: deep groupwise registration of multispectral images. In: Chung, A.C.S., Gee, J.C., Yushkevich, P.A., Bao, S. (eds.) IPMI 2019. LNCS, vol. 11492, pp. 706–717. Springer, Cham (2019). https://doi.org/10.1007/978-3-030-20351-1_55

7. Dalca, A.V., Balakrishnan, G., Guttag, J., Sabuncu, M.R.: Unsupervised learning of probabilistic diffeomorphic registration for images and surfaces. Med. Image Anal. **57**, 226–236 (2019)
8. Geng, X., Christensen, G.E., Gu, H., Ross, T.J., Yang, Y.: Implicit reference-based group-wise image registration and its application to structural and functional MRI. Neuroimage **47**(4), 1341–1351 (2009)
9. He, Z., Chung, A.C.: Unsupervised end-to-end groupwise registration framework without generating templates. In: 2020 IEEE International Conference on Image Processing (ICIP), pp. 375–379. IEEE (2020)
10. Hering, A., et al.: Learn2Reg: comprehensive multi-task medical image registration challenge, dataset and evaluation in the era of deep learning. IEEE Trans. Med. Imaging **42**(3), 697–712 (2022)
11. Hu, Y., et al.: Weakly-supervised convolutional neural networks for multimodal image registration. Med. Image Anal. **49**, 1–13 (2018)
12. Joshi, S.C., Davis, B.C., Jomier, M., Gerig, G.: Unbiased diffeomorphic atlas construction for computational anatomy. Neuroimage **23**, S151–S160 (2004)
13. Kavur, A.E., Selver, M.A., Dicle, O., Barı, M., Gezer, N.S.: CHAOS - Combined (CT-MR) Healthy Abdominal Organ Segmentation Challenge Data, April 2019. https://doi.org/10.5281/zenodo.3362844
14. Kingma, D.P., Ba, J.: Adam: a method for stochastic optimization. In: 3rd International Conference on Learning Representations (2015)
15. Learned-Miller, E.G.: Data driven image models through continuous joint alignment. IEEE Trans. Pattern Anal. Mach. Intell. **28**(2), 236–250 (2005)
16. Liao, S., Jia, H., Wu, G., Shen, D.: A novel framework for longitudinal atlas construction with groupwise registration of subject image sequences. Neuroimage **59**(2), 1275–1289 (2012)
17. Lorenzen, P., Prastawa, M., Davis, B., Gerig, G., Bullitt, E., Joshi, S.: Multi-modal image set registration and atlas formation. Med. Image Anal. **10**(3), 440–451 (2006)
18. Luo, X., Zhuang, X.: X-metric: an n-dimensional information-theoretic framework for groupwise registration and deep combined computing. IEEE Trans. Pattern Anal. Mach. Intell. (2022)
19. Menze, B.H., et al.: The multimodal brain tumor image segmentation benchmark (brats). IEEE Trans. Med. Imaging **34**(10), 1993–2024 (2014)
20. Metz, C.T., Klein, S., Schaap, M., van Walsum, T., Niessen, W.J.: Nonrigid registration of dynamic medical imaging data using nD+ t B-splines and a groupwise optimization approach. Med. Image Anal. **15**(2), 238–249 (2011)
21. Oktay, O., et al.: Attention u-net: learning where to look for the pancreas. In: Medical Imaging with Deep Learning (2018)
22. Orchard, J., Mann, R.: Registering a multisensor ensemble of images. IEEE Trans. Image Process. **19**(5), 1236–1247 (2009)
23. Paszke, A., et al.: Automatic differentiation in pytorch (2017)
24. Polfliet, M., Klein, S., Huizinga, W., Paulides, M.M., Niessen, W.J., Vandemeulebroucke, J.: Intrasubject multimodal groupwise registration with the conditional template entropy. Med. Image Anal. **46**, 15–25 (2018)
25. Shi, Y., Paige, B., Torr, P., et al.: Variational mixture-of-experts autoencoders for multi-modal deep generative models. Adv. Neural Inf. Process. Syst. **32**, 15718–15729 (2019)
26. Vahdat, A., Kautz, J.: Nvae: a deep hierarchical variational autoencoder. arXiv preprint arXiv:2007.03898 (2020)

27. Wachinger, C., Navab, N.: Simultaneous registration of multiple images: similarity metrics and efficient optimization. IEEE Trans. Pattern Anal. Mach. Intell. **35**(5), 1221–1233 (2012)
28. Zhuang, X., et al.: Cardiac segmentation on late gadolinium enhancement MRI: a benchmark study from multi-sequence cardiac MR segmentation challenge. Med. Image Anal. **81**, 102528 (2022)

Learning Probabilistic Piecewise Rigid Atlases of Model Organisms via Generative Deep Networks

Amin Nejatbakhsh[1]([✉]), Neel Dey[2], Vivek Venkatachalam[3], Eviatar Yemini[4], Liam Paninski[1], and Erdem Varol[5]

[1] Departments of Neuroscience and Statistics, Columbia University, New York, USA
mn2822@cumc.columbia.edu
[2] Computer Science and Artificial Intelligence Lab, MIT, Massachusetts, USA
[3] Department of Physics, Northeastern University, Boston, USA
[4] Department of Neurobiology, University of Massachusetts Chan Medical School, Worcester, USA
[5] Department of Computer Science and Engineering, New York University, New York, USA

Abstract. Atlases are crucial to imaging statistics as they enable the standardization of inter-subject and inter-population analyses. While existing atlas estimation methods based on fluid/elastic/diffusion registration yield high-quality results for the human brain, these deformation models do not extend to a variety of other challenging areas of neuroscience such as the anatomy of *C. elegans* worms and fruit flies. To this end, this work presents a general probabilistic deep network-based framework for atlas estimation and registration which can flexibly incorporate various deformation models and levels of keypoint supervision that can be applied to a wide class of model organisms. Of particular relevance, it also develops a deformable piecewise rigid atlas model which is regularized to preserve inter-observation distances between neighbors. These modeling considerations are shown to improve atlas construction and key-point alignment across a diversity of datasets with small sample sizes including neuron positions in *C. elegans* hermaphrodites, fluorescence microscopy of male *C. elegans*, and images of fruit fly wings. Code is accessible at https://github.com/amin-nejat/Deformable-Atlas.

1 Introduction

Constructing biological atlases via image registration helps summarize normative patterns and variability within a target population. An atlas also provides a common coordinate system for image registration and segmentation, which can help decouple and quantify different sources of variability observed in the data [15,18,22,23]. However, while atlas estimation of structures such as the

Supplementary Information The online version contains supplementary material available at https://doi.org/10.1007/978-3-031-34048-2_26.

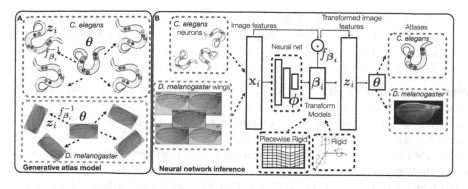

Fig. 1. Schematic of generative model of atlas construction. A: Each observation (Z_i) is modeled as a random draw from an atlas parametrized by θ and perturbed by transformation $f_{\beta_i}^{-1}$. **B**: We infer atlas parameters (θ) from observations (X_i) by optimizing a neural network loss function that penalizes the distance of each transformed observation (Z_i) to the latent atlas (θ) and also learn the transformation model parameters (β_i) that minimizes the loss.

human brain are well served with existing registration techniques [4,9,10,12,19], this work argues that other domains of neuroscience require alternative models.

Motivation. While fluid, elastic, and diffusion-based deformation models are well-motivated for the human brain, other structures may benefit from piecewise rigid deformations. For example, fluorescence microscopy of the nematode *C. elegans* nervous system is of high interest for atlas construction [7,20,28,31,33], where the shape and position of each neuron may individually deform rigidly while the number and function of neurons are conserved across individuals. Another model organism whose morphometry is of interest to neuroscience is the fruit fly *Drosophila melanogaster* [17]. In the fly, one suitable structure for atlas building is the wing whose inter-fly deformation can be well approximated by piece-wise rigid motion. However, probabilistic atlases that explore the structural variability amongst wings of different phenotypes and sexes has not yet been established. Further, atlas-building methods are *typically* hand-tailored to accommodate the specifications of a single organism, a single experimental condition, or the specifics of a developing/degenerating population [24], and are thus cannot be generally repurposed for experimentalists that require atlases for novel biological datasets that they curate [16].

Contributed Methods. This paper provides a general probabilistic framework for building atlases for any model organism using point clouds and/or images. Herein, we demonstrate the utility for two disparate organisms: nematodes and the fruit flies. We model individual observations, e.g., neural positions in a particular worm or the wing shape in an individual fly, as drawing from a generative atlas perturbed by a deformation jointly estimated via registration networks. The proposed framework allows for incorporating arbitrary transformations and atlas distributions with an emphasis on piece-wise rigid transformations well-suited to the considered model organisms. Further, we develop regularizers that encourage the conservation of inter-keypoint distances within an observation to prevent self-intersections. Lastly, it allows for flexible supervision-levels, ranging from full manual annotations of neural positions (supervised) to partial annotations (semi-supervised), and no annotations (unsupervised).

Experimental Results. We show three applications of our framework to model organisms under varying levels of supervision. First, we construct rigid and piecewise-rigid atlases of *C. elegans* neural positions from fully supervised datasets [33]. Then we build a semi-supervised atlas of male *C.elegans* using pixel intensities and partial annotations of neural positions [29]. Lastly, we show fully unsupervised atlas building of fruit fly wings from natural images [26]. To our knowledge, this work presents the first piecewise rigid atlases of *C. elegans* neural point clouds, *C. elegans* images, and *D. melanogaster* wing images.

The proposed procedure can provide valuable insights into the appropriateness of the transformation models. For example, the contained analyses address whether worm posture can be modeled using rigid or piecewise rigid motion and the constructed fly wing atlases enable morphometric comparisons across genotypes.

1.1   Related Work

C. Elegans Statistical Atlases. A number of atlases of neural positions in the *C. elegans hermaphrodite* have been introduced, utilizing a variety of shape and pose models [6,21,25,30,31]. However, atlas construction of the *male C. elegans* nervous system introduces further challenges due to having more neurons and higher density [14], additional ganglia enclosing these neurons [27], and greater variability in their neuronal and gangliar positions [29]. Therefore, existing atlas models of hermaphrodite neuron positions do not necessarily generalize to males.

Piecewise Rigid Registration. Piecewise rigid deformations are of key interest when modeling articulated structures (such as bones [32], joints, or *C. elegans* neurons) where local movements may be linear and global motion may be deformable. Theoretical treatments for such registration models have appeared in [1,2,8], yet to our knowledge, this paper is the first to construct atlases with such models, especially in a probabilistic deep network framework.

Atlas Building. In practice, several atlas building methods alternate between registering observations to a template and updating the template with a pointwise intensity and/or shape average of the aligned observations [3,4,19]. More recent deep network-based methods [9,12,13] instead explicitly synthesize templates via regularized registration objectives without averaging. This latter method typically yield sharper and more interpretable estimates, and this is the approach that we follow and modify towards probabilistic atlas construction with deformation models well-suited for model organisms.

Most relevant to our work, atlas construction via statistical inference with deep networks has been done for diffusion-regularized registration for large-scale 3D neuroimages in [9]. The proposed framework is distinct in that: (1) We specifically use piecewise rigid deformations for atlas construction on model organisms.

As our flow fields are explicitly constrained, we do not require diffusion regularization; (2) Our framework constructs atlases across varying levels of keypoint supervision (un/semi/fully-supervised) which is crucial for model organisms; (3) We develop an inter-keypoint distance conserving regularization to avoid self-intersections; (4) The datasets considered here are of much lower sample sizes and do not admit training the large registration and synthesis networks of [9].

2   Methods

Setup. We denote the atlas as a latent variable $Z \in \mathbb{R}^D$ following the distribution $P_\theta(Z)$ (Fig. 1A). Both X (the observations) and Z random variables can be high-dimensional or low dimensional depending on the application. For example, an atlas constructed using point clouds is lower dimensional than an atlas that is constructed using image intensities [31]. Given the atlas, i.e. a distribution over the random variable Z, the observations X_i are samples from the prior Z_i that are warped by transformation $f_{\beta_i} \in \mathcal{F}$ where $\mathcal{F}$ is a function class of feasible deformations between the atlas and observations and β_i are the parameters of the deformation for the ith observation. In this work, $\mathcal{F}$ is the space of rigid or piecewise rigid transformations.

Inference and Optimization. With $\mathcal{F}$ and the functional form of P_θ prespecified, our goal is to solve the inverse problem for parameters $\theta, \beta_{1:n}$. In general terms, we write a probabilistic cost function informed by our statistical model and optimize it w.r.t. $\theta, \beta_{1:n}$ as:

$$\mathcal{L}(\theta, \beta_{1:n}) = \log P_{\theta,\beta_{1:n}}(X_{1:n}, Z_{1:n}) = \sum_{i=1}^{n} \log P_{\beta_i}(X_i|Z_i) + \log P_\theta(Z_i)$$

$$\text{where } Z_i \sim P_\theta(Z) \quad \text{and} \quad X_i|Z_i \sim P(X|Z_i) = f_{\beta_i}(Z_i) + \epsilon_i$$

We take an alternating approach for optimizing $\mathcal{L}$ where we iteratively optimize $\mathcal{L}$ w.r.t. θ and $\beta_{1:n}$. Given our estimate of the values $\beta_{1:n}$ denoted by $\hat{\beta}_{1:n}$ we find the best fit $\hat{\theta}$ to the data in the following way:

$$\hat{\theta} = \max_\theta \mathcal{L}(\theta|\hat{\beta}_{1:n}) = \max_\theta \sum_{i=1}^{n} \log P_\theta(f_{\hat{\beta}_i}^{-1}(X_i)) \tag{1}$$

Notice that here we are trying to find the sufficient statistics of P_θ from known observations $Z_{1:n}$. For the case of multivariate normal distributions where $\theta = \{\mu, \Sigma\}$, the maximum likelihood estimate (MLE) is the empirical mean and covariance. However, our formulation allows for incorporating arbitrarily complex distributions where we solve the MLE problem using stochastic variational inference in the parameter space. This is facilitated by probabilistic programming where generic algorithms for MLE and MAP estimation are provided.

On the other hand if we have a reasonable estimate of θ then in order to update our estimates of $\beta_{1:n}$ we need to solve the following for each i:

$$\hat{\beta}_i = \max_{\beta_i \in \mathcal{F}} \mathcal{L}(\beta_i|\hat{\theta}) = \max_{\beta_i \in \mathcal{F}} \log P_{\hat{\theta}}(f_{\beta_i}^{-1}(X_i)) \tag{2}$$

Analytical solutions might exist for specific $\mathcal{F}$, but in general specific algorithms are required for particular choices of $\mathcal{F}$ as demonstrated in the experiments.

Amortized Learning of Transformations. Instead of optimizing this loss function with respect to the parameters $\{\beta_{1:n}\}$ directly, we solve an alternative amortized optimization problem. We parameterize $\beta_i = \text{nn}_\phi(X_i)$ where nn_ϕ is a neural network with weights ϕ and optimize the loss w.r.t. the parameters of the neural net. We parameterize rigid and piecewise rigid transformations using 3 angles and 3 translation parameters per piece. The optimization is performed using the Adam optimizer. The updates of $\beta_{1:n}$ are performed by backpropagating the gradients of ϕ while the maximum likelihood estimation of θ is performed with the stochastic variational inference module of Pyro [5].

A full instantiation of the model requires an observation model $P(X|Z)$, transformation parameters $\beta_{1:n}$, transformation function f_β, transformation regularization $\mathcal{R}(\beta)$, and a prior model $P_\theta(Z)$. In the experiments section, we provide various instantiations of the model consistent with the assumption of their corresponding datasets, as summarized in Table 1. For each instantiation, the loss is determined by the choice of its components and is used to update the transformation $(\beta_{1:n})$ and prior parameters (θ).

3   Experiments

Our experiments are presented below and are split into fully-supervised, semi-supervised, and unsupervised settings acting on point cloud and/or image representations. To benchmark registration error when ground truth landmarks are available, we use the commonly used target to registration error (TRE) which measures the l_2 distance between moved keypoints and target keypoints.

3.1   Supervised Atlas of Hermaphrodite *C. elegans* Neuron Positions

We used a public dataset of five point clouds of hermaphrodite *C. elegans* tail neurons with 42 neurons per worm [33]. Each neuron in worm i has a 3D location denoted by $p_{i,n} \in \mathbb{R}^3$ and an RGB color denoted by $c_{i,n} \in \mathbb{R}^3$. Therefore, the observation for each worm consists of positions and colors of all its neurons $X_i = \{(p_{i,n}, c_{i,n}), n = 1, \ldots, 42\}$. Every neuron in the worm body corresponds to a ganglion, therefore for every neuron in the dataset we also have a label determining which ganglion that neuron corresponds to. The ganglia provide a natural grouping of the neurons that move together in space, making our proposed PR model suitable for the registration of the point clouds. Our regularization ensures that the distances between neighboring ganglia are approximately preserved after alignment (Eq. 14) as depicted in Fig. 2B.

Statistical Model: We experiment with two different deformation classes for the spatial component of the point clouds, namely rigid (R) and regularized piecewise rigid (PR) warps (Eq. 9). The function class for the color component is a simple

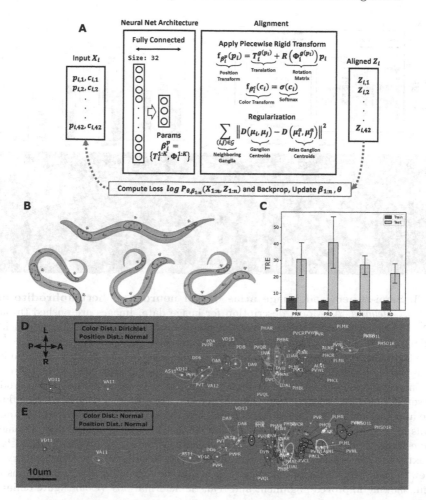

Fig. 2. Supervised positional and color atlas of tail neurons of hermaphrodite *C. elegans*. A: Model description for point clouds of neuron positions and colors in 3 dimensions. An MLP processes the input data and outputs piecewise rigid (PR) parameters β_i per piece. **B:** Motivation of PR for worm point clouds. PR aligns ganglia (denoted by g_1, g_2, g_3) between the atlas (gray worm) and observed point clouds (blue worms) while maintaining the distances between the neighboring ganglia. **C:** Train errors and cross-validated TRE ($\pm$ standard error) for 4 different models (PR: piecewise rigid, R: rigid, D: Dirichlet, N: normal). Notice that all models exhibit overfitting due to the small sample size. **D, E:** Learned atlas using PR-Dirichlet (**D**) and PR-Normal (**E**) models. Small dots indicate individuals' neural positions, larger dots indicate mean positions in the atlas and ellipses indicate one standard deviation of mass. The PRD model projects the mean colors into 3 distinct ones while the PRN model preserves a more detailed description of mean colors. (Color figure online)

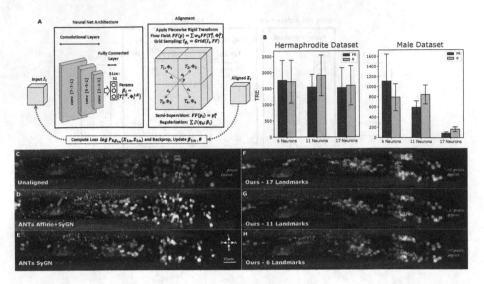

Fig. 3. Semi-supervised image atlas of tail neurons of hermaphrodite and male *C. elegans*. A: Model description for image data. Images are pushed through a CNN which regresses PR transformation parameters and flow field. The flow field is regularized using the Jacobian of the transformation on a grid of points and the position of the landmark points. Transformations $(\beta_{1:n})$ and atlas (θ) parameters are optimized under the log-likelihood cost (red box). **B:** Out-of-sample alignment error decreases with increasing number of neural annotations for the male dataset. The hermaphrodite dataset exhibits less deformations and hence the TRE does not improve by adding more annotations. **C:** Superposition of unaligned NeuroPAL [33] strain hermaphrodite worms used as input. **D, E:** Atlases constructed using a widely-adopted neuroimaging atlas estimation method (SyGN [4]), highlighting how human brain-specific models do not extend to the considered model organisms. **F-H:** Results from our semi-supervised piecewise rigid atlas estimation framework with a varying number of annotations per worm, showing improved alignment and realistic deformations. (Color figure online)

$\texttt{softmax}$ operator normalizing the transformed RGB colors to sum to one (Eq. 11). For the prior distribution over the positions and colors, we considered two statistical models. We chose the prior distribution over the positions to be multivariate normal as suggested by previous work [6,31]. However, for the color distribution we experimented with Normal and Dirichlet distributions[1] (Eq. 16, 17). For registration amortization, we use a fully connected architecture for ϕ (Fig. 2A).

Results. In Fig. 2, we illustrate the atlas parameters θ and aligned point clouds $Z_{1:n}$ as well the uncertainties and the training and testing errors. We learned the atlas and the registration network using 4 worms and tested whether the registration model is capable of aligning the test image to the learned atlas. In Fig. 2 the training TRE refers to the error between the atlas points and aligned

---

[1] Dirichlet is an appropriate choice for color distribution as its samples sum to one and prior information about expected color can be encoded in its parameters α.

points corresponding to the training worms while the test TRE measures this error for the test worms when aligned using the trained registration model. As we are using a very small sample size ($n = 4$) we expect the test error to be higher than the training error. Our results suggest that the rigid model with Dirichlet color distribution (RD) achieves the best test error (5-fold cross-validated) in the fully-supervised point cloud setting but all 4 models achieve comparable performances in terms of training error.

3.2   Semi-supervised Atlas of *C. elegans* images w/partial keypoints

We now showcase the flexibility of our framework in a semi-supervised setting by applying it to images (instead of point clouds) and using partial landmark annotations to guide the transformation. We use a dataset of 5 images of hermaphrodite *C. elegans* and 12 images of male *C. elegans* along with landmarks corresponding to the locations of neurons in those images.

Statistical Model. Formally, $\boldsymbol{X}_i = \{\boldsymbol{I}_i, \boldsymbol{p}_{i,n}\}$ where $\boldsymbol{I}_i \in \mathbb{R}^{W \times H \times D \times C}$, $n = 1, \ldots, 42$, $\boldsymbol{p}_{i,n} \in \mathbb{R}^3$, and where W, H, D, C are the width, height, depth, and the number of channels. Here, PR transformation operates both on images and landmarks (Eq. 10). The semi-supervision is achieved by regularizing the transformation parameters to align the landmarks (Eq. 12). We have further regularization using the Jacobian of the transformation to ensure the feasibility of the transformation (Eq. 15). The prior distribution is the image space is pixel-wise standard normal (Eq. 18). Instead of a fully connected architecture for ϕ here we used a convolutional neural network to extract image features that are useful for registration. The pictorial description of the model is illustrated in Fig. 3A.

Results. We use a subset of keypoints for registration and atlas construction and hold out other key points for benchmarking generalization. The test keypoints are chosen to cover different parts of the image to provide a full picture of the registration quality (Fig. 3). We vary the level of semi-supervision by using {6, 11, 17} landmarks per image and observed that test TRE drops with more landmarks as expected (Fig. 3B). Further, the alignment parameters lead to more biologically feasible transformations when we include more landmarks (Fig. 3F-H). We then compare our method against the *de facto* standard deformable atlas construction technique for human brains (SyGN) [4] with and without affine prealignment. As SyGN does not make use of landmarks or piecewise rigid deformations, it fails to align the neurons (especially in lower density posterior regions) and does not yield biologically plausible atlases (Fig. 3D,E). We observe that for the hermaphrodite dataset the PR and R models achieve comparable performances. However, for the male dataset, where images contain denser subsets of neurons in smaller regions, the PR model outperforms R when more landmarks are included as expected.

3.3   Unsupervised Atlas of Transgenic *D. melanogaster* wings

We used a public dataset [26] of 128 2D fruit fly images from 4 genotypes (egfr, samw, star, tkv) and 2 sexes to infer a latent atlas image that represents an

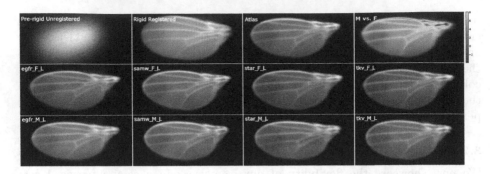

Fig. 4. Unsupervised atlas of fruit fly wing: Atlas: We infer a latent canonical atlas wing in the pixel space without the use of any markers or annotations using 128 example images of fruit fly wings in varying poses. **Pre-rigid Unregistered, Unregistered:** pre-aligned using rigid transformation. The unregistered image is shown in panels before and after rigid alignment. **egfr_F_L - tkv_M_L:** Averaging wing images of different genotypes and sexes enables a visual comparison of morphological differences between these groups. **M vs. F:** Pointwise t-statistics (q<0.05) between males and females yields a heatmap that shows that females have more mass in the the medial part the wing than males.

average wing corrected for postural differences by a piecewise rigid motion model. This dataset can be organized into pairs $X_i = (I_i, Y_i)$ where $I_i \in \mathbb{R}^{W \times H}$ and $Y_i \in \{\text{efgr}, \text{samw}, \text{star}, \text{tkv}\} \times \{\text{male}, \text{female}\}$.

Statistical Model. The observation model is the same as Sect. 3.2 with landmarks removed, i.e., we only rely on image intensities for learning the atlas (Eq. 13). Similar to [9] our framework allows for incorporating genotype-dependent parameters and learning conditional atlases (Eq. 19). To do this, we chose a prior of the form $P_\theta(Z|Y_i) = \mathcal{N}(Z; \mu(Y_i))$.

Results. The resulting image atlas is illustrated in Fig. 4. Using the atlas coordinate framework, we performed a pixelwise t-test on the aligned wings of females and males to observe statistically significant differences in the wing tip density in the medial part of the wing. Furthermore, our results show morphological differences between genotypes which matches domain knowledge.

4   Discussion

Limitations. Some limitations exist in the presented work and will be addressed in the future: (1) The sample sizes considered in the *C. elegans* experiments are small and cannot yield broadly generalizable templates. As more data is publicly released, we will retrain our models to lower atlas bias. (2) Unlike our point cloud experiments, our image experiments do not yield pixelwise uncertainties. We will modify our atlas prior in future work to remedy this.

Table 1. Instantiations of our model in Fully Supervised (`Full`), Semi-Supervised (`Semi`), and Unsupervised (`None`) settings. Each statistical model consists of the dataset (`Data`), observation model (`Obs.`), warp parameters (`Par.`), transformation function (`Func.`), warp regularization (`Reg.`), and prior/atlas distribution (`Prior`). **Notation:** I is an image, p is a 3D point, c is an RGB color, Y is a discrete label, δ is the delta distribution, β denotes the transformation parameters, Φ is a vector of rotation angles and R creates a rotation matrix with Φ, T is the set of translations, FF is a flow field parameterized by β, $w_{1:K}$ are scalar weights summing to one, μ is the centroid of positions within a ganglion, $\mathcal{G}$ is the ganglion neighborhood graph, and $\mathcal{J}$ is the Jacobian. We use $g(p)$ to denote the ganglion that the neuron positioned at point p corresponds to. The positional parameters described above determine a piecewise rigid transformation in the spatial domain. The prior $\mathcal{R}(\beta)$ regularizes warps and ensures smoothness across neighboring ganglia.

	Setting (EqNo.)	Description
Data	Full (3)	$\{(p_{i,n}, c_{i,n}), n = 1, \ldots, 42\}$
	Semi (4)	$\{I_i, p_{i,n}, n = 1, \ldots, 42\}$ $I_i \in \mathbb{R}^{W \times H \times D \times C}$ $p_{i,n} \in \mathbb{R}^3$
	None (5)	$(I_i, Y_i), I_i \in \mathbb{R}^{W \times H}, Y_i \in \{\text{efgr, samw, star, tkv}\} \times \{\text{male, female}\}$
Obs.	Full (6)	$\delta(p_i; f_{\beta_i^p}(p_i)) \delta(c_i; f_{\beta_i^c}(c_i)) \exp(-\mathcal{R}(\beta_i))$
	Semi (7)	$\delta(I_i, p_i; f_{\beta_i}(I_i, p_i)) \exp(-\mathcal{R}(\beta_i))$
	None (8)	$\delta(I_i; f_{\beta_i}(I_i)) \exp(-\mathcal{R}(\beta_i))$
Par.	Full (9)	$\beta_i^c = \{\}$ $\beta_i^p = \{T_i^{1:K}, \Phi_i^{1:K}\}$, $T_i^k \in [-1,1]^3, \Phi_i^k \in [-\pi, \pi]^3$
	Semi/None (10)	$\beta_i = \{T_i^{1:K}, \Phi_i^{1:K}\}$, $T_i^k \in [-1,1]^3, \Phi_i^k \in [-\pi, \pi]^3$
Func.	Full (11)	$f_\beta(p, c) = \left[f_{\beta_p}(p), f_{\beta_c}(c)\right] = \left[\text{R}(\Phi^{g(p)})p + T^{g(p)}, \texttt{softmax}(c)\right]$
	Semi (12)	$f_\beta(I, p) = \left[\text{FF}_\beta \circ I, \text{FF}_\beta(p)\right]$ $\text{FF}(p) = \sum_{k=1}^{K} w_k(p)(\text{R}(\Phi_k)p + T_k)$
	None (13)	$f_\beta(I) = \text{FF}_\beta \circ I$, $\text{FF}(p) = \sum_{k=1}^{K} w_k(p)(\text{R}(\Phi_k)p + T_k)$
Reg.	Full (14)	$\mathcal{R}(\beta_i) = \sigma_\beta \sum_{i,j \in \mathcal{G}} \left\| D(\mu_i, \mu_j) - D(\mu_i^a, \mu_j^a) \right\|^2$
	Semi/None (15)	$\mathcal{R}(\beta) = \sigma_\beta \sum_{k=1}^{K} \det \mathcal{J}_\beta(q_k)$
Prior	Full-1 (16)	$P_\theta(Z) = P(p)P(c) = \mathcal{N}(p; \mu_p, \sigma_p I) \mathcal{N}(c; \mu_c, \sigma_c I)$ MVN-MVN
	Full-2 (17)	$P_\theta(Z) = P(p)P(c) = \mathcal{N}(p; \mu_p, \sigma_p I) \text{Dir}(c; \mu_c)$ MVN-Dir
	Semi (18)	$P_\theta(I_i, p_i) = \mathcal{N}(I_i; \mu^I, \sigma_I I) \mathcal{N}(p_i; \mu^p, \sigma_p I)$
	None (19)	$P_\theta(Z|Y_i) = \mathcal{N}(Z; \mu(Y_i))$

Conclusions. This work developed a general probabilistic framework to compute piecewise rigid point cloud and image atlases in novel imaging datasets of model organisms such as *C. elegans* and fruit flies. As new imaging modalities emerge to capture different views of the nervous systems of model animals, we expect that the flexibility of our framework will be valuable for standardizing downstream analyses. We also expect piecewise rigid atlas construction to apply to applications such as motion correction [11,34] and kinematics modeling [32].

Acknowledgements. Paninski: NSF NeuroNex DBI-1707398, Gatsby Charitable Foundation, DMS 1912194, Simons Foundation Collab. on the Global Brain. Yemini: Klingenstein-Simons Fellowship in Neuroscience, Hypothesis Fund. Dey: NIH NIBIB NAC P41EB015902, NIBIB 5R01EB032708. Varol: 1K99MH128772-01A1. Venkatachalam: Burroughs Wellcome Fund and NIH R01 NS126334.

References

1. Arsigny, V., Commowick, O., Ayache, N., Pennec, X.: A fast and log-Euclidean polyaffine framework for locally linear registration. J. Math. Imaging Vis. **33**(2), 222–238 (2009)
2. Arsigny, V., Pennec, X., Ayache, N.: Polyrigid and polyaffine transformations: a new class of diffeomorphisms for locally rigid or affine registration. In: Ellis, R.E., Peters, T.M. (eds.) MICCAI 2003. LNCS, vol. 2879, pp. 829–837. Springer, Heidelberg (2003). https://doi.org/10.1007/978-3-540-39903-2_101
3. Avants, B., Gee, J.C.: Geodesic estimation for large deformation anatomical shape averaging and interpolation. Neuroimage **23**, S139–S150 (2004)
4. Avants, B.B., et al.: The optimal template effect in hippocampus studies of diseased populations. Neuroimage **49**(3), 2457–2466 (2010)
5. Bingham, E., et al.: Pyro: Deep universal probabilistic programming. J. Mach. Learn. Res. **20**, 28:1–28:6 (2019)
6. Bubnis, G., Ban, S., DiFranco, M.D., Kato, S.: A probabilistic atlas for cell identification (2019)
7. Choe, K.P., Strange, K.: Molecular and genetic characterization of osmosensing and signal transduction in the nematode Caenorhabditis elegans. FEBS J. **274**(22), 5782–5789 (2007)
8. Commowick, O., et al.: An efficient locally affine framework for the smooth registration of anatomical structures. Med. Image Anal. **12**(4), 427–441 (2008)
9. Dalca, A., Rakic, M., Guttag, J., Sabuncu, M.: Learning conditional deformable templates with convolutional networks. In: Advances in Neural Information Processing Systems. vol. 32 (2019)
10. Davis, B.C., Fletcher, P.T., Bullitt, E., Joshi, S.: Population shape regression from random design data. Int. J. Comput. Vis. **90**(2), 255–266 (2010)
11. Dey, N., Messinger, J., Smith, R.T., Curcio, C.A., Gerig, G.: Robust non-negative tensor factorization, diffeomorphic motion correction, and functional statistics to understand fixation in fluorescence microscopy. In: Shen, D., et al. (eds.) MICCAI 2019. LNCS, vol. 11764, pp. 658–666. Springer, Cham (2019). https://doi.org/10.1007/978-3-030-32239-7_73
12. Dey, N., Ren, M., Dalca, A.V., Gerig, G.: Generative adversarial registration for improved conditional deformable templates. In: Proceedings of the IEEE/CVF International Conference on Computer Vision, pp. 3929–3941 (2021)
13. Ding, Z., Niethammer, M.: Aladdin: Joint atlas building and diffeomorphic registration learning with pairwise alignment. In: Proceedings of the IEEE/CVF Conference on Computer Vision and Pattern Recognition, pp. 20784–20793 (2022)
14. Emmons, S.W., Sternberg, P.W.: Male development and mating behavior (2011)
15. Greitz, T., Bohm, C., Holte, S., Eriksson, L.: A computerized brain atlas: construction, anatomical content, and some applications. J. Comput. Assist. Tomogr. **15**(1), 26–38 (1991)
16. Heckscher, E.S., et al.: Atlas-builder software and the eNeuro atlas: resources for developmental biology and neuroscience. Development **141**(12), 2524–2532 (2014)

17. Houle, D., Govindaraju, D.R., Omholt, S.: Phenomics: the next challenge. Nat. Rev. Genet. **11**(12), 855–866 (2010)
18. Jones, A.R., Overly, C.C., Sunkin, S.M.: The Allen brain atlas: 5 years and beyond. Nat. Rev. Neurosci. **10**(11), 821–828 (2009)
19. Joshi, S., Davis, B., Jomier, M., Gerig, G.: Unbiased diffeomorphic atlas construction for computational anatomy. Neuroimage **23**, S151–S160 (2004)
20. Kaiser, M., Hilgetag, C.C.: Nonoptimal component placement, but short processing paths, due to long-distance projections in neural systems. PLoS Comput. Biol. **2**(7), e95 (2006)
21. Long, F., Peng, H., Liu, X., Kim, S.K., Myers, E.: A 3D digital atlas of C. elegans and its application to single-cell analyses. Nature Meth. **6**(9), 667–672 (2009)
22. Roland, P., et al.: Human brain atlas: for high-resolution functional and anatomical mapping. Hum. Brain Mapp. **1**, 137–184 (1994)
23. Scheffer, L.K., Meinertzhagen, I.A.: The fly brain atlas. Annu. Rev. Cell Dev. Biol. **35**, 637–653 (2019)
24. Schuh, A., et al.: Unbiased construction of a temporally consistent morphological atlas of neonatal brain development. bioRxiv, p. 251512 (2018)
25. Skuhersky, M., Wu, T., Yemini, E., Boyden, E., Tegmark, M.: Toward a more accurate 3D atlas of c. elegans neurons. bioRxiv (2021)
26. Sonnenschein, A., VanderZee, D., Pitchers, W.R., Chari, S., Dworkin, I.: An image database of drosophila melanogaster wings for phenomic and biometric analysis. GigaScience **4**(1), s13742-015 (2015)
27. Sulston, J.E., Horvitz, H.R.: Post-embryonic cell lineages of the nematode, Caenorhabditis elegans. Dev. Biol. **56**(1), 110–156 (1977)
28. Szigeti, B., et al.: OpenWorm: an open-science approach to modeling Caenorhabditis elegans. Front. Comput. Neurosci. **8**, 137 (2014)
29. Tekieli, T., et al.: Visualizing the organization and differentiation of the male-specific nervous system of C. elegans. Development, **148**, dev199687 (2021)
30. Toyoshima, Y., et al.: An annotation dataset facilitates automatic annotation of whole-brain activity imaging of C. elegans. bioRxiv (2019). https://doi.org/10.1101/698241
31. Varol, E., et al.: Statistical atlas of *C. elegans* neurons. In: Martel, A.L., et al. (eds.) MICCAI 2020. LNCS, vol. 12265, pp. 119–129. Springer, Cham (2020). https://doi.org/10.1007/978-3-030-59722-1_12
32. Wustenberg, R.: Carpal bone rigid-body kinematics by log-euclidean polyrigid estimation (2022)
33. Yemini, E., et al.: Neuropal: a multicolor atlas for whole-brain neuronal identification in C. elegans. Cell **184**(1), 272–288 (2021)
34. Yu, J., et al.: Versatile multiple object tracking in sparse 2D/3D videos via diffeomorphic image registration. bioRxiv (2022)

Harmonization/Federated Learning

Harmonizing Flows: Unsupervised MR Harmonization Based on Normalizing Flows

Farzad Beizaee[1,2]([✉]), Christian Desrosiers[1], Gregory A. Lodygensky[2,3], and Jose Dolz[1]

[1] École de Technologie Supérieure (ETS), Montreal H3C 1K3, Canada
farzad.beizaee.1@ens.etsmtl.ca
[2] CHU Sainte-Justine, University of Montreal, Montreal H3T 1C5, Canada
[3] Canadian Neonatal Brain Platform, Montreal, Canada

Abstract. In this paper, we propose an unsupervised framework based on normalizing flows that harmonizes MR images to mimic the distribution of the source domain. The proposed framework consists of three steps. First, a shallow harmonizer network is trained to recover images of the source domain from their augmented versions. A normalizing flow network is then trained to learn the distribution of the source domain. Finally, at test time, a harmonizer network is modified so that the output images match the source domain's distribution learned by the normalizing flow model. Our unsupervised, source-free and task-independent approach is evaluated on cross-domain brain MRI segmentation using data from four different sites. Results demonstrate its superior performance compared to existing methods. The code is available at https://github.com/farzad-bz/Harmonizing-Flows.

Keywords: Harmonization · Normalizing flows · Test-time adaptation

1 Introduction

Deep learning models have become the *de facto* solution for most image-based problems, including those in the medical domain. Despite significant progress, these models still suffer under distributional drift, and their performance largely degrades when they are applied to data obtained in different conditions.

Clinical studies using magnetic resonance imaging (MRI) often have to deal with such large domain shifts. Due to the qualitative nature of the MRI acquisition process, generated images are sensitive to imaging devices, acquisition protocols, scanner artifacts, as well as to patient populations [27]. For instance, images from the same modality (e.g., T1-w) acquired from two different scanners with separate configurations will likely present noticeable differences, which can be considered a domain shift. Consequently, collecting a multi-center MRI dataset to address a particular clinical question does not guarantee a greater statistical power, as the increase in variance comes from a non-clinical source.

© The Author(s), under exclusive license to Springer Nature Switzerland AG 2023
A. Frangi et al. (Eds.): IPMI 2023, LNCS 13939, pp. 347–359, 2023.
https://doi.org/10.1007/978-3-031-34048-2_27

Furthermore, this data heterogeneity can also hamper the generalizability of deep learning models, preventing their large dissemination. In particular, when trained on a specific site, such models are typically unable to provide similar performance for other centers.

To alleviate this issue, image harmonization addresses the distributional shift problem from an image-to-image mapping perspective, where the objective is to transfer image contrasts across different domains. Nevertheless, most harmonization methods in the literature make strong assumptions that might hamper their scalability and usability in real-life scenarios. First, some methods must have access to source images during the adaptation, which may no longer be available. Labels associated with the downstream task may also be required in other approaches. Finally, most harmonization techniques need to know the target domains during training, while these domains are often unknown.

In this work, we make the following contributions:

- We relax all these assumptions and present a novel MR harmonization method that is *source-free* ($\mathcal{SF}$), *task-agnostic* ($\mathcal{TA}$) and can handle *unknown-domains* ($\mathcal{UD}$) without requiring to be retrained for each target distribution. Indeed, our method only needs one domain and modality at training time, as opposed to existing approaches.
- In particular, we propose to use a novel family of generative models, i.e., normalizing flows, which have shown to be a powerful method to model data distributions in generative tasks. We stress that leveraging normalizing flows to guide the adaptation of a harmonizer network has not been explored.
- In addition to the methodological novelty, our empirical results demonstrate that our approach brings substantial improvements compared to existing techniques, while alleviating their weaknesses.
- Furthermore, due to its task-agnostic nature and its capability to work under the *unknown-domains* scenario, the proposed method can also be employed in the task of test-time adaptation. In this setting, our method largely outperforms a popular task-agnostic test-time adaptation strategy.

2   Related Work

Image Harmonization. Several techniques have been proposed for the harmonization of images in the medical domain, and particularly for MRI data. Classical post-processing steps, such as intensity histogram matching [24,26], reduce the influence of biases across scanners, but may also remove informative local variations in intensity. Statistical approaches can model image intensity and dataset bias at the voxel level [2,11,12], however they must often be adjusted each time images from new sites are provided. Modern strategies for image harmonization, which are based on deep learning models, have shown to be a promising alternative for this problem [4,5,8,21,34,35]. Nevertheless, they make unrealistic assumptions that hamper the scalability of existing approaches to large scale multi-site harmonization tasks. First, images of the same target anatomy across multiple sites, commonly referred to as *traveling subjects* are employed to identify intensity transformations between different sites [5]. This involves that a given number of subjects are scanned at every site or scanner required for training, a condition rarely met in practice. Second, another group of methods is limited to two domains [34] and requires target domains to be known at training time [21,34]. In addition, each time a new domain is added, these approaches must be fine-tuned in order to accommodate the characteristics of each domain. Calamity [21] further needs paired multi-modal MR sequences, limiting even more its applicability to single modality scenarios. Last, task-dependent approaches leverage labels associated to each image for a given down-stream task [4,8], thus optimizing the harmonization for this specific problem. Nevertheless, having access to large labeled datasets might be impractical due to the underlying labeling cost.

Test-Time Adaptation. Our method also relates to the problem of test-time domain adaptation (TTA) [3,20,23,29] which aims to quickly adapt a pre-trained deep network to domain shifts during inference on test examples. One key difference between TTA and the well-known unsupervised domain adaption (UDA) problem is that, in TTA, the source examples are no longer available. One of the earliest TTA approaches, called TENT [29], updates the affine transformation parameters of normalization layers by minimizing the Shannon entropy of predictions for test examples. In [23], this strategy is improved by optimizing a log-likelihood ratio instead of entropy, as well as by considering the normalization statistics of the test batch. The method named SHOT [20] fine-tunes the entire feature extractor with a mutual information loss and uses pseudo-labels to provide additional test-time guidance. Instead of updating the network parameters, LAME [3] uses Laplacian regularization to do a post-hoc adaptation of the softmax predictions.

Normalizing Flows. Recently, normalizing Flows (NFs) have emerged as a popular approach for constructing probabilistic and generative models with tractable distributions [19]. NFs aim at transforming unknown complex distributions into simpler ones, for instance, a standard normal distribution. This is achieved by applying a sequence of invertible and differentiable transformations. While most existing literature has leveraged NFs for generative tasks (e.g., image generation [15,17], noise modeling [1], graph modeling [32]) and anomaly detection [14,18], recent evidence also suggests their usefulness for aligning a given set of source domains [13,28]. To our knowledge, a single work has investigated NFs in the context of harmonization [31]. However, it aimed at performing causal inference on pre-extracted features (brain ROI volume measures), and not image harmonization as in our work. Moreover, since extracting ROIs requires pixel-wise labels, the method in [31] is not task-agnostic.

3   Methodology

We first define the problem addressed in our work. Let $\mathcal{X}_{\mathcal{S}} = \{\mathbf{x}_n\}_{n=1}^N$ be a set of unlabeled images in the source domain $\mathcal{S}$, where a given image i is represented by

$\mathbf{x}_i \in \mathbb{R}^{|\Omega|}$ and Ω denotes its spatial domain (i.e., $W \times H$). Similarly, we denote as $\mathcal{X}_T = \{\mathbf{x}_n\}_{n=1}^M$ the set of unlabeled images in a potential target domain T^1. The goal of unsupervised data harmonization is to find a mapping function $f_\theta : \mathcal{S} \to \mathcal{T}$ without having access to labeled images for any of the domains. In what follows, we present our NF-based solution for this problem, whose framework is depicted in Fig. 1.

3.1 Learning the Source Domain Distribution

We leverage Normalizing Flows (NFs) [7] to model the distribution of the source domain. NFs are a recent family of generative methods that can model a complex probability density $p_x(\mathbf{x})$ (i.c., the source) as a series of transformation functions, denoted as $g_\phi = g_1 \circ g_2 \circ \ldots g_T$, applied on simpler and tractable probability density $p_u(\mathbf{u})$ (e.g., a standard multi-variate Gaussian distribution). We can express a source image as $\mathbf{x} = g_\phi(\mathbf{u})$, where $\mathbf{u} \sim p_u(\mathbf{u})$ and $p_u(\mathbf{u})$ is the base distribution of the flow model. An important requirement of the transformation function g_ϕ is that it must be *invertible*, and both g_ϕ and g_ϕ^{-1} should be *differentiable*. Under these conditions, the density of the original variable $\mathbf{x}$ is well-defined and its likelihood can be computed exactly using the change of variables rule as:

$$
\begin{aligned}
\log p_{\mathbf{x}}(\mathbf{x}) &= \log p_{\mathbf{z}}\left(g_\phi^{-1}(\mathbf{x})\right) + \log\left|\det\left(\mathbf{J}_{g_\phi^{-1}}(\mathbf{x})\right)\right| \\
&= \log p_{\mathbf{z}}\left(g_\phi^{-1}(\mathbf{x})\right) + \sum_{t=1}^T \log\left|\det\left(\mathbf{J}_{g_t^{-1}}(\mathbf{u}_{t-1})\right)\right|
\end{aligned}
\tag{1}
$$

where the first term on the right-hand side is the log-likelihood under the simple distribution, and $\mathbf{J}_{g_t^{-1}}(\mathbf{u}_{t-1})$ is the Jacobian matrix of the inverse transformation g_t. To train the NF model and learn the source data distribution, the model parameters ϕ are typically optimized so to minimize the negative log-likelihood in Eq. 1. This results in the following loss function:

$$
\mathcal{L}_{NF} = -\log p_{\mathbf{x}}(\mathbf{x})
\tag{2}
$$

Building the Normalizing Flow. To build a bijective transformation function for the NF model, stacking a sequence of affine coupling layers [7,17] has been demonstrated to be an efficient strategy. Because flows based on coupling layers are computationally symmetric, i.e., equally fast to evaluate or invert, they can overcome the usability issues of asymmetric flows such as masked autoregressive flows, making them a popular choice. Let us consider $\mathbf{z} \in \mathbb{R}^D$ as the input to the coupling layer, which is split into a disjoint partition: $(\mathbf{z}^A, \mathbf{z}^B) \in \mathbb{R}^d \times \mathbb{R}^{D-d}$. The transformation function $g(\cdot) : \mathbb{R}^D \to \mathbb{R}^D$ can then be defined as:

$$
\mathbf{y}^A = \mathbf{z}^A, \quad \mathbf{y}^B = \mathbf{z}^B \odot \exp\left(s\left(\mathbf{z}^A\right)\right) + t\left(\mathbf{z}^A\right)
\tag{3}
$$

---

[1] Note that for simplicity, we assume here that there exists only a single domain. Nevertheless, our formulation is directly applied to T different domains.

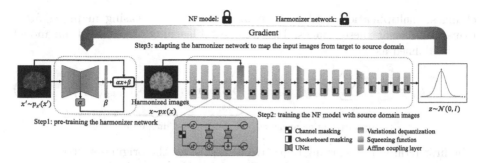

Fig. 1. Pipeline of the proposed Harmonizing Flows method. Our approach consists of two steps. First, we employ normalizing flows (NFs) to capture the distribution of the source domain. During the second stage, the trained NFs are leveraged to update the parameters of a harmonizer network, which are updated in order to maximize the similarity between the harmonized outputs and the distribution learned by the NF. Note that steps 1 and 2 are not dependent on each other, and can therefore be performed in any order.

where $\odot$ is element-wise multiplication. This setting offers simplicity for calculating the Jacobian determinant, which makes it possible to use complex neural networks as shift $s(\cdot)$ and scale $t(\cdot)$ networks. Note that the transformation in Eq. 3 is invertible and therefore allows for efficient Jacobian computation in Eq. 1. The work in [7] presented coupling flows on simpler tasks and datasets, e.g., CIFAR, which required less enriched representations. In contrast, the problem at hand requires pixel-to-pixel mappings on more challenging images. Thus, we replace the simple convolutional blocks in [7] with shallow U-shaped convolutional neural networks to find the shift and scale parameters of the affine transformation, as they capture more global context and provide higher representation power. Furthermore, as NFs are based on the change of variables rule, which is defined in continuous space, it is crucial to make the input continuous. Dequantization of the input can be achieved by adding a uniform noise $u \in U[0, 1]$ to the discrete values. However, it might result in a hypercube representation of the images with sharp borders. These sharp borders are hard to model for a flow as it uses smooth transformations. Recently, a variational framework was proposed [15] to extend dequantization to more sophisticated distributions, by replacing the uniform distribution with a learnable distribution.

Constraining the Source-Distribution Learning. Optimizing the objective in Eq. 2 with only source images might bias the model to focus on characteristics of subjects, such as age and gender, rather than on source-specific features like contrast and brightness. To overcome this issue, we propose a strategy that facilitates the learning of the source-domain distribution. This technique consists in randomly selecting N' images from the original dataset $\mathcal{X}_S$ and applying a series of augmentations $f_{aug}(\cdot)$ such that the resulting image has a dissimilarity to the original image (measured by mean squared distance) higher than a specified threshold. In particular, we employ contrast augmentation, brightness changes, multiplication, and random monotonically increasing mapping functions to augment these images. Then, the total learning objective of our model can be defined as:

$$\mathcal{L}_T = \underbrace{- \sum_{n=1}^{N-N'} \log p_\mathbf{x}(\mathbf{x}_n)}_{\text{Source distribution modeling}} \underbrace{- \sum_{n=1}^{N'} \min\left(c, -\log p_\mathbf{x}(f_{aug}(\mathbf{x}_n))\right)}_{\text{Guiding term}}. \quad (4)$$

The first term is the learning objective in Eq. 2 over the original source images, whereas the second one forces the NF model to decrease the likelihood on the augmented images, which facilitates the learning of domain-specific characteristics (e.g., contrast or brightness) instead of subject-related features (e.g., sex or age). Furthermore, we use a constant margin c in the second term to prevent the negative log-likelihood of an augmented sample from diverging to infinity.

3.2   Achieving Image Harmonization

Harmonizer Network. A simple solution to perform image-to-image translation is to employ a harmonizer network $h_\theta(\cdot)$, such that MRIs from the target domain are translated to the source domain. This can be expressed as $p_\mathbf{x}(\mathbf{x}) = p_{\mathbf{x}'}(h_\theta(\mathbf{x}'))$, where θ is the set of learnable parameters of the harmonizer network, and $\mathbf{x}$ and $\mathbf{x}'$ are images from the source and target domains, respectively. To train this model, we can simply use a standard reconstruction loss over images across different domains. However, we want the proposed method to follow a *domain-free* paradigm, where target domains remain unknown at training time. Toward this goal, we train the harmonizer network to reconstruct the original source images from their augmented versions. As in the previous step, we augment the original images by using different types of contrast augmentation, brightness changes, multiplication, or random monotonically increasing mapping functions. Contrary to the first step, there is no constraint on the magnitude of the augmentations. The learning objective for the harmonizer network thus becomes:

$$\theta^{init} = \underset{\theta}{\arg\min} \frac{1}{N} \sum_{n=1}^{N} \|(\mathbf{x}_n - h_\theta(f_{aug}(\mathbf{x}_n))\|^2 \quad (5)$$

We stress that the performed augmentations are not reliable representations of potential unseen target domains. Consequently, the direct application of the learned parameters θ^{init} for image-to-image mapping will result in suboptimal domain transformations. Nevertheless, they can serve as the initial model for the subsequent step. A simple UNet is considered for the harmonizer network, which learns two values. First, the last layer of the network (β) is employed as a bias value having the same dimension as the input image. Second, a scalar α from the middle layer of the network is used as a coefficient value. In this way, the output of the harmonizer can be defined as $h_\theta(\mathbf{x}) = \alpha * \mathbf{x} + \beta$.

Guiding the Harmonizer Network with the Normalizing Flow. The final step involves updating the harmonizer network so that images from the target domain are mapped into the source domain distribution. To achieve this, we propose to leverage the trained NF, which is stacked at the output of the harmonizer network. Note that the NF model has already learned the distribution of source data, and therefore its parameters remain frozen during the adaptation of the harmonizer. Thus, the learning objective of the adaptation stage consists in increasing the likelihood of the harmonizer outputs for images from the target domain, based on the NF model's density estimation. This loss function can be formally defined as follows:

$$\mathcal{L}_{Adap} = - \sum_{m=1}^{M} \log p_{\mathbf{x}}\big(g_\phi\big(h_\theta(\mathbf{x}_m)\big)\big) \tag{6}$$

As stopping criterion for updating the harmonizer, we evaluate two possible alternatives. First, we measure the Shannon entropy of the predictions for the target task (e.g., segmentation or classification), stopping the adaptation when the entropy plateaus. We also consider the bits per dimension (bpd), a scaled version of the *negative log-likelihood* widely used for evaluating generative models: $bpd = - \log p_{\mathbf{x}}(\mathbf{x}) \cdot (\log 2 \cdot \prod_i \Omega_i)^{-1}$ where Ω_1, ..., Ω_T, is the spatial dimension of the input images. More concretely, we can stop updating the harmonizer parameters when the reached bpd value is the same as the one observed for the source domain using the NF model. In practice, this value can be obtained at training time using a validation set.

4   Experiments

4.1   Experimental Setting

We evaluate the proposed method on the task of brain MRI segmentation across multiple sites. The reason behind this choice stems from the fact that the segmentation performance is a reliable indicator of whether the structural information is well preserved during the mapping.

Datasets. Four sites of the Autism Brain Imaging Data Exchange (ABIDE) [6] dataset are employed: California Institute of Technology (CALTECH), Kennedy Krieger Institute (KKI), University of Pittsburgh School of Medicine (PITT) and NYU Langone Medical Center (NYU). The selection of these sites is based on their cross-site difference, as these datasets present the most distinct histogram from each other, which better highlights the impact of harmonization. These sites are denoted as $\mathcal{D}_1$, $\mathcal{D}_2$, $\mathcal{D}_3$, and $\mathcal{D}_4$, respectively. From each site, we selected 20 T1-weighted MRIs from the healthy control population (19 from CALTECH), which were skull-stripped, motion-corrected, and quantized to 256 levels of intensity. 2D coronal slices of 60% of these images are used for training, 15% for validation, and the remaining 25% for testing. Furthermore, the segmentation labels are obtained from FreeSurfer [10], following other large-scale studies

[9], and grouped into 15 labels: background, cerebellum gray matter, cerebellum WM, cerebral GM, cerebral WM, thalamus, hippocampus, amygdala, ventricles, caudate, putamen, pallidum, ventral DC, CSF, and brainstem.

Harmonization Baselines. The proposed approach is benchmarked against a set of relevant harmonization and image-to-image translation methods. We first consider a simple *Baseline* applying the segmentation network directly on non-harmonized images, in order to assess the impact of each harmonization approach. Our comparison also includes: Histogram Matching [24], aleatoric uncertainty estimation (AUE) [30], Combat [25], BigAug [33] (which uses heavy augmentations for generalization of the segmentation networks), and two popular generative-based approaches, i.e., Cycle-GAN [22] and Style-Transfer [21].

Evaluation Protocol. To assess the performance of our harmonization approach, we resort to a segmentation task as it requires the preservation of fine-grained structural details. First, a segmentation network $S_\Phi(\cdot)$ is trained on the images from the source domain, whose parameters remain frozen thereafter. The harmonized images from each method are then employed to evaluate segmentation performance, which is measured with the Dice Similarity Coefficient (DSC) and modified Hausdorff distance (HD). To evaluate the robustness of tested methods, we repeat the experiments four times, each employing a different source and set of target domains. These different settings are denoted as $\mathcal{A} : \mathcal{D}_1 \to \{\mathcal{D}_2, \mathcal{D}_3, \mathcal{D}_4\}$; $\mathcal{B} : \mathcal{D}_2 \to \{\mathcal{D}_1, \mathcal{D}_3, \mathcal{D}_4\}$; $\mathcal{C} : \mathcal{D}_3 \to \{\mathcal{D}_1, \mathcal{D}_2, \mathcal{D}_4\}$; $\mathcal{D} : \mathcal{D}_4 \to \{\mathcal{D}_1, \mathcal{D}_2, \mathcal{D}_3\}$.

Implementation Details. The Normalizing flow model is trained for 1600 epochs using Adam optimizer with an initial learning rate of 1×10^{-3}, a weight decay of 0.5 every 200 epochs and a batch-size of 32. We use a U-shaped network inside the coupling layers, which consists of four levels of different scales with a scaling factor of 2. Each level includes a modified version of the ELU activation function, i.e., concat(ELU(x), ELU($-x$)), and a convolutional layer followed by a normalizing layer. To construct the NF model, we first cascade four coupling layers with checkerboard masking to learn the noise distribution using variational dequantization. After applying four of the same coupling layers, features are squeezed as explained in [7] to have a lower spatial dimension and more channels. We then add four coupling layers using a channel-masking strategy, another feature squeezing function, and a final set of four coupling layers with channel-masking. The overall architecture of the flow model is shown in Fig. 1. The margin c used for guiding the flow is set empirically to 1.2. **The harmonizer** has five levels of different scales with a scaling factor of 2, each level including two layers of the modified ELU activation function followed by a convolutional layer. The number of kernels of each level is 16, 32, 48, 64, and 64, respectively. The harmonizer is trained for 200 epochs using Adam optimizer with a learning rate starting at 1×10^{-3}, a weight decay of 0.5 every 30 epochs and a batch-size of 32. **The segmentation network** is trained for 200 epochs using Adam optimizer with an initial learning rate of 4×10^{-3}, a weight decay of 0.5 every 30 epochs and a batch-size of 32. All the models were implemented in PyTorch and were run on NVIDIA RTX A6000 GPU cards.

Table 1. Performance overview. Main results for the compared methods across different settings ($\mathcal{A}, \mathcal{B}, \mathcal{C}, \mathcal{D}$). The best results are highlighted in bold.

	$\mathcal{SF}$	$\mathcal{TA}$	$\mathcal{UD}$	$\mathcal{A}$	$\mathcal{B}$	$\mathcal{C}$	$\mathcal{D}$	Average
				DSC (%)				
Baseline	–	–	–	54.6±7.5	60.8±4.6	62.9±5.8	72.6±4.5	62.7±5.6
AUE [30]	✓	✗	✓	54.7±7.4	60.7±4.7	62.6±5.7	72.4±4.5	62.6±5.6
Hist matching [24]	✓	✓	✓	55.7±8.6	58.1±5.1	62.2±4.8	69.5±4.9	61.4±5.9
Combat [25]	✓	✓	✓	75.7±9.2	79.9±6.0	79.5±8.1	79.9±7.8	78.7±7.8
BigAug [33]	✓	✗	✓	54.2±7.6	67.9±3.6	61.5±4.5	78.0±3.7	65.4±4.8
Cycle-GAN [22]	✗	✓	✗	74.5±3.0	78.8±2.9	80.1±2.2	83.1±2.0	79.1±2.5
Style-transfer [21]	✓	✓	✗	56.9±7.1	80.0±1.7	67.8±4.9	73.4±4.0	69.5±4.4
Ours	✓	✓	✓	**80.8±3.2**	**82.3±2.2**	**83.2±3.3**	**85.2±1.5**	**82.9±2.6**
				HD (mm)				
Baseline	–	–	–	18.20±8.27	9.57±3.23	9.07±2.78	5.73±1.81	10.64±4.03
AUE [30]	✓	✗	✓	17.57±8.18	9.67±3.39	9.03±2.87	5.57±1.85	10.46±4.08
Hist matching [24]	✓	✓	✓	17.40±8.10	10.47±3.77	12.00±4.56	6.73±2.40	11.65±4.71
Combat [25]	✓	✓	✓	5.23±3.87	3.67±2.47	3.30±1.80	3.17±2.14	3.84±2.57
BigAug [33]	✓	✗	✓	19.53±10.51	8.43±3.40	18.87±7.76	3.70±1.07	12.63±5.69
Cycle-GAN [22]	✗	✓	✗	4.63±2.89	3.63±1.93	2.63±0.62	2.30±0.55	3.30±1.50
Style-transfer [21]	✓	✓	✗	14.23±7.20	2.93±0.78	7.53±2.38	4.27±1.35	7.24±2.92
Ours	✓	✓	✓	**3.10±1.63**	**2.77±0.87**	**2.37±0.77**	**2.30±0.50**	**2.63±0.94**

4.2 Results

Comparison to State-of-the-Art. Segmentation results obtained on the images harmonized by different methods are reported in Table 1. We can observe that the proposed approach consistently outperforms compared methods by a noticeable margin, across datasets and for both segmentation metrics. In particular, the average improvement gain is nearly 4% in terms of DSC, and 0.7 mm in terms of HD, compared to the second best performing method, CycleGAN.

Impact of Normalizing Flows. This section assesses the impact of each component of the proposed method. In particular, we evaluate the segmentation performance when images are: *i)* not normalized, *ii)* normalized with the pre-trained harmonizer θ^{init}, or *iii)* normalized with the proposed harmonizing flow. The results from this ablation study (Table 2) empirically motivate the proposed NF model as a powerful mechanism to guide the harmonizer network. First, the strategy proposed to pre-train the harmonizer brings a substantial improvement over non-harmonized images, yet it is very simple and does not require access to images from the target domain. Secondly, driving the adaptation of the harmonizer with the proposed NF further improves the segmentation results by a large margin, demonstrating the benefits of our model.

Table 2. Ablation study on the different components in terms of DSC.

	$\mathcal{A}$	$\mathcal{B}$	$\mathcal{C}$	$\mathcal{D}$	Average
Without harmonization	54.6 ±7.5	60.8 ±4.6	62.9 ±5.8	72.6 ±4.5	62.7 ±5.6
Pre-trained harmonizer	71.9 ±5.1	77.0 ±3.0	76.0 ±5.2	75.3 ±4.5	75.1 ±4.5
Adapting using NF	80.8 ±3.2	82.3 ±2.2	83.2 ±3.3	85.2 ±1.5	82.9 ±2.6

Adaptation Stopping Criterion. In this section, we address the important question of when to stop the adaptation. The first alternative is to stop it when the Shannon entropy of the segmentation predictions reaches its minimum point. As this objective does not require any labeled data, it gives a valid stopping point for adapting the harmonizer. As a second criterion, we stop adapting when the output *bpd* of the NF model reaches the observed source domain *bpd*. As opposed to entropy, this criterion is not task-dependent and is suitable for unsupervised tasks or tasks where entropy is not applicable. Last, we resort to the segmentation performance, and stop the adaptation when it reaches the best DSC score, which we define as *Oracle*. Note that this criterion is unrealistic, and its purpose is just to demonstrate how a good stopping criterion can improve harmonization. As shown in Table 3, although minimum entropy is a better criterion compared to *bpd*, both achieve comparable performances. In addition, both stopping criteria are a suitable choice, as their results are very close to the *Oracle*.

Table 3. Impact of the adaptation stopping criterion (in terms of DSC).

	$\mathcal{A}$	$\mathcal{B}$	$\mathcal{C}$	$\mathcal{D}$	Average
Minimum Entropy	80.8 ±3.2	82.3 ±2.2	83.2 ±3.3	85.2 ±1.5	82.9 ±2.6
Source BPD	80.5 ±3.1	82.6 ±2.3	82.7 ±3.6	84.8 ±1.6	82.6 ±2.6
Oracle (best epoch)	81.0 ±3.0	82.7 ±2.3	84.0 ±2.9	85.2 ±1.5	83.2 ±2.4

Qualitative Results. Fig. 2 depicts several examples of harmonized images produced by the proposed approach. These results illustrate that, regardless of the target domain, our method produces reliable image-to-image mappings to the source distribution.

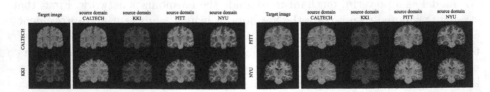

Fig. 2. Examples of harmonized images produced by the proposed method.

Table 4. Quantitative results on bias corrected images (in terms of DSC).

	$\mathcal{A}$	$\mathcal{B}$	$\mathcal{C}$	$\mathcal{D}$	Average
Baseline	71.9 ±3.7	78.2 ±3.6	82.9 ±1.7	82.6 ±2.9	78.9 ±3.0
AUE [30]	71.4 ±3.7	78.3 ±3.6	82.7 ±1.7	82.7 ±2.8	78.8 ±2.9
Hist matching [24]	78.7 ±1.9	80.6 ±2.4	83.1 ±2.0	81.3 ±2.6	80.9 ±2.2
Combat [25]	77.4 ±1.6	80.2 ±1.3	79.8 ±2.0	78.6 ±2.0	79.0 ±1.7
BigAug [33]	81.7 ±2.1	82.4 ±1.5	85.2 ±0.9	84.6 ±1.3	83.5 ±1.4
Cycle-GAN [22]	78.3 ±2.0	79.9 ±1.0	82.9 ±1.3	83.4 ±1.0	81.1 ±1.3
Style-transfer [21]	74.1 ±3.0	80.0 ±1.3	80.6 ±1.5	81.1 ±1.5	79.0 ±1.8
Ours	**83.0 ±1.8**	**84.4 ±1.5**	**85.4 ±1.2**	**85.6 ±1.3**	**84.6 ±1.5**

Results When N4 Bias Correction is Applied. In previous sections, we used the original MRIs of the ABIDE dataset without bias correction to evaluate the proposed harmonization method on more challenging scenarios, where pre-processing steps to enhance the images might not be applicable. Compared to bias-corrected MRIs, original MRIs have arguably more complex distributions, which makes it more difficult for harmonization methods to map MRIs from a target domain to the source one. To demonstrate that our method also achieves satisfactory performance when the initial domain shifts are reduced, we repeated the previous steps with N4-biased corrected MRIs. These results, shown in Table 4, also showcase the advantage of our method in this different setting. For conciseness, we report here the average results for the HD metric (in mm) across different methods: Baseline (5.04 ± 1.41), Hist matching (4.45 ± 1.33), Combat (3.28 ± 0.75), BigAUG (2.64 ± 0.47), Cycle-GAN (2.55 ± 0.31), Style-transfer (3.77 ± 1.07) and ours (2.36 ± 0.62).

Experiments on Test-Time Adaptation (TTA). Our model can also be employed in a TTA scenario, where the network needs to be updated at inference for a given image, or set of images. To motivate this assumption, we compare the performance of our approach to the popular TENT model [29]. Adapting the segmentation network $S_\Phi(\cdot)$ with TENT yields 65.1 ± 5.0 of DSC, which represents a considerable gap compared to our model, i.e., 82.9 ± 2.6. Note that there exist other TTA methods for segmentation in the medical field, e.g., [16], however, they require segmentation masks for the adaptation.

5   Conclusion

In this work, we proposed a novel harmonization method that leverages Normalizing Flows to guide the adaptation of a harmonizer network. Our approach is source-free, task-agnostic, and works with unseen domains. These characteristics make our model applicable in real-life problems where the source domain is not accessible during adaptation, target domains are unknown at training time and harmonization is task-independent. The proposed method yields state-of-the-art harmonization performance based on the segmentation task, yet relaxes the strong assumptions made by existing harmonization strategies. Thus, we believe that our model is a powerful alternative for MRI multi-site harmonization.

References

1. Abdelhamed, A., Brubaker, M.A., Brown, M.S.: Noise flow: noise modeling with conditional normalizing flows. In: ICCV, pp. 3165–3173 (2019)
2. Beer, J.C., et al.: Longitudinal combat: a method for harmonizing longitudinal multi-scanner imaging data. Neuroimage **220**, 117129 (2020)
3. Boudiaf, M., et al.: Parameter-free online test-time adaptation. In: CVPR, pp. 8344–8353 (2022)
4. Delisle, P.L., et al.: Realistic image normalization for multi-Domain segmentation. Med. Image Anal. **74**, 102191 (2021)
5. Dewey, B.E., et al.: Deepharmony: a deep learning approach to contrast harmonization across scanner changes. Magn. Reson. Imaging **64**, 160–170 (2019)
6. Di Martino, A., et al.: The autism brain imaging data exchange: towards a large-scale evaluation of the intrinsic brain architecture in autism. Mol. Psychiatry **19**(6), 659–667 (2014)
7. Dinh, L., et al.: Density estimation using real NVP. In: ICLR (2017)
8. Dinsdale, N.K., et al.: Deep learning-based unlearning of dataset bias for MRI harmonisation and confound removal. Neuroimage **228**, 117689 (2021)
9. Dolz, J., Desrosiers, C., Ayed, I.B.: 3D fully convolutional networks for subcortical segmentation in MRI: a large-scale study. Neuroimage **170**, 456–470 (2018)
10. Fischl, B.: Freesurfer. Neuroimage **62**(2), 774–781 (2012)
11. Fortin, J.P., et al.: Removing inter-subject technical variability in magnetic resonance imaging studies. Neuroimage **132**, 198–212 (2016)
12. Fortin, J.P., et al.: Harmonization of multi-site diffusion tensor imaging data. Neuroimage **161**, 149–170 (2017)
13. Grover, A., et al.: Alignflow: cycle consistent learning from multiple domains via normalizing flows. In: AAAI, pp. 4028–4035 (2020)
14. Gudovskiy, D., et al.: Cflow-ad: real-time unsupervised anomaly detection with localization via conditional normalizing flows. In: WACV, pp. 98–107 (2022)
15. Ho, J., et al.: Flow++: improving flow-based generative models with variational dequantization and architecture design. In: ICML, pp. 2722–2730 (2019)
16. Karani, N., et al.: Test-time adaptable neural networks for robust medical image segmentation. Med. Image Anal. **68**, 101907 (2021)
17. Kingma, D.P., Dhariwal, P.: Glow: generative flow with invertible 1×1 convolutions. NeurIPS **31** (2018)
18. Kirichenko, P., Izmailov, P., Wilson, A.G.: Why normalizing flows fail to detect out-of-distribution data. NeurIPS **33**, 20578–20589 (2020)
19. Kobyzev, I., Prince, S.J., Brubaker, M.A.: Normalizing flows: an introduction and review of current methods. IEEE PAMI **43**(11), 3964–3979 (2020)
20. Liang, J., et al.: Do we really need to access the source data? Source hypothesis transfer for unsupervised domain adaptation. In: ICML, pp. 6028–6039 (2020)
21. Liu, M., et al.: Style transfer using generative adversarial networks for multi-site MRI harmonization. In: de Bruijne, M., et al. (eds.) MICCAI 2021. LNCS, vol. 12903, pp. 313–322. Springer, Cham (2021). https://doi.org/10.1007/978-3-030-87199-4_30

22. Modanwal, G., et al.: MRI image harmonization using cycle-consistent generative adversarial network. In: SPIE Medical Imaging 2020, vol. 11314, pp. 259–264 (2020)
23. Mummadi, C.K., et al.: Test-time adaptation to distribution shift by confidence maximization and input transformation. In: ICLR (2022)
24. Nyúl, L.G., Udupa, J.K., Zhang, X.: New variants of a method of MRI scale standardization. IEEE Trans. Med. Imaging **19**(2), 143–150 (2000)
25. Pomponio, R., et al.: Harmonization of large MRI datasets for the analysis of brain imaging patterns throughout the lifespan. Neuroimage **208**, 116450 (2020)
26. Shinohara, R., et al.: Statistical normalization techniques for magnetic resonance imaging. NeuroImage Clin. **6**, 9–19 (2014)
27. Takao, H., et al.: Effect of scanner in longitudinal studies of brain volume changes. J. Magn. Reson. Imaging **34**(2), 438–444 (2011)
28. Usman, B., et al.: Log-likelihood ratio minimizing flows: towards robust and quantifiable neural distribution alignment. NeurIPS **33**, 21118–21129 (2020)
29. Wang, D., et al.: TENT: fully test-time adaptation by entropy minimization. In: ICLR (2020)
30. Wang, G., et al.: Aleatoric uncertainty estimation with test-time augmentation for medical image segmentation with convolutional neural networks. Neurocomputing **338**, 34–45 (2019)
31. Wang, R., Chaudhari, P., Davatzikos, C.: Harmonization with flow-based causal inference. In: de Bruijne, M., et al. (eds.) MICCAI 2021. LNCS, vol. 12903, pp. 181–190. Springer, Cham (2021). https://doi.org/10.1007/978-3-030-87199-4_17
32. Zang, C., Wang, F.: Moflow: an invertible flow model for generating molecular graphs. In: Proceedings of the 26th ACM SIGKDD, pp. 617–626 (2020)
33. Zhang, L., et al.: Generalizing deep learning for medical image segmentation to unseen domains via deep stacked transformation. In: IEEE TMI, pp. 2531–2540 (2020)
34. Zhu, J.Y., et al.: Unpaired image-to-image translation using cycle-consistent adversarial networks. In: ICCV, pp. 2223–2232 (2017)
35. Zuo, L., et al.: Information-based disentangled representation learning for unsupervised MR harmonization. In: Feragen, A., Sommer, S., Schnabel, J., Nielsen, M. (eds.) IPMI 2021. LNCS, vol. 12729, pp. 346–359. Springer, Cham (2021). https://doi.org/10.1007/978-3-030-78191-0_27

Vicinal Feature Statistics Augmentation for Federated 3D Medical Volume Segmentation

Yongsong Huang[1,2], Wanqing Xie[1,3], Mingzhen Li[1,4], Mingmei Cheng[1,2], Jinzhou Wu[1], Weixiao Wang[1], Jane You[5], and Xiaofeng Liu[1(✉)]

[1] Harvard Medical School, Harvard University, Boston, MA 02114, USA
xliu61@mgh.harvard.edu
[2] Graduate School of Engineering, Tohoku University, Sendai 980-8579, Japan
[3] Department of Intelligent Medical Engineering, School of Biomedical Engineering, Anhui Medical University, Hefei 230032, China
[4] Department of Mathematics, Washington University in St. Louis, St. Louis, MO 63130, USA
[5] Department of Computing, The Hong Kong Polytechnic University, Hung Hom, Hong Kong

Abstract. Federated learning (FL) enables multiple client medical institutes collaboratively train a deep learning (DL) model with privacy protection. However, the performance of FL can be constrained by the limited availability of labeled data in small institutes and the heterogeneous (i.e., non-i.i.d.) data distribution across institutes. Though data augmentation has been a proven technique to boost the generalization capabilities of conventional centralized DL as a "free lunch", its application in FL is largely underexplored. Notably, constrained by costly labeling, 3D medical segmentation generally relies on data augmentation. In this work, we aim to develop a vicinal feature-level data augmentation (VFDA) scheme to efficiently alleviate the local feature shift and facilitate collaborative training for privacy-aware FL segmentation. We take both the inner- and inter-institute divergence into consideration, without the need for cross-institute transfer of raw data or their mixup. Specifically, we exploit the batch-wise feature statistics (e.g., mean and standard deviation) in each institute to abstractly represent the discrepancy of data, and model each feature statistic probabilistically via a Gaussian prototype, with the mean corresponding to the original statistic and the variance quantifying the augmentation scope. From the vicinal risk minimization perspective, novel feature statistics can be drawn from the Gaussian distribution to fulfill augmentation. The variance is explicitly derived by the data bias in each individual institute and the underlying feature statistics characterized by all participating institutes. The added-on VFDA consistently yielded marked improvements over six advanced FL methods on both 3D brain tumor and cardiac segmentation.

Y. Huang and W. Xie—Contribute equally.

A. Frangi et al. (Eds.): IPMI 2023, LNCS 13939, pp. 360–371, 2023.
https://doi.org/10.1007/978-3-031-34048-2_28

1   Introduction

Federated learning (FL) [25] for medical image analysis, which follows a privacy-aware decentralized paradigm to learn a global model on several local medical institutes, has recently been the center of much attention. In practice, however, the participating institutes may have diverse data collection schemes, which can lead to inefficient global collaboration [17,25] in two ways. First, some of the small institutes may have limited training data to support effective local updating. This is especially evident in the medical segmentation task, which is often constrained by the availability of costly labeled training data [7]. Second, the data collected from different institutes with heterogeneous vendors, doses, and populations can result in biased gradient uploading to hinder the convergence of the federal model. Though in conventional centralized learning, the widely used data augmentation [27] can be a straightforward and unified solution to the challenges of limited sample and poor generalized data distribution [37], it remains largely underexplored in FL segmentation.

It is of great importance to explore the proper global data augmentation scheme under the strict privacy restriction of FL, in which each client cannot access the data across institutes [25]. Simply utilizing the data augmentation methods in centralized learning without injecting global information is sub-optimal, inheriting the institute's bias. Early attempts [13,15,28] require accessing a global dataset to achieve balanced training, which is restricted in standard FL settings. Astraea [6] addresses this issue with several mediators between global servers and institutes. However, since each mediator group is a set of institutes, privacy among them is not protected. On the other hand, FedMix [34] and XORMix [26] propose to transmit the averaged images across institutes, which still take risk of privacy breaches and are inherently weak at constructing semantic transform with image-level MixUp [18]. In addition, the above methods focus on classification and do not apply to segmentation–an important medical image analysis task with great demand for data augmentation [7].

In this work, we propose to take both the inner- and inter-institute divergence into consideration without the need for any cross-institute transmission of raw data or their MixUp [26,34]. To mitigate the aforementioned limitations, we resort to the feature-level vicinal risk minimization (VRM) [4] to expand an example point in the feature space to a probabilistically modeled flexible distribution, e.g., a conceptually simple Gaussian prototype, centered at the original point, with the variation reflecting the local and global data divergence. Novel feature statistics can be drawn from the Gaussian prototype distribution to fulfill augmentation. Moreover, inspired by the previous image style generation and domain adaptation works [3,16,20], we propose to exploit the batch-wise feature statistics (e.g., mean and standard deviation) in each institute to abstractly represent the discrepancy of data, which are shared among institutes without any raw data transmission.

With our VFDA framework, the core issue is to properly associate the variance with the data bias in each individual institute and the underlying feature statistics characterized by all participating institutes. For effective augmentation, a proper variance is determined based on variances of feature statistics within each institute, regulated by the global variance of feature statistics characterized by all participating institutes. The augmentation in VFDA allows the local model to be trained over samples drawn from diverse feature distributions, alleviating local distribution shifts and facilitating institute-invariant representation learning, eventually leading to a better global model.

The main contributions of this work can be summarized as follows:

- To our knowledge, this is the first attempt to investigate an efficient data augmentation scheme in FL segmentation, which is especially pertinent for 3D medical data analysis.
- We propose a vicinal feature-level data augmentation (VFDA) scheme to model each feature statistic probabilistically via a conceptually simple Gaussian prototype and use the statistics drawn from the Gaussian distribution to implement augmentation. The transmission of raw images or their MixUp is not needed, meeting requirements for strict privacy protection.
- Both local and global divergence are taken into account in the abstractly represented batch feature statistics to quantify the augmentation scope.
- We evaluated our VFDA scheme on 3D brain tumor and cardiac anatomical segmentation tasks with 6 advanced FL methods to demonstrate its general efficacy and superiority.

2   Methodology

In FL segmentation, we are given N local client institutes, in which the n-th institute has M^n pairs of input volume $x_m^n \in \mathcal{X}$ and the corresponding segmentation label $y_m^n \in \mathcal{Y}$. We are expected to learn a global segmentation model $f_g(w_g) : \mathcal{X} \to \mathcal{Y}$ parameterized by w_g across all institutes, generally enforced by the global empirical risk minimization (ERM) objectives:

$$\mathcal{L}_{ERM}^g(w_g) = \frac{1}{N} \sum_{n=1}^{N} \mathbb{E}_{(x_m^n, y_m^n) \sim \mathcal{P}^n}[\mathcal{L}_{ERM}^n(x_m^N, y_m^N; w_n)], \qquad (1)$$

where $\mathcal{P}^n$ is the underlying distribution of institute n (i.e., $\{x_m^n, y_m^n\}_{m=1}^{M^n} \sim \mathcal{P}^n$), and $\mathcal{L}_{ERM}^n$ is the institutes-wise empirical risk with local network parameter w_n. Specifically, in the segmentation task, $\mathcal{L}_{ERM}^n$ typically follows the empirical form of voxel-wise cross-entropy or Dice loss. With a privacy-aware decentralized setting, directly calculating $\mathcal{L}_{ERM}^g(w_g)$ with all data across institutes as conventional centralized training is infeasible. In contrast, FL [21,25] relies on local training of models $f_n(w_n), n \in \{1, \cdots, N\}$ in each institute in parallel with the local data only. In each round, the trained local models are aggregated to a global model $f_g(w_g)$, which is then further distributed to each institute for the next round of local training. Therefore, the local training objective in FL is equivalent to empirically approximating the local distribution $\mathcal{P}^n$ by a finite M^n number of samples, i.e., $\mathcal{P}_{ERM}^n(x, y) = \frac{1}{M^n} \sum_{m=1}^{M^n} \delta(x = x_m^n, y = y_m^n)$, where $\delta(x = x_m^n, y = y_m^n)$ is a Dirac delta distribution with a point mass at (x_m^n, y_m^n).

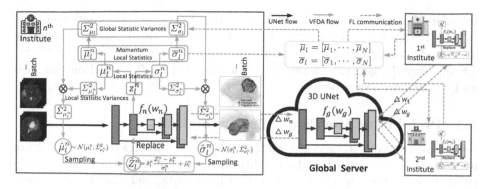

Fig. 1. Illustration of our VFDA over an encoder layer in the classical FL 3D segmentation framework with heterogeneous clients.

Although the ERM objectives have been widely adopted in deep FL frameworks as the training scheme, the underlying assumption is that the empirical local distributions $\mathcal{P}_{ERM}^n$ are homologous to the underlying global distribution $\mathcal{P}^g$, which is unrealistic in actual clinical scenarios [16,24]. In practice, there can be a significant performance drop since each $\mathcal{P}_{ERM}^n$ exhibits diverse data drifts with respect to $\mathcal{P}^n$ and $\mathcal{P}^g$, which lead to inconsistency of local and global empirical objectives and difficulties of generalization to testing distribution [1,9]. In addition, each local model can be data starved, given only access to institute samples.

2.1 Label Consistent Vicinal Feature Distribution Extrapolation

Based on the above concerns, we propose to expand the Dirac delta distribution inherent in $\mathcal{P}_{ERM}^n(x, y)$ to a more expressive one to approximate the true distribution following the idea of VRM [4,36]. Therefore, we are able to examine the vicinal region of each sample (x_m^n, y_m^n) for infinite data augmentation. Specifically, we can define a institute-wise vicinity distribution $\mathcal{V}^n(\hat{x}_m^n, \hat{y}_m^n | x_m^n, y_m^n)$ and apply it to each sample (x_m^n, y_m^n) to generate numerous pseudo samples $(\hat{x}_m^n, \hat{y}_m^n)$ and support the local institute training. More formally, we construct the vicinal local distribution as $\mathcal{P}_{VRM}^n = \frac{1}{M^n} \sum_{m=1}^{M^n} \mathcal{V}^n(\hat{x}_m^n, \hat{y}_m^n | x_m^n, y_m^n)$, and expect a better mimic of $\mathcal{P}^n$ with a proper $\mathcal{V}^n(\cdot)$.

There are numerous successful candidates of $\mathcal{V}^n(\cdot)$ in conventional centralized learning, e.g., MixUp [36] and CutMix [35]. Though simply utilizing MixUp [36] or CutMix locally can potentially improve the performance (as shown in experiments), it can be sub-optimal as there are no global cues injected. In these cases, $\mathcal{P}_{VRM}^n$ gives a better mimic of the true local distribution $\mathcal{P}^n$ rather than the global distribution $\mathcal{P}^g$.

Instead of the raw voxel level MixUp [35,36], VFDA estimates a vicinity distribution $\mathcal{V}_l^n$ at each encoder layer l to augment its batch-wise intermediate feature Z_l^n in n-th institute for more flexible expansion [18]. Of note, $Z_l^n \in \mathbb{R}^{B \times C \times H \times W \times S}$ denotes the intermediate feature representation of B minibatch volumes, with height H, width W, slices S, and channels C. In addition, to achieve consistency training [32], we define $\mathcal{V}_l^n$ to be label consistent, i.e., $\mathcal{V}_l^n(\hat{Z}_l^n, \hat{Y}^n | Z_l^n, Y^n) = \mathcal{V}_l^n(\hat{Z}_l^n | Z_l^n)\delta(\hat{Y}^n = Y^n)$, which only extrapolate the latent feature Z_l^n while maintaining the consistency of the label $Y^n \in \mathbb{R}^{B \times H \times W \times S}$. Then, a key challenge is adaptively configuring $\mathcal{V}_l^n(\hat{Z}_l^n | Z_l^n)$ based on the local and global data divergence.

2.2 Probabilistic Modeling of Feature Statistics

As opposed to explicitly modeling $\mathcal{V}_l^n(\hat{Z}_l^n | Z_l^n)$, VFDA resorts to implicit feature augmentation by exploiting the batch-wise feature statistics in each institute to abstractly represent the discrepancy of input data, and model each feature statistic probabilistically via a Gaussian prototype. Specifically, we utilize the channel-wise feature statistics of mean μ_l^n and standard deviation σ_l^n. For Z_l^n, its channel-wise statistics of $\mu_l^n \in \mathbb{R}^{B \times C}$ and $\sigma_l^n \in \mathbb{R}^{B \times C}$ can be formulated as:

$$\mu_l^n = \frac{1}{H \times W \times S} \sum_{h=1}^{H} \sum_{w=1}^{W} \sum_{s=1}^{S} Z_l^n; \quad \sigma_l^n = \sqrt{\frac{1}{H \times W \times S} \sum_{h=1}^{H} \sum_{w=1}^{W} \sum_{s=1}^{S} (Z_l^n - \mu_l^n)}. \tag{2}$$

Recent works [3,19,20,30] demonstrated that these low-order batch statistics are domain-specific, owing to the divergence of feature representations. As the abstraction of latent features, the feature statistics among local institutes will also exhibit inconsistency and follow shifts from the statistics of the true distribution.

We propose to capture such shifts via probabilistic modeling. We hypothesize that each feature statistic follows a multi-variate Gaussian prototype, i.e., $\mu_l^n \sim \mathcal{N}(\mu_l^n, \hat{\Sigma}_{\mu_l^n}^2)$ and $\sigma_l^n \sim \mathcal{N}(\sigma_l^n, \hat{\Sigma}_{\sigma_l^n}^2)$, where each Gaussian prototype's center corresponds to the original statistic, and the variance is an estimation of the potential feature statistic shift from the true distribution.

2.3 Local and Global Statistic Variances Quantification

Determining an appropriate variant range is challenging since each institute has only access to the data itself but has no sense of its statistical biases. Recent research [29,31] has shown that deep feature space contains many semantic directions, and feature variances provide a reasonable measurement of potential meaningful semantic changes along the directions. This motivates us to estimate the variance of the Gaussian prototype from the variance of feature statistics.

Local Statistic Variances. In each institute, we compute local variances of feature statistics based on the information within each mini-batch:

$$\Sigma_{\mu_l^n}^2 = \frac{1}{B} \sum_{b=1}^{B} (\mu_l^n - \mathbb{E}[\mu_l^n])^2 \in \mathbb{R}^C; \quad \Sigma_{\sigma_l^n}^2 = \frac{1}{B} \sum_{b=1}^{B} (\sigma_l^n - \mathbb{E}[\sigma_l^n])^2 \in \mathbb{R}^C, \tag{3}$$

where $\Sigma_{\mu_l^n}^2$ and $\Sigma_{\sigma_l^n}^2$ denote the variance of feature mean μ_l^n and standard deviation σ_l^n that are specific to each institute. Each value in $\Sigma_{\mu_l^n}^2$ and $\Sigma_{\sigma_l^n}^2$ is the variance of feature statistics in a particular channel, and its magnitude captures the potential change in that particular channel at that specific institute.

Global Statistic Variances. The institute-specific statistic variances are solely computed based on the data in each institute and are thus likely biased due to the bias in the local dataset. To resolve this, we further estimate the variances of the institute-sharing feature statistics, taking information from all institutes into account. Particularly, we propose a momentum version of feature statistics for each institute, which is updated online with an exponential momentum decay (EMD) strategy:

$$\overline{\mu}_l^n \leftarrow (1-\eta)\frac{1}{B}\sum_{b=1}^{B}\mu_l^n + \eta\overline{\mu}_l^n \in \mathbb{R}^C; \quad \overline{\sigma}_l^n \leftarrow (1-\eta)\frac{1}{B}\sum_{b=1}^{B}\sigma_l^n + \eta\overline{\sigma}_l^n \in \mathbb{R}^C, \quad (4)$$

where the momentum factor $\eta = \eta^0 \exp(-r)$ follows an exponential decay over round r. Notably, $\overline{\mu}_l^n$ and $\overline{\sigma}_l^n$ are the momentum updated feature statistics of encoder layer l in institute n. η^0 is empirically initialized to 10.

In each communication round, these accumulated local feature statistics are sent to the server along with model parameters. Let $\overline{\mu}_l = [\overline{\mu}_1, \cdots, \overline{\mu}_N] \in \mathbb{R}^{N\times C}$ and $\overline{\sigma}_l = [\overline{\sigma}_1, \cdots, \overline{\sigma}_N] \in \mathbb{R}^{N\times C}$ denote the collections of accumulated feature statistics of all institutes, then the global, institute sharing statistic variances are calculated as:

$$\Sigma_{\mu_l}^2 = \frac{1}{N}\sum_{n=1}^{N}(\overline{\mu}_l^n - \mathbb{E}[\overline{\mu}_l])^2 \in \mathbb{R}^C; \quad \Sigma_{\sigma_l}^2 = \frac{1}{N}\sum_{n=1}^{N}(\overline{\sigma}_l^n - \mathbb{E}[\overline{\sigma}_l])^2 \in \mathbb{R}^C. \quad (5)$$

Along with the aggregated model parameters as in classical FL methods, e.g., FedAvg [21], these variances are distributed back to each institute to inform a global estimation of feature statistic variances. Note that $\Sigma_{\mu_l}^2$ and $\Sigma_{\sigma_l}^2$ are shared by all participating institutes.

Institute sharing estimations $\Sigma_{\mu_l}^2$ and $\Sigma_{\sigma_l}^2$ provide a quantification of distribution divergence among institutes, and larger values imply potentials of more significant changes of the corresponding channels in the true feature statistic space. Therefore, for each institute, we weight the institute-specific statistic variances $\Sigma_{\mu_l^n}^2$, $\Sigma_{\sigma_l^n}^2$ with $\Sigma_{\mu_l}^2$, $\Sigma_{\sigma_l}^2$, so that each institute has a sense of such global divergence, i.e., $\hat{\Sigma}_{\mu_l^n}^2 = \Sigma_{\mu_l}^2 \Sigma_{\mu_l^n}^2$ and $\hat{\Sigma}_{\sigma_l^n}^2 = \Sigma_{\sigma_l}^2 \Sigma_{\sigma_l^n}^2$.

2.4   Implementation of Vicinal Feature-Level Data Augmentation

After establishing the Gaussian prototype, we calculate novel feature $\hat{Z}_l^n$ in the vicinity of Z_l^n as follows:

$$\hat{Z}_l^n = \hat{\sigma}_l^n \frac{Z_l^n - \mu_l^n}{\sigma_l^n} + \hat{\mu}_l^n; \quad \hat{\mu}_l^n \sim \mathcal{N}(\mu_l^n, \hat{\Sigma}_{\mu_l^n}^2), \quad \hat{\sigma}_l^n \sim \mathcal{N}(\sigma_l^n, \hat{\Sigma}_{\sigma_l^n}^2), \quad (6)$$

where Z_l^n is first normalized with its original statistics by $\frac{Z_l^n - \mu_l^n}{\sigma_l^n}$, and further scaled with novel statistics $\hat{\mu}_l^n$ and $\hat{\sigma}_l^n$ that are randomly sampled from the

Table 1. Comparisons of different FL methods with or without our VFDA framework in FeTS2021, and the ablation study of VFDA modules.

Method	Additional augmentation	Dice	Score	[%] ↑	
		ET	TC	WT	Mean
FedAvg [21]	–	75.63	62.01	76.55	71.14
FedAvg [21]	MixUp	75.80	62.74	76.95	71.83
FedAvg [21]	VFDA (Ours)	**76.59**	**64.10**	**77.86**	**72.85**
FedAvg [21]	VFDA w/o EMD	76.08	63.47	77.29	72.28
FedAvg [21]	VFDA w/o Global Statistic Variances	75.92	63.15	77.03	72.03
FedNorm [33]	–	75.60	63.76	76.98	72.11
FedNorm [33]	MixUp	75.82	63.96	77.05	72.28
FedNorm [33]	VFDA (Ours)	**76.49**	**64.82**	**78.17**	**73.16**
FedNorm [33]	VFDA w/o EMD	76.14	64.55	77.83	72.84
FedNorm [33]	VFDA w/o Global Statistic Variances	76.01	64.34	77.50	72.62

corresponding Gaussian distribution [12]. To make the sampling differentiable, we apply the re-parameterization trick [10]:

$$\hat{\mu}_l^n = \mu_l^n + \epsilon_\mu \hat{\Sigma}_{\mu_l^n}^2, \quad \hat{\sigma}_l^n = \sigma_l^n + \epsilon_\sigma \hat{\Sigma}_{\sigma_l^n}^2), \qquad (7)$$

where $\epsilon_\mu \sim \mathcal{N}(0, 1)$ and $\epsilon_\sigma \sim \mathcal{N}(0, 1)$ follow normal distribution.

The proposed VFDA in Eq. (6) works in a plug-and-play fashion, which can be inserted at arbitrary positions of the model to facilitate latent semantic augmentation. In our implementation, we add a VFDA after each encoder layer of UNet. Of note, we explore the batch-wise feature statistics for vicinal expansion while not relying on the network with batch normalization layer [14,16]. Our FDA can be widely generalized to modern deep learning models with mini-batch training. During testing, no augmentation is performed.

3   Experiments and Results

To show the effectiveness of our framework, we experimented on both 3D federated brain tumor and cardiac anatomical segmentation tasks and added VFDA to numerous advanced FL segmentation models, e.g., FedAvg [21], FedProx [8], FedBN [14], FedNorm [33], PRRF [5], and FedCRLD [23]. Of note, our VFDA can be directly applied to 2D segmentation by setting $S = 0$.

We implemented all modules on a server with an NVIDIA A100 GPU and used the PyTorch toolbox. For the evaluation metrics, we employed the widely accepted Dice similarity coefficient (DSC), which measures the overlap between the predicted segmentation mask and the label.

3.1   Federated Brain Tumor Segmentation

The Federated Tumor Segmentation (FeTS) 2021 Challenge Task-1 [22] incorporates the magnetic resonance imaging (MRI) volumes and segmentation label of

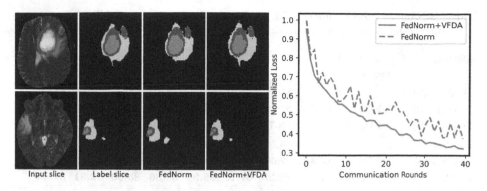

Fig. 2. Left: The qualitative comparisons of 2 slices from two subjects with FedNorm with or without our VFDA. CT: red; ET: purple; WT: red+purple+green; background: black (best viewed in color). Right: the normalized training loss of FedNorm with or without our VFDA.

three brain tumor structures, i.e., tumor core (TC), enhanced tumor (ET) and whole tumor (WT), from 341 subjects. For FL, we followed the standard dataset partition-2 to separate 22 local clients based on their original institutions and further split the large institutes into subsets according to the tumor sizes, which are subject-independent. MRI scans can be highly heterogeneous among participating clients in the FeTS challenge as various scanners and image protocols were employed. The size-based subset also involves the divergence of different tumor grades. With the fine-grained partition in FeTS partition-2, the subjects in each local client are limited.

According to the standard evaluation protocol, the segmentation model of ResNet-based 3D UNet is fixed for all participants [22]. We adopted the successful solutions in FeTS, i.e., FedNorm [33] and FedAvg [21] with different model aggregation schemes as our baselines. For fair comparisons, we followed the detailed setting of federated aggregation with tensor normalization (FedNorm) [33] based on the open-fl framework. All of the experiments were conducted under a fixed train validation split and random seed to make our results convincing and deterministic. Specifically, in each communication round, we performed one epoch for local client training and initialized the learning rate to 5e-4 with a polynomial decaying factor of 0.9 consistently. For all methods, we added the vanilla data augmentation of rotation, scaling, elastic deformation, brightness, and aggressiveness adjustment as in [33].

The quantitative evaluation results with respect to Dice score are presented in Table 1. Simply adding the MixUp augmentation improved the performance of both FedAvg and FedNorm significantly. By taking a more flexible feature-level expansion based on the local and global divergence described by the abstract feature statistic variances, our VFDA can achieve remarkable results. For the ablation study, we removed the global statistic variances or EMD module and denoted them as VFDA w/o EMD or VFDA w/o Global Statistic Variances,

Table 2. Comparisons of different FL methods with or without our VFDA framework in cardiac segmentation task, and the ablation study of VFDA modules.

Method	Additional augmentation	Dice Score [%] ↑						
		A	B	C	D	E	F	Mean
FedAvg [21]	–	85.84	85.39	89.08	79.77	84.42	18.36	73.81
FedAvg [21]	VFDA w/o Global	87.42	86.25	88.30	82.59	85.45	45.08	79.18
FedAvg [21]	VFDA (Ours)	**88.72**	**87.53**	**89.91**	**84.15**	**86.05**	**52.47**	**81.4**
FedProx [8]	–	86.70	84.41	88.81	82.85	83.66	27.26	75.62
FedProx [8]	VFDA w/o Global	87.65	86.40	89.28	84.91	85.13	49.72	80.51
FedProx [8]	VFDA (Ours)	**88.23**	**87.13**	**89.95**	**84.86**	**85.60**	**52.81**	**81.43**
FedBN [14]	–	86.98	85.87	89.58	81.91	84.73	27.49	76.09
FedBN [14]	VFDA w/o Global	88.05	86.85	89.74	83.62	85.82	45.65	79.96
FedBN [14]	VFDA (Ours)	**88.62**	**87.28**	**90.20**	**85.07**	**86.15**	**48.30**	**80.94**
PRRF [5]	–	87.04	86.11	88.05	84.65	83.85	53.09	80.47
PRRF [5]	VFDA w/o Global	87.86	87.03	89.65	85.12	85.44	61.05	82.69
PRRF [5]	VFDA (Ours)	**88.92**	**88.75**	**90.26**	**86.07**	**86.29**	**62.38**	**83.79**
FedCRLD [23]	–	88.06	87.28	90.88	86.96	86.40	76.15	85.96
FedCRLD [23]	VFDA w/o Global	89.02	88.34	91.24	87.35	87.23	77.80	86.83
FedCRLD [23]	VFDA (Ours)	**89.74**	**89.05**	**92.11**	**87.83**	**87.62**	**78.53**	**87.48**

respectively. Their inferior performance compared to VFDA demonstrates the contribution of global statistic variances or EMD training.

In Fig. 2, some example slices are shown, in which the VFDA achieves more accurate delineations compared to FedNorm. In addition, adding VFDA contributes to a stabler continuous optimization process than FedNorm.

3.2   Federated Cardiac Anatomical Segmentation

To further demonstrate the generalizability of VFDA, we also evaluated it on multi-center multi-sequence cardiac MRI segmentation as in [23]. Specifically, a real-world FL task is constructed using the publicly available M&M [2] and Emidec [11] datasets. Of note, M&M dataset incorporates the cine-MRI of the subjects from five centers in three countries and is scanned with four different scanner vendors, and the Emidec is a delayed enhancement (DE) MRI dataset. As in [23], the M&M dataset is split into five institutes/centers, i.e., client A-E, while Emidec is configured as the sixth institute, i.e., client F. There are notable appearance shifts across institute due to the diverge centers, devices, and contrast agents. We chose the subject-independent 7/1/2 split for each client institute and adopted the 3D UNet as the segmentation model backbone.

For fair comparisons, we adopted the vanilla augmentation of rotation, translation, scale, and mirror as [23]. Notably, the compared baselines of FedProx [8], FedBN [14], PRRF [5], and FedCRLD [23] are designed for non-IID cases, which explicitly target the data bias among local institutes using different techniques.

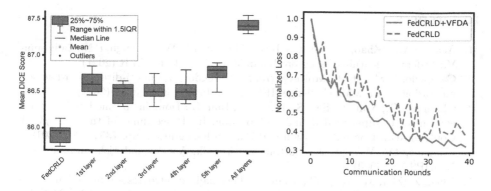

Fig. 3. Left: Sensitivity analysis of adding VFDA to each layer of the 3DUNet encoder in FedCRLD with five times random trails. Right: the normalized training loss of FedCRLD with or without our VFDA.

However, how to efficiently induce the global data divergence for FL debias is a long-lasting challenge. Our VFDA can be a unified and general solution to simply add on these methods to efficiently improve the segmentation performance, as shown in Table 2. In the ablation study, we also demonstrated that taking the global statistic variances into consideration is important for VFDA for all these baselines.

In Fig. 3 left, we investigated the effect of VFDA in different encoder layers. Of note, adding VFDA to the decoder layer does not improve the performance. In Fig. 3 right, the FedCRLD training in client F (i.e., Emidec with different DE-MRI) is relatively unstable, while adding VFDA smooths the optimization and leads to a lower loss.

4 Conclusion

In this work, we proposed a novel and efficient data augmentation methodology for federated learning in 3D medical volume segmentation, which suffers from the imbalance of clients in small institutes, and the inner- and inter-institute heterogeneous data shift. We resort to batch-wise feature statistics as an abstract quantification of the local and global statistic variances and utilize them probabilistically via a Gaussian prototype. We utilize the mean corresponding to the original statistic and the variance to define the proper augmentation scope in a label-preserving vicinal risk minimization framework on the feature space to expand the feature by simply sampling the Gaussian distribution with the re-parameterization trick. The experiments in both 3D brain tumor and cardiac anatomical structure segmentation FL tasks with six advanced FL methods consistently demonstrate its efficiency and generality. It has the potential to be widely adapted to different FL scenarios with low additional community costs.

Funding. This work is partly supported by JSPS KAKENHI JP23KJ0118.

References

1. Acar, D.A.E., Zhao, Y., Navarro, R.M., Mattina, M., Whatmough, P.N., Saligrama, V.: Federated learning based on dynamic regularization. ICLR (2021)
2. Campello, V.M., et al.: Multi-centre, multi-vendor and multi-disease cardiac segmentation: the M&Ms challenge. IEEE TMI **40**(12), 3543–3554 (2021)
3. Chang, W.G., You, T., Seo, S., Kwak, S., Han, B.: Domain-specific batch normalization for unsupervised domain adaptation. In: Proceedings of the IEEE/CVF Conference on Computer Vision and Pattern Recognition, pp. 7354–7362 (2019)
4. Chapelle, O., Weston, J., Bottou, L., Vapnik, V.: Vicinal risk minimization. NeurIPS **13** (2000)
5. Chen, Z., Zhu, M., Yang, C., Yuan, Y.: Personalized retrogress-resilient framework for real-world medical federated learning. In: de Bruijne, M., et al. (eds.) MICCAI 2021. LNCS, vol. 12903, pp. 347–356. Springer, Cham (2021). https://doi.org/10.1007/978-3-030-87199-4_33
6. Duan, M., Liu, D., Chen, X., Liu, R., Tan, Y., Liang, L.: Self-balancing federated learning with global imbalanced data in mobile systems. IEEE Trans. Parallel Distrib. Syst. **32**(1), 59–71 (2020)
7. Eaton-Rosen, Z., Bragman, F., Ourselin, S., Cardoso, M.J.: Improving data augmentation for medical image segmentation. MIDL (2018)
8. Gudur, G.K., Perepu, S.K.: Resource-constrained federated learning with heterogeneous labels and models for human activity recognition. In: Li, X., Wu, M., Chen, Z., Zhang, L. (eds.) DL-HAR 2021. CCIS, vol. 1370, pp. 57–69. Springer, Singapore (2021). https://doi.org/10.1007/978-981-16-0575-8_5
9. Karimireddy, S.P., Kale, S., Mohri, M., Reddi, S., Stich, S., Suresh, A.T.: Scaffold: stochastic controlled averaging for federated learning. In: International Conference on Machine Learning, pp. 5132–5143. PMLR (2020)
10. Kingma, D.P., Welling, M.: Auto-encoding variational bayes. arXiv (2013)
11. Lalande, A., et al.: Emidec: a database usable for the automatic evaluation of myocardial infarction from delayed-enhancement cardiac MRI. Data **5**(4), 89 (2020)
12. Li, B., Wu, F., Lim, S.N., Belongie, S., Weinberger, K.Q.: On feature normalization and data augmentation. In: Proceedings of the IEEE/CVF Conference on Computer Vision and Pattern Recognition, pp. 12383–12392 (2021)
13. Li, D., Wang, J.: Fedmd: heterogenous federated learning via model distillation. arXiv preprint arXiv:1910.03581 (2019)
14. Li, X., Jiang, M., Zhang, X., Kamp, M., Dou, Q.: Fedbn: federated learning on non-iid features via local batch normalization. ICLR (2021)
15. Lin, T., Kong, L., Stich, S.U., Jaggi, M.: Ensemble distillation for robust model fusion in federated learning. NeurIPS **33**, 2351–2363 (2020)
16. Liu, X., Xing, F., El Fakhri, G., Woo, J.: Memory consistent unsupervised off-the-shelf model adaptation for source-relaxed medical image segmentation. Med. Image Anal. 102641 (2022)
17. Liu, X., Yoo, C., Xing, F., Oh, H., El Fakhri, G., Kang, J.W., Woo, J., et al.: Deep unsupervised domain adaptation: a review of recent advances and perspectives. APSIPA Trans. Signal Inf. Process. **11**(1) (2022)
18. Liu, X., et al.: Data augmentation via latent space interpolation for image classification. In: ICPR, pp. 728–733. IEEE (2018)
19. Mancini, M., Porzi, L., Bulo, S.R., Caputo, B., Ricci, E.: Boosting domain adaptation by discovering latent domains. In: CVPR, pp. 3771–3780 (2018)

20. Maria Carlucci, F., Porzi, L., Caputo, B., Ricci, E., Rota Bulo, S.: Autodial: automatic domain alignment layers. In: ICCV, pp. 5067–5075 (2017)
21. McMahan, B., Moore, E., Ramage, D., Hampson, S., y Arcas, B.A.: Communication-Efficient learning of deep networks from decentralized data. In: Artificial Intelligence and Statistics, pp. 1273–1282. PMLR (2017)
22. Pati, S., Bakas, S.: The federated tumor segmentation (fets) challenge (2021)
23. Qi, X., Yang, G., He, Y., Liu, W., Islam, A., Li, S.: Contrastive re-localization and history distillation in federated CMR segmentation. In: Wang, L., Dou, Q., Fletcher, P.T., Speidel, S., Li, S. (eds.) Medical Image Computing and Computer Assisted Intervention–MICCAI 2022. MICCAI 2022. LNCS, vol. 13435, pp. 256–265. Springer, Cham (2022). https://doi.org/10.1007/978-3-031-16443-9_25
24. Qu, Z., Li, X., Duan, R., Liu, Y., Tang, B., Lu, Z.: Generalized federated learning via sharpness aware minimization. arXiv preprint arXiv:2206.02618 (2022)
25. Rieke, N., et al.: The future of digital health with federated learning. NPJ Digit. Med. 3(1), 1–7 (2020)
26. Shin, M., Hwang, C., Kim, J., Park, J., Bennis, M., Kim, S.L.: Xor mixup: privacy-preserving data augmentation for one-shot federated learning. arXiv preprint arXiv:2006.05148 (2020)
27. Shorten, C., Khoshgoftaar, T.M.: A survey on image data augmentation for deep learning. J. Big Data 6(1), 1–48 (2019)
28. Tuor, T., Wang, S., Ko, B.J., Liu, C., Leung, K.K.: Overcoming noisy and irrelevant data in federated learning. In: ICPR, pp. 5020–5027. IEEE (2021)
29. Upchurch, P., et al.: Deep feature interpolation for image content changes. In: CVPR (2017)
30. Wang, X., Jin, Y., Long, M., Wang, J., Jordan, M.: Transferable normalization: towards improving transferability of deep neural networks. arXiv:2019 (2019)
31. Wang, Y., Pan, X., Song, S., Zhang, H., Huang, G., Wu, C.: Implicit semantic data augmentation for deep networks. NeurIPS 32 (2019)
32. Xie, Q., Dai, Z., Hovy, E., Luong, T., Le, Q.: Unsupervised data augmentation for consistency training. NeurIPS 33, 6256–6268 (2020)
33. Yin, Y., et al.: Efficient federated tumor segmentation via normalized tensor aggregation and client pruning. In: Crimi, A., Bakas, S. (eds.) Brainlesion: Glioma, Multiple Sclerosis, Stroke and Traumatic Brain Injuries. BrainLes 2021. LNCS, vol. 12963, pp. 433–443. Springer, Cham (2022). https://doi.org/10.1007/978-3-031-09002-8_38
34. Yoon, T., Shin, S., Hwang, S.J., Yang, E.: Fedmix: approximation of mixup under mean augmented federated learning. arXiv preprint arXiv:2107.00233 (2021)
35. Yun, S., Han, D., Oh, S.J., Chun, S., Choe, J., Yoo, Y.: Cutmix: regularization strategy to train strong classifiers with localizable features. In: ICCV, pp. 6023–6032 (2019)
36. Zhang, H., Cisse, M., Dauphin, Y.N., Lopez-Paz, D.: mixup: beyond empirical risk minimization. ICML (2018)
37. Zhang, L., et al.: Generalizing deep learning for medical image segmentation to unseen domains via deep stacked transformation. IEEE TMI 39(7), 2531–2540 (2020)

Image Synthesis

S2DGAN: Generating Dual-energy CT from Single-energy CT for Real-time Determination of Intracerebral Hemorrhage

Caiwen Jiang[1], Yongsheng Pan[1], Tianyu Wang[3], Qing Chen[4], Junwei Yang[1], Li Ding[4], Jiameng Liu[1], Zhongxiang Ding[3], and Dinggang Shen[1,2,5(✉)]

[1] School of Biomedical Engineering, ShanghaiTech University, Shanghai, China
{jiangcw,panysh,dgshen}@shanghaitech.edu.cn
[2] Shanghai United Imaging Intelligence Co., Ltd., Shanghai, China
[3] Department of Radiology, Affiliated Hangzhou First People's Hospital,
Zhejiang University School of Medicine, Hangzhou, China
xiang@shanghaitech.edu.cn, tianyuwang1221@zju.edu.cn
[4] Zhejiang Chinese Medical University, Hangzhou, China
[5] Shanghai Clinical Research and Trial Center, Shanghai 201210, China

Abstract. Timely determination of whether there is intracerebral hemorrhage after thrombectomy is essential for follow-up treatment. But, this is extremely challenging with standard single-energy CT (SECT), because blood and contrast agents (injected during thrombectomy) have similar CT values under a single energy spectrum. In contrast, dual-energy CT (DECT) employs two different energy spectra, thus allowing to differentiate between hemorrhage and contrast extravasation in real time, based on energy-related attenuation characteristics between blood and contrast. However, compared to SECT scanners, DECT scanners have limited popularity due to high price. To address this dilemma, in this paper we first attempt to generate pseudo DECT images from a SECT image for real-time diagnosis of hemorrhage. More specifically, we propose a SECT-to-DECT generative adversarial network (S2DGAN), which is a 3D transformer-based multi-task learning framework equipped with a shared attention mechanism. Among them, the *transformer-based architecture* can guide S2DGAN to focus more on high-density areas (crucial for hemorrhage diagnosis) during the generation. Meanwhile, the introduced *multi-task learning strategy* and *shared attention mechanism* enable S2DGAN to model dependencies between interconnected generation tasks, improving generation performance while significantly reducing model parameters and computational complexity. Validated on clinical data, S2DGAN can generate DECT images better than state of-the-art methods and achieve an accuracy of 90% in hemorrhage diagnosis based only on SECT images.

Keywords: Intracerebral hemorrhage · Single-energy CT (SECT) · Dual-energy CT (DECT) · Generation · Transformer · Shared attention

© The Author(s), under exclusive license to Springer Nature Switzerland AG 2023
A. Frangi et al. (Eds.): IPMI 2023, LNCS 13939, pp. 375–387, 2023.
https://doi.org/10.1007/978-3-031-34048-2_29

1   Introduction

Thrombectomy is a common treatment option for acute stroke, but it may damage blood-brain barrier and lead to postinterventional cerebral hyper-density (PCHD), which is an essential concern in the follow-up treatments [1]. Depending on the damage level of blood-brain barrier, PCHD may be caused by contrast extravasation, hemorrhage, or even both. However, since blood and contrast agents have similar attenuation coefficients, it is difficult to distinguish them by standard single-energy CT (SECT) which uses only a single energy spectrum. In clinic, twice or more SECT scans with time lag are necessary to diagnose PCHD depending on the body's different absorption rates of blood and contrast agents, which greatly delays the treatment time. To avoid such delay, dual-energy CT (DECT), which uses two energy spectra to scan patients, is recommended. Two typical cases of SECT and DECT are provided in Fig. 1. Superior to SECT, DECT can indirectly calculate the virtual non-contrast images (VNCs) and iodine overlay maps (IOM), by observing which we can differentiate contrast extravasation from hemorrhage in real time [2].

However, the accessibility of DECT scanners is greatly limited due to their expensive costs. This results in a demand to develop cheap and efficient techniques as alternatives of DECT to perform real-time PCHD diagnosis. Based on current clinical situation, a possible alternative is to generate VNC and IOM from widely-existing SECT by the rapidly developing image translation techniques. Moreover, we choose to generate pseudo VNC and IOM rather than directly predict diagnostic results from SECT considering the following facts: 1) The use of VNC and IOM is more clinically natural in terms of diagnosis and thus more subjectively convincing. 2) The generation of VNC and IOM can better exploit in the information entailed SECT, thus more likely to produce a correct diagnosis. 3) The generated VNC and IOM can provide additional information, such as the extent of hemorrhage, and the degree of vascular damage.

Among the existing image translation techniques, generative adversarial networks (GANs) [3–6] have shown large potential and obtained great success. For example, Frid et al. used GAN to synthesize high-quality lesion ROIs for liver lesion classification [7]. Pan et al. proposed to use GAN-based techniques to impute the missing PET images from MRI for Alzheimer's disease diagnosis [8], where PET image and MRI are registered with certain registration method [9–11]. Moreover, the transformer [12] technique has played an increasingly critical role in image translation tasks, especially those focusing on ROIs (i.e., high-density areas in our task). Gradually, more and more GANs evolved the transformer components in their network architectures [13–17]. For example, Luo et al. proposed a CNN+transformer based GAN for high-quality PET image reconstruction [13]. Jiang et al. proposed a GAN constructed purely on transformer-based generators and discriminators, and achieved state-of-the-art performance [16]. Peiris et al. proposed a volumetric transformer for predicting segmentation maps from 3D images [17]. Based on these successful applications, it is convincing to generate pseudo VNC and IOM from SECT for real-time diagnosis of PCHD.

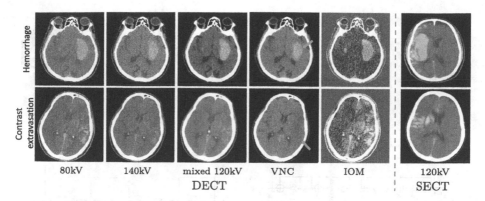

Fig. 1. Two typical cases of DECT and SECT. Each DECT case (first 5 columns) contains five images including 80 kV, 140 kV, mixed 120 kV, VNC, and IOM, where 80 kV and 140 kV are the original scanned images while mixed 120 kV, VNC, and IOM are indirectly calculated from 80 kV and 140 kV. Each SECT case (last column) only contains a 120 kV image. The key for diagnosis is whether there are high-density areas in VNC images, as indicated by the red arrows. (Color figure online)

To this end, we design the SECT-to-DECT generative adversarial network (S2DGAN), by incorporating three strategies, namely *transformer-based architecture*, *multi(two)-task learning framework*, and *shared attention mechanism*, to generate VNC and IOM simultaneously from one SECT (mixed 120 kV /120 kV) image for real-time diagnosis of PCHD. Specifically, our S2DGAN adopts a pure transformer-based generator, thus can focus more on high-density regions, which are key to diagnosing PCHD. Meanwhile, the multi-task learning strategy allows the generation tasks of VNC and IOM to share the same encoder, which models the connections between these interconnected tasks while effectively limiting the model size. Moreover, we share attention across VNC and IOM generation, further capturing their dependencies and reducing the computational complexity of S2DGAN. Extensive experiments conducted on clinical data demonstrate that S2DGAN-generated VNC and IOM have superior image quality and better fidelity, compared to those generated by state-of-the-art methods.

The main contributions of our work include: 1) This is the first attempt to achieve real-time diagnosis of PCHD by generating DECT from SECT. 2) Our novel S2DGAN achieves superior performance over the state-of-the-art methods on both VNC and IOM generation. 3) We use a large clinical dataset (consisting of 200 DECT and 30 SECT samples from individuals with PCHD) for validation of our approach.

2   Method

As shown in Fig. 2 (a), our proposed S2DGAN adopts vision transformer based on encoder-decoder architecture [17,18]. S2DGAN consists of a transformer encoder, a bottleneck layer, two task-specific transformer decoders, and two discriminators. When given a 3D SECT image, its features that contain both global

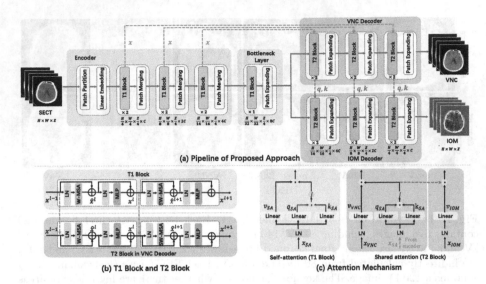

Fig. 2. Overview of the proposed approach. (a) Architecture of proposed S2DGAN; (b) Details of T1 Block and T2 Block; (c) Sketches of shared attention and self-attention.

and contextual information are first extracted by the encoder, and then pass through the bottleneck layer and enter into two decoders. These two decoders, equipped with a *shared attention* mechanism, progressively decode and upsample features to obtain the VNC and IOM predictions, respectively. Finally, the predictions are fed into two discriminators to determine their correctness. Below, we will describe the details of each component of S2DGAN.

2.1 Transformer Encoder

The transformer encoder is a pyramidal backbone [19] that contains four blocks, where the $\{1^{st}\}$ block consists of a patch partitioning layer along with a linear embedding layer while the $\{2^{nd}, 3^{rd}, 4^{th}\}$ blocks consist of two successive T1 blocks followed by a patch merging layer. These blocks successively decrease the resolution and double the channel dimensionality of input, jointly producing a hierarchical representation.

Patch Partitioning. Since transformer-based models work with a sequence of tokens, we use the patch partition layer to create a set of tokens by splitting the $H \times W \times Z$ SECT image into multiple non-overlapping $4 \times 4 \times 4$ patches, where each patch is treated as a token. The feature of each patch is the concatenation of raw voxel values, thus having a feature dimension of $4 \times 4 \times 4 = 64$. These raw-valued features are then projected to dimension C (set as 48 in the implementation) by a linear embedding layer, and finally we can obtain $\frac{H}{4} \times \frac{Z}{4} \times \frac{W}{4}$ tokens with the dimension of C.

T1 Block. To reduce the complexity of attention computation, the transformer blocks (T1 and T2 blocks) in S2DGAN adopt the same windowing operation as Swin Transformer [19], i.e., computing attention in the partitioned windows, instead of the whole images or feature maps. As illustrated in Fig. 2 (b), the T1 block contains a regular window-based multi-head self-attention (W-MSA) module and a shifted window-based MSA (SW-MSA) module, followed by a 2-layer multi-layer perceptron (MLP). Meanwhile, layer normalization (LN) is employed before each MSA module and MLP layer, and a residual connection is applied after each module.

In the W-MSA module, we split the tokens evenly into non-overlapping windows with a size of $4 \times 4 \times \frac{Z}{4}$ and compute self-attention within each local window. It may limit the modeling power due to lacking connections across windows, thus we add a SW-MSA module after the W-MSA module to establish cross-window connections. In the SW-MSA module, the partitioned windows are shifted by half of their size $(4/2 = 2)$ along the height (H) and width (W) dimensions, and self-attention is then calculated in these shifted windows.

Patch Merging. The patch merging operation is performed after each T1 block to produce hierarchical representation, which is essential to generate finer details in the output for the dense prediction tasks [20]. Each patch merging operation merges adjacent tokens along the height and width dimensions in a non-overlapping manner to produce new tokens, i.e., each group of 2×2 neighboring tokens is concatenated along the channel dimension. Thus, the height and width dimensionalities are halved and the channel dimension is increased by four times. Then, the channel dimensionality of concatenated tokens is halved by a linear mapping. For example, the input size of $\frac{H}{4} \times \frac{Z}{4} \times \frac{W}{4} \times C$ will become $\frac{H}{8} \times \frac{Z}{8} \times \frac{W}{4} \times 2C$ after going through a patch merging layer.

2.2 Bottleneck Layer

The bottleneck layer connects the encoder and decoders, consisting of a T1 block followed by a patch expanding layer.

Patch Expanding. The patch expanding, which increases the resolution of features, can be considered as the inverse operation of patch merging. For the input feature map with size of $\frac{H}{32} \times \frac{W}{32} \times \frac{Z}{4} \times 8C$, the operation first doubles its channel dimensionality (with size becoming $\frac{H}{32} \times \frac{W}{32} \times \frac{Z}{4} \times 16C$) by a linear mapping, and then reshapes it to double the height and weight dimensionalities while reducing the channel dimensionality (with size becoming $\frac{H}{16} \times \frac{W}{16} \times \frac{Z}{4} \times 4C$). Namely, after going through the patch expanding operation, $\frac{H}{16} \times \frac{W}{16} \times \frac{Z}{4}$ tokens with the dimensionality of $4C$ are created.

After passing through the bottleneck layer, the common features extracted by the encoder are fed into two task-specific transformer-based decoders.

2.3 Transformer Decoder

There are two transformer decoders with the same structure to predict VNC and IOM, respectively. Between those two decoders, a shared attention mechanism

380 C. Jiang et al.

is employed, i.e., sharing query and key vectors derived from the encoder. The decoder contains three cascaded blocks, each of which consists of two successive T2 blocks and a patch expanding layer. These blocks increase the resolution and halve the channel dimension of the features, producing a hierarchical representation corresponding to encoder. The output of the last block is passed to a linear layer to obtain the final $H \times W \times Z \times 1$ prediction.

T2 Block. T2 block has the same architecture as T1 block, but adopts a shared attention mechanism instead of the self-attention mechanism adopted in T1 blocks. In this way, we can achieve attention sharing between generation tasks of VNC and IOM, where details of the shared attention mechanism will be discussed in Sect. 2.4.

2.4 Shared Attention

Considering that VNC and IOM generation tasks focus on similar regions, i.e., high-density regions, during the generation, we design a shared attention mechanism to further capture dependencies between these two interconnected tasks beyond sharing encoder. The steps involved in the mechanism are described as follows, and such procedures are applied for all the T2 blocks.

Formally, for a particular T2 block in the VNC generation decoder, let x_{VNC} denote the upsampled output of the previous block, and x_{SA} denote the output of the corresponding T1 block in the encoder. The standard way to compute self-attention for VNC generation is to obtain the key, query, and value vectors from its previous output x_{VNC} only. By contrast, our shared attention uses linear layers in the T2 block to compute the query q_{SA} and key k_{SA} from x_{SA} (from the encoder), as shown in Fig. 2 (c). Meanwhile, the value v_{VNC} is still computed using the previous block output x_{VNC} since the output of the T2 block should be related to the VNC generation task. Thus, the attention value is calculated as:

$$A_{SA} = softmax(\frac{q_{SA}k_{SA}}{\sqrt{C_{SA}}} + B), \qquad (1)$$

where C_{SA} is the number of channels and B is the position bias. Then the output of a particular attention head is $A_{SA} \cdot v_{VNC}$.

For the IOM generation decoder, we adopt the same scheme as above, but we only calculate A_{SA} in VNC generation decoder and then feed it to the IOM generation decoder directly. Empirically, we find that sharing the attention calculated in the VNC decoder is more effective to improve performance compared to the IOM decoder, thus we choose VNC generation task to share attention.

2.5 Discriminator and Loss

The final VNC and IOM predictions $\hat{y}_{VNC}$ and $\hat{y}_{IOM}$ are fed to two discriminators with the same architecture of four convolutional layers and a fully connected layer. The loss functions $\mathcal{L}_{VNC}$ and $\mathcal{L}_{IOM}$ for VNC and IOM generation are defined as follows:

$$\mathcal{L}_{VNC} = \mathcal{L}_{l1}(\hat{y}_{VNC}, y_{VNC}) + \mathcal{L}_{d1}(\hat{y}_{VNC}, y_{VNC}),$$
$$\mathcal{L}_{IOM} = \mathcal{L}_{l1}(\hat{y}_{IOM}, y_{IOM}) + \mathcal{L}_{d2}(\hat{y}_{IOM}, y_{IOM}),$$
$$(2)$$

Table 1. Quantitative results of ablation analysis, in terms of PSNR and SSIM.

Generator	VNC		IOM	
	PSNR↑	SSIM↑	PSNR↑	SSIM↑
S-UNet	22.37	0.864	20.88	0.822
D-UNet	25.43	0.892	22.38	0.835
S-S2DGAN	25.36	0.907	23.23	0.846
D-S2DGAN	26.85	0.921	24.52	0.853
S2DGAN	**28.54**	**0.932**	**24.67**	**0.876**

where $\mathcal{L}_{l1}$ is $L1$ loss, $\mathcal{L}_{d1}$ and $\mathcal{L}_{d2}$ are adversarial losses, while y_{VNC} and y_{IOM} are the ground truths (GTs).

During the training, we employ the GradNorm [21] technique to dynamically adjust the training weights of $\mathcal{L}_{VNC}$ and $\mathcal{L}_{IOM}$ according to their gradients to avoid overfitting.

3   Experiments

3.1   Dataset and Implementation

We collected 200 DECT samples and 30 SECT samples from individuals with PCHD. Each DECT sample contains five spatially-aligned images, including 80 kV, 140 kV, mixed 120 kV, VNC, and IOM images, while each SECT sample only contains one 120 kV image. All of the samples were collected within 24 h post-operatively, and there were only two cases of contrast extravasation and contrast extravasation accompanied by hemorrhage. It has been verified by [22] that the mixed 120 kV image is approximately equivalent to the actual 120 kV image. Thus, we reserve all the SECT samples for test and only employ the DECT samples to train S2DGAN. Herein, the training input is the mixed 120 kV image while the output is VNC and IOM. To investigate the bias between mixed 120 kV images and actual 120 kV images, 40 DECT samples are also reserved for test.

In our implementation, experiments were conducted on the PyTorch platform using two NVIDIA Tesla A100 GPUs and an Adam optimizer with initial learning rate of 0.001. All images were resampled to a voxel spacing of $1 \times 1 \times 1\,\mathrm{mm}^3$ with the size of $256 \times 256 \times 128$, and their intensity was normalized within [0, 1] by min-max normalization. To augment the training samples and reduce the usage of GPU memory, the original image was randomly cropped to the size of $96 \times 96 \times 96$ as input.

To evaluate the generated VNC and IOM, we first calculated the peak signal to noise ratio (PSNR) and structural similarity index (SSIM) for quantitative assessment, and then evaluated these generated images by downstream tasks, including the PCHD diagnosis and segmentations of hemorrhage and contrast extravasation, to demonstrate their clinical practicability. Moreover, to exclude randomness, five-fold cross-validation was also performed during evaluation.

Table 2. Quantitative comparison of our S2DGAN with several state-of-the-art generation methods, in terms of PSNR, SSIM, model size ($^\#$param), and GFLOPs.

Method	VNC		IOM		Model Efficiency	
	PSNR↑	SSIM↑	PSNR↑	SSIM↑	$^\#$param↓	GFLOPs↓
AutoCNN [3]	22.32	0.854	20.84	0.812	22M	274
cGAN [4]	22.97	0.858	21.31	0.817	28M	312
MedGAN [5]	24.33	0.866	22.28	0.821	52M	478
Auto-GAN [6]	23.46	0.873	22.54	0.829	56M	536
3D Transformer-GAN [13]	24.87	0.883	22.93	0.844	148M	1270
SwinGAN [14]	25.13	0.887	23.12	0.837	84M	964
TarGAN [16]	25.56	0.894	23.45	0.843	56M	723
ViTGAN [15]	26.89	<u>0.925</u>	24.54	0.857	182M	1570
TransGAN [23]	<u>27.32</u>	0.918	**24.83**	<u>0.863</u>	197M	1812
S2DGAN	**28.54**	**0.932**	<u>24.67</u>	**0.876**	96M	1059

3.2 Ablation Analysis

To evaluate the effectiveness of each network component in S2DGAN, we designed another four variant generators, including: 1) S-UNet, consisting of a CNN encoder and a CNN decoder; 2) D-UNet, consisting of a CNN encoder and two CNN decoders; 3) S-S2DGAN, consisting of a transformer encoder and a transformer decoder; 4) D-S2DGAN, consisting of a transformer encoder and two transformer decoders. D-S2DGAN has the same architecture as S2DGAN, but without adopting the shared attention mechanism in decoders. In addition, two separate models of S-UNet and S-S2DGAN need to be trained to generate the VNC and IOM, respectively, since their design only allows to output one type of image.

The quantitative results are provided in Table 1, from which, we can find the following observations. (1) D-UNet/D-S2DGAN achieves better generation results than S-UNet/S-S2DGAN. This proves that a multi-task framework is more appropriate than a single-task framework for interrelated VNC and IOM generation. (2) The transformer-based S-S2DGAN and D-S2DGAN, respectively, achieve better results than the CNN-based S-UNet and D-UNet. This may be because the transformer can capture global information, and thus can benefit the generation of significant high-density areas in VNC and IOM. (3) S2DGAN achieves better results on both VNC and IOM generation tasks than D-S2DGAN and other variants. This demonstrates that the shared attention mechanism can strengthen the connection between VNC and IOM generations, thus resulting in better performance. These three comparisons conjointly verify the effective design of S2DGAN, where the *multi-task learning strategy, transformer-based architecture*, and *shared attention mechanism* all can benefit generations of VNC and IOM.

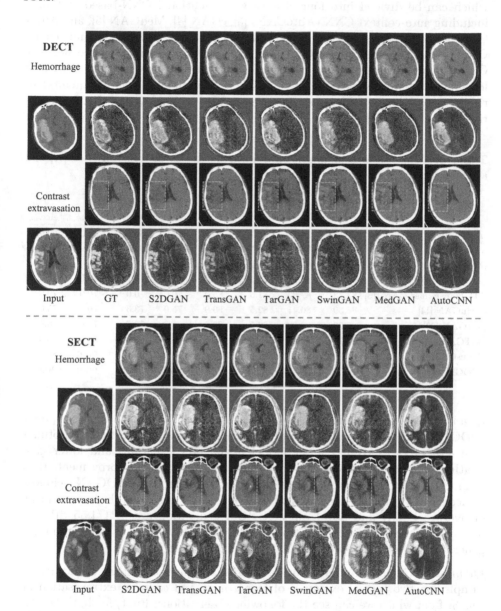

Fig. 3. Visual comparison of DECT and SECT samples on two typical cases. In each case, the first row and second row show the VNC and IOM, respectively, and from left to right are the input (mixed 120 kV images for DECT and 120 kV images for SECT), GT (SECT without GT), and six images generated by our S2DGAN and five other methods. Red boxes show areas for detailed comparison. (Color figure online)

3.3   Comparison with State-of-the-art Methods

We further compare S2DGAN with several state-of-the-art generation methods, which can be divided into four classes: a) conventional CNN-based methods, including auto-context CNN (AutoCNN) [3], cGAN [4], MedGAN [5], and Auto-GAN [6]; b) transformer-based methods, including ViTGAN [15] and Trans-GAN [16]; c) combinations of CNN and transformer, including 3D Transformer-GAN [13] and Swin-Transformer-based GAN (SwinGAN) [14]; and d) multi-task learning frameworks, including TarGAN [23]. The quantitative and qualitative results of VNC and IOM generation are provided in Table 2 and Fig. 3, respectively.

Table 3. Quantitative comparison on downstream tasks, in terms of Dice [%] and accuracy [%].

Method	DECT					SECT		
	$Dice_{hem}$	$Dice_{con}$	Acc_{whole}	Acc_{hem}	Acc_{con}	Acc_{whole}	Acc_{hem}	Acc_{con}
AutoCNN [3]	74.12	75.72	62.5	70.0	55.0	60.00	60.0	60.0
cGAN [4]	75.28	77.74	72.5	80.0	65.0	63.3	60.0	65.0
MedGAN [5]	78.54	82,54	75.0	85.0	65.0	66.7	70.0	65.0
Auto-GAN [6]	80.42	83.12	77.5	85.0	70.0	66.7	60.0	70.0
3D Transformer-GAN [13]	83.73	87.63	77.5	85.0	70.0	70.0	70.0	70.0
SwinGAN [14]	84.39	86.41	82.5	90.0	75.0	73.3	70.0	75.0
TarGAN [23]	88.26	89.32	82.5	90.0	75.0	76.7	70.0	80.0
ViTGAN [15]	91.11	91.57	85.0	90.0	80.0	83.3	80.0	85.0
TransGAN [16]	90.53	92.11	87.5	95.0	80.0	83.3	90.0	80.0
S2DGAN	**94.21**	**96.03**	**92.5**	**100.0**	**85.0**	**90.0**	**90.0**	**90.0**

Quantitative Comparison: It can be noticed from Table 2 that our proposed S2DGAN achieves generally the optimal results. Compared to the sub-optimal TransGAN, S2DGAN has fewer model parameters (#param) and lower giga floating point of operations (GFLOPs) but achieves slight improvement, thus is more efficient. Meanwhile, with similar model size, our S2DGAN achieves noticeable performance improvement to SwinGAN. Specifically, the improvement of average PSNR and SSIM on VNC/IOM generation are 3.41/1.55 dB and 0.045/0.039, respectively. This implies that our S2DGAN can achieve promising generation results while maintaining a reasonable model size.

Qualitative Comparison: We provide a visual comparison of DECT and SECT samples on two typical examples of hemorrhage and contrast extravasation in Fig. 3, from which we can see the following observations. First, for both DECT and SECT samples, the VNC and IOM generated by our S2DGAN have less noise, fewer artifacts but clearer structural details, especially the in high-density areas. Second, in the contrast extravasation cases, only the VNC images generated by our S2DGAN can lead to correct diagnosis, while others still retain high-density areas (see red box in Fig. 3) that would lead to wrong diagnosis (i.e., diagnosed as hemorrhage). Such key observations demonstrate that our S2DGAN is superior to these state-of-the-art methods.

3.4 Downstream Task Evaluation

We further resort to two downstream tasks to evaluate the generated images more comprehensively, including: 1) Segmentation of hemorrhage and contrast extravasation on the VNC and IOM, respectively. Taking the segmentation on real images as GT, we can calculate $Dice_{hem}$ and $Dice_{con}$ of different methods for quantification. 2) Diagnosis of PCHD, i.e., determining whether any contrast extravasation or hemorrhage will be determined. We use Acc_{whole}, Acc_{hem}, and Acc_{con} to denote classification accuracies of the whole, hemorrhage, and contrast extravasation, respectively. All downstream tasks were performed with the assistance of two experienced imaging physicians, where only 10 hemorrhage DECT samples were used for segmentation due to the high workload, while 40 DECT (20 hemorrhage and 20 contrast extravasation) samples and 30 SECT (10 hemorrhage and 20 contrast extravasation) samples were used for PCHD diagnosis.

With the quantitative results provided in Table 3, we can see our S2DGAN achieves the highest Dice coefficients [%] for segmentation of hemorrhage and contrast extravasation. This verifies that the high-density areas in S2DGAN-generated VNC and IOM are the closest to the real images. Moreover, our S2DGAN achieves the highest accuracy of PCHD diagnosis on both DECT and SECT samples. This proves that the VNC and IOM generated by S2DGAN are more realistic in diagnosic semantics. These pieces of evidence prove the reliable usability and great potential of the IOM and VNC images generated by S2DGAN for downstream tasks.

4   Conclusion and Limitations

In this paper, to achieve SECT-based real-time determination of hemorrhage, we propose a SECT-to-DECT adversarial generative network (S2DGAN) to generate VNC and IOM simultaneously from one SECT (120 kV/ mixed 120 kV) image. Specifically, we adopt three strategies including *transformer-based architecture*, *multi-task learning strategy*, and *shared attention mechanism* in S2DGAN, to improve the generation performance while efficiently limiting the model size and computational complexity. Validated on the collected clinical dataset, we demonstrate that S2DGAN is designed effectively and can achieve superior performance quantitatively and qualitatively over the state-of-the-art methods.

Despite the advance of S2DGAN, our current work still suffers from a few limitations as follows: (1) Lacking real paired data since there are no corresponding VNC and IOM for actual SECT (120 kV) images. Although mixed 120 kV images can be approximated as 120 kV images, there still exists domain gap between them, which will be explored in our future work. 2) Focusing only on the diagnosis of PCHD in the 0–24 h postoperative period. Actually, multiple diagnoses within postoperative 0–72 h are needed, and it will be more complex in 24–72 h due to the absorption of contrast agents. In our future work, we will collect data in 24–72 h to validate our approach for the whole PCHD diagnostic process.

Acknowledgment. This work was supported in part by National Natural Science Foundation of China (No. 62131015), Science and Technology Commission of Shanghai Municipality (STCSM) (No. 21010502600), The Key R&D Program of Guangdong Province, China (No. 2021B0101420006), and the China Postdoctoral Science Foundation (Nos. BX2021333, 2021M703340). This work was completed under the close collaboration between C. Jiang and Y. Pan, and they contributed equally to this work.

References

1. Shao, Y., Xu, Y., Li, Y., Wen, X., He, X.: A new classification system for postinterventional cerebral hyperdensity: the influence on hemorrhagic transformation and clinical prognosis in acute stroke. Neural Plasticity, 2021 (2021)
2. Lyu, T., et al.: Estimating dual-energy CT imaging from single-energy CT data with material decomposition convolutional neural network. Med. Image Anal. **70**, 102001 (2021)
3. Xiang, L., et al.: Deep auto-context convolutional neural networks for standard-dose PET image estimation from low-dose PET/MRI. Neurocomputing **267**, 406–416 (2017)
4. Wang, Y., et al.: 3D conditional generative adversarial networks for high-quality PET image estimation at low dose. Neuroimage **174**, 550–562 (2018)
5. Armanious, K., et al.: MedGAN: medical image translation using GANs. Comput. Med. Imaging Graph. **79**, 101684 (2020)
6. Cao, B., Zhang, H., Wang, N., Gao, X., Shen, D.: Auto-GAN: self-supervised collaborative learning for medical image synthesis. In: Proceedings of the AAAI Conference on Artificial Intelligence, vol. 34, no. 07, pp. 10486–10493 (2020)
7. Frid-Adar, M., Diamant, I., Klang, E., Amitai, M., Goldberger, J., Greenspan, H.: GAN-based synthetic medical image augmentation for increased CNN performance in liver lesion classification. Neurocomputing **321**, 321–331 (2018)
8. Pan, Y., Liu, M., Xia, Y., Shen, D.: Disease-image-specific learning for diagnosis-oriented neuroimage synthesis with incomplete multi-modality data. IEEE Trans. Pattern Anal. Mach. Intell. **27**(5), 1675–1686 (2021)
9. Wu, G., Jia, H., Wang, Q., Shen, D.: SharpMean: groupwise registration guided by sharp mean image and tree-based registration. Neuroimage **56**(4), 1968–1981 (2011)
10. Jia, H., Wu, G., Wang, Q., Shen, D.: ABSORB: atlas building by self-organized registration and bundling. Neuroimage **51**(3), 1057–1070 (2010)
11. Jia, H., Yap, P., Shen, D.: Iterative multi-atlas-based multi-image segmentation with tree-based registration. Neuroimage **59**(1), 422–430 (2012)
12. Vaswani, A., et al.: Attention is all you need. Adv. Neural Inf. Process. Syst. **30**, 5999–6009 (2017)
13. Luo, Y., et al.: 3D transformer-GAN for high-quality PET reconstruction. In: de Bruijne, M., et al. (eds.) MICCAI 2021. LNCS, vol. 12906, pp. 276–285. Springer, Cham (2021). https://doi.org/10.1007/978-3-030-87231-1_27

14. Pan, K., Cheng, P., Huang, Z., Lin, L., Tang, X.: Transformer-based T2-weighted MRI synthesis from T1-weighted images. In: 2022 44th Annual International Conference of the IEEE Engineering in Medicine & Biology Society (EMBC), pp. 5062–5065 (2022)
15. Lee, K., Chang, H., Jiang, L., Zhang, H., Tu, Z., Liu, C.: VitGAN: training GANs with vision transformers. arXiv preprint arXiv:2107.04589 (2021)
16. Jiang, Y., Chang, S., Wang, Z.: TransGAN: two transformers can make one strong GAN. arXiv preprint arXiv:2102.07074, 1(3) (2021)
17. Peiris, H., Hayat, M., Chen, Z., Egan, G., Harandi, M.: A robust volumetric transformer for accurate 3D tumor segmentation. In: Wang, L., Dou, Q., Fletcher, P.T., Speidel, S., Li, S. (eds.) Medical Image Computing and Computer Assisted Intervention – MICCAI 2022. MICCAI 2022. LNCS, vol. 13435, pp. 162–172. Springer, Cham (2022). https://doi.org/10.1007/978-3-031-16443-9_16
18. Bhattacharjee, D., Zhang, T., Sustrunk, S., Salzmann, M.: MulT: an end-to-end multitask learning transformer. In: Proceedings of the IEEE/CVF Conference on Computer Vision and Pattern Recognition, pp. 12031–12041 (2022)
19. Liu, Z., et al.: Swin transformer: hierarchical vision transformer using shifted windows. In: Proceedings of the IEEE/CVF International Conference on Computer Vision, pp. 10012–10022 (2021)
20. Chen, L., Papandreou, G., Kokkinos, I., Murphy, K., Yuille, A.: Deeplab: semantic image segmentation with deep convolutional nets, Atrous convolution, and fully connected CRFs. IEEE Trans. Pattern Anal. Mach. Intell. **40**, 834–848 (2017)
21. Chen, Z., Badrinarayanan, V., Lee, C.Y., Rabinovich, A.: Gradnorm: gradient normalization for adaptive loss balancing in deep multitask networks. In: International Conference on Machine Learning, pp. 794–803 (2018)
22. Bodanapally, U., et al.: Dual-energy CT in hemorrhagic progression of cerebral contusion: overestimation of hematoma volumes on standard 120-kv images and rectification with virtual high-energy monochromatic images after contrast-enhanced whole-body imaging. Am. J. Neuroradiol. **39**(4), 658–662 (2018)
23. Chen, J., Wei, J., Li, R.: TarGAN: target-aware generative adversarial networks for multi-modality medical image translation. In: International Conference on Medical Image Computing and Computer-Assisted Intervention, pp. 24–33 (2021)

SADM: Sequence-Aware Diffusion Model for Longitudinal Medical Image Generation

Jee Seok Yoon[1,3] , Chenghao Zhang[2] , Heung-Il Suk[1] , Jia Guo[2] ,
and Xiaoxiao Li[3(✉)]

[1] Korea University, Seoul 02841, Republic of Korea
{wltjr1007,hisuk}@korea.ac.kr
[2] Columbia University, New York, NY 10027, USA
{cz2639,jg3400}@columbia.edu
[3] The University of British Columbia, Vancouver, BC V6T 1Z4, Canada
xiaoxiao.li@ece.ubc.ca

Abstract. Human organs constantly undergo anatomical changes due
to a complex mix of short-term (*e.g.*, heartbeat) and long-term (*e.g.*,
aging) factors. Evidently, prior knowledge of these factors will be beneficial when modeling their future state, *i.e.*, via image generation. However, most of the medical image generation tasks only rely on the input
from a single image, thus ignoring the sequential dependency even when
longitudinal data is available. Sequence-aware deep generative models,
where model input is a sequence of ordered and timestamped images,
are still underexplored in the medical imaging domain that is featured
by several unique challenges: 1) Sequences with various lengths; 2) Missing data or frame, and 3) High dimensionality. To this end, we propose
a sequence-aware diffusion model (SADM) for the generation of longitudinal medical images. Recently, diffusion models have shown promising results in high-fidelity image generation. Our method extends this
new technique by introducing a sequence-aware transformer as the conditional module in a diffusion model. The novel design enables learning longitudinal dependency even with missing data during training and
allows autoregressive generation of a sequence of images during inference.
Our extensive experiments on 3D longitudinal medical images demonstrate the effectiveness of SADM compared with baselines and alternative
methods. The code is available at https://github.com/ubc-tea/SADM-
Longitudinal-Medical-Image-Generation.

Keywords: Diffusion model · Sequential image generation ·
Autoregressive conditioning

1 Introduction

Iconic advancements of generative models in the medical domain have been possible due to several factors, such as state-of-the-art computational hardware and,

© The Author(s), under exclusive license to Springer Nature Switzerland AG 2023
A. Frangi et al. (Eds.): IPMI 2023, LNCS 13939, pp. 388–400, 2023.
https://doi.org/10.1007/978-3-031-34048-2_30

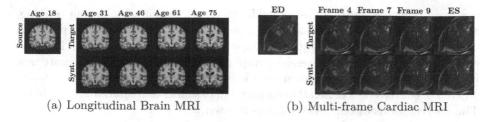

(a) Longitudinal Brain MRI   (b) Multi-frame Cardiac MRI

Fig. 1. Examples of longitudinal medical images synthesized by our proposed SADM.

more importantly, the availability of medical datasets (both open and in-house). Hence, promising solutions for medical image synthesis, restoration, acceleration, and many other tasks have been proposed over the past few years [19]. Recent efforts to generate longitudinal medical images were mainly proposed for the following two tasks: 1) generation of longitudinal brain image [16], which takes a source brain image and generates a new image with respect to chronological age (*i.e.*, normal) progression or disease (*i.e.*, abnormal) progression [11]; and 2) generation of multi-frame cardiac image [4,12], which typically take a starting frame of a cardiac cycle (*i.e.*, end-diastolic or ED phase) and generates the final frame of the cycle (*i.e.*, end-systolic or ES phase). An illustration showing examples of these tasks is presented in Fig. 1. Generative adversarial networks (GANs) have been a *de facto* standard for these tasks in the past few years, but recent advances in diffusion models have shown promising results. For example, the latent diffusion model, which uses the latent embedding of an image as input to the diffusion model to improve computational efficiency, has been used to synthesize high-quality 3D brain MRI [17]. Similarly, a diffusion model was combined with a deformation image registration model, namely VoxelMorph [2], to synthesize the end-systolic frame of cardiac MRI [12,13]. However, most of these works rely on the input from a single image to generate its longitudinal images. Even when longitudinal samples are available, these methods often ignore the sequential dependency in the medical domain.

Sequence-aware [18] deep generative models are a class of generative models that can learn the sequential or temporal dependency of the longitudinal input data. A sequence is defined as an ordered and timestamped set [18], and sequence-aware generative models take input from a sequence of images and output a generated image (formal definition in *Problem settings* in Sect. 3). Although such generative models for video datasets have been studied that often take a sequence of frames as input and learn their temporal dependence [1,10], the existing solutions are not feasible for longitudinal medical data generation tasks. Because video datasets rarely have issues very common in the medical domain, such as *1) longitudinal data scarcity, 2) missing frames or data, 3) high dimensionality, and 4) low temporal resolution.*

To this end, we explore ways to address these issues and propose a novel generative model for longitudinal medical image generation that can learn the temporal dependency given a sequence of medical images. Our proposed method is named sequence-aware diffusion model (SADM). Specifically, during training, SADM learns to estimate attentive representations in the longitudinal positions of given tokens based on sequential input, even with missing data. In inference time, we use an autoregressive sampling scheme to effectively generate new images. Our extensive experiments on longitudinal 3D medical images demonstrate the effectiveness of SADM compared to baselines and alternative methods. The contributions of our SADM are as follows:

1. To the best of our knowledge, we are one of the first to explore the temporal dependency of sequential data and use it as a prior in diffusion models for medical image generation.
2. Our proposed SADM can work in various real-world settings, such as single image input, longitudinal data with missing frames, and high-dimensional images via the essential transformer module design.
3. We present state-of-the-art results in longitudinal image generation and missing data imputation for a multi-frame cardiac MRI and a longitudinal brain MRI dataset.

2 Preliminary

2.1 Diffusion Models

Diffusion models consist of a forward process, which starts with the data $\mathbf{x} \sim p(\mathbf{x})$ and gradually adds noise to obtain a noisy version of the data $\mathbf{z} = \{\mathbf{z}_t | t \in [0, 1]\}$, and a reverse process, which reverts the forward process by predicting and subtracting the noise in the reverse direction (*i.e.*, from $t = 1$ to $t = 0$).

Formally, following [9], we define the forward process $q(\mathbf{z}|\mathbf{x})$ specified in continuous time $0 \le s < t \le 1$ as:

$$q(\mathbf{z}_t|\mathbf{x}) = \mathcal{N}(\alpha_t \mathbf{x}, \sigma_t^2 \mathbf{I}), \quad q(\mathbf{z}_t|\mathbf{z}_s) = \mathcal{N}((\alpha_t/\alpha_s)\mathbf{z}_s, \sigma_{t|s}^2 \mathbf{I}) \tag{1}$$

where $\alpha_t^2 = 1/(1 + e^{-t})$ and $\sigma_t^2 = 1 - \alpha_t^2$ are the continuous-time noise schedules, $\sigma_{t|s}^2 = (1 - e^{\lambda_t - \lambda_s}) \sigma_t^2$ is the variance term of the s to t transition, and $\lambda_t = \log[\alpha_t^2/\sigma_t^2]$ is the signal-to-noise-ratio of the noise schedules that is monotonically decreasing [14]. This forward process can be reformulated in the reverse direction as $q(\mathbf{z}_s|\mathbf{z}_t, \mathbf{x}) = \mathcal{N}(\tilde{\boldsymbol{\mu}}_{s|t}(\mathbf{z}_t, \mathbf{x}), (\tilde{\sigma}_{s|t}^2)\mathbf{I})$, where $\tilde{\boldsymbol{\mu}}_{s|t}(\mathbf{z}_t, \mathbf{x}) = e^{\lambda_t - \lambda_s}(\alpha_s/\alpha_t)\mathbf{z}_t + (1 - e^{\lambda_t - \lambda_s})\alpha_s \mathbf{x}$ and $\tilde{\sigma}_{s|t}^2 = (1 - e^{\lambda_t - \lambda_s})\sigma_s^2$.

The reverse process is parameterized by a generative model $\hat{\mathbf{x}}_\theta$ in the form:

$$p_\theta(\mathbf{z}_s|\mathbf{z}_t) = \mathcal{N}(\tilde{\boldsymbol{\mu}}_{s|t}(\mathbf{z}_t, \hat{\mathbf{x}}_\theta(\mathbf{z}_t, \lambda_t)), \tilde{\Sigma}_{s|t}\mathbf{I}) \tag{2}$$

where the variance $\tilde{\Sigma}_{s|t} = (\tilde{\sigma}_{s|t}^2)^{1-v}(\sigma_{t|s}^2)^v$ is an interpolation between $\tilde{\sigma}_{s|t}^2$ and $\sigma_{t|s}^2$ [14], and v is the hyperparameter that controls the stochasticity of the sampler [15]. We use the ancestral sampler [8] where $\lambda_0 < ... < \lambda_T = \lambda_1$ for discrete T time steps:

$$\mathbf{z}_s = \tilde{\boldsymbol{\mu}}_{s|t}(\mathbf{z}_t, \hat{\mathbf{x}}_\theta(\mathbf{z}_t, \lambda_t)) + \sqrt{\tilde{\Sigma}_{s|t}}\boldsymbol{\epsilon} \quad \text{where } \boldsymbol{\epsilon} \sim \mathcal{N}(\mathbf{0}, \mathbf{I}). \tag{3}$$

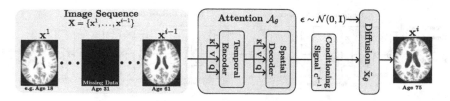

Fig. 2. An overview of our proposed sequence-aware diffusion model (SADM).

2.2 Classifier-Free Guidance

There are two branches of conditioning methods for diffusion models: 1) Classifier guided [5]; and 2) Classifier-free guidance [9]. However, it is often hard to define the problem setting for a classifier in the medical domain, and even the state-of-the-art classifiers often do not have the performance suitable for classifier-guided models. Thus, we opt to use classifier-free guidance for conditioning the diffusion model. The classifier-free guided diffusion model takes the conditioning signal $\mathbf{c}$ as an additional input, and is defined as

$$\tilde{\mathbf{x}}_\theta\left(\mathbf{z}_t, \mathbf{c}, \lambda_t\right) = (1+w)\hat{\mathbf{x}}_\theta\left(\mathbf{z}_t, \mathbf{c}, \lambda_t\right) - w\hat{\mathbf{x}}_\theta\left(\mathbf{z}_t, \emptyset, \lambda_t\right), \tag{4}$$

which is the weighted sum of the model with condition $\mathbf{c}$ and model with zero tensor $\emptyset$, *i.e.*, unconditional model. The guidance strength w controls the trade-off between sample quality and diversity, *i.e.*, the higher the guidance strength the lower the diversity. Equation (4) can also be performed in ϵ-space $\tilde{\epsilon}_\theta\left(\mathbf{z}_t, \mathbf{c}, \lambda_t\right) = (1+w)\hat{\epsilon}_\theta\left(\mathbf{z}_t, \mathbf{c}, \lambda_t\right) - w\hat{\epsilon}_\theta\left(\mathbf{z}_t, \emptyset, \lambda_t\right)$.

During training, we can randomly replace the conditioning signals $\mathbf{c}$ by a zero tensor with probability p_{uncond}. Then, the noise-prediction loss term [8] for the reverse process conditional generative model is:

$$\mathcal{L}(\mathbf{x}) = \mathbb{E}_{\epsilon \sim \mathcal{N}(0,\mathbf{I}), t \sim U(0,1), \hat{\mathbf{I}} \sim Be(p_{uncond})}\left[\left\|\hat{\epsilon}_\theta\left(\mathbf{z}_t, \mathbf{c}\hat{\mathbf{I}}, \lambda_t\right) - \epsilon\right\|_2^2\right] \tag{5}$$

where $\hat{\mathbf{I}} \sim Be(p_{uncond})$ is either a zero or an identity tensor sampled from a Bernoulli distribution, $\mathbf{z}_t = \alpha_t\mathbf{x} + \sigma_t\epsilon$, and $\hat{\epsilon}_\theta\left(\mathbf{z}_t, \mathbf{c}, \lambda_t\right) = \sigma_t^{-1}(\mathbf{z}_t - \alpha_t\hat{\mathbf{x}}_\theta(\mathbf{z}_t, \mathbf{c}, \lambda_t))$.

3   SADM: Sequence-Aware Diffusion Model

We propose a sequence-aware diffusion model (SADM) for longitudinal medical image generation. Specifically, our proposed SADM uses a transformer-based attention module for conditioning a diffusion model. The attention module is a 4D generalization of the video vision transformer (ViViT) [1], and it is specifically used to generate the conditioning signals for the diffusion model. In this section, we will briefly explain the problem setting and the details of SADM. An overview of SADM is illustrated in Fig. 2.

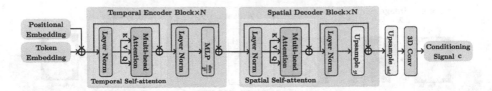

Fig. 3. An illustration of the attention module $\mathcal{A}_\theta$. The temporal encoder performs temporal self-attention followed by dimension reduction using an MLP, while the spatial decoder performs a spatial self-attention followed by an upsampling operation.

Problem Setting. Let $\mathbf{X} \sim p(\mathbf{X}) \in \mathbb{R}^{L \times W \times H \times D}$ be a longitudinal 3D medical image with temporal length L. We partition $\mathbf{X}$ into conditioning images $\mathbf{X}^\mathcal{C} \in \mathbb{R}^{n_\mathcal{C} \times W \times H \times D}$, missing images $\mathbf{X}^\mathcal{M} \in \mathbb{R}^{n_\mathcal{M} \times W \times H \times D}$, and future images $\mathbf{X}^\mathcal{F} \in \mathbb{R}^{n_\mathcal{F} \times W \times H \times D}$, where $\mathcal{C} = \{c_1, \ldots, c_{n_\mathcal{C}}\}$, $\mathcal{M} = \{m_1, \ldots, m_{n_\mathcal{M}}\}$, and $\mathcal{F} = \{f_1, \ldots, f_{n_\mathcal{F}}\}$ are a sequence of scalar indices for tensor indexing. We define these sequences as ordered, timestamped, and non-intersecting sets [18] such that $\mathcal{C} \cap \mathcal{M} = \mathcal{C} \cap \mathcal{F} = \mathcal{M} \cap \mathcal{F} = \emptyset$, $\mathcal{C} \cup \mathcal{M} \cup \mathcal{F} = \{1, \ldots, L\}$, and $n_\mathcal{C} + n_\mathcal{M} + n_\mathcal{F} = L$. We assume that indices of $\mathcal{F}$ are always in future of $\mathcal{C}$ and $\mathcal{M}$, *i.e.*, $c < f$ and $m < f$ for all $c \in \mathcal{C}, m \in \mathcal{M}$, and $f \in \mathcal{F}$. Also, we assume that the first image of the sequence is known, *i.e.*, $c_1 = 1$. The objective is to maximize the posteriors $p(\mathbf{X}^\mathcal{M} | \mathbf{X}^\mathcal{C})$ and $p(\mathbf{X}^\mathcal{F} | \mathbf{X}^\mathcal{C})$, *i.e.*, synthesize the missing and future images given a sequence of conditioning images.

3.1 Attention Module $\mathcal{A}_\theta$

Unlike many other longitudinal vision datasets (*e.g.*, video data), longitudinal medical images have the following unique properties: various sequence lengths, missing data or frames, and high dimensionality. Existing generative solutions used in common computer vision fields are not optimized for these properties, thus longitudinal medical image generation requires a specialized architecture. Inspired by the success of transformers for vision datasets and their ability to calculate attention for long-distance spatio-temporal representations [1], we propose a transformer-based attention module for generating conditioning signals for the diffusion model. This attention module will direct which frames of the conditioning image $\mathbf{X}^\mathcal{C}$ will be beneficial to generate the future or missing frames. An overview of our attention module is illustrated in Fig. 3.

Token Embedding. A common approach to embedding an image or video into tokens is by running a fusing window over the input with non-overlapping strides [1]. Given $\mathbf{X} \in \mathbb{R}^{L \times W \times H \times D}$, we run a non-overlapping linear projection window of dimension $\mathbb{R}^{l \times w \times h \times d \times dim}$ over the image, *i.e.*, a 4D convolution. The resulting unflattened token $\mathbf{h}$ have the shape of $\mathbb{R}^{\lfloor \frac{L}{l} \rfloor \times \lfloor \frac{W}{w} \rfloor \times \lfloor \frac{H}{h} \rfloor \times \lfloor \frac{D}{d} \rfloor \times dim}$. However, as we are dealing with longitudinal medical images that may contain missing frames, fusing through the temporal axis is not feasible. Thus, we set $l = 1$ for experiments conducted in this paper. Also, the temporal resolution of

---

Algorithm 1: SADM Training

---

1 **repeat**
2   Initialize an empty sequence of images $\tilde{\mathbf{X}}$
3      $(\mathbf{X}, \mathcal{C}, \mathcal{M}, \mathcal{F}) \sim p(\mathbf{X})$ // Sample data and indices
4      $i \sim U(\mathcal{M} \cup \mathcal{F})$ // Randomly select target index
5   **for** $k = 1, ..., i - 1$ **do**
6   **if** $k \in \mathcal{C}$ **then**
7              $\tilde{\mathbf{X}} \leftarrow \tilde{\mathbf{X}} + [deque(\mathbf{X}^{\mathcal{C}})]$ // Append conditioning image
8   **else**
9              $\tilde{\mathbf{X}} \leftarrow \tilde{\mathbf{X}} + [\emptyset]$ // Append zero tensor
10      $\mathbf{c}^{i-1} \leftarrow \mathcal{A}_{\theta}(\tilde{\mathbf{X}})$ // Generate conditioning signal
11      $(\epsilon, t, \hat{\mathbf{I}}) \sim (\mathcal{N}(\mathbf{0}, \mathbf{I}), U(0, 1), Be(p_{uncond}))$
12      $\mathbf{z}_t \leftarrow \alpha_t \mathbf{X}^i + \sigma_t \epsilon$ // Sample noisy image
13   Take gradient step on $\nabla_{\theta} \| \hat{\epsilon}_{\theta} \left(\mathbf{z}_t, \mathbf{c}^{i-1} \hat{\mathbf{I}}, \lambda_t \right) - \epsilon \|_2^2$
14 **until** converged;

---

longitudinal images is typically very low compared to its spatial resolution, so setting $l = 1$ will only slightly affect the computational efficiency.

Temporal Encoder. Inspired by factorized transformers [1], we factorize our attention module into a temporal encoder and a spatial decoder that can benefit from long-range spatio-temporal attention with high computational efficiency. The temporal encoder computes the self-attention temporally among all tokens in the same spatial index. Specifically, it takes the unflattened token and reshapes them into $\mathbf{h}_{\text{tmp}} \in \mathbb{R}^{\lfloor \frac{W}{w} \rfloor \cdot \lfloor \frac{H}{h} \rfloor \cdot \lfloor \frac{D}{d} \rfloor \times L \times dim}$, where the leading dimension is the batch dimension. Then, it computes the self-attention along the temporal dimensions, and an MLP reduces the token's dimension by a factor of 2^3 and reshapes them to $\mathbf{h}_{\text{tmp}}^{\ell} \in \mathbb{R}^{\lfloor \frac{W}{2^{\ell}w} \rfloor \cdot \lfloor \frac{H}{2^{\ell}h} \rfloor \cdot \lfloor \frac{D}{2^{\ell}d} \rfloor \times L \times dim}$ for each temporal transformer block ℓ.

Spatial Decoder. The output of the temporal encoder is reshaped into a spatial token $\mathbf{h}_{\text{spt}} \in \mathbb{R}^{L \times \lfloor \frac{W}{2^N w} \rfloor \cdot \lfloor \frac{H}{2^N h} \rfloor \cdot \lfloor \frac{D}{2^N d} \rfloor \times dim}$, and the spatial decoder calculates self-attention to spatial dimensions between all tokens in the same temporal index. Then, it is upsampled by a factor of 2 for each spatial dimension to obtain $\mathbf{h}_{\text{spt}}^{\ell} \in \mathbb{R}^{L \times \lfloor \frac{W}{2^{N-\ell}w} \rfloor \cdot \lfloor \frac{H}{2^{N-\ell}h} \rfloor \cdot \lfloor \frac{D}{2^{N-\ell}d} \rfloor \times dim}$. Finally, we unflatten and reshape the output of the last block into $\mathbb{R}^{\lfloor \frac{W}{w} \rfloor \times \lfloor \frac{H}{h} \rfloor \times \lfloor \frac{D}{d} \rfloor \times L \cdot dim}$, and perform upsampling and a 3D convolution operation to obtain the conditioning signal $\mathbf{c} \in \mathbb{R}^{W \times H \times D}$.

Since transformers can mask specific indexes of a token, we can train and infer even when there are missing frames in the longitudinal images. However, we have found that using zero tensors for missing frames with non-zero positional encoding performs better than masking missing frames.

Algorithm 2: Autoregressive Sampling

---

1 **Input:** Conditioning indices $\mathcal{C} = \{c_1, ..., c_{n_c}\}$ and images $\mathbf{X}^{\mathcal{C}}$
2 **Output:** Missing images $\mathbf{X}^{\mathcal{M}}$ and future images $\mathbf{X}^{\mathcal{F}}$
3 Initialize an empty sequence of images $\tilde{\mathbf{X}}$
4 **for** $i = 1, ..., L$ **do**
5   **if** $i \in \mathcal{C}$ **then** // We assume $c_1 = 1$
6       $\tilde{\mathbf{X}} \leftarrow \tilde{\mathbf{X}} + [deque(\mathbf{X}^{\mathcal{C}})]$
7   **else**
8       $\mathbf{c}^{i-1} \leftarrow \mathcal{A}_\theta(\tilde{\mathbf{X}})$ // $\tilde{\mathbf{X}} = [\mathbf{x}^1, ..., \mathbf{x}^{i-1}]$
9       $\mathbf{z}_1 \leftarrow \mathcal{N}(\mathbf{0}, \mathbf{I})$
10   **for** $t = 1, ..., \frac{2}{T}, \frac{1}{T}$ **do**
11         $s \leftarrow t - \frac{1}{T}$
12         $\epsilon \sim \mathcal{N}(\mathbf{0}, \mathbf{I})$
13         $\mathbf{z}_s \leftarrow \tilde{\mu}_{s|t}\left(\mathbf{z}_t, \tilde{\mathbf{x}}_\theta\left(\mathbf{z}_t, \mathbf{c}^{i-1}, \lambda_t\right)\right) + \sqrt{\tilde{\Sigma}_{s|t}}\epsilon$ // Eq. (3) and (4)
14       $\tilde{\mathbf{x}}^i \leftarrow \mathbf{z}_0$
15       $\tilde{\mathbf{X}} \leftarrow \tilde{\mathbf{X}} + [\tilde{\mathbf{x}}^i]$ // $\tilde{\mathbf{X}} = [\mathbf{x}^1, ..., \mathbf{x}^{i-1}] + [\tilde{\mathbf{x}}^i]$
16 $\mathbf{X}^{\mathcal{C}}, \mathbf{X}^{\mathcal{M}}, \mathbf{X}^{\mathcal{F}} \leftarrow partition(\tilde{\mathbf{X}})$
17 **return** $\mathbf{X}^{\mathcal{M}}, \mathbf{X}^{\mathcal{F}}$

3.2 Conditional Diffusion Model

Our proposed SADM follows the formulation of classifier-free diffusion model [9] defined in Sect. 2. We extend this diffusion model by using a sequence-aware conditioning signal explained in the previous section. Furthermore, we use an autoregressive sampling scheme that can effectively capture the long-distance temporal dependency during inference.

Training SADM. During training, the input to the diffusion model is a randomly selected target image from unobserved indices $\mathcal{M} \cup \mathcal{F}$ and a conditioning signal from previous indices, *i.e.*, $\mathbf{x}^i \in \mathbb{R}^{W \times H \times D}$ and $\mathbf{c} = \mathcal{A}(\{\mathbf{x}^1, ..., \mathbf{x}^{i-1}\})$, respectively. The attention module and the conditional diffusion model can be pretrained separately and finetuned together or trained end-to-end from scratch with the loss term defined in Eq. 5. The attention module can be pretrained by minimizing the ℓ_2 loss between the target image and the conditioning signal $\mathbf{c}$, and the diffusion model can be pretrained with a zero or random-valued tensor as a conditioning signal. However, we have found that training end-to-end from scratch performs better. Our training pipeline is defined in Algorithm 1.

Autoregressive Sampling. Our SADM samples the next-frame image $\mathbf{x}^i$ given the conditional signals of its previous images, *i.e.*, $\mathbf{c}^{i-1} = \mathcal{A}_\theta(\{\mathbf{x}^1, ..., \mathbf{x}^{i-1}\})$. However, real-world data often have missing data or only a single image per subject. Thus, we use an autoregressive sampling scheme that imputes missing images with synthesized images autoregressively. This autoregressive sampling scheme

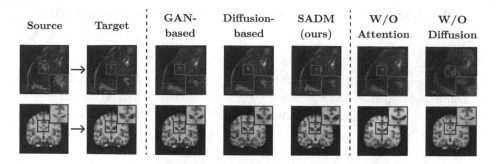

Fig. 4. Qualitative comparison between baselines, proposed SADM, and ablated models for single image setting. GAN-based model [4] uses a UNet-based GAN model to synthesize MRI frame, and diffusion-based model [12] uses a diffusion model with a deep registration model to register one frame to another. W/O Attention model is SADM with only the diffusion model, while W/O diffusion model is SADM with only the attention module. Red boxes indicate blood pool regions and blue boxes indicate ventricle regions. (Color figure online)

has been shown to improve the generative performance of diffusion models [7]. An overview of the autoregressive sampling scheme is shown in Algorithm 2.

4   Experiments

In this section, we show the effectiveness of our proposed SADM in medical image generation on one public 3D longitudinal cardiac MRI dataset and one simulated 3D longitudinal brain MRI dataset. We compare our work with GAN-based [4] and diffusion-based [12] baselines quantitatively and qualitatively. Finally, an ablation study of our model components with various settings for the input sequence is presented.

4.1   Dataset and Implementation

Cardiac Dataset. We use the multi-frame cardiac MRI curated by ACDC (Automated Cardiac Diagnosis Challenge) organizers [3]. A common task for cardiac image generation is to synthesize the final frame of a cardiac cycle (*i.e.*, end-systolic or ES), given a starting frame of the cycle (*i.e.*, end-diastolic or ED). The ACDC dataset consists of cardiac MRI from 100 training subjects and 50 testing subjects. We take intermediate frames from ED to ES and resize them to $\mathbf{X} \in \mathbb{R}^{12 \times 128 \times 128 \times 32}$, where each dimension is the length of the frame, the width, the height, and the depth, respectively. Then we Min-Max normalize the dataset subject-wise. Although MRI resizing results in uneven resolution, we opt for this approach, as it is the most reproducible preprocessing method. For training, we randomly select conditioning, missing, and future indices. During inference, we experiment with three settings: 1) Single image, where only the ED frame is given as input; 2) Missing data, where the input sequence has randomly

Table 1. Quantitative comparison between baselines and our propose SADM.

Method	Cardiac MRI			Brain MRI		
	SSIM ↑	PSNR ↑	NRMSE ↓	SSIM ↑	PSNR ↑	NRMSE ↓
GAN-based [4]	0.788	27.394	0.176	0.955	30.502	0.138
Diffusion-based [12]	0.842	28.863	0.154	0.961	31.229	0.121
Ours	**0.851**	**28.992**	**0.153**	**0.978**	**31.699**	**0.090**

missing frames; and 3) Full sequence, where the input sequence is fully loaded with conditioning images.

Brain Dataset. Simulating healthy subjects' brain changes over time is essential for understanding human aging [6]. The in-house synthesis of longitudinal brain MRI was carried out in two main steps, using 2,851 subject scans evenly distributed in age. We first divided these subject scans into five age groups (18–30, 31–45, 46–60, 61–74, and 75–97 years old) and generated five age-specific templates following [20]. Then we used a GAN-based registration model to register each subject scan to these five templates, respectively, to simulate the longitudinal images of the same person at different ages [21]. Templates and registered images were divided into ten cross-validation folds.

Implementation. We follow the classifier-free diffusion model [9] architecture and hyperparameters, and modify the model into a 3D model. For the transformer, we use the spatial and temporal transformer blocks introduced in ViViT [1] (specifically, Model 3 of ViViT). We trained the model for 3 million iterations, which took about 150 GPU hours using Nvidia V100 32GB GPUs. For inference, we use diffusion time steps of $T = 1,000$ and classifier-free guidance of $w = 0.1$.

4.2 Comparison with Baseline Methods

We choose two state-of-the-art baseline models for comparison: 1) GAN-based model [4], which uses a UNet-based GAN model to synthesize an ES frame given an ED frame; and 2) Diffusion-based model [12], which uses a diffusion model with a deep registration model to register an ED frame into an ES frame. We follow the same training and inference pipeline for cardiac and brain image generation, so we will explain the settings only for a cardiac dataset as follows. For training, SADM uses the intermediate frames between ED and ES, so we augment the baselines with pairs of intermediate frames and its ES frame for a fair comparison. Since these models can only perform single image synthesis (*i.e.*, ED to ES translation), we follow their inference pipeline using only the ED frame as input. We used the source code provided in each respective paper, only modifying the data loader for preprocessing and augmentation.

A qualitative comparison is presented in Fig. 4, and a quantitative comparison is presented in Table 1. For cardiac image generation, our proposed SADM

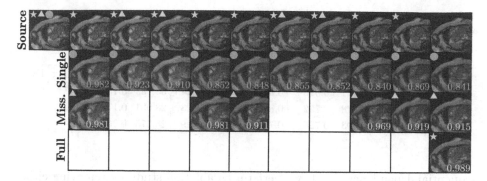

Fig. 5. An illustration of different settings for input sequence. The first row shows the ground-truth progression from ED frame to ES frame. The symbols in upper left corner of images in the first row is the conditioning image $\mathbf{X}^C$ (red star for full sequence, yellow triangle for missing data, and green circle for single image). The remaining rows show the synthesized images with single image, missing data, and full sequence settings, respectively. The numbers on bottom right corner of each image is the SSIM between the ground truth and the synthesized frame. (Color figure online)

shows a better depiction of the blood pool (red box) compared to the baseline methods. Also, other areas surrounding the blood pool, such as the myocardium and ventricles, are also synthesized with higher fidelity. For brain image generation, the ventricular regions (blue box) synthesized by SADM are more crisp compared to baselines, and the cortical surface is synthesized more accurately. Then, we perform a quantitative comparison by calculating the structural similarity index (SSIM), peak signal-to-noise ratio (PSNR), and normalized root-mean-square deviation (NRMSE) between the target and the synthesized ES frame. Our model outperforms the GAN-based method [4] by $3 - -13\%$ in each metric while slightly outperforming the diffusion-based model [12]. It is worth noting that the diffusion-based baseline uses a source image and a reference image for registration, whereas we only use the source image as input. Although our proposed SADM is capable of working with a single image, it is designed to perform even better with a sequence of images, as shown in the next section.

4.3 Ablation Study

In this section, we perform an ablation study on the components of our model with various settings for the input sequences. First, we experiment with different settings for the input sequence defined in the first paragraph of Sect. 4.1, *i.e.*, single image, missing data, and full sequence settings. As presented in Fig. 5, synthesis using the full sequence and missing data settings show a higher SSIM compared to a single input setting. Also, as observed by the high peak in SSIM for frames in the vicinity of conditioning frames, our SADM is learning which frames of the input sequence are important in generating future frames, *i.e.*, the

398   J. S. Yoon et al.

Table 2. An ablation study of SADM components with single image, missing data, and full sequence settings using the ACDC cardiac dataset.

Method		SSIM ↑			PSNR ↑			NRMSE ↓		
Diffusion	Attention	Single	Missing	Full	Single	Missing	Full	Single	Missing	Full
✓	✓	0.851	0.916	0.977	28.992	30.314	33.733	0.153	0.143	0.087
✓		0.778	0.792	0.802	26.955	26.223	28.031	0.179	0.175	0.167
	✓	0.703	0.726	0.727	25.533	25.629	25.275	0.267	0.234	0.217

sequential dependency. Next, we perform an ablation study by removing either the attention module or the diffusion model.

The diffusion-only model can be trained with the raw pixel values of the sequential image as conditioning signals, and the attention-only model can be trained by minimizing the ℓ_2 loss between the target image and the output of the transformer. As shown in Table 2, evidently, the attention-only module has the worst performance as the transformers are not designed for image generation (typically due to flattening operations [22]). The diffusion-only model performs on par with GAN-based baseline [4], but it is unable to learn the sequential dependency, as observed by the minimal performance increase in full sequence setting compared to single input settings.

5   Conclusion

To this end, we propose a sequence-aware diffusion model for the generation of longitudinal medical images. Specifically, our model consists of a transformer-based attention module that can learn the sequential or temporal dependence of longitudinal data input and a diffusion model that can synthesize high-fidelity medical images. We tested our proposed SADM on longitudinal cardiac and brain MRI generation and presented state-of-the-art performance quantitatively and qualitatively. Our approach to learning the temporal dependence of sequential data and using it as a prior in diffusion models is an exciting new research topic in the field of medical image generation. However, the limitations of our model's computational efficiency for large medical datasets suggest that further work is needed to improve sampling efficiency. We hope our research inspires researchers to pursue this newly found topic and find solutions to these challenges.

Acknowledgments. This work is supported in part by the Natural Sciences and Engineering Research Council of Canada (NSERC), and NVIDIA Hardware Award, and Institute of Information & Communications Technology Planning & Evaluation (IITP) grant funded by the Korea government (MSIT) No. 2022-0-00959 ((Part 2) Few-Shot Learning of Causal Inference in Vision and Language for Decision Making), and the MOTIE (Ministry of Trade, Industry, and Energy) in Korea, under Human Resource Development Program for Industrial Innovation (Global) (P0017311) supervised by the Korea Institute for Advancement of Technology (KIAT).

References

1. Arnab, A., et al.: ViViT: a video vision transformer (2021)
2. Balakrishnan, G., et al.: VoxelMorph: a learning framework for deformable medical image registration. IEEE Trans. Med. Imaging **38**(8), 1788–1800 (2019)
3. Bernard, O., et al.: Deep learning techniques for automatic MRI cardiac multi-structures segmentation and diagnosis: is the problem solved? IEEE Trans. Med. Imaging **37**(11), 2514–2525 (2018)
4. Campello, V.M., et al.: Cardiac aging synthesis from cross-sectional data with conditional generative adversarial networks. Front. Cardiovasc. Med. 9 (2022)
5. Dhariwal, P., Nichol, A.: Diffusion models beat GANs on image synthesis. In: Advances in Neural Information Processing Systems, vol. 34, pp. 8780–8794 (2021)
6. Feng, X., et al.: Estimating brain age based on a uniform healthy population with deep learning and structural magnetic resonance imaging. Neurobiol. Aging **91**, 15–25 (2020)
7. Harvey, W., et al.: Flexible diffusion modeling of long videos. In: Advances in Neural Information Processing Systems (2022)
8. Ho, J., Jain, A., Abbeel, P.: Denoising diffusion probabilistic models. In: Advances in Neural Information Processing Systems, vol. 33, pp. 6840–6851 (2020)
9. Ho, J., Salimans, T.: Classifier-free diffusion guidance (2022)
10. Ho, J., et al.: Imagen video: high definition video generation with diffusion models (2022)
11. Kaddour, J., et al.: Causal machine learning: a survey and open problems (2022)
12. Kim, B., Han, I., Ye, J.C.: DiffuseMorph: unsupervised deformable image registration using diffusion model. In: Avidan, S., Brostow, G., Cissé, M., Farinella, G.M., Hassner, T. (eds.) Computer Vision (ECCV 2022). LNCS, vol. 13691, pp. 347–364. Springer, Cham (2022). https://doi.org/10.1007/978-3-031-19821-2_20
13. Kim, B., Ye, J.C.: Diffusion deformable model for 4D temporal medical image generation. In: Wang, L., Dou, Q., Fletcher, P.T., Speidel, S., Li, S. (eds.) Medical Image Computing and Computer Assisted Intervention (MICCAI 2022). LNCS, vol. 13431, pp. 539–548. Springer, Cham (2022). https://doi.org/10.1007/978-3-031-16431-6_51
14. Kingma, D., et al.: Variational diffusion models. In: Advances in Neural Information Processing Systems, vol. 34, pp. 21696–21707 (2021)
15. Nichol, A.Q., Dhariwal, P.: Improved denoising diffusion probabilistic models. In: ICML, vol. 139, pp. 8162–8171 (2021)
16. Oh, K., Yoon, J.S., Suk, H.I.: Learn-explain-reinforce: counterfactual reasoning and its guidance to reinforce an Alzheimer's disease diagnosis model. IEEE Trans. Pattern Anal. Mach. Intell. 1–15 (2022)
17. Pinaya, W.H.L., et al.: Brain imaging generation with latent diffusion models. In: Deep Generative Models (2022)
18. Quadrana, M., Cremonesi, P., Jannach, D.: Sequence-aware recommender systems. ACM Comput. Surv. **51**(4), 1–36 (2019)
19. Yi, X., Walia, E., Babyn, P.: Generative adversarial network in medical imaging: a review. Med. Image Anal. **58**, 101552 (2019)

20. Zhang, C., et al.: Constructing age-specific MRI brain templates based on a uniform healthy population across life span with transformer. In: 2023 ISMRM and SMRT Annual Meeting and Exhibition (2023)
21. Zhang, C., et al.: Cycle inverse consistent deformable medical image registration with transformer. In: ISMRM and SMRT Annual Meeting and Exhibition (2023)
22. Zhang, X., et al.: RSTNet: captioning with adaptive attention on visual and non-visual words. In: CVPR. IEEE (2021)

Image Enhancement

An Unsupervised Framework for Joint MRI Super Resolution and Gibbs Artifact Removal

Yikang Liu, Eric Z. Chen, Xiao Chen, Terrence Chen, and Shanhui Sun$^{(\boxtimes)}$

United Imaging Intelligence, Cambridge, MA, USA
{yikang.liu,zhang.chen,xiao.chen01,terrence.chen,shanhui.sun}@uii-ai.com

Abstract. The k-space data generated from magnetic resonance imaging (MRI) is only a finite sampling of underlying signals. Therefore, MRI images often suffer from low spatial resolution and Gibbs ringing artifacts. Previous studies tackled these two problems separately, where super resolution methods tend to enhance Gibbs artifacts, whereas Gibbs ringing removal methods tend to blur the images. It is also a challenge that high resolution ground truth is hard to obtain in clinical MRI. In this paper, we propose an unsupervised learning framework for both MRI super resolution and Gibbs artifacts removal without using high resolution ground truth. Furthermore, we propose regularization methods to improve the model's generalizability across out-of-distribution MRI images. We evaluated our proposed methods with other state-of-the-art methods on eight MRI datasets with various contrasts and anatomical structures. Our method not only achieves the best SR performance but also significantly reduces the Gibbs artifacts. Our method also demonstrates good generalizability across different datasets, which is beneficial to clinical applications where training data are usually scarce and biased.

Keywords: Super resolution · Gibbs artifact · Unsupervised learning

1 Introduction

The super resolution (SR) for magnetic resonance imaging (MRI) images is different from natural images due to the distinct image generation process. Due to various constraints such as hardware limitations and acquisition time, the k-space data generated from MRI is only a finite sampling of spatial frequencies, which leads to the loss of high spatial frequencies. The generated low resolution (LR) images make subtle structures and boundaries hardly distinguishable. Another common MRI image artifact called Gibbs ringing often arises in practice due to abrupt cutoff in spatial frequency (Fig. 1a), which may appear as anatomical structures and cause misinterpretation. Therefore, SR for MRI images is more complex than natural images since solving these two problems individually often leads to a dilemma. SR tends to enhance Gibbs ringing artifacts (Fig. 1b) while Gibbs ringing artifact removal (deGibbs) algorithms might

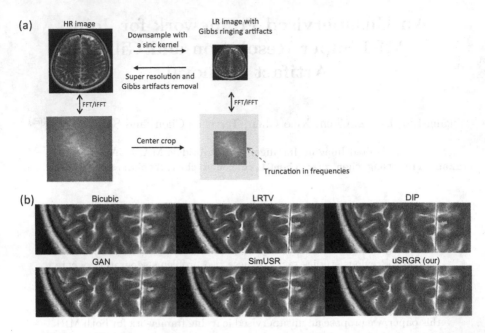

Fig. 1. (a) An illustration of LR MRI data generation process. The acquired LR MRI image can be viewed as a result of a center crop of the underlying HR k-space, which leads to the common Gibbs ringing artifacts. Here HR image is used for demonstration purposes and is rarely available in practice. (b) Examples of SR (2x) and Gibbs artifacts (red arrows). No ground truth of HR image is available. (Color figure online)

generate blurry results (Fig. 4d). Furthermore, the artifact removal algorithms are often performed on the original resolution [13,22,32]. Since both problems are caused by finite k-space signals, we propose to solve them simultaneously with a unified deep learning framework. However, there are several challenges. First, supervised learning becomes a nontrivial task due to the lack of high resolution (HR) ground-truth MRI images. Second, there is usually a distribution mismatch between training and testing data, since it is impractical to collect training images of all contrasts, anatomies, diseases, etc. Third, it is crucial for clinical diagnosis that the model should not introduce additional artifacts into SR images.

To address these problems, we propose an **U**nsupervised learning framework for joint MRI **S**uper **R**esolution and **G**ibbs artifact **R**emoval (uSRGR), which utilizes specifically designed loss constraints for image fidelity and model generalizability. The model learns the inverse process of downsampling MRI images, which is to crop out the peripheral (high-frequency) part of the frequency space (obtained by applying Fourier transform to MRI images) with a box window and leads to reduced image resolution and Gibbs artifacts (Fig. 1a).

To avoid using HR images, LR images I_{LR} are downsampled into images of lower resolution $I_{LR'}$, and the model is trained to fit $I_{LR'}$ to I_{LR}. At the same time, the model predicts HR images $\hat{I}_{\mathrm{HR}}$ from I_{LR} images. To ensure the fidelity of prediction, the difference between downsampled $\hat{I}_{\mathrm{HR}}$ and I_{LR} is minimized. Moreover, since cropping peripheral parts of frequency domain (hereinafter referred to as 'f-cropping') has a more global effect than the downsampling process in natural images (e.g. convolution with a kernel of a finite size), we utilize large 1D convolutional kernels to model such an effect. Furthermore, to improve the model generalizability, we regularize the model to match a Sinc convolution process that downsamples I_{HR} to I_{LR}, since Sinc convolution in the spatial domain is equivalent to f-cropping. In summary, our contributions are as follows:

- We propose a novel unsupervised deep learning framework for joint SR and Gibbs artifact removal without using HR MRI images.
- We demonstrate an innovative training paradigm that regularizes the model to a Sinc deconvolution process to improve learning generalization.

2   Related Works

Many supervised deep learning methods have been proposed for image SR [6,9,18,19,23,24,28]. Some unsupervised deep learning methods have also been proposed [14,21,24,30], where paired LR and HR images are not required for training but HR images are still needed to learn the HR image manifold. Without using HR images, several unsupervised or self-supervised SR methods have been proposed for natural images. Deep image prior (DIP) [16,20] uses a network structure as a prior to constrain HR prediction. Zero-shot SR (ZSSR) [27] is another online-learning method that is trained in a self-supervised manner using a pair of the downsampled and original LR images. Following ZSSR, SimUSR [1] is trained offline on many image pairs of downsampled and original LR images to reduce inference time. GAN based methods have also been proposed for unsupervised SR [5,14,21,24,30]. However, GAN tends to generate new artifacts [33]. In the absence of a supervised loss from HR images, there is a risk that the GAN model may generate artifacts that result in clinical misdiagnosis.

As for Gibbs removal, traditional methods aim to minimize oscillation in images with image filtering [8,11], Gegenbauer polynomial reconstruction [2], frequency space extrapolation regularized by total variation [4], and sub-voxel shift resampling [13]. Since image details can also be smoothed as oscillation, these methods blur images to varying degrees [13]. Recently proposed deep learning models [22,32] are trained on pairs of artifactual and clean images, where artifactual images were synthesized by f-cropping HR images [22,32]. Similar to our method, [32] also tried to model the inverse of the f-cropping process. However, it is different from our work in multiple ways: first, it is a supervised learning method; second, distribution mismatch between training and test data is not considered; third, [32] crops out 50% of the original frequency space as the downsampling operation whereas we crop out 75%.

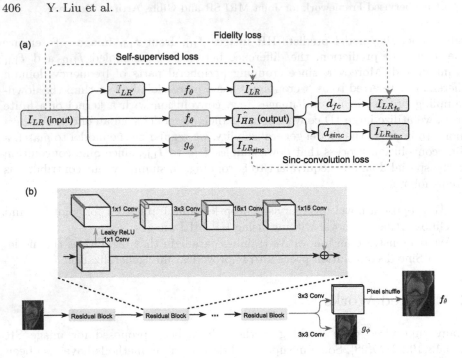

Fig. 2. (a) The proposed unsupervised framework uSRGR for MRI SR and Gibbs removal. (b) The network architecture used in uSRGR. The networks f_θ and g_ϕ share the same architecture except for the last pixel shuffle layer due to different output sizes.

3   Methods

3.1   Problem Formulation

A LR image $\boldsymbol{I}_{LR} \in \mathbb{R}^{n \times m}$ can be generated by downsampling a HR image $\boldsymbol{I}_{HR} \in \mathbb{R}^{N \times M}$ as $\boldsymbol{I}_{LR} = d(\boldsymbol{I}_{HR}) + \epsilon$, where d is a downsampling operation, and ϵ is the noise. For MRI, d is f-cropping d_{fc}, and $\boldsymbol{I}_{LR}$ often bears Gibbs ringing artifacts. Therefore, the joint task of SR and deGibbs can be formulated as an inverse problem where we need to approximate the inverse operation of the downsampling operation d_{fc} with a deep neural network f_θ: $\boldsymbol{I}_{HR} = f_\theta(\boldsymbol{I}_{LR})$.

3.2   An Unsupervised Learning Framework

Without HR ground truth, we propose to regularize SR neural network f_θ with three constraints given only LR images (Fig. 2a): a fidelity constraint, a self-supervision constraint, and a Sinc-convolution constraint.

Self-supervision Constraint: Based on the assumption that the inverse operation of downsampling d_{fc} is locally consistent across multiple scales [1], we downsampled the LR image I_{LR} with d_{fc} to a lower resolution $I_{LR'}$ and define the self-supervision loss $\mathcal{L}_{ss}$ on pairs of $I_{LR'}$ and I_{LR}:

$$\mathcal{L}_{ss}(\theta) = \sum_i \mathcal{L}_d(f_\theta(I_{LR'}^{(i)}), I_{LR}^{(i)}) \tag{1}$$

Fidelity Constraint: Signal fidelity in medical images is crucial in clinical settings, especially for parametric mappings where the intensities bear physical and physiological meanings. We downsample the predicted HR image $\hat{I}_{HR}$ with d_{fc} and minimize its difference with the original LR image I_{LR} as

$$\mathcal{L}_f(\theta) = \sum_i \mathcal{L}_d(d_{fc}(f_\theta(I_{LR}^{(i)})), I_{LR}^{(i)}). \tag{2}$$

Sinc-Convolution Constraint: We leverage the knowledge about the generation of LR MRI images and introduce another regularization on the f-cropping process d_{fc}. It enforces the model f_θ to learn the inverse of the underlying downsampling process d_{fc} and prevents model overfitting on spurious image features. Thus the model can generalize better on out-of-distribution data.

The central cropping in frequency domain is mathematically equivalent to convolution with a Sinc kernel of an infinite size in the spatial domain. Therefore, if we downsample $f_\theta(I_{LR})$ by convolution with a Sinc kernel of a *finite* size, it should be close to I_{LR}, which provides another supervision on the mapping from I_{LR} to I_{HR}. To compensate the difference between d_{fc} and d_{sinc} caused by the finite Sinc kernel, we first pretrain another network g_ϕ that maps images downsampled with f-cropping $d_{fc}(I_{LR})$ to the ones downsampled with Sinc convolution $d_{sinc}(I_{LR})$) on a dataset of LR images I_{LR} with the loss function:

$$\mathcal{L}(\phi) = \sum_i \mathcal{L}_d(g_\phi(d_{fc}(I_{LR})), d_{sinc}(I_{LR})) \tag{3}$$

Then the Sinc-convolution constraint on the model f_θ is

$$\mathcal{L}_{sinc}(\theta) = \sum \max(\mathcal{L}_d(d_{sinc}(f_\theta(I)), g_\phi(I)), a), \tag{4}$$

where $a = 0.001$ is a small number to compensate the prediction error in g_ϕ and I is sampled from I_{LR} and $I_{LR'}$ images.

Taken together, the total loss function for the SR model f_θ is

$$\mathcal{L}(\theta) = \mathcal{L}_{ss} + \beta\mathcal{L}_f + \gamma\mathcal{L}_{sinc}, \tag{5}$$

where $\beta = 1$ and $\gamma = 0.5$ are weights. We use a combination of L1 loss and multi-scale SSIM loss [34] for $\mathcal{L}_d$. The model g_ϕ is also fined tuned with the loss function Eq. 3 during the training of the SR model f_θ, where both I_{LR} and $\hat{I}_{HR}$ (i.e. $f_\theta(I_{LR})$) are used. In this way, the models f_θ and g_ϕ need to come to agreements on both HR and LR scales, which provides extra regularization for f_θ in addition to Eq. 2.

Network: For efficient computation, we adopt a simple residual network architecture with wide activation [29] for both f_θ and g_ϕ (Fig. 2b). To model the global effects of Gibbs artifacts, we utilized two 1D large-kernel convolutions to approximate a 2D large-kernel convolution. We employ a pixel shuffle layer in f_θ that takes LR images as inputs, which is more efficient than taking interpolated LR images as inputs [26].

4   Experiment Setup

Datasets. We included four datasets (T1, T1pre, T1post, T2) from the public (under a non-exclusive, royalty-free license) fastMRI brain images [31] and two datasets (PDFS and PD) from fastMRI knee images [31]. The fastMRI dataset[1] [31] contains MRI k-space data of brain and knee images. We reconstructed images from k-space data and cropped them to 320×320. The brain images have four contrasts: T1 (denoted as *brain-T1*), T1 before and after injection of contrast agent (denoted as *brain-T1pre* and *brain-T1post*, respectively), T2 (denoted as *brain-T2*), and T2 FLAIR. T2-FLAIR images were not used due to high motion artifacts. The knee images have two contrasts: proton-density weighting with and without fat suppression, denoted as *knee-PDFS* and *knee-PD*, respectively. The imaging parameters can be found in [31]. brain-T2, brain-T1post, brain-T1pre, brain-T1, knee-PD, knee-PDFS contains 21500, 123600, 32700, 9200, 5000, and 9900 images for testing. 267800 brain-T2 and 48400 knee-PD images were used for training separately, with 60000 brain-T2 and 5000 knee-PD images used respectively for validation.

We also included two private cardiac cine MRI datasets, where retroCine dataset contains 2,955 images from 51 subjects and rtCine dataset contains 57 images from two subjects. retroCine data of 51 subjects (total 2955 images) were acquired on a 3T scanner (uMR 790 United Imaging Healthcare, Shanghai, China) and phased-arrayed coils using a bSSFP sequence. rtCine data of 57 images were acquired from two patients using the same equipment and a bSSFP sequence using variable Latin Hypercube undersampling [17]. Imaging parameters were: matrix size = 192×180, TR = 2.8 ms, TE = 1.3 ms, spatial resolution = $1.82 \times 1.82 \, \text{mm}^2$, and temporal resolution = 34 ms and 42 ms for retroCine and rtCine. Acquisition of both retroCine and rtCine was approved by a local institutional review board.

Compared Methods. We compared uSRGR with four state-of-the-art SR methods. LRTV [25] is a MRI SR method using low-rank and total variation regularization[2]. Three unsupervised deep learning methods are included: DIP [16], SimUSR [1], and a GAN-based method extended from [3,7,10]. In the GAN-based method, we used a patch discriminator to distinguish patches derived from LR inputs and HR predictions and the objective function is $f = \arg\min_f \{ (\max_D \mathbb{E}_{x \sim patches(I_{LR})} |D(x) - 1| + |D(f(x))|) + \lambda \mathcal{L}_d(ds(f(I_{LR})), I_{LR}) \}$, where f is a network with the same architecture as in Fig. 2.

---

[1] https://fastmri.org/dataset/.
[2] https://bitbucket.org/fengshi421/superresolutiontoolkit.

Table 1. Residual blocks with (left) and without (right) 1D convolutions.

Layer Name	Parameter Dimension
conv2d	$6N \times N \times 1 \times 1$
LeakyReLU	
conv2d	$\text{int}(4.8N) \times 6N \times 1 \times 1$
conv2d	$\text{int}(4.8N) \times \text{int}(4.8N) \times 3 \times 3$
conv1d	$\text{int}(4.8N) \times \text{int}(4.8N) \times 15$
conv1d	$N \times \text{int}(4.8N) \times 15$

Layer Name	Parameter Dimension
conv2d	$6N \times N \times 1 \times 1$
LeakyReLU	
conv2d	$\text{int}(4.8N) \times 6N \times 1 \times 1$
conv2d	$N \times \text{int}(4.8N) \times 3 \times 3$

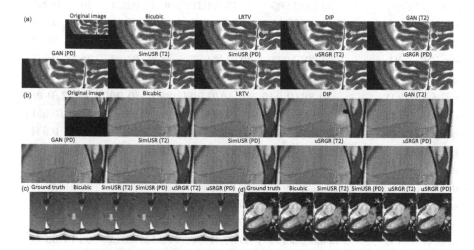

Fig. 3. Comparison of different methods tested on (a), (b) the original datasets (brain-T2 and knee-PD) and (c), (d) synthetic LR datasets (brain-T1post and retroCine). T2 and PD in figure titles indicate training datasets. Red arrows indicate Gibbs artifacts. (Color figure online)

Implementation Details. The uSRGR model included 12 residual blocks ($N = 80$) with 1D convolutional layers (Table 1 and Fig. 2) and two 3×3 2D convolutional layers before and after the residual blocks. For uSRGR-1Dconv and other baselines, we removed the 1D convolutions in the residual blocks (Table 1) and increased the number of channels N to 135, so that they had comparable number of parameters (uSRGR: 4.2M parameters; others: 4.4M parameters).

All models were implemented with PyTorch 1.5 and trained on NVIDIA V100. We used Adam optimizer [15] with a learning rate of 10^{-4}. Batch size was 4 and images were cropped into 160×160 patches. All models were trained for 200 epochs. Training losses converged well and no overfitting was observed.

5   Results and Conclusion

Quantitative Comparisons: Since HR MRI images are not available, we used the original resolution images as the evaluation reference for quantitative comparisons. We synthesized LR images ($2\times$ downsampling) from original images

Table 2. Evaluation on synthetic LR images with original images as ground truth. Rows indicate the methods (training datasets) and columns indicate test datasets.

PSNR/SSIM	brain-T2	brain-T1	brain-T1post	brain-T1pre
Bicubic	32.08/0.912	35.01/0.940	34.94/0.942	34.85/0.933
LRTV	29.64/0.898	32.88/0.931	32.65/0.934	32.67/0.925
DIP	34.52/0.919	37.93/0.943	37.82/0.946	37.72/0.929
GAN (T2)	33.56/0.905	36.75/0.931	36.39/0.934	36.16/0.925
GAN (PD)	33.24/0.905	36.73/0.932	36.25/0.936	36.15/0.928
simUSR (T2)	36.39/0.938	39.67/0.952	39.12/0.956	38.60/0.948
simUSR (PD)	35.80/0.936	39.91/0.957	38.92/0.960	38.38/0.951
uSRGR (T2)	**36.87/0.949**	**40.48/0.963**	**39.65/0.962**	**38.98/0.959**
uSRGR (PD)	**36.86/0.945**	**40.48/0.964**	**39.63/0.960**	**38.97/0.960**
uSRGR-fid (T2)	36.43/0.939	40.05/0.951	39.17/0.955	38.42/0.951
uSRGR-fid (PD)	36.40/0.935	39.99/0.956	39.25/0.959	38.60/0.954
uSRGR-sinc (T2)	36.46/0.941	40.25/0.953	39.43/0.959	38.54/0.953
uSRGR-sinc (PD)	35.68/0.937	40.06/0.957	38.91/0.960	38.39/0.952
uSRGR-1Dconv (T2)	36.73/0.942	40.27/0.955	39.538/0.960	38.813/0.955
uSRGR-1Dconv (PD)	36.70/0.923	40.31/0.959	39.63/0.960	38.97/0.958
deG+SR (T2)	36.12/0.910	39.02/0.941	38.64/0.952	38.20/0.944
deG+SR (PD)	36.08/0.913	38.87/0.936	38.55/0.953	38.13/0.941
PSNR/SSIM	knee-PD	knee-PDFS	retroCine	rtCine
Bicubic	34.07/0.888	34.32/0.823	30.97/0.896	31.49/0.914
LRTV	32.28/0.874	33.09/0.807	27.97/0.877	28.69/0.895
DIP	34.81/0.897	34.14/0.834	33.91/0.913	34.27/0.932
GAN (T2)	33.90/0.884	33.84/0.815	32.27/0.896	32.87/0.913
GAN (PD)	33.78/0.883	33.88/0.815	31.97/0.896	32.58/0.913
simUSR (T2)	36.98/0.916	36.13/0.854	35.23/0.944	36.06/0.954
simUSR (PD)	37.29/0.922	36.15/0.856	35.47/0.948	36.42/0.960
uSRGR (T2)	**37.59/0.930**	**36.50/0.863**	**35.81/0.953**	**36.74/0.965**
uSRGR (PD)	**37.61/0.932**	**36.52/0.863**	**35.81/0.954**	**36.75/0.966**
uSRGR-fid (T2)	37.09/0.919	36.14/0.852	35.32/0.945	36.16/0.956
uSRGR-fid (PD)	37.00/0.918	36.05/0.854	35.30/0.946	36.23/0.957
uSRGR-sinc (T2)	37.24/0.922	36.16/0.852	35.57/0.949	36.37/0.959
uSRGR-sinc (PD)	37.28/0.923	36.26/0.856	35.54/0.947	36.41/0.960
uSRGR-1Dconv (T2)	37.35/0.923	36.33/0.856	35.62/0.949	36.55/0.961
uSRGR-1Dconv (PD)	37.41/0.923	36.33/0.858	35.65/0.950	36.56/0.961
deG+SR (T2)	36.54/0.905	35.93/0.850	34.97/0.932	35.78/0.939
deG+SR (PD)	36.69/0.910	35.88/0.851	35.12/0.935	35.91/0.938

using f-cropping to mimic the MRI downsampling process. Furthermore, to test the generalizability of different methods, we trained models on brain-T2 and knee-PD datasets respectively and tested on all eight datasets, where brain-T2 and knee-PD were split into training, validation, and testing sets. The test data with the same imaging sequence and organ as the training data is referred to as the *within-distribution*(WD) data and otherwise as the *out-of-distribution*(OD) data. Table 2 shows quantitative results for different SR methods. Our method performed significantly better than other methods on both WD and OD data ($p < 0.05$). No significant difference ($p > 0.2$) between uSRGR models trained on brain-T2 and knee-PD was found on any test dataset, indicating its good generalizability.

Qualitative Comparisons: For qualitative evaluation, besides the synthetic LR images, we also trained all models on the original images to mimic the real MRI SR setting, where no HR ground truth is available. Similarly, all models were trained on brain-T2 and knee-PD datasets respectively and tested on all eight datasets to evaluate their generalizability. Figure 3 shows that LRTV produces over-smoothed images while DIP generates hole-like artifacts and GAN also predicts dot and stripe artifacts. Moreover, these methods all enhance Gibbs artifacts. These observations are consistent with low PSNR/SSIM in Table 2. simUSR removes Gibbs artifacts on WD data (simUSR(T2) on brain-T2) but fails to generalize on OD data (simUSR(PD) on brain-T2), also indicated by the significant difference ($p < 0.05$) in PSNR/SSIM between simUSR(T2) and simUSR(PD) tested on WD and OD data in Table 2. simUSR(PD) also failed to remove Gibbs artifacts in brain-T1post and retroCine images. In contrast, uSRGR generated sharper images with significantly ($p < 0.05$) reduced Gibbs artifacts consistently across WD and OD data.

Ablation Studies: We removed individual components in uSRGR: the fidelity constraint, the Sinc-convolution constraint, and the 1D convolutional layers, denoted as uSRGR-fid, uSRGR-sinc and uSRGR-1Dconv, respectively. As shown in Table 2, removing those components individually leads to significantly lower PSNR/SSIM across all datasets ($p < 0.05$). Removing the Sinc-convolution constraint resulted in more Gibbs artifacts (Fig. 4a), especially when trained on knee-PD and tested on brain-T2. This is consistent with our motivation that regularization based on the f-cropping process helps the model to learn the inverse process and thus improves the model generalizability. This also boosts the performance on WD data, since the models in Eq. 1 and Eq. 2 operate on different image scales. Removing the fid loss significantly increases the fidelity error between the downsampled HR and the original image as in Fig. 4c (PSNR on all datasets: 49.2 vs. 44.7 with and without fid loss, $p < 0.001$). Removing the 1D convolutions resulted in more blurry SR images (Fig. 4b).

To demonstrate the benefit of joint SR and deGibbs, we trained separate models for deGibbs and SR (denoted as deG+SR). The same network g_ϕ was trained for deGibbs on image pairs downsampled with d_{cubic} and d_{fc}. Then the separately trained network f_θ was applied for SR. Table 2 shows that deG+SR

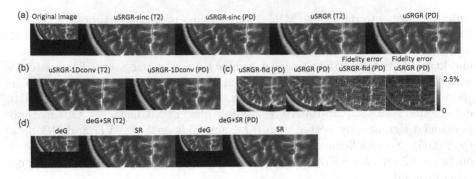

Fig. 4. Ablation studies tested on brain-T2.

has lower PSNR/SSIM than uSRGR on all datasets ($p < 0.05$). Figure 4d shows that the separate deGibbs model blurred LR images and resulted in blurry SR images, indicates the advantage of combining deGibbs and SR in one framework.

6 Limitations and Future Work

This paper's aim is to develop an unsupervised deep learning framework for joint MRI SR and Gibbs artifact removal with good generalizability. Therefore, we did not put much effort into testing various loss functions that measure difference between two images (e.g. perceptual loss [12]), which may improve quality of predicted SR images. In addition, since Gibbs artifact has heavy effects on certain applications such as fiber tracking in diffusion tensor imaging, it would be beneficial to test the proposed method on such applications.

7 Summary

We proposed an unsupervised deep learning framework for MRI SR and Gibbs artifacts removal without using HR images as ground truth. The proposed method shows superior performance and generalizability than other unsupervised SR methods across various MRI datasets.

References

1. Ahn, N., Yoo, J., Sohn, K.A.: SimUSR: a simple but strong baseline for unsupervised image super-resolution. In: Proceedings of the IEEE/CVF Conference on Computer Vision and Pattern Recognition Workshops, pp. 474–475 (2020)
2. Archibald, R., Gelb, A.: A method to reduce the Gibbs ringing artifact in MRI scans while keeping tissue boundary integrity. IEEE Trans. Med. Imaging **21**(4), 305–319 (2002)
3. Bell-Kligler, S., Shocher, A., Irani, M.: Blind super-resolution kernel estimation using an internal-GAN. Adv. Neural Inf. Process. Syst. **32**, 284–293 (2019)

4. Block, K.T., Uecker, M., Frahm, J.: Suppression of MRI truncation artifacts using total variation constrained data extrapolation. Int. J. Biomed. Imaging **2008**, 1–8 (2008). https://doi.org/10.1155/2008/184123, http://www.hindawi.com/journals/ijbi/2008/184123/

5. Chen, S., et al.: Unsupervised image super-resolution with an indirect supervised path. In: Proceedings of the IEEE/CVF Conference on Computer Vision and Pattern Recognition (CVPR) Workshops (2020)

6. Chen, Y., Xie, Y., Zhou, Z., Shi, F., Christodoulou, A.G., Li, D.: Brain MRI super resolution using 3D deep densely connected neural networks. In: 2018 IEEE 15th International Symposium on Biomedical Imaging (ISBI 2018), no. Isbi, pp. 739–742. IEEE (2018). https://doi.org/10.1109/ISBI.2018.8363679, https://ieeexplore.ieee.org/document/8363679/

7. Goodfellow, I.J., et al.: Generative adversarial nets. In: Proceedings of the 27th International Conference on Neural Information Processing Systems, NIPS 2014, vol. 2, pp. 2672–2680. MIT Press, Cambridge (2014)

8. Gottlieb, D., Shu, C.W.: On the Gibbs phenomenon and its resolution. SIAM Rev. **39**(4), 644–668 (1997). https://doi.org/10.1137/S0036144596301390

9. Huang, X., Zhang, Q., Wang, G., Guo, X., Li, Z.: Medical image super-resolution based on the generative adversarial network. In: Jia, Y., Du, J., Zhang, W. (eds.) CISC 2019. LNEE, vol. 593, pp. 243–253. Springer, Singapore (2020). https://doi.org/10.1007/978-981-32-9686-2_29

10. Isola, P., Zhu, J.Y., Zhou, T., Efros, A.A.: Image-to-image translation with conditional adversarial networks. In: Proceedings of the IEEE Conference on Computer Vision and Pattern Recognition, pp. 1125–1134 (2017)

11. Jerri, A.J.: Lanczos-like σ-factors for reducing the Gibbs phenomenon in general orthogonal expansions and other representations. J. Comput. Anal. Appl. **2**, 111–127 (2000)

12. Johnson, J., Alahi, A., Fei-Fei, L.: Perceptual losses for real-time style transfer and super-resolution. In: Leibe, B., Matas, J., Sebe, N., Welling, M. (eds.) ECCV 2016. LNCS, vol. 9906, pp. 694–711. Springer, Cham (2016). https://doi.org/10.1007/978-3-319-46475-6_43

13. Kellner, E., Dhital, B., Kiselev, V.G., Reisert, M.: Gibbs-ringing artifact removal based on local subvoxel-shifts. Magn. Reson. Med. **76**(5), 1574–1581 (2016). https://doi.org/10.1002/mrm.26054, https://onlinelibrary.wiley.com/doi/10.1002/mrm.26054

14. Kim, H., Kim, J., Won, S., Lee, C.: Unsupervised deep learning for super-resolution reconstruction of turbulence. J. Fluid Mech. **910**, A29 (2021)

15. Kingma, D.P., Ba, J.: Adam: a method for stochastic optimization. In: Bengio, Y., LeCun, Y. (eds.) 3rd International Conference on Learning Representations, ICLR 2015, San Diego, CA, USA, 7–9 May 2015, Conference Track Proceedings (2015). http://arxiv.org/abs/1412.6980

16. Lempitsky, V., Vedaldi, A., Ulyanov, D.: Deep image prior. In: Proceedings of the IEEE Computer Society Conference on Computer Vision and Pattern Recognition, pp. 9446–9454 (2018). https://doi.org/10.1109/CVPR.2018.00984

17. Lyu, J., et al.: Toward single breath-hold whole-heart coverage compressed sensing MRI using VAriable spatial-temporal LAtin hypercube and echo-sharing (VALAS). In: ISMRM (2019)

18. Lyu, Q., Shan, H., Wang, G.: MRI super-resolution with ensemble learning and complementary priors. IEEE Trans. Comput. Imaging **6**, 615–624 (2020). https://doi.org/10.1109/tci.2020.2964201

19. Mahapatra, D., Bozorgtabar, B., Garnavi, R.: Image super-resolution using progressive generative adversarial networks for medical image analysis. Comput. Med. Imaging Graph. **71**, 30–39 (2019). https://doi.org/10.1016/j.compmedimag.2018.10.005

20. Mataev, G., Milanfar, P., Elad, M.: DeepRED: deep image prior powered by RED. In: Proceedings of the IEEE/CVF International Conference on Computer Vision Workshops (2019)

21. Menon, S., Damian, A., Hu, S., Ravi, N., Rudin, C.: PULSE: self-supervised photo upsampling via latent space exploration of generative models. In: Proceedings of the IEEE/CVF Conference on Computer Vision and Pattern Recognition, pp. 2437–2445 (2020)

22. Muckley, M.J., et al.: Training a neural network for gibbs and noise removal in diffusion MRI, pp. 1–18 (2019). https://doi.org/10.1002/mrm.28395, http://arxiv.org/abs/1905.04176

23. Pham, C.H., et al.: Multiscale brain MRI super-resolution using deep 3D convolutional networks. Comput. Med. Imaging Graph. **77** (2019). https://doi.org/10.1016/j.compmedimag.2019.101647

24. Ravì, D., Szczotka, A.B., Pereira, S.P., Vercauteren, T.: Adversarial training with cycle consistency for unsupervised super-resolution in endomicroscopy. Med. Image Anal. **53**, 123–131 (2019). https://doi.org/10.1016/j.media.2019.01.011

25. Shi, F., Cheng, J., Wang, L., Yap, P.T., Shen, D.: LRTV: MR image super-resolution with low-rank and total variation regularizations. IEEE Trans. Med. Imaging **34**(12), 2459–2466 (2015). https://doi.org/10.1109/TMI.2015.2437894

26. Shi, W., et al.: Real-time single image and video super-resolution using an efficient sub-pixel convolutional neural network. In: Proceedings of the IEEE Computer Society Conference on Computer Vision and Pattern Recognition, vol. 2016-Decem, pp. 1874–1883 (2016). https://doi.org/10.1109/CVPR.2016.207

27. Shocher, A., Cohen, N., Irani, M.: "Zero-shot" super-resolution using deep internal learning. In: Proceedings of the IEEE Computer Society Conference on Computer Vision and Pattern Recognition, pp. 3118–3126 (2017). https://doi.org/10.1109/CVPR.2018.00329, http://arxiv.org/abs/1712.06087

28. Umehara, K., Ota, J., Ishida, T.: Application of super-resolution convolutional neural network for enhancing image resolution in chest CT. J. Digit. Imaging **31**(4), 441–450 (2017). https://doi.org/10.1007/s10278-017-0033-z

29. Yu, J., Fan, Y., Huang, T.: Wide activation for efficient image and video super-resolution. In: 30th British Machine Vision Conference 2019, BMVC 2019, pp. 1–13 (2020)

30. Yuan, Y., Liu, S., Zhang, J., Zhang, Y., Dong, C., Lin, L.: Unsupervised image super-resolution using cycle-in-cycle generative adversarial networks. In: 2018 IEEE/CVF Conference on Computer Vision and Pattern Recognition Workshops (CVPRW), pp. 814–81409 (2018). https://doi.org/10.1109/CVPRW.2018.00113

31. Zbontar, J., et al.: fastMRI: An open dataset and benchmarks for accelerated MRI. CoRR abs/1811.08839 (2018). http://arxiv.org/abs/1811.08839

32. Zhang, Q., et al.: MRI Gibbs-ringing artifact reduction by means of machine learning using convolutional neural networks. Magn. Reson. Med. **82**(6), 2133–2145 (2019). https://doi.org/10.1002/mrm.27894

33. Zhang, X., Karaman, S., Chang, S.F.: Detecting and simulating artifacts in GAN fake images. In: 2019 IEEE International Workshop on Information Forensics and Security, WIFS 2019 (2019). https://doi.org/10.1109/WIFS47025.2019.9035107

34. Zhao, H., Gallo, O., Frosio, I., Kautz, J.: Loss functions for neural networks for image processing, pp. 1–11 (2015). http://arxiv.org/abs/1511.08861

OTRE: Where Optimal Transport Guided Unpaired Image-to-Image Translation Meets Regularization by Enhancing

Wenhui Zhu[1]($\boxtimes$), Peijie Qiu[2], Oana M. Dumitrascu[3], Jacob M. Sobczak[3], Mohammad Farazi[1], Zhangsihao Yang[1], Keshav Nandakumar[1], and Yalin Wang[1]

[1] School of Computing and Augmented Intelligence, Arizona State University, Tempe, AZ, USA
wzhu59@asu.edu
[2] McKeley School of Engineering, Washington University in St. Louis, St. Louis, MO, USA
[3] Department of Neurology, Mayo Clinic, Scottsdale, AZ, USA

Abstract. Non-mydriatic retinal color fundus photography (CFP) is widely available due to the advantage of not requiring pupillary dilation, however, is prone to poor quality due to operators, systemic imperfections, or patient-related causes. Optimal retinal image quality is mandated for accurate medical diagnoses and automated analyses. Herein, we leveraged the *Optimal Transport (OT)* theory to propose an unpaired image-to-image translation scheme for mapping low-quality retinal CFPs to high-quality counterparts. Furthermore, to improve the flexibility, robustness, and applicability of our image enhancement pipeline in the clinical practice, we generalized a state-of-the-art model-based image reconstruction method, regularization by denoising, by plugging in priors learned by our OT-guided image-to-image translation network. We named it as *regularization by enhancing (RE)*. We validated the integrated framework, OTRE, on three publicly available retinal image datasets by assessing the quality after enhancement and their performance on various downstream tasks, including diabetic retinopathy grading, vessel segmentation, and diabetic lesion segmentation. The experimental results demonstrated the superiority of our proposed framework over some state-of-the-art unsupervised competitors and a state-of-the-art supervised method.

Keywords: Retinal color fundus photography · Image enhancement · Optimal transport · Regularization by enhancing · Unsupervised learning

1 Introduction

Retinal color fundus photography (CFP) is widely and routinely used to diagnose various ocular diseases. Automated analyses are being developed for

---

W. Zhu and P. Qiu—The two authors contributed equally to this paper.

© The Author(s), under exclusive license to Springer Nature Switzerland AG 2023
A. Frangi et al. (Eds.): IPMI 2023, LNCS 13939, pp. 415–427, 2023.
https://doi.org/10.1007/978-3-031-34048-2_32

point-of-care disease screening, based on non-mydriatic CFP [1], Furthermore, research is conducted to unlock the CFPs potential to screen for neurodegenerative disorders such as Alzheimer's disease [2]. Both human and computer-aided analysis methods prefer operating on high-quality retinal CFPs. Whereas patient and provider-friendly, non-mydriatic retinal CFP is prone to noise, e.g., shading artifacts and blurring because of light transmission disturbance, defocusing, abnormal pupils, or suboptimal human operations [3], resulting in low-quality CFPs. CFP degradation such as obscuration of blood vessels, and missing or artifactual new lesions, leads to inaccurate diagnostic interpretation. Enhancing low-quality retinal CFPs into high-quality counterparts is of key importance for many downstream tasks, e.g., diabetic retinopathy (DR), blood vessel segmentation, DR lesion segmentation, etc. Shen et al. [3] proposed a clinically oriented fundus enhancement network (cofe-Net) by inputting pairs of degraded images synthesized by a fundus degradation model and clean, high-quality images. However, collecting paired noisy-clean retinal training data is difficult and expensive in reality. Lehtinen et al. [4] relaxed the paired noisy-clean images as noisy image pairs obtained from the same condition by arguing that the noisy data approaches the clean data on expectation. Krull et al. [5] extended [4] to a self-supervised training scheme by predicting surrounding pixels around a blind-spot pixel. This family of self-supervised methods strongly assumed that the "noise" causing the degradation of images was pixel/image-independent. However, unlike natural images, the composition of noise in retinal fundus images was more complicated, therefore, more challenging to model.

Unsupervised methods have recently attracted much attention. Such a process was usually modeled as an end-to-end image-to-image translation task. Many previous explorations [6–9] in this task were built on top of generative adversarial networks(GANs) to map a source domain Y to a target domain X. Such a mapping became more challenging when the input and target were unpaired due to extensive mapping ambiguities. To reduce the large searching space, [6,9] regularized the GAN by a task-specific regularization, while the CycleGAN [8] proposed a generalized regularization called cycle consistency. However, the expensive computation, destruction of lesions, and introduction of non-existing vessel structures limited the application of CycleGAN to retinal fundus images. To reduce the computational complexity of the CycleGAN, Wang et al. [10] proposed an OT-guided GAN (OTTGAN) for unsupervised image denoising with a single generator and discriminator. Although it achieved very significant results in natural image denoising, its adoption of mean squared error cost as the metric and its lack of high-quality image consistency led to the destruction or over-tampering of the vessel and lesion structures in our tasks.

Here, we proposed an integrated unsupervised end-to-end image enhancement framework based on optimal transport (OT) and regularization by denoising [11] methods. Our novel OT formulation maximally preserves structural consistency (e.g., lesions, vessel structures, optical discs) between enhanced and low-quality images to prevent over-tampering of important structures. To further improve flexibility and robustness to images from different distributions and applicability in real clinical practice where no sufficient data is available to

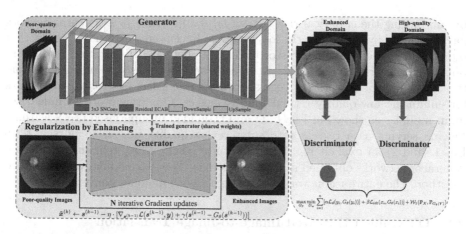

Fig. 1. The framework of our proposed method includes OT-guided GAN-based fundus enhancing network and regularization by enhancing framework where SNConv denotes the spectral normalized convolutional layer, and Residual ECAB denotes the residual block with efficient channel attention [14]. The architecture of the discriminator is adopted from [15].

train the model [11–13], we refined the enhanced images by our proposed regularization by enhancing (RE), a variant of regularization by denoising (RED) method [11], whose priors were learned by the OT-guided network. The contributions of this paper are three folded: (1) We proposed a novel OT-guided GAN-based unsupervised end-to-end retinal image enhancement training scheme, where a maximal information-preserving consistency mechanism was adopted to prevent lesion and structure over-tampering. (2) An RE module was introduced to refine the output of the OT module, improving the flexibility, robustness, and applicability of the system. Our study was the first of its kind to bridge the gap between OT-guided generative models and model-based enhancement frameworks. It is a general approach, adaptable to other structure-preserving medical image enhancement research. (3) Our extensive experimental results on three large retinal imaging cohorts demonstrated the superiority of our proposed method over unsupervised and state-of-the-art (SOTA) supervised methods.

2 Methods

Restoring clean images $x \sim X$ from their corruptions $y \sim Y$ can be formulated as a variational regularization in the Bayesian framework

$$\hat{x} = \underset{x}{\arg\min} f(x) + R(x), \tag{1}$$

where f is the data fidelity measuring the consistency between the restoration and the corrupted data and R is the regularization/prior term. The modern deep learning-based image restoration seeks to train an end-to-end regressor by minimizing the empirical risk $\mathbb{E}_{x,y}[\mathcal{L}(f_\theta(y), x)]$, where f_θ is a neural network parameterized by θ, and $\mathcal{L}$ is the loss function. Recent advances [11–13] show that plugging a learned image regressor into the model-based restoration framework boosts its performance. This work unified the model-based regularization and the learned image restoration regressor to provide a flexible, robust framework for enhancing low-quality retinal fundus images. Our framework included two main modules as shown in Fig. 1: 1) an OT-guided unsupervised GAN learning scheme serving as a regressor to enhance low-quality images to pursue f_θ in Eq. 1, and 2) an explicit regularization term RE as $R(x)$, refining the trained generator networks obtained in the first module. The two modules were cascaded together. The entire framework iterated until both modules converged.

2.1 OT-Guided Unpaired Image-to-Image Translation

Let $\mu \sim \mathbb{P}_X$ and $\nu \sim \mathbb{P}_Y$ be two probability measures on the target and source probability manifolds, respectively. The *Monge's* optimal transport problem of transporting masses from domain Y to X $(Y \rightarrow X)$ can be defined as

$$\inf \int_Y C(y, T(y))dv(y) \tag{2}$$

where $C(\cdot, \cdot)$ is the cost of transporting y to $T(y)$. The minimal cost among all possible ν-mensurable mappings T yields the optimal transport $u = T^*(\nu)$. Intuitively, the transport defined in Eq. 2 matches the objective of Image-to-Image translation which seeks an optimal mapping from the source domain to the target domain, which we define as *Domain* transport. We turn the proposed OT-guided Image-to-Image translation scheme into an optimization problem.

Definition 1. *The Image-to-Image translation from a source to a target domain* $Y \rightarrow X$ *suggested by the optimal mass transport can be expressed as*

$$\inf \int_Y C(y, T(y))dv(y), \quad \textit{subject to} \quad u = T^*(\nu) \tag{3}$$

By further parameterizing the optimal transport map T as a neural network T_θ, the Eq. 3 can be discretized as

$$\min_\theta \mathbb{E}_{y \sim \mathbb{P}_Y}[C(y, T_\theta(y))], \quad \textbf{subject to} \quad \mathbb{P}_{T_\theta(Y)} = \mathbb{P}_X \tag{4}$$

By applying the *Lagrange Multiplier*, Eqn. 4 is relaxed to a constrained optimization given by

$$\min_\theta \mathbb{E}_{y \sim \mathbb{P}_Y}[C(y, T_\theta(y))] + \lambda d(\mathbb{P}_{T_\theta(Y)}, \mathbb{P}_X), \tag{5}$$

Likewise, transporting a given measurement in the target domain will also produce another measurement in the target domain X, However, we do not desire discrepancies between the measurements on the target. An *Identity* cost constraint is introduced to prevent the network from over-learning or generating

---

Algorithm 1. OT-Guided Unpaired Image-to-Image Translation.

Require: The learning rate η, the batch size m, the gradient penalty weight λ, the consistency loss weight $\alpha \leq 1$, the identity loss weight β.
Require: Initial discriminator parameters w_0, initial generator parameters θ_0.
  while not converge **do**
    Sample a batch of low-quality images $y = \{y_i\}_{i=1}^m \sim \mathbb{P}_Y$ with $\{g_i\}_{i=1}^m$.
    Sample a batch of high-quality images $x = \{x_i\}_{i=1}^m \sim \mathbb{P}_X$ with $\{g_i\}_{i=1}^m$.
    for $i = 1, \ldots, m$ **do**
      Sample a random $\epsilon \sim U[0,1]$.
      $\tilde{x}_i \leftarrow G_\theta(y_i)$
      $\hat{x}_i \leftarrow \epsilon x_i + (1-\epsilon)\tilde{x}_i$
      $\mathcal{L}_{D_w}(i) \leftarrow D_w(\tilde{x}_i) - D_w(x_i) + \lambda(\|\nabla_{\hat{x}_i} D_w(\hat{x}_i)\|_2 - 1)_+^2$
    end for
    $w \leftarrow w + \eta \cdot \text{RMSProp}(w, \nabla_w \frac{1}{m}\sum_{i=1}^m \mathcal{L}_{D_w}(i))$
    $\mathcal{L}_{G_\theta} \leftarrow \frac{1}{m}\sum_{i=1}^m -D_w(G_\theta(y)) + \alpha\mathcal{L}_d(y, G_\theta(y)) + \beta\mathcal{L}_{idt}(x, G_\theta(x))$
    $\theta \leftarrow \theta - \eta \cdot \text{RMSProp}(\theta, \nabla_\theta \mathcal{L}_{G_\theta})$
  end while

---

unexpected measurements. Meanwhile, it is utilized for maintaining consistency in the target domain. Adding this term to Eq. 5 can be expressed as:

$$\min_\theta \mathbb{E}_{y \sim \mathbb{P}_Y}[C(y, T_\theta(y))] + \mathbb{E}_{x \sim \mathbb{P}_X}[C(x, T_\theta(x))] + \lambda d(\mathbb{P}_{T_\theta(Y)}, \mathbb{P}_X), \qquad (6)$$

which is defined as *Identity* constrain. where $d(\cdot, \cdot)$ measures the divergence of $\mathbb{P}_X$ and $\mathbb{P}_{T_\theta(Y)}$, and λ is a weight parameter. It is noteworthy that we use the same cost representation, but the *Identity* term is utilized as a constraint in the target domain X and is not related to the optimal transport between the source and target domain.

Proposition 1. *Supposing Wasserstein-1 distance $\mathcal{W}_1(\cdot, \cdot)$ is applied to measure the divergence between $\mathbb{P}_X$ and $\mathbb{P}_{T_\theta(Y)}$, Eq. 6 suggests an adversarial training scheme of unpaired Image-to-Image translation from $Y \to X$, given by*

$$\max_{G_\theta} \min_{D_w} \mathbb{E}_Y[\mathcal{L}_d(y, G_\theta(y))] + \mathbb{E}_X[\mathcal{L}_{idt}(x, G_\theta(x))] + \lambda\mathcal{W}_1(\mathbb{P}_X, \mathbb{P}_{G_\theta(Y)})$$

$$\mathcal{W}_1(\mathbb{P}_X, \mathbb{P}_{G_\theta(Y)}) = \sup_{\|D_w\|_L \leq 1} \mathbb{E}_X[D_w(x)] - \mathbb{E}_Y[D_w(G_\theta(y))] \qquad (7)$$

where G_θ is the generator parameterized by θ, D_w, the discriminator, is a 1-Lipschitz function parameterized by w, and $\mathcal{L}_d$ and $\mathcal{L}_{idt}$ denotes the domain transport cost and identity constraint cost, respectively. To better preserve the important structure (e.g., lesions), we ensured that the unpaired input with the matched disease labels g_i if there were any, where g denotes the disease type.

The *1-Lipshcitz* constraint is approached by the gradient penalty [16] in our experiment. The *Domain* transport and the *Identity* constraint shares the same cost function, as detailed below.

Information-Preserving Consistency Mechanism. There are two main concerns of the proposed OT-guided unpaired image-to-image translation in our task: 1) maintaining the underlying information, e.g., optical discs, lesions, and vessels, consistency before and after the translation; 2) minimizing the duality gap between the primal problem (Eq. 4) and the dual problem (Eq. 5). We will introduce our information-preserving consistency mechanism centered on addressing those two main concerns.

CycleGAN addresses the first concern by introducing the L_1 norm as the loss function to enforce low-frequency consistency leading us to the optimal median. In addition, a Patch Discriminator is incorporated to capture high-frequency components by enforcing local structural consistency at a patch level. The Patch Discriminator needs to specify architecture with a pre-defined receptive field usually resulting in a "shallow" discriminator. Our early experiments with CycleGAN, however, indicated that it destroyed the lesion structures and introduced non-existing vessels. Inspired by the Patch Discriminator, we used the multi-scale structural similarity index measure (SSIM) [17] as our consistency loss function $\mathcal{L}_d$ given by $\mathcal{L}_d(y, G_\theta(y)) = 1 - \text{SSIM}_{MS}(y, G_\theta(y))$. Followed by the CycleGAN, we also incorporated the identity loss $\mathcal{L}_{idt}$ to make sure that a high-quality input would result in a high-quality enhancement given by $\mathcal{L}_{idt}(x, G_\theta(x)) = 1 - \text{SSIM}_{MS}(x, G_\theta(x))$. We also used the UNet [18] as our generator to help the low-level semantics flow from the poor-quality domain to the high-quality domain. The following theorem [17] provides a theoretical guarantee to our loss function definition.

Theorem 1 ([17]). *The Structural Similarity Index Measure is proven to be locally Quasi-Convex which minimizes the duality gap between the primal and dual problem and weak duality holds.*

To better balance identity loss, domain loss, and the divergence between $\mathbb{P}_X$ and $\mathbb{P}_{G_\theta(y)}$, the final objective function was rewritten as

$$\max_{G_\theta} \min_{D_w} \sum_{i=1}^{n} [\alpha\mathcal{L}_d(y_i, G_\theta(y_i)) + \beta\mathcal{L}_{idt}(x_i, G_\theta(x_i))] + \mathcal{W}_1(\mathbb{P}_X, \mathbb{P}_{G_\theta(Y)}), \quad (8)$$

where α, β are weight parameters of the domain loss and identity loss, respectively. The algorithm of our OT-guided unpaired image-enhancing training scheme is given by Algorithm 1.

2.2 Regularization by Enhancing

Regularization by Denoising(*RED*) [11] is an off-the-shelf model-based framework that can take advantage of a variety of existing CNN priors without modifying the model's architecture to guide image restoration. We generalized the denoiser-centered RED idea to a more generic one that leveraged our image prior learned from the proposed OT-guided enhancing networks. We formulated the enhancement as an image prior to guiding the restoration of any test images

---

Algorithm 2. Regularization by Enhancing.

Require: The step size η, regularization strength γ, tolerance **tol**, Generator G_θ
Require: Initial $\tilde{x}^{(0)}$, $s^{(0)} = \tilde{x}^{(0)}$, $t^{(0)} = 1$
  while not converge **do**
    $t^{(k)} = \frac{1}{2}(1 + \sqrt{1 + 4(t^{(k-1)})^2})$
    $\mathbf{Der}(s^{(k-1)}) = \nabla_{s^{(k-1)}}\mathcal{L}(s^{(k-1)}, y) + \gamma(s^{(k-1)} - G_\theta(s^{(k-1)}))$
    $\tilde{x}^{(k)} \leftarrow s^{(k-1)} - \eta \cdot \mathbf{Der}(s^{(k-1)})$
    $s^{(k)} \leftarrow \tilde{x}^{(k)} + \frac{t^{(k-1)}-1}{t^{(k)}}(\tilde{x}^{(k)} - \tilde{x}^{(k-1)})$
    if $\|\tilde{x}^{(k)} - \tilde{x}^{(k-1)}\| \leq \mathbf{tol} \cdot \|\tilde{x}^{(k-1)}\|$ **then**
      break
    end if
  end while

---

whenever there are not enough samples for the end-to-end training. The objective of our proposed regularization by Enhancing (*RE*) is given by

$$\hat{x} = \underset{x}{\mathrm{argmin}}\, \mathbb{E}_x[\mathcal{L}(x, y)] + \gamma R(x) \text{ with } R(x) = \frac{1}{2}x^T(x - G_\theta(x)), \qquad (9)$$

where γ controls the regularization strength, and $\mathcal{L}$ denotes the multi-scale structural similarity loss. The gradient of the *RE* prior has a simple form

$$\nabla_x R(x) = x - G_\theta(x), \qquad (10)$$

under the condition that G_θ is locally homogeneous and has a symmetric Jacobian. The 1-Lipschitz constraint of G_θ can further guarantee the passivity of G_θ resulting in a convex objective function. We regularized the spectral radius of the weight of each convolutional layer in our generator G_θ via spectral normalization [19] to approximate the 1-Lipschitz constraint. In the optimization phase, the accelerated gradient descent was chosen to iteratively approach the optimum. The iterative optimization of the *RE* is given by Algorithm 2.

3 Experimental Results

We conducted extensive experiments in scenarios where the ground-truth clean images are available (*full-reference assessment*) and unavailable (*no-reference assessment*). Three downstream tasks including DR grading, vessel segmentation, and lesion segmentation, were studied to further evaluate the performance of our proposed method. Visual inspection was conducted by human ophthalmologists to evaluate the performance of no-reference assessment. The vanilla ResNet-50 [20] and UNet [18] were used to train and test the downstream tasks.

3.1 Datasets

Our proposed method was extensively evaluated on three publicly available retinal CFP datasets: the EyeQ dataset [21], the DRIVE dataset [22], and the

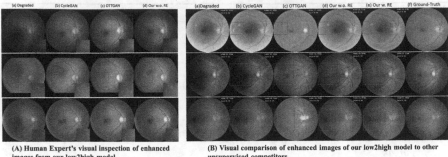

(A) Human Expert's visual inspection of enhanced images from our low2high model.

(B) Visual comparison of enhanced images of our low2high model to other unsupervised competitors.

Fig. 2. (A).The highlight blocks denote that contrast the structure of the lesion and vessel. It is very obvious that all other methods changed the structure of the lesion or vessel. (B). The red blocks denote generate extra structure compared with our method and areas where noise reduction is not obvious. (Color figure online)

IDRID dataset [23]. The EyeQ dataset was manually labeled into three quality levels: good, usable, and reject. We used 7886 training images and 8161 testing images (good & reject) in our training and evaluation. The DRIVE dataset evaluated our proposed method on the vessel segmentation task with 40 subjects. The IDRID dataset containing 81 subjects with pixel-level annotation of microaneurysms (MA), soft exudates (SE), hemorrhages (HE), and hard exudates (EX) were used to evaluate our method on DR lesion segmentation. All images are center-cropped and resized to a size of 256×256.

3.2   Experimental Design

For the no-reference assessment, our proposed OT-guided Image-to-Image translation GAN was trained with 7886 training images on the EyeQ dataset by unpaired low-quality images and high-quality images. It was defined as low-quality to high-quality (*low2high*) model. For the full-reference assessment, the model was trained on the subset of the high-quality EyeQ training dataset with degraded images obtained by [3] and unpaired high-quality images. It was defined as degradation to high-quality (*deg2high*) model. The disease label g in Algorithm 1 was the DR grading label from the EyeQ. Data augmentation, including random horizontal/vertical flips, random crops, and random rotations, was performed to prevent over-fitting during training. All models were trained with the RMSprop optimizer for 200 epochs with an initial learning rate of 1×10^{-4} for the discriminator and 5×10^{-5} for the generator with a decay of 10 by every 100 epochs. The optimal hyperparameters were $\alpha = 60$, $\beta = 20$ for both low2high and deg2high models. In the testing phase, the optimal hyperparameter γ was grid-searched within a range from 1×10^{-3} to 1×10^{-4} with the number of iterations equal to 400 for all experiments. All methods were implemented in PyTorch, and the code will be available on GitHub after the paper acceptance.

Table 1. Evaluation metrics of the DR grading task with enhancements from different methods in the Lesion structure changed ratio (LCR), background-color changed ratio (BCR), and generated extra structures ratio (GESR).

Method	DR Grading			Experts Evaluation		
	Accuracy	Kappa	ROC	LCR	BCR	GESR
CycleGAN	0.7148	0.5378	0.9083	0.449	**0.0**	0.347
OTTGAN	0.6996	0.5105	0.8995	0.429	0.102	0.490
OTRE	**0.7767**	**0.6814**	**0.9403**	**0.020**	0.040	**0.326**

3.3 No-Reference Quality Assessment

Evaluating the quality of the enhancement without knowing the ground-truth clean images is challenging. We considered combining the DR grading task with visual inspection by human experts to assess the performance of the enhancement. The DR grading task can be viewed as a criterion to judge whether lesion information is preserved after the enhancement. A ResNet-50 model was trained on high-quality images following the experimental setup in [24] and evaluated by the low-quality images and their enhancements from different methods. The performance of the enhancement will be indicated by the classification accuracy, Area under Receiver Operating Characteristic Curve (AU-ROC), and Cohen's Kappa Coefficient (kappa).

For visual inspection by human experts, 50 low-quality images were randomly chosen and processed by different enhancement methods. Visual inspection was done to measure 1) the ratio of changing lesion structure (LCR), 2) the ratio of changing the main background color (BCR), and 3) the ratio of generating non-existing structures (GESR). For the fairness of our experiments, the assessment was first conducted by three volunteers who were pretrained with the designed protocol and then finalized by the ophthalmologist.

Figure 2 (A) illustrates some results from different unsupervised enhancement methods. All methods can enhance image quality. However, our method can better maintain the lesion and vessel structure while reducing the noise. We introduced two experiments to verify that our method can preserve the maximal information (Table 1). First we applied DR grading algorithm to the enhanced images (low2high) and evaluated their grading accuracy. As shown in Table 1, the OTRE outperformed other methods in all three measures, expecially by more than 20% in the kappa measure. The human expert inspection also verified that our method could maximize information preservation. Our method performed best in LCR and GESR, with a dramatic improvement in LCR, even over 40% improvement over the other two methods.

3.4 Full-Reference Quality Assessment

For full-reference assessment, we degraded high-quality images following the degradation model introduced by Shen et al. [3] to synthesize the low-quality images for each dataset. The training dataset consisted of 6500 high-quality images selected from the EyeQ training dataset and other 6500 synthesized low-quality images degraded from non-overlapping 6500 high-quality images from the EyeQ training dataset. The testing dataset was made up of 500 images from the EyeQ testing dataset, the entire DRIVE dataset, and the entire IDRID dataset. The commonly used Peak-Signal-to-Noise Ratio (PSNR) and Structural Similarity Index Measure (SSIM) were used to evaluate the quality of the enhanced low-quality images. To further validate our proposed method on the downstream tasks, we evaluated the performance of our proposed method on the blood vessel segmentation and DR lesion segmentation tasks.

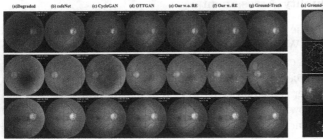 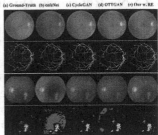

(A) deg2high model evaluation based on the full-reference experiment.

(B) The visualization of downstream segmentation tasks.

Fig. 3. (A). The highlighted red block shows the comparison of structure-preserving and extra structure generation. Visually the image enhancement is good, but the PSNR is not high, so further downstream task evaluation is essential. (B). The highlighted red blocks denote the comparison of fine vessel bifurcation segmentation results. (Color figure online)

First, we evaluated the consistency of the enhanced images and their high-quality counterparts. We performed two different experiments by applying both no-reference trained models (Sect. 3.3, low2high) and full-reference trained models (deg2high). We also tested whether the inclusion of RE module improved our results. Figure 2(B) and Fig. 3(A) show some image examples and Table 2 reports the numerical results. As shown in Table 2, except for EyeQ's SSIM measure, our OTRE outperformed all other supervised and unsupervised methods in three different datasets, and the PSNR achieved respectively the highest 24.63, 22.81, and 22.05. Remarkably, the OTRE beat the SOTA supervised method (cofe-Net) given that the cofe-Net was trained with paired images, but the OTRE was not. Interestingly, our method no-reference trained model (*low2high*) achieved competitive results for the unseen degradation noises. We also learned that the inclusion of RE module gained improved performance. Figure 2(B), Fig. 3(A) also provided stronger support of effectiveness that our method preserved the structure and achieved better noise reduction.

Table 2. Result comparison of unsupervised methods when trained with the no-reference training data (Sec. 3.3, low2high) and full-reference training data (deg2high) on the current degrading testing dataset. The OTRE frameworks with/without RE module were investigated on both datasets. The supervised method (coef-Net) was trained/evaluated with the degrading dataset only.

	Method	EyeQ		DRIVE		IDRID	
		PSNR	SSIM	PSNR	SSIM	PSNR	SSIM
Supervised	cofe-Net	23.11	**0.910**	21.87	0.767	20.25	0.825
Unsupervised	CycleGAN (low2high)	18.57	0.836	18.78	0.705	19.13	0.799
	CycleGAN (deg2high)	22.75	0.895	21.92	0.766	21.56	0.855
	OTTGAN (low2high)	18.93	0.859	19.20	0.723	19.70	0.828
	OTTGAN (deg2high)	23.69	0.894	21.61	0.750	21.93	0.839
	OTRE without RE (low2high)	20.39	0.878	20.03	0.733	20.50	0.837
	OTRE with RE (low2high)	21.08	0.880	20.61	0.740	20.55	0.836
	OTRE without RE (deg2high)	24.29	0.906	22.40	0.772	21.51	**0.860**
	OTRE with RE (deg2high)	**24.63**	0.905	**22.81**	**0.794**	**22.05**	0.852

Table 3. Result comparison of the segmentation of blood vessels on the DRIVE cohort [22] and diabetic lesions (EX and HE) on the IDRID dataset [23]. The OTRE compared favorably to other supervised and unsupervised methods. (ROC: Area under Receiver Operating Characteristic Curve, PR: Area under the Precision-Recall, F1: F1 score, SE: Sensitivity, SP: Specificity).

Method	Vessel Segmentation					EX			HE		
	ROC	PR	F1	SE	SP	ROC	PR	F1	ROC	PR	F1
cofe-Net	0.923	0.787	0.714	0.644	**0.977**	0.926	0.442	0.469	0.807	0.103	0.090
CycleGAN	0.910	0.762	0.696	0.622	0.975	0.900	0.474	0.347	0.845	0.155	0.141
OTTGAN	0.900	0.739	0.667	0.581	0.976	0.912	0.507	**0.512**	0.855	0.107	0.145
OTRE	**0.927**	**0.796**	**0.726**	**0.672**	0.975	**0.934**	**0.529**	0.441	**0.894**	**0.233**	**0.273**

To further confirm the superiority of our method, two downstream segmentation tasks were studied using the groundtruth data from DRIVE and IDRID datasets. Since the training and testing of our segmentation task were based entirely on enhanced images, without adding any preprocessing and additional tricks, in the lesion segmentation task, we only considered large blocks of lesions which were easy to train, such as EX and HE. As shown in Table 3, the OTRE method achieved excellent results in three segmentation tasks. It achieved the highest ROC and PR results in all segmentation results, 2 out of 3 bests results in F1 measure. From some image examples shown in Fig. 3(B), it is easy to see that other methods have the problem of insignificant enhancement performance, resulting in the altered vessel and lesion structures.

4   Conclusion and Future Work

This work integrated OT-guided GAN-based enhancing network with the RE module and achieved promising results on three datasets, surpassing or on a par with SOTA unsupervised and supervised methods. A limitation of the current system is that it assumed all input data were usable but in real clinical applications, there exist some images that are completely corrupted. A screening procedure to classify whether the input images are usable will make our work more practical. We will study it in our future work.

Acknowledgement. This work was partially supported by grants from NIH (R21AG065942, R01EY032125, and R01DE030286).

References

1. Wolf, R.M., Channa, R., Abramoff, M.D., Lehmann, H.P.: Cost-effectiveness of autonomous point-of-care diabetic retinopathy screening for pediatric patients with diabetes. JAMA Ophthalmol. **138**(10), 1063–1069 (2020)
2. Cheung, C.Y., et al.: A deep learning model for detection of Alzheimer's disease based on retinal photographs: a retrospective, multicentre case-control study. Lancet Digit. Health **4**(11), e806–e815 (2022)
3. Shen, Z., Fu, H., Shen, J., Shao, L.: Modeling and enhancing low-quality retinal fundus images. IEEE Trans. Med. Imaging **40**(3), 996–1006 (2021)
4. Lehtinen, J., et al.: Noise2Noise: learning image restoration without clean data. ICML **80**, 2965–2974 (2018)
5. Krull, A., et al.: Noise2void-learning denoising from single noisy images. In: Proceedings of the IEEE Computer Society Conference Computer Vision and Pattern Recognition, pp. 2129–2137 (2019)
6. Bousmalis, K., Silberman, N., et al.: Unsupervised pixel-level domain adaptation with generative adversarial networks. In: CVPR (2016)
7. Isola, P., Zhu, J.Y., Zhou, T., Efros, A.A.: Image-to-image translation with conditional adversarial networks. In: CVPR (2017)
8. Zhu, J., Park, T., Isola, P., Efros, A.A.: Unpaired image-to-image translation using cycle-consistent adversarial networks. In: CVPR, pp. 2242–2251 (2017)
9. Liu, M.Y., Tuzel, O.: Coupled generative adversarial networks. In: Advances in Neural Information Processing Systems (2016)
10. Wang, W., Wen, F., Yan, Z., Liu, P.: Optimal transport for unsupervised denoising learning. IEEE PAMI 1 (2022)
11. Romano, Y., Elad, M., Milanfar, P.: The little engine that could: regularization by denoising (RED). SIAM J. Imag. Sci. **10**(4), 1804–1844 (2017)
12. Ryu, E., Liu, J., Wang, S., Chen, X., Wang, Z., Yin, W.: Plug-and-play methods provably converge with properly trained denoisers. PMLR **97**, 5546–5557 (2019)
13. Lucas, A., Iliadis, M., Molina, R., Katsaggelos, A.K.: Using deep neural networks for inverse problems in imaging: beyond analytical methods. IEEE Signal Process. Mag. **35**(1), 20–36 (2018)
14. Wang, Q., Wu, B., Zhu, P., Li, P., Zuo, W., Hu, Q.: ECA-net: efficient channel attention for deep convolutional neural networks. In: The IEEE Conference on Computer Vision and Pattern Recognition (2020)

15. Ledig, C., Theis, L., et al.: Photo-realistic single image super-resolution using a generative adversarial network. In: CVPR (2016)
16. Gulrajani, I., Ahmed, F., Arjovsky, M., et al.: Improved training of wasserstein GANs. In: Advances in Neural Information Processing Systems, vol. 30 (2017)
17. Brunet, D., Vrscay, E.R., Wang, Z.: On the mathematical properties of the structural similarity index. IEEE Trans. Image Process. **21**(4), 1488–1499 (2012)
18. Ronneberger, O., Fischer, P., Brox, T.: U-net: convolutional networks for biomedical image segmentation. In: Navab, N., Hornegger, J., Wells, W.M., Frangi, A.F. (eds.) MICCAI 2015. LNCS, vol. 9351, pp. 234–241. Springer, Cham (2015). https://doi.org/10.1007/978-3-319-24574-4_28
19. Miyato, T., et al.: Spectral normalization for generative adversarial networks. In: International Conference on Learning Representations (2018)
20. He, K., Zhang, X., Ren, S., Sun, J.: Deep residual learning for image recognition. In: In: Proceedings of the IEEE Computer Society Conference on Computer Vision and Pattern Recognition (2016)
21. Fu, H., et al.: Evaluation of retinal image quality assessment networks in different color-spaces. In: Shen, D., et al. (eds.) MICCAI 2019. LNCS, vol. 11764, pp. 48–56. Springer, Cham (2019). https://doi.org/10.1007/978-3-030-32239-7_6
22. Staal, J., et al.: Ridge-based vessel segmentation in color images of the retina. IEEE Trans. Med. Imaging **23**(4), 501–509 (2004)
23. Porwal, P., et al.: IDRID: a database for diabetic retinopathy screening research. Data **3**(3) (2018)
24. Zhu, W., Qiu, P., Lepore, N., Dumitrascu, O., Wang, Y.: Self-supervised equivariant regularization reconciles multiple instance learning: joint referable diabetic retinopathy classification and lesion segmentation. In: 18th International Symposium on Medical Information Processing and Analysis (SIPAIM) (2022)

Super-Resolution Reconstruction of Fetal Brain MRI with Prior Anatomical Knowledge

Shijie Huang[1], Geng Chen[2], Kaicong Sun[1], Zhiming Cui[1], Xukun Zhang[3], Peng Xue[1,4], Xuan Zhang[5], He Zhang[6], and Dinggang Shen[1,7,8(✉)]

[1] School of Biomedical Engineering, ShanghaiTech University, Shanghai 201210, China
dgshen@shanghaitech.edu.cn

[2] National Engineering Laboratory for Integrated Aero-Space-Ground-Ocean Big Data Application Technology, School of Computer Science and Engineering, Northwestern Polytechnical University, Xi'an, China

[3] Academy for Engineering and Technology, Fudan University, Shanghai, China

[4] School of Mechanical, Electrical and Information Engineering, Shandong University, Weihai 264209, China

[5] Department of Radiology, The First Affiliated Hospital of Nanjing Medical University, Nanjing 210000, China

[6] Department of Radiology, Obstetrics and Gynecology Hospital of Fudan University, Shanghai, China

[7] Shanghai United Imaging Intelligence Co., Ltd., Shanghai, China

[8] Shanghai Clinical Research and Trial Center, Shanghai 201210, China

Abstract. Super-resolution reconstruction (SRR) of fetal brain MRI from motion-corrupted thick-slice stacks can provide high-resolution isotropic 3D images that are vital for prenatal examination and quantification of brain development. Existing fetal brain SRR methods generally rely on a two-stage optimization procedure by performing rigid *slice-to-volume registration* and *volumetric reconstruction* in an alternating manner. Despite their advantages, these methods have not considered additional guidance from external anatomical priors, resulting in unsatisfactory performance in various challenging cases. To address this issue, we propose a novel **P**rior **A**natomical **K**nowledge guided fetal brain **S**uper-**R**esolution **R**econstruction method, namely **PAK-SRR**. In PAK-SRR, we consider two key kinds of prior anatomical information. First, we integrate the anatomical prior provided by tissue segmentation into both the *slice-to-volume registration* and *volumetric reconstruction* to enforce registration consistency on boundaries, effectively alleviating misregistration caused by blurry tissue boundaries of brain. Second, to enrich the structural details of the reconstructed images, we further employ longitudinal fetal brain atlases to guide *volumetric reconstruction*. Extensive experiments on multi-site clinical datasets demonstrate that our PAK-SRR significantly outperforms the state-of-the-art SRR methods for fetal brain MRI, quantitatively and qualitatively. Our code is publicly available at https://github.com/sj-huang/PAK-SRR for reproducibility and further research.

S. Huang and G. Chen— Contributed equally.

Keywords: Fetal Brain · Prior Anatomical Knowledge ·
Super-Resolution Reconstruction · Brian Tissue Segmentation

1   Introduction

Acquiring isotropic high-resolution (HR) magnetic resonance (MR) images from fetal brain is essential for prenatal examination and brain development studies. However, it is clinically difficult to acquire such isotropic HR volumetric images due to large irregular motion of fetal subjects. In this way, fetal brains are often scanned in a short time with fast imaging protocols (e.g., SSFSE: single-shot fast spin echo [4]), which acquire multiple thick-slice stacks from different views, as shown in Fig. 1. These different-view stacks are further integrated using super-resolution reconstruction (SRR) techniques at the post-acquisition stage to obtain desired isotropic HR volumetric images. However, it is challenging to conduct fetal brain SRR due to 1) poor slice image quality caused by large irregular motion, 2) misalignment among intra- and inter-stack slices, and 3) large variation of fetal brains in the acquired MR images [13].

To tackle these challenges, Rousseau *et al.* [13] proposed an approach to reconstruct 3D isotropic HR volumetric images for fetal brain from multi-view (direction) and multi-resolution 2D slices, by iteratively performing *slice-to-volume registration* and *volumetric reconstruction*, after preprocessing of acquired data. Kuklisova-Murgasova *et al.* [9] used expectation-maximization statistics to discard the potential outliers for more robust SRR. Moreover, they applied intensity matching to compensate for the inconsistency of multi-slices. Rousseau *et al.* [14] later improved the regularization procedure associated with SRR to preserve better boundaries. However, these approaches are semi-automatic since fetal brain region needs to be predefined [2]. To resolve this issue, Ebner *et al.* [2] proposed an automatic approach to integrate localization, segmentation, and SRR into a unified framework. Besides, deep learning has been employed to improve the performance of fetal brain SRR [7,15,18]. However, deep learning-based methods rely on large-scale high-quality training data, and suffer from generalization issues, thus limiting their practical applications. Despite the progress of existing methods [2,5,9,13,14,19], they all ignore the use of anatomical prior, which can provide vital information to improve the SRR of fetal brain MRI. This leads to unsatisfactory performance for various challenging cases with large irregular motions and low image quality, as shown in Fig. 1.

To this end, we propose a novel method, namely **PAK-SRR**, to employ rich prior anatomical knowledge (PAK), including tissue segmentation maps (i.e., *seg-priors*) and longitudinal fetal brain atlases (i.e., *atlas-priors*), as strong guidance to improve the SRR of fetal brain MRI (Fig. 1). Specifically, we employ *seg-priors* from tissue segmentation maps to guide more accurate *slice-to-volume registration*, which can improve the quality of reconstructed HR volumetric images. Furthermore, we employ *atlas-priors* (with rich anatomical priors) to address image degradation caused by motion artifacts and thick-slice acquisition. These two kinds of prior anatomical information are integrated into a unified regularization framework, which is optimized by an effective linear least-squares solver,

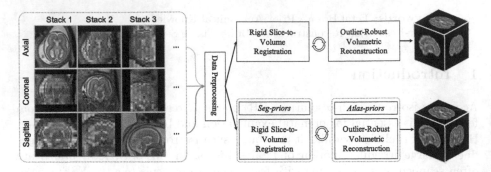

Fig. 1. Comparison of the existing framework (right top) [2] with our PAK-SRR (right bottom).

called LSMR [3]. Experimental results on multi-site clinical datasets demonstrate that our PAK-SRR achieves state-of-the-art reconstruction performance along with strong robustness against motion artifacts.

2   Background

The SRR of fetal brain MRI utilizes stacks of acquired 2D slices to reconstruct 3D super-resolution image and is usually formulated by the slice acquisition model [5] as below:

$$\mathbf{y}_k = \mathbf{A}_k(\mathbf{x}) + \mathbf{e}_k, \tag{1}$$

where $\mathbf{y}_k$ is the acquired k_{th} slice and $\mathbf{x}$ is the tentative 3D HR image. $\mathbf{A}_k$ is an operator including *slice-to-volume* transformation $\boldsymbol{T}_k$ and blurring operator $\boldsymbol{B}$ (based on a certain point spread function (PSF) [11]). Generally, we can formulate $\mathbf{A}_k(\cdot) := [\boldsymbol{T}_k \circ \boldsymbol{B}(\cdot)]_k$, where the symbol $\circ$ denotes resampling operation, and $[\cdot]_k$ indicates the corresponding plane of the slice $\mathbf{y}_k$ in the 3D image. The vector $\mathbf{e}_k$ represents an additive noise.

Due to the ill-posedness of the above problem in Eq. (1), the commonly used methods for SRR of fetal brain MRI are based on a two-stage iterative registration-reconstruction framework, including 1) *slice-to-volume registration* and 2) *volumetric reconstruction*. In each iteration i, *slice-to-volume registration* is performed to align each slice $\mathbf{y}_k$ with the tentatively reconstructed volumetric image $\mathbf{x}$ usually by minimizing the following objective function:

$$\boldsymbol{T}_k^{(i)} = \underset{\boldsymbol{T}_k}{\operatorname{argmin}} \left(\mathcal{S}(\mathbf{x}^{(i-1)}, \mathbf{y}_k; \boldsymbol{T}_k) + \mathcal{R}_{\mathcal{T}}(\boldsymbol{T}_k) \right), \tag{2}$$

where $\mathcal{S}$ represents the similarity measure between slice $\mathbf{y}_k$ and the transformed volume $\mathbf{x}$, and $\mathcal{R}_{\mathcal{T}}(\cdot)$ is the regularization term for the transformation $\boldsymbol{T}_k$. It should be noted that transformation is usually applied on volume $\mathbf{x}$ instead of slice $\mathbf{y}_k$, since resampling a 2D slice in 3D space is not as well defined as sampling a volume. Rigid transformation is usually employed for registration [2,9,13,14], which includes six Degrees of Freedom (DoF) for each individual

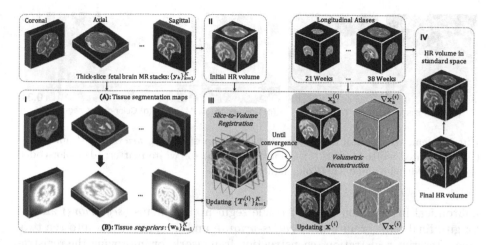

Fig. 2. Overview of our proposed PAK-SRR for fetal brain MRI. (I) Extracting brain tissue maps and the corresponding *seg-priors* (i.e., $\{\mathbf{w}_k\}_{k=1}^{K}$). (II) Aligning multi-view stacks (i.e., $\{\mathbf{y}_k\}_{k=1}^{K}$) into the same space for obtaining an initial coarse HR volumetric image. (III) The two-stage iterative registration-reconstruction stage, including a *seg-priors* guided *slice-to-volume registration* for updating $\{\boldsymbol{T}_k^{(i)}\}_{k=1}^{K}$ and a *seg-priors* and *atlas-priors* jointly guided (with $\{\mathbf{w}_k\}_{k=1}^{K}$ and $\nabla \mathbf{x}_a^{(i)}$) *volumetric reconstruction* for updating $\mathbf{x}^{(i)}$. (IV) Reorienting the final reconstructed 3D image into standard anatomical space defined by longitudinal atlases.

slice. Mathematically, we omit subscript k for clarity and can formulate the transformation as a 4×4 matrix $\boldsymbol{T} = \begin{bmatrix} \boldsymbol{R}(\boldsymbol{\theta}) & \boldsymbol{d} \\ 0 & 1 \end{bmatrix}$, where $\boldsymbol{R}(\boldsymbol{\theta})$ denotes a rotation matrix parameterized by three rotational parameters $\boldsymbol{\theta} = (\theta_u, \theta_v, \theta_w)^T$, and $\boldsymbol{d} = (d_u, d_v, d_w)^T$ is a vector of three translation parameters in the 3D space.

Volumetric reconstruction is performed when the transformations $\{\boldsymbol{T}_k^{(i)}\}_{k=1}^{K}$ are updated in each iteration. Assuming $\mathbf{e}_k$ being an additive white Gaussian noise, the image $\mathbf{x}$ can be updated as below:

$$\mathbf{x}^{(i)} = \underset{\mathbf{x}}{\operatorname{argmin}} \left(\sum_k \frac{1}{2} \left\| \mathbf{y}_k - \mathbf{A}_k^{(i)}(\mathbf{x}) \right\|_2^2 + \mathcal{R}_{\mathbf{x}}(\mathbf{x}) \right), \tag{3}$$

where $\mathcal{R}_{\mathbf{x}}(\cdot)$ regularizes the solution space of the latent 3D image $\mathbf{x}$. The optimization problem can be solved by popularly used least-square solver. The above-described two-stage framework performs alternating update of transformations $\{\boldsymbol{T}_k^{(i)}\}_{k=1}^{K}$ and 3D image $\mathbf{x}^{(i)}$ until convergence.

3   Method

An overview of our proposed PAK-SRR is shown in Fig. 2. Different from the existing methods for SRR of fetal brain MRI, we consider two key kinds of prior

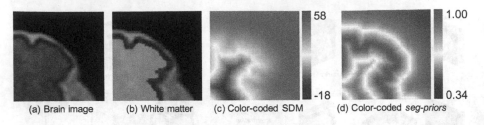

(a) Brain image   (b) White matter   (c) Color-coded SDM   (d) Color-coded *seg-priors*

Fig. 3. Demonstration of the stages for calculation of the *seg-priors*. (a) Brain image; (b) Segmented white matter; (c) Color-coded SDM for white matter; (d) Color-coded *seg-priors* for white matter. (Color figure online)

anatomical information from **1)** tissue segmentation maps (*seg-priors*) and **2)** longitudinal fetal brain atlases (*atlas-priors*), and integrate them into the two-stage iterative registration-reconstruction framework for improving the registration accuracy and reconstruction performance. The effectiveness of our *seg-priors* and *atlas-priors* will be demonstrated in Sect. 4.

3.1   *Slice-to-Volume Registration* Guided by *Seg-priors*

The *slice-to-volume registration* stage aligns multi-resolution and multi-view input slices with the tentatively-reconstructed volume by rigid registration. However, accurate registration is challenging due to large irregular motion and blurry tissue boundaries in the input stacks, which may cause misalignment and result in unsatisfactory SRR. To resolve this issue, we perform tissue segmentation and employ the *seg-priors* to assist the *slice-to-volume registration*. Specially, we perform segmentation for six types of tissues, including cerebrospinal fluid, grey matter, white matter, ventricles, cerebellum, and brainstem. Based on the segmentation results, we calculate the individual Signed Distance Map (SDM) [1] for each of six types of tissues, and then calculate the mean of six SDMs ($N = 6$) as the *seg-priors* $\{\mathbf{w}_k\}_{k=1}^{K}$ as follows:

$$\mathbf{w}_k = \frac{1}{N} \sum_{t=1}^{N} c^{\frac{-\left|SDM_k^{(t)}\right|}{\left\|SDM_k^{(t)}\right\|_\infty}}, \qquad (4)$$

where c is a constant used to control the contrast of the map (the larger c, the larger the contrast), and in the experiments, we set $c = 1e^3$. The operators $|\cdot|$ and $\|\cdot\|_\infty$ calculate the absolute value and ℓ_∞-norm, respectively. The calculation of *seg-priors* for white matter is demonstrated in Fig. 3.

We incorporate the *seg-priors* $\{\mathbf{w}_k\}_{k=1}^{K}$ into the calculation of similarity measure, i.e., normalized cross-correlation coefficient (NCC), and obtain our proposed weighted NCC as

$$\mathrm{NCC}_\mathbf{w}(I_1, I_2) := \frac{\sum_j \mathbf{w}^{(j)}\left(I_1^{(j)} - \bar{I}_1\right)\left(I_2^{(j)} - \bar{I}_2\right)}{\sqrt{\sum_j \left(I_1^{(j)} - \bar{I}_1\right)^2}\sqrt{\sum_j \left(I_2^{(j)} - \bar{I}_2\right)^2}}, \qquad (5)$$

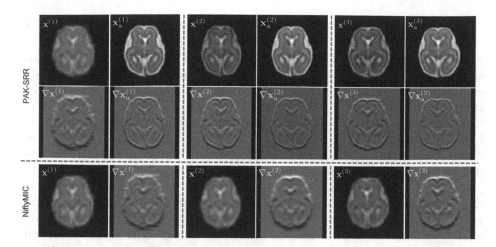

Fig. 4. Visualization of the reconstructed volume ($\mathbf{x}^{(i)}$), the registered atlas ($\mathbf{x}_a^{(i)}$), and the corresponding gradient maps ($\nabla \mathbf{x}^{(i)}$ and $\nabla \mathbf{x}_a^{(i)}$) by both methods (shown in first two rows for PAR-SRR, and third row for NiftyMIC) at three update iterations (shown horizontally with two columns for each iteration).

where I_1 and I_2 denote the vectorized images and $\mathbf{w}$ is the vectorized *seg-priors*. The index j denotes the j_{th} element of the vector. Therefore, we can update the transformation $\boldsymbol{T}_k^{(i)}$ of the k_{th} slice at iteration i with

$$\boldsymbol{T}_k^{(i)} = \underset{\boldsymbol{T}_k}{\arg\min} \left(-\text{NCC}_{\mathbf{w}_k}(\mathbf{y}_k, [\boldsymbol{T}_k \circ \boldsymbol{B}(\mathbf{x}^{(i-1)})]_k) \right). \tag{6}$$

In this way, we penalize more on boundary misalignment, and can obtain improved registration performance on boundary areas, which usually convey more structural information and are thus critical for downstream applications.

3.2 *Volumetric Reconstruction* Jointly Guided by *Seg-priors* and *Atlas-priors*

The *volumetric reconstruction* stage aims to reconstruct a super-resolution volumetric image based on the aligned thick-slice stacks. In this stage, we intend to employ the *seg-priors* and *atlas-priors* jointly. First, we integrate the calculated *seg-priors* $\{\mathbf{w}_k\}_{k=1}^K$ into the maximum a posteriori (MAP)-based *volumetric reconstruction* stage:

$$\mathbf{x}^{(i)} = \underset{\mathbf{x}}{\arg\min} \left(\sum_{k \in \mathscr{K}_\delta^{(i)}} \frac{1}{2} \left\| \mathbf{y}_k - \mathbf{A}_k^{(i)}(\mathbf{x}) \right\|_{\mathbf{w}_k}^2 + \frac{\alpha}{2} \|\nabla \mathbf{x}\|_2^2 \right), \tag{7}$$

where $\|\cdot\|_{\mathbf{w}_k}^2$ represents the weighted ℓ_2-norm with the weights $\mathbf{w}_k$ being diagonal matrix for the k_{th} slice. Following the work of [2], we eliminate the outlier slices which are severely corrupted by large motion according to the following rule:

$$\mathscr{K}_\delta^{(i)} := \left\{ 1 \leq k \leq K : \mathrm{NCC}(\mathbf{y}_k, \mathbf{A}_k^{(i)}(\mathbf{x}^{(i-1)})) \geq \delta \right\}, \tag{8}$$

where $\mathscr{K}_\delta^{(i)}$ denotes a set of indices of inlier slices. $\mathrm{NCC}(\cdot)$ measures the slice similarity such that the slices with similarity less than a predefined threshold δ are considered as outliers and will be discarded in the *volumetric reconstruction* stage.

Second, to compensate for the information loss due to motion artifacts, we resort to longitudinal fetal brain atlases and use gradient maps of certain time-point atlases as auxiliary information for *volumetric reconstruction*. In Fig. 4, we demonstrate the gradient maps of a reconstructed volumetric image and the atlas [6]. As can be observed, the atlas usually has sharper boundaries and sharper gradient maps than the initially reconstructed volumetric image, implying that atlas information can be utilized to compensate for blurry boundaries and improve quality of the reconstructed images. Therefore, we register the atlas to the tentatively reconstructed volumetric image, denoted as $\mathbf{x}_a$, and impose similarity match between the gradient of the volume $\nabla \mathbf{x}$ and the corresponding one of the *atlas-priors* (i.e., $\nabla \mathbf{x}_a$) to obtain clearer boundaries in the reconstructed images. Mathematically, we can use the following formulation:

$$\mathbf{x}^{(i)} = \underset{\mathbf{x}}{\mathrm{argmin}} \left(\sum_{k \in \mathscr{K}_\delta^{(i)}} \frac{1}{2} \left\| \mathbf{y}_k - \mathbf{A}_k^{(i)}(\mathbf{x}) \right\|_{\mathbf{w}_k}^2 + \frac{\alpha}{2} \| \nabla \mathbf{x} - \varepsilon \nabla \mathbf{x}_a \|_2^2 \right), \tag{9}$$

where ε and α are scalar parameters. The prior knowledge from the atlases can effectively guide the volumetric reconstruction from low-quality slices. Those slices with severe quality degradation usually decrease reconstruction performance. It is worth noting that, when introducing *atlas-priors* into the regularization, the overall intensities of the reconstructed volumetric image $\mathbf{x}$ might slightly deviate from the input slices during the optimization, as shown in Fig. 4. To alleviate this effect, we perform histogram matching between the tentatively reconstructed volumetric image and its corresponding input stacks during *slice-to-volume registration* in each iteration.

4   Experiments

4.1   Dataset

We have evaluated our proposed PAK-SRR on fetal brain MRI of 66 subjects covering 20 to 38 gestational weeks (GWs) from multi-site clinic centers, with totally 213 stacks. Each subject was scanned under at least three different-view (direction) stacks by the SSFSE sequence with different spatial resolutions including $1.3 \times 1.3 \times 5$, $1.5 \times 1.5 \times 4.2$, $0.625 \times 0.625 \times 3$, and $0.55 \times 0.55 \times 4.4$ mm. We use bias field correction [8] to preprocess all different-view stacks.

4.2   Experimental Settings

We do the following settings for our experiments. **1)** For selecting a target stack used initially to build a tentative HR 3D MR image of each fetal brain, we first

Table 1. Quantitative comparison between PAK-SRR and other methods on multi-site datasets of 66 subjects. The best results are in **bold**.

	Rousseau *et al.* [13]	SVRTK [9]	BTK [14]	NiftyMIC [2]	PAK-SRR
PSNR	18.68 ± 3.81	19.42 ± 4.55	19.65 ± 4.31	22.88 ± 5.44	**24.60 ± 5.19**
SSIM	0.7303 ± 0.1909	0.7074 ± 0.1907	0.7527 ± 0.1950	0.7935 ± 0.2027	**0.8401 ± 0.1412**
NCC	0.78 ± 0.24	0.78 ± 0.24	0.80 ± 0.24	0.85 ± 0.23	**0.91 ± 0.09**

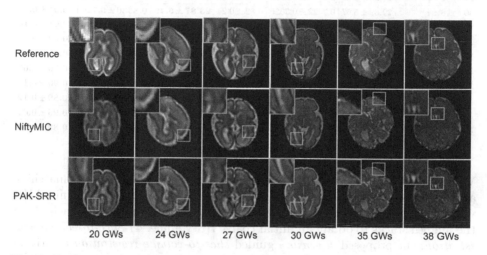

Fig. 5. Qualitative results by NiftyMIC and PAK-SRR for subjects at 20, 24, 27, 30, 35, and 38 GWs, shown in 6 columns, respectively.

use U-Net to segment brain tissue map for each different-view stack, and then convert them to *seg-priors* as shown in Fig. 3. Next, we calculate an overlap ratio for each stack with all other stacks according to their respective brain tissue maps. In this way, we can select a stack with the maximum overlap ratio with all other stacks. Finally, by using this selected stack as a target space and aligning all other stacks to it, we can build a tentative HR 3D MR image (for each fetal brain). **2)** For parameters used in Eq. 9, we set the outlier-threshold δ as 0.6, 0.7, and 0.8, as well as the hyper-parameter ε as 0.8, 0.7, and 0.6, for the three iterations (for our case of using three iterations), respectively. We set the regularization parameter α to 0.01 for all iterations. **3)** The isotropic resolution for the final HR MR image can be set to a range of 0.5 mm to 1.0 mm. We choose 0.8 mm since the resolution of our employed longitudinal atlases is 0.8 mm (Fig. 5).

4.3   Implementation Details

The following pipeline is conducted for all the experiments. **1)** Intensity correction: following the work in [2], we employ linear regression to make intensities of all other stacks consistent with the intensities of the selected target stack. **2)**

Table 2. Quantitative results by NiftyMIC and PAK-SRR for normal (N) and abnormal (ABN) subjects from 20 to 38 GWs.

GWs	# Subjects	NiftyMIC			PAK-SRR		
	N:ABN	PSNR	SSIM	NCC	PSNR	SSIM	NCC
20	1 : 0	20.76±5.10	0.7778±0.2013	0.84±0.20	**24.80±5.12**	**0.9201±0.0467**	**0.96±0.03**
24	2 : 0	21.87±5.80	0.8001±0.2062	0.87±0.20	**26.07±3.61**	**0.9226±0.0624**	**0.97±0.02**
27	2 : 0	21.86±5.26	0.7975±0.2125	0.83±0.25	**23.08±4.90**	**0.8658±0.0909**	**0.93±0.06**
30	8 : 1	22.39±5.67	0.7922±0.2025	0.82±0.24	**23.87±5.49**	**0.8536±0.1201**	**0.91±0.10**
31	11 : 1	20.29±6.13	0.7197±0.2383	0.77±0.29	**22.90±4.30**	**0.8166±0.1361**	**0.91±0.10**
32	4 : 1	24.32±3.87	0.8691±0.1000	0.90±0.16	**25.02±5.11**	**0.8700±0.1078**	**0.92±0.08**
33	2 : 2	23.06±5.47	0.8094±0.1853	0.85±0.23	**24.78±4.89**	**0.8709±0.1076**	**0.93±0.07**
34	4 : 2	23.24±5.48	0.8072±0.1985	0.84±0.25	**23.39±4.91**	**0.8333±0.1282**	**0.90±0.09**
35	4 : 3	23.90±4.57	0.8342±0.1566	0.87±0.20	**24.36±5.39**	**0.8455±0.1331**	**0.90±0.11**
36	7 : 4	23.26±5.19	0.7618±0.2340	0.86±0.21	**23.80±5.54**	**0.8006±0.1979**	**0.89±0.12**
37	2 : 2	26.22±3.39	0.8761±0.1062	0.92±0.14	**27.25±4.26**	**0.8988±0.0751**	**0.95±0.04**
38	1 : 2	23.88±5.44	0.8106±0.1605	0.86±0.21	**24.00±6.08**	**0.8124±0.1479**	**0.88±0.10**

Volume-to-volume rigid registration: we employ the symmetric block-matching algorithm REGALADIN [12] to align all stacks with the target stack. **3)** Initial HR volume Estimation: we apply the scattered data approximation approach [2] on the aligned stacks to obtain an initial coarse HR volume. **4)** *Slice-to-volume registration*: our proposed *seg-priors* guided *slice-to-volume registration* algorithm (as described in Sect. 3.1) is used to improve registration accuracy (especially in challenging areas) by introducing tissue *seg-priors*. **5)** *Volumetric reconstruction*: we employ the longitudinal atlases from Gholipour *et al.* [6] and register the corresponding time-point atlas (according to GWs) to the tentatively-estimated HR volume. Based on the guidance of *atlas-priors* (as described in Sect. 3.2), we update the reconstructed HR volume using LSMR algorithm [3]. **6)** Alternatingly optimizing **4)** and **5)** until convergence: the above-described two-stage framework performs the alternating update of transformation $T_k^{(i)}$ and 3D image $\mathbf{x}^{(i)}$ until convergence. **7)** Reorienting HR image: the reconstructed HR volumetric image is further reoriented into the standard anatomical space (defined by longitudinal atlases) for downstream analysis.

4.4   Experimental Results

We compare our PAK-SRR with four state-of-the-art methods [2,9,13,14] in terms of peak-signal-to-noise ratio (PSNR), structural similarity index measure [17] (SSIM), and NCC. As shown in Table 1, our PAK-SRR obtains the best mean performance and the best standard deviation in SSIM and NCC. The second-best method, NiftyMIC, outperforms the other three methods also by a large margin. Therefore, in the following experiments, we will focus on the comparison between our PAK-SRR and NiftyMIC.

Table 2 summarizes the statistical results of different GWs for all 66 subjects. We divide the data into normal (clinically defined normal fetal brain growth and

Table 3. Quantitative results of NiftyMIC and PAK-SRR for a normal 31 GWs subject under different combinations of input stacks. Note that A, C, and S denote stacks that are acquired with high-resolution in axial, coronal, and sagittal views, respectively.

Stacks	NiftyMIC			PAK-SRR		
	PSNR	SSIM	NCC	PSNR	SSIM	NCC
3A	15.34 ± 4.96	0.5618 ± 0.2598	0.55 ± 0.31	$\mathbf{20.72 \pm 3.96}$	$\mathbf{0.7341 \pm 0.1506}$	$\mathbf{0.77 \pm 0.12}$
3A+2C	16.22 ± 4.80	0.5832 ± 0.2358	0.62 ± 0.31	$\mathbf{20.80 \pm 3.19}$	$\mathbf{0.7485 \pm 0.1367}$	$\mathbf{0.80 \pm 0.10}$
1A+1C+1S	18.19 ± 4.95	0.6966 ± 0.2094	0.75 ± 0.24	$\mathbf{21.56 \pm 3.23}$	$\mathbf{0.7941 \pm 0.1088}$	$\mathbf{0.84 \pm 0.12}$
3A+2C+2S	15.58 ± 4.47	0.5481 ± 0.3114	0.59 ± 0.31	$\mathbf{20.50 \pm 2.78}$	$\mathbf{0.7358 \pm 0.1284}$	$\mathbf{0.79 \pm 0.09}$

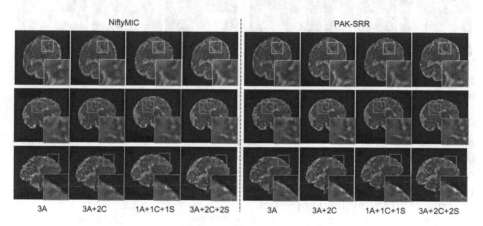

Fig. 6. Qualitative results by NiftyMIC and PAK-SRR on 31-GWs subject. A, C, and S denote stacks that are acquired with high-resolution in axial, coronal, and sagittal views, respectively.

development, with appearance close to the atlases) and abnormal (as opposed) cases for better evaluation. As can be observed, for all the cases, our PAK-SRR achieves promising performance in all the evaluation metrics. For the cases mainly containing normal subjects, our proposed PAK-SRR obtains significant improvement over NiftyMIC, especially in SSIM and NCC, which is attributed to the use of additional information from the *atlas-priors*.

For in-depth analysis, we employ different combinations of axial (A), coronal (C), and sagittal (S) views as input to evaluate the effectiveness of our proposed methods. The qualitative performance is illustrated in Fig. 6. It can be seen that NiftyMIC generates blurring boundaries and gets worse when the number of input views reduces. In contrast, PAK-SRR obtains high-quality isotropic images, even using only the axial view from three angles (3A). In fact, in Table 3, we show that, using only 3A views, PAK-SRR generates much better quantitative performance than NiftyMIC using the combinations of all views. Interestingly, the result of 1A+1C+1S outperforms 3A+2C+2S for both methods, indicating that the redundant or possibly-damaged inlier slices may degrade the performance due to misregistration, and another reason might be that the use of more

438   S. Huang et al.

Table 4. Ablation study of key components of our PAK-SRR, ($\mathcal{P}$) denotes the paired
t-test p-values for comparison of group means before and after intervention.

	NiftyMIC	SP-SRR ($\mathcal{P}$)	AP-SRR ($\mathcal{P}$)	PAK-SRR ($\mathcal{P}$)
PSNR	22.88 ± 5.44	24.26 ± 5.51 (2.8e–8)	24.18 ± 5.35 (3.7e–8)	**24.60 ± 5.19 (1.5e–8)**
SSIM	0.7935 ± 0.2027	0.8386 ± 0.1442 (7.8e–8)	0.8391 ± 0.1434 (6.2e–8)	**0.8401 ± 0.1412 (5.1e–8)**
NCC	0.85 ± 0.23	0.89 ± 0.12 (1.3e–6)	0.89 ± 0.12 (1.2e–6)	**0.91 ± 0.09 (1.1e–6)**

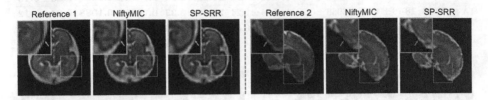

Fig. 7. Qualitative performance of NiftyMIC and SP-SRR with the assistance of tissue
segmentation in two cases (Left: 27 GWs; Right: 31 GWs).

angular stacks also causes errors in *slice-to-volume registration*. Overall, based
on both qualitative and quantitative evaluations, our PAK-SRR achieves much
clearer anatomical structures, especially for the challenging samples, and outperforms NiftyMIC for all cases of different GWs and different acquisition views,
in terms of all the evaluation metrics.

4.5   Ablation Study

In PAK-SRR, we propose two key components, i.e., *seg-priors* (SP) and *atlas-
priors* (AP). To demonstrate their effectiveness, we conduct ablation study on
two variants, SP-SRR and AP-SRR, for evaluating *seg-priors* and *atlas-priors*
(as shown in Table 4), respectively.

Effectiveness of *Seg-priors*: The *seg-priors* are obtained by tissue segmentation which is one of the key contributions. Figure 7 demonstrates the results of
NiftyMIC and SP-SRR for two fetal brains of 27 GWs and 31 GWs. We can see
that both methods obtain reasonable HR images for two fetal brains. However,
NiftyMIC reconstructs HR images not only with blurry boundaries but also with
less fidelity compared to the reference images. On the contrary, SP-SRR achieves
successful reconstruction in these challenging regions as masked and zoomed-up
in Fig. 7, demonstrating that the *seg-priors* extracted from the tissue segmentation maps indeed improve both the registration accuracy and the reconstruction
performance.

Effectiveness of *Atlas-priors*: To show the effectiveness of AP-SRR, we provide a challenging case with large motion and artifacts in Fig. 8. In this case, the
brain cannot be identified in the other two views. The NiftyMIC fails to reconstruct the k_{th} slice in the sagittal view, and shows severe streak artifacts in the
other two views. In contrast, with the guidance of longitudinal atlases (i.e., *atlas-
priors*), AP-SRR achieves promising reconstruction in all three views. It is worth

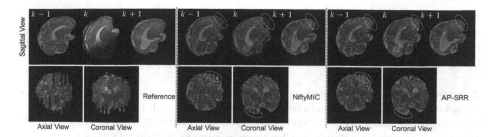

Fig. 8. Qualitative performance of NiftyMIC (middle three columns) and AP-SRR (right three columns) for a 35 GWs subject. The k_{th} reference slice of the sagittal view is corrupted with large motion and artifacts (left three column).

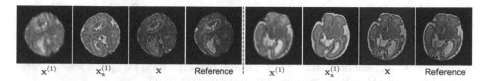

Fig. 9. Visualization of the initial image ($\mathbf{x}^{(1)}$), initially-registered atlas ($\mathbf{x}_a^{(1)}$), reconstructed HR image ($\mathbf{x}$), and reference.

noting that since the super-resolved sagittal view provided by AP-SRR intends to compensate for the motion artifact and generates more plausible images than the corrupted input slice, which is usually considered as a reference, it may reduce quantitative assessment for this particular case. In some cases, there might be significant differences between the appearances of the atlas and the reference. Even in this way, the anatomical structures of the reconstructed slices using *atlas-priors* can still be consistent with the references as shown in Fig. 9. This demonstrates the robustness of our proposed method.

5 Conclusion

In this paper, we proposed a novel prior-anatomical-knowledge assisted method for super-resolution reconstruction (SRR) of fetal brain MRI, namely PAK-SSR. Our method is built on a two-stage iterative registration-reconstruction scheme. Different from the existing methods, we introduce the *seg-priors*, derived from the segmentation maps of six tissue types in fetal brains, to guide the *slice-to-volume registration* and *volumetric reconstruction*. Moreover, we propose to employ the *atlas-priors*, derived from the longitudinal fetal brain atlases, to exploit developmental characteristics for assisting the *volumetric reconstruction* stage. We performed extensive experiments to evaluate our PAK-SRR. Experimental results show that PAK-SRR outperforms the state-of-the-art SRR methods by a large margin in terms of both quantitative and qualitative evaluations. In the ablation study, we further show that our proposed *seg-priors* and *atlas-priors* can effectively improve *slice-to-volume registration* and *volumetric reconstruction*, eventually providing reconstructed 3D images with high fidelity and clear tissue details. In the future, we will explore to leverage these high-quality fetal brain images to improve existing brain development studies [10,16] by extending the temporal dimension to pro-gestation period.

Acknowledgement. This work was supported in part by National Natural Science Foundation of China (grant number 62131015), Science and Technology Commission of Shanghai Municipality (STCSM) (grant number 21010502600), and The Key R&D Program of Guangdong Province, China (grant number 2021B0101420006).

References

1. Danielsson, P.E.: Euclidean distance mapping. Comput. Graph. Image Process. **14**(3), 227–248 (1980)
2. Ebner, M., et al.: An automated framework for localization, segmentation and super-resolution reconstruction of fetal brain MRI. Neuroimage **206**, 116324 (2020)
3. Fong, D.C.L., Saunders, M.: LSMR: an iterative algorithm for sparse least-squares problems. SIAM J. Sci. Comput. **33**(5), 2950–2971 (2011)
4. Garel, C.: MRI of the Fetal Brain: Normal Development and Cerebral Pathologies. Springer, Heidelberg (2004). https://doi.org/10.1007/978-3-642-18747-6
5. Gholipour, A., Estroff, J.A., Warfield, S.K.: Robust super-resolution volume reconstruction from slice acquisitions: application to fetal brain MRI. IEEE Trans. Med. Imaging **29**(10), 1739–1758 (2010)
6. Gholipour, A., et al.: A normative spatiotemporal MRI atlas of the fetal brain for automatic segmentation and analysis of early brain growth. Sci. Rep. **7**(1), 1–13 (2017)
7. Hou, B., et al.: 3-D reconstruction in canonical co-ordinate space from arbitrarily oriented 2-D images. IEEE Trans. Med. Imaging **37**(8), 1737–1750 (2018)
8. Kim, K., et al.: Bias field inconsistency correction of motion-scattered multislice MRI for improved 3D image reconstruction. IEEE Trans. Med. Imaging **30**(9), 1704–1712 (2011)
9. Kuklisova-Murgasova, M., Quaghebeur, G., Rutherford, M.A., Hajnal, J.V., Schnabel, J.A.: Reconstruction of fetal brain MRI with intensity matching and complete outlier removal. Med. Image Anal. **16**(8), 1550–1564 (2012)
10. Li, G., Nie, J., Wang, L., Shi, F., Lyall, A.E., Lin, W., Gilmore, J.H., Shen, D.: Mapping longitudinal hemispheric structural asymmetries of the human cerebral cortex from birth to 2 years of age. Cereb. Cortex **24**(5), 1289–1300 (2014)
11. Liang, Z.P., Lauterbur, P.C.: Principles of Magnetic Resonance Imaging: A Signal Processing Perspective. "The" Institute of Electrical and Electronics Engineers Press (2000)
12. Modat, M., Cash, D.M., Daga, P., Winston, G.P., Duncan, J.S., Ourselin, S.: Global image registration using a symmetric block-matching approach. J. Med. Imaging **1**(2), 024003 (2014)
13. Rousseau, F., et al.: Registration-based approach for reconstruction of high-resolution in utero fetal MR brain images. Acad. Radiol. **13**(9), 1072–1081 (2006)
14. Rousseau, F., et al.: BTK: an open-source toolkit for fetal brain MR image processing. Comput. Methods Programs Biomed. **109**(1), 65–73 (2013)
15. Shi, W., et al.: AFFIRM: affinity fusion-based framework for iteratively random motion correction of multi-slice fetal brain MRI. IEEE Trans. Med. Imaging (2022)

16. Wang, L., Shi, F., Lin, W., Gilmore, J.H., Shen, D.: Automatic segmentation of neonatal images using convex optimization and coupled level sets. Neuroimage **58**(3), 805–817 (2011)
17. Wang, Z., Bovik, A.C., Sheikh, H.R., Simoncelli, E.P.: Image quality assessment: from error visibility to structural similarity. IEEE Trans. Image Process. **13**(4), 600–612 (2004)
18. Xu, J., Moyer, D., Grant, P.E., Golland, P., Iglesias, J.E., Adalsteinsson, E.: SVoRT: iterative transformer for slice-to-volume registration in fetal brain MRI. In: Wang, L., Dou, Q., Fletcher, P.T., Speidel, S., Li, S. (eds.) Medical Image Computing and Computer Assisted Intervention (MICCAI 2022). LNCS, vol. 13436, pp. 3–13. Springer, Cham (2022). https://doi.org/10.1007/978-3-031-16446-0_1
19. Zhao, C., Dewey, B.E., Pham, D.L., Calabresi, P.A., Reich, D.S., Prince, J.L.: SMORE: a self-supervised anti-aliasing and super-resolution algorithm for MRI using deep learning. IEEE Trans. Med. Imaging **40**(3), 805–817 (2020)

Multimodal Learning

Q2ATransformer: Improving Medical VQA via an Answer Querying Decoder

Yunyi Liu[1], Zhanyu Wang[1], Dong Xu[2], and Luping Zhou[1(✉)]

[1] The University of Sydney, Sydney, NSW, Australia
{yunyi.liu,zhanyu.wang,luping.zhou}@sydney.edu.au
[2] The University of Hong Kong, Hong Kong, Hong Kong SAR
dongxu@cs.hku.hk

Abstract. Medical Visual Question Answering (VQA) systems play a supporting role to understand clinic-relevant information carried by medical images. The questions to a medical image include two categories: close-end (such as Yes/No question) and open-end. To obtain answers, the majority of the existing medical VQA methods rely on classification approaches, while a few works attempt to use generation approaches or a mixture of the two to process the two kinds of questions separately (classification for the close-end and generation for the open-end). The classification approaches are relatively simple but perform poorly on long open-end questions, while the generation approaches face the challenge of generating many non-existent answers, resulting in low accuracy rates. To bridge this gap, in this paper, we propose a new Transformer based framework for medical VQA (named as Q2ATransformer), which integrates the advantages of both the classification and the generation approaches and provides a unified treatment for the close-end and open-end questions. Specifically, we introduce an additional Transformer decoder with a set of learnable candidate answer embeddings to query the existence of each answer class to a given image-question pair. Through the Transformer attention, the candidate answer embeddings interact with the fused features of the image-question pair to make the decision. In this way, despite being a classification-based approach, our method provides a mechanism to interact with the answer information for prediction like the generation-based approaches. On the other hand, by classification, we mitigate the task difficulty by reducing the search space of answers. Our method achieves new state-of-the-art performance on two medical VQA benchmarks. Especially, for the open-end questions, we achieve 79.19% on VQA-RAD and 54.85% on PathVQA, with 16.09% and 41.45% absolute improvements, respectively.

Keywords: Medical VQA · Attention Mechanism · Classification

1 Introduction

Visual question answering (VQA) is known to be a challenging AI task that answers image-related questions based on image content. This process involves

A. Frangi et al. (Eds.): IPMI 2023, LNCS 13939, pp. 445–456, 2023.
https://doi.org/10.1007/978-3-031-34048-2_34

both image and natural language processing techniques and usually comprises of four key components: extracting image features, extracting question features, integrating features, and answering. Recent years have witnessed significant progress in this field [8,15]. Medical VQA is a natural extension of VQA to medical images accompanied by clinic-relevant questions. Through questioning and answering, it offers a user-friendly way to assist clinic decisions. The questions in medical VQA could be either close-end, such as Yes/No questions, or open-end.

Medical VQA is still in its early stage of development and the current performance is far from being satisfying. Most existing methods [4–7,13] could be referred to as closed-type approaches, as illustrated in Fig. 1(a), which treat each answer as a class and apply a classification model directly to the fused features of the input image-question pair to predict answers. The advantage of such approaches is that by treating VQA as classification tasks, they reduce the complexity of the task and make the answer search space smaller. Despite the good performance on Yes/No questions, closed-type approaches are difficult to accurately predict the answer for open-end questions that are much longer and more varied than the close-end ones. On the other hand, a few works [2,9] treat VQA as a generation task and employ generation-based approaches to produce answers word by word. They are referred to as the open-type approaches in Fig. 1(b). In these approaches, current word generation usually depends on previous words of the answer. Therefore, these approaches allow the image-question features to interact with the answer information for the prediction, potentially improving the long answer prediction. However, due to the tremendous search space of the generated answers, these approaches tend to produce many nonexistent answers, leading to low accuracy rates, therefore are not currently the mainstream of medical VQA. Although there are attempts to combine these two types of approaches [14], they straightforwardly treat close-end and open-end questions separately, e.g., classification for close-end ones and generation for open-end ones.

To bridge the research gap and promote medical VQA, we introduce a new model framework Q2ATransformer and refer it to as semi-open type, as shown in Fig. 1(c). By semi-open, we keep adopting classification-based approaches to make the answer search space small, and at the same time introduce the learning of answer semantics so that the fused image-question features and the answer semantics could interact for better prediction, like the generation-based approaches. Our model mitigates the shortcomings of the classification-based closed-type framework while enjoying the advantages of the generation-based open-type framework. To achieve this, we introduce a set of learnable candidate answer embeddings and let the image-question feature interact with the candidate answer embeddings by sending them through a transformer decoder. In the decoder, the candidate answer embeddings work as a query to calculate their relationships with the fused image-question features to decide the existence of the answer classes. By this, our classification considers the interaction of answer information and the fused image-question features, which is different from the

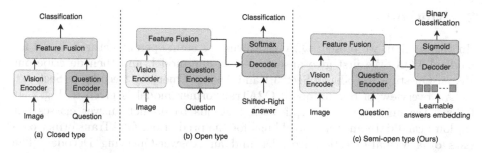

Fig. 1. Paradigms for medical VQA frameworks. (a) Closed-type framework treats VQA as predicting answer classes, where a classifier is built directly on top of the fused image-question features. (b) Open-type framework is generation-based, where the fused image-question features interact with the previous words of the answer to generate the next word of the answer through a text decoder. (c) Our proposed semi-open framework learns candidate answer embeddings through a decoder, where they interact with the fused image-question features to improve the prediction of answer classes.

existing classification-based approaches. Compared with the generation-based open-type approaches, our model reduces the task difficulty and significantly improves the accuracy rates.

Last but not the least, our model provides a uniform treatment for both the close-end and the open-end questions.

The main contribution of this paper could be summarized as follows.

First, we proposed a framework of semi-open type for medical VQA, which bridges the advantages of both the classification-based closed-type framework and the generation-based open-type frameworks in medical VQA literature. This is achieved by a designed mechanism to learn and make use of candidate answer embedding through a transformer decoder while limiting the search scope of answers through classification.

Second, we proposed a Cross-modality Fusion Network (CMAN) to effectively fuse the image and question features. It directly concatenates the two modal features instead of conducting matrix multiplication or summation for feature fusion to mitigate information loss. Then the relations between the image and question features are captured through computing self-attention on the concatenated features to produce the fused features. CMAN outperforms the commonly used image-question fusion methods in medical VQA as shown in our ablation study.

Third, our model demonstrates superior performance on two large medical VQA benchmarks for both close-end and open-end questions. Especially, our improvement on open-end question answering is overwhelming, with 16% and 41% absolute improvements on VQA-RAD and PathVQA, respectively, verifying the effectiveness of our proposed semi-open framework.

2   Method

In this section, we present Q2ATransformer, a semi-open structured model for medical VQA. We first give an overview of our model, and then describe our Visual-Question Encoder in Sect. 3.1 and Answer Querying Decoder in Sect. 3.2.

An overview of our proposed Q2ATransformer model is given in Fig. 2. It follows the majority of medical VQA methods to predict answer classes but exploits candidate answer embeddings for the prediction. Q2ATransformer consists of a Visual-Question Encoder and an Answer-Querying Decoder. The Visual-Question Encoder takes a medical image and a clinic-relevant question as the input and outputs a fused feature with both image and question information. It consists of three parts: vision encoder, question encoder, and fusion network. We use Swin transformer as our vision encoder and BERT as the question encoder. For the fusion network, we propose a Cross-modality Attention Network (CMAN) to integrate image and question features. The Answer Querying Decoder takes the fused image-question feature and learnable candidate answer embeddings as the input and outputs the probability of each candidate answer. Our Answer Querying Decoder consists of two layers of transformer decoders and a classifier to make predictions.

2.1   Visual-Question Encoder

The Visual-Question Encoder consists of an image encoder, a question encoder, and a feature fusion module, elaborated as follows.

Image Encoder. Our encoder uses the Swin Transformer [12] rather than CNN-based model as our image feature extractor. The advantages of Swin Transformer are three-fold. First, Swin Transformer makes a vision transformer to a hierarchical structure as CNN, which can make the vision transformer more flexible at various scales and has linear computational complexity with the increase of image size. Second, Swin Transformer considers cross-window connection through window shift to obtain long-range dependencies, which introduces more interactions between grids. Therefore, it can provide more regional features and interactions compared with CNN, which is more suitable for the fine-grained nature of medical images. Third, Swin Transformer was pretrained on a large dataset, so it is a very robust feature extractor. Based on these characteristics, we choose Swin Transformer to encode our input image.

Given an input image $\mathbf{I} \in \mathbb{R}^{H \times W \times C}$, where C is the number of channels and H and W stand for image height and width, respectively, the image embeddings $\mathbf{F}_i \in \mathbb{R}^{N \times D_f}$ can be expressed as $\mathbf{F}_i = \mathbf{W}_i \times \text{SwinTransformer}(\mathbf{I}) + \mathbf{b}_i$, where $\mathbf{W}_i$ and $\mathbf{b}_i$ are learnable parameters to project the output of Swin Transformer into the same dimension D_f as the question embeddings. They also provide certain flexibility to adapt Swin Transformer to the datasets in our task. Here N is the number of the extracted image regional features.

Question Encoder. For the input question, we use the pre-trained BERT model [3] as the encoder to extract text features.

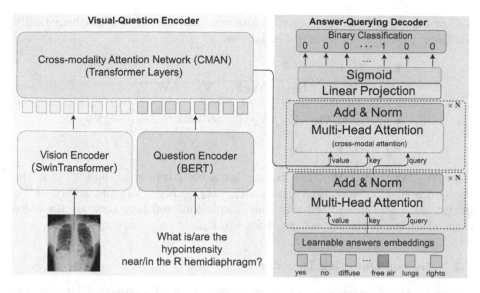

Fig. 2. Overview of Q2ATransformer. The input image-question pair is sent to a Visual-Question Encoder to extract and fuse image and question features. The Visual-Question Encoder consists of a Swin-Transformer-based vision encoder, a BERT-based question encoder, and a proposed Cross-modality Attention Network for feature fusion. The fused feature proceeds to the Answer-Querying Decoder, where the input learnable candidate answer embeddings are utilized as the query to compute the attention map and refined according to the attended fused image-question features to predict the presence of the queried answers.

BERT [3] is a successful NLP model. It incorporates context from both directions of a sentence when embedding questions. It has been applied to question answering tasks with the state-of-the-art results, and is therefore chosen in our task as the question encoder. The question embeddings $\mathbf{F}_q \in \mathbb{R}^{M \times D_f}$ is obtained by $\mathbf{F}_q = \text{BERT}(\mathbf{Q}_e)$, where $\mathbf{Q}_e$ denotes the input question and M is the question feature number and D_f the question feature dimension.

Feature Fusion Mechanism. After the image and question features are extracted, respectively, we propose the Cross-modality Attention Network (CMAN) to fuse the information from these two modalities. As medical images are fine-grained and the visual differences of clinical importance are often subtle, we explore a sophisticated way for feature fusion by investigating the interactions between image regional features and question features. In our proposed fusion module CMAN, we first integrate the image features $\mathbf{F}_i$ and the question features $\mathbf{F}_q$ by concatenating them together. Compared with the commonly used matrix multiplication or summation for feature fusion, concatenation could mitigate information loss and facilitate the subsequent computation of image-question interaction in our module. After that, the concatenated features are passed to two transformer encoder layers to calculate the relationship between every pair of image question features through the self-attention mechanism of the Transformer. In this way, we could obtain the fused feature carrying the relation of image question features with minimal information loss. Mathematically, the fused feature $\mathbf{F}_f$ is obtained as follows.

$$
\begin{aligned}
\mathbf{F}_c &= [\mathbf{F}_i; \mathbf{F}_q] \\
\mathbf{Q}_{F_c} &= \mathbf{W}_q\mathbf{F}_c, \quad \mathbf{K}_{F_c} = \mathbf{W}_k\mathbf{F}_c, \quad \mathbf{V}_{F_c} = \mathbf{W}_v\mathbf{F}_c \\
\mathbf{F}_{att} &= \mathbf{Att}(\mathbf{Q}_{\mathbf{F}_c}, \mathbf{K}_{\mathbf{F}_c}, \mathbf{V}_{\mathbf{F}_c}) = \mathbf{softmax}(\frac{\mathbf{Q}_{\mathbf{F}_c}\mathbf{K}_{\mathbf{F}_c}^{\mathbf{T}}}{\sqrt{\mathbf{d}_{\mathbf{k}}}})\mathbf{V}_{\mathbf{F}_c} \\
\mathbf{F}_f &= \mathbf{W}_f\mathbf{F}_{att} + \mathbf{b}_f
\end{aligned}
\tag{1}
$$

Here $\mathbf{W}_q$, $\mathbf{W}_k$, $\mathbf{W}_v$, $\mathbf{W}_f$, and $\mathbf{b}_f$ are learnable parameters, and ";" indicates the concatenation operation. The matrices $\mathbf{Q}_{F_c}$, $\mathbf{K}_{F_c}$, $\mathbf{V}_{F_c}$ are known as the *query*, *key*, and *value* in self-attention calculation, and here they are the linear transformation of the concatenated feature $\mathbf{F}_c$.

2.2 Answer Querying Decoder

Given an input image question pair, among a set of answers of interest, our Answer Querying Decoder predicts whether each candidate answer matches the corresponding image question pair and uses the candidate with the highest probability as the final answer. For this purpose, we employ a two-layer transformer decoder followed by a linear projector as our classifier, and introduce a set of learnable candidate answer embeddings together with the fused image-question feature $\mathbf{F}_f$ as the input of the decoder. Assuming there are C answer classes in total, we need C candidate answer embeddings with one-to-one correspondence to the C answer classes. These answer embeddings, collectively represented by a matrix $\mathbf{A}$, are randomly initialised and will be updated during training through a self-attention module, a cross-attention module, and a feed-forward network(FFN) in order. Both the self-attention module and the cross-attention module implement the multi-head self-attention ($MSA(query, key, value)$) but with different *key*, *query*, and *value*. The self-attention module computes the relation between different answer embeddings by using $\mathbf{A}$ to construct all the *key*, *query*, and *value* matrices. The cross-attention module cares about the relation between the answer embeddings $\mathbf{A}$ and the fused image-question feature $\mathbf{F}_f$. It thus uses the answer embedding $\mathbf{A}$ as the *query* and the fused image-question feature $\mathbf{F}_f$ as the *key* and *value* to compute the attention and further updates the answer embeddings by combining the attended image-question features. Mathematically, denoting the answer embeddings at the l-th layer as $\mathbf{A}_l$, it will be updated from the output of the previous layer $\mathbf{A}_{l-1}$ as follows:

$$
\begin{aligned}
\mathbf{A}_l &= MSA(\mathbf{A}_{l-1}, \mathbf{A}_{l-1}, \mathbf{A}_{l-1}) \\
\mathbf{A}_l &= MSA(\mathbf{A}_l, \mathbf{F}_f, \mathbf{F}_f) \\
\mathbf{A}_l &= FFN(\mathbf{A}_l),
\end{aligned}
\tag{2}
$$

where $l = 1 \cdots L$ and L is the number of Transformer decoder layers. Through this process, the image-question features are injected into the answer embeddings and used to refine the latter. The refined C answer embeddings are sent to the final linear projection layer followed by a sigmoid function $\sigma(\cdot)$ to predict the probabilities of answer classes. That is:

$$\mathbf{p} = \sigma(\mathbf{W}_A \mathbf{A}_L + \mathbf{b}), \tag{3}$$

where $\mathbf{W}_A$ and $\mathbf{b}$ are learnable parameters, and $\mathbf{p}$ is a vector comprising of C probabilities corresponding to C answer classes. The answer class with the highest probability is chosen as the predicted answer.

2.3 Loss Function

Medical VQA faces a significant class imbalance problem: Yes/No answer classes are much larger than long open answer classes. In order to address the sample imbalance problem more effectively, we choose a simplified asymmetric loss, which is a variant of focal loss while the hyper-parameter γ is set differently for positive and negative classes, as shown in Eq. 4:

$$\mathcal{L} = \frac{1}{C} \sum_{c=1}^{C} \begin{cases} (1 - p_c)^{\gamma+} \log(p_c), \ y_c = 1 \\ (p_c)^{\gamma-} \log(1 - p_c), \ y_c = 0 \end{cases} \tag{4}$$

where y_c is the ground-truth binary label, indicating if the input image-question pair has the answer class c, while p_c is the predicted probability for the class c. The total loss is computed by averaging this loss over all samples in the training data set. We set the hyper-parameters $\gamma+ = 1$ and $\gamma- = 4$ by default.

3 Experiments and Results

3.1 Datasets

We conduct our experiments on two medical VQA benchmarks: VQA-RAD [11] and PathVQA [7], which are described as follows.

VQA-RAD is the most commonly used radiology dataset seen to date, containing 315 images and 3515 question-answer pairs, each corresponding to at least one question-answer pair. The types of questions include 11 categories: "anomalies", "properties", "color", "number", "morphology", "organ type", "other", and "section". 58% of the questions are close-end questions and the rest are open-end questions. The images are of the body's head, chest, and abdomen. Manual division of the training and test sets is required. For comparability, we divide the data set according to the MMQ method [4].

PathVQA is a dataset for exploring pathology VQA. Images with captions were extracted from digital resources (electronic textbooks and online libraries). Open-end questions account for 50.2% of all questions. For the closed-end yes/no questions, the answers are balanced with 8,145 yes and 8,189 no questions. PathVQA consists of 32,799 question-answer pairs, 1,670 pathology images collected from two pathology textbooks, and 3,328 pathology images collected from the PEIR digital library [1]. For comparability, we also divide the data set according to the MMQ method [4].

3.2   Comparison with the State-of-the-Art Methods

We compare our proposed model with 7 state-of-the-art (SOTA) Medical VQA approaches, including StAn [7], BiAn [7], MAML [6], MEVF [13], MMQ [4], Pub-MedCLIP [5], and MMBERT [9]. The first 6 methods are classification-based approaches. They are chosen because they are among the best performers on the two benchmarks VQA-RAD and PathVQA. The last method MMBERT [9] is chosen as a representative of generation-based approaches, which has the reported performance on VQA-RAD. Except PubMedCLIP [5] and MMBERT [9] whose results are quoted from their original papers, the results of other methods are quoted from MMQ [4]. It is noted that same as our Q2ATransformer, PubMedCLIP [5] and MMBERT [9] employ the same data split as MMQ [4]. Therefore these results are strictly comparable.

As shown in Table 1, on both datasets, our Q2ATransformer consistently outperforms the compared models. Specifically, compared with the second best performer, on VQA-RAD, we achieve an accuracy of 79.19% (16.09% absolute improvement) on Open-end questions, 81.2% (1.2% absolute improvement) on close-end questions, and 80.48% (8.48% absolute improvemen) across all questions; on PathVQA, we achieve an accuracy of 54.85% (41.45% absolute improvement) on open-end questions, 88.85% (4.85% absolute improvement) on Yes/No questions, and 74.61% (25.81% absolute improvement) across all questions. The results could be even better if we increase the dimension of the candidate answer embeddings, as shown in our ablation experiments. From these results, we can see our Q2ATransformer demonstrates overwhelming advantages on open-end questions, which supports our analysis that by interacting answer information with fused image-question features, our model could better tackle long answer questions. Our model also outperforms the generation-based method MMBERT [9], since we reduce the search space of answers while MMBERT [9] could generate non-existent answers.

Table 1. Performance comparison of different methods. † and ‡ indicate the methods are classification-based(closed-type) or generation-based(open-type), respectively.

References Methods	Fusion Methods	PathVQA			VQA-RAD		
		Free-form	Yes/No	Over-all	Open-ended	Close-ended	Over-all
StAn† [7]	SAN	1.6	59.4	30.5	24.2	57.2	44.2
BiAn† [7]	BAN	2.9	68.2	35.6	28.4	67.9	52.3
MAML† [6]	SAN	5.4	75.3	40.5	38.2	69.7	57.1
	BAN	5.9	79.5	42.9	40.1	72.4	59.6
MEVF† [13]	SAN	6.0	81.0	43.6	40.7	74.1	60.7
	BAN	8.1	81.4	44.8	43.9	75.1	62.7
MMQ† [4]	SAN	11.2	82.7	47.1	46.3	75.7	64.0
	BAN	13.4	84.0	48.8	53.7	75.8	67.0
PubMedCLIP† [5]	-	-	-	-	60.1	80	72.1
MMBERT‡ [9]	-	-	-	-	63.1	77.9	72.0
Ours		**54.85**	**88.85**	**74.61**	**79.19**	**81.2**	**80.48**

Table 2. Ablation Studies. BAN, SAN, and CMAN stand for Bilinear Attention Network [10], Stacked Attention Network [16] and ours Cross-modality Attention Network, respectively; Decoder refers to our Answer-Querying Decoder.

#	BAN	SAN	CMAN	Decoder	VQA-RAD			PathVQA		
					open	closed	overall	free-form	yes/no	overall
1	✓				43.62	75.56	64.1	15.03	78.24	51.69
2	✓			✓	54.36	80.07	70.84	44.78	88.29	70.09
3		✓			61.07	77.07	71.33	44.58	86.29	68.88
4		✓		✓	73.83	80.08	77.83	52.88	88.44	73.51
5			✓		69.13	76.32	73.73	47.53	86.73	70.31
6			✓	✓	79.19	81.2	80.48	54.85	88.85	74.61

3.3 Ablation Study

To investigate the contributions of our proposed feature fusion module CMAN and the decoder for answer querying, we conduct extensive ablation studies to compare different configurations of our model, as presented in Table 2. Here BAN, SAN, and CMAN are three attention networks to fuse image and question features, representing Bilinear Attention Network [10], Stacked Attention Network [16] and ours Cross-modality Attention Network, respectively; Decoder represents our Answer-Querying Decoder. The symbol ✓ indicates the inclusion of the corresponding component. All the experiments in Table 2 are performed based on the same image and question encoders.

Impact of the CMAN. The benefit of using CMAN can be well reflected by the improvement from #1 to #5 or from #3 to #5 in Table 2, indicating the effectiveness of our proposed CMAN over BAN and SAN for image and question feature fusion. This is because compared with BAN which multiplies the image and question features or SAN which does a direct matrix summation for fusion, our CMAN directly concatenates the two channels of features together and then calculates attention for fusion. Through this, our CMAN mitigates the information loss due to the multiplication or summation operation during feature fusion in BAN or SAN.

Contribution of the Decoder. As shown, the inclusion of our answer querying decoder could boost the model performance. To verify the robustness of our decoder, we incorporate it with three different attention modules shown in #2, #4 and #6. By comparing #2 to #1, #4 to #3, or #6 to #5, it can be observed that our answer querying decoder can bring significant performance gain with all three attention mechanisms. Especially, when combining our CMAN and decoder, we can achieve the new SOTA results.

Impact of Answer Embedding Size. The experimental results in Fig. 3 show that as the dimension of answer embedding increases, the model's performance improves while the best result is obtained when the embedding size is around

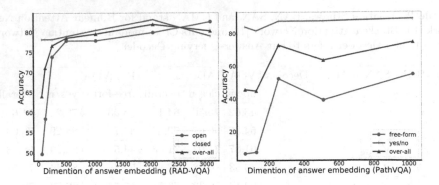

Fig. 3. Ablation study about different dimensions of answer embeddings.

2048. However, increasing the embedding size will also increase the computational cost, while the performance improvement becomes saturated. As a trade-off, our model adopts 1024-dimensional answer embeddings.

3.4 Qualitative Results

Example results from PathVQA and VQA-RAD datasets are given in Fig. 4 and Fig. 5, respectively. As can be seen, for these examples where MMQ using BAN for feature fusion fails, our Q2ATransformer w/o decoder has been able to correct most of them using the proposed CMAN fusion module. The performance could be further improved with our Answer Querying Decoder by learning candidate answer embedding through their interactions with the fused image-question features.

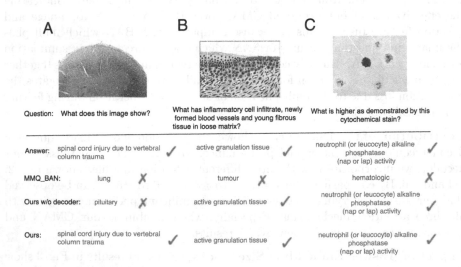

Fig. 4. Example results from PathVQA dataset.

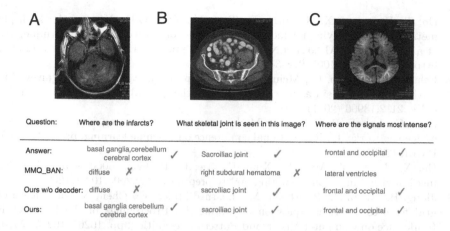

Fig. 5. Example results from VQA-RAD dataset.

3.5   Limitation and Discussion

As described in Sect. 2, we treat each answer as a learnable embedding and use all embeddings as the query to compute the attention map in our decoder. Since we compute the global self-attention, this may increase computation overhead when the number of answer classes is very large. This problem has been encountered in NLP when processing long sequences. Some solutions have been proposed, such as dynamically computing sparse attention , which can significantly reduce computational overhead and will be explored in our future work.

4   Conclusion

In this paper, we propose a semi-open framework for medical VQA, which successfully enrolls answer semantic information into the answer class prediction process through our designed mechanism to correlate the answering embeddings with the fused image-question features, which improves the accuracy significantly. It enriches the existing closed-type and open-type medical VQA frameworks and refreshes the SOTA performance on the two benchmarks, especially for the open-end questions.

References

1. Peir digital library. http://peir.path.uab.edu/library/index.php?/category/2
2. Ambati, R., Dudyala, C.R.: A sequence-to-sequence model approach for imageclef 2018 medical domain visual question answering. In: 2018 15th IEEE India Council International Conference (INDICON), pp. 1–6. IEEE (2018)
3. Devlin, J., Chang, M.W., Lee, K., Toutanova, K.: Bert: Pre-training of deep bidirectional transformers for language understanding. arXiv preprint arXiv:1810.04805 (2018)

4. Do, T., Nguyen, B.X., Tjiputra, E., Tran, M., Tran, Q.D., Nguyen, A.: Multiple meta-model quantifying for medical visual question answering. In: de Bruijne, M., et al. (eds.) MICCAI 2021. LNCS, vol. 12905, pp. 64–74. Springer, Cham (2021). https://doi.org/10.1007/978-3-030-87240-3_7

5. Eslami, S., de Melo, G., Meinel, C.: Does clip benefit visual question answering in the medical domain as much as it does in the general domain? arXiv preprint arXiv:2112.13906 (2021)

6. Finn, C., Abbeel, P., Levine, S.: Model-agnostic meta-learning for fast adaptation of deep networks. In: International conference on machine learning, pp. 1126–1135. PMLR (2017)

7. He, X., Zhang, Y., Mou, L., Xing, E., Xie, P.: Pathvqa: 30000+ questions for medical visual question answering. arXiv preprint arXiv:2003.10286 (2020)

8. Jiang, H., Misra, I., Rohrbach, M., Learned-Miller, E., Chen, X.: In defense of grid features for visual question answering. In: Proceedings of the IEEE/CVF Conference on Computer Vision and Pattern Recognition, pp. 10267–10276 (2020)

9. Khare, Y., Bagal, V., Mathew, M., Devi, A., Priyakumar, U.D., Jawahar, C.: Mmbert: multimodal bert pretraining for improved medical VQA. In: 2021 IEEE 18th International Symposium on Biomedical Imaging (ISBI), pp. 1033–1036. IEEE (2021)

10. Kim, J.H., Jun, J., Zhang, B.T.: Bilinear attention networks. In: Advances in Neural Information Processing Systems, vol. 31 (2018)

11. Lau, J.J., Gayen, S., Ben Abacha, A., Demner-Fushman, D.: A dataset of clinically generated visual questions and answers about radiology images. Sci. Data 5(1), 1–10 (2018)

12. Liu, Z., et al.: Swin transformer: Hierarchical vision transformer using shifted windows. In: Proceedings of the IEEE/CVF International Conference on Computer Vision, pp. 10012–10022 (2021)

13. Nguyen, B.D., Do, T.-T., Nguyen, B.X., Do, T., Tjiputra, E., Tran, Q.D.: Overcoming data limitation in medical visual question answering. In: Shen, D., et al. (eds.) MICCAI 2019. LNCS, vol. 11767, pp. 522–530. Springer, Cham (2019). https://doi.org/10.1007/978-3-030-32251-9_57

14. Ren, F., Zhou, Y.: CGMVQA: a new classification and generative model for medical visual question answering. IEEE Access 8, 50626–50636 (2020)

15. Wu, C., Liu, J., Wang, X., Li, R.: Differential networks for visual question answering. In: Proceedings of the AAAI Conference on Artificial Intelligence, vol. 33, pp. 8997–9004 (2019)

16. Yang, Z., He, X., Gao, J., Deng, L., Smola, A.: Stacked attention networks for image question answering. In: Proceedings of the IEEE Conference on Computer Vision and Pattern Recognition, pp. 21–29 (2016)

Using Multiple Instance Learning to Build Multimodal Representations

Peiqi Wang[1]($\boxtimes$), William M. Wells[1], Seth Berkowitz[2], Steven Horng[2], and Polina Golland[1]

[1] CSAIL, MIT, Cambridge, MA, USA
wpq@mit.edu, polina@csail.mit.edu
[2] BIDMC, Harvard Medical School, Boston, MA, USA

Abstract. Image-text multimodal representation learning aligns data across modalities and enables important medical applications, e.g., image classification, visual grounding, and cross-modal retrieval. In this work, we establish a connection between multimodal representation learning and multiple instance learning. Based on this connection, we propose a generic framework for constructing permutation-invariant score functions with many existing multimodal representation learning approaches as special cases. Furthermore, we use the framework to derive a novel contrastive learning approach and demonstrate that our method achieves state-of-the-art results in several downstream tasks.

Keywords: representation learning · multiple instance learning

1 Introduction

In this paper, we propose a framework for designing multimodal representation learning methods that encompasses previous approaches as special cases and implies a new algorithm for multimodal learning that advances the state of the art. Specifically, we establish a connection between self-supervised representation learning based on contrastive learning and multiple instance learning [3] and show that they share similar assumptions and goals. We bring insights from multiple instance learning to offer a fresh perspective on self-supervised representation learning and ideas for performance improvements. With this connection in mind, we derive a novel algorithm for learning image-text representations that capture the structure shared between the two modalities and generalize well in a variety of downstream tasks.

We aim to establish alignment between images and associated text to improve clinical workflow. For example, an image model that mimics the radiologists' interpretation could retroactively label images to select relevant patients for a clinical trial. Further, local alignment between image regions and text fragments (e.g., sentences) promises to benefit many downstream tasks. For example, cross-modal retrieval can provide description of an image region for automated documentation or enable comparisons with similar previously imaged patients for

A. Frangi et al. (Eds.): IPMI 2023, LNCS 13939, pp. 457–470, 2023.
https://doi.org/10.1007/978-3-031-34048-2_35

better interpretation based on local anatomy or pathology. Similarly, radiologists documenting findings can verify the accuracy of the report by noting if the referred location (i.e., visual grounding of the text) is consistent with their impression of the image.

Self-supervised representation learning is a useful tool for reducing annotation burden for machine learning models in medical imaging. Despite the need and opportunities for automation, development of robust machine learning methods is held back by the lack of annotations that serve as the supervision signal for learning. Self-supervised representation learning on paired image-text data offers two advantages: (i) learning requires no further annotations and (ii) treating text as "labels" enables us to use natural language to reference visual concepts and vice versa [30]. Thus, we focus on learning image-text multimodal representations but the proposed framework is broadly applicable to representation learning on other multimodal data.

Learning joint representations involves training image and text encoders to perform self-supervised tasks on paired image-text data [5, 22, 25] and evaluating on relevant downstream tasks. We focus on contrastive learning, i.e., classifying image-text pairs as matched (i.e., corresponding to the same imaging event), or mismatched. Contrastive learning has been applied to the medical domain, demonstrating impressive transfer capabilities on a diverse set of tasks [2, 4, 13, 23, 28, 36]. The biggest improvements come from addressing challenges unique to this domain, e.g., the use of cross attention to deal with the lack of effective pathology detectors [13] and adaptation of language models to address linguistic challenges in clinical notes [2]. Training the models has involved increasingly complex contrastive loss functions that treat image and text symmetrically [2, 4, 13, 36] and on multiple scales [2, 13, 23, 28]. In contrast to previous work that relies on many loss terms, our proposed contrastive loss is simple to implement and yields superior performance.

Borrowing ideas from multiple instance learning, we treat local image region features as "data" and sentence features as (complex) "labels". Multiple instance learning is a type of weakly supervised learning that is effective for problems that lack fine-grain annotations [3]. For example, it can help to locate tumor cells in whole slide images with just image-level labels [21]. Central to multiple instance learning is the construction of permutation-invariant score functions [14], and the choice of how the instance scores or features are aggregated to be evaluated against an image-level label. Effective instance aggregators leverage domain knowledge [8], e.g., the Noisy-OR aggregator for drug activity prediction [26], the Noisy-AND aggregator for cellular phenotype classification [19]. In our work, we extend multiple instance classification to contrastive learning by constructing permutation-invariant image-text score functions. Drawing on insights from multiple instance classification with correlated instances [21], our proposed instance aggregator exploits correlation among instances to build representations that perform well in downstream tasks.

Many prior multiple instance learning methods focused on one particular task of interest, e.g., detection [35], region classification [7], or retrieval [17].

Some investigated the choices of instance aggregators for more than one downstream task [10,27] but are limited in generality (i.e., not intended for other applications) and scope (i.e., explored a few simple instance aggregators). In contrast, our proposed framework for constructing permutation-invariant score functions can be readily applied to other applications. We systematically investigate instance aggregators and their effect on representation learning, leading to a novel approach for learning joint representations. We evaluate the resulting image-text representations on a diverse set of downstream tasks and demonstrate state-of-the-art performance across all tasks in the context of a large set of chest X-ray images and associated radiological reports.

2 Method

We first introduce notation and discuss the local and global approaches for constructing permutation-invariant image-document score functions at the core of the learning procedure. We then instantiate the framework for a specific choice of aggregators for contrastive learning.

2.1 Problem Setup

A local D-dimensional representation of an image with N proposed regions is a collection of N features vectors $x_n \in \mathcal{X} \subset \mathbb{R}^D$, $n \in \{1, \cdots, N\}$. In our experiments, we use regular tiling to generate image regions and leave more sophisticated proposal methods (e.g., [31]) for future work. A local representation of a M-sentence document (e.g., a radiology report) is a collection of sentence feature vectors $y_m \in \mathcal{Y} \subset \mathbb{R}^D$, $m \in \{1, \cdots, M\}$.

Function $h : \mathcal{X} \times \mathcal{Y} \to \mathbb{R}$ measures the similarity between representations, e.g., $h(x_n, y_m)$ is the similarity between a region and a sentence. In our experiments, we use cosine similarity $h(x, y) = \langle x, y \rangle / (\|x\| \|y\|)$, though the formulation accepts any differentiable similarity function.

For any vector space $\mathcal{U}$, aggregator function $\pi : \mathcal{P}(\mathcal{U}) \to \mathcal{U}$ aggregates elements in the input set into a "representative". $\mathcal{P}(\mathcal{U})$ is the set of all finite subsets of $\mathcal{U}$. For example, $\pi(\{x_n\}) = \frac{1}{N} \sum_n x_n$ aggregates N region features $x_n \in \mathcal{X}$ by averaging them, while $\pi(\{h_n\}) = \max_n h_n$ aggregates N similarity scores into a single score by computing the maximum score. We restrict our attention to aggregators that are permutation-invariant, i.e., they treat their input as an unordered set rather than an ordered vector.

Permutation-invariant image-document score function $S : \mathcal{P}(\mathcal{X}) \times \mathcal{P}(\mathcal{Y}) \to \mathbb{R}$ measures the similarity between an image and a document based on region features $\{x_n\}$ and sentence features $\{y_m\}$.

2.2 Local and Global Permutation-Invariant Score Functions

Contrastive representation learning can be seen as maximizing the likelihood of correctly classifying image-text pairs as matched or mismatched. Since supervision is provided at the image-document level, we define a framework to build permutation-invariant image-document score functions.

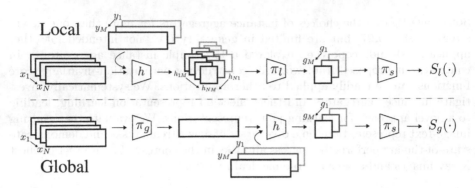

Fig. 1. Local (top) and global (bottom) image-document score functions.

The local approach aggregates region-sentence scores into an image-sentence score. The image-sentence score g_m for sentence m in the document is obtained by applying a local aggregator function π_l to region-sentence scores, i.e., $g_m = \pi_l(\{h(x_n, y_m)\}_n) \triangleq \pi_l(\{h(x_1, y_m), \cdots, h(x_N, y_m)\})$.

The global approach first aggregates local region features $\{x_n\}$ into a single image feature vector $\pi_g(\{x_n\})$ using a global aggregator function π_g. The image-sentence score g_m is computed using the similarity function h on the image feature vector $\pi_g(\{x_n\})$ and sentence feature vector y_m, i.e., $g_m = h(\pi_g(\{x_n\}), y_m)$.

In both approaches, the image-document score S is obtained by aggregating image-sentence scores with another aggregator function π_s, i.e., $S(\{x_n\}, \{y_m\}) = \pi_s(\{g_m\})$. Figure 1 illustrates the framework for constructing S. To summarize, the local and global image-document scores S_l and S_g are computed as follows:

$$S_l(\{x_n\}, \{y_m\}) = \pi_s(\{\pi_l(\{h(x_n, y_m)\}_n)\}_m), \tag{1}$$
$$S_g(\{x_n\}, \{y_m\}) = \pi_s(\{h(\pi_g(\{x_n\}), y_m)\}_m). \tag{2}$$

As the aggregator functions are permutation-invariant, the image-document score function S is naturally permutation-invariant as well. We emphasize that S treats image features and text features differently, and that the order of application of similarity evaluation $h(\cdot)$ and aggregators $\pi(\cdot)$ is empirically relevant. This design decision is motivated by the fact that each sentence in a radiology report represent a concept and its location in the image, i.e., it is akin to a label for some region in the image. The converse is not necessarily true as some parts of the image are not described in the report.

2.3   Representation Learning with LSE+NL Aggregators

In this section, we introduce our method LSE+NL for learning multimodal representations that relies on a combination of local and global image-document score functions and an asymmetric text-to-image contrastive loss.

Inspired by [21], we use a soft maximum function to identify the most relevant region for a sentence, i.e., the critical region, and attend more to regions that are similar to the critical region. Specifically, the local aggregator π_l is the log-sum-exp (LSE) function

$$\pi_l(\{h_n\}) = \frac{1}{\gamma_l} \log \sum_{n=1}^{N} \exp(\gamma_l \, h_n), \tag{3}$$

where γ_l is a scale parameter that controls how well the LSE function approximates the max function. The global aggregator π_g linearly combines the region features using the distance to the critical region as weights, i.e.,

$$\pi_g(\{x_n\}) = \sum_{n=1}^{N} \frac{\exp(\gamma_g \langle Ax_n, Ax_k \rangle)}{\sum_{n'=1}^{N} \exp(\gamma_g \langle Ax_{n'}, Ax_k \rangle)} x_n, \tag{4}$$

where k is the index of the critical region, i.e., $k = \arg\max_n h(x_n, y_m)$, A is a learned weight matrix, and γ_g is the scale parameter for the softmax function. We can interpret π_g as a form of attention where regions that are more similar to the critical region are given a higher attention weight. In effect, π_g exploits the correlation between each region and the critical region using attention. In addition, π_g can be seen as a form of non-local (NL) network [34]. Both π_l and π_g are permutation-invariant functions. We choose π_s to be the average function.

We use the local and global image-document scores in (1) and (2) computed with our choice of π_l and π_g for contrastive learning. Given a document, we form an image-document score vector $s \triangleq (s^+, s_1^-, \cdots, s_K^-)$ where $s^+ \in \mathbb{R}$ is the image-document score with its matched image and $s_k^- \in \mathbb{R}$ for $k = 1, \cdots, K$ is the image-document score with K mismatched images. We use s_l and s_g to denote $(K+1)$-length score vectors defined above computed using the local and the global score functions respectively. The image and text encoders are trained to minimize $\mathcal{L}(s_l) + \mathcal{L}(s_g)$ over documents in the training set where $\mathcal{L}$ is the text-to-image contrastive loss [29, 36]

$$\mathcal{L}(s) \triangleq -\log \frac{\exp(\gamma \, s^+)}{\exp(\gamma \, s^+) + \sum_{k=1}^{K} \exp(\gamma \, s_k^-)} \tag{5}$$

with scale parameter γ. In the equation above, s is either vector s_l computed using (1) with π_l defined in (3) or vector s_g computed using (2) with π_g defined in (4). The image-to-text contrastive loss where the negative scores are computed for an image with K different mismatched documents is often used alongside $\mathcal{L}$ in prior work [2,4,13,36]. We choose to treat images and text asymmetrically and show that the simple text-to-image contrastive loss is sufficient to induce representations that generalize well.

3   Connection to Multiple Instance Learning

In multiple instance learning [3], a set that contains many instances $\{x_1, \cdots, x_N\}$ is referred to as a bag. The training set consists of bags and their associated

Table 1. Taxonomy of related methods for image-language representation learning in our multiple instance learning inspired framework. For each method, we report image segments captured by x_n (region or video), language segments captured by y_m (word, sentence, or audio), local aggregator π_l if used (Max or LSE), global aggregator π_g if used (Avg, NN for generic non-linear functions, cross attention (CA) $\pi_l(\{x_n\}, y_m) = \sum_n \exp(\langle x_n, y_m \rangle)/\sum_{n'} \exp(\langle x_{n'}, y_m \rangle)x_n$, or NL in (4)), and the final score aggregator π_s (Sum, Max, LSE, Id, Avg).

Methods	x_n	y_m	π_l	π_g	π_s
NeuralTalk [16]	region	word	Max	–	Sum
DAVEnet-MISA [10]	region	audio	Max	–	Sum
MIML [9]	video	audio	Max	–	Max
MIL-NCE [27]	video	sentence	–	Avg	LSE
ConVIRT/CLIP [30,36]	region	sentence	–	NN ∘ Avg	Id
GLoRIA/BioViL [2,13]	region	word	–	CA	LSE
	region	sentence	–	Avg	Id
LSE+NL (Ours)	region	sentence	LSE	–	Avg
	region	sentence	–	NL	Avg

bag labels y while the instance labels are not provided. For binary bag labels, a positive bag is guaranteed to include at least one positive instance, while a negative bag includes no positive instances. The bag-level labels are used to train classifier to assign instance-level and bag-level labels in new, unseen bags.

Existing image-text representation learning algorithms that are either predictive [6] or contrastive [30] can be seen as a form of multiple instance learning. Specifically, we can view an image as a bag of region features and the corresponding sentence that describes the image as the bag label. Instead of taking on binary values, the bag labels can represent arbitrary categories via natural language. Although the exact region that corresponds to the sentence is unknown, the matched image contains at least one region that corresponds to the text while a randomly sampled image most likely does not. Similar to multiple instance learning, self-supervised representation learning methods use these assumptions for learning.

More generally, we consider the text label as a bag of sentences. For example, sentences describing findings within a chest X-ray image most likely can be permuted without changing the overall meaning. Therefore, representation learning can be interpreted as predicting the label bag $\{y_m\}$ given the input bag $\{x_m\}$. This setup corresponds to multi-instance multi-label learning [37].

Moreover, multiple instance learning and multimodal representation learning share comparable goals. Multiple instance learning aims to align instances and bags with labels such that the pre-trained model performs well in classification tasks. Multimodal representation learning aims to align images and their subregions with text such that the pre-trained model perform well on tasks that rely on such alignment, e.g., image classification relies on image-sentence alignment, visual grounding and cross-modal retrieval rely on region-sentence alignment.

There are two main multiple instance learning approaches, instance-level and embedding-level approaches [1]. The instance-level approach computes the bag score by aggregating the instance scores, while the embedding-level approach computes the bag score based on a bag feature that is aggregated from the instance features. The local and global approaches in Sect. 2.2 are extensions of the instance and embedding approaches to contrastive learning.

This parallel enables us to analyze prior methods as instances of the framework defined in Sect. 2.2 that is inspired by multiple instance learning (Table 1). We make one generalization to the formulation in Sect. 2.2 to accommodate cross attention [20]: the local aggregator function π_l can potentially rely on label features y_m to multiplex its behavior, i.e., $\pi_l : \mathcal{P}(\mathcal{X}) \times \mathcal{Y} \to \mathcal{X}$. In summary, a diverse set of aggregators π_l, π_g, π_s have been demonstrated on multimodal representation learning at varying scales, implying there may not be a single set of aggregators that works well for every problem. More realistically, the best aggregator functions are the ones that fit application-specific assumptions well.

4 Experiments

We illustrate the proposed approach by building a representation of frontal chest X-ray images and associated radiology reports and using it in downstream tasks. In all of the experiments, the data used for representation learning is disjoint from the test sets used to evaluate the downstream tasks.

We normalize the images and resize them to 512×512 resolution. We apply random image augmentations, i.e., 480×480 random crops, brightness and contrast variations, and random affine transforms (only for image model fine-tuning during evaluation). We use PySBD [32] for sentence tokenization.

We employ ResNet-50 [12] as the image region encoder and CXR-BERT [2] as the sentence encoder. Each encoder is followed by a linear projection to a 128 dimension embedding space. In particular, the projected ResNet-50 conv-5 activations act as the region features $\{x_n\}$ and the projected mean-pooled contextualized word embeddings acts as the sentence features $\{y_m\}$.

4.1 Representation Learning

We use a subset of 234,073 chest X-ray images and report from MIMIC-CXR [15] for representation learning. We randomly initialize the image encoder and use the CXR-BERT model [2] pre-trained on a biomedical corpus (i.e., the stage II model) as the sentence encoder. We use the AdamW optimizer [24] and decay the initial learning rate of 5e−5 using a cosine schedule with 2k warmup steps. we initialize γ to 14 and optimize this hyperparameter alongside the encoder parameters. We set other scale parameters as follows: $\gamma_l = 0.1, \gamma_g = e$. We use a batch size of 64. For each image in the batch, we sample 5 sentences, with replacement if needed, to make up the label bag. Here, $N = 225$ and $M = 5$.

4.2   Downstream Tasks

Image Classification. To evaluate zero-shot (ZS) and fine-tuned (FT) classification performance, we use the same split of RSNA Pneumonia (RSNA) [33] as in [13], specifically, 18,678/4,003/4,003 for training/validation/testing. To evaluate in-distribution fine-tuned classification performance in the ablation study, we use 5 CheXpert labels (Atelectasis, Cardiomegaly, Edema, Pleural Effusion, Pneumothorax) on the MIMIC-CXR data set [15] that we denote MIMIC-CheXpert (CheX). There are roughly 1k images in the test set associated with each CheXpert label. To evaluate the data efficiency of representation learning approaches, we use different amounts of training data (1% and 100%).

For zero-shot image classification, we first tokenize and encode the class-specific text prompts (e.g., "Findings suggesting pneumonia." and "No evidence of pneumonia."). For each image, we assign a binary label that corresponds to the prompt with the higher image-sentence score. We find it important to normalize the scores to $[0, 1]$ for each class before applying the softmax. For fine-tuned image classification, we use the Adam optimizer [18] with a learning rate of 3e-3 to optimize the randomly initialized weights and a bias over the mean-pooled region features while keeping the encoder weights fixed. For RSNA Pneumonia, we report accuracy and AUC. For MIMIC-CheXpert, we report the average AUC over five binary classification tasks.

Visual Grounding. We evaluate visual grounding performance using the MS-CXR region-sentence annotations [2]. This data set consists of 1,448 bounding boxes over 1,162 images, where each bounding box is associated with a sentence that describes its dominant radiological feature. We compute region-sentence scores to quantify how well the sentence is localized in the image. We report a measure of discrepancy between region-sentence scores inside and outside the bounding box, i.e., contrast-to-noise ratio (CNR) [2], and how well the thresholded region-sentence scores overlap with the bounding box on average, i.e., mean intersection over union (mIoU). In contrast to [2], we pick thresholds that span $[-1, 1]$ in 0.05 increments to compute the mIoU for a fair comparison.

Cross-Modal Retrieval. We evaluate cross-modal retrieval performance using the MS-CXR data set as well. We compute the bounding box features from the region features with RoIAlign [11]. We compute box-sentence scores and sort them to retrieve items in one modality given a query from the other modality. The correctly retrieved item is the one that is paired with the query item. We report the fraction of times the correct item was found in the top K results (R@K) and the median rank of the correct item in the ranked list (MedR).

4.3   Results

Comparison with State-of-the-Art Methods. We compare the proposed approach LSE+NL with the state-of-the-art methods GLoRIA [13] and BioViL

Table 2. Image classification performance on the RSNA Pneumonia data set. We report accuracy and AUC on zero-shot and fine-tuned classification (fine-tuned on 1% and 100% labels). Our approach compares favorably to BioViL [2].

Method	Zero-Shot		1%		100%	
	ACC↑	AUC↑	ACC↑	AUC↑	ACC↑	AUC↑
BioViL	0.73	0.83	0.81	**0.88**	0.82	**0.89**
LSE+NL	**0.80**	**0.84**	**0.84**	0.87	**0.85**	**0.89**

Table 3. Visual grounding performance. We report contrast-to-noise ratio (CNR) and mean intersection-over-union (mIoU). mIoU measures mean IoU of a thresholded region-sentence map and the ground truth bounding box over a set of thresholds. Our approach outperforms BioViL [2] on both measures.

Method	CNR↑	mIoU↑
BioViL	1.14	0.17
LSE+NL	**1.44**	**0.19**

Table 4. Cross-modal retrieval performance. We report recall for the top 10, 50 and 100 answers returned by the method, as well as the median rank of the ground truth element for sentence retrieval based on region queries and for region retrieval based on sentence queries. Our method outperforms the baselines on all measures.

Method	Region → Sentence				Sentence → Region			
	R@10↑	R@50↑	R@100↑	MedR↓	R@10↑	R@50↑	R@100↑	MedR↓
GLoRIA	0.06	0.21	0.37	162	0.06	0.21	0.34	183
BioViL	0.07	0.26	0.40	151	0.08	0.26	0.40	146
LSE+NL	**0.11**	**0.29**	**0.45**	**119**	**0.11**	**0.36**	**0.51**	**97**

[2]. GLoRIA is a representation learning method that learns based on image-sentence and region-word pairs. BioViL improves upon GLoRIA by using a better text encoder, relying on a symmetric contrastive loss and masked language modeling for representation learning. We omit reporting GLoRIA's classification and visual grounding performance for GLoRIA as [2] showed that BioViL is better than GLoRIA on these tasks. Our simple model provides consistently better performance than these state-of-the-art algorithms.

Table 2 reports image classification accuracy based on the learned representations for different amounts of data used to fine-tune the representation for the downstream task (zero-shot, 1%, and 100%). Our method is competitive or better than the baseline, especially in the zero-shot setup, underscoring its promise for limited annotation scenarios. Table 3 and Table 4 report the methods' performance on visual grounding and cross-modal retrieval respectively. Our method significantly outperforms the baseline.

466   P. Wang et al.

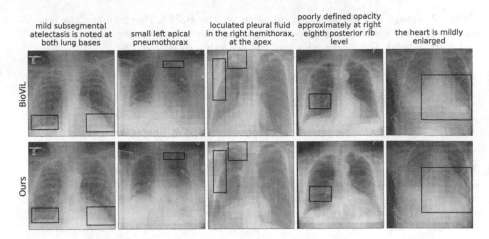

Fig. 2. Example visual grounding results for several challenging cases for BioVil [2] (top row) and our method (bottom row). Text queries and the corresponding ground truth bounding boxes are shown for each image. Colormap overlay visualizes region-sentence scores (blue corresponds to low scores, red highlights regions with high scores). Our method provides maps that align better with the ground truth bounding boxes. (Color figure online)

Table 5. Ablation study results. For each variant of the method, performance statistics are reported for each downstream task consistently with Tables 2, 3, and 4. RSNA is RSNA Pneumonia. CheX is MIMIC-CheXpert. FT is fine-tuned classification using 100% of the labels. ZS is zero-shot classification. We report AUC for image classification. Local representations perform well for image classification, while visual grounding and cross-modal retrieval benefit from integration of local and global representations.

Method	Classification			Grounding	Cross-Modal Retrieval	
	RSNA-ZS↑	RSNA-FT↑	CheX-FT↑	CNR↑	MedR(I → T)↓	MedR(T → I)↓
LSE	**0.856**	**0.892**	**0.874**	1.308	146	137
NL	0.636	0.871	0.854	0.836	264	272
LSE+Average	0.851	0.889	0.868	0.915	191	161
LSE+NL	0.846	0.891	0.870	1.403	**110**	102
w. ResNet-50	0.844	0.890	0.870	**1.438**	119	**97**

Figure 2 illustrates examples of visual grounding. Unlike [2], we do not smooth the region-sentence scores produced by our model. Our method yield qualitatively better region-sentence scores than BioViL on a few challenging failure cases discussed in [2]. In particular, our pre-trained model captures location specifications more effectively, e.g., recognizing "at both lung bases" in the first image and "right" in the third image. Both our method and BioViL are prone to false positives, i.e., regions outside the ground-truth bounding box with high region-sentence scores, which highlights the need for further improvements.

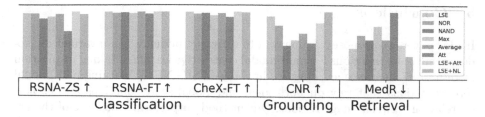

Fig. 3. Effects of aggregator choice on the performance. Performance of models trained with local aggregators (shades of blue), global aggregators (shades of orange) and combinations of local and global aggregators (shades of green) is shown for image classification (AUC), visual grounding (CNR) and cross-modality retrieval (MedR averaged for both directions). The metrics are normalized to unit interval for easier comparisons across tasks. The choice of aggregators effects image classification performance much less than that of visual grounding and cross-modality retrieval. There is high performance variations within each group. Combination approaches do well on all tasks. (Color figure online)

Ablation. In the ablation study (Table 5), we compare our method LSE+NL with using either the local LSE or the global NL approach only, as well as replacing the NL with average as the region aggregator, i.e., LSE+Average. To enable extensive experimentation, we use ResNet-18 as the image encoder. LSE+NL provides good trade-off between region-sentence and image-sentence alignment. LSE+NL has comparable performance to LSE for image classification tasks while significantly outperforming all alternatives in visual grounding and cross-modal retrieval. Using a larger image encoder model ResNet-50 provides only a modest improvement in visual grounding.

Aggregator Choices. Figure 3 compares a few instance aggregators' performance on downstream tasks. We compare the local approach (e.g., LSE, NOR [26], NAND [19]) the global approach (e.g., Max, Average, Att [14]) and a combination of local and global approaches (e.g., LSE+Att, LSE+NL). Aggregators within each approach exhibits high performance variations. The best local aggregator is superior to the best global aggregators we explored on all downstream tasks. Combining local and global approaches yields the best performing method.

4.4 Limitations

Though empirically useful, our framework doesn't provide theoretical guarantees on downstream task performance. We did not investigate what properties of an aggregator determine its transfer behaviors. In addition, our proposed method LSE+NL is sensitive to the value of scaling parameters; Finding the optimal hyperparameters automatically is crucial for model scaling.

5   Conclusions

In this paper, we propose a framework to construct permutation-invariant image-document score functions for multimodal contrastive learning. Taking inspiration from multiple instance learning, we introduce LSE+NL for learning multimodal representations that rely on both local and global score functions and exploit correlation between image regions. Our method outperforms the state-of-the-art approaches on image classification, visual grounding, and cross-modal retrieval. In addition, we show that contrastive representation learning is a form of multiple instance learning, providing us with valuable insights from a related field for solving shared challenges to learn representations that generalized well.

Acknowledgements. Work supported by MIT JClinic, Philips, and Wistron.

References

1. Amores, J.: Multiple instance classification: review, taxonomy and comparative study. Artif. Intell. **201**, 81–105 (2013)
2. Boecking, B., et al.: Making the Most of Text Semantics to Improve Biomedical Vision-Language Processing. In: Avidan, S., Brostow, G., Cissé, M., Farinella, G.M., Hassner, T. (eds.) ECCV 2022. LNCS, vol. 13696, pp. 1–21. Springer, Cham (2022). https://doi.org/10.1007/978-3-031-20059-5_1
3. Carbonneau, M.A., Cheplygina, V., Granger, E., Gagnon, G.: Multiple instance learning: a survey of problem characteristics and applications. Pattern Recognit. **77**, 329–353 (2018)
4. Chauhan, G., et al.: Joint modeling of chest radiographs and radiology reports for pulmonary edema assessment. In: Martel, A.L., et al. (eds.) MICCAI 2020. LNCS, vol. 12262, pp. 529–539. Springer, Cham (2020). https://doi.org/10.1007/978-3-030-59713-9_51
5. Chen, Y.-C., et al.: UNITER: UNiversal image-TExt representation learning. In: Vedaldi, A., Bischof, H., Brox, T., Frahm, J.-M. (eds.) ECCV 2020. LNCS, vol. 12375, pp. 104–120. Springer, Cham (2020). https://doi.org/10.1007/978-3-030-58577-8_7
6. Desai, K., Johnson, J.: VirTex: learning visual representations from textual annotations. In: CVPR (2021)
7. Fang, H., et al.: From captions to visual concepts and back. In: CVPR (2015)
8. Foulds, J., Frank, E.: A review of multi-instance learning assumptions. Knowl. Eng. Rev. **25**, 1–25 (2010)
9. Gao, R., Feris, R., Grauman, K.: Learning to separate object sounds by watching unlabeled video. In: Ferrari, V., Hebert, M., Sminchisescu, C., Weiss, Y. (eds.) ECCV 2018. LNCS, vol. 11207, pp. 36–54. Springer, Cham (2018). https://doi.org/10.1007/978-3-030-01219-9_3
10. Harwath, D., Recasens, A., Surís, D., Chuang, G., Torralba, A.: Jointly discovering visual objects and spoken words from raw sensory input. In: IJCV (2020)
11. He, K., Gkioxari, G., Dollár, P., Girshick, R.: Mask R-CNN. In: ICCV (2017)

12. He, K., Zhang, X., Ren, S., Sun, J.: Deep residual learning for image recognition. In: CVPR (2016)
13. Huang, S.C., Shen, L., Lungren, M.P., Yeung, S.: GLoRIA: a multimodal global-local representation learning framework for label-efficient medical image recognition. In: ICCV (2021)
14. Ilse, M., Tomczak, J., Welling, M.: Attention-based deep multiple instance learning. In: ICML (2018)
15. Johnson, A.E.W., et al.: MIMIC-CXR, a de-identified publicly available database of chest radiographs with free-text reports. Sci. Data **6**, 317 (2019)
16. Karpathy, A., Fei-Fei, L.: Deep visual-semantic alignments for generating image descriptions. TPAMI (2017)
17. Karpathy, A., Joulin, A., Fei-Fei, L.: Deep fragment embeddings for bidirectional image sentence mapping. In: NIPS (2014)
18. Kingma, D., Ba, J.: Adam: a method for stochastic optimization. arXiv:1412.6980 (2014)
19. Kraus, O.Z., Ba, J.L., Frey, B.J.: Classifying and segmenting microscopy images with deep multiple instance learning. Bioinformatics **32**, i52–i59 (2016)
20. Lee, K.-H., Chen, X., Hua, G., Hu, H., He, X.: Stacked cross attention for image-text matching. In: Ferrari, V., Hebert, M., Sminchisescu, C., Weiss, Y. (eds.) ECCV 2018. LNCS, vol. 11208, pp. 212–228. Springer, Cham (2018). https://doi.org/10.1007/978-3-030-01225-0_13
21. Li, B., Li, Y., Eliceiri, K.W.: Dual-stream multiple instance learning network for whole slide image classification with self-supervised contrastive learning. In: CVPR (2021)
22. Li, L.H., Yatskar, M., Yin, D., Hsieh, C.J., Chang, K.W.: VisualBERT: a simple and performant baseline for vision and language. arXiv:1908.0355 (2019)
23. Liao, R., et al.: Multimodal representation learning via maximization of local mutual information. In: de Bruijne, M., et al. (eds.) MICCAI 2021. LNCS, vol. 12902, pp. 273–283. Springer, Cham (2021). https://doi.org/10.1007/978-3-030-87196-3_26
24. Loshchilov, I., Hutter, F.: Decoupled weight decay regularization. In: ICLR (2019)
25. Lu, J., Batra, D., Parikh, D., Lee, S.: ViLBERT: pretraining task-agnostic visiolinguistic representations for vision-and-language tasks. In: NeurIPS (2019)
26. Maron, O., Lozano-Pérez, T.: A framework for multiple-instance learning. In: NIPS (1998)
27. Miech, A., Alayrac, J.B., Smaira, L., Laptev, I., Sivic, J., Zisserman, A.: End-to-end learning of visual representations from uncurated instructional videos. In: CVPR (2020)
28. Müller, P., Kaissis, G., Zou, C., Rueckert, D.: Joint learning of localized representations from medical images and reports. In: Avidan, S., Brostow, G., Cissé, M., Farinella, G.M., Hassner, T. (eds.) ECCV 2022. LNCS, vol. 13686, pp. 685–701. Springer, Cham (2022). https://doi.org/10.1007/978-3-031-19809-0_39
29. van den Oord, A., Li, Y., Vinyals, O.: Representation learning with contrastive predictive coding. arXiv:1807.03748 (2018)
30. Radford, A., et al.: Learning transferable visual models from natural language supervision. In: ICML (2021)
31. Ren, S., He, K., Girshick, R., Sun, J.: Faster R-CNN: towards real-time object detection with region proposal networks. In: NIPS (2015)
32. Sadvilkar, N., Neumann, M.: PySBD: pragmatic sentence boundary disambiguation. In: NLP-OSS (2020)

33. Shih, G., et al.: Augmenting the national institutes of health chest radiograph dataset with expert annotations of possible pneumonia. Radiol. Artif. Intell. **1**, e180041 (2019)
34. Wang, X., Girshick, R., Gupta, A., He, K.: Non-local neural networks. In: CVPR (2018)
35. Zhang, C., Platt, J., Viola, P.: Multiple instance boosting for object detection. In: NIPS (2005)
36. Zhang, Y., Jiang, H., Miura, Y., Manning, C.D., Langlotz, C.P.: Contrastive learning of medical visual representations from paired images and text. In: MLHC (2022)
37. Zhou, Z.H., Zhang, M.L., Huang, S.J., Li, Y.F.: Multi-instance multi-label learning. Artif. Intell. **176**, 2291–2320 (2012)

X-TRA: Improving Chest X-ray Tasks with Cross-Modal Retrieval Augmentation

Tom van Sonsbeek[(✉)] and Marcel Worring

University of Amsterdam, Amsterdam, The Netherlands
{t.j.vansonsbeek,m.worring}@uva.nl

Abstract. An important component of human analysis of medical images and their context is the ability to relate newly seen things to related instances in our memory. In this paper we mimic this ability by using multi-modal retrieval augmentation and apply it to several tasks in chest X-ray analysis. By retrieving similar images and/or radiology reports we expand and regularize the case at hand with additional knowledge, while maintaining factual knowledge consistency. The method consists of two components. First, vision and language modalities are aligned using a pre-trained CLIP model. To enforce that the retrieval focus will be on detailed disease-related content instead of global visual appearance it is fine-tuned using disease class information. Subsequently, we construct a non-parametric retrieval index, which reaches state-of-the-art retrieval levels. We use this index in our downstream tasks to augment image representations through multi-head attention for disease classification and report retrieval. We show that retrieval augmentation gives considerable improvements on these tasks. Our downstream report retrieval even shows to be competitive with dedicated report generation methods, paving the path for this method in medical imaging.

Keywords: Information Retrieval · Medical Image Classification · Multi-modal Learning

1 Introduction

The promise of automated deep learning systems to assist radiologists is enormous. At the moment, important milestones, such as better consistency or even better performance have been achieved on an increasing number of use-cases [18,37]. A source of inspiration in further improvement of these efforts is the way humans register and analyze images, which for deep learning has shown to be effective in the past [17,37].

In any analysis, a doctor provides the memory and knowledge to place what is currently seen in the context of what has been seen before. In principle this can be compared to what implicitly happens at scale in any deep learning method. A doctor's analysis is not implicit though. Their analysis process can be described

A. Frangi et al. (Eds.): IPMI 2023, LNCS 13939, pp. 471–482, 2023.
https://doi.org/10.1007/978-3-031-34048-2_36

and verified. We wonder whether (medical) deep learning methods could benefit from an explicit memory/knowledge infusion.

Making deep learning methods more explicit in terms of using past observations has already been studied in Natural Language Processing (NLP), in the form of retrieval augmentation [14,21]. Supplementing data by retrieving relevant retrieved information can lead to performance gains [4]. This process can be thought of to work as both an enrichment and regularization process. A benefit of retrieval augmentation is that context from a trusted knowledge source is used as a supplement [13,29]. The versatility of retrieval augmentation, which essentially provides a non-parametric memory expansion, is gaining traction in the multi-modal field [4,28].

Multi-modal data modalities typically have different strengths leading to a strong and a weak data modality [37]. For instance, radiology reports generally contain richer and more complete information than X-rays, since the report is essentially a clinician's annotation [24]. With retrieval augmentation information can be transferred explicitly from the strong to the weak modality.

A reason retrieval augmentation methods are not yet adopted for medical applications lies in the weakness of retrieval methods for the medical domain. Retrieval in the general domain is focused on global image regions [8,16] whereas in medical images global features, such as body/organ structure are similar across patients. Meanwhile more fine-grained aspects are more discriminating as disease indicators, but are easily overlooked. The need for fine-grained results makes medical image retrieval magnitudes more complex.

We propose X-Ray Task Retrieval Augmentation (X-TRA), a framework for retrieval augmentation in a multi-modal medical setting, specifically designed for X-ray and radiology report analysis. To do so we introduce a cross-modal retrieval model and retrieval augmentation method. We make the following contributions.

- We propose a CLIP-based multi-modal retrieval framework with a dedicated fine-tuning component for efficient content alignment of medical information which improves state-of-the-art results in multi- and single-modal retrieval on radiology images and reports.
- We introduce a multi-modal retrieval augmentation component for disease classification and report retrieval pipelines.
- We show that our method (1) reaches state-of-the-art performance both in multi-label disease classification and report retrieval. (2) Our report retrieval is competitive with dedicated report generation methodologies. (3) We show the cross-dataset versatility and the limitations of our method.

2   Related Work

Multi-modal Alignment. The introduction of Transformers for natural language processing (NLP) accelerated the development of integrated vision-language (VL) alignment models suitable for various VL-tasks, such as ViLBERT [19],

LXMERT [30] and SimVLM [33]. These methods provide alignment on region to sentence- or word-level scale. The next step in multi-modal alignment was made by methods using contrastive learning combined with substantially larger datasets. Examples are CLIP [27] and ALIGN [10] which significantly outperform existing methods by using datasets for training consisting of 400M and 1.8B VL-pairs respectively. Domain-specific versions of CLIP, which is open-source, have been fine-tuned with additional data, such as PubMedCLIP [3].

Retrieval Augmentation. The origin of retrieval augmentation lies in the NLP field. It was created to fully utilise the power of large datasets. With retrieval augmentation we are not only dependent on a parametric model, but can also supplement data as a non-parametric component. Previous methods have shown the simple yet effective and versatility working of retrieval augmentation in a number of applications [5,13,29].

Retrieval in Medical Imaging. Up until recently the only retrieval methods in medical imaging were tailored hand-crafted methods [16]. With access to large datasets and pre-trained methods the balance shifted towards making automated retrieval methods [6,26]. Especially in the histopathology and radiology domain major strides were made with retrieval methods [2,8]. The use of text to improve image retrieval has been adopted for improving chest X-ray retrieval. Yu *et al.* [35] use CNN and word2vec features for multi-modal alignment and retrieval. Zhang *et al.* [36] approach this problem with a hash-based retrieval method.

Retrieval for Chest X-ray Analysis. Common tasks in chest X-ray analysis are disease classification and report generation [1,11,15]. Using retrieval for report generation has been a common approach. The approaches often entail the use of retrieved information as an input or template for a decoder which crafts a custom report [23,32,34]. Augmentation of chest X-ray tasks with synthetically generated diffusion-based images was shown to be possible [1], however the clinical use of non-genuine images can lead to complications and is not undisputed [37].

3 Methods

Our method is composed of two separate parts (Fig. 1). The first part is the alignment of the two modalities and construction of the retrieval model. The second part uses the output of the retriever as a non-parametric component in (cross-modal) retrieval augmentation to enhance the downstream tasks.

We consider a dataset $\Theta^N_{\{\mathbf{x},\mathbf{r}\}}$ consisting of pairs containing an X-ray ($\mathbf{x}_i$) and radiology report ($\mathbf{r}_i$). To align these modalities we make use of the powerful CLIP vision-language aligner. Our objective is to minimize the distance between $\mathbf{x}$ and $\mathbf{r}$, to make cross-modal tasks possible. These aligned features will be used for retrieval augmentation to do multi-label classification and report retrieval as downstream tasks.

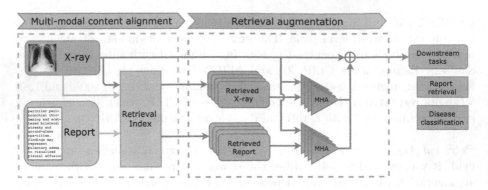

Fig. 1. Architecture overview of X-TRA.

3.1 Stage I: Multi-modal Content Alignment

We leverage the pre-trained features from CLIP for initial feature representations. However, there is a domain shift between the natural image data CLIP is trained on and medical images we want to use in our method. Medical images can be visually very similar, while holding drastically different information. Small localized markers can be indicators for disease. In natural images global representations are more decisive and thus more suitable for unsupervised contrastive alignment. Alignment in CLIP goes as follows [27],

$$\mathcal{L}_{CLIP} = -\frac{1}{N} \sum_{z \in Z} \sum_{i=1}^{N} \log \frac{e^{\left(\text{sim}\left(z_i^0, z_i^1\right)/\tau\right)}}{\sum_{j=1}^{N} e^{\left(\text{sim}\left(z_i^0, z_j^1\right)/\tau\right)}} \quad with \ Z = \{(\mathbf{x}, \mathbf{r}), (\mathbf{r}, \mathbf{x})\}. \quad (1)$$

We need to overcome the obvious domain shift between medical images and the natural images on which CLIP is trained. Therefore, we require a more specific type of fine-tuning that is especially geared towards content-based extraction. We introduce the following loss, requiring a global class label for each dataset. With this fine-tuning step we are creating a supervised content-based alignment method with content classifier C:

$$\mathcal{L}_{ours} = -\frac{1}{N} \sum_{z \in Z} \sum_{i=1}^{N} y_i log_e(\widehat{C(z_i)}) \qquad with \ Z = \{\mathbf{x}, \mathbf{r}, (\mathbf{x}, \mathbf{r})\}. \quad (2)$$

This content based alignment loss should improve the alignment of detailed content-level details over the global visual appearance of the image.

Creating a Retrieval Index. At retrieval time we need to retrieve images that have a high similarity with query images. To efficiently do so we make use of Facebook AI Similarity Search (FAISS) [12]. This retrieval tool efficiently performs nearest-neighbour similarity search. After multi-modal alignment we encode our data to a FAISS index I conditioned on the training dataset. We can construct indices that only retrieve images (I^x), only reports (I^r), or both (I^{xr}).

Given a query Q_s in source modality s, we can obtain its k neighbours of target modality t through:

$$\mathcal{N}_{s \to t}^k = I^t(Q_s, k), \tag{3}$$

this can be either $\mathbf{x}$, $\mathbf{r}$ or both. Once retrieval index I is trained based on the newly aligned training dataset we can consider the retriever as a non-parametric component which retrieves information from a fixed dataset in the subsequent retrieval augmentation steps. Note that during testing time, a query from the test set will be used to retrieve neighbours from the training set.

3.2 Stage II: Retrieval Augmentation

The purpose of retrieval augmentation is to effectively leverage similar representations to adopt a more informative representation of a given input, with our already trained retrieval index we retrieve similar representations.

To obtain a richer representation of $\mathbf{x}_i$, we retrieve intra- $\mathcal{N}_{x \to x}^k$ and inter-modal neighbours $\mathcal{N}_{x \to r}^k$ from I^x and I^r respectively. To integrate the retrieved neighbouring samples, we can use various fusion methods [25]. The simplest one is concatenation: $(\mathbf{x}_i, \mathcal{N}_{x \to x}^k, \mathcal{N}_{x \to r}^k)$. A more suitable method is multi-head attention (MHA) which is able to capture the long range dependencies between the original image and the retrieved information [31]:

$$\mathbf{x}_i^{TRA} = (\mathbf{x}_i, \text{MHA}(\mathcal{N}_{x \to x}^k, \mathbf{x}_i), \text{MHA}(\mathcal{N}_{x \to r}^k, \mathbf{x}_i)). \tag{4}$$

3.3 Downstream Tasks

We are tackling two common tasks in chest X-ray analysis. These are multi-label disease classification and report retrieval. For this last task our objective is to show how well a retriever can perform on the report generation task. We measure performance by comparing task performance of $\mathbf{x}^{TRA}$ in comparison to $\mathbf{x}$.

A useful property of our retrieval index would be usability of an pre-trained model across datasets. Three clinically relevant scenarios for this are: From scratch training on the new dataset, frozen usage of the trained retrieval model and fine-tuning of the existing retrieval model with another image-report dataset.

3.4 Datasets

The primary dataset to which our method is applied is **MIMIC-CXR** *(200k image-report pairs)* [11]. Disease labels for each pair are extracted from the report through a rule-based extraction method [9]. To evaluate the versatility and cross-domain capabilities of our method, we use the small **openI** *(4k image-report pairs)* [20] and image-only **CheXpert** *(200k images)* [9] datasets. Official train-test splits are used.

3.5   Experimental Setup

As pre-processing step, the X-ray images are normalized and standardized by rescaling with center-cropping to scale 256 × 256, from which images of size 224 × 224 are sampled. The maximum number of tokens for representing radiology reports in the text encoder is set to 256. Three different VL models are used as encoders. At first a CNN-BERT model, composed of a DenseNet121 image encoder and a ClinicalBERT [7] text encoder. Given the strong performance of large vision-language models we also use CLIP (ViT-32 image encoder and text encoder) [27] and its medically fine-tuned equivalent PubMedCLIP [3]. This model is fine-tuned using the Radiology Objects in COntext (ROCO) dataset [22].

Multi-modal alignment is implemented as a single pass through a two-layer ReLu activated MLP, with dimension z_{enc}, a dropout rate of 0.5, and layer normalization. z_{enc} is the output dimension of the encoder. We implement C as a three layer classifier head with dimensions $\{z_{enc}, 256, 14\}$. During retrieval we make use of $k = 10$ retrieved neighbours. To prevent overfitting, early stopping with a tolerance of 3 is applied to all training operations.

4   Results

4.1   Cross-Modal Retrieval

We are comparing the performance of our retrieval method against previous methods in Table 1 in terms of class-based mean average precision (mAP). Due to the powerful alignment of CLIP and tailor made fine-tuning we are outperforming all existing retrieval approaches for radiology images and/or reports by a large margin. The performance difference with similarly fine-tuned encoder-decoder combination DenseNet121 and ClinicalBERT further underwrites the power of CLIP in building a strong retrieval method, specifically on cross-domain retrieval. Interestingly, we observe that PubMedCLIP is not outperforming CLIP. This can be explained by a domain shift between MIMIC-CXR and ROCO, together with the ability of CLIP to generalize well out-of-domain [27]. In our downstream tasks image-based retrieval is most important, which is performing similar on inter- and intra-modal retrieval tasks.

4.2   Multi-label Disease Classification

Disease classification results in terms of AUC in Table 2 show that retrieval augmentation gives a clear improvement across different disease classes. It is interesting to see that we find a positive, albeit weak, correlation (R≈0.60) between the increase in class AUC performance and retrieval mAP. Moreover, the performance gain from retrieval augmentation (0.80 → 0.85) is similar to additional training with synthetic diffusion-generated X-rays (0.80 → 0.84) [1]. The benefit of our method is that the supplemented information originates from the trusted dataset itself and is not synthetically generated.

Table 1. Class-based retrieval performance (source → target) for images (**x**) and reports (**r**) in terms of mAP on MIMIC-CXR on our content alignment method, compared against other methods.

		No Finding	Enl. Cardiomed.	Cardiomegaly	Lung Opacity	Lung Lesion	Edema	Consolidation	Pneumonia	Atelectasis	Pneumothorax	Pleural Effusion	Pleural other	Fracture	Support Devices	wAvg	Avg
Yu et al. [35]		-	.65	.75	.72	.43	.80	.73	.60	.76	.76	.85	.43	.16	.86	-	.680
CLIP ($\mathcal{L}_{CLIP}$)		.71	.52	.74	.78	.39	.79	.39	.40	.76	.42	.67	.44	.43	.64	.578	.761
CNN+BERT	x → x	.87	.63	.88	.90	.49	.90	.57	.60	.85	.85	.83	.29	.47	.82	.678	.769
PubmedCLIP		.90	.63	.82	.83	.39	.86	.45	.63	.87	.53	.90	.48	.51	.79	.685	.795
CLIP		.84	.62	.89	.89	.56	.91	.55	.59	.89	.60	.86	.49	.57	.84	.713	.840
Zhang et al. [36]		-	-	-	-	-	-	-	-	-	-	-	-	-	-	-	.498
CLIP ($\mathcal{L}_{CLIP}$)		.50	.60	.73	.81	.53	.70	.73	.87	.85	.59	.78	.55	.31	.77	.666	.762
CNN+BERT	x → r	.61	.74	.89	.95	.45	.69	.76	.82	.71	.71	.77	.32	.67	.84	.713	.756
PubmedCLIP		.74	.65	.67	.62	.38	.70	.13	.76	.72	.51	.83	.51	.90	.60	.623	.728
CLIP		.64	.71	.91	.92	.73	.87	.89	.94	.94	.67	.95	.61	.48	.84	.793	.793
CLIP ($\mathcal{L}_{CLIP}$)		.76	.66	.81	.88	.61	.73	.67	.53	.84	.54	.79	.57	.74	.70	.679	.739
CNN+BERT	x → xr	.85	.85	.76	.75	.51	.83	.64	.66	.82	.58	.95	.53	.62	.84	.728	.766
PubmedCLIP		.90	.77	.71	.89	.81	.86	.57	.44	.81	.59	.93	.65	.64	.82	.742	.824
CLIP		.85	.86	.91	.90	.68	.84	.54	.66	.90	.64	.87	.68	.78	.81	.780	.857
Zhang et al. [36]		-	-	-	-	-	-	-	-	-	-	-	-	-	-	-	.485
CLIP ($\mathcal{L}_{CLIP}$)		.62	.52	.93	.88	.50	.60	.29	.44	.75	.54	.85	.50	.36	.71	.606	.723
CNN+BERT	r → x	.77	.54	.73	.91	.52	.83	.39	.87	.77	.63	.74	.23	.61	.73	.662	.735
PubmedCLIP		.75	.65	.92	.99	.23	.79	.21	.51	.59	.72	.81	.56	.43	.67	.645	.720
CLIP		.63	.62	.96	.94	.62	.69	.47	.61	.85	.69	.91	.57	.46	.82	.703	.779
CLIP ($\mathcal{L}_{CLIP}$)		.77	.88	.86	.92	.59	.75	.67	.70	.87	.70	.93	.54	.28	.77	.731	.852
CNN+BERT	r → r	.83	.63	.86	.98	.60	.84	.68	.66	.88	.64	.96	.47	.54	.80	.741	.843
PubmedCLIP		.99	.75	.90	.98	.67	.84	.83	.60	.95	.82	.98	.38	.28	.84	.772	.887
CLIP		.93	.93	.87	.96	.73	.94	.77	.79	.87	.85	.95	.55	.42	.84	.814	.895
CLIP ($\mathcal{L}_{CLIP}$)		.90	.77	.80	.87	.77	.72	.61	.74	.86	.77	.69	.31	.28	.80	.707	.828
CNN+BERT	r → xr	.74	.49	.91	.98	.42	.77	.68	.79	.87	.64	.78	.26	.61	.84	.734	.836
PubmedCLIP		.92	.81	.94	.99	.68	.86	.93	.82	.99	.76	.79	.30	.27	.84	.793	.903
CLIP		.91	.91	.96	.97	.76	.84	.76	.77	.92	.96	.84	.46	.37	.80	.803	.909

Table 2. Chest X-ray classification on MIMIC-CXR with and without retrieval augmentation. The results show the beneficial effect of retrieval augmentation on classification performance.

	X-TRA	No Finding	Enl. Cardiomed.	Cardiomegaly	Lung Opacity	Lung Lesion	Edema	Consolidation	Pneumonia	Atelectasis	Pneumothorax	Pleural Effusion	Pleural other	Fracture	Support Devices	wAvg	Avg
CNN+BERT	✗	.81	.63	.73	.67	.62	.83	.69	.59	.68	.75	.83	.70	.58	.84	.71	.79
	✓	.81	.74	.75	.69	.63	.81	.72	.63	.75	.75	.83	.69	.63	.85	.73	.82
	Δ	.00	.11	.02	.02	.01	.02	.03	.04	.07	.00	.00	-.01	.05	.01	.02	.03
PubmedCLIP	✗	.78	.65	.72	.66	.61	.82	.70	.61	.73	.76	.81	.62	.54	.84	.70	.78
	✓	.84	.76	.78	.69	.64	.83	.73	.64	.76	.75	.82	.75	.67	.85	.75	.83
	Δ	.06	.11	.06	.03	.03	.01	.03	.03	.03	-.01	.01	.13	.13	.01	.05	.05
CLIP	✗	.77	.65	.71	.67	.62	.85	.73	.61	.72	.75	.80	.59	.51	.83	.70	.80
	✓	.82	.78	.74	.70	.71	.82	.75	.63	.79	.78	.86	.74	.72	.91	.77	.85
	Δ	.05	.13	.03	.03	.09	-.03	.02	.02	.07	.03	.06	.15	.21	.08	.07	.05

Table 3. Chest X-ray report retrieval on MIMIC-CXR with and without X-TRA retrieval augmentation. Compared to dedicated report generation methods.

		BLEU-1	BLEU-2	BLEU-3	BLEU-4	ROUGE-L	METEOR	BERTScore
Report generation	Pino *et al.* [23]	–	–	–	.094	.185	–	–
	Wang *et al.* [32]	.344	.215	.146	.105	.279	.138	–
	Yang *et al.* [34]	.438	.297	.216	.164	.332	–	–
	Li *et al.* [15]	**.467**	**.334**	**.261**	**.215**	**.415**	.201	–
Report retrieval	Chambon *et al.* [1]	–	–	–	–	–	–	.432
	Yang *et al.* [34]	.306	.179	.116	.076	.232	–	–
	CNN+BERT	.268 (↑.025)	.193 (↑.064)	.106 (↑.036)	.072 (↑.029)	.288 (↑.042)	.248 (↑.027)	.572(↑.17)
	PubmedCLIP	.308 (↑.031)	.206 (↑.021)	.111 (↑.021)	.074 (↑.006)	.330 (↑.022)	.286 (↑.025)	.610(↑.29)
	CLIP	.318 (↑.041)	.226 (↑.041)	.121 (↑.024)	.085 (↑.023)	.339 (↑.044)	**.296 (↑.055)**	**.617(↑.31)**

4.3 Report Generation

In retrieval augmented report retrieval we show interesting performance on the report generation metrics compared to a selection of previous methods. While it should not be expected that simple retrieval outperforms dedicated report generation methods we are able to provide a result that can be considered competitive (Table 3). On the METEOR and ROUGE metric we are even outperforming most existing methods. The metrics reflect that the strength of report retrieval is in the global representation of the report. Our retriever is fine-tuned to retrieve samples with equivalent label spaces, hence good results on metrics that reward global similarity. An interesting outlook is the application of this method in a dedicated report generation framework which could boost performance further.

4.4 Cross-Dataset

By evaluating the cross-dataset scenarios (Table 4) with the CheXpert and openI datasets we can conclude that transferability to images from other domains is limited. However we do see that if retrieval augmentation is not useful, it can be ignored by the model and will not be detrimental for performance. The domain shift between different chest X-rays is a remaining problem [24]. Currently the most practical solution for this problem is the addition of a fine-tuning step.

Cross-domain results on open-I show that learning across modalities is possible with fine-tuning. When adding the openI dataset to the existing retrieval index, we can integrate the existing index with this new dataset. We can see that X-TRA benefits openI in this setting. In the updated retrieval index 23% of the retrieved information originates from openI and 77% from MIMIC-CXR.

4.5 Ablation Studies

We study the effect of the components in our retrieval augmentation method in Fig. 2. Specifically we look at the influence of each component in content- and CLIP based alignment. Interestingly, the composition of data modalities in retrieval augmentation does not have a big effect, since the retriever has similar results in inter- and intra-modal retrieval. In case randomly selected data is used

Table 4. Cross-domain result on downstream tasks: Report retrieval (RR) and multi-label classification (MLC) with and without X-TRA.

Target ↓ dataset	Retrieval source→	Target		MIMIC-CXR		MIMIC-CXR	
	Setting→	From scratch		Frozen		Finetuning	
		RR	MLC	RR	MLC	RR	MLC
CheXpert	CNN+BERT	–	–	–	.81(↓.01)	–	–
	PubmedCLIP	–	–	–	**.82**(↑.01)	–	–
	CLIP	–	–	–	.81(.00)	–	–
OpenI	CNN+BERT	.31(↑.05)	.88(↑.01)	.31(↑.03)	.87(↑.01)	.33(↑.05)	.90(↑.05)
	PubmedCLIP	.26(↑.05)	.86(↓.01)	.34(↑.04)	.89(↑.03)	.38(↑.05)	.91(↑.05)
	CLIP	**.29**(↑.04)	**.90**(↑.04)	**.35**(↑.02)	**.90**(↑.02)	**.38**(↑.07)	**.93**(↑.05)

instead of retrieved information, we achieve comparable results compared to our method without X-TRA. This is in accordance with cross-modal results, showing that if X-TRA supplemented information is not useful, it can be ignored. Using a partial retrieval index we can conclude that X-TRA can be useful with a small retrieval index, however performance reaches optimal levels when $N > 100k$.

4.6 Insight and Limitations

Qualitative results from our retrieval method for 2 different query images is shown in Fig. 3. We retrieve from the image index and report index. The retrieved

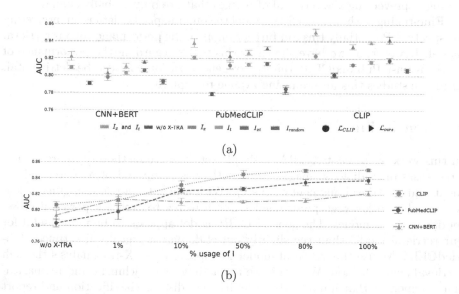

(a)

(b)

Fig. 2. Ablation studies on X-TRA on disease classification, for five different random seeds, with (a) different compositions of the retrieval index for $\mathcal{L}_{CLIP}$ and $\mathcal{L}_{ours}$ and (b) partial usage of the retrieval index.

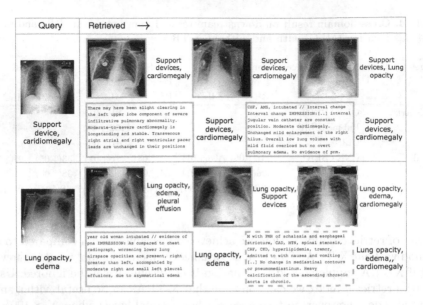

Fig. 3. Examples of image-image and image-text retrieval including disease class labels. A green outline means a correct retrieval, orange or dashed means a missed or extra disease label respectively. (Color figure online)

images match well in terms of labels attributed to them, showing that our fine-tuning is preventing the retrieval of images that are only globally similar.

Fine-tuning of the entire CLIP model to domain-specific data is an interesting prospective. Potentially this can further improve the performance of our retrieval model. However, as we have shown in this paper regarding the performance of CLIP against PubMedCLIP, the loss of generalization can also be detrimental. In future studies this an promising avenue to explore.

5   Conclusion

In this work we present X-TRA, a simple yet effective method to improve multiple tasks on radiology images. Our method is composed of a content alignment and a retrieval augmentation step. With a new label-based alignment loss we are able to leverage pre-trained CLIP features to create a powerful cross-modal retrieval model. The general CLIP model appears to be more useful for our retrieval model than the slightly out-of-domain medically fine-tuned Pub-MedCLIP. We use this retrieval model to improve chest X-ray analysis through retrieval augmentation. With this we are adding an enrichment and regularization component that improves both multi-label disease classification and report retrieval by up to over 5%. On this last task we are even showing to be competitive with dedicated report retrieval methods. It opens up possibilities for retrieval augmentation as a generic tool in medical imaging.

References

1. Chambon, P., et al.: RoentGen: vision-language foundation model for chest x-ray generation. arXiv preprint arXiv:2211.12737 (2022)
2. Endo, M., Krishnan, R., Krishna, V., Ng, A.Y., Rajpurkar, P.: Retrieval-based chest x-ray report generation using a pre-trained contrastive language-image model. In: Machine Learning for Health, pp. 209–219. PMLR (2021)
3. Eslami, S., de Melo, G., Meinel, C.: Does CLIP benefit visual question answering in the medical domain as much as it does in the general domain? arXiv e-prints arXiv:2112.13906 (Dec 2021)
4. Gur, S., Neverova, N., Stauffer, C., Lim, S.N., Kiela, D., Reiter, A.: Cross-modal retrieval augmentation for multi-modal classification. In: Findings of EMNLP 2021, pp. 111–123 (2021)
5. Guu, K., Lee, K., Tung, Z., Pasupat, P., Chang, M.: Retrieval augmented language model pre-training. In: ICML, pp. 3929–3938 (2020)
6. Hu, B., Vasu, B., Hoogs, A.: X-MIR: explainable medical image retrieval. In: WACV, pp. 440–450 (2022)
7. Huang, K., Altosaar, J., Ranganath, R.: ClinicalBERT: modeling clinical notes and predicting hospital readmission. arXiv preprint arXiv:1904.05342 (2019)
8. Ionescu, B., et al.: Overview of the ImageCLEF 2022: multimedia retrieval in medical, social media and nature applications. In: CLEF, pp. 541–564 (2022)
9. Irvin, J., et al.: CheXpert: a large chest radiograph dataset with uncertainty labels and expert comparison. In: AAAI, vol. 33, pp. 590–597 (2019)
10. Jia, C., et al.: Scaling up visual and vision-language representation learning with noisy text supervision. In: ICML, pp. 4904–4916 (2021)
11. Johnson, A.E., et al.: MIMIC-CXR, a de-identified publicly available database of chest radiographs with free-text reports. Sci. Data 6(1), 317 (2019)
12. Johnson, J., Douze, M., Jégou, H.: Billion-scale similarity search with GPUs. IEEE Trans. Big Data 7(3), 535–547 (2019)
13. Komeili, M., Shuster, K., Weston, J.: Internet-augmented dialogue generation. In: ACL, pp. 8460–8478 (2022)
14. Lewis, P., et al.: Retrieval-augmented generation for knowledge-intensive NLP tasks. NeurIPS 33, 9459–9474 (2020)
15. Li, J., Li, S., Hu, Y., Tao, H.: A self-guided framework for radiology report generation. In: International Conference on Medical Image Computing and Computer-Assisted Intervention, pp. 588–598 (2022)
16. Li, Z., Zhang, X., Müller, H., Zhang, S.: Large-scale retrieval for medical image analytics: a comprehensive review. Med. Image Anal. 43, 66–84 (2018)
17. Litjens, G., et al.: A survey on deep learning in medical image analysis. Med. Image Anal. 42, 60–88 (2017)
18. Liu, X., et al.: A comparison of deep learning performance against health-care professionals in detecting diseases from medical imaging: a systematic review and meta-analysis. Lancet Digit. Health 1(6), e271–e297 (2019)
19. Lu, J., Batra, D., Parikh, D., Lee, S.: ViLBERT: pretraining task-agnostic visiolinguistic representations for vision-and-language tasks. NeurIPS 32 (2019)
20. OpenI: Indiana University - chest x-rays (PNG images). https://openi.nlm.nih.gov/faq.php
21. Pasupat, P., Zhang, Y., Guu, K.: Controllable semantic parsing via retrieval augmentation. In: EMNLP, pp. 7683–7698 (2021)

22. Pelka, O., Koitka, S., Rückert, J., Nensa, F., Friedrich, C.M.: Radiology Objects in COntext (ROCO): a multimodal image dataset. In: Stoyanov, D., et al. (eds.) LABELS/CVII/STENT -2018. LNCS, vol. 11043, pp. 180–189. Springer, Cham (2018). https://doi.org/10.1007/978-3-030-01364-6_20

23. Pino, P., Parra, D., Besa, C., Lagos, C.: Clinically correct report generation from chest x-rays using templates. In: International Workshop on Machine Learning in Medical Imaging, pp. 654–663 (2021)

24. Pooch, E.H., Ballester, P.L., Barros, R.C.: Can we trust deep learning models diagnosis? The impact of domain shift in chest radiograph classification. arXiv preprint arXiv:1909.01940 (2019)

25. Priyasad, D., Fernando, T., Denman, S., Sridharan, S., Fookes, C.: Memory based fusion for multi-modal deep learning. Inf. Fusion **67**, 136–146 (2021)

26. Qayyum, A., Anwar, S.M., Awais, M., Majid, M.: Medical image retrieval using deep convolutional neural network. Neurocomputing **266**, 8–20 (2017)

27. Radford, A., et al.: Learning transferable visual models from natural language supervision. In: ICML, pp. 8748–8763 (2021)

28. Ramos, R., Martins, B., Elliott, D., Kementchedjhieva, Y.: SmallCap: lightweight image captioning prompted with retrieval augmentation. arXiv preprint arXiv:2209.15323 (2022)

29. Siriwardhana, S., Weerasekera, R., Wen, E., Kaluarachchi, T., Rana, R., Nanayakkara, S.: Improving the domain adaptation of retrieval augmented generation (RAG) models for open domain question answering. arXiv preprint arXiv:2210.02627 (2022)

30. Tan, H., Bansal, M.: LXMERT: learning cross-modality encoder representations from transformers. In: EMNLP, pp. 5100–5111 (2019)

31. Vaswani, A., et al.: Attention is all you need. NeurIPS 30 (2017)

32. Wang, J., Bhalerao, A., He, Y.: Cross-modal prototype driven network for radiology report generation. In: Avidan, S., Brostow, G., Cissé, M., Farinella, G.M., Hassner, T. (eds.) Computer Vision (ECCV 2022). LNCS, vol. 13695, pp. 563–579. Springer, Cham (2022). https://doi.org/10.1007/978-3-031-19833-5_33

33. Wang, Z., Yu, J., Yu, A.W., Dai, Z., Tsvetkov, Y., Cao, Y.: SimVLM: simple visual language model pretraining with weak supervision. In: ICLR (2021)

34. Yang, X., Ye, M., You, Q., Ma, F.: Writing by memorizing: hierarchical retrieval-based medical report generation. In: ACL, pp. 5000–5009 (2021)

35. Yu, Y., Hu, P., Lin, J., Krishnaswamy, P.: Multimodal multitask deep learning for x-ray image retrieval. In: de Bruijne, M., et al. (eds.) MICCAI 2021. LNCS, vol. 12905, pp. 603–613. Springer, Cham (2021). https://doi.org/10.1007/978-3-030-87240-3_58

36. Zhang, Y., Ou, W., Zhang, J., Deng, J.: Category supervised cross-modal hashing retrieval for chest x-ray and radiology reports. Comput. Electr. Eng. **98**, 107673 (2022)

37. Zhou, S.K., et al.: A review of deep learning in medical imaging: imaging traits, technology trends, case studies with progress highlights, and future promises. Proc. IEEE **109**(5), 820–838 (2021)

Optimization

Differentiable Gamma Index-Based Loss Functions: Accelerating Monte-Carlo Radiotherapy Dose Simulation

Sonia Martinot[1,2,3]($\boxtimes$), Nikos Komodakis[4], Maria Vakalopoulou[2], Norbert Bus[1], Charlotte Robert[3], Eric Deutsch[3], and Nikos Paragios[1]

[1] Therapanacea, Paris, France
sonia.martinot@hotmail.fr
[2] CentraleSupélec, Gif-sur-Yvette, France
[3] Institut Gustave Roussy, Villejuif, France
[4] University of Heraklion, Crete, Heraklion, Greece

Abstract. The Gamma index Passing Rate (GPR) is considered the preferred metric to evaluate dose distributions in order to deliver safe radiotherapy treatments. For this reason, in the context of accelerating Monte-Carlo dose simulations using deep neural networks, the GPR remains the default clinical metric used to validate the predictions of the models. However, the optimization criterion that is used for training these neural networks is based on loss functions that are different than GPR. To address this important issue, in this work we introduce a new class of GPR-based loss functions for deep learning. These functions allow us to successfully train neural networks that can directly yield the best dose predictions from a clinical standpoint. Our approach overcomes the mathematical non-differentiability of the GPR, thus allowing a successful application of gradient descent. Moreover, it brings the GPR computation time down to milliseconds, therefore enabling fast trainings. We demonstrate that models trained with our GPR-based loss functions outperform models trained with other commonly used loss functions with respect to several metrics and display a 15% improvement of the GPR over the test data. Code is available at https://rb.gy/vf5jwv.

Keywords: Deep Learning · Gamma index · Monte-Carlo · Radiotherapy

1 Introduction

In photon radiation therapy, accurate dose modeling is paramount to ensure treatment plans safely target the tumour. Existing algorithms such as Collapsed Cone Convolution algorithm [1] or Pencil Beam [11] fail to match the Monte-Carlo (MC) radiation transport calculations in terms of precision of the deposited dose [4,6]. Yet, MC generation of radiotherapy dose distributions remains too time-consuming for clinical adoption. Recent deep learning accelerated MC dose

A. Frangi et al. (Eds.): IPMI 2023, LNCS 13939, pp. 485–496, 2023.
https://doi.org/10.1007/978-3-031-34048-2_37

calculation methods [12,16] offer a solution for this problem, utilizing common computer vision loss functions. Even if such methods provide a good trade-off between time and performance, training on such loss functions amounts to solving a proxy problem, with no strict assurance to conjointly optimize the clinical validation of the generated dose, which is performed using the Gamma index Passing Rate (GPR).

The GPR is one of the most essential and commonly used clinical evaluation metric for verification of complex radiotherapy dose delivery such as Intensity Modulated Radiation Therapy or Volumetric Modulated Arc Therapy (VMAT) [13]. As such, the GPR provides a clinical criterion to assess the quality of the model's predictions. Therefore, training directly with the GPR as primary objective would yield more accurate training from a clinical standpoint. However, the GPR has two main limitations that deter from using it as loss function. First, training neural networks in a supervised setting requires a differentiable loss function to allow backpropagation. Yet, the GPR is non-differentiable , thus jeopardizing gradient descent. Secondly, despite efforts to bridge the gap, current Gamma index and GPR computations remain time-consuming, especially when comparing high dimensional dose distributions.

By taking a medical imaging perspective, we circumvent these challenges to incorporate the GPR as an optimization criterion during training of neural networks. According to our knowledge, this is the first study to create a new class of loss functions based on the GPR and to bring the speed of gamma index computations down to milliseconds, both for 2D and 3D dose distributions. We provide a proof-of-concept showcasing deep learning acceleration of MC dose simulations with models trained to optimize the presented GPR-based loss functions. Finally, we study the behavior of the GPR-based loss functions and benchmark them against the Structural Similarity Index Measure (SSIM), the Mean Absolute Error (MAE), and the Mean Squared Error (MSE). Our code and models will be publicly released.

2   Related Work

Loss Functions: When training a neural network on a task, the choice of the loss function is crucial. Loss functions such as the Dice Loss [15], the Focal Loss [7], or the Structural Similarity Index Measure (SSIM) [17] have revolutionized, respectively, segmentation, object detection, and image processing tasks. Moreover, all loss functions do not yield the same impact on the training and inference, as explained in the study introducing the Multiscale-SSIM [19].

This problem becomes even more evident in the medical field, in which models need to ensure reliable performance. For this reason, integrating mathematical objectives that train the models to optimize clinically relevant properties is of utmost importance for their integration into clinical practice.

In light of these considerations, we overcome the mathematical challenge of the GPR and turn this clinical metric into a viable loss function for our task of accelerating the simulation of MC radiotherapy dose distributions. We provide a family of GPR-based criteria that are therefore in adequacy with clinical requirements.

Gamma Index: The main challenges of computing the gamma index matrix reside in the pixel-wise computation of gamma index values that can be time-consuming proportionally to the dimensionality of the evaluated dose distribution. Prohibitive calculation time hinders the potential of the GPR as loss function. Many works propose ways to decrease the computation complexity, either by changing the mathematical formalism or accelerating the calculations. In [5], Gu et al. use a geometric method with a GPU-accelerated radial pre-sorting technique to speed up calculations. Chen et al. [3] consider reducing the search distance by using a fast Euclidean distance transform.

In this paper, we present an acceleration approach adequate for deep learning frameworks that significantly reduces the calculation speed and enables fast training with our GPR-based loss functions.

3   Methods

3.1   The Gamma Passing Rate

Gamma Index: Let D_r and D_e be two dose distributions ($\mathbb{R}^k \to \mathbb{R}$), respectively the reference and the evaluated. In our case, the evaluated dose distribution is the model's prediction. To each of them corresponds a grid of points in which each point, P_r of D_r, and P_e of D_e has a coordinate vector, respectively $\vec{d}(P_r)$ and $\vec{d}(P_e)$, and a dose value, $D_r(P_r)$ and $D_e(P_e)$.

Let us consider a point P_r in D_r and the points P_e in a vicinity $V(P_r)$ around P_r. Then the gamma index Γ is defined as a function of real values such that for all $P_r \in D_r$, $\Gamma(P_r)$ writes as follows:

$$\Gamma(P_r) = \min_{P_e \in V(P_r)} \sqrt{\frac{||\vec{d}(P_e) - \vec{d}(P_r)||^2}{DTA^2} + \frac{(D_e(P_e) - D_r(P_r))^2}{\Delta^2}} \qquad (1)$$

where DTA is the tolerance on the Distance-To-Agreement (DTA), commonly in mm, and Δ is the tolerance on the relative dose difference expressed as a percentage of the reference dose value $D_r(P_r)$. This definition entails that each point P_r has its own gamma index value in Γ, which indicates how close neighbouring points P_e are, both spatially and dose-wise.

GPR: Let us introduce a dose threshold δ and consider a point P_r of the reference distribution such that $D_r(P_r) \geq \delta$. Then, given a DTA and dose tolerance Δ, the evaluated distribution matches the reference at P_r, if the passing criterion is satisfied, i.e. if:

$$\text{Passing criterion:} \quad \Gamma(P_r) \leq 1 \qquad (2)$$

The GPR is defined as the percentage of points P_r that satisfy the condition in Eq. 2 while $D_r(P_r) \geq \delta$.

Let $\mathbb{1}_{D_r \geq \delta}$ and $\mathbb{1}_{\Gamma \leq 1}$ be the indicator functions defined such that:

$$\mathbb{1}_{D_r \geq \delta}(P_r) = \begin{cases} 1 & \text{if } D_r(P_r) \geq \delta. \\ 0 & \text{otherwise.} \end{cases} \qquad \mathbb{1}_{\Gamma \leq 1}(P_r) = \begin{cases} 1 & \text{if } \Gamma(P_r) \leq 1. \\ 0 & \text{otherwise.} \end{cases} \quad (3)$$

Then we can write the GPR as follows:

$$GPR(D_r, D_e) = \frac{\sum_{P_r \in D_r} \mathbb{1}_{D_r \geq \delta}(P_r) \cdot \mathbb{1}_{\Gamma \leq 1}(P_r)}{\sum_{P_r \in D_r} \mathbb{1}_{D_r \geq \delta}(P_r)} \quad (4)$$

Minimization Problem: With the GPR formulation in Eq. 4, maximizing the GPR amounts to minimizing the corresponding loss function L_{GPR}^{δ} which draws values in $[0, 1]$:

$$L_{GPR}^{\delta}(D_r, D_e) = 1 - \frac{\sum_{P_r \in D_r} \mathbb{1}_{D_r \geq \delta}(P_r) \cdot \mathbb{1}_{\Gamma \leq 1}(P_r)}{\sum_{P_r \in D_r} \mathbb{1}_{D_r \geq \delta}(P_r)} = 1 - GPR \quad (5)$$

Due to the fact that the indicator function $\mathbb{1}_{D_r \geq \delta}$ does not depend on Γ, the gradient of L_{GPR}^{δ} (with respect to the trainable parameters) can be written as follows:

$$\frac{\partial L_{GPR}^{\delta}}{\partial w} = \frac{1}{\sum_{P_r \in D_r} \mathbb{1}_{D_r \geq \delta}(P_r)} \cdot \sum_{P_r \in D_r} \mathbb{1}_{D_r \geq \delta}(P_r) \frac{\partial \mathbb{1}_{\Gamma \leq 1}(P_r)}{\partial \Gamma} \frac{\partial \Gamma(P_r)}{\partial w}, \quad (6)$$

where w represents any of the trainable parameters of the neural network.

The problem with the above definition of L_{GPR}^{δ} is that it generates zero gradients, which is a direct consequence of the fact that the indicator function $\mathbb{1}_{\Gamma \leq 1}(\cdot)$ is stepwise constant with respect to Γ, preventing SGD training. To address this issue, in the following we propose the use of a soft approximation of the objective function L_{GPR}^{δ} with non-zero gradients.

3.2   Soft Counting with Sigmoid-GPR

To avoid the propagation of null gradients, we propose to use the sigmoid function, $\sigma(x) = (1 + \exp^{-\beta x})^{-1}$, to approximate counting passing voxels. The slope of the sigmoid depends on the value of its sharpness β that we consider as a hyperparameter.

Moreover, we note that for all $P_r \in D_r$, it stands that:

$$\lim_{\beta \to +\infty} \sigma(\beta \cdot (1 - \Gamma(P_r)) = \mathbb{1}_{\Gamma \leq 1}(P_r) \quad (7)$$

Hence, the asymptotic behaviour of the sigmoid function combined with shifting the gamma index values can provide an estimate of the count of passing voxels by summation over all points P_r. The accuracy of the estimation then depends on the value of β: the bigger the β, the more precise the estimation will be.

Thus, we approximate the loss L_{GPR}^{δ} in Eq. 5 with $L_{\sigma-GPR}^{\delta}$ defined using the sigmoid function:

$$L_{\sigma-GPR}^{\delta} = 1 - \frac{\sum_{P_r \in D_r} \sigma(\beta \cdot (1 - \Gamma(P_r))) \mathbb{1}_{D_r \geq \delta}(P_r)}{\sum_{P_r \in D_r} \mathbb{1}_{D_r \geq \delta}(P_r)} \tag{8}$$

$L_{\sigma-GPR}^{\delta}$ is differentiable everywhere and, provided the sharpness β is not too high, gradients are non-zero and allow gradient descent to update the model's weights during backpropagation.

Given Eq. 7, we remark that $L_{\sigma-GPR}^{\delta}$ accurately approximates the true GPR loss function, i.e., $L_{\sigma-GPR}^{\delta} \to L_{GPR}^{\delta}$ as $\beta \to +\infty$.

Annealing Schedule of β: In light of the equations above, we propose to consider β as a hyperparameter. At the beginning of training, the model usually predicts poorly and the majority of voxels fail to satisfy the gamma index passing criterion. This implies that the corresponding loss computed with $L_{\sigma-GPR}^{\delta}$ generates zero gradients everywhere if the value of β is set too high. To avoid this behaviour, we propose an annealing schedule for β that starts with low initial values and progressively increases β over the training. Moreover, when $\beta \sim 0^+$, the Taylor series expansion of the sigmoid function yields:

$$\sigma(\beta \cdot (1 - \Gamma(P_r))) \sim \beta \cdot (1 - \Gamma(P_r)) \tag{9}$$

Given Eq. 9, we can write the Taylor expansion of $L_{\sigma-GPR}^{\delta}$ when $\beta \sim 0^+$:

$$L_{\sigma-GPR}^{\delta} \sim 1 - \beta + \frac{\sum_{P_r \in D_r} \Gamma(P_r) \mathbb{1}_{D_r \geq \delta}(P_r)}{\sum_{P_r \in D_r} \mathbb{1}_{D_r \geq \delta}(P_r)} \tag{10}$$

Consequently, we introduce the loss function $L_{\Gamma}^{\delta}(D_r, D_e)$ to model the linear behaviour of $L_{\sigma-GPR}^{\delta}$ at the start of the annealing schedule:

$$L_{\Gamma}^{\delta}(D_r, D_e) = \frac{\sum_{P_r \in D_r} \Gamma(P_r) \cdot \mathbb{1}_{D_r \geq \delta}(P_r)}{\sum_{P_r \in D_r} \mathbb{1}_{D_r \geq \delta}(P_r)} \tag{11}$$

As the training continues and the loss decreases, the annealing scheme proceeds in progressively increasing β in order to improve the approximation of the GPR loss L_{GPR}^{δ} defined in Eq. 5. As β increases and is acquiring larger values, minimizing the loss amounts to getting failing voxels (voxels with $\Gamma > 1$) to satisfy the passing criterion.

To model the behaviour of $L_{\sigma-GPR}^{\delta}$ at this stage (i.e. as β is acquiring larger values), we modify L_{Γ}^{δ} in Eq. 11 by introducing the loss function $L_{\Gamma>1}^{\delta}$. To prevent backpropagation of zero gradients with respect to $\mathbb{1}_{\Gamma>1}$, we use the *stopgrad* operation:

$$L_{\Gamma>1}^{\delta}(D_r, D_e) = \frac{\sum_{P_r \in D_r} \Gamma(P_r) \cdot \mathbb{1}_{D_r \geq \delta}(P_r) \cdot stopgrad(\mathbb{1}_{\Gamma>1}(P_r))}{\sum_{P_r \in D_r} \mathbb{1}_{D_r \geq \delta}(P_r)} \tag{12}$$

For the sake of characterizing the behaviour of $L_{\sigma-GPR}^{\delta}$ in Eq. 8, we also study trainings that involve the use of loss functions L_{Γ}^{δ} and $L_{\Gamma>1}^{\delta}$ in the following experiments.

3.3 Accelerating Gamma Index Matrix Computations

Having a differentiable GPR loss does not make it directly applicable for neural network training since, by definition, it requires iterating over all voxels in the given distributions, therefore leading to prohibitive computation time when considering high-resolution distributions. To deal with this issue, we propose an accelerated version of GPR for faster calculations.

To avoid physical incoherence when computing gamma index values, we sample the evaluated and reference distributions to the resolution $1\,mm^3$ with bilinear interpolation. By the definition in Eq. 1, one can observe that the evaluated voxels located farther than $DTA\,mm$ from P_r automatically yield a gamma index superior to 1. Thus we limit the search within an invariant vicinity defined by the chosen DTA. More precisely, the gamma index value of a reference point then stems from comparing gamma values computed with voxels in a cube comprising $(2 \times DTA + 1)^k$ voxels, in the case of k dimensional dose distributions. We then use unfolding to extract sliding local blocks of the evaluated distribution generated by the model. This operation creates one channel per voxel in the vicinity defined by the DTA. We then apply the minimum operation over the channel dimension to get the minimal gamma index value.

The approach enables fast computation of the gamma index distribution Γ, which is necessary for the calculations of $L^\delta_{\sigma-GPR}$, L^δ_Γ an $L^\delta_{\Gamma>1}$ presented in Eq. 8, 11, 12. Computation times are discussed in Sect. 5.

4 Dataset and Experimental Design

Dataset: We carried out the experiments on the publicly available dataset presented in [10] comprising 50 patients treated with VMAT plans. Each patient has a reference dose distribution computed from 1×10^{11} particles and a low precision simulation computed from 1×10^9 particles. The main goal of the methods benchmarked on this task is to generate the high precision simulation of the dose from the available low precision one. More details about the dataset can be found in the original publication. For our experiments, we split patients to 35-5-10 for respectively, the train, validation and test sets. The cases in the dataset correspond to various anatomies and therefore, we split them as equally as possible between sets to avoid biases.

Even though our approach enables training on 3D dose distributions, the dataset comprises a small number of samples. Thus, we decided to carry out the experiments in 2D to favour significant experiments and a relevant benchmark. In this setting, a training sample corresponds to an axial slice of a patient's dose volume. The 2D training dataset therefore comprises around 11k training samples, where a sample is a pair of corresponding slices of low precision and high precision dose simulation.

Preprocessing: We normalized both low precision and reference distributions using the average dose maximum computed over the reference dose volumes from the training set. We then applied the same normalization on the validation and test sets. To enable batch training, we padded each training sample with zeros in order to match a fixed size of 256 × 256. To further help the model generate accurate dose predictions, we added the corresponding CT slice as second input channel to incorporate the corresponding anatomy. We applied minimax normalization to CT volumes so voxel values remain in $[0,1]$ range.

Model: In all experiments, the model is a standard UNet architecture [14] with skip connections between the encoder and the decoder. The encoder part of the model performs downsampling twice with convolutional layers using 4×4 filters and a stride of 2. Symmetrically, transposed convolutions upsample feature maps in the decoder. Each stage of the UNet comprises two convolution blocks before downsampling or upsampling. Much like the convolutional block presented by Liu et al. in [8], a convolution block first applies a convolution with 7×7 filters and 3×3 padding, and then two convolutions with 3×3 filters to further process the features maps. Each convolution is followed by Gaussian Error Linear Units (GELU) activation units. The block ends with a residual connection to keep high frequency details from the block's input. Overall, the model has around 10 million trainable parameters.

Optimization Set-Up: In all trainings, we trained the model using AdamW optimizer [9]. We set the initial learning rate to $3e^{-4}$ and decreased it progressively during the training when the validation loss stagnated. Weight decay was set to $5e^{-4}$ and batch size to 16. We trained for 20k iterations on a NVidia GeForce RTX 3090 GPU. The trainings were stopped when overfitting appeared by adopting the early stopping strategy. With this training scheme, early stopping occurred after around 15k iterations, when the validation loss fails to decrease 2% after 500 iterations.

Loss Functions: To train with the GPR-based loss using sigmoid count $L^\delta_{\sigma-GPR}$ presented in Eq. 8, we designed the following annealing schedule for the sharpness parameter β. We set the inital value of β to 2×10^{-2} for the first 150 iterations. Then, β increased by a factor of 5% every 50 iterations until it reached an intermediate value of 3 where updates slowed down to 5% every 100 iterations. Increasing updates stopped when β reached a chosen ceiling value of $\beta_{max} = 5$. Setting β_{max} prevented the slope of the sigmoid from getting too sharp and the loss from encountering a vanishing gradient problem, which would stop the updates of gradient descent. Additional benchmarks with the approximating loss functions L^δ_Γ and $L^\delta_{\Gamma>1}$ have been also conducted, in order to better characterize the behaviour of $L^\delta_{\sigma-GPR}$ for small values of β.

For all GPR-based functions, we set the dose threshold δ to 20%. This means that, while loss functions compute the gamma index distribution by considering all voxels, the computed approximated GPR value takes into account only voxels P_r for which the dose value is superior to 20% of the maximum dose of the reference distribution, i.e. $D_r(P_r) \geq 20\% \cdot \max_{P_r \in D_r} D_r(P_r)$.

Table 1. Evaluation metrics over the dose distributions comprised in the test set. Different benchmarks over the considered loss functions for different metrics are highlighted with their mean and standard deviation. With bold we indicate the best performing methods per metric.

Loss function	GPR 2%/2mm	GPR 3%/2mm	GPR 3%/3mm	SSIM (%)	MAE	MSE
MAE	47.6 ± 7.6	53.8 ± 8.4	66.5 ± 10.0	83.3 ± 9.2	0.39 ± 0.10	0.51 ± 0.28
MSE	50.2 ± 7.8	56.7 ± 8.7	69.5 ± 10.3	83.9 ± 11.5	0.34 ± 0.09	0.43 ± 0.29
SSIM+MAE	46.3 ± 8.8	52.3 ± 9.8	64.6 ± 11.9	**94.0 ± 8.8**	0.35 ± 0.09	0.74 ± 0.36
SSIM+MSE	49.1 ± 10.0	55.3 ± 11.1	67.5 ± 13.2	93.3 ± 3.1	0.30 ± 0.07	0.30 ± 0.15
L_{Γ}^{δ}	57.7 ± 3.3	65.0 ± 3.5	79.2 ± 4.3	87.5 ± 7.6	0.28 ± 0.07	0.30 ± 0.23
$L_{\Gamma>1}^{\delta}$	57.2 ± 3.9	64.6 ± 4.1	79.0 ± 4.4	86.7 ± 8.1	0.27 ± 0.07	0.25 ± 0.16
$L_{\sigma-GPR}^{\delta}$	**59.3 ± 3.3**	**66.8 ± 3.6**	**81.4 ± 4.0**	88.2 ± 7.5	**0.24 ± 0.06**	**0.22 ± 0.16**

To benchmark against our proposed GPR-based loss functions, we considered several other loss functions commonly used in computer vision. The benchmark includes the MAE and the MSE for a comparison with pixel-wise errors. Finally, we considered the combination of pixel-wise errors with the SSIM. More precisely, the benchmark includes SSIM-MAE and SSIM-MSE, which are the equally weighted sum of respectively the SSIM and MAE, and the SSIM and MSE. For each training on the loss functions considered above, we used the exact same model architecture and optimization strategy, in order to promote the reliability and fairness of the comparison.

5   Results

5.1   Training with GPR-Based Loss Functions

Extensive quantitative comparison on the test set for each training, using the MAE, MSE, SSIM and GPR with various values of DTA and dose tolerance δ are summarised in Table 1. As the test set comprises 10 patients, we computed the metrics over each slice of each patient's volume, and then average over the test set per considered metric. Results point out that models trained with GPR-based loss functions tend to outperform others with respect to the GPR, the MAE and MSE. In contrast, models trained with SSIM-MAE and SSIM-MSE show the highest SSIM scores. With a closer look however, one can observe that they report among the lowest performance for the rest of the metrics. This result indicates that the SSIM may not be a well-suited metric to evaluate the quality of dose distributions, since it seems to be biased.

To assert statistical significance of the results, we take an in-depth look at each patient in the test set to explain the high standard deviation values observed in Table 1. Boxplots a), b) and d) in Fig. 1 point out the presence of an outlier patient case on which models tend to fail with respect to the GPR, SSIM and MSE. In contrast with SSIM, MSE and MAE-trained models, we observe that models trained with GPR-based loss functions not only display robustness to this outlier, but also show smaller standard deviation over the whole test set.

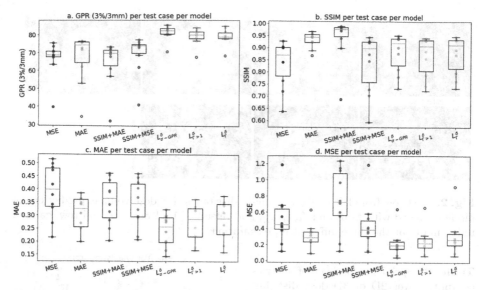

Fig. 1. Boxplots representing the evaluation metrics achieved by trained models for each case in the test set depending on the loss function used for training. The y axis indicates the values of the considered metric. The x axis spcifies the loss function with which the corresponding model was trained.

Figure 1 also allows to compare discrepancies within the family of GPR-based loss functions. While all of them produce better performing models with respect to all evaluation metrics except the SSIM, the loss function with sigmoid counting $L^\delta_{\sigma-GPR}$ outperforms L^δ_Γ and $L^\delta_{\Gamma>1}$. We explain this behaviour by the fact that both L^δ_Γ and $L^\delta_{\Gamma>1}$ focus only on minimizing gamma index values, and not directly maximizing the number of voxels satisfying the passing criterion. We conclude that $L^\delta_{\sigma-GPR}$ yields better maximization of the GPR and is therefore the better approximation of the true GPR loss function L^δ_{GPR}.

The MSE-trained model outperforms other models trained with non GPR-based loss functions with respect to the GPR, so we chose to display its dose prediction conjointly with the dose generated by the $L^\delta_{\sigma-GPR}$ trained model in Fig. 2. Although both trainings achieved convergence, the prediction of the MSE-trained model manifests important artefacts at the bottom of the generated dose. Additionally, the dose itself seems to be smoother than the dose predicted with the $L^\delta_{\sigma-GPR}$ training. Finally, the MSE-trained model appears to overestimate the dose in low-dose regions to a greater extent than the $L^\delta_{\sigma-GPR}$-trained model.

5.2   Speed-Up of GPR-Acceleration Approach

In an effort to promote the GPR-based loss functions as viable deep learning optimization criteria that allow fast error computations and training, we had to accelerate gamma index computations. To quantify the extent of our accel-

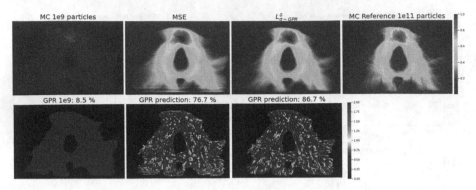

Fig. 2. First row from left to right: a single slice of the 1e9 dose volume, predictions of models trained with MSE and $L^\delta_{\sigma-GPR}$, and reference 1e11 dose. Second row: gamma index maps for the three different representations.

Table 2. Speed comparison of metrics computed over 2D or 3D dose distributions.

Metric	3D dose Time(ms)	2D dose Time(ms)
MSE	0.14 ± 0.02	0.13 ± 0.03
MAE	0.23 ± 0.06	0.19 ± 0.04
SSIM	27.49 ± 7.73	5.06 ± 3.32
$L^\delta_{\sigma-GPR}$, $L^\delta_{\Gamma>1}$, L^δ_Γ	**30.54 ± 0.01**	**4.51 ± 0.00**
Exhaustive	985 ± 515	8.01 ± 2.28
PyMedPhys	> 1 second	> 100 ms

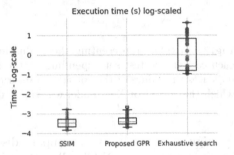

Fig. 3. Boxplots of execution times of the SSIM, our proposed approach and the exhaustive search method on 3D dose distributions.

eration approach, we benchmark against two methods. The first one is a GPU-accelerated exhaustive search approach in a limited vicinity of $3\,\text{mm}^3$ around the considered reference voxel. The second is an open-source tool from PyMedPhys [2] which makes use of acceleration ideas from Wendling et al. [18] and executes on CPU and is single-threaded. Regarding the latter, we limit the interpolation ratio to 2 to have a fair comparison.

The time estimation was twofold. We timed each evaluation metric and GPR-based loss functions on 3D or 2D distributions stemming from the MC dataset used for the experiments. 3D dose distributions were interpolated to resolution $1\,\text{mm}^3$ and of shape $128 \times 200 \times 200$, comprising around 5×10^6 voxels. The 2D dose distributions comprised axial slices of the 3D dose distributions and were interpolated to a size 400×400. For the GPR calculations, we set the DTA and Δ to respectively $2\,\text{mm}$ and 2%. Execution times are displayed in Table 2.

Figure 3 and Table 2 highlight that our approach has equivalent speed to that of the SSIM. Compared to the exhaustive search method, our approach improves the speed of gamma index computations by a factor of at least 30 in the case of 3D dose distributions and twofold for 2D distributions. Consistently with these results, we note that trainings took around 24 h for SSIM-MAE, SSIM-MSE and GPR-based loss function, whereas they lasted for 15 h for experiments with the MAE and the MSE. Results therefore validate our GPR-based loss functions to efficiently train deep neural networks. Nonetheless, our comparison is limited to speed assessment and does not encompass RAM usage and precision considerations. Although our approach highlights significant speed gain in the computation of the GPR metric and, by extension, of the GPR-based loss functions presented in this study, it comes at the price of an increased RAM usage caused by the unfolding operation.

We make the remark that for all loss functions, the obtained GPRs do not meet the 95% GPR threshold indicating clinical validation. Nevertheless, the goal of the experiments was to show the benefits of optimizing directly the clinical metric during training and results support that statement.

6   Conclusion

Adopting the correct optimization criterion is essential to train deep learning models adequately with the task they are designed to solve. For the task of accelerating MC radiotherapy dose simulation with deep learning, this work proves that directly optimizing models with the clinical validation metric yields significant improvement in predicted dose quality when compared to other loss functions. We provide a fast computation of the GPR to enable such results. Moreover, the GPR is a similarity metric for distributions in general, and may be applied to other tasks such as radiotherapy dose generation or even finding adversarial examples for generative adversarial networks. Future work will focus on addressing the remaining limitations of our approach and assessing the potential of our new class of loss functions in solving other deep learning tasks.

References

1. Ahnesjö, A.: Collapsed cone convolution of radiant energy for photon dose calculation in heterogeneous media. Med. Phys. **16**(4), 577–592 (1989)
2. Biggs, S., et al.: Pymedphys: a community effort to develop an open, python-based standard library for medical physics applications. J. Open Source Softw. **7**(78), 4555 (2022)
3. Chen, M., Lu, W., Chen, Q., Ruchala, K., Olivera, G.: Efficient gamma index calculation using fast Euclidean distance transform. Phys. Med. Biol. **54**(7), 2037 (2009)
4. Deng, Y., et al.: Comparison of pencil beam and Monte Carlo calculations with ion chamber array measurements for patient-specific quality assurance. Radiat. Med. Prot. **3**(3), 115–122 (2022)
5. Gu, X., Jia, X., Jiang, S.B.: GPU-based fast gamma index calculation. Phys. Med. Biol. **56**(5), 1431 (2011)

6. Lee, B.I., Boss, M., LaRue, S.M., Martin, T.W., Leary, D.: Comparative study of the collapsed cone convolution and monte carlo algorithms for radiation therapy planning of canine sinonasal tumors reveals significant dosimetric differences. Veterinary Radiol. Ultrasound Official J. Am. Coll. Veterinary Radiol. Int. Veterinary Radiol. Assoc. **63** (2021)
7. Lin, T.Y., Goyal, P., Girshick, R.B., He, K., Dollár, P.: Focal loss for dense object detection. In: 2017 IEEE ICCV, pp. 2999–3007 (2017)
8. Liu, Z., Mao, H., Wu, C.Y., Feichtenhofer, C., Darrell, T., Xie, S.: A convnet for the 2020s. In: Proceedings of the IEEE/CVF Conference on Computer Vision and Pattern Recognition, pp. 11976–11986 (2022)
9. Loshchilov, I., Hutter, F.: Decoupled weight decay regularization. arXiv preprint arXiv:1711.05101 (2017)
10. Martinot, S., Bus, N., Vakalopoulou, M., Robert, C., Deutsch, E., Paragios, N.: High-particle simulation of monte-carlo dose distribution with 3D convlstms. In: MICCAI, pp. 499–508 (2021)
11. Mohan, R., Chui, C., Lidofsky, L.: Differential pencil beam dose computation model for photons. Med. Phys. **13**(1), 64–73 (1986)
12. Neph, R., Lyu, Q., Huang, Y., Yang, Y.M., Sheng, K.: Deepmc: a deep learning method for efficient monte carlo beamlet dose calculation by predictive denoising in magnetic resonance-guided radiotherapy. Phys. Med. Biol. **66**(3), 035022 (2021)
13. Quan, E., et al.: A comprehensive comparison of IMRT and VMAT plan quality for prostate cancer treatment. Int. J. Radiat. Oncol. Biol. Phys. **83**, 1169–78 (2012)
14. Ronneberger, O., Fischer, P., Brox, T.: U-net: convolutional networks for biomedical image segmentation. In: MICCAI, pp. 234–241 (2015)
15. Sudre, C.H., Li, W., Vercauteren, T., Ourselin, S., Jorge Cardoso, M.: Generalised dice overlap as a deep learning loss function for highly unbalanced segmentations. In: Deep Learning in Medical Image Analysis and Multimodal Learning for Clinical Decision Support, pp. 240–248 (2017)
16. Vasudevan, V., Huang, C., Simiele, E., Yu, L., Xing, L., Schuler, E.: Combining monte carlo with deep learning: Predicting high-resolution, low-noise dose distributions using a generative adversarial network for fast and precise monte carlo simulations. Int. J. Radiat. Oncol. Biol. Phys. **108**(3), S44–S45 (2020)
17. Wang, Z., Bovik, A.C., Sheikh, H.R., Simoncelli, E.P.: Image quality assessment: from error visibility to structural similarity. IEEE Trans. Image Process. **13**(4), 600–612 (2004)
18. Wendling, M., et al.: A fast algorithm for gamma evaluation in 3D. Med. Phys. **34**(5), 1647–1654 (2007)
19. Zhao, H., Gallo, O., Frosio, I., Kautz, J.: Loss functions for image restoration with neural networks. IEEE Trans. Comput. Imaging **3**(1), 47–57 (2016)

Scalable Orthonormal Projective NMF via Diversified Stochastic Optimization

Abdalla Bani[1]([✉]), Sung Min Ha[1], Pan Xiao[1], Thomas Earnest[1], John Lee[1], and Aristeidis Sotiras[1,2]

[1] Department of Radiology, Washington University School of Medicine in St. Louis, St. Louis, MO 63108, USA
{a.bani,sungminha,panxiao,tom.earnest,jjlee,aristeidis.sotiras}@wustl.edu
[2] Institute for Informatics, Washington University School of Medicine in St. Louis, St. Louis, MO 63108, USA

Abstract. The increasing availability of large-scale neuroimaging initiatives opens exciting opportunities for discovery science of human brain structure and function. Data-driven techniques, such as Orthonormal Projective Non-negative Matrix Factorization (opNMF), are well positioned to explore multivariate relationships in big data towards uncovering brain organization. opNMF enjoys advantageous interpretability and reproducibility compared to commonly used matrix factorization methods like Principal Component Analysis (PCA) and Independent Component Analysis (ICA), which led to its wide adoption in clinical computational neuroscience. However, applying opNMF in large-scale cohort studies is hindered by its limited scalability caused by its accompanying computational complexity. In this work, we address the computational challenges of opNMF using a stochastic optimization approach that learns over mini-batches of the data. Additionally, we diversify the stochastic batches via repulsive point processes, which reduce redundancy in the mini-batches and in turn lead to lower variance in the updates. We validated our framework on gray matter tissue density maps estimated from 1000 subjects part of the Open Access Series of Imaging (OASIS) dataset. We demonstrated that operations over mini-batches of data yield significant reduction in computational cost. Importantly, we showed that our novel optimization does not compromise the accuracy or interpretability of factors when compared to standard opNMF. The proposed model enables new investigations of brain structure using big neuroimaging data that could improve our understanding of brain structure in health and disease.

Keywords: NMF · Stochastic optimization · MRI · Big data

1 Introduction

Maturation of *in vivo* neuroimaging technology has led to large-scale studies of healthy and diseased populations that are often augmented with new

A. Frangi et al. (Eds.): IPMI 2023, LNCS 13939, pp. 497–508, 2023.
https://doi.org/10.1007/978-3-031-34048-2_38

data [5,11,28]. (Such) Large scale datasets provide increased statistical power that may enable a better understanding of brain structure in health and disease. Data-driven analytical techniques are well-positioned to explore multivariate relationships in large sample size settings towards extracting data representation in unbiased, hypothesis-free way, thus complementing conventional representations that rely on predefined regions of of interests. Unsupervised machine learning techniques, such as Principal Component Analysis (PCA) and Independent Component Analysis (ICA) have been commonly used to derive data-driven representations of brain structure and function in neuroimaging studies. However, PCA and ICA model the data through complex mutual cancellation between component regions of opposite sign, leading to representations that are difficult to interpret, lack regional specificity and do not generalize well in unseen datasets.

In contrast to PCA and ICA, non-negative matrix factorization (NMF) factorizes the data under a non-negativity constraint. This has been shown to lead to a parts-based representation of the data, where parts are combined in additive way to form a whole. More recently, an NMF variant termed orthonormal projective NMF (opNMF) [31] has been proposed and applied to neuroimaging data [26]. opNMF demonstrated improved interpretability and reproducibility compared to PCA and ICA [26], which has led to its increasing adoption in neuroimaging studies. Without attempting to be exhaustive, opNMF has been used to investigate normative brain organization across a wide range of age groups, including neonate [16,20,29], adolescents [27], and adults [6,21,23,26]. Additionally, it has been used to examine brain alterations associated with psychopathology [9,17,18]. Importantly, it has been used to analyze neuroimaging data derived from T1- and T2-weighted Magnetic Resonance Imaging (MRI) [9,16,17,26,27], Fluid Attenuated Inversion Recovery (FLAIR) imaging [6], diffusion MRI [20,29], and multi-modal data [18,21,23].

However, applying opNMF in large scale neuroimaging studies is challenging due to its high memory demands. The factorization is quadratically driven by the number of features and requires the storage and multiplication of high dimensional and dense matrices that are unable to reside in memory. Current implementations of opNMF have bypassed this problem by changing the order of matrix multiplication while trading off running time [26]. Unfortunately, this does not scale in big data scenarios because the high sample size presents an additional bottleneck. Stochastic variants of opNMF that operate on mini-batches of the data have been proposed as a solution to handle large-scale data and have been shown to numerically reduce memory requirements [32] but with no theoretical convergence guarantees. Furthermore, current stochastic opNMF variants utilize random sampling and do not take into account information about how representative of the population the stochastic samples are. This might be problematic as there is mounting evidence suggesting that non-representative sampling may bias results [12,24].

Accordingly, we propose in the present work a scalable stochastic algorithm for opNMF in big neuroimaging data that learns over mini-batches of the data and provide theoretical considerations for its convergence. Additionally, we enhance the solution of the algorithm by reducing redundancy in the mini-batches using repulsive point processes. We demonstrate that our approach yields a significant decrease in the computational complexity of opNMF without a significant compromise in performance. Furthermore, we show that optimizing over diverse mini-batches that are representative of features of the data leads to faster convergence and smaller error compared to uniformly sampled mini-batches.

2   Method

Vanilla opNMF requires reading all data into the memory at once. Our key idea to reduce memory requirements is twofold. First, we adopt stochastic optimization to approximate the opNMF solution over mini-batches of the data that require less memory to store. Second, we adopt repulsive point processes that create diversified mini-batches, leading to less redundant and more representative mini-batches. In the remainder of the section, we provide the theoretical framework supporting the convergence of the stochastic variant, which we will subsequently use as basis for detailing the proposed variant that optimizes over diversely sampled mini-batches of the data, termed d-sopNMF.

2.1   Orthonormal Projective Non-negative Matrix Factorization

Let $\mathbf{x}_1, \mathbf{x}_2, ..., \mathbf{x}_n \in \mathbb{R}_+^m$ be vectorized images of n different subjects and m be the number of non-negative voxels of each image. A data matrix $\mathbf{X}$ is constructed by arraying the images column-wise $\mathbf{X} = [\mathbf{x}_1, \mathbf{x}_2, ..., \mathbf{x}_n] \in \mathbb{R}_+^{m \times n}$. Our goal is to meaningfully decompose $\mathbf{X}$ into a number of components $r \ll \min(m, n)$ that represent the patterns of structural covariation across the subjects. Towards this goal, we solve the orthonormal non-negative factorization [31]

$$\min_{\mathbf{W} \geq 0} \left\| \mathbf{X} - \mathbf{W}\mathbf{W}^T\mathbf{X} \right\|_F^2 \quad \text{s.t.} \quad \mathbf{W}^T\mathbf{W} = \mathbf{I}, \tag{1}$$

where $\mathbf{W} \in \mathbb{R}_+^{m \times r}$ is the components matrix, $\| \cdot \|_F$ is Frobenius norm, and $\mathbf{I}$ denotes the identity matrix. The previous minimization problem can be solved using the iterative multiplicative update (MU) rule [31]

$$\mathbf{W}_{t+1} \leftarrow \mathbf{W}_t \frac{(\mathbf{X}\mathbf{X}^T\mathbf{W})_t}{(\mathbf{W}\mathbf{W}^T\mathbf{X}\mathbf{X}^T\mathbf{W})_t}. \tag{2}$$

Compared to vanilla NMF ($\min \| \mathbf{X} - \mathbf{W}\mathbf{H} \|_F^2$), opNMF estimates the loading coefficients $\mathbf{H} \in \mathbb{R}_+^{r \times n}$ as a projection of the data to the approximated components $\mathbf{H} = \mathbf{W}^T\mathbf{X}$ [31]. The orthonormality and projectivity constraints result in components that enjoy increased spatial specificity as they greatly decrease the amount of overlapping components [26,27,31]. However, in high-resolution settings (e.g., magnetic resonance (MR) imaging where the features are in the order of millions) the product $\mathbf{X}\mathbf{X}^T$ in Eq. 2 becomes too large to store in memory.

Implementations of opNMF have avoided this issue by using a memory-optimized version of the update rule, where $\mathbf{X}^T\mathbf{W}$ in Eq. 2 is calculated first in every iteration at the cost of redundant calculations and additional runtime. The previous problem is further exasperated in large sample size settings because as the number of samples grows, the entire product $\mathbf{X}\mathbf{X}^T\mathbf{W}$ in Eq. 2 becomes prohibitive.

2.2   Stochastic Orthonormal Projective Non-negative Matrix Factorization

sopNMF is a stochastic variant of opNMF that aims to approximate the solution opNMF over mini-batches of the data. We use the stochastic approach to alleviate the effects of the costly term $\mathbf{X}\mathbf{X}^T\mathbf{W}$ that appears twice in Eq. 2 by replacing it with $\mathbf{X}_p\mathbf{X}_p^T\mathbf{W}$, where $\mathbf{X}_p \in \mathbb{R}_+^{m \times p}$ ($p \ll n$) is a small matrix constructed by randomly sampling subsets of the columns of the data matrix $\mathbf{X}$ [32]. The MU rule in Eq. 2 can be rewritten as

$$\mathbf{W}_{t+1} \leftarrow \mathbf{W}_t \frac{\left(\mathbf{X}_p\mathbf{X}_p^T\mathbf{W}\right)_t}{\left(\mathbf{W}\mathbf{W}^T\mathbf{X}_p\mathbf{X}_p^T\mathbf{W}\right)_t}. \tag{3}$$

Theoretical Considerations. We first derive a form for the stochastic sequence of surrogate functions that model the sopNMF objective. Second, we show that the surrogate functions converge to the opNMF objective almost surely (a.s.), given reasonable assumptions. We closely follow the framework proposed in [14] with the main distinction being the opNMF objective, and due to space limitations we omit the details. It is important to note that finding a global minimum for the opNMF objective is generally unattainable due to the non-convexity of the problem. Hence, all the provided convergence results are with respect to local minima. The solution of the matrix factorization amounts to optimizing an empirical cost function given a finite set of samples $\mathbf{X} = [\mathbf{x}_1, \mathbf{x}_2, ..., \mathbf{x}_n]$, $\mathbf{x}_i \in \mathbb{R}_+^m$

$$f_n(\mathbf{W}) \triangleq \frac{1}{n} \sum_{i=1}^n \ell(\mathbf{x}_i, \mathbf{W}), \tag{4}$$

where $\ell(\mathbf{x}_i, \mathbf{W})$ is a loss function. For example, in the case of a single subject, we can write $\ell(\mathbf{x}, \mathbf{W})$ as $\ell(\mathbf{x}, \mathbf{W}) \triangleq \min_{\mathbf{W} \geq 0} \|\mathbf{x} - \mathbf{W}\mathbf{W}^T\mathbf{x}\|_2^2$. We constrain the solution to prevent large values of $\mathbf{W}$ and enforce the orthonormality of $\mathbf{W}$ (or approximate orthonormality since all of the elements of $\mathbf{W} \geq 0$) by constraining its columns, $\mathcal{C} \triangleq \{\mathbf{W} \in \mathbb{R}_+^{m \times r} \text{ s.t. } \forall j = 1, ..., r, \quad \mathbf{w}_j^T\mathbf{w}_j \leq 1\}$. Considering all of the subjects, the optimization of the empirical cost function can be written as $\min_{\mathbf{W} \in \mathcal{C}} \sum_{i=1}^n \left(\|\mathbf{x}_i - \mathbf{W}\mathbf{W}^T\mathbf{x}_i\|_2^2\right)$, where $\|\cdot\|_2$ is the ℓ_2-norm, or in matrix factorization format $\min_{\mathbf{W} \in \mathcal{C}} \|\mathbf{X} - \mathbf{W}\mathbf{W}^T\mathbf{X}\|_F^2$.

In machine learning applications, we often do not seek the minimization of the empirical cost function $f_n(\mathbf{W})$ with high accuracy, but instead we aim to minimize the expected cost $f(\mathbf{W}) \triangleq \mathbb{E}_\mathbf{x}[\ell(\mathbf{x}, \mathbf{W})] = \lim_{n \to \infty} f_n(\mathbf{W})$ a.s. , where the previous expectation is taken relative to the probability distribution of the data $p(\mathbf{x})$ [1, 14]. Considering a data set composed of subjects from a distribution $p(\mathbf{x})$, we draw a single subject (or a mini-batch) and update $\mathbf{W}_{t+1}$ by minimizing over $\mathcal{C}$ the function

$$\hat{f}_t(\mathbf{W}) \triangleq \frac{1}{t} \sum_{i=1}^{t} \left(\|\mathbf{x}_i - \mathbf{W}\mathbf{W}^T \mathbf{x}_i\|_2^2 \right).$$

In proposition 1 we show that $\hat{f}$ acts a surrogate and upperbounds the empirical cost. Given that $\hat{f}$ acts as a surrogate of f, we can iterate over min-batches of the data (Algorithm 1).

Proposition 1 (*The sequence of surrogate objective functions converges a.s.*). $\hat{f}_t$ acts as a converging surrogate of f, and $f(\mathbf{W}_t) - \hat{f}_t(\mathbf{W}_t)$ converges a.s. to zero under the assumptions: (1) the distribution of the imaging data is bounded and admits a compact support $\mathcal{V}$, (2) for $(\mathbf{x}, \mathbf{W}) \in \mathcal{V} \times \mathcal{C}$, $\ell(\mathbf{x}, \mathbf{W})$ is strongly convex for some constant, and (3) the surrogate functions are strongly convex with lower bounded Hessians on $\mathcal{C}$.

Proof sketch. We first show that the iterates $\mathbf{W}_{t+1} - \mathbf{W}_t = O\left(\frac{1}{t}\right)$. This is done by upper bounding $\|\mathbf{W}_{t+1} - \mathbf{W}_t\|_F$ using strong convexity and Lipschitz continuity of the surrogate $\hat{f}$. Then, we show that the series of positive stochastic processes defined as $u_t \triangleq \hat{f}_t(\mathbf{W}_t) \geq 0$ is quasi-martingale [14]. This is done by showing that the expected sum of the variations of the stochastic process conditioned on the filtration of past information $\mathcal{F}_t$, $\sum_{t=1}^{\infty} \mathbb{E}\left[\max\left(\mathbb{E}\left[u_{t+1} - u_t \mid \mathcal{F}_t\right], 0\right)\right]$ converges, indicating that u_t is quasi-martingale, which converges with probability 1. We then bound each summand of the series using Donsker's theorem [14]. The convergence of u_t indicates that $\hat{f}_t(\mathbf{W}_t) - \hat{f}_t(\mathbf{W}_t) \xrightarrow{\text{a.s.}} 0$. We then show that $\ell(\mathbf{x}, \mathbf{D})$ is continuously differentiable and uniformly Lipschitz on a compact set using perturbation theory [14]. Using that fact, we invoke results from empirical processes to show that $f(\mathbf{W}_t) - \hat{f}_t(\mathbf{W}_t) \xrightarrow{\text{a.s.}} 0$ and hence $f(\mathbf{W}_t) \xrightarrow{\text{a.s.}} 0$.

2.3 Diversified Stochastic Orthonormal Projective Non-negative Matrix Factorization

Random sampling may lead to batches that contain redundant data that do not represent well the population, which may lead to biased results [10, 34]. To account for this, we propose a variant, termed diversified sopNMF (d-sopNMF), which utilizes repulsive point processes to create diversified mini-batches.

Determinantal Point Processes (DPP). DPP is a sampling technique that aims to obtain samples that are as diverse as possible by encoding negative correlations [10]. The strength of negative correlations can be modeled using a real, symmetric and positive semi-definite kernel matrix $\mathbf{L} \in \mathbb{R}^{n \times n}$, where element $\mathbf{L}_{ij}$ is a measure of similarity between points i and j in the data set [10]. The probability of subsampling any subset $\mathcal{X}$ of $\{1, \ldots, n\}$ in a DPP is proportional to the determinant of the submatrix that indexes the subset $\mathbf{L}_{\mathcal{X}}$ of $\mathbf{L}$:

$$P(\mathcal{X}) = \frac{\det(\mathbf{L}_{\mathcal{X}})}{\det(\mathbf{L} + \mathbf{I})} \propto \det(\mathbf{L}_{\mathcal{X}}),$$

we further use DPPs with fixed sample size (kDPPs) and condition the inclusion probability on the size of the subset $\mathcal{P}_{\mathbf{L}}^k(\mathcal{X}) = \frac{\det(\mathbf{L}_{\mathcal{X}})}{\sum_{|\mathcal{X}'|} \det(\mathbf{L}_{\mathcal{X}'})}$, where $|\mathcal{X}| = p$ [10]. In this work, we construct a linear kernel that induces diversity in the batches given how similar the subjects' images in the study are $\mathbf{L}_{ij} = \mathbf{x}_i^T \mathbf{x}_j$.

d-sopNMF. The diversification approach is beneficial in two ways: (1) DPPs allow us to obtain mini-batches that are representative of the population, and (2) allow for faster decay in the error by reducing varaince in the stochastic gradients. For a DPP mini-batch $\mathbf{X}_{p \sim kDPP}$, the sopNMF MU in Eq. 3 becomes

$$\mathbf{W}_{t+1} \leftarrow \mathbf{W}_t \frac{\left(\mathbf{X}_{p \sim kDPP} \mathbf{X}_{p \sim kDPP}^T \mathbf{W}\right)_t}{\left(\mathbf{W}\mathbf{W}^T \mathbf{X}_{p \sim kDPP} \mathbf{X}_{p \sim kDPP}^T \mathbf{W}\right)_t}. \tag{5}$$

Diversified Loss. The opNMF multiplicative update can be viewed as a classical gradient additive update with a non-negativity preserving step γ^+, $\mathbf{W}_{t+1} = \mathbf{W}_t - \gamma_t^+ \frac{\partial f(\mathbf{W})}{\partial \mathbf{W}}$ [31]. Using that fact, we can obtain a diversified loss function by taking the expectation with respect to the kDPP batches $f^d(\mathbf{W}) \triangleq \frac{1}{p}\mathbb{E}_{\mathbf{x} \sim kDPP}[\ell(\mathbf{x}, \mathbf{W})]$ [34]. For B kDPP batches of size p, we minimize

$$\mathbf{W}_{t+1} = \mathbf{W}_t - \gamma_t^+ \frac{1}{p} \sum_{i \in B \sim kDPP} \frac{\partial \ell(\mathbf{x}_i, \mathbf{W})}{\partial \mathbf{W}}.$$

Next, we closely follow [34] to show that the diversified loss is a weighted empirical risk, which can lead to lower variance in the gradients. With a smart kernel choice, the weighting procedure can result in lower error.

Proposition 2. *(f^d is a re-weighted empirical risk with kDPP weights π_i)* [34]:

$$f^d(\mathbf{W}) = \frac{1}{p} \sum_{i=1}^{n} \pi_i \ell(\mathbf{x}_i, \mathbf{W}).$$

Proof. Let $v_i \in \{0, 1\}$ denote if subject i was sampled according to a kDPP or not, and let the expectation $\mathbb{E}[\sum_{i=1}^{n} v_i f(\mathbf{x}_i)] \triangleq \mathbb{E}_{\mathbf{x} \sim k-DPP}[F(\mathbf{X})]$. Let the marginal probability for a subject $\mathbf{x}_i$ being sampled according to a DPP as $\mathbb{E}[v_i] \triangleq \pi_i$, then $pf^d(\mathbf{W}) = \mathbb{E}_{\mathbf{x} \sim kDPP}[\ell(\mathbf{x}; \mathbf{W})]$ is equal to

$$\mathbb{E}\left[\sum_{i=1}^{n} v_i \ell(\mathbf{x}_i; \mathbf{W})\right] = \sum_{i=1}^{n} \mathbb{E}[v_i] \ell(\mathbf{x}_i; \mathbf{W}) = \sum_{i=1}^{n} \pi_i \ell(\mathbf{x}_i; \mathbf{W}).$$

Proposition 3. (*f^d results in variance reduction in the empirical gradient*) [34]: the product of the gradients $\mathbf{x}_i$ and $\mathbf{x}_j$, $\left[\frac{\partial\ell(\mathbf{x}_i,\mathbf{W})}{\partial\mathbf{W}}\right]^T\left[\frac{\partial\ell(\mathbf{x}_j,\mathbf{W})}{\partial\mathbf{W}}\right]$ is always positive whenever the correlation Σ_{ij} is negative

$$\forall_{i\neq j}: \Sigma_{ij}\left[\frac{\partial\ell(\mathbf{x}_i,\mathbf{W})}{\partial\mathbf{W}}\right]^T\left[\frac{\partial\ell(\mathbf{x}_j,\mathbf{W})}{\partial\mathbf{W}}\right] < 0.$$

Proof. The correlation between $\mathbf{x}_i$ and $\mathbf{x}_j$ is $\Sigma_{ij} = \frac{\mathbb{E}[(v_i-\pi_i)(v_j-\pi_j)]}{\mathbb{E}[v_i]\mathbb{E}[v_j]} = \frac{\mathbb{E}[v_iv_j]}{\pi_i\pi_j} - 1$ where $\mathbb{E}[v_iv_j] = \mathbb{E}[v_i^2]\delta_{ij}+\mathbb{E}[v_iv_j](1-\delta_{ij}) = \mathbb{E}[v_i]\delta_{ij}+(\Sigma_{ij}+1)\pi_i\pi_j(1-\delta_{ij})$, δ_{ij} is the Kronecker delta, and $v_{ij}^2 = v_{ij}$. Define the diversified gradient g_d and the full gradient of the diversified loss $g_F = \mathbb{E}[g_d]$

$$g_d(\mathbf{x}_i,\mathbf{W}) = \frac{1}{p}\sum_{i=1}^n v_i\frac{\partial\ell(\mathbf{x}_i,\mathbf{W})}{\partial\mathbf{W}} \quad\text{and}\quad g_F(\mathbf{x}_i,\mathbf{W}) = \frac{1}{p}\sum_{i=1}^n \pi_i\frac{\partial\ell(\mathbf{x}_i,\mathbf{W})}{\partial\mathbf{W}},$$

and obtaining the difference $\Delta = g_d - g_F = \frac{1}{p}\sum_{i=1}^n(\pi_i - v_i)\frac{\partial\ell(\mathbf{x}_i,\mathbf{W})}{\partial\mathbf{W}}$. Taking the variance of the diversified gradient, $\mathbb{V}(g_d) = \mathrm{Tr}(\mathrm{Cov}(g_d)) = \mathbb{E}[\Delta^T\Delta]$ and inserting $\mathbb{E}[v_iv_j]$ we get

$$\mathbb{V}(g_d) = \frac{1}{p^2}\sum_{i,j=1}^n \underbrace{\mathbb{E}[(v_i-\pi_i)(v_j-\pi_j)]}_{\mathbb{E}[v_iv_j]-\pi_i\pi_j}\left[\frac{\partial\ell(\mathbf{x}_i,\mathbf{W})}{\partial\mathbf{W}}\right]^T\left[\frac{\partial\ell(\mathbf{x}_j,\mathbf{W})}{\partial\mathbf{W}}\right]$$

$$= \underbrace{\frac{1}{p^2}\sum_{i=1}^n(\pi_i-\pi_i^2)\left\|\frac{\partial\ell(\mathbf{x}_i,\mathbf{W})}{\partial\mathbf{W}}\right\|_2^2}_{(\mathrm{I})} + \underbrace{\frac{1}{p^2}\sum_{i\neq j}\Sigma_{ij}\pi_i\pi_j\left[\frac{\partial\ell(\mathbf{x}_i,\mathbf{W})}{\partial\mathbf{W}}\right]^T\left[\frac{\partial\ell(\mathbf{x}_j,\mathbf{W})}{\partial\mathbf{W}}\right]}_{(\mathrm{II})},$$

(II) is always negative due to the choice of the kernel and may reduce the variance of the regular stochastic gradient (I) which is always positive (since $\pi_i < 1$).

3   Experiments and Results

We compared sopNMF variants to opNMF in terms of reconstruction error, sparsity, and memory consumption. We also compared the similarity of the sopNMF components and coefficients to those of opNMF. We chose a factorization rank that is often used in neuroimaging analysis to provide a typical low dimensional representation for the brain [3,22,25–27,33]. For sopNMF based on random sampling, we considered sampling with replacement (batch-randomization (BR)) and sampling without replacement (shuffle (SH)). We considered mini-batch sizes $\{50, 125, 250\}$ for d-sopNMF, sopNMF$_{\mathrm{BR}}$, and sopNMF$_{\mathrm{SH}}$. Since there is no constraint in d-sopNMF and sopNMF$_{\mathrm{BR}}$ on the number of times a subject is visited, we revisit every subject at the end of the learning to ensure that all of the subjects are passed through the model. We have used NNDSVD as initialization because it allows for deterministic initialization and leads to faster

---

Algorithm 1: (d-)sopNMF

Data: data matrix $\mathbf{X} \in \mathbb{R}_+^{m \times n}$, target rank $r \in \mathbb{N}^+$, batch size $p \ll n$, the
number of batches $B = \frac{n}{p}$, initial guess of $\mathbf{W} \in \mathbb{R}_+^{m \times r} = \mathbf{W}_o$, stopping
criterion, max epochs, **Bool:** DPP

Result: factorization matrix $\mathbf{W} \in \mathbb{R}_+^{m \times r}$

while *stopping $==$ False or epoch $<$ max epoch* **do**

    for $t = 1$ **to** B **do**

        if DPP $==$ *True* **then**

          | Read DPP mini-batch $\mathbf{X}_p \in \mathbb{R}_+^{m \times p}$;

        else

          | Read random mini-batch $\mathbf{X}_p \in \mathbb{R}_+^{m \times p}$;

        end

        Update $\mathbf{W}_{t+1}$ using MU;

        Record batch loss $\left\| \mathbf{X}_p - \mathbf{W}_{t+1}\mathbf{W}_{t+1}^T\mathbf{X}_p \right\|_F^2$;

    end

    calculate loss $= \frac{1}{B}\sum_{i=1}^{B}\left\| \mathbf{X}_i - \mathbf{W}_{t+1}\mathbf{W}_{t+1}^T\mathbf{X}_i \right\|_F^2$;

    if *Stopping $==$ True* **then**

        | Stop;

    else

        | $\mathbf{W} = \mathbf{W}_{t+1}$;

    end

end

---

convergence [2]. opNMF was run to completion with 50,000 iterations. For sop-NMF variants, the iterations were divided into epochs based on the batch size, i.e. $\frac{50,000}{n/p}$, with stopping criterion based on the consecutive decrease of the loss function in the last 100 epochs. All experiments had access to 8 threads on Xeon Gold 6226R CPUs and 32 GB of DDR4 RAM. All implementations are in Python and codes will be made available.

Data. We benchmarked the performance of sopNMF on baseline T1-weighted MR images of 1000 subjects (69.78±9.53 years, 550 female) from the Open Access Series of Imaging Studies (OASIS) dataset [11]. We chose OASIS because it is one of the largest publicly available neuroimaging datasets (compared to median size of 23 available on openneuro.org [15]). Gray matter (GM) segmentations were preprocessed using FSL FLIRT [8] and DRAMMS [19] to register to MNI152, modulated with Jacobian determinant, and smoothed with 6mm full width half max Gaussian kernel. Then, smoothed GM tissue density maps were vectorized and arrayed column-wise to construct the input data matrix $\mathbf{X} \in \mathbb{R}_+^{1187150 \times 1000}$.

Memory Usage. We assessed the scalability of sopNMF by comparing its maximum memory footprint during the MU against that of opNMF. At the largest p, sopNMF variants used approximately a third of the memory of opNMF, and the footprint was further alleviated at smaller p (Fig. 1.h).

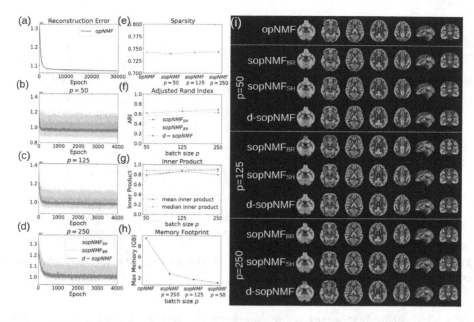

Fig. 1. (a) Reconstruction error of opNMF. **(b-d)** Batch-wise reconstruction error of sopNMF variants (sopNMF$_{BR}$ and sopNMF$_{SH}$ are overlapping). **(e)** Sparsity of **W**. **(f)** ARI between **W** of each sopNMF variant and opNMF. **(g)** Mean and median inner products between **W** of each sopNMF variant and opNMF. **(h)** Max memory usage during MU. **(i)** **W** visualizations for opNMF and the sopNMF variants at different batch size.

Approximation Error. Our goal is to alleviate the computational challenges of opNMF in the analysis of large-scale neuroimaging data while retaining its accuracy as quantified by the reconstruction error ($\|\mathbf{X} - \mathbf{WH}\|_F$). sopNMF variants retained the accuracy of opNMF across sampling methods, with d-sopNMF achieving lower reconstruction error across batch sizes (Fig. 1.a-d).

Sparsity. Importantly, we preserved the interpretability of opNMF as quantified by the sparsity (spatial specificity) of the components. Sparsity was calculated using the definition: $s(\mathbf{W}) = \frac{\sqrt{m} - (\sum |\mathbf{W}(i)|) / \sqrt{\sum \mathbf{W}(i)^2}}{\sqrt{m} - 1} \in \{0, 1\}$ [7], where $\mathbf{W}(i)$ is the ith element of **W**. Similar to opNMF, sopNMF variants produced highly sparse and well-localized components (Fig. 1.e).

Components and Loading Coefficients Similarity. The stochastic optimization is approximating the objective of opNMF, hence we want to ensure that the generated components do not changes. This is important because the components is what we interpret. That is why we evaluate the similarity of the components using the *Adjusted Rand Index* (ARI), as well as the mean and median inner products between matched components. ARI ranges between $[-1, 1]$, with higher values indicating better similarity between two clusters and 0 indicating random assignment. Mean and median inner products range between $[0, 1]$ with higher

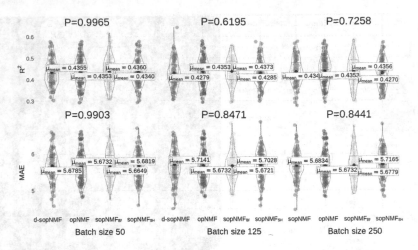

Fig. 2. R^2 and MAE for all folds in the age prediction task on 10 realizations of 10-fold cross validated OASIS data.

values indicating better correspondence between clusters. We adopted a winner-takes-all approach to derive a clustering solution from the components of each method. The components produced by sopNMF variants demonstrate a high degree of similarity to those of opNMF (Fig. 1.f-g). Further visual inspection of typical components of sopNMF variants supports the high degree of agreement as shown by the similarity measures across different batch sizes (Fig. 1.i). The loading coefficients **H** inform us of how the components contribute to the formation of the data. The NMF coefficients are used often for brain age predictions [30] and in a brain-aging study such as OASIS, we expect **H** to capture the aging pattern of the cohort. We used **H** in support vector regression (SVR) to predict the age of the subjects. We performed 10 realizations of 10-fold cross-validation and report the resulting R-squared (R^2) and mean absolute error (MAE). The approximated coefficients replicate the performance of opNMF coefficients in the task of brain age prediction (Fig. 2).

4   Discussion

Given constraints on computational resources, we focused on a single decomposition rank that is widely used to understand brain organization in low resolutions. Future work will examine the reproducibility of the components at higher granularity, which could uncover different insights regarding underlying brain networks. The choice of the DPP kernel gives us the freedom to consider different characteristics of the data when constructing our mini-batches. For example, we can create Gaussian kernels that diversely sample across imaging meta-data to prevent the learning, and hence the produced factors, from being dominated by certain demographics. Thus the choice of the kernel depends on the data and the research question. Additionally, repulsive sampling comes with computational overhead compared to uniform sampling. However, it is important to note that the sampling is independent of the learning process and can be parallelized. Further, the benefits of diversification are particularly handy for opNMF where the updates are expensive. Finally, there are recent fast implementations of repeated DPP sampling utilizing Nystrom approximations and tree-based methods [4,13]. In summary, we presented a solution for the computational demands of opNMF in its application to big neuroimaging data. The methods were able to replicate the performance of opNMF while significantly decreasing the memory demand. As a consequence, it can enable data-driven analyses of large-scale neuroimaging studies, leading to unique insights on brain organization. Accordingly, we plan to apply it to large-scale multi-modal clinical populations studies to better understand how the brain is affected in disease.

Acknowledgments. This work was supported by NIH R01-AG067103 grant. Computations were performed on Washington University Center for High Performance Computing.

References

1. Bottou, L.: Large-scale machine learning with stochastic gradient descent. In: Proceedings of COMPSTAT 2010 (2010)
2. Boutsidis, C., Gallopoulos, E.: SVD based initialization: a head start for nonnegative matrix factorization. Pattern Recognit. **41**(4), 1350–1362 (2008)
3. Damoiseaux, J.S., et al.: Consistent resting-state networks across healthy subjects. Proc. Natl. Acad. Sci. U.S.A **103**(37), 13848–13853 (2006)
4. Gillenwater, J., et al.: A tree-based method for fast repeated sampling of determinantal point processes. In: Proceedings of the 36th ICML '19 (2019)
5. Glasser, M.F., et al.: The minimal preprocessing pipelines for the human connectome project. NeuroImage **80**, 105–124 (2013). Mapping the Connectome
6. Habes, M., et al.: White matter lesions: spatial heterogeneity, links to risk factors, cognition, genetics, and atrophy. Neurology **91**(10), 964–975 (2018)
7. Hoyer, P.O.: Non-negative matrix factorization with sparseness constraints. J. Mach. Learn. Res. **5**, 1457–1469 (2004)
8. Jenkinson, M., et al.: FSL. NeuroImage. 20 YEARS OF fMRI 62.2 (2012)
9. Kaczkurkin, A.N., et al.: Evidence for dissociable linkage of dimensions of psychopathology to brain structure in youths. Am. J. Psychiatry **176**(12), 1000–1009 (2019)
10. Kulesza, A., Taskar, B.: Determinantal point processes for machine learning. Found. Trends® Mach. Learn. **5**(2-3), 123–286 (2012)
11. LaMontagne, P.J., et al.: OASIS-3: longitudinal neuroimaging, clinical, and cognitive dataset for normal aging and Alzheimer disease. Technical report. medRxiv (2019)
12. LeWinn, K.Z., et al.: Sample composition alters associations between age and brain structure. Nat. Commun. **8**(1) (2017)
13. Li, C., Jegelka, S., Sra, S.: Fast DPP sampling for Nystrom with application to kernel methods. In: Proceedings of the 33rd ICML (2016)
14. Mairal, J., et al.: Online learning for matrix factorization and sparse coding. J. Mach. Learn. Res. **11**(2), 19–60 (2010)

15. Markiewicz, C.J., et al.: OpenNeuro: an open resource for sharing of neuroimaging data (2021)
16. Nazeri, A., et al.: Neurodevelopmental patterns of early postnatal white matter maturation represent distinct underlying microstructure and histology. Neuron 0.0 (2022)
17. Neufeld, N.H., et al.: Structural brain networks in remitted psychotic depression. Neuropsychopharmacol. **45**(7), 1223–1231 (2020)
18. Ochi, R., et al.: Investigating structural subdivisions of the anterior cingulate cortex in schizophrenia, with implications for treatment resistance and glutamatergic levels. J. Psychiatry Neurosci. **47**(1), 1–10 (2022)
19. Ou, Y., et al.: DRAMMS: deformable registration via attribute matching and mutual-saliency weighting. Med. Image Anal. **15**(4), 622–639 (2011)
20. Ouyang, M., et al.: Differential cortical microstructural maturation in the preterm human brain with diffusion kurtosis and tensor imaging. Proc. Natl. Acad. Sci. U.S.A. **116**(10), 4681–4688 (2019)
21. Patel, R., et al.: Investigating microstructural variation in the human hippocampus using non-negative matrix factorization. NeuroImage **207**, 116348 (2020)
22. Pizzo, F., et al.: Deep brain activities can be detected with magnetoencephalography. Nat. Commun. **10**(1), 971 (2019)
23. Robert, C., et al.: Analyses of microstructural variation in the human striatum using non-negative matrix factorization. NeuroImage **246**, 118744 (2022)
24. Sankar, A., et al.: Diagnostic potential of structural neuroimaging for depression from a multi-ethnic community sample. BJ Psych. Open **2**(4), 247–254 (2016)
25. Smith, S.M., et al.: Correspondence of the brain's functional architecture during activation and rest. Proc. Natl. Acad. Sci. U.S.A **106**(31), 13040–13045 (2009)
26. Sotiras, A., Resnick, S.M., Davatzikos, C.: Finding imaging patterns of structural covariance via non-negative matrix factorization. NeuroImage **108**, 1–16 (2015)
27. Sotiras, A., et al.: Patterns of coordinated cortical remodeling during adolescence and their associations with functional specialization and evolutionary expansion. Proc. Natl. Acad. Sci. U.S.A **114**(13), 3527–3532 (2017)
28. Sudlow, C., et al.: UK Biobank: an open access resource for identifying the causes of a wide range of complex diseases of middle and old age. PLOS Med. **12**(3) (2015)
29. Thompson, E., et al.: Non-negative data-driven mapping of structural connections with application to the neonatal brain. NeuroImage **222**, 117273 (2020)
30. Varikuti, D.P., et al.: Evaluation of non-negative matrix factorization of grey matter in age prediction. NeuroImage **173**, 394–410 (2018)
31. Yang, Z., Oja, E.: Linear and nonlinear projective nonnegative matrix factorization. IEEE Trans. Neural Netw. **21**(5), 734–749 (2010)
32. Yang, Z., Zhang, H., Oja, E.: Online projective nonnegative matrix factorization for large datasets. In: Huang, T., Zeng, Z., Li, C., Leung, C.S. (eds.) ICONIP 2012. LNCS, vol. 7665, pp. 285–290. Springer, Heidelberg (2012). https://doi.org/10.1007/978-3-642-34487-9_35
33. Yeo, B.T.T., et al.: Estimates of segregation and overlap of functional connectivity networks in the human cerebral cortex. NeuroImage **88**, 212–227 (2014)
34. Zhang, C., Kjellstrom, H., Mandt, S.: Determinantal Point Processes for Mini-Batch Diversification (2017)

Reconstruction

Deep Physics-Informed Super-Resolution of Cardiac 4D-Flow MRI

Fergus Shone[1,2,3,4(✉)], Nishant Ravikumar[1,2], Toni Lassila[1,2,3],
Michael MacRaild[1,2,3], Yongxing Wang[1,2], Zeike A. Taylor[1,2], Peter Jimack[3],
Erica Dall'Armellina[1,2,4], and Alejandro F. Frangi[1,2,3,4]

[1] Centre for Computational Imaging and Simulation Technologies in Biomedicine,
Schools of Computing, Mechanical Engineering and Medicine, University of Leeds,
Leeds, UK
mm16f2s@leeds.ac.uk
[2] National Institute for Health and Care Research (NIHR) Leeds Biomedical
Research Centre (BRC), Leeds, UK
[3] EPSRC Centre for Doctoral Training in Fluid Dynamics, School of Computing,
University of Leeds, Leeds, UK
[4] Leeds Institute of Cardiovascular and Metabolic Medicine, School of Medicine,
University of Leeds, Leeds, UK

Abstract. 4D-flow magnetic resonance imaging (MRI) provides non-invasive blood flow reconstructions in the heart. However, low spatio-temporal resolution and significant noise artefacts hamper the accuracy of derived haemodynamic quantities. We propose a physics-informed super-resolution approach to address these shortcomings and uncover hidden solution fields. We demonstrate the feasibility of the model through two synthetic studies generated using computational fluid dynamics. The Navier-Stokes equations and no-slip boundary condition on the endocardium are weakly enforced, regularising model predictions to accommodate network training without high-resolution labels. We show robustness to each type of data degradation, achieving normalised velocity RMSE values of under 16% at extreme spatial and temporal upsampling rates of 16× and 10× respectively, using a signal-to-noise ratio of 7.

Keywords: Physics-informed machine learning · 4D-flow MRI · Super-resolution

1 Introduction

4D-Flow MRI. Phase-contrast magnetic resonance imaging (PC-MRI) is the modality of choice for quantifying cardiovascular blood flow across a range of target domains. For visualising cardiac flow, 4D-flow MRI [8] is the preferred option, due to its ability to provide time-resolved, 3D reconstructions of the

---

Erica Dall'Armellina and Alejandro F Frangi: Joint senior authors.

© The Author(s), under exclusive license to Springer Nature Switzerland AG 2023
A. Frangi et al. (Eds.): IPMI 2023, LNCS 13939, pp. 511–522, 2023.
https://doi.org/10.1007/978-3-031-34048-2_39

velocity field. However, its direct application in clinic has been hampered by several shortcomings: low spatio-temporal resolution, significant noise artefacts and long scan times. Moreover, clinically relevant derived quantities like pressure, vorticity and wall shear stresses are not directly measured, and thus susceptible to effects of corruption in the velocity data. Thus, efforts have been made to increase the spatio-temporal resolution and denoise 4D-flow MRI data to improve the accuracy of predictive haemodynamic quantities [3,4,11].

Super-Resolution of 4D-Flow MRI. Unlike in structural MRI, super-resolution of 4D-flow MRI is an emerging field, with few publications to date. Machine learning approaches have dominated recent efforts, all of which have focused entirely on flow in the vasculature. Residual networks are used in [4] and [11], increasing the spatial resolution by factors of 2× and 4× respectively whilst denoising the data. Both approaches here used architectures designed for single image super-resolution, and therefore address only spatial upsampling. This not only precludes temporal super-resolution, but also ignores rich information in flow data at neighbouring time points, which could also improve spatial upsampling capabilities. Furthermore, both approaches require paired low- and high-resolution data during training, which is not readily available in 4D-flow MRI studies. In contrast, super-resolution in both space and time was achieved in [3] using a physics-informed neural network (PINN), without high-resolution target data. Upsampling factors of 100× and 5× in space and time respectively were reported. However, the network architecture and hyperparameters used in [3] are not capable of capturing complex, small-scale flow features, like those present in cardiac flow. The model used in [3] belongs to a class of supervised machine learning methods called PINNs [10] which are designed to model physical systems for which only sparse and potentially noisy data are available. Weak imposition of symmetries, physical laws and other *a priori* domain knowledge via a multi-component loss function provides regularisation during network training to heavily restrict the space of possible solutions, and render this small-data setting feasible.

Our Contributions. We present the first application of PINNs to fluid flow problems in moving domains, and the first instance of flow super-resolution in the left ventricle (LV). We demonstrate the feasibility of our model using two synthetic cases, generated using computational fluid dynamics (CFD): a 2D idealised ventricle, with a focus on exploring network parameter configurations, including a comparative study with various activation functions; and a 3D patient-specific LV geometry, for which we complete a further network hyperparameter study. In both cases, we evaluate robustness to different levels of image degradation, testing performance at different spatial/temporal downsampling rates and noise levels.

2 Methods

PINN Model. The proposed PINN model uses a fully connected neural network to approximate a function mapping spatio-temporal coordinates on a domain to corresponding velocity and pressure fields, as seen in Fig. 1. Predicted solutions are constrained through a multi-component loss function involving low-resolution measurements, boundary conditions, and known physical laws. Namely: 4D-flow MRI data, the no-slip condition on the endocardium, and residuals of the Navier-Stokes equations, respectively. Unlike purely data-driven machine learning approaches, the physics-based constraints used in PINNs restrict the space of possible solutions by penalising non-physical predictions, facilitating efficient training with sparse and noisy data whilst ensuring predicted solution fields obey the underlying physics. With this model, we aim to only use information that would be present in a real cardiac 4D-flow MRI study, where the boundary motion of the endocardium is extracted through segmentation and registration of structural cine-MR images.

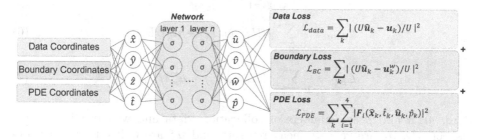

Fig. 1. Basic overview of the physics-informed neural network used in this paper, displaying network inputs and outputs, and the three loss components.

In large vessels and cardiac chambers, blood is considered a Newtonian, incompressible fluid. Therefore, the flow is governed by the incompressible Navier-Stokes equations. After nondimensionalising our input and output variables using characteristic length and velocity scales [7] and standardising our input variables, they are expressed as

$$F_{1-3} = \frac{1}{\sigma_{t^*}}\partial_{\hat{t}}\hat{u} + (\hat{u}\cdot\hat{\nabla})\hat{u} + \hat{\nabla}\hat{p} - \frac{1}{\text{Re}}\hat{\nabla}^2\hat{u} = 0 \tag{1}$$

$$F_4 = \hat{\nabla}\cdot\hat{u} = 0, \tag{2}$$

where

$$\hat{\nabla} = \left(\frac{1}{\sigma_{x_1^*}}\partial_{\hat{x}_1}, \frac{1}{\sigma_{x_2^*}}\partial_{\hat{x}_2}, \frac{1}{\sigma_{x_3^*}}\partial_{\hat{x}_3}\right)^T, \quad \hat{\nabla}^2 = \frac{1}{\sigma_{x_1^*}^2}\partial_{\hat{x}_1}^2 + \frac{1}{\sigma_{x_2^*}^2}\partial_{\hat{x}_2}^2 + \frac{1}{\sigma_{x_3^*}^2}\partial_{\hat{x}_3}^2, \tag{3}$$

for Reynolds number Re $= \rho U L/\mu$, with dimensionless velocity $\hat{u} = u/U$ and pressure $\hat{p} = p/\rho U^2$, density ρ and dynamic viscosity μ. σ_i values are standard deviations of each input variable. We assume $\rho = 1066$ kg m^{-3} and $\mu = 0.0035$ Pa s [7]. For wall velocity u^w, the no-slip boundary condition is enforced as $U\hat{u} = u^w$.

Loss Function. The loss function consists of three components: *Data loss*, $\mathcal{L}_{data}$, in which the error between velocity predictions and 4D-flow MRI measurements is minimised, *PDE (partial differential equation) loss*, $\mathcal{L}_{PDE}$, containing residuals of the Navier-Stokes equations, and *BC (boundary condition) loss*, $\mathcal{L}_{BC}$, where predictions on the domain wall are constrained to obey the no-slip condition. In all cases, the $L2$ loss is used, as per [10]. The total loss to be minimised during training is given by

$$\mathcal{L} = \mathcal{L}_{PDE} + \alpha\mathcal{L}_{data} + \beta\mathcal{L}_{BC}, \tag{4}$$

where

$$\mathcal{L}_{PDE} = \frac{1}{m}\sum_{k=1}^{m}\sum_{i=1}^{4}||F_i(\hat{\boldsymbol{x}}_k, \hat{t}_k, \hat{\boldsymbol{u}}_k, \hat{p}_k)||^2 \tag{5}$$

$$\mathcal{L}_{data} = \frac{1}{n}\sum_{k=1}^{n}||(U\hat{\boldsymbol{u}}(\hat{\boldsymbol{x}}_k, \hat{t}_k) - \boldsymbol{u}_k)/U||^2 \tag{6}$$

$$\mathcal{L}_{BC} = \frac{1}{p}\sum_{k=1}^{p}||(U\hat{\boldsymbol{u}}(\hat{\boldsymbol{x}}_k, \hat{t}_k) - \boldsymbol{u}_k^w)/U||^2. \tag{7}$$

Here, m, n and p are numbers of collocation, data and wall training points, respectively, $\boldsymbol{u}_k$ are velocity measurements, and $\boldsymbol{u}_k^w$ are wall velocity measurements.

In Eq. 4, α and β are weighting coefficients selected dynamically during training, as per [5]. Incorporating a loss weighting scheme such as this is highly important when using PINNs to balance the gradient contributions from each loss component during training. At training iteration k, we calculate

$$\hat{\alpha}^{k+1} = \frac{|\nabla_\theta\mathcal{L}_{PDE}|}{|\nabla_\theta\mathcal{L}_{data}|}, \qquad \hat{\beta}^{k+1} = \frac{|\nabla_\theta\mathcal{L}_{PDE}|}{|\nabla_\theta\mathcal{L}_{BC}|}, \tag{8}$$

where $\nabla_\theta\mathcal{L}_i$ is the gradient of loss component i w.r.t. the network weights, θ. This operation is performed layer-wise and we take the mean value. We then calculate weights for epoch $k+1$ as

$$\alpha^{k+1} = (1 - \lambda)\alpha^k + \lambda\hat{\alpha}^{k+1}, \qquad \beta^{k+1} = (1 - \lambda)\beta^k + \lambda\hat{\beta}^{k+1}, \tag{9}$$

for constant λ, chosen to be 0.1 as per [5].

Network Architecture. To determine optimal network architecture and hyperparameter choices, we performed individual ablation studies for the 2D and 3D cases. Details of these studies can be found in the Experiments section. The networks used in both cases are fully connected, where hidden layers use Siren activation functions [12] and the output layer uses a linear activation. For both cases, the optimal number of layers was 9. For the 2D case, 900 neurons per layer with a dropout rate of 0.55 performed best, whereas in the 3D case these values were 750 and 0.35, respectively. For hidden layer i, the Siren activation is formulated as

$$\phi_i(x_i) = \sin(\mathbf{W}_i x_i + \mathbf{b}_i), \tag{10}$$

for input x_i, weights $\mathbf{W}_i$ and biases $\mathbf{b}_i$, where weights in each hidden layer, w_i, are drawn from $w_i \sim \mathcal{U}(-\sqrt{6/n}, \sqrt{6/n})$ at initialisation. This activation function is designed specifically to capture high-frequency solution information and model higher-order derivatives, both of which are relevant to this study. Further details about this activation function can be found in [12].

ADAM [6] is used as the optimiser with an initial learning rate of 1×10^{-5} for the 2D case, and 1×10^{-4} for the 3D case, which is selected to be the highest possible value without divergence. Plateau-based annealing is used, where the learning rate decays by a factor of 0.1 if the validation loss plateaus, within a tolerance of 10^{-6}. The annealing initiates after 5 epochs, and has a cool down period of 3 epochs following a decay action. All code is written in Python, primarily using the package TensorFlow 2 [1].

Error Metrics. We evaluate performance using max-normalised root mean square error (NRMSE) between predicted solution fields and full-resolution ground truth fields, with velocity NRMSE given by:

$$\text{vNRMSE} := \frac{1}{\max|\boldsymbol{u}|} \sqrt{\frac{1}{N} \sum_{k=1}^{N} (\boldsymbol{u}_{pred} - \boldsymbol{u})_k^2},$$

where $\boldsymbol{u}_{pred} = U\hat{\boldsymbol{u}}$, and $\boldsymbol{u}$ is the ground truth velocity, at N data locations. We also calculate pressure NRMSE by zero-centring the mean value for prediction and ground truth data at each time step. The pressure NRMSE then assumes the same form as velocity NRMSE.

3 Experiments and Results

In this section, we describe our two experiments used to demonstrate the feasibility of our model. Both cases are synthetic, with solution ground truth data generated from CFD simulations. Use of synthetic data is of particular importance in this study, as such higher resolution ground truth data are not available for *in vivo* cases.

3.1 Case 1: 2D Idealised Ventricle

Computational Fluid Dynamics. Case 1 consists of a 2D idealised ventricle, as seen on the left in Fig. 2. The simulation was computed in ANSYS Fluent (ANSYS Inc., Canonsburg, PA). Boundary motion across diastole and systole was prescribed analytically with a sinusoidal function, driving the flow in both phases. The period of one complete cardiac cycle was 0.5 s, resulting in a peak $Re = 5600$, in line with expectations for true LV flow. In diastole, the inlet was opened and the outlet closed, and vice versa in systole. Two small protrusions were placed at the inlet to represent the open leaflets of the mitral valve. Zero-normal stress conditions were applied to the inlet and outlet when opened, with the no-slip condition applied on the walls and closed openings. We used a highly-resolved mesh of ~330k elements and error-based adaptive time-stepping to capture flow features across a range of scales, and to emulate complex flow patterns observed in a real LV. The minimum time step was 10^{-6} s. Adaptive meshing was used with diffusion-based smoothing, remeshing every 3 time steps based on minimum and maximum length scales determined by the initial mesh sizing.

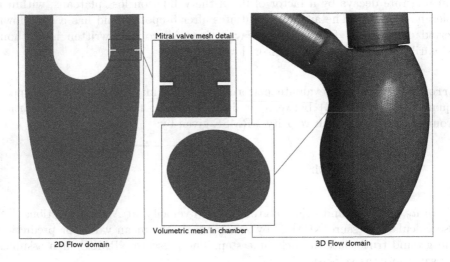

Fig. 2. Geometry and mesh details for the 2D idealised left ventricle (left) and 3D patient-specific left ventricle (right). In end-diastole, the 2D mesh consisted of ~0.33 million elements, whilst the 3D mesh consisted of ~1.3 million elements.

Training Data Generation. Temporal downsampling of the data was performed by simply removing requisite frames from the cardiac cycle. The spatial downsampling process was designed to reflect real 4D-flow MRI acquisition processes: 1) CFD solution data were interpolated on a high-resolution Cartesian grid at each time step; 2) a square region of interest (ROI) was extracted, removing inflow and outflow channels; 3) four-point balanced encoding [9] was applied to the velocity data, with the encoding velocity set to 10% higher than the maximum velocity; 4) velocity 'images' were converted to k-space images using the fast Fourier transform (FFT); and 5) spatial downsampling was achieved by cropping the outer edges of the k-space images, with the effect of truncating high-frequency modes [4]. Zero-mean Gaussian noise was added in the frequency domain to both real and imaginary signals, matching the true noise distribution of 4D-Flow MRI data [3,4]. To control noise levels, we altered the standard deviation, which was based on a signal power calculated using the encoding velocity. Finally, inverse FFT (IFFT) and balanced four-point decoding were applied to recover the downsampled velocity data in the spatial domain. The maximum downsampling rates and noise levels simulated exceed what would be expected in real 4D-flow MRI data to demonstrate the effective range of the model.

In studies with real data, the boundary motion and collocation point clouds are generated using segmentation and registration of structural MRI images acquired alongside the 4D-flow MRI data. However, in the case of the synthetic studies discussed in this paper, we simply used the CFD results at the boundary to mask the flow domain and provide the wall coordinates and associated velocities. The collocation point cloud was chosen to be the cell centres of the underlying CFD mesh, since this was already highly resolved and evenly distributed in space and time. This produced a set of $\sim$6,000,000 collocation points throughout space and time.

The complete synthetic 4D-flow MRI dataset was split 80:20 for training and validation. We constructed our test set using 10% of the ground truth CFD data evenly distributed in space and time. We confirmed that batch size has a minimal effect on accuracy, and thus selected the maximum possible batch size for each case for computational efficiency. Since the size of each training set differs significantly, we continuously looped through the batches of the smaller sets without shuffling them, while the largest set completed the epoch. Then, all training sets were shuffled before generating new batches.

Results. In Fig. 3, we see network predictions in late diastole at spatial upsampling rates of 8× and 16×, using a temporal upsampling rate of 5×, and zero noise. To ensure pressure colour bars are matched, we shifted the predicted pressure values by the difference in predicted and ground truth means. This is required because pressure predictions are only unique up to a constant, given that we have no pressure measurements and pressure appears only as a derivative in the Navier-Stokes equations. We see good qualitative agreement in both velocity and pressure fields, even at the higher spatial upsampling rate, with much of the fine-scale flow captured well. The results far outperform the cubic spline interpolation used for comparison, which achieves vNRMSE values of 0.115 and 0.181 for the two spatial upsampling rates. Pressure NRMSE remains low (0.0459) even at 16× upsampling, while the velocity NRMSE is rather higher, at 0.141. This could partly be due to small misalignment of high-velocity regions in the flow, e.g. in the left region of the ventricle.

We assess the denoising capabilities of our model in Fig. 4, where we add noise with a standard deviation of 0.15 to our 4D-flow MRI training data, and compare the performance against two competing methods: cubic spline interpolation and the PINN configuration used in Sect. 4.2 of [3]. Qualitatively, our model shows robustness to noise and is able to capture most of the solution features present in the ground truth data. Conversely, the cubic spline algorithm is heavily corrupted

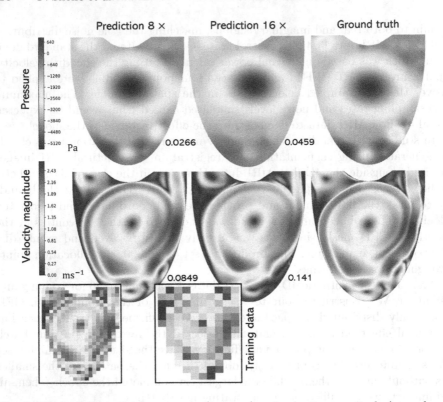

Fig. 3. 2D idealised ventricle: pressure (top row) and velocity magnitude (second row) contour plots in late diastole, using noise-free training data, temporal upsampling rate of 5×, and spatial upsampling rates of 8× (left) and 16× (centre), alongside the ground truth (right). Inserts show the resolution of training data. The values reported next to each figure are NRMSE for the given case.

by the presence of noise and is clearly unsuitable in this context, while the PINN model used in [3] is able to remove the data noise but eradicates much of the finer flow detail, which is also shown quantitatively with increased NRMSE values. In the presence of noise, the spatial upsampling capabilities of our model are more limited, where NRMSE values at the spatial upsampling rate of 4× with noise are comparable to those at 8× with noise-free data. However, in this work we have not deployed any specific component to negate the effects of noise, such as label smoothing, and the inclusion of such components should improve robustness to these effects.

3.2   Case 2: Patient-Specific Left Ventricle

CFD Setup. As before, results for this case were produced using ANSYS Fluent. The mean of the principle component analysis (PCA) shape model in [13] was used as a 3D LV geometry. Inflow and outflow channels were added at mitral

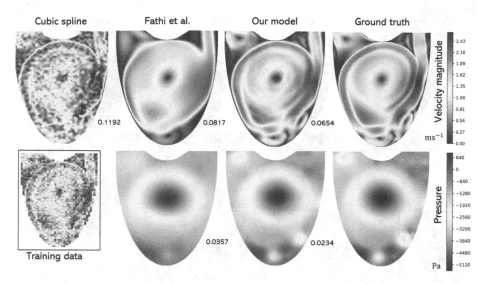

Fig. 4. 2D idealised ventricle: velocity magnitude (top row) and pressure (bottom row) contour plots in late diastole, comparing our model (middle right) with that used in [3] (middle left) and cubic spline (left). Here we have used noisy (standard deviation: 15%) training data with temporal and spatial upsampling rates of 5× and 4× respectively. Insert displays the resolution of training data. Again, reported values are NRMSE.

and aortic openings to ensure fully developed inlet and outlet flows, which were again modelled using zero-normal stress conditions. Aortic and mitral valves opened and closed at corresponding times in the cardiac cycle, whose period was set to 1 s. For simplicity, we assumed either fully open or closed valves, with no opening/closing phase in between. In diastole, we approximated the open mitral valve shape using the mitral valve plane generated in [13], with some intrusion of the valve leaflets into the flow domain. Endocardial motion was sinusoidal as for the 2D case, with the ventricular base fixed and lower regions toward the apex moving, and with the same time-stepping configuration. In end-diastole, the mesh consisted of ~1.3M elements. Figure 2 (right) depicts the geometry and mesh.

Training Data Generation. The downsampling procedure was similar to the 2D case, with the corresponding 3D algorithms used for FFT and IFFT operations. The ROI was selected to remove most of the inflow and outflow channels, and the CFD mesh cell centres were again used as coordinates for the collocation point cloud, producing ~25M points in space and time.

Results. A selection of results is shown in Fig. 5, where we compare the effect of varying noise levels at temporal and spatial upsampling rates of 5× and 8× respectively. Pressure NRMSE values are mean-normalised, as the maximum pressures in the ground truth data were unrealistically large due to some of

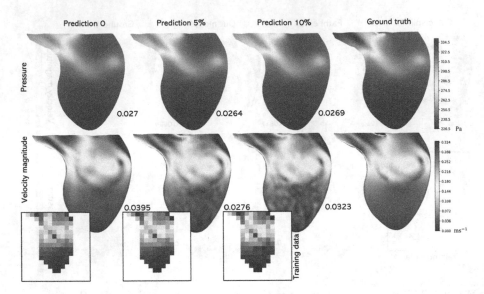

Fig. 5. 3D patient-specific ventricle: pressure (top row) and velocity magnitude (second row) contour plots across central x-z plane in early systole, using training data at a temporal upsampling rate of 5× and spatial upsampling rate of 8×, in the presence of noise with a standard deviation of 5% (centre left), 10% (centre right) and no noise (left), alongside the ground truth (right). Inserts display resolution of training data. Again, reported values are NRMSE, although the pressure NRMSE values are normalised using the mean instead of the maximum.

the outflow channel being included in the dataset. Visually, noise has a clear effect in low velocity regions toward the apex, but flow structures toward the base and valve region are well captured. This occurs because velocity scales with signal intensity in 4D-flow MRI, and thus higher velocity voxels have a higher signal-to-noise ratio. In contrast, pressure predictions appear far more robust to noise levels, reflected qualitatively and quantitatively in the contour plots and NRMSE values. This again supports the hypothesis that the PDE regularisation component is robust to uncertainty generated from noise in the velocity measurements. Some discrepancy is apparent in the low pressure regions near the aortic valve. However, this could be due to inconsistency in the ground truth data. Part of this region belongs to the outflow channel on the other side of the aortic valve plane, which experiences unrealistically extreme pressures when the valve is closed.

3.3 Hyperparameter Optimisation Studies

To perform our hyperparameter optimisation studies we used the Bayesian tree-structured parzen estimator algorithm from the Python package Hyperopt [2]. The ranges set in the search space were based on maximum and minimum values associated with PINN models seen in the literature. Complete search spaces for both studies can be found in Table 1. We performed 20 complete training runs per hyperparameter option, resulting in 140 training runs in total. For the 3D case, we performed a slightly restricted study based on the assumption that the Siren activation function would remain the optimal choice. Although optimal network depth remained the same for both 2D and 3D cases, the optimal width was found to be smaller for the 3D case. This, perhaps unexpectedly, suggests that network size scales more closely with complexity of flow rather than number of spatial dimensions, with results in the 2D case exhibiting more complicated flow patterns.

Table 1. Search spaces defined for the 2D and 3D hyperparameter optimisation studies

Case	Hidden layers	Neurons per layer	Dropout rate	Activation function
2D	5–12	100–1000	0–0.7	Tanh, Swish, Siren
3D	5–12	400–1000	0–0.7	-

4 Discussion and Conclusion

In this work, we propose a PINN-based method for super-resolution of 4D-flow MRI data in the LV and demonstrate its feasibility through two synthetic studies. The two cases presented were constructed using specific modelling and downsampling procedures to ensure that the data generated were representative of true cardiac 4D-flow MRI. We performed individual network ablation studies for both 2D and 3D cases, and then evaluated model performance at varying levels of data degradation, demonstrating robustness of the model to both noise artefacts and low spatio-temporal resolution at levels of data corruption beyond what would be expected in real cases. We found greater robustness to image degradation in the 3D case, although this is likely due to the simpler flow patterns simulated in the ground truth data, with a longer cardiac cycle and relatively coarser mesh used to produce the CFD results. In the presence of significant levels of noise, the spatial upsampling capabilities of the model in the 2D case are evidently restricted, however, the inclusion of specific denoising components could alleviate these issues. In both cases, pressure predictions were shown to be less affected by the presence of noise in the velocity data.

The capability of the model to super-resolve in both space and time, whilst also uncovering the hidden pressure fields without the use of any additional data, presents a clear advantage over previous 4D-flow MRI super-resolution approaches that are based on single image super-resolution models. Further, during training the model does not rely on the availability of paired low- and high-resolution data, which are not readily available in most 4D-flow MRI studies. The main disadvantage when compared with these methods is the inability of our model, and PINNs in general, to generalise across cases, meaning the network must be re-trained each time it is presented with a new patient or volunteer. However, accelerating training cycles using transfer learning or similar techniques could be a useful avenue of investigation here to compensate for the inherent lack of generalisation capabilities.

Acknowledgements. This work was partially supported by the EPSRC Centre for Doctoral Training in Fluid Dynamics (EP/L01615X/1) and the Royal Academy of Engineering Chair in Emerging Technologies (CiET1919/19). The computational work was undertaken on the UK National Tier-2 high performance computing service JADE-2 (EP/T022205/1).

References

1. Abadi, M., Agarwal, A., Barham, P., Brevdo, E., Chen, Z., Citro, C., G.S, Corrado.: TensorFlow: large-scale machine learning on heterogeneous distributed systems. In: 12th USENIX Symposium on OSDI (2016)
2. Bergstra, J., Yamins, D., Cox, D.D.: Making a science of model search: hyperparameter optimization in hundreds of dimensions for vision architectures. Comput. Sci. Discov. 8 (2015)
3. Fathi, M.F., et al.: Super-resolution and denoising of 4D-flow MRI using physics-informed deep neural nets. Compu.t Methods Programs Biomed. **197**, 105729 (2020)
4. Ferdian, E., et al.: 4DFlowNet: super-resolution 4D flow MRI using deep learning and computational fluid dynamics. Front. Phys. 8(183) (2020)
5. Jin, X., Cai, S., Li, H., Karniadakis, G.E.: NSFnets (Navier-Stokes flow nets): physics-informed neural networks for the incompressible Navier-Stokes equations. J. Comput. Phys. **426**, 109951 (2021)
6. Kingma, D.P., Ba, J.: Adam: A Method for Stochastic Optimization. 3rd ICLR (2015)
7. Kissas, G., Yang, Y., Hwuang, E., Witschey, W.R., Detre, J.A., Perdikaris, P.: Machine learning in cardiovascular flows modeling: predicting arterial blood pressure from non-invasive 4D flow MRI data using physics-informed neural networks. Comput. Methods Appl. Mech. Eng. **358**, 112623 (2020)
8. Markl, M., Frydrychowicz, A., Kozerke, S., Hope, M., Wieben, O.: 4D flow MRI. J. Magn. Reson. Imaging **36**(5), 1015–1036 (2012)
9. Pelc, N.J., Bernstein, M.A., Shimakawa, A., Glover, G.H.: Encoding strategies for three-direction phase-contrast MR imaging of flow. J. Magn. Reson. Imaging **1**, 405–413 (1991)
10. Raissi, M., Perdikaris, P., Karniadakis, G.E.: Physics-informed neural networks: a deep learning framework for solving forward and inverse problems involving nonlinear partial differential equations. J. Comput. Phys. **378**, 686–707 (2019)
11. Shit, S., Zimmermann, J.: SRflow: Deep learning based super-resolution of 4D-flow MRI data. Front. Artif. Intell. (2022)
12. Sitzmann, V., Martel, J.N.P., Bergman, A.W., Lindell, D.B., Wetzstein, G.: Implicit neural representations with periodic activation functions. In: 34th Conference on NeurIPS (2020)
13. Xia, Y., et al.: Automatic 3D+t four-chamber CMR quantification of the UK biobank: integrating imaging and non-imaging data priors at scale. Med. Image Anal. **80**, 102498 (2022)

Fast-MC-PET: A Novel Deep Learning-Aided Motion Correction and Reconstruction Framework for Accelerated PET

Bo Zhou[1]([✉]), Yu-Jung Tsai[2], Jiazhen Zhang[1], Xueqi Guo[1], Huidong Xie[1], Xiongchao Chen[1], Tianshun Miao[2], Yihuan Lu[3], James S. Duncan[1,2], and Chi Liu[1,2]

[1] Department of Biomedical Engineering, Yale University, New Haven, US
bo.zhou@yale.edu
[2] Department of Radiology and Biomedical Imaging, Yale University, New Haven, US
[3] United Imaging Healthcare, Shanghai, China

Abstract. Patient motion during PET is inevitable. Its long acquisition time not only increases the motion and the associated artifacts but also the patient's discomfort, thus PET acceleration is desirable. However, accelerating PET acquisition will result in reconstructed images with low SNR, and the image quality will still be degraded by motion-induced artifacts. Most of the previous PET motion correction methods are motion type specific that require motion modeling, thus may fail when multiple types of motion present together. Also, those methods are customized for standard long acquisition and could not be directly applied to accelerated PET. To this end, modeling-free universal motion correction reconstruction for accelerated PET is still highly under-explored. In this work, we propose a novel deep learning-aided motion correction and reconstruction framework for accelerated PET, called Fast-MC-PET. Our framework consists of a universal motion correction (UMC) and a short-to-long acquisition reconstruction (SL-Reon) module. The UMC enables modeling-free motion correction by estimating quasi-continuous motion from ultra-short frame reconstructions and using this information for motion-compensated reconstruction. Then, the SL-Recon converts the accelerated UMC image with low counts to a high-quality image with high counts for our final reconstruction output. Our experimental results on human studies show that our Fast-MC-PET can enable 7-fold acceleration and use only 2 min acquisition to generate high-quality reconstruction images that outperform/match previous motion correction reconstruction methods using standard 15 min long acquisition data.

Keywords: Accelerated PET · Universal Motion Correction · Deep Reconstruction

© The Author(s), under exclusive license to Springer Nature Switzerland AG 2023
A. Frangi et al. (Eds.): IPMI 2023, LNCS 13939, pp. 523–535, 2023.
https://doi.org/10.1007/978-3-031-34048-2_40

1   Introduction

Positron Emission Tomography (PET) is a commonly used functional imaging modality with wide applications in oncology, cardiology, neurology, and biomedical research. However, patient motion during the PET scan, including both involuntary motions (i.e. respiratory, cardiac, and bowel motions) and voluntary motions (i.e. body and head motions), can lead to significant motion artifacts, degrading the downstream clinical tasks. Moreover, the long acquisition time that easily exceeds 15 min, will lead to increased patient motion, patient discomfort, and low patient throughput.

In previous works of PET motion correction (MC), a variety of external device-aided and data-driven MC methods have been developed for correcting specific motion types. For example, in respiratory MC, Chan et al. [4] developed a non-rigid event-by-event continuous MC list-mode reconstruction method. Lu et al. [10] further improved their method by generating matched attenuation-corrected gate PET for respiratory motion estimation. In body MC, Andersson et al. [1] proposed to divide the PET list-mode data into predefined temporal frames for reconstructions, where the reconstructions of each frame are registered to a reference frame for body MC. Later, Lu et al. [11] further developed a reconstruction-free center-of-distribution-based body motion detection and correction method. In cardiac MC, cardiac cycle tracking/gating using electrocardiography (ECG) is still the gold-standard [12]. While providing efficient MC solutions to reduce motion artifacts for different motion types, these methods usually require prior knowledge of the motion type and need motion-type-specific modeling. Thus, these previous MC methods may lead to sub-optimal image quality or fail when multiple motion types are present simultaneously. There are also recent attempts in using ultra-fast list-mode reconstruction of short PET frames to estimate motion during the PET scan [18,21]. However, these methods may not adapt well to many motion types with non-rigid motion [18], and extending to non-rigid motion is computationally infeasible, i.e. requiring non-rigid registration of thousands of frames for a single scan using traditional registration algorithms [21]. In addition, it still requires the standard long acquisition to collect sufficient events to achieve a reasonable signal-to-noise ratio (SNR) in the final reconstruction. On the other hand, previous works have also investigated the feasibility of reducing the PET acquisition time. Lindemann et al. [9] and Lasnon et al. [8] found that one can reasonably maintain the PET image quality and lesion detectability with two-fold acquisition time reduction using traditional reconstructions. Weyts et al. [19] show that a deep learning-based denoising model can enable two-fold PET acquisition time reduction and provide image quality that matches with the full acquisition. However, these works only show the feasibility of a 2-fold time reduction and did not consider the residual motions during the accelerated acquisition.

In this work, we aim to address these challenges by developing a PET reconstruction framework that can 1) reduce the acquisition time, i.e. 7-fold acceleration, and 2) correct the residual motion, regardless of the motion type, in the accelerated acquisition. Specifically, we propose a novel deep learning-aided data-driven motion reduction and accelerated PET reconstruction framework, called Fast-MC-PET. In the Fast-MC-PET, we first design a universal motion correction method aided by deep learning to reconstruct a motion-reduced image from the short acquisition. While reducing the motion artifacts given the accelerated acquisition and our motion correction, the reconstructed image still suffers from high noise levels due to low event counts. Thus, in the second step of Fast-MC-PET, we also deploy a deep generative network to convert the low-counts images to high-counts images. Our experimental results on real human data demonstrate the Fast-MC-PET can generate high-quality images with reduced motion-induced errors while enabling 7-fold accelerated PET acquisition.

2   Methods

Our Fast-MC-PET consists of two key components, including a universal motion correction (UMC) module and a short-to-long acquisition reconstruction (SL-Recon) module. In UMC, we first partition the list-mode data into ultra-short list-mode data, i.e. every 500 ms, and estimate a quasi-continuous motion over the short acquisition. Given the motion and the original list-mode data, a motion-corrected short-acquisition image is then reconstructed by a motion-compensated OSEM list-mode reconstruction. Finally, a deep generative model is devised to transform the motion-corrected short-acquisition image into a high-count long-acquisition image, thus providing a motion-corrected high-count image using only accelerated short-acquisition. In the following sections, we will describe these steps in detail (Fig. 1).

2.1   Universal Motion Correction

With the short acquisition data, the UMC aims to generate a motion reduced low-count reconstruction. The UMC consists of three steps, including point cloud image (PCI) & paired gated image generation, quasi-continuous motion estimation, and motion-compensated OSEM list-mode reconstruction.

Point Cloud and Paired Gated Image Generation. To estimate a continuous motion, the list-mode data is first partitioned into a series of ultra-short list-mode data, i.e. every 500 ms. For every 500 ms list-mode data, we back-project the Line-of-Response (LOR) of each event within the time-of-flight (TOF) bin, and all the back-projected LORs form a PCI for this short time frame. The PCI reconstruction can be formulated as

$$P_{j,t} = \sum_i \frac{c_{i,j,t} L_{i,t}}{Q_j}, \tag{1}$$

where $c_{i,j,t}$ is the system matrix that represents the contribution of an annihilation originating from pixel j being detected on LOR i at time t, accounting for geometry, resolution, and solid angle effects. $L_{i,t}$ is the decay correction factor. Q_j is the sensitivity of voxel j that is pre-computed via $Q_j = \sum_i c_{i,j}$, and $P_{j,t}$ is the back-projected value of voxel j at time t with sensitivity correction.

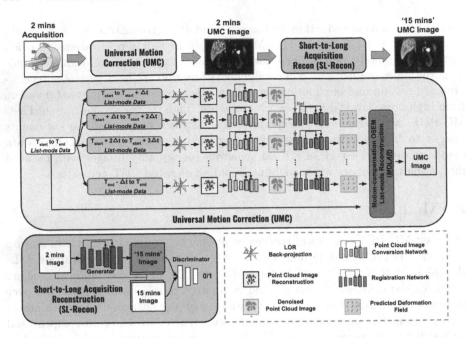

Fig. 1. The overall pipeline of Fast-MC-PET. The Universal Motion Correction (UMC) module (grey box) reconstructs motion-reduced image from the short acquisition data. The Short-to-Long Acquisition Reconstruction (SL-Recon) module (pink box) converts the UMC image from short acquisition to long acquisition. (Color figure online)

Due to the ultra-low-counts level, the signal-to-noise ratio (SNR) of PCI is low and is unsuitable for motion estimation tasks, as demonstrated in Fig. 2's 1st row. Thus, we deploy a deep learning-based denoising network, i.e. UNet [16], that aims to convert PCIs to gated OSEM images with high SNR. To train the denoising network, we first reconstruct the amplitude-based respiratory gated OSEM images [15] using the body motion free list-mode data, extracted by the Centroid-Of-Distribution (COD)-based body motion detection method [11]. Then, within each gate, we randomly extract 10% PCIs to construct the training pairs of PCI and the corresponding gated image. $\mathcal{L}_2$ loss is used for the network training, and can be formulated as

$$\mathcal{L}_{dn} = ||\gamma_g - f_{dn}(P_g)||_2^2 \qquad (2)$$

where γ_g is the gated OSEM image and P_g is the randomly extracted PCI that lies in the same gate. With a trained denoising model $f_{dn}(\cdot)$, the series of PCIs can then be converted to a series of high-quality denoised PCI (dPCI) via:

$$\gamma_t = f_{dn}(P_t) \qquad \qquad . \qquad (3)$$

where γ_t is the denoised images with $t = (0 \sim \Delta t, \Delta t \sim 2\Delta t, ..., T - \Delta t \sim T)$. Here, we set $\Delta t = 0.5s$ and $T = 120s$ here, thus generating 240 3D images. Examples of dPCIs are illustrated in Fig. 2's 2nd row.

Quasi-Continuous Motion Estimation. A quasi-continuous motion can be estimated using the series of dPCIs from the previous step. Within the first 5 s, the dPCI in the expiration phase, i.e. with the highest COD coordinates in the z-direction, is chosen as the reference frame γ_{ref} for all the other frames γ_t, resulted in 239 dPCI pairs requiring registration. Conventional registration methods [14, 20] are time-consuming, and it is prohibitively long to register hundreds of 3D pairs here. Thus, we propose to use a deep learning-based registration method for fast motion estimation [3] in our framework. Given the reference dPCI image γ_{ref} and the source dPCI image γ_t, we use a motion estimation network, i.e. UNet [16], to predict the motion deformation $M_t = f_m(\gamma_{ref}, \gamma_t)$. The network is trained by optimizing the following loss function:

$$\mathcal{L}_m = ||\gamma_{ref} - M_t \circ \gamma_t||_2^2 + \beta||\nabla M_t||_2^2 \qquad (4)$$

where the first term measures the image similarity after applying the motion prediction M_t, and the second term is a deformation regularization that adopts a L2-norm of the gradient of the deformation. The regularization's weight is set as $\beta = 0.001$. During training, γ_{ref} and γ_t are randomly selected from the gated images. With a trained motion estimation network $f_m(\cdot)$, we can then estimate the quasi-continuous motion using $M_t = f_m(\gamma_{ref}, \gamma_t)$ with $t = (0 \sim \Delta t, \Delta t \sim 2\Delta t, ..., T - \Delta t \sim T)$.

Motion-compensated OSEM List-mode Reconstruction. To reconstruct a single image λ at the reference location γ_{ref} using all the coincidence events, we can deform the system matrix at each time t to the reference location, generating new deformed system matrixs $c_{i,j}^{t \rightarrow ref}$ using M_t from the previous step. Deforming the system matrix can be seen as "bending" the LORs into curves of response (CORs), where both forward and back-projections are traced along the CORs. In list-mode notation, for event k occurring on LOR $i(k)$ at time $t(k)$, we replace indexes i by k, and substitute $c_{k,j}$ in the previous TOF-MOLAR [6] by $c_{k,j,\tau_k}^{t \rightarrow ref}$. The OSEM updating equation can thus be formulated as:

$$\lambda_j^{n+1} = \frac{\lambda_j^n}{Q_j} \sum_{k=1}^{K} \frac{c_{k,j,\tau_k}^{t \rightarrow ref} L_k A_k N_k}{T(\sum_{j'} c_{k,j',\tau_k}^{t \rightarrow ref} L_k A_k N_k \lambda_{j'}^n + R_{k,\tau_k} + S_{k,\tau_k})} \qquad (5)$$

$$Q_j = \frac{1}{n_T} \sum_{t'=1}^{n_T} \sum_{i=1}^{I} \sum_{\tau=1}^{n_\tau} c_{i,j,t',\tau}^{t \rightarrow ref} L_{i,t'} A_{i,t'} N_i \qquad (6)$$

where n is the number of iteration, k is the index of each detected event, $c_{k,j,\tau_k}^{t \rightarrow ref}$ is the deformed system matrix element with τ_k denoting the TOF bin for event k. L_k is the decay factor and A_k is the attenuation factor derived from CT. N_k is the sensitivity term, R_{k,τ_k} is the randoms rate estimate, and S_{k,τ_k} is the scatter rate estimate in counts per second in TOF bin τ_k. The random events are estimated from the product of the singles rates of the two detectors for each LOR, and then uniformly distributed across all TOF bins. Here, Q is the sensitivity image that is pre-computed by back-projecting randomly sampled events along the CORs to account for the motion on voxel sensitivity. When calculating Q, each time frame of duration T is divided into n_T short time bins, i.e. t'. Moreover, n_τ denotes the total number of TOF bins ($n_\tau = 13$ for the Siemens mCT PET scanner used in this study). Here, we set the number of iteration to 2 and the number of subsets to 21 for our UMC reconstructions.

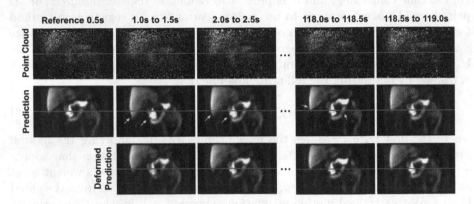

Fig. 2. Examples of the Point Cloud Images (PCIs), the denoised PCIs (dPCIs), and the deformed dPCIs using estimated motion fields.

2.2   Short-to-Long Acquisition Reconstruction

Even though the UMC reduces the motion effects in the reconstruction, the UMC image still suffers from low SNR due to the limited counts from the short acquisition, as compared to the long acquisition. Thus, we propose to use a short-to-long acquisition reconstruction (SL-Recon) to convert the UMC image from a short-acquisition to a long-acquisition one. Here, we use a conditional generative adversarial network for this reconstruction. Given a UMC image λ_s from the short acquisition, we can use a generative network, i.e. UNet [16], that directly predicts the UMC image λ_l from a long acquisition from it. The SL-Recon network is trained using both a pixel-wise L2 loss and an adversarial loss defined as:

$$\mathcal{L}_2 = ||G(\lambda_s) - \lambda_l||_2^2 \tag{7}$$

$$\mathcal{L}_{adv} = -log(D_{gan}(\lambda_l|\lambda_s)) - log(1 - D_{gan}(G(\lambda_s)|\lambda_s)) \tag{8}$$

where G is the SL-Recon generative network and D is the discriminator network. Here, we simply use OSEM reconstructions from long acquisitions (15 min), paired with OSEM reconstructions from short acquisitions (2 min in the center period), for the network's training.

2.3   Evaluation on Human Data

We included 26 pancreatic ^{18}F-FPDTBZ [13] PET/CT patient studies. All PET data were obtained in list mode using the 4-ring Siemens Biograph mCT scanners equipped with the AZ-733V respiratory gating system (Anzai Medical, Tokyo, Japan). The Anzai respiratory trace was recorded 40 Hz for all subjects. The average dose administered to the patients is 9.13±1.37 mCi. 15 min of the list-mode acquisition were used for each patient study. We used 23 patients to generate the training data for the PCI denoising model, the motion estimation model, and the SL-Recon model. Extensive evaluations were performed on the remaining 3 patients with different motion types. For training the PCI denoising model and the motion estimation model, we generated 5 gated images for each patient using OSEM (21 subsets and 2 iterations). For training the SL-Recon model, the training pairs of long/short acquisition images were reconstructed using the same OSEM protocol without gating. All the images were reconstructed into $200 \times 200 \times 109$ 3D volumes with a voxel size of $2.032 \times 2.032 \times 2.027 \ mm^3$.

2.4 Implementation Details

We implemented our deep learning modules using Pytorch. We used the ADAM optimizer [7] with a learning rate of 10^{-4} for training the PCI denoising network, motion estimation network, and the SL-Recon network. We set the batch size to 3 for all networks' training. All of our models were trained on an NVIDIA Quadro RTX 8000 GPU. The PCI denoising network was trained for 200 epochs, and then fine-tuned for 10 epochs on the patient-specific gated images of the test patient during the test time. The motion estimation network was trained for 250 epochs, and the SL-Recon network was trained for 200 epochs. To prevent overfitting, we also implemented 'on-the-fly' data augmentation for the PCI denoising and SL-Recon networks. During training, we performed $64 \times 64 \times 64$ random cropping, and then randomly flip the cropped volumes along the x, y, and z-axis.

3 Results

The qualitative comparison of Fast-MC-PET reconstructions is shown in Fig. 3. As we can observe, the 2 min reconstruction with no motion correction (NMC) suffers from both motion blurring and high-noise levels due to low counts. The first patient has both body/torso motion and respiratory motion during the 2 min PET scan, thus introducing heavy blurring for major organ boundaries, i.e. liver and kidneys. The 2 min UMC image recovers the sharp organ boundaries by correcting those motions during the short acquisition. Based on the UMC image from 2 min acquisition, the final Fast-MC-PET image further reduces the noise thus providing a near motion-free and high-count image, matching the 15 min UMC image quality. The second patient with respiratory and bowel motion introduces significant image blurring for the pancreas (view 1) and intestines (view 2). The 2 min UMC image can recover the diminished details inside these organ regions. The final Fast-MC-PET image further reduces the noise, thus generating a high-quality image with motion correction and high counts. On the other hand, by reducing the acquisition time from 15 min to 2 min, we can see that the diminished organ structures, especially the intestine structure (view 2) in 15 min

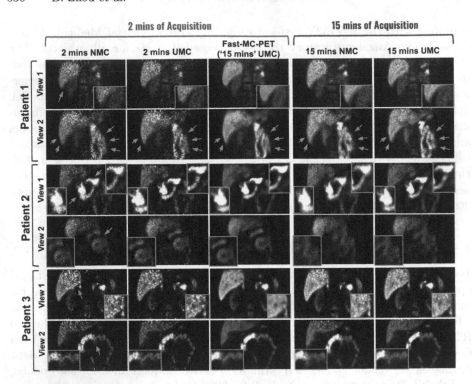

Fig. 3. Visualization of Fast-MC-PET reconstructions. The 2min UMC images (2nd column) contain less motion blurring, as compared to the no motion correction (NMC) images (1st column). The virtual 15 min UMC images (3rd column) predicted from 2 min UMC images (2nd column) provide image-quality that match the true 15 min images (last column).

NMC, can be preliminarily restored in 2 min NMC. Complex motion, e.g. bowel motion, in a 15 min long acquisition is extremely challenging to correct even with UMC. Thus, based on 2 min acquisition, the Fast-MC-PET here shows better reconstruction quality with better structural recovery. Similar observations can be found for the third patient with respiratory and bowel motion, where the 2 min-based Fast-MC-PET provides reconstruction quality matched the 15 min UMC reconstruction.

We compared our 2 min-based Fast-MC-PET reconstructions to previous correction methods that are long acquisition based, i.e. 15 min. The visual comparison is shown in Fig. 4. First, we compared with the classic respiratory motion correction method [2] that reduces the motion and noise by averaging the aligned amplitude-gated images, where non-rigid registration [14] is used for alignments. Then, we compared our method with the NR-INTEX [4] that compensates for the respiratory motion by estimating the continuous deformation field using internal-external motion correlation which is considered the current state-of-the-art method. Both previous methods require specific motion-type modeling, and

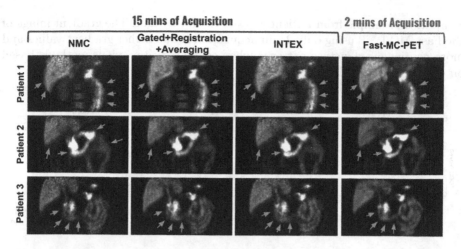

Fig. 4. Comparisons to previous motion correction methods. Our Fast-MC-PET with 2 min acquisition show improved structural details recovery (orange arrows), as compared to previous methods with 15 min acquisition.

thus fail when additional motion types are present, e.g. body motion (Patient 1) and bowel motion (Patient 3). The UMC module in the Fast-MC-PET is not specific to any motion type and thus can correct different types of motion together. Therefore, our Fast-MC-PET can provide consistently better results when multiple types of motion co-exist (Patients 1 and 3), and generate comparable reconstruction quality when respiratory motion is dominating (Patient 2).

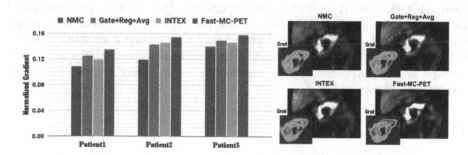

Fig. 5. Comparison of the gradient of reconstructions. Left: quantitative evaluation using the mean gradient value. Right: visual comparison of the reconstruction and the gradient.

For quantitative evaluation, we computed the mean normalized gradient of the reconstructions, where better reconstruction with sharper structure will have higher gradient values. The results are summarized in Fig. 5. The normalized gradient values of Fast-MC-PET are 0.159, 0.154, and 0.132 for Patients 1, 2, and 3, respectively, which are consistently higher than all previous methods. A comparison example from Patient 2 is shown on the right. The gradient image of the Fast-MC-PET using only 2 min acquisition shows higher gradient values and more continuous structure patterns when compared to previous methods based on 15 min acquisition.

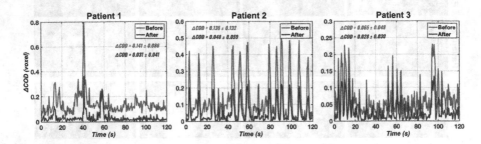

Fig. 6. The difference of COD trace between the reference frame and the current frame (ΔCOD) over the 2 min acquisition. The ΔCOD before (red) and after (blue) UMC correction are plotted for all three patients. The mean ΔCODs are reported in the plots. (Color figure online)

Ablative evaluation of motion correction is shown in Fig. 6. The difference of COD between the reference frame and the current frame (ΔCOD) over the 2 min acquisition is visualized. For Patient 1 with body motion and irregular breathing pattern, the ΔCOD curve before correction contains irregular steep changes leading to a mean ΔCOD of 0.141 ± 0.086. With the UMC in our Fast-MC-PET, the curve after correction is much more stable with a reduced mean ΔCOD of 0.031 ± 0.041 with significance ($p < 0.001$). For Patients 2 and 3 with more stable and regular motion patterns, the UMC can also reduce the mean ΔCOD from 0.135 ± 0.132 to 0.048 ± 0.059 and from 0.065 ± 0.048 to 0.028 ± 0.030, respectively. Both with significance ($p < 0.001$). A patient example of PCIs over the 2 min acquisition before and after applying the UMC correction is shown in Fig. 2.

4   Discussion

In this work, we propose a novel deep learning-aided data-driven motion correction and reconstruction framework for accelerated PET (Fast-MC-PET). The proposed method can accelerate the PET acquisition by nearly 7-fold and use only 2 min acquisition while providing high-quality reconstruction with motion correction. In this framework, we first devise a UMC module that estimates continuous motion based on PCIs and use this information to reconstruct motion-compensated images. Instead of using 15 min long acquisition that 1) inherits more motion due to long scanning time and 2) requires registrations of 1800 PCI pairs in UMC, we use 2 min accelerated acquisition with less motion and only requires registrations of 240 PCI pairs. The averaged registration inference time for one pair is 0.41s, thus needing about 98.5s for all registration in UMC which is more manageable. The UMC reconstruction from accelerated acquisition can then be inputted into the SL-Recon module to directly generate the 15 min long acquisition motion-corrected reconstruction. With this simple yet efficient pipeline, we can generate high-quality motion corrected accelerated PET reconstruction that potentially outperforms previous methods with the standard long acquisition.

There are a few limitations and opportunities that are the subject of our ongoing work. First, our pilot study only tested on ^{18}F-FPDTBZ patients who were all scanned using Siemens mCT. The trained model may not directly generalize well to a different PET tracer/scanner. However, if the training data of different tracers/scanners is available, the Fast-MC-PET can be fine-tuned and potentially adapted to these distributions. Multi-institutional federated learning [22] may also be used to improve the adaptation. In the future, we will further evaluate the performance using patients scanned with different PET tracers/scanners. Second, we used a temporal resolution of 500 ms for PCI in UMC with a focus on abdominal region motion correction in this work. A higher temporal resolution, e.g. 100 ms, may be needed for cardiac motion correction in the chest region, which is an important direction in our future investigation. Third, the UMC correction performance is still not perfect, as shown in Fig. 6 blue curves, where the ΔCOD values are non-zero. The current implementation uses a simple 3-level UNet for motion prediction. Deploying a more advanced registration network, e.g. transformer-based network [5] and temporal registration networks [23], may potentially further reduce the registration error and improve the final reconstruction quality. Lastly, the PCI denoising step requires supervised training from paired gated images, which is time-consuming to prepare. In the future, we will also investigate self-supervised denoising methods, e.g. Noise2Void [17], for PCI denoising in our Fast-MC-PET.

5   Conclusion

This paper presents a deep learning-aided motion correction and reconstruction framework for accelerated PET, called Fast-MC-PET. The Fast-MC-PET consisting of UMC and SL-Recon, uses only 2 min accelerated PET acquisition data for high-quality reconstruction. The UMC reconstructs motion-corrected short acquisition image, regardless of the motion type in the abdominal region. The SL-Recon then converts the 2 min UMC image into virtual 15 min UMC image. The experimental results demonstrate that our proposed method can accelerate acquisition by nearly 7-fold and generate high-quality motion-corrected reconstruction for patients with different motions.

References

1. Andersson, J.L.: How to obtain high-accuracy image registration: application to movement correction of dynamic positron emission tomography data. Eur. J. Nucl. Med. **25**(6), 575–586 (1998)

2. Bai, W., Brady, M.: Regularized b-spline deformable registration for respiratory motion correction in pet images. Phys. Med. Biol. **54**(9), 2719 (2009)
3. Balakrishnan, G., Zhao, A., Sabuncu, M.R., Guttag, J., Dalca, A.V.: VoxelMorph: a learning framework for deformable medical image registration. IEEE Trans. Med. Imaging **38**(8), 1788–1800 (2019)
4. Chan, C., et al.: Non-rigid event-by-event continuous respiratory motion compensated list-mode reconstruction for pet. IEEE Trans. Med. Imaging **37**(2), 504–515 (2017)
5. Chen, J., Frey, E.C., He, Y., Segars, W.P., Li, Y., Du, Y.: TransMorph: transformer for unsupervised medical image registration. Med. Image Anal. **82**, 102615 (2022)
6. Jin, X., et al.: List-mode reconstruction for the biograph MCT with physics modeling and event-by-event motion correction. Phys. Med. Biol. **58**(16), 5567 (2013)
7. Kingma, D.P., Ba, J.: Adam: a method for stochastic optimization. arXiv preprint arXiv:1412.6980 (2014)
8. Lasnon, C., et al.: How fast can we scan patients with modern (digital) PET/CT systems? Eur. J. Radiol. **129**, 109144 (2020)
9. Lindemann, M.E., Stebner, V., Tschischka, A., Kirchner, J., Umutlu, L., Quick, H.H.: Towards fast whole-body PET/MR: investigation of pet image quality versus reduced pet acquisition times. PLoS ONE **13**(10), e0206573 (2018)
10. Lu, Y., et al.: Respiratory motion compensation for PET/CT with motion information derived from matched attenuation-corrected gated pet data. J. Nucl. Med. **59**(9), 1480–1486 (2018)
11. Lu, Y., et al.: Data-driven voluntary body motion detection and non-rigid event-by-event correction for static and dynamic pet. Phys. Med. Biol. **64**(6), 065002 (2019)
12. Lu, Y., Liu, C.: Patient motion correction for dynamic cardiac pet: current status and challenges. J. Nucl. Cardiol. **27**(6), 1999–2002 (2020)
13. Normandin, M.D., et al.: In vivo imaging of endogenous pancreatic β-cell mass in healthy and type 1 diabetic subjects using 18f-fluoropropyl-dihydrotetrabenazine and pet. J. Nucl. Med. **53**(6), 908–916 (2012)
14. Papademetris, X., et al.: Bioimage suite: an integrated medical image analysis suite: an update. Insight J. **2006**, 209 (2006)
15. Ren, S., et al.: Data-driven event-by-event respiratory motion correction using TOF PET list-mode centroid of distribution. Phys. Med. Biol. **62**(12), 4741 (2017)
16. Ronneberger, Olaf, Fischer, Philipp, Brox, Thomas: U-Net: convolutional networks for biomedical image segmentation. In: Navab, Nassir, Hornegger, Joachim, Wells, William M.., Frangi, Alejandro F.. (eds.) MICCAI 2015. LNCS, vol. 9351, pp. 234–241. Springer, Cham (2015). https://doi.org/10.1007/978-3-319-24574-4_28
17. Song, T.A., Yang, F., Dutta, J.: Noise2void: unsupervised denoising of pet images. Phys. Med. Biol. **66**(21), 214002 (2021)
18. Spangler-Bickell, M.G., Deller, T.W., Bettinardi, V., Jansen, F.: Ultra-fast list-mode reconstruction of short pet frames and example applications. J. Nucl. Med. **62**(2), 287–292 (2021)
19. Weyts, K., et al.: Artificial intelligence-based pet denoising could allow a two-fold reduction in [18f] FDG PET acquisition time in digital PET/CT. Eur. J. Nucl. Med. Mol. Imaging **49**, 1–11 (2022). https://doi.org/10.1007/s00259-022-05800-1
20. Xu, Z., et al.: Evaluation of six registration methods for the human abdomen on clinically acquired CT. IEEE Trans. Biomed. Eng. **63**(8), 1563–1572 (2016)
21. Zhang, J., Fontaine, K., Carson, R., Onofrey, J., Lu, Y.: Deep learning-aided data-driven quasi-continous non-rigid motion correction in PET. In: 2021 28th IEEE Nuclear Science Symposium and Medical Imaging Conference (2021)

22. Zhou, B., et al.: Federated transfer learning for low-dose pet denoising: a pilot study with simulated heterogeneous data. IEEE Trans. Rad. Plasma Med. Sci. (2022)

23. Zhou, B., Tsai, Y.J., Chen, X., Duncan, J.S., Liu, C.: MDPET: a unified motion correction and denoising adversarial network for low-dose gated PET. IEEE Trans. Med. Imaging **40**(11), 3154–3164 (2021)

MeshDeform: Surface Reconstruction of Subcortical Structures via Human Brain MRI

Junjie Zhao[1], Siyuan Liu[2], Sahar Ahmad[3,4], and Pew-Thian Yap[3,4]($\boxtimes$)

[1] Department of Computer Science, University of North Carolina,
Chapel Hill, USA
[2] College of Marine Engineering, Dalian Maritime University, Dalian, China
[3] Department of Radiology, University of North Carolina, Chapel Hill, USA
ptyap@med.unc.edu
[4] Biomedical Research Imaging Center, University of North Carolina,
Chapel Hill, USA

Abstract. Surface reconstruction of cortical and subcortical structures is crucial for brain morphological studies. Existing deep learning surface reconstruction methods, such as DeepCSR and Vox2Surf, learn an implicit field function for computing the isosurface, but do not consider mesh topology. In this paper, we propose a novel and efficient deep learning mesh deformation network, called MeshDeform, to reconstruct topologically correct surfaces of subcortical structures using brain MR images. MeshDeform combines features extracted from a U-Net encoder with mesh deformation blocks to predict surfaces of subcortical structures by deforming spherical mesh templates. MeshDeform is able to reconstruct in less than 10 s the surfaces of a left-right pair of subcortical structures with subvoxel accuracy. Reconstruction of all 17 subcortical structures takes less than one and a half minutes. By contrast, Vox2Surf takes about 20–30 min for all subcortical structures. Visual and quantitative evaluation on the Human Connectome Project (HCP) dataset demonstrate that MeshDeform generates accurate subcortical surfaces in limited time while preserving mesh topology.

Keywords: Surface reconstruction · Subcortical structures · MRI

1 Introduction

Morphometric analysis of gray matter (GM) structures in the brain subcortex is essential for studying cognitive dysfunction and neurodegeneration [15,18]. To quantify morphological alterations, surfaces of the subcortical structures need to be accurately reconstructed. Traditional cortical surface reconstruction methods like FreeSurfer [6] often require a series of image processing steps, including brain tissue segmentation, surface fitting, mesh tessellation, and topology correction. These reconstruction approaches usually take an extensive amount of computational time, limiting its application in time-sensitive tasks.

Deep learning methods can be employed to accurately reconstruct 3D structures in a much shorter time. There are two categories of deep learning surface

© The Author(s), under exclusive license to Springer Nature Switzerland AG 2023
A. Frangi et al. (Eds.): IPMI 2023, LNCS 13939, pp. 536–547, 2023.
https://doi.org/10.1007/978-3-031-34048-2_41

reconstruction approaches: (i) implicit surface representation learning; and (ii) explicit surface representation learning. Implicit learning methods [3] such as DeepCSR [5] and Vox2Surf [9] learn signed distance functions (SDFs) as implicit surface representations and then compute 3D shapes using the marching cubes algorithm [13]. Potential topological defects can be fixed with a topology correction algorithm [1]. Topology correction typically takes a significant amount of time and, in practice, the corrected meshes can still contain some topological errors.

Explicit representation learning methods deform pre-defined mesh templates to target meshes. Pixel2Mesh [16] and Voxel2Mesh [17] combine features extracted from volumetric images with a mesh deformation process to predict target shapes. These methods implicitly preserve the original topology of the mesh template; therefore, no further topology correction is needed.

In this paper, we propose an end-to-end deep learning pipeline that directly predicts subcortical structures from T1-weighted (T1w) and T2-weighted (T2w) brain MR images. Considering the fact that the vertices on the surface mesh can be nonuniformly distributed, we use a graph convolutional neural network [2,7] to predict vertex-wise deformation vectors.

Each mesh deformation block receives mesh features and image feature maps extracted from a 3D U-Net [14] encoder, predicts mesh deformation, and passes the updated result to the next block. Since subcortical structures are homeomorphic to a sphere, we use a spherical mesh template that has almost the same number of vertices as the ground truth mesh. A segmentation module is added for additional supervision of the encoder for further performance improvement. Experimental results demonstrate that our approach can reconstruct the left-right surface pair of each subcortical structure from T1w and T2w images in less than 10 s.

The key advantages of our method are summarized as follows:

1. An end-to-end graph convolutional neural network that directly predicts brain subcortical surfaces from T1w and T2w images.
2. An explicit surface representation method that deforms a given mesh template to the target space, while preserving the spherical topology.
3. Efficient prediction of subcortical structures, taking less than 10 s to generate a bilateral surface pair of each subcortical structure.

2 Methods

Our surface reconstruction network (Fig. 1) takes T1w and T2w MR images as input and processes them through a 3D U-Net encoder to extract features at different levels. A binary image segmentation decoder is utilized to guide the encoder in extracting features, taking into account tissue boundary information. The deformation blocks concatenate image features pooled in multiple layers with mesh coordinate features obtained through graph convolution. Each mesh deformation block predicts the displacement of each vertex to eventually generate the final surface.

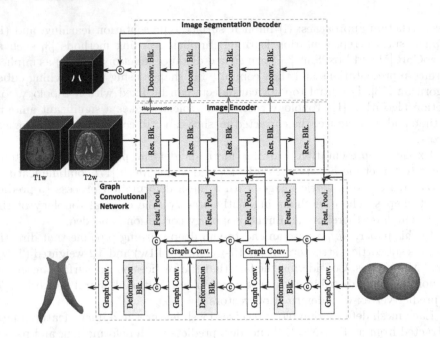

Fig. 1. Network architecture of MeshDeform.

2.1 Graph Deformation Informed by Multi-scale Spatial Features

Given an initial spherical mesh for each subcortical structure (e.g., lateral ventricle in Fig. 1), MeshDeform utilizes multi-scale spatial features extracted from T1w and T2w MR images to guide a graph convolutional network (GCN) to deform in a cascaded fashion an initial spherical mesh to construct the mesh of a subcortical structure. Specifically, we use an image encoder to extract multi-scale spatial features, and an auxiliary segmentator is incorporated to ensure that the extracted features are sensitive to tissue boundaries. We employ a multi-step deformation network to gradually deform each vertex by merging spatial features at multiple scales to reconstruct surfaces of subcortical structures.

2.2 Network Architecture

Our network was trained for 17 subcortical structures, including the left and right nuclei accumbentes, amygdalae, caudate nuclei, hippocampi, globi pallidi, putamina, thalami, lateral ventricles, and the third ventricle. It contains (i) an image encoder for feature extraction, (ii) a segmentation decoder to facilitate the training process, and (iii) mesh deformation blocks for vertex displacements.

Image Encoder. We used a 3D U-Net encoder [10] to extract feature maps at several resolutions to represent increasingly abstract information. The T1w and T2w input images were cropped to $96 \times 144 \times 128$ voxel dimensions each for the left and right hemispheres. The encoder contains five residual blocks, followed by a $3 \times 3 \times 3$ convolution kernel with stride 2 to downsample the feature maps.

Image Segmentation Decoder. The image segmentation decoder provides additional information to improve surface reconstruction. Since the goal is not to actually predict segmentations, the decoder produces only a binary segmentation of the subcortical surfaces for additional supervision for the image encoder. The segmentation decoder is derived from [10]. It contains the same number of decoder blocks as the image encoder residual blocks. Each decoder block output is upsampled and passed to a $3 \times 3 \times 3$ convolutional layer, then concatenated with feature maps extracted from the corresponding encoder block to predict a segmentation map. The final prediction is formed by combining the outputs of the three decoder blocks.

Graph Convolution Network. The MeshDeform network (Fig. 1) utilizes graph convolutions [2] to predict vertex-wise displacements from an initial spherical mesh template. The network contains three mesh deformation blocks. Each deformation block takes image feature maps (extracted from the image encoder) and mesh features (processed by a graph convolutional layer) as input. Since high-level image feature maps encode a more abstract representation, we project these features to the initial mesh deformation block to learn the rough shape of a subcortical structure. The lower level image features are then projected onto the subsequent deformation blocks, helping the model learn fine details of the target mesh.

A mesh deformation block contains three graph residual blocks, which consists of two graph convolutional layers. After being processed by a deformation block, the mesh features are passed to an additional graph convolutional layer to predict the displacement for each vertex. We then add the predicted displacements to the coordinates of the corresponding vertices in the mesh template to obtain an intermediate mesh. Image features for the next layer are projected onto this deformed mesh.

2.3   Mesh Initialization

The proposed network deforms an initial spherical mesh template to reconstruct the surface of a subcortical structure. Since the subcortical structure and the sphere are homeomorphic, deforming a sphere inherently preserves the correct topology. We also found that matching the rough number of vertices between ground truth mesh and spherical mesh template is essential for predicting an accurate surface. Thus, we used two spherical mesh templates, one fine-grained for more complicated structures and one coarse-grained spherical mesh for simpler structures. The mesh templates are generated by subdividing an icosphere, followed by decimation to control the number of vertices.

To initialize the location and scale of the mesh template, we compute the average centroid coordinates and the average surface areas for each subcortical structure based on the training dataset. This prevents irregular deformations and improves prediction accuracy.

2.4   Loss Function

The network loss function is a combination of four loss terms: (i) point loss and (ii) normal loss to encourage accurate geometry of the subcortical structures, and (iii) edge length loss and (iv) Laplacian loss for regularization of the predicted meshes. The binary segmentation loss is added in the end for extra supervision in training.

Point Loss. We use the Chamfer distance to evaluate the vertex distance between a predicted mesh and the ground truth:

$$L_c = \frac{1}{N} \left(\sum_{p \in P} \min_{q \in Q} \|p - q\|_2^2 + \sum_{q \in Q} \min_{p \in P} \|p - q\|_2^2 \right), \tag{1}$$

where p represents a vertex in the predicted mesh P and q represents a vertex in the ground truth mesh Q.

Normal Loss. Normal loss is used to ensure smooth surface meshes. This loss measures the difference between a predicted normal vector and the ground truth normal vector:

$$L_n = \sum_{p} \sum_{q = \arg\min_{q \in Q} \|p - q\|_2^2} \|(p_1 - p) \times (p_2 - p) - \mathbf{n}_q\|_2^2, \tag{2}$$

where p_1 and p_2 are neighboring vertices that share the same face with p. $(p_1 - p) \times (p_2 - p)$ is the cross product between these two vectors and $\mathbf{n}_q$ is the surface normal vector of q in the ground truth mesh.

Edge Length Loss. To penalize spikes (long edges) in the predicted mesh, we add an edge length regularization loss. It regularizes the differences between all edge lengths in p's neighborhood and an average edge length μ:

$$L_e = \sum_{p \in P} \sum_{k \in \mathcal{N}(p)} \left| \|p - k\|_2^2 - \mu^2 \right|, \tag{3}$$

where $\mathcal{N}(p)$ are the neighbors of p.

Laplacian Loss. The Laplacian term encourages surface smoothness. It regularizes the difference between a given vertex p and the average location of all its neighbors:

$$L_{\text{lap}} = \sum_{p \in P} \left\| p - \sum_{k \in \mathcal{N}(p)} \frac{1}{\mathcal{N}(p)} k \right\|_2^2. \tag{4}$$

Total Loss. Since multiple loss terms with different scales are hard to tune, we use weighted geometric mean of all losses to obtain the total loss for meshes [4]:

$$L_{\text{mesh}} = \sum_N L_c^{\lambda_1} L_n^{\lambda_2} L_e^{\lambda_3} L_{\text{lap}}^{\lambda_4}, \tag{5}$$

where N is the number of subcortical structures to predict. This loss combination is invariant to the scale of individual losses [4]. It enables us to tune four hyperparameters based on their significance. We generated several sets of hyperparameters and chose the one that yielded the best results. The hyperparameters adopted were $\lambda_1 = 0.55$, $\lambda_2 = 0.25$, $\lambda_3 = 0.12$ and $\lambda_4 = 0.12$.

For segmentation loss, we use the loss function defined in [10], containing cross-entropy and dice-score losses. The final loss is calculated as the sum of the mesh losses from the three deformation blocks and the binary segmentation loss:

$$L_{\text{total}} = L_{\text{mesh}}^{(1)} + L_{\text{mesh}}^{(2)} + L_{\text{mesh}}^{(3)} + L_{\text{seg}}. \tag{6}$$

3 Experimental Results

3.1 Dataset Preprocessing

Experiments were conducted using data collected via the Human Connectome Project (HCP) [8,11], consisting of minimally preprocessed T1w and T2w images and anatomical segmentation maps of 22 subjects. We used 15 subjects for training, 3 for validation, and 4 for testing. The T1w and T2w images were aligned by rigid registration using FMRIB's Linear Image Registration Tool (FLIRT) [11,12] and were cropped to $96 \times 144 \times 128$ voxel dimensions to fit in the GPU memory and to speed up training. Ground truth meshes were generated from anatomical segmentation maps via the marching cubes algorithm, followed by mesh smoothing.

3.2 Implementation Details

The proposed method was implemented using Tensorflow 2.8. Evaluation was based on a machine with an Intel E5-2650 v4 CPU and an NVIDIA TITAN Xp. In the training process, we used the Adam optimizer to minimize the total loss function, with an initial learning rate of 0.001. The learning rate decreased when no loss improvement was detected for 10 epochs. We used the point loss on the validation set as the metric to monitor training progress.

3.3 Evaluation Metrics

We evaluated MeshDeform using the average symmetric surface distance (ASSD):

$$\text{ASSD} = \sum_{p \in P} \min_{q \in Q} \frac{\|p - q\|}{|P|} + \sum_{q \in Q} \min_{p \in P} \frac{\|p - q\|}{|Q|},$$

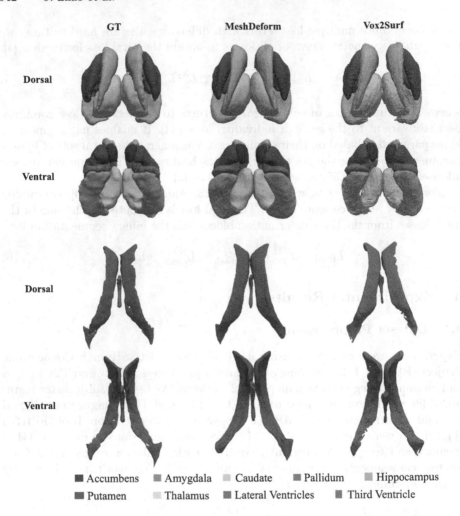

Fig. 2. Comparison of subcortical surfaces reconstructed via MeshDeform and Vox2Surf with the ground truth (GT).

where P is the predicted mesh and Q is the ground truth (GT) mesh. The ASSD computes the average distance of all points from the predicted mesh to the GT mesh. We also evaluated the predicted surface meshes with the modified Hausdorff distance (MHD):

$$\text{MHD} = \max\Big(\underset{p \in P}{\text{median}} \min_{q \in Q} \|p - q\|, \underset{q \in Q}{\text{median}} \min_{p \in P} \|p - q\| \}\Big),$$

which measures the spatial distance between the vertices of the predicted mesh and those of the GT mesh.

Lastly, we projected the predicted mesh to an image volume to generate a segmentation map for evaluation using the dice similarity coefficient (DSC):

$$\mathrm{DSC} = \frac{2|S_P \cap S_Q|}{|S_P| + |S_Q|},$$

where S_P and S_Q are the voxel sets in the predicted and ground truth segmentation maps, respectively. The DSC ranges from 0 to 1, with 1 indicating the greatest agreement in segmentation and 0 otherwise.

To evaluate the spherical topology of the reconstructed surfaces, we computed the Euler characteristic: $V - E + F$, where V, E and F are the number of vertices, edges and faces, respectively. For a spherical polyhedron, the Euler characteristic should always be 2.

Computational efficiency was measured via GPU runtime.

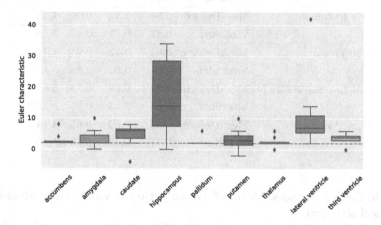

Fig. 3. Boxplots of Euler characteristic values given by the surfaces reconstructed using Vox2Surf. The Euler characteristic value for MeshDeform is 2 for all the structures (blue dashed line). (Color figure online)

3.4   Performance Comparison

Figure 2 shows the comparison results of the reconstructed surfaces of subcortical structures using Vox2Surf and our method MeshDeform. We can see that MeshDeform preserves topology since the meshes are obtained from deforming spherical mesh templates. On the other hand, Vox2Surf does not explicitly ensure spherical topology, leading to topological defects that are visible on the edges of the lateral ventricles. MeshDeform also generates smoother surfaces that preserve detailed shapes of the subcortical structures.

Quantitative comparison results based on ASSD, MHD, and DSC, summarized in Table 1, indicate that the surfaces reconstructed with MeshDeform are

Table 1. Quantitative evaluation of the compared surface reconstruction methods.

		ASSD (mm)	MHD (mm)	DSC	Runtime (sec)
L/R Accumbens	MeshDeform	0.64	0.57	0.75	4.29
	Vox2Surf	0.91	0.84	0.68	N/A
L/R Amygdala	MeshDeform	0.74	0.66	0.83	4.36
	Vox2Surf	1.15	1.15	0.72	N/A
L/R Caudate	MeshDeform	0.59	0.50	0.87	4.48
	Vox2Surf	0.71	0.61	0.85	N/A
L/R Hippocampus	MeshDeform	0.75	0.61	0.84	4.39
	Vox2Surf	0.98	0.80	0.79	N/A
L/R Pallidum	MeshDeform	0.85	0.77	0.79	4.42
	Vox2Surf	1.15	1.08	0.70	N/A
L/R Putamen	MeshDeform	0.64	0.52	0.90	4.36
	Vox2Surf	0.87	0.75	0.85	N/A
L/R Thalamus	MeshDeform	0.64	0.52	0.93	4.31
	Vox2Surf	0.93	0.80	0.88	N/A
L/R Lateral Ventricle	MeshDeform	0.73	0.59	0.76	5.87
	Vox2Surf	0.66	0.53	0.85	N/A
Third Ventricle	MeshDeform	0.68	0.60	0.65	4.13
	Vox2Surf	0.71	0.63	0.66	N/A

similar to the GT surfaces. MeshDeform outperforms Vox2Surf for almost every subcortical structure.

Figure 3 shows the Euler characteristics of Vox2Surf reconstructions. The values are not consistent across the subcortical structures, indicating there are holes or discontinuities in the reconstructed surfaces. By contrast, MeshDeform gives a constant Euler characteristic value of 2 (blue dashed line in Fig. 3), indicating perfect spherical topology.

MeshDeform is computationally efficient (Table 1) and takes less than 10 s to generate the bilateral surfaces of a single structure in both cerebral hemispheres. In comparison, Vox2Surf takes over 20 min for point sampling and another 20 min for predicting 22 subcortical and cortical surfaces.

3.5 Ablation Study

We conducted an ablation study to verify the effectiveness of the image decoder in providing additional supervision in training. The results shown in Table 2 indicate consistent degradation in surface reconstruction accuracy when the image decoder is removed.

Table 2. ASSD and MHD evaluation results for ablation study.

		ASSD (mm)	MHD (mm)
L/R Accumbens	MeshDeform	0.64	0.57
	No decoder	0.70	0.64
L/R Amygdala	MeshDeform	0.74	0.66
	No decoder	0.77	0.66
L/R Caudate	MeshDeform	0.59	0.50
	No decoder	0.58	0.50
L/R Hippocampus	MeshDeform	0.75	0.61
	No decoder	0.69	0.55
L/R Pallidum	MeshDeform	0.85	0.77
	No decoder	0.88	0.82
L/R Putamen	MeshDeform	0.64	0.52
	No decoder	0.69	0.57
L/R Thalamus	MeshDeform	0.64	0.52
	No decoder	0.65	0.55
L/R Lateral Ventricle	MeshDeform	0.73	0.59
	No decoder	0.76	0.57
Third Ventricle	MeshDeform	0.68	0.60
	No decoder	0.46	0.39

4   Conclusion

In this paper, we proposed MeshDeform—a deep neural network to efficiently and accurately reconstruct surfaces of the subcortical structures from brain MRI. The method utilizes image features and mesh spatial features to predict mesh deformation via graph convolution. This end-to-end pipeline is efficient and rapidly reconstructs subcortical surfaces directly from T1- and T2-weighted MR images. Evaluation results show that the reconstructed surfaces are close to the ground truth, while at the same time preserve spherical topology.

Further work entails overcoming some limitations of MeshDeform. One such improvement involves creating a more lightweight version of MeshDeform to allow it to process images with higher resolution, circumventing GPU memory constraint. Another improvement entails making MeshDeform robust to images of varying quality, including those acquired in clinical settings.

Acknowledgments. This work was supported in part by the United States National Institutes of Health (NIH) through grants MH125479 and EB008374.

References

1. Bazin, P.L., Pham, D.L.: Topology correction of segmented medical images using a fast marching algorithm. Comput. Methods Programs Biomed. **88**(2), 182–190 (2007)
2. Bronstein, M.M., Bruna, J., LeCun, Y., Szlam, A., Vandergheynst, P.: Geometric deep learning: going beyond Euclidean data. IEEE Signal Process. Mag. **34**(4), 18–42 (2017)
3. Chen, Z., Zhang, H.: Learning implicit fields for generative shape modeling. In: Proceedings of the IEEE/CVF Conference on Computer Vision and Pattern Recognition, pp. 5939–5948 (2019)
4. Chennupati, S., Sistu, G., Yogamani, S., A Rawashdeh, S.: Multinet++: multi-stream feature aggregation and geometric loss strategy for multi-task learning. In: Proceedings of the IEEE/CVF Conference on Computer Vision and Pattern Recognition Workshops (2019)
5. Cruz, R.S., Lebrat, L., Bourgeat, P., Fookes, C., Fripp, J., Salvado, O.: DeepCSR: a 3D deep learning approach for cortical surface reconstruction. In: Proceedings of the IEEE/CVF Winter Conference on Applications of Computer Vision, pp. 806–815 (2021)
6. Dale, A.M., Fischl, B., Sereno, M.I.: Cortical surface-based analysis: I. Segmentation and surface reconstruction. NeuroImage **9**(2), 179–194 (1999)
7. Defferrard, M., Bresson, X., Vandergheynst, P.: Convolutional neural networks on graphs with fast localized spectral filtering. In: Advances in Neural Information Processing Systems, vol. 29 (2016)
8. Glasser, M.F., et al.: The minimal preprocessing pipelines for the human connectome project. Neuroimage **80**, 105–124 (2013)
9. Hong, Y., Ahmad, S., Wu, Y., Liu, S., Yap, P.-T.: Vox2Surf: implicit surface reconstruction from volumetric data. In: Lian, C., Cao, X., Rekik, I., Xu, X., Yan, P. (eds.) MLMI 2021. LNCS, vol. 12966, pp. 644–653. Springer, Cham (2021). https://doi.org/10.1007/978-3-030-87589-3_66
10. Isensee, F., Kickingereder, P., Wick, W., Bendszus, M., Maier-Hein, K.H.: Brain tumor segmentation and radiomics survival prediction: contribution to the BRATS 2017 challenge. In: Crimi, A., Bakas, S., Kuijf, H., Menze, B., Reyes, M. (eds.) BrainLes 2017. LNCS, vol. 10670, pp. 287–297. Springer, Cham (2018). https://doi.org/10.1007/978-3-319-75238-9_25
11. Jenkinson, M., Beckmann, C.F., Behrens, T.E., Woolrich, M.W., Smith, S.M.: FSL. NeuroImage **62**(2), 782–790 (2012)
12. Jenkinson, M., Smith, S.: A global optimisation method for robust affine registration of brain images. Med. Image Anal. **5**(2), 143–156 (2001)
13. Lorensen, W.E., Cline, H.E.: Marching cubes: a high resolution 3D surface construction algorithm. ACM SIGGRAPH Comput. Graph. **21**(4), 163–169 (1987)
14. Ronneberger, O., Fischer, P., Brox, T.: U-Net: convolutional networks for biomedical image segmentation. In: Navab, N., Hornegger, J., Wells, W.M., Frangi, A.F. (eds.) MICCAI 2015. LNCS, vol. 9351, pp. 234–241. Springer, Cham (2015). https://doi.org/10.1007/978-3-319-24574-4_28
15. Tang, X., et al.: Regional subcortical shape analysis in premanifest Huntington's disease. Hum. Brain Mapp. **40**(5), 1419–1433 (2019)
16. Wang, N., et al.: Pixel2Mesh: 3D mesh model generation via image guided deformation. IEEE Trans. Pattern Anal. Mach. Intell. **43**(10), 3600–3613 (2020)

17. Wickramasinghe, U., Remelli, E., Knott, G., Fua, P.: Voxel2Mesh: 3D mesh model generation from volumetric data. In: Martel, A.L., et al. (eds.) MICCAI 2020. LNCS, vol. 12264, pp. 299–308. Springer, Cham (2020). https://doi.org/10.1007/978-3-030-59719-1_30

18. Zheng, F., et al.: Age-related changes in cortical and subcortical structures of healthy adult brains: a surface-based morphometry study. J. Magn. Reson. Imaging **49**(1), 152–163 (2019)

Neural Implicit k-Space for Binning-Free Non-Cartesian Cardiac MR Imaging

Wenqi Huang[1]([✉]), Hongwei Bran Li[1,2], Jiazhen Pan[1], Gastao Cruz[3], Daniel Rueckert[1,4], and Kerstin Hammernik[1,4]

[1] Technical University of Munich, Munich, Germany
wenqi.huang@tum.de
[2] University of Zurich, Zurich, Switzerland
[3] University of Michigan, Michigan, USA
[4] Department of Computing, Imperial College London, London, UK

Abstract. In this work, we propose a novel image reconstruction framework that directly learns a neural implicit representation in k-space for ECG-triggered non-Cartesian Cardiac Magnetic Resonance Imaging (CMR). While existing methods bin acquired data from neighboring time points to reconstruct one phase of the cardiac motion, our framework allows for a *continuous*, *binning-free*, and *subject-specific* k-space representation. We assign a unique coordinate that consists of time, coil index, and frequency domain location to each sampled k-space point. We then learn the subject-specific mapping from these unique coordinates to k-space intensities using a multi-layer perceptron with frequency domain regularization. During inference, we obtain a complete k-space for Cartesian coordinates and an arbitrary temporal resolution. A simple inverse Fourier transform recovers the image, eliminating the need for density compensation and costly non-uniform Fourier transforms for non-Cartesian data. This novel imaging framework was tested on 42 radially sampled datasets from 6 subjects. The proposed method outperforms other techniques qualitatively and quantitatively using data from four and one heartbeat(s) and 30 cardiac phases. Our results for one heartbeat reconstruction of 50 cardiac phases show improved artifact removal and spatio-temporal resolution, leveraging the potential for real-time CMR. (Code available: https://github.com/wenqihuang/NIK_MRI).

Keywords: Image Reconstruction · Non-Cartesian MRI · Cardiac MRI · Neural Implicit Functions · Deep Learning · k-Space Interpolation

1 Introduction

Cardiac Magnetic Resonance Imaging (CMR) plays an important role in the clinical assessment of cardiac morphology and function. However, due to the requirement of breath-holds during the data acquisition and the rapid motion of the heart, it is challenging to obtain images with both high temporal and spatial

© The Author(s), under exclusive license to Springer Nature Switzerland AG 2023
A. Frangi et al. (Eds.): IPMI 2023, LNCS 13939, pp. 548–560, 2023.
https://doi.org/10.1007/978-3-031-34048-2_42

resolution. Considering the limited ability of cardiac patients to perform breath-holds, fast CMR data acquisition with high undersampling rates has attracted great interest in the field of reconstruction.

To reconstruct images from undersampled k-space data, Parallel Imaging (PI) and Compressed Sensing (CS) have been introduced to Magnetic Resonance Imaging (MRI) reconstruction with great success from both the hardware and algorithmic side. PI introduced multiple receiver coils to MRI scanners. The coherence among coils enables a higher undersampling rate for data acquisition. In CS, sparse priors are widely used in cardiac MRI as regularization terms due to their effective ability to reduce the solution space and avoid local minima of the reconstruction problem. Many of the PI and CS reconstruction algorithms are nowadays available directly on the MRI scanners [5,11,13,14,21].

Besides advances in the reconstruction algorithms for MRI, the sampling patterns of MRI have evolved greatly. In clinical practice, Cartesian sampling patterns are commonly used, which can be implemented efficiently on the MRI scanner and facilitate image reconstruction. Here, each data point is equally spaced on a Cartesian grid and can be used directly to perform an inverse fast Fourier transform (FFT) calculation. Although Cartesian sampling is widely used in CMR, it is not robust to motion artifacts. Non-Cartesian coordinate sampling trajectories, such as radial sampling and spiral sampling, were introduced to alleviate this problem. Unlike Cartesian sampling, non-Cartesian sampling can have changeable phase encoding directions during acquisition. Therefore, the noise from moving anatomical structures does not propagate as discrete ghosting along a fixed phase encoding direction but is more distributed over the entire image [26], resulting in more incoherent artifacts. This advantage makes non-Cartesian sampling increasingly popular in CMR. However, the data points are not sampled on a regular grid and therefore require an inverse non-uniform fast Fourier transform (NUFFT) to transform between the k-space and image domains. In NUFFT, the sampled points are re-gridded onto a Cartesian grid using an interpolation kernel. To compensate for the inhomogeneous distribution of sampled points, a density compensation function (DCF) is additionally required, which is time-consuming and challenging to estimate [12,26].

Recent developments in deep learning have enabled MRI reconstruction of highly undersampled data. Existing works have shown that exploiting the acquisition physics and the raw k-space data can improve the reconstruction performance substantially [6,7,15–17,24]. These approaches usually require large training databases with fully-sampled k-space data for training. Akçakaya et al. proposed a database-free approach to learn k-space interpolation from a fully-sampled k-space center [1]. However, this approach can only be realized for Cartesian sampling patterns. Yaman et al. proposed a self-supervised learning approach for Cartesian MRI [27]. Deep image prior was investigated for radially sampled data in [28]. However, this method requires data binning.

As an emerging technique in computer vision, neural implicit functions (NIF) can directly model physical properties from spatial coordinates [3,4,10]. Such modeling is proven to be effective in shape representation [4] and scene rendering [10]. NIF have attracted increasing attention in medical image analysis tasks, such as radiation therapy [22], super-resolution [23], shape reconstruction [2], image registration [25], image segmentation [8] and view reconstruction [29,30]. In the context of MRI, Shen *et al.* [18] proposed a sparsely-sampled image reconstruction framework with NIF, which learns a mapping from coordinate to image intensity in the image domain. This approach requires artifact-free images from fully-sampled k-space data for training. In practice, fully-sampled k-space data are hardly available. Additionally, images from undersampled k-space data contain unavoidable global artifacts, which are challenging to remove when NIF operates in the image domain. To overcome these challenges, we directly learn a mapping *from coordinate to k-space signal value* with NIF.

This paper proposes a framework based on neural implicit functions, named neural implicit k-space (NIK), to solve the challenging non-Cartesian CMR reconstruction problem with only a few heartbeats. The proposed NIK framework makes the following three main contributions:

- We present a novel k-space point estimation scheme using neural implicit functions for MRI reconstruction, which takes a completely different technical route from the traditional image domain and k-space domain approaches;
- We overcome typical challenges in CMR reconstruction: We achieve non-Cartesian reconstruction without NUFFT or conventional gridding, and we can reconstruct an arbitrary number of cardiac phases without loss of temporal information and without data binning.
- Our method achieves a single-shot reconstruction, only requiring the imaged object and no additional training data. We compare our method to several single-shot reconstruction methods and it outperforms the baselines with superior qualitative and quantitative results.

2   Methods

Our proposed method learns a continuous implicit representation of k-space from undersampled non-Cartesian spokes. The representation is optimized with tailored techniques, including frequency domain regularization and high dynamic range loss. After training, the k-space intensities on the Cartesian grid can be generated and converted to the final image by an inverse Fourier transform.

2.1   Non-Cartesian MR Data Acquisition

The signal of an MR image is acquired in a frequency domain, the so-called k-space. Limitations of MR physics, organ motion, and patient compliance allow us to acquire only a limited number of k-space lines (spokes) in a single heartbeat. However, this is often insufficient to reconstruct a high-quality image x with high temporal resolution. To achieve adequate image quality, multiple heartbeat acquisitions during breath-holds and data binning are usually used. While holding the breath virtually eliminates the effect of respiratory motion, capturing the cardiac motion with high temporal resolution remains challenging. For acquisitions across multiple heartbeats, the combined data acquired from different heartbeats need to be sufficiently similar. Otherwise, motion correction is required to account for, e.g., arrhythmia. Figure 1(a) demonstrates the acquisition and mapping across multiple heartbeats. Time-stamped k-space lines and the ECG signal are acquired, and the ECG signal is used to map all k-space lines to one heartbeat for further reconstruction. Most existing methods divide these k-space lines y into a fixed number of cardiac phases according to their relative timestamps and obtain the binned k-space data $\tilde{y}$. The neighbouring lines in the time dimension will be clustered together and considered to be the same time point. The reconstruction problem can be formulated as the following standard MR reconstruction problem:

$$\hat{x} = \arg \min_{x} \|Ax - \tilde{y}\|_2^2 + \lambda R(x), \tag{1}$$

where A is a non-Cartesian encoding matrix. When only one receiver coil is considered (single coil imaging), A applies NUFFT only to the image x. For multi-coil imaging, the different receiver coils are sensitive in different spatial regions as encoded by the coil sensitivity maps. $R(x)$ imposes regularization on x to reduce the solution space and avoid local minima and λ is the regularization weight. However, instead of preserving the original sampling timestamps of the k-space lines, the data binning scheme merges the adjacent k-space lines, assuming they occurred at the same moment, and then reconstructs the cardiac phases at that time point (see Fig. 1(*)). This results in a loss of temporal resolution and lower spatial-temporal coherence exploitation. Notably, data binning will not be required in our proposed method.

2.2   Neural Implicit k-Space Representation

For 2D dynamic imaging, the coordinate of each k-space point has four elements: the spatial frequencies k_x and k_y, the time t, and a coil index c. In our setting, the alignment of the data across different heartbeats is still required. However, instead of binning the adjacent k-space lines, we keep their exact timestamps for each point on the k-space lines, which maintains temporal resolution. In this way, each sampled point has its k-space coordinate $v_i = [t_i, k_{x_i}, k_{y_i}, c_i]^T$ and the corresponding k-space signal value y_i, $i = 1, 2, \ldots, N$, and N is the total number of sampled points. Our goal is to model the implicit representation of the continuous k-space by learning a mapping $f : V \to K$, where $V \subseteq \mathbb{R}^4$ is the coordinate space, and $K \subseteq \mathbb{C}$ is the corresponding complex signal value space. The sampled k-space signal y_i is on a non-Cartesian lattice on V. We train a neural network $G_\theta : V \to K$ with parameters θ on the data pairs $\{(v_i, y_i)|i = 1, 2, \ldots, N\}$ to approximate the underlying mapping f. The optimization of parameters θ is formulated as follows:

$$\theta^* = \arg \min_{\theta} \|G_\theta(v) - y\|_2^2. \tag{2}$$

Since the coordinate space is a subset of $\mathbb{R}^4$, the coordinates from any kind of sampling trajectory can be used for network training. Importantly, we can query

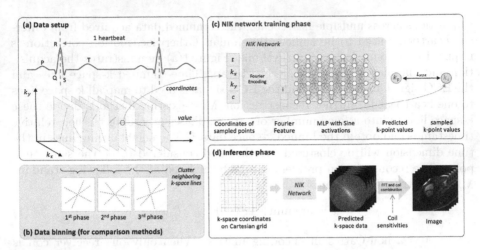

Fig. 1. Schematic illustration of neural implicit k-space (NIK). (a) The k-space lines (spokes) are sorted and mapped to one heartbeat. Instead of the traditional data binning (*), we train the MLP to learn the implicit representation of the k-space with the k-space coordinate-intensity pairs (b). t, k_x, k_y, and c refer to time point, k-space coordinates, and coil channel, respectively. (c) In the inference phase, we feed a set of coordinates from the Cartesian grid and obtain the corresponding k-space signal value. The final image can be easily reconstructed by applying the inverse fast Fourier transform and coil combination.

any coordinates during the inference phase. Hence, it is possible to avoid NUFFT by simply feeding the coordinates $\bar{v}$ from a Cartesian grid to the network and obtaining the value of each point on that grid. The reconstructed image $\hat{x}$ can then be obtained by applying the inverse Fourier transform $\mathcal{F}^{-1}$ once:

$$\hat{x} = \mathcal{F}^{-1}(G_{\theta*}(\bar{v})). \tag{3}$$

2.3   Frequency Domain Regularization

To avoid local minima and overfitting to the noisy measurements, we introduce the point-wise frequency-domain regularization to Eq. (2):

$$\arg\min_{\theta} \|G_\theta(v) - y\|_2^2 + \lambda \|G_\theta(v) - D(G_\theta(v))\|_2^2, \tag{4}$$

where D denotes a frequency-domain denoiser. As the output of the network $G_\theta(v)$ gets closer to the clean signal y^*, the denoiser will play a negligible role to ensure $D(G_\theta(v))$ closer to $G_\theta(v)$. However, the choice of D is not trivial. Kernel-based denoising methods have been proposed in the image domain using a set of convolution kernels g to denoise the image x. However, these techniques cannot be directly incorporated into our method because the object $G_\theta(v)$ is no longer in the image domain. We leverage the convolution theorem and take advantage of that the image domain convolution $g * x$ equals the frequency domain multiplication $\mathcal{F}(g) \cdot \mathcal{F}(x)$. Therefore, Eq. (4) can be rewritten as:

$$\arg\min_{\theta} \|G_\theta(v) - y\|_2^2 + \lambda \|G_\theta(v) - \mathcal{F}(g, v) \cdot G_\theta(v)\|_2^2, \tag{5}$$

where $\mathcal{F}(g, v)$ denotes the value of position v for the filter kernel g in frequency domain. In this work, we choose the Gaussian kernel as a denoiser kernel. The frequency domain filter value is given as $\mathcal{F}(g, v) = e^{-|v|/2\sigma^2}$, in which $|v|$ is the distance to the k-space center of coordinate v and σ is a parameter to control the strength of denoising.

2.4 High Dynamic Range Loss

Another optimization challenge in k-space is the large range of k-space values with an imbalanced distribution. The magnitude values around the k-space center which denote the low-frequency components are usually thousands of times higher than the others. If one uses the traditional L2 norm loss in Eq. (5), the high-magnitude points will be dominant in the training phase and produce an inferior reconstruction result. Inspired by [9], we introduce a log transform $\phi(z) = \log(z + \epsilon)$ to the k-space data magnitudes (where ϵ is a small constant), which compresses the high-magnitude signal to be in comparable magnitude as the low-magnitude ones:

$$\min_{\theta} \|\phi(G_\theta(v)) - \phi(y)\|_2^2 + \lambda \|\phi(G_\theta(v)) - \phi(\mathcal{F}(g, v) \cdot G_\theta(v))\|_2^2. \tag{6}$$

However, the non-linear transform will also change the noise distribution in the measurement data y. To enforce the training to converge to an unbiased result, we approximate the log transform by linearizing it at $G_\theta(v)$, with $\phi'(z) = \frac{1}{z+\epsilon}$:

$$\min_{\theta} \|\phi'(G_\theta(v))(G_\theta(v) - y)\|_2^2 + \lambda \|\phi'(G_\theta(v))(G_\theta(v) - \mathcal{F}(g, v) \cdot G_\theta(v))\|_2^2. \tag{7}$$

2.5 Training and Inference

During the training phase, we use the acquired data pair $\{(v_i, y_i)|i = 1, 2, \ldots, N\}$ from one single scan to train the network with the loss defined in Eq. (7). Once the training is finished, the network has learned the implicit representation of the data distribution in k-space. In inference, we can generate a set of coordinates $\bar{v}$ on the Cartesian grid with an arbitrary temporal resolution. As shown in Eq. (3), we feed these generated coordinates to the trained network to predict the complex value of these coordinates. Then, a simple inverse Fourier transform is applied to attain the individual coil images. The final image can be obtained by coil combination with coil sensitivity maps.

3   Experimental Setup

Dataset. The proposed NIK framework was evaluated in 6 healthy subjects at a 1.5T scanner (Ingenia, Philips, Best, The Netherlands) with 28 receive coils. Each subject has 7 short-axis slices. To reduce the training time we only used the data from 6 selected coils, discarding coils which are not sensitive in the heart region. All in vivo experiments were conducted with IRB approval and informed consent. The following sequence parameters were used: short axis slice; FOV = 256×256 mm^2 (considering inherent 1.6× frequency encoding oversampling of radial trajectories); 8 mm slice thickness; resolution = 2×2 mm^2; TE/TR = 1.16/2.3 ms; b-SSFP readout; radial tiny golden angle of ∼23.6°; flip angle 60°; 8960 radial spokes acquired; nominal scan time ∼20 s; breath-hold acquisition.

Coordinate Settings. The coordinate system for the NIK is flexible and can contain an arbitrary number of continuous or discrete dimensions. In this work, we included only four dimensions: the time dimension t, which represents the time point within the averaged heartbeat; the spatial frequencies k_x and k_y, which denote the 2D k-space location within the slice plane; the coil index c indicates from which coil the point data originates. All of these coordinates are normalized to $[-1, 1]$ to guarantee their equal contributions to the optimization.

Model Details. The proposed NIK framework is based on a multi-layer perceptron (MLP) with Fourier features and periodical activation functions. For each of the input coordinate element $v = [t, k_x, k_y, c]^T$ we generate a random Gaussian matrix B, where each entry is drawn independently from a normal distribution $\mathcal{N}(0, 1)$. The Fourier features position encoding of the coordinate element v is represented as

$$\gamma(v) = [\cos(2\pi B v), \sin(2\pi B v)] \tag{8}$$

according to [20]. The following MLP has 8 layers with a hidden feature dimension of 512. We use Sine activation functions in our MLP according to its effectiveness demonstrated in [19]. The output layer has two channels, representing the real and imaginary part of the predicted complex k-space value.

Training Settings. During the training, we feed all of the k-space coordinate-value pairs from one or four heartbeats of a single scan to the network. The models were trained by the Adam optimizer with parameters $\beta_1 = 0.9$, $\beta_2 = 0.999$, $\varepsilon = 10^{-8}$, learning rate $3 \cdot 10^{-5}$, and batch size 1. The training procedure stopped after 50000 steps when the loss did not decrease any further. The models were implemented with PyTorch 1.12. The training and testing were performed on an NVIDIA A6000 GPU. The regularization weight λ was empirically selected as 0.4 and 0.5 for four heartbeats and one heartbeat reconstruction, respectively. The ϵ in training loss was set to 0.01, and $\sigma = 1$ for frequency-domain regularization. The network training took approximately 10 h with four heartbeats

Table 1. The average MSE, PSNR and SSIM of INUFFT, CG-SENSE, L+S and the proposed NIK for 4 heartbeats and 1 heartbeat (mean±std).

Heartbeat	Methods	NRMSE	PSNR (dB)	SSIM
4	INUFFT	0.38 ± 0.08	28.18 ± 2.07	0.58 ± 0.08
	CG-SENSE [13]	0.23 ± 0.04	31.38 ± 1.84	0.75 ± 0.05
	L+S [11]	0.22 ± 0.04	32.06 ± 1.69	0.76 ± 0.04
	NIK (proposed)	$\mathbf{0.19 \pm 0.04}$	$\mathbf{33.52 \pm 2.08}$	$\mathbf{0.84 \pm 0.05}$
1	INUFFT	0.79 ± 0.14	21.53 ± 1.52	0.34 ± 0.06
	CG-SENSE [13]	0.43 ± 0.06	25.48 ± 1.28	0.54 ± 0.06
	L+S [11]	0.35 ± 0.04	27.77 ± 1.44	0.65 ± 0.04
	NIK (proposed)	$\mathbf{0.25 \pm 0.05}$	$\mathbf{30.67 \pm 1.94}$	$\mathbf{0.77 \pm 0.05}$

data, and 3 h for one heartbeat. The reconstruction of 30 cardiac phases with a resolution of 256×256 after training is less than 30 s.

Comparison of Methods. Data were reconstructed for 30 cardiac phases using inverse NUFFT (INUFFT), Conjugate Gradient SENSE (CG-SENSE) [13], Low Rank plus Sparse (L+S) [11], and the proposed NIK with one heartbeat and four heartbeats. For comparison methods, there are about 14 and 55 spokes for each cardiac phase for one heartbeat and four heartbeats, respectively. Reference CG-SENSE reconstructions with twenty heartbeats (~270 spokes per cardiac phase) were considered to evaluate the performance of all methods.

Evaluation Metrics. Both visual comparison and quantitative evaluation were used for performance evaluation. For a quantitative evaluation, the normalized root mean squared error (NRMSE), peak signal-to-noise ratio (PSNR) and structural similarity index (SSIM) were calculated. A higher PSNR and SSIM and lower NRMSE indicate better quantitative performance.

4 Results

Table 1 summarized the mean and standard deviation of the reconstruction results for 42 slices from 6 subjects. For the four heartbeats and one heartbeat reconstructions, we can see that our proposed NIK consistently outperforms the comparison methods. A visual comparison of the four heartbeat reconstruction can be found in Fig. 2, which supports our quantitative findings. From the zoomed images in the second row, we notice that the comparison methods suffer from the loss of small anatomical structures (red arrows), while NIK is able to recover them. x-t profiles show larger errors for the comparison methods with respect to the temporal dimension. This is related to the striking artifacts produced by radial sampling in the 2D plane, which are challenging to eliminate if the conventional NUFFT is used. In contrast, our proposed method reconstructs the temporal information coherently and accurately. The reconstruction of one heartbeat is showcased in Fig. 3. Due to the extremely small number of spokes, INUFFT, CG-SENSE and L+S end up either with noisy images or blurred images. In contrast, NIK is competent in recovering sharp images with high temporal resolution.

Figure 4(a) shows the results for 50 cardiac phases of one heartbeat reconstruction. The reconstruction performance of INUFFT, CG-SENSE and L+S is much inferior compared to Fig. 3 and the images are overwhelmed by the artifacts caused by radial sampling. Because of the data-binning scheme, the number of spokes available for each cardiac phase dropped to 3/5 of the original when we increase from 30 to 50 phases ($\sim$8 spokes per cardiac phase). While the proposed NIK still has similar image quality as 30 phase reconstruction. We also demonstrated the effectiveness of frequency domain regularization in Fig. 4(b). The result with frequency domain regularization shows slightly reduced noise.

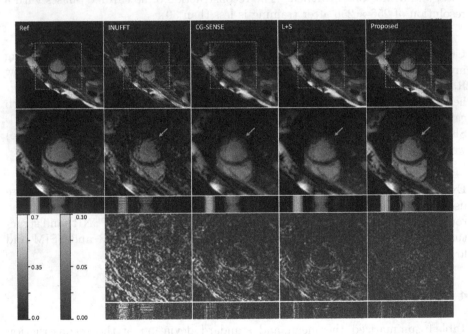

Fig. 2. Reconstructions with data from four heartbeats. The 2nd and 4th rows show the zoomed views of the yellow boxes and their error maps. The border of the myocardium is marked by red arrows. The 3rd and 5th row show the x-t profile and their error maps. (Color figure online)

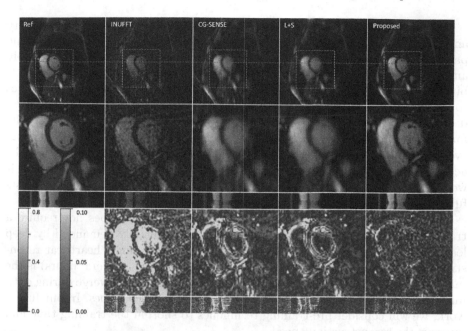

Fig. 3. Reconstructions with data from one heartbeat. The 2nd and 4th rows show the zoomed views of the yellow boxes and their error maps. The 3rd and 5th rows show the x-t profile and their error maps. (Color figure online)

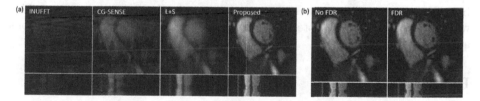

Fig. 4. (a) Reconstructions for 50 cardiac phases and (b) influence of frequency-domain regularization (FDR) with 30 cardiac phases, with data from one heartbeat.

5 Discussion and Conclusion

In this work, we present the concept of neural implicit k-space (NIK) for non-Cartesian CMR. NIK is a new database-free learning framework to resolve temporal information in a continuous non-Cartesian acquisition paradigm. Instead of merging adjacent k-space lines in the temporal dimension to reconstruct each cardiac phase, a continuous implicit representation of k-space can be learned so that k-space points on a Cartesian grid can be generated and transformed into clean images by inverse Fourier transform. In this paper, we focus on radially sampled CMR data. We believe that NIK can be easily extended to arbitrary sampling patterns and N-dimensional datasets by modifying the input coordinate. More sampling patterns and modalities will be investigated.

Our experiments illustrate that NIK can reconstruct images in high quality and shows superior performance in removing artifacts caused by radial sampling patterns. NIK outperforms all reference methods by a large margin qualitatively and quantitatively. As NIK learns a continuous k-space representation and is binning-free, the image quality is not affected by the number of cardiac phases to be reconstructed. This is not possible for conventional methods based on data binning, as some phases might have too few k-space lines. The results for one heartbeat reconstruction show promise for real-time cardiac imaging, even though some small details are lost as they might not be encoded in the acquired k-space. Noise can still remain in the image reconstructions, which is well-known from Cartesian k-space interpolation techniques [1,5]. This suggests future research in frequency-domain filtering beyond Gaussian filtering.

Our method achieves a single-shot reconstruction that does not require a training dataset from a large population, as it is commonly demanded by deep learning-based reconstruction methods. While enabling single heartbeat reconstruction, we acknowledge that the long training time (i.e., several hours) is the main limitation of our approach. This is caused by the slow converge during optimization given the large range nature of sampled k-space values. In our future work, we will consider pre-training techniques to shorten the training time and improve the reconstruction speed.

In summary, we introduce a novel concept of neural implicit k-space (NIK) for database-free k-space interpolation of non-Cartesian CMR data. The k-space pointwise estimation scheme comes naturally with the advantages of binning-free, sampling pattern-independent, and NUFFT-free imaging. This work demonstrates great potential to leverage NIK in further CMR applications and beyond.

References

1. Akçakaya, M., Moeller, S., Weingärtner, S., Uğurbil, K.: Scan-specific robust artificial-neural-networks for k-space interpolation (RAKI) reconstruction: database-free deep learning for fast imaging. Magn. Reson. Med. **81**(1), 439–453 (2019)
2. Amiranashvili, T., Lüdke, D., Li, H.B., Menze, B., Zachow, S.: Learning shape reconstruction from sparse measurements with neural implicit functions. In: International Conference on Medical Imaging with Deep Learning, pp. 22–34. PMLR (2022)
3. Barron, J.T., Mildenhall, B., Tancik, M., Hedman, P., Martin-Brualla, R., Srinivasan, P.P.: Mip-NeRF: a multiscale representation for anti-aliasing neural radiance fields. In: Proceedings of the IEEE/CVF International Conference on Computer Vision, pp. 5855–5864 (2021)
4. Chibane, J., Alldieck, T., Pons-Moll, G.: Implicit functions in feature space for 3D shape reconstruction and completion. In: Proceedings of the IEEE/CVF Conference on Computer Vision and Pattern Recognition, pp. 6970–6981 (2020)
5. Griswold, M.A., et al.: Generalized autocalibrating partially parallel acquisitions (GRAPPA). Magn. Reson. Med.: Official J. Int. Soc. Magn. Reson. Med. **47**(6), 1202–1210 (2002)

6. Hammernik, K., et al.: Learning a variational network for reconstruction of accelerated MRI data. Magn. Reson. Med. **79**(6), 3055–3071 (2018)
7. Huang, W., et al.: Deep low-rank plus sparse network for dynamic MR imaging. Med. Image Anal. **73**, 102190 (2021)
8. Kuang, K., et al.: What makes for automatic reconstruction of pulmonary segments. In: Wang, L., Dou, Q., Fletcher, P.T., Speidel, S., Li, S. (eds.) MICCAI 2022. LNCS, pp. 495–505. Springer, Cham (2022). https://doi.org/10.1007/978-3-031-16431-6_47
9. Mildenhall, B., Hedman, P., Martin-Brualla, R., Srinivasan, P.P., Barron, J.T.: NeRF in the dark: high dynamic range view synthesis from noisy raw images. In: Proceedings of the IEEE/CVF Conference on Computer Vision and Pattern Recognition, pp. 16190–16199 (2022)
10. Mildenhall, B., Srinivasan, P.P., Tancik, M., Barron, J.T., Ramamoorthi, R., Ng, R.: NeRF: representing scenes as neural radiance fields for view synthesis. Commun. ACM **65**(1), 99–106 (2021)
11. Otazo, R., Candes, E., Sodickson, D.K.: Low-rank plus sparse matrix decomposition for accelerated dynamic MRI with separation of background and dynamic components. Magn. Reson. Med. **73**(3), 1125–1136 (2015)
12. Pipe, J.G., Menon, P.: Sampling density compensation in MRI: rationale and an iterative numerical solution. Magn. Reson. Med.: Official J. Int. Soc. Magn. Reson. Med. **41**(1), 179–186 (1999)
13. Pruessmann, K.P., Weiger, M., Börnert, P., Boesiger, P.: Advances in sensitivity encoding with arbitrary k-space trajectories. Magn. Reson. Med.: Official J. Int. Soc. Magn. Reson. Med. **46**(4), 638–651 (2001)
14. Pruessmann, K.P., Weiger, M., Scheidegger, M.B., Boesiger, P.: SENSE: sensitivity encoding for fast MRI. Magn. Reson. Med.: Official J. Int. Soc. Magn. Reson. Med. **42**(5), 952–962 (1999)
15. Qin, C., Schlemper, J., Caballero, J., Price, A.N., Hajnal, J.V., Rueckert, D.: Convolutional recurrent neural networks for dynamic MR image reconstruction. IEEE Trans. Med. Imaging **38**(1), 280–290 (2018)
16. Ramzi, Z., Chaithya, G., Starck, J.L., Ciuciu, P.: NC-PDNet: a density-compensated unrolled network for 2D and 3D non-Cartesian MRI reconstruction. IEEE Trans. Med. Imaging **41**, 1625–1638 (2022)
17. Schlemper, J., Caballero, J., Hajnal, J.V., Price, A., Rueckert, D.: A deep cascade of convolutional neural networks for MR image reconstruction. In: Niethammer, M., et al. (eds.) IPMI 2017. LNCS, vol. 10265, pp. 647–658. Springer, Cham (2017). https://doi.org/10.1007/978-3-319-59050-9_51
18. Shen, L., Pauly, J., Xing, L.: NeRP: implicit neural representation learning with prior embedding for sparsely sampled image reconstruction. IEEE Trans. Neural Netw. Learn. Syst. (2022)
19. Sitzmann, V., Martel, J., Bergman, A., Lindell, D., Wetzstein, G.: Implicit neural representations with periodic activation functions. In: Advances in Neural Information Processing Systems, vol. 33, pp. 7462–7473 (2020)
20. Tancik, M., et al.: Fourier features let networks learn high frequency functions in low dimensional domains. In: Advances in Neural Information Processing Systems, vol. 33, pp. 7537–7547 (2020)
21. Uecker, M., et al.: ESPIRiT-an eigenvalue approach to autocalibrating parallel MRI: where SENSE meets GRAPPA. Magn. Reson. Med. **71**(3), 990–1001 (2014)
22. Vasudevan, V., et al.: Neural representation for three-dimensional dose distribution and its applications in precision radiation therapy. Int. J. Radiat. Oncol. Biol. Phys. **114**(3), e552 (2022)

23. Vasudevan, V., et al.: Implicit neural representation for radiation therapy dose distribution. Phys. Med. Biol. **67**(12), 125014 (2022)
24. Wang, S., et al.: Accelerating magnetic resonance imaging via deep learning. In: 2016 IEEE 13th International Symposium on Biomedical Imaging (ISBI), pp. 514–517. IEEE (2016)
25. Wolterink, J.M., Zwienenberg, J.C., Brune, C.: Implicit neural representations for deformable image registration. In: Medical Imaging with Deep Learning (2021)
26. Wright, K.L., Hamilton, J.I., Griswold, M.A., Gulani, V., Seiberlich, N.: Non-cartesian parallel imaging reconstruction. J. Magn. Reson. Imaging **40**(5), 1022–1040 (2014)
27. Yaman, B., Hosseini, S.A.H., Moeller, S., Ellermann, J., Uğurbil, K., Akçakaya, M.: Self-supervised learning of physics-guided reconstruction neural networks without fully sampled reference data. Magn. Reson. Med. **84**(6), 3172–3191 (2020)
28. Yoo, J., Jin, K.H., Gupta, H., Yerly, J., Stuber, M., Unser, M.: Time-dependent deep image prior for dynamic MRI. IEEE Trans. Med. Imaging **40**(12), 3337–3348 (2021)
29. Zang, G., Idoughi, R., Li, R., Wonka, P., Heidrich, W.: IntraTomo: self-supervised learning-based tomography via sinogram synthesis and prediction. In: Proceedings of the IEEE/CVF International Conference on Computer Vision, pp. 1960–1970 (2021)
30. Zha, R., Zhang, Y., Li, H.: NAF: neural attenuation fields for sparse-view CBCT reconstruction. In: Wang, L., Dou, Q., Fletcher, P.T., Speidel, S., Li, S. (eds.) MICCAI 2022. LNCS, pp. 442–452. Springer, Cham (2022). https://doi.org/10.1007/978-3-031-16446-0_42

Registration

Geometric Deep Learning for Unsupervised Registration of Diffusion Magnetic Resonance Images

Jose J. Bouza[1], Chun-Hao Yang[2] (ORCID), and Baba C. Vemuri[3](✉)

[1] Intuitive Surgical, 1020 Kifer Road, Sunnyvale, CA, USA
[2] Institute of Applied Mathematical Sciences, National Taiwan University, Taipei, Taiwan
`chunhaoy@ntu.edu.tw`
[3] Department of CISE, University of Florida, Gainesville, FL, USA
`vemuri@ufl.edu`

Abstract. Deep learning based models for registration predict a transformation directly from moving and fixed image appearances. These models have revolutionized the field of medical image registration, achieving accuracy on-par with classical registration methods at a fraction of the computation time. Unfortunately, most deep learning based registration methods have focused on scalar imaging modalities such as T1/T2 MRI and CT, with less attention given to more complex modalities such as diffusion MRI. In this paper, to the best of our knowledge, we present the first end-to-end geometric deep learning based model for the non-rigid registration of fiber orientation distribution fields (fODF) derived from diffusion MRI (dMRI). Our method can be trained in a fully-unsupervised fashion using only input fODF image pairs, i.e. without ground truth deformation fields. Our model introduces several novel differentiable layers for local Jacobian estimation and reorientation that can be seamlessly integrated into the recently introduced manifold-valued convolutional network in literature. The results of this work are accurate deformable registration algorithms for dMRI data that can execute in the order of seconds, as opposed to dozens of minutes to hours consumed by their classical counterparts.

Keywords: Registration · Geometric Deep Learning · Diffusion MRI

1 Introduction

Image registration is a fundamental operation in medical image analysis. Broadly speaking, the image registration problem involves finding the correspondence (match) between images in different coordinate systems. This correspondence can be established by finding an appropriate geometric transformation between the coordinate systems. The nature of this transformation can vary from a simple global affine transformation to a full non-rigid transformation yielding a dense deformation field. In this paper, we develop the first end-to-end deep learning based model for full-deformable registration of fiber orientation distribution function (fODF) fields derived from diffusion MRI (dMRI) data. Our method can be easily ported (after some simple modifications) to other commonly used derived representations from dMRI such as, the diffusion tensor (DT) or the ensemble average propagator (EAP) fields.

© The Author(s), under exclusive license to Springer Nature Switzerland AG 2023
A. Frangi et al. (Eds.): IPMI 2023, LNCS 13939, pp. 563–575, 2023.
https://doi.org/10.1007/978-3-031-34048-2_43

This paper firstly builds in novel ways on the existing manifold valued convolution operations presented in [5] and then develops a novel architecture suitable for the non-rigid registration of fODFs derived from dMRI. Our novel contributions include: 1) An efficient CUDA implementation of the core operations presented in [5] 2) Several differentiable layers required for diffusion MRI registration (Jacobian estimation layer, reorientation layer) 3) Design and implementation of end-to-end deformable dMRI registration networks 4) A detailed experimental analysis of these models.

1.1   Prior Work: Classical dMRI Registration

Image registration is a fundamental problem in medical image analysis with numerous applications. We now present a brief note on dMRI registration applied to derived representations such as: diffusion tensor images (DTI), ensemble average propagator (EAP) fields or fiber orientation density function (fODF) fields. All of these images are manifold-valued images in that, at each voxel, we have a manifold-valued 'object'. In the case of DTI, this object is an element of the manifold of (n, n) symmetric positive definite matrices denoted by P_n, for EAP (fODF) fields/images, it is the manifold of probability density functions.

Early work on DTI registration used a registration cost function based on either fractional anisotropy (FA) or some rotation invariant features computed from the DTs [14, 26] These methods are however not applicable to higher order tensor field representation which might be needed to cope with crossing fibers in the data. In this context, groupwise registration of fourth order tensor representations of dMRI data was presented in [3]. In [29], a non-rigid registration and reorientation algorithm applied directly to the raw dMRI data was presented. Their algorithm performs the reorientation via the use of pre-specified fiber basis functions.

Several approaches to register the EAP (fODF) fields have been proposed in literature. These methods first compute EAPs (fODFs) at each voxel and then register these derived manifold-valued fields/images [7, 17]. For an extensive literature review, we refer the reader to [10].

1.2   Prior Work: Deep Learning Based Registration

Modern classical registration algorithms are relatively accurate, but require substantial computation time due to their iterative nature. It is not uncommon for even well-optimized software such as ANTs [1] to take upwards of 30 min to register a pair of high-resolution brain MRI volumes. With the introduction of deep learning based registration methods the possibility of registration with accuracy on-par with classical methods, but with a runtime on the order of seconds is within reach.

The first deep learning based registration methods [27] required ground truth deformation fields and processed image patches instead of full images. Later methods used fully unsupervised methods for training but still using a path-based approach [16, 23]/With the introduction of the VoxelMorph architecture [2], Balakrishnan et al. showed that full-resolution image registration is possible within a deep network. Further work has extended the VoxelMorph architecture to guarantee diffeormorphic registrations [8] and learning contrast-invariant registrations [11]. These recent models achieve accuracy on-par with classical registration algorithms, with several orders of magnitude improvements in runtime. All the above methods are however designed for registration of scalar-valued images.

In this paper, we focus on the task of dMRI registration. Registration of dMRI data is more challenging than traditional modalities for several reasons. First, because dMRI contains directional information, a reorientation step must follow the application of a deformation field. Second, dMRI data has substantially more information at each voxel than most other modalities, and thus requires more memory and computation to process. Finally, in the context of deep learning based methods, there has been little attention given to developing network architectures that respect the manifold geometry of the dMRI derived image representations such as DTI, fODF images etc., which form the input to the network or can be estimated within the network.

To the best of our knowledge, the only existing deep-learning based dMRI registration method in literature is DDMReg [28]. This model extracts FA images and several tract orientation maps (TOMs) from the dMRI data. Each FA and TOM image is passed through a separate registration subnetwork (each of which is a VoxelMorph style architecture [2]). Each subnetwork outputs a proposal deformation field, and a multi-deformation fusion subnetwork combines these fields to generate a final predicted deformation field. This approach has a few pitfalls. First, the registration subnetworks and fusion subnetworks are all trained separately, thus not achieving the performance of end-to-end trained models. Second, the FA and TOM inputs are hand-crafted features extracted from the dMRI data which consumes preprocessing time during inference. Further, we show that we can achieve improved performance by building a model that can directly process dMRI derived (fODF) data in a way that respects the underlying geometry.

1.3 Paper Organization

In Sect. 2 we briefly review the Manifold Valued Convolution (MVC) and Manifold Valued Volterra Series (MVVS) operations, the core layers used to process the fODF fields derived from dMRI. In Sect. 3, we present efficient CUDA implementations of the MVC and MVVS operations. Section 4 contains a description of our deep network architectures for deformable registration, including a differentiable implementation of a dMRI reorientation method. Finally, Sect. 5 contains an extensive set of experimental results demonstrating the performance of our geometric deep network.

2 Manifold Valued Volterra Series

We will now very briefly review the manifold valued convolution (MVC) and manifold valued Volterra series (MVVS) operations introduced in [5]. A manifold-valued image is a map $F : \mathbb{Z}^n \to \mathcal{M}$ and this image modality naturally arises in various dMRI data representations. Recall that for $x_1, x_2 \in \mathbb{R}^n$, the Hadamard product of x_1 and x_2 is $x_1 \odot x_2 = [x_{11}x_{21}, \ldots, x_{1n}x_{2n}]$. The Nth order Volterra series for $g : \mathbb{R}^n \to \mathbb{R}$ and $f : \mathbb{R} \to \mathbb{R}$ is given by $h(x) = \sum_{n=1}^{N} \int \cdots \int g_n(x - \tau_1, \ldots, x - \tau_n) \prod_{i=1}^{n} f(\tau_i) d\tau_i$.

In this work, we only consider the first and second order MVVS, which are defined by

$$MVC(F, w)(\mathbf{y}) = \mathbf{Exp}_{m(\mathbf{y})} \left(\sum_{\mathbf{z}=1}^{K} w(\mathbf{z} - \mathbf{y}) \mathbf{Log}_{m(\mathbf{y})} F(\mathbf{z}) \right)$$

$$MVVS(F, w_1, w_2)(\mathbf{y}) = \mathbf{Exp}_{m(\mathbf{y})} \left[\sum_{\mathbf{z}=1}^{K} w_1(\mathbf{z} - \mathbf{y}) \mathbf{Log}_{m(\mathbf{y})} F(\mathbf{z}) \right.$$

$$\left. + \sum_{\mathbf{z}_1, \mathbf{z}_2=1}^{K} w_2(\mathbf{z}_1 - \mathbf{y}, \mathbf{z}_2 - \mathbf{y}) \left(\mathbf{Log}_{m(\mathbf{y})} F(\mathbf{z}_1) \right) \odot \left(\mathbf{Log}_{m(\mathbf{y})} F(\mathbf{z}_2) \right) \right]$$

where $F : \mathbb{Z}^d \to \mathcal{M}$ is a manifold-valued image and $w_j : (\mathbb{Z}^d)^j \to \mathbb{R}$ is the j-th kernel with size K. For $\mathbf{y} \in Z^d$ where $m(\mathbf{y}) = \mathsf{FM}(F(\mathbf{z}))$ where each $\mathbf{z}$ ranges over the support of the Volterra masks w_j centered at $\mathbf{y}$ and FM is the unweighted Frechet mean. The N-th order MVVS is a straightforward generalization and we refer the reader to [5].

In [4], it was reported that using the sample in the middle voxel of the moving window gives similar performance to using the Frechet mean. This modification significantly improves performance and ease of implementation, thus we opt to use the mid point of the moving window as the base point throughout this paper. In the interest of computational and parameter efficiency, we only use MVC and second-order MVVS layers in our experiments.

3   Implementation

In this section, we present a novel CUDA implementation of the MVVS layer. This implementation allows us to build networks for processing the fODF field derived from the full dMRI volume.

3.1   Data of Interest

In this work we limit our focus to the fODF representation of dMRI data [18]. An fODF describes the distribution of fiber orientations in a voxel. The diffusion signal is modeled as the convolution of the fODF and the response function characterizing diffusion along coherent fiber bundles. Thus, given the diffusion signal and the response function, the fODF can be solved for via deconvolution [20]. The space of all fODFs can be defined by the set $\Phi \colon \left\{ \phi : \hat{\mathbf{u}} \in S^2 \to R_0^+ \,\middle|\, \phi(\hat{\mathbf{u}}) \geq 0, \int_{S^2} \phi(\hat{\mathbf{u}}) d\hat{\mathbf{u}} = 1 \right\}$ where, R_0^+ is the set of nonnegative reals. Using a square root density parameterization, this distribution can be identified with a point on the unit Hilbert sphere, a Riemannian manifold whose geometry is fully known, and has been used in literature for EAP (fODF) estimation [22]. We represent a sampled fODF as a point on the unit hypersphere $\mathbb{S}^M$ following the convention in [5], where M is the number of sample points. The unit hypersphere is a Riemannian manifold, thus this representation fits well into the MVC/MVVS framework. By virtue of the fact that this representation of fODF leads to elements in the space of probability distributions, unlike the spherical harmonic representation of fODFs, it does not require explicit enforcement of non-negativity and integration to one constraints.

3.2 CUDA Implementation

We now present an optimized CUDA implementation of the MVC operation. This allows us for the first time to use these operations at the scale of full dMRI volumes, unlike the original MVC work presented in [5]. For clarity, we first present a naive CUDA implementation and then briefly describe several optimizations added to the naive implementation. We only present the MVC operation details in this section, but we also implemented forward and backward passes for the MVVS operation. The code is made public.

The input manifold-valued image will be represented by a tensor of shape $C \times D \times W \times H \times M$, where C, D, W, H are the channels, depth, width and height respectively. The output manifold-valued image will be a tensor of shape $C_{\text{out}} \times S(D) \times S(W) \times S(H) \times M$ where $S(x) = \frac{x-K}{T} + 1$, K is the filter size and T the stride. The data at each voxel is a point on the hypersphere embedded in Euclidean space, $S^{M-1} \subset \mathbb{R}^M$, and is thus an M dimensional vector. The weight filter will be represented as a tensor of shape $C \times C_{\text{out}} \times K^3$ where C_{out} is the number of output channels and K is the filter size.

In the naive CUDA implementation, each CUDA thread will compute one output voxel. Suppose a thread is assigned to compute voxel $\mathbf{v} = [c, d, w, h]$ in the output image. It will perform the following steps:

1. Compute the input voxels coordinates in the receptive field of the output voxel $\mathbf{v}$. Let $R_i(\mathbf{v})$ denote the ith voxel in the receptive field of $\mathbf{v}$.
2. Compute the Riemannian Log of each voxel value in the receptive field with base point set to the midpoint of the receptive field. $\mathbf{Log}_{R_m(\mathbf{v})}(R_i(\mathbf{v}))$ for $i = 1, \ldots, K^3$ where m is the index of the midpoint.
3. Perform the following weighted sum, $T = \sum_{i=1}^{K^3} w_i \mathbf{Log}_{R_m(\mathbf{v})}(R_i(\mathbf{v}))$ where w_i are the weight filter values.
4. Perform the Riemannian exponential $\mathbf{Exp}_{R_m(\mathbf{v})}(T)$ and write it to the output manifold-valued image at voxel coordinate $\mathbf{v}$.

The closed form expressions for the **Log** and **Exp** maps on the sphere are given by the following expression, where $U = X - \langle X, Y \rangle Y$ [19]. $\mathbf{Exp}_Y(X) = \cos(\|X\|)Y + \sin(\|X\|)\frac{X}{\|X\|}$ and, $\mathbf{Log}_Y(X) = U \cos^{-1}(\langle X, Y \rangle)/\langle U, U \rangle$.

We optimize the naive CUDA implementation with two strategies: 1) remove temporary memory allocations by performing the Riemannian Log, weighted sum and Riemannian Exp in-place (steps 2–4). 2) Use tiling [25] to reduce redundant global memory reads by using shared memory to do spatial caching.

Performance Analysis. We perform a benchmark analysis of the following implementations of the MVC operation: 1. PyTorch CPU Implementation 2. PyTorch GPU Implementation 3. Naive CUDA Implementation 4. Memory Optimized CUDA Implementation 5. Tiled CUDA Implementation. The metrics of interest are runtime and peak global GPU memory usage. All experiments are run on an RTX 2080 Ti GPU with CUDA version 11.1. A random S^{M-1}-valued image was generated with spatial dimensions 80^3 and $M = 45$. A random weight kernel was generated with kernel size $K = 3$.

The input and output channels are both 1. A CUDA thread block size of 5^3 is used. All reported metrics are averaged over 10 runs.

Table 1. Performance of tested MVC Implementations. Memory usage is measured as peak GPU global memory usage.

Method	Runtime (s)	Memory Usage (MB)
PyTorch CPU	15.646	N/A
PyTorch GPU	0.378	9627
Naive CUDA	0.162	452
Memory-Efficient (M-E) CUDA	0.154	187
Tiled M-E CUDA	0.129	187

Results are reported in Table 1. We can see that even the naive CUDA implementation offers a substantial improvement in both runtime and memory usage. Beyond this, the memory efficient CUDA implementation achieves the goal of peak memory usage no greater than that required to store the input and output tensors. Finally, the tiled memory efficient CUDA implementation further improves runtime. For all experiments we utilize the tiled memory efficient CUDA implementation.

4   An MVC/MVVS Architecture for Deformable Diffusion MRI Registration

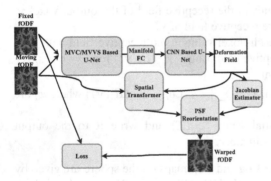

Fig. 1. Deformable registration network architecture.

We now present architectures for unsupervised registration of dMRI data represented using fODF images (fODFs at each voxel). We present architectures for deformable (non-rigid) registration tasks. Several layers must be introduced. The first is the previously presented MVC/MVVS layer, which extract features from the fODF images. The second is a differentiable estimator of the local Jacobian of the deformation field. The third is a differentiable point spread function fODF (defined subsequently) reorientation layer. Finally, we utilize the spatial transformer [12] layer for resampling of the fODF field. By using differentiable versions of the Jacobian estimation, reorientation and resampling operations we can perform the transformation of the moving fODF image inside the network, and compute a loss directly between the input fixed image and output warped image. This allows us to train in a fully unsupervised manner. This strategy has been used in several other methods for neural network based registration (e.g. [2, 28]). We release all of the code required for inference and training using these models.

4.1 Differentiable Jacobian Estimator

When registering fODF images, a vital step is reorientation. When a deformation field is applied to an fODF image, the underlying spatial grid will be warped. Since the fODF functions at each voxel represent directional information, they must be transformed in accordance with the deformation field. We implement the reorientation method utilized in [17], which uses the local Jacobian of the deformation field to reorient the fODF functions.

Our model will be trained in an unsupervised manner, with the deformation and reorientation of the moving fODF image occurring inside the network during the training stage. Thus, we must compute the local Jacobian of the deformation field during training. To this end, we implement an efficient second order central difference based approximator for computing the Jacobian of the deformation field at each voxel. The 3D deformation field is represented as a tensor of shape $B \times D \times W \times H \times 3$, where B, D, W, and H are the batch, depth, width and height respectively. The Jacobian estimator computes a second order central difference estimation of the partial derivatives along each direction of the deformation vectors to output a $B \times D \times W \times H \times 3 \times 3$ tensor, i.e. a field of local Jacobian matrices.

4.2 Differentiable PSF Reorientation

Given the local Jacobian of the deformation field at every voxel, the next step is to reorient the fODF. Recall that the fODF at each voxel is represented as a density function on the sphere $f : \mathbb{S}^2 \to \mathbb{R}$. But we sample the fODF density along M sample points on the sphere to represent it as an M-dimensional probability vector. Thus our reorientation method must operate on this representation. We opt to implement a differentiable version of the method presented in [17]. In short, this method approximates the fODF function as a weighted sum of spherical point spread functions (PSFs), reorients the PSFs, and then resamples the weighted sum of the reoriented PSFs to return to the original representation, an M-directional probability vector. This method was shown to give improved results for fODF reorientation relative to previous methods, and satisfies some useful properties (e.g. the fODF integral and the partial volume fractions are guaranteed to be preserved).

We implement an efficient, batch-mode, differentiable version of this operation in PyTorch which can run on GPUs. To the best of our knowledge, this is the first differentiable implementation of an fODF reorientation method.

4.3 Registration Architecture

We now have all the building blocks for an fODF registration network. A schematic of the proposed architecture is presented in Fig. 1. The moving and fixed images are concatenated along a channel axis and passed through a feature extraction head which consists of an MVC or MVVS based UNet, a ManifoldFC block and a traditional CNN based UNet. The ManifoldFC operation, originally introduced in [6], allows us to map manifold-valued features to scalar-valued features.

The output of the UNet heads is a deformation field (deformation vectors are stored across the channel dimension as in e.g. [2]). This deformation field is used to resample the input moving image. Simultaneously, the deformation field is passed through a Jacobian estimator block. The Jacobian matrices are then used to reorient the resampled moving image.

The MVC/MVVS UNet block consist of 4 layers mapping across the following channel sizes: $2 \rightarrow 8 \rightarrow 16 \rightarrow 32 \rightarrow 32$, each layer with a kernel size of 3. The traditional U-Net block consists of 3 encoder and 3 decoder layers, with a maximum of 1024 channels and skip connections between encoder and decoder layers.

In the deformable registration network a loss function is applied to the output warped image, with the input fixed image as the target as in [2]. By performing the resampling and reorientation inside the network, we can train in a fully unsupervised manner, in contrast to previous approaches such as [27] which required ground truth deformation fields to train.

5   Experiments

We now present several experiments demonstrating the performance of our MVC and MVVS registration networks on deformable registration tasks.

5.1   Dataset

We train and evaluate on dMRI data from the Human Connectome Project (HCP) Young Adult dataset. HCP consists of dMRI scans of the brain for 1200 subjects aged 22–35. For details about acquisition parameters, subject criteria, preprocessing etc. we refer the reader to the HCP study [21]. We randomly selected 400 subjects from the HCP dataset and run them through an fODF generation pipeline which consists of response function estimation using the technique in [9] to generate subject specific white matter, grey matter and CSF response functions. We then use multi-shell multi-tissue constrained spherical deconvolution [13] to reconstruct the white-matter fODF functions from the diffusion signal and estimated response function.

5.2   Evaluation Strategy

We evaluate registration accuracy by computing the DICE score between known fixed and warped structures. We limit our evaluation to white matter tracts, since this is a structure well captured by dMRI. Optimally, one would use expert labeled segmentation's of white matter tracts to perform the registration evaluation, but no such segmentation's exist for the HCP dataset. Instead, we opt to use a well-validated automatic segmentation algorithm to generate segmentation masks for white matter tracts in the moving and fixed image. Specifically, we utilize the TractSeg segmentation model [24]. A total of 72 white matter tracts were segmented. For subjects in the validation set, segmentation accuracy was visually reviewed and poor segmentation's were discarded. At evaluation time, a moving and fixed image pair are registered. The transformation is then used to transform each of the moving image white matter tract segmentation

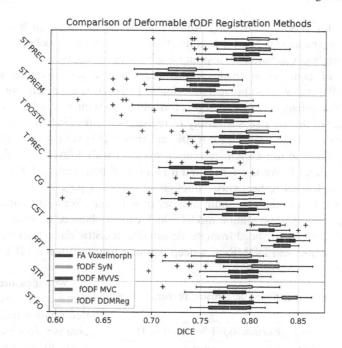

Fig. 2. Comparison of deformable fODF registration methods on a subset of white matter tracts. The shortened tract names correspond to the following structures. ST PREC: Striato-precentral. ST PREM: Striato-premotor. T POSTC: Thalamo-postcentral. T PREC: Thalamo-precentral. CG: Cingulum left. CST: Corticospinal. FPT: Fronto-pontine. STR: Superior Thalamic Radiation. ST FO: Striato-fronto-orbital

masks. Finally, the DICE score is computed between the warped tract segmentation map and fixed tract segmentation mask.

For deformable registration, we focus on a template registration task. A randomly chosen subject from our 400 subject dataset is selected as a reference fixed image. All other samples from the dataset will be registered to this subject. Thus we have a dataset of 399 moving, fixed image pairs where the fixed image is the same for all samples. We split this dataset into 349 training samples, 25 validation samples and 25 test samples. Deformable registration algorithms generally require an accurate initialization to perform well. Thus, moving and fixed image pairs are first pre-aligned using an affine registration algorithm.

For this experiment, we take the additional step of downsampling the moving and fixed fODF images by a scale factor of $1/2$ to obtain images of spatial dimensions $64 \times 40 \times 64$. In spite of the optimizations made in Sect. 2, fODF registration is a memory intensive task, thus downsampling was necessary to allow our registration networks to run on a single GPU. Future advances in hardware and algorithms will allow us to run our method on full resolution images.

We train for 1000 epochs using Adam optimizer with an initial learning rate of 0.0001 with the mean square error loss function. We select the model parameters at the epoch with the best validation set loss and use it for our evaluation.

Our set of comparison methods includes classical and deep learning based approaches. We start with a classical deformable registration approach. For this we use Symmetric Normalization (SyN) based deformable registration, which has been shown to be the state of the art across a variety of registration tasks [15]. We use the mrtrix3 implementation of SyN designed for fODF registration [18] We configure the SyN registration algorithm to use a multi-resolution pyramid with 4 levels at scale factors of 1/8, 1/4, 1/2, and 1. We use a maximum number of iterations of 1000, 1000, 1000 and 100 at each level respectively, although an early stoppage criteria usually stops the registration before the maximum number of iterations. We also evaluate a Voxel-Morph style UNet FA registration model which can be obtained by removing the initial MVC/MVVS encoder head from the deformable registration architecture presented previously. We also test DDMReg, a deep learning based diffusion MRI registration method [28].

Table 2. Performance of tested deformable registration methods.

Method	DICE	Runtime (s)	% Voxels $\det(J) \leq 0$
FA Voxelmorph	0.7126	0.9	0.007%
fODF Classical	0.7601	839	0.012%
fODF MVVS	0.7493	2.6	0.006%
fODF MVC	0.7317	2.1	0.005%
fODF DDMReg	0.7417	12.2	0.003%

We measure the percentage of voxels in the deformation field with non-positive Jacobian determinants to determine regions where the deformation field is non-diffeomorphic. We found that for our methods, explicit regularization of generated deformation fields with a loss function penalty term was not necessary. Indeed, our model achieves a very low percentage of voxels with non-positive Jacobians without an explicit regularization term.

5.3   Results

All evaluation results are computed across test sets unseen during training and not used for model checkpoint selection or hyperparameter optimization.

The deformable registration experiment results are presented in Table 2. Figure 2 shows the DICE performance of all compared methods on a subset of the 72 white matter tract structures. Again, DICE scores are averaged over all 72 white matter tract structures generated in our evaluation pipeline. Among the learning based (non-classical) techniques we again see the MVVS based model achieving the best DICE overlap score. The FA based VoxelMorph style model still lags behind all methods which use the full fODF. The MVVS based model again outperforms the MVC based model. The DDM-Reg based model is not trained end-to-end, instead opting for independently training several registration proposal networks and a registration fusion head. It also does not take the fODF as input directly, instead extracting hand-engineered features from the fODF image and inputting those features into the network (see Sect. 1.2 for details).

Optimally, the network should have the capacity to learn features useful for the registration task internally. We attribute the improved performance of our MVVS based model relative to the DDMReg model to these limitations of the DDMReg model, which are not present in our MVVS based fODF registration technique. Again, we approach the accuracy of the classical registration method at just 0.3% of the runtime. We see that the MVVS based model also outperforms DDMReg on registration time, in large part due to the optimizations made in our custom CUDA implementation of the MVVS layer Sect. 2. The runtime results in this table does not include preprocessing time required to generate the FA and TOM features required for DDMReg, which can take several additional minutes [28], thus our results underestimate the true runtime improvement of our MVVS based model over DDMReg. Finally, all methods achieve a very small percentage of deformation field voxels with non-positive Jacobian, indicating that generated deformation fields are close to being diffeomorphic.

6 Conclusions

In this paper we presented a novel geometric deep neural network for registration of fODF images. We presented a registration model that respects the underlying geometry of fODF (manifold-valued) images. We also presented an efficient CUDA implementation of the vital manifold-valued image processing layer (MVC/MVVS) and introduced a novel Jacobian estimation and reorientation layer. Overall, our method is the first end-to-end trained model for fODF (dMRI) image registration. Finally, we presented several experiments demonstrating that our MVVS (MVC) for deformable registration achieve accuracy in par with classical methods but at a fraction of the processing time.

Acknowledgements. This research was in part supported by the NSF grant IIS-1724174, the NIH NINDS and NIA via RF1NS121099 to Vemuri and the MOST grant 110-2118-M-002-005-MY3 to Yang.

References

1. Avants, B.B., Tustison, N., Song, G.: Advanced normalization tools (ANTS). Insight j **2**(365), 1–35 (2009)
2. Balakrishnan, G., Zhao, A., Sabuncu, M.R., et al.: An unsupervised learning model for deformable medical image registration. In: Proceedings of the IEEE CVPR, pp. 9252–9260 (2018)
3. Barmpoutis, A., Vemuri, B.C.: Groupwise registration and atlas construction of 4th-order tensor fields using the R+ Riemannian metric. In: Yang, G.-Z., Hawkes, D., Rueckert, D., Noble, A., Taylor, C. (eds.) MICCAI 2009. LNCS, vol. 5761, pp. 640–647. Springer, Heidelberg (2009). https://doi.org/10.1007/978-3-642-04268-3_79
4. Bouza, J.J.: Manifold valued Volterra CNNs with apps. Med. Img. Ph.D. thesis, University of Florida (2022)
5. Bouza, J.J., Yang, C.-H., Vaillancourt, D., Vemuri, B.C.: A higher order manifold-valued convolutional neural network with applications to diffusion MRI processing. In: Feragen, A., Sommer, S., Schnabel, J., Nielsen, M. (eds.) IPMI 2021. LNCS, vol. 12729, pp. 304–317. Springer, Cham (2021). https://doi.org/10.1007/978-3-030-78191-0_24

6. Chakraborty, R., Bouza, J., Manton, J., Vemuri, B.C.: ManifoldNet: a deep neural network for manifold-valued data with applications. IEEE TPAMI **44**(2), 799–810 (2022)

7. Cheng, G., Vemuri, B.C., Carney, P.R., Mareci, T.H.: Non-rigid registration of high angular resolution diffusion images represented by Gaussian mixture fields. In: Yang, G.-Z., Hawkes, D., Rueckert, D., Noble, A., Taylor, C. (eds.) MICCAI 2009. LNCS, vol. 5761, pp. 190–197. Springer, Heidelberg (2009). https://doi.org/10.1007/978-3-642-04268-3_24

8. Dalca, A.V., Balakrishnan, G., Guttag, J., Sabuncu, M.R.: Unsupervised learning of probabilistic diffeomorphic registration for images and surfaces. Med. Img. Anal. **57**, 226–236 (2019)

9. Dhollander, T., Raffelt, D., Connelly, A.: Unsupervised 3-tissue response function estimation from single-shell or multi-shell diffusion MR data without a co-registered T1 image. In: ISMRM Workshop on Breaking the Barriers of Diffusion MRI, vol. 5. ISMRM (2016)

10. Henrik, J.G.: Density-based similarity in the registration of diffusion-weighted images. Ph.D. thesis, University of Copenhagen (2018)

11. Hoffmann, M., Billot, B., Greve, D.N., Iglesias, J.E., Fischl, B., Dalca, A.V.: SynthMorph: learning contrast-invariant registration without acquired images. arXiv preprint arXiv:2004.10282 (2020)

12. Jaderberg, M., Simonyan, K., Zisserman, A., et al.: Spatial transformer networks. In: Advances in NeurIPS, pp. 2017–2025 (2015)

13. Jeurissen, B., Tournier, J.D., Dhollander, T., Connelly, A., Sijbers, J.: Multi-tissue constrained spherical deconvolution for improved analysis of multi-shell diffusion MRI data. Neuroimage **103**, 411–426 (2014)

14. Jones, D., Lewis, D., Alexander, D., et al.: Spatial normalization and averaging of diffusion tensor MRI data sets. Neuroimage **17**(2), 592–617 (2002)

15. Klein, A., Andersson, J., Ardekani, B.A., et al.: Evaluation of 14 nonlinear deformation algorithms applied to human brain MRI registration. Neuroimage **46**(3), 786–802 (2009)

16. Li, H., Fan, Y.: Non-rigid image registration using fully convolutional networks with deep self-supervision. arXiv preprint arXiv:1709.00799 (2017)

17. Raffelt, D., Tournier, J.D., Crozier, S., Connelly, A., Salvado, O.: Reorientation of fiber orientation distributions using apodized point spread functions. MRM **67**(3), 844–855 (2012)

18. Raffelt, D., Tournier, J.D., Fripp, J., et al.: Symmetric diffeomorphic registration of fibre orientation distributions. Neuroimage **56**(3), 1171–1180 (2011)

19. Srivastava, A., Jermyn, I., Joshi, S.: Riemannian analysis of probability density functions with applications in vision. In: 2007 IEEE CVPR, pp. 1–8. IEEE (2007)

20. Tournier, J.D., Calamante, F., Connelly, A.: Robust determination of the fibre orientation distribution in diffusion MRI: non-negativity constrained super-resolved spherical deconvolution. Neuroimage **35**(4), 1459–1472 (2007)

21. Van Essen, D.C., Ugurbil, K., Auerbach, E., et al.: The human connectome project: a data acquisition perspective. Neuroimage **62**(4), 2222–2231 (2012)

22. Vemuri, B.C., Sun, J., Banerjee, M., et al.: A geometric framework for ensemble average propagator reconstruction from diffusion MRI. Med. Img. Anal. **57**, 89–105 (2019)

23. de Vos, B.D., Berendsen, F.F., Viergever, M.A., Staring, M., Išgum, I.: End-to-end unsupervised deformable image registration with a convolutional neural network. In: Cardoso, M.J., et al. (eds.) DLMIA/ML-CDS 2017. LNCS, vol. 10553, pp. 204–212. Springer, Cham (2017). https://doi.org/10.1007/978-3-319-67558-9_24

24. Wasserthal, J., Neher, P., Maier-Hein, K.H.: TractSeg-fast and accurate white matter tract segmentation. Neuroimage **183**, 239–253 (2018)

25. van Werkhoven, B., Maassen, J., Bal, H.E., Seinstra, F.J.: Optimizing convolution operations on GPUs using adaptive tiling. Future Gener. Comput. Syst. **30**(C), 14–26 (2014)

26. Yang, J., Shen, D., Davatzikos, C., Verma, R.: Diffusion tensor image registration using tensor geometry and orientation features. In: Metaxas, D., Axel, L., Fichtinger, G., Székely, G. (eds.) MICCAI 2008. LNCS, vol. 5242, pp. 905–913. Springer, Heidelberg (2008). https://doi.org/10.1007/978-3-540-85990-1_109

27. Yang, X., Kwitt, R., Styner, M., Niethammer, M.: Quicksilver: fast predictive image registration-a deep learning approach. Neuroimage **158**, 378–396 (2017)

28. Zhang, F., Wells, W.M., O'Donnell, L.J.: Deep diffusion MRI registration (DDMReg): a deep learning method for diffusion MRI registration. IEEE Trans. Med. Img. (2021)

29. Zhang, P., Niethammer, M., Shen, D., Yap, P.-T.: Large deformation diffeomorphic registration of diffusion-weighted images with explicit orientation optimization. In: Mori, K., Sakuma, I., Sato, Y., Barillot, C., Navab, N. (eds.) MICCAI 2013. LNCS, vol. 8150, pp. 27–34. Springer, Heidelberg (2013). https://doi.org/10.1007/978-3-642-40763-5_4

MetaMorph: Learning Metamorphic Image Transformation with Appearance Changes

Jian Wang[1](✉), Jiarui Xing[2], Jason Druzgal[3], William M. Wells III[4,5],
and Miaomiao Zhang[1,2]

[1] Computer Science, University of Virginia, Charlottesville, VA, USA
jw4hv@virginia.edu
[2] Electrical and Computer Engineering, University of Virginia,
Charlottesville, VA, USA
[3] Radiology and Medical Imaging, University of Virginia, Charlottesville, VA, USA
[4] Brigham and Women's Hospital, Harvard Medical School, Boston, MA, USA
[5] Computer Science and Artificial Intelligence Laboratory, MIT,
Cambridge, MA, USA

Abstract. This paper presents a novel predictive model, MetaMorph, for metamorphic registration of images with appearance changes (i.e., caused by brain tumors). In contrast to previous learning-based registration methods that have little or no control over appearance-changes, our model introduces a new regularization that can effectively suppress the negative effects of appearance changing areas. In particular, we develop a piecewise regularization on the tangent space of diffeomorphic transformations (also known as initial velocity fields) via learned segmentation maps of abnormal regions. The geometric transformation and appearance changes are treated as joint tasks that are mutually beneficial. Our model MetaMorph is more robust and accurate when searching for an optimal registration solution under the guidance of segmentation, which in turn improves the segmentation performance by providing appropriately augmented training labels. We validate MetaMorph on real 3D human brain tumor magnetic resonance imaging (MRI) scans. Experimental results show that our model outperforms the state-of-the-art learning-based registration models. The proposed MetaMorph has great potential in various image-guided clinical interventions, e.g., real-time image-guided navigation systems for tumor removal surgery.

1 Introduction

Deformable image registration is an important tool in a variety of medical image analysis tasks, such as multi-modality image alignment [12,18,25], statistical analysis for population image studies [26,32,35], atlas-guided image segmentation or classification [27,30,33], and object tracking with anomaly detection [11,24]. In many clinical applications, it is desirable that the estimated transformations are diffeomorphisms (i.e., bijective, smooth, and inverse smooth mappings) because they produce anatomically plausible images [7]. Despite recent achievements in treating the problem of diffeomorphic image registration as a fast learning task, current approaches oftentimes have an assumption that the topology of objects presented in images is intact [6,10,17,31]. Existing algorithms fail badly in cases where appearance changes occur (e.g., missing data caused by pathology, such as tumors, myocardial scars, multiple sclerosis, and etc.) because they have little to no control over these unknown variables.

To address this issue, a few algorithms of image metamorphosis have been developed to incorporate the modeling of appearance changes in registration functions [8,14,16,21,23]. Existing metamorphic image registration methods mainly fall into two categories: (i) exclude appearance changes via manually delineated segmentations of abnormal regions [21,23], and (ii) treat the appearance changes as unknown variables estimated out from images [8,14]. These approaches either heavily depend on manually segmented labels of 3D volumetric data that are time and labor-consuming, or struggle with balancing between the effects of appearance vs. geometric changes. A recent work [8] has developed a metamorphic autoencoder that estimates the deformation and appearance variations by decoupling the geometric and appearance representations in latent spaces. However, such a model is highly sensitive to parameter-tuning due to its difficulty in differentiating changes caused by geometric transformations vs. appearances.

In this paper, we develop a novel learning-based model of metamorphic image registration, named as MetaMorph, that provides more robust and accurate registration results in images with appearance changes. In contrast to previous approaches [8,14,21,23], we incorporate a new *appearance-aware regularization* in the network loss function that enforces a piecewise constraint on geometric transformation fields. Such a constraint will be learned simultaneously from a jointly optimized segmentation task. In addition, we effectively augment the segmentation labels by utilizing the learned transformations in the training process. This not only substantially improves the segmentation performance, but also reduces the requirement for massive ground truth segmentation labels. The main contributions of our proposed MetaMorph are summarized in three folds:

- To the best of our knowledge, MetaMorph is the first predictive registration algorithm that utilizes jointly learned segmentation maps to model appearance changes.
- MetaMorph learns a new appearance-aware regularization that piecewisely constrains the variations of image intensities caused by geometric transformations separately from appearance changes.
- The joint learning scheme of MetaMorph maximizes the mutual benefits of metamorphic image registration and segmentation.

To demonstrate the effectiveness of our model, we validate MetaMorph on real 3D human brain tumor MRIs. Experimental results show that MetaMorph outperforms the state-of-the-art learning-based registration models [6,8] with substantially increased accuracy. The developed MetaMorph has great potential in various image-guided clinical interventions, e.g., real-time image-guided navigation systems for tumor removal surgery.

2   Background: Diffeomorphic Image Registration

In this section, we briefly review the concept of the diffeomorphic image registration in the setting of large deformation diffeomorphic metric mapping (LDDMM) with a geodesic shooting algorithm [7,20,29].

Let S be the source image and T be the target image defined on a d-dimensional torus domain $\Gamma = \mathbb{R}^d/\mathbb{Z}^d$ ($S(x), T(x) : \Gamma \to \mathbb{R}$). The problem of diffeomorphic image registration is to find the geodesic (a.k.a. shortest path) to generate time-varying diffeomorphisms $\{\psi_t(x)\} : t \in [0,1]$ such that $S \circ \psi_1$ is similar to T, where $\circ$ is an interpolation operation that deforms S by the smooth deformation field ψ_1. This is typically formulated as an optimization problem by minimizing an explicit energy function over the transformation fields ψ_t as

$$E(v_t) = \text{Dist}[S \circ \psi_1(v_t), T] + \text{Reg}[\psi_t(v_t)], \tag{1}$$

where the distance function $\text{Dist}(\cdot, \cdot)$ measures the image dissimilarity between the source and the deformed image. Commonly used distance functions include a sum-of-squared difference of image intensities [7], normalized cross correlation [4], and mutual information [34,36]. The regularization term $\text{Reg}(\cdot)$ is a constraint that enforces the spatial smoothness of transformations, arising from a distance metric on the tangent space V of diffeomorphisms, i.e., an integral over the norm of time-dependent velocity fields $\{v_t(x)\} \in V$,

$$\text{Reg}(\psi_t) = \int_0^1 (Lv_t, v_t)\, dt, \quad \text{with} \quad \frac{d\psi_t}{dt} = -D\psi_t \cdot v_t, \tag{2}$$

where $L : V \to V^*$ is a symmetric, positive-definite differential operator that maps a tangent vector $v_t \in V$ into its dual space as a momentum vector $m_t \in V^*$. We typically write $m_t = Lv_t$, or $v_t = Km_t$, with K being an inverse operator of L. The notation $(\cdot, \cdot)$ denotes the pairing of a momentum vector with a tangent vector, which is similar to an inner product. Here, the operator D denotes a Jacobian matrix and $\cdot$ represents element-wise matrix multiplication.

A geodesic curve with a fixed endpoint is characterized by an extremum of the energy function (2) that satisfies the Euler-Poincaré differential (EPDiff) equation [2,20],

$$\frac{\partial v_t}{\partial t} = -K\left[(Dv_t)^T \cdot m_t + Dm_t \cdot v_t + m_t \cdot \text{div}\, v_t\right], \tag{3}$$

where div is the divergence. This process in Eq. (3) is known as *geodesic shooting*, stating that the geodesic path $\{\psi_t\}$ can be uniquely determined by integrating a given initial velocity v_0 forward in time by using the rule (3).

Therefore, we rewrite the optimization of Eq. (1) equivalently as

$$E(v_0) = \text{Dist}[S \circ \psi_1(v_0), T] + (Lv_0, v_0), \text{ s.t. Eq. (2)\&Eq. (3).} \tag{4}$$

3   Our Model: MetaMorph

The objective function of diffeomorphic image registration in Eq. (4) works well under the condition that images are ideally of good quality with preserved topology. This assumption breaks when corruptions such as appearance changes or occlusions occur. In this section, we first define an objective function of the metamorphic image registration that considers the modeling of appearance changes. An appearance-aware regularization is developed to effectively suppress the negative influences of appearance changes in typical diffeomorphic image registration algorithms. We then develop a joint learning framework that includes i) a segmentation network for appearance change detection, and ii) a metamorphic registration network incorporating the newly formulated objective function as part of the network loss.

Appearance-Aware Regularization. The purpose of metamorphic image registration is to find an optimal transformation $\psi(v_0, \delta)$ that is composed of two variables: the optimal initial velocity field v_0, and the appearance change δ. A recent work proposed to learn these variables via disentangled latent representations in an encoder-decoder neural network [8]. However, it is extremely challenging for this algorithm to differentiate the variations of image intensities caused by geometric transformations from appearance changes since they unavoidably compensate for each other. The ambiguity introduced by optimizing two compensating variables without any guidance fails to search for accurate registration solutions. Additionally, this makes the algorithm highly sensitive to network parameters with an increased risk of poor convergence. To alleviate this issue, we introduce an appearance-aware regularization in the registration framework, guided by learned segmentations of the appearance-changing areas.

Assume U is a union of the learned segmentations of appearance-changing areas from the source image S and the target image T. Analogous to Eq. (4), we define the appearance-aware regularization $\mathbf{Reg}^*(\cdot)$ in the space of initial velocity fields. To suppress the effects of appearance variations, we piecewisely constrain the initial velocity fields through a segmentation indicator, i.e.,

$$\mathbf{Reg}^*(v_0) = (L(v_0 \odot (1 - U)), v_0 \odot (1 - U)), \text{ s.t. Eq. (3)}, \qquad (5)$$

where $\odot$ represents an element-wise multiplication between a vector field and a scalar field. For the purpose of notation simplicity, we define $\hat{v}_0 \triangleq v_0 \odot (1 - U)$ in the following sections.

With the newly defined regularization in Eq. (5), we arrive at the objective function of metamorphic image registration as

$$\mathbf{E}^*[\hat{\psi}(\hat{v}_0)] = \mathbf{Dist}^*[\hat{S} \circ \hat{\psi}_1(\hat{v}_0), \hat{T}] + \mathbf{Reg}^*(\hat{v}_0), \qquad (6)$$

where $\hat{S}$ and $\hat{T}$ denotes the source and target images with appearance changes masked out, i.e., $\hat{S} = S \odot (1 - U)$, and $\hat{T} = T \odot (1 - U)$. Here, the $\mathbf{Dist}^*[\cdot, \cdot]$ is the image dissimilarity term that measures the dissimilarity between the consistent area between the deformed image and target.

3.1   Predictive Metamorphic Image Registration

We develop a deep learning framework to jointly learn the segmentation for appearance change and the masked-out velocity field $\hat{v}_0$. An overview of our proposed MetaMorph architecture is shown in Fig. 1.

Appearance change can be masked by a fixed foreground segmentation via pre-running image segmentation algorithms [21,23]. However, performing manual annotations of segmentation labels is time and labor-consuming. In this work, instead of using a fixed mask, we treat the appearance change as a variable from the segmentation network and jointly optimize with the optimal registration solution. We utilize an encoder-decoder based neural network to learn the segmentation masks and then apply them to the associate image pairs for masking out the appearance change. Although we adopt UNet-based architecture for segmentation in this work [28], other networks such as recurrent residual neural networks [1], transformer-based networks [9,15] can also be easily plugged into the proposed method.

With the developed segmentation network, now we are ready to formulate the loss function of MetaMorph,

$$\ell = \mathbf{Dist}^*[\hat{S} \circ \hat{\psi}_1(\hat{v}_0), \hat{T}] + \mathbf{Reg}^*(\hat{v}_0) + \gamma \cdot \ell_{seg}, \;\; \text{s.t. Eq. (5).} \qquad (7)$$

Here, γ is a weighting parameter that balances the segmentation and registration loss, ℓ_{seg} is a segmentation loss that maximizes the Sørensen-Dice coefficient [13] between ground truth y and the predicted $\hat{y}$,

$$\ell_{seg} = 1 - \text{Dice}(y, \hat{y}), \qquad (8)$$

where $\text{Dice}(y, \hat{y}) = 2(|y| \cap |\hat{y}|)/(|y| + |\hat{y}|)$.

We adopt an approximated region-based mutual information (RMI) [36], which is a broadly-used distance metric for images from different domains. For simplicity, we let $\hat{S}_\psi$ denote the deformed image. Let $f(\hat{S}_\psi)$ and $f(\hat{T})$ denote the probability density functions for the deformed image and target respectively, and their joint probability density function is $f(\hat{S}_\psi, \hat{T})$. The image dissimilarity with RMI can be formulated as

$$\mathbf{Dist}^*[\hat{S}_\psi, \hat{T}] = \text{RMI}(\hat{S}_\psi, \hat{T}) = \int_{\hat{S}_\psi} \int_{\hat{T}} f(\hat{S}_\psi, \hat{T}) \log \frac{f(\hat{S}_\psi, \hat{T})}{f(\hat{S}_\psi)f(\hat{T})}$$

$$\approx l_{ce}(\hat{S}_\psi, \hat{T}) - \frac{1}{B} \sum_{b=1}^{B} I_b(\hat{T}; \hat{S}_\psi), \qquad (9)$$

where $L_{ce}(\cdot, \cdot)$ is a cross entropy loss between two images. The $I_b(\cdot; \cdot)$ is a batch-wise lower bound that $I_b(\hat{T}; \hat{S}_\psi) = \frac{1}{2} \log[\det(\Sigma_{\hat{T}|\hat{S}_\psi})]$, where $\Sigma_{\hat{T}|\hat{S}_\psi}$ is the posterior covariance matrix of $\hat{T}$ (a symmetric positive semi-definite matrix), given $\hat{S}_\psi$. Here B denotes the number of images in a mini-batch b. Please refer to [36] for more derivation details.

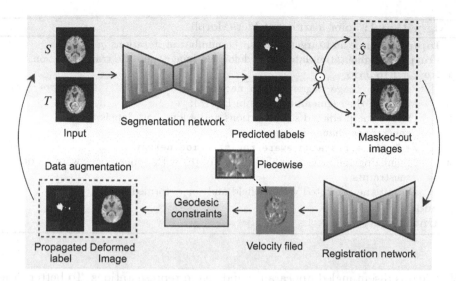

Fig. 1. An illustration of the network architecture for MetaMorph. Top left to right: input a pair of images into a segmentation network, and apply predicted labels onto images to mask out the appearance change. Bottom right to left: input a pair of images (with masked-out appearance change) to the registration network and predict a piecewise velocity field, integrate geodesic constraints, and produce a deformed image and transformation-propagated segmentation. The deformed images and labels are circulated into the segmentation network as augmented data.

We develop an alternating optimization scheme [22] to minimize the network loss defined in Eq. (7). All network parameters are optimized jointly by alternating between the training of segmentation and image registration. A summary of our joint learning of MetaMorph is in Algorithm 1.

4   Experimental Evaluation

To demonstrate the effectiveness of the proposed model, we compare both segmentation and registration tasks with state-of-the-arts.

Data. For 3D brain tumor MRI scans with tumor segmentation labels, we include 100 public T1-weighted brain scans of different subjects from Brain Tumor Segmentation (BraTS) [5,19] challenge 2021. We also include 28 landmarks (16 for brain ventricle and 12 for corpus callosum) that are annotated by clinicians to better evaluate the image registration performance. All MRIs are $155 \times 240 \times 240$, $1.25\,\mathrm{mm}^3$ isotropic voxels. As a preprocessing step, we run affine registration, intensity normalization, and bias field correction on all images.

Experiments. We compare our metamorphic image registration method with two registration baselines, an unsupervised predictive diffeomorphic registration method (VoxelMorph as VM) [6], and a metamorphic autoencoder (MAE) [8]

Algorithm 1: Joint learning of MetaMorph.

Input   : Source and target images, the number of iterations q.
Output: Segmentation labels, the deformed image, and the transformation.

1 **for** $i = 1$ to q **do**

　　/* Train image segmentation network   */

2　　Minimize the segmentation loss in Eq. (8);

3　　Output the predicted segmentations and adopt both labels to mask
　　　appearance change in images;

　　/* Train appearance-aware registration network   */

4　　Minimizing the metamorphic loss in Eq. (6) with appearance-aware geodesic
　　　constraints;

5　　Output the predicted velocity field and the deformed image;

6 **end**

7 **Until** convergence

---

that learns disentangled appearance and shape representations. To better visualize the deformations, we show predicted transformation grids and deformed images with transformation-propagated landmarks for all methods. Quantitatively, we compute the L_2 distance of landmarks as registration error between the propagated and the target frames over 60 pairs.

We evaluate the brain tumor segmentation via computing Dice score [13] by comparing MetaMorph with three segmentation backbones, U-Net architecture [28], U-Net based on recurrent residual convolutional neural network (R2-Unet) [1], and transformer-based Unet (UnetR) [15]. We also show the performance of MetaMorph by replacing the segmentation module in our model with all backbones (named MetaMorph:Unet, MetaMorph:R2-Unet, and MetaMorph:UnetR). We visualize the predicted segmentations overlaid with testing images across all methods.

Parameter Settings. We set parameter $\alpha = 3$ for the operator L, the number of time steps for Euler integration in EPDiff (Eq. (3)) as 10. We set the weight parameter $\gamma = 0.5$ and the batch size as 4. We use an adaptive cosine annealing learning rate scheduler that starts from an initial value at $\eta = 5e - 4$ for network training. We run all models for 100 epochs with Adam optimizer and save the networks with the best validation performance. The training and prediction procedure of all learning-based methods are performed on two Nvidia GTX 2070Ti GPUs. We run five-fold cross validation and split the images by using 70% as training images, 20% as validation images, and 10% as testing images.

Results. Figure 2 visualizes the image registration prediction of two 3D brain MRIs of study across all methods. It shows MetaMorph significantly outperforms both VM and MAE. General diffeomorphic registration models (e.g., VM) without an appearance-control mechanism may fail and produce less satisfied deformed images without sufficient deformations. MAE offers accurate deformations to a certain level while it produces artifacts. By excluding the appearance change, MetaMorph more accurately deforms all regions (e.g., ventricles and

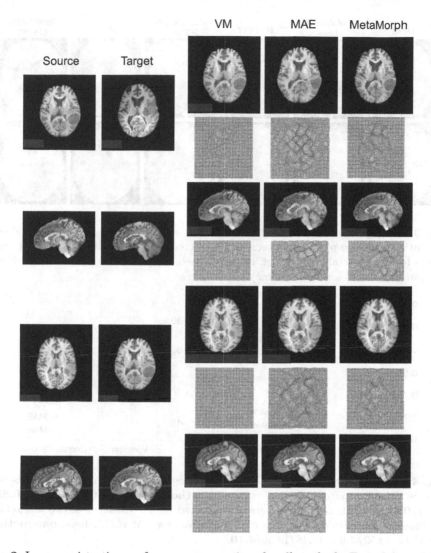

Fig. 2. Image registration performance comparison for all methods. From left to right, source, target, deformed images by VoxelMorph (VM), metamorphic autoencoder (MAE), and our method. All images are overlaid with annotated landmarks (red circle for ventricle and blue cross for corpus callosum). (Color figure online)

corpus callosum). It also shows that our propagated landmarks align best with the target.

Figure 3 shows two examples of image segmentation performance comparison for all methods. It indicates that MetaMorph-based models predict better segmentation labels (closer to ground truth) than original backbones. The predicted labels by MetaMorph have slightly better segmentations of the brain tumor boundary. This is because we use deformed images and labels that are

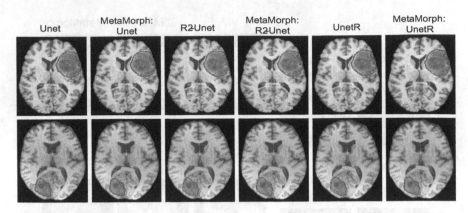

Fig. 3. Image segmentation visualization for all methods. Left to right: overlaid segmentation map comparison between the predicted label (red) and the ground truth (blue) for Unet, MetaMorph: Unet, R2-Unet, MetaMorph: R2-Unet, UnetR and MetaMorph: UnetR. (Color figure online)

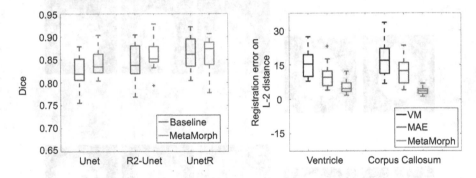

Fig. 4. Left: Dice comparison on brain tumor segmentation across all methods over images. The means of baseline vs. **our method** are $0.815/\mathbf{0.834}$, $0.835/\mathbf{0.856}$, $0.861/\mathbf{0.874}$; Right: registration error (computed on L_2 distance) of two anatomical landmarks for 60 brain pairs. The means of errors for VM vs. MAE vs. **our method** are $15.02/10.53/\mathbf{4.64}$, $16.48/13.59/\mathbf{4.10}$.

produced by a joint registration framework as augmented data for each subject; thus learning a broader spectrum for appearance variation in data and offering more accurate prediction when new testing data arrives.

Figure 4 (left panel) statistical reports the Dice coefficient comparison. It shows that MetaMorph consistently achieves a higher segmentation accuracy than backbones. Transformer-based methods (UnetR-based) produce the highest Dice for all methods. Figure 4 (right panel) reports the landmark-based registration error between the target image and the deformed image. MetaMorph outperforms other methods with the lowest error, indicating our proposed method finishes the metamorphic image registration task with higher accuracy.

5   Conclusion

We present a predictive metamorphic image registration model, MetaMorph, via deep neural networks in this paper. Different from existing models that have limited control over appearance change, we develop a joint learning framework that adopts a segmentation module to accurately guide the registration network to learn diffeomorphic transformation fields. The developed segmentation module maximally excludes the disadvantageous effect caused by appearance change for learned deformations; thus enabling more precise correspondence alignment between deformed and target frames. Experimental results on 3D brain MRIs with real tumors show that our proposed framework yields a better registration as well as a segmentation model. While our algorithm is presented in the setting of LDDMM with geodesic shooting, the theoretical development is generic to other deformation models, e.g., stationary velocity fields [3]. Our model has great clinical potential on solving one of the most challenging registration problems, e.g., real-time brain shift estimation between preoperative and intraoperative MRI scans with missing data values. Interesting future works of MetaMorph will be i) building a probabilistic model to quantify the registration uncertainty along the boundary of tumor areas and ii) extending the proposed method to more advanced clinical scenarios that appearance changes are difficult to detect, e.g., real-time automated image registration for ultrasound images.

Acknowledgments. This work was supported by NSF CAREER Grant 2239977.

References

1. Alom, M.Z., Hasan, M., Yakopcic, C., Taha, T.M., Asari, V.K.: Recurrent residual convolutional neural network based on U-Net (R2U-Net) for medical image segmentation. arXiv preprint arXiv:1802.06955 (2018)
2. Arnold, V.: Sur la géométrie différentielle des groupes de lie de dimension infinie et ses applications à l'hydrodynamique des fluides parfaits. In: Annales de l'institut Fourier, vol. 16, pp. 319–361 (1966)
3. Arsigny, V., Commowick, O., Pennec, X., Ayache, N.: A log-Euclidean framework for statistics on diffeomorphisms. In: Larsen, R., Nielsen, M., Sporring, J. (eds.) MICCAI 2006. LNCS, vol. 4190, pp. 924–931. Springer, Heidelberg (2006). https://doi.org/10.1007/11866565_113
4. Avants, B.B., Epstein, C.L., Grossman, M., Gee, J.C.: Symmetric diffeomorphic image registration with cross-correlation: evaluating automated labeling of elderly and neurodegenerative brain. Med. Image Anal. **12**(1), 26–41 (2008)
5. Baid, U., et al.: The RSNA-ASNR-MICCAI brats 2021 benchmark on brain tumor segmentation and radiogenomic classification. arXiv preprint arXiv:2107.02314 (2021)
6. Balakrishnan, G., Zhao, A., Sabuncu, M.R., Guttag, J., Dalca, A.V.: VoxelMorph: a learning framework for deformable medical image registration. IEEE Trans. Med. Imaging **38**(8), 1788–1800 (2019)
7. Beg, M.F., Miller, M.I., Trouvé, A., Younes, L.: Computing large deformation metric mappings via geodesic flows of diffeomorphisms. Int. J. Comput. Vis. **61**(2), 139–157 (2005)

8. Bône, A., Vernhet, P., Colliot, O., Durrleman, S.: Learning joint shape and appearance representations with metamorphic auto-encoders. In: Martel, A.L., et al. (eds.) MICCAI 2020. LNCS, vol. 12261, pp. 202–211. Springer, Cham (2020). https://doi.org/10.1007/978-3-030-59710-8_20

9. Chen, J., et al.: TransUNet: transformers make strong encoders for medical image segmentation. arXiv preprint arXiv:2102.04306 (2021)

10. Chen, J., Frey, E.C., He, Y., Segars, W.P., Li, Y., Du, Y.: TransMorph: transformer for unsupervised medical image registration. Med. Image Anal. **82**, 102615 (2022)

11. Chen, M., Kanade, T., Pomerleau, D., Rowley, H.A.: Anomaly detection through registration. Pattern Recogn. **32**(1), 113–128 (1999)

12. Chung, A.C.S., Wells, W.M., Norbash, A., Grimson, W.E.L.: Multi-modal image registration by minimising Kullback-Leibler distance. In: Dohi, T., Kikinis, R. (eds.) MICCAI 2002. LNCS, vol. 2489, pp. 525–532. Springer, Heidelberg (2002). https://doi.org/10.1007/3-540-45787-9_66

13. Dice, L.R.: Measures of the amount of ecologic association between species. Ecology **26**(3), 297–302 (1945)

14. François, A., Gori, P., Glaunès, J.: Metamorphic image registration using a semi-Lagrangian scheme. In: Nielsen, F., Barbaresco, F. (eds.) GSI 2021. LNCS, vol. 12829, pp. 781–788. Springer, Cham (2021). https://doi.org/10.1007/978-3-030-80209-7_84

15. Hatamizadeh, A., et al.: UNETR: transformers for 3D medical image segmentation. In: Proceedings of the IEEE/CVF Winter Conference on Applications of Computer Vision, pp. 574–584 (2022)

16. Holm, D., Trouvé, A., Younes, L.: The Euler-Poincaré theory of metamorphosis. Q. Appl. Math. **67**(4), 661–685 (2009)

17. Kim, B., Han, I., Ye, J.C.: DiffuseMorph: unsupervised deformable image registration along continuous trajectory using diffusion models. arXiv preprint arXiv:2112.05149 (2021)

18. Maes, F., Collignon, A., Vandermeulen, D., Marchal, G., Suetens, P.: Multimodality image registration by maximization of mutual information. IEEE Trans. Med. Imaging **16**(2), 187–198 (1997)

19. Menze, B.H., et al.: The multimodal brain tumor image segmentation benchmark (BRATS). IEEE Trans. Med. Imaging **34**(10), 1993–2024 (2014)

20. Miller, M.I., Trouvé, A., Younes, L.: Geodesic shooting for computational anatomy. J. Math. Imaging Vis. **24**(2), 209–228 (2006)

21. Niethammer, M., et al.: Geometric metamorphosis. In: Fichtinger, G., Martel, A., Peters, T. (eds.) MICCAI 2011. LNCS, vol. 6892, pp. 639–646. Springer, Heidelberg (2011). https://doi.org/10.1007/978-3-642-23629-7_78

22. Nocedal, J., Wright, S.J.: Numerical Optimization. Springer, Heidelberg (1999)

23. Patenaude, B., Smith, S.M., Kennedy, D.N., Jenkinson, M.: A Bayesian model of shape and appearance for subcortical brain segmentation. Neuroimage **56**(3), 907–922 (2011)

24. Prastawa, M., Bullitt, E., Ho, S., Gerig, G.: A brain tumor segmentation framework based on outlier detection. Med. Image Anal. **8**(3), 275–283 (2004)

25. Qin, C., Shi, B., Liao, R., Mansi, T., Rueckert, D., Kamen, A.: Unsupervised deformable registration for multi-modal images via disentangled representations. In: Chung, A.C.S., Gee, J.C., Yushkevich, P.A., Bao, S. (eds.) IPMI 2019. LNCS, vol. 11492, pp. 249–261. Springer, Cham (2019). https://doi.org/10.1007/978-3-030-20351-1_19

26. Rao, A., Aljabar, P., Rueckert, D.: Hierarchical statistical shape analysis and prediction of sub-cortical brain structures. Med. Image Anal. **12**(1), 55–68 (2008)

27. Riklin-Raviv, T., Van Leemput, K., Menze, B.H., Wells, W.M., III., Golland, P.: Segmentation of image ensembles via latent atlases. Med. Image Anal. **14**(5), 654–665 (2010)
28. Ronneberger, O., Fischer, P., Brox, T.: U-Net: convolutional networks for biomedical image segmentation. In: Navab, N., Hornegger, J., Wells, W.M., Frangi, A.F. (eds.) MICCAI 2015. LNCS, vol. 9351, pp. 234–241. Springer, Cham (2015). https://doi.org/10.1007/978-3-319-24574-4_28
29. Vialard, F.X., Risser, L., Rueckert, D., Cotter, C.J.: Diffeomorphic 3D image registration via geodesic shooting using an efficient adjoint calculation. Int. J. Comput. Vis. **97**(2), 229–241 (2012)
30. Wachinger, C., Golland, P.: Atlas-based under-segmentation. In: Golland, P., Hata, N., Barillot, C., Hornegger, J., Howe, R. (eds.) MICCAI 2014. LNCS, vol. 8673, pp. 315–322. Springer, Cham (2014). https://doi.org/10.1007/978-3-319-10404-1_40
31. Wang, J., Zhang, M.: DeepFLASH: an efficient network for learning-based medical image registration. In: Proceedings of the IEEE/CVF Conference on Computer Vision and Pattern Recognition, pp. 4444–4452 (2020)
32. Wang, J., Zhang, M.: Bayesian atlas building with hierarchical priors for subject-specific regularization. In: de Bruijne, M., et al. (eds.) MICCAI 2021. LNCS, vol. 12904, pp. 76–86. Springer, Cham (2021). https://doi.org/10.1007/978-3-030-87202-1_8
33. Wang, J., Zhang, M.: Geo-SIC: learning deformable geometric shapes in deep image classifiers. In: The Conference on Neural Information Processing Systems (2022)
34. Wells, W.M., III., Viola, P., Atsumi, H., Nakajima, S., Kikinis, R.: Multi-modal volume registration by maximization of mutual information. Med. Image Anal. **1**(1), 35–51 (1996)
35. Zhang, M., Wells, W.M., Golland, P.: Low-dimensional statistics of anatomical variability via compact representation of image deformations. In: Ourselin, S., Joskowicz, L., Sabuncu, M.R., Unal, G., Wells, W. (eds.) MICCAI 2016. LNCS, vol. 9902, pp. 166–173. Springer, Cham (2016). https://doi.org/10.1007/978-3-319-46726-9_20
36. Zhao, S., Wang, Y., Yang, Z., Cai, D.: Region mutual information loss for semantic segmentation. In: Advances in Neural Information Processing Systems, vol. 32 (2019)

NeurEPDiff: Neural Operators to Predict Geodesics in Deformation Spaces

Nian Wu[1,2(✉)] and Miaomiao Zhang[2,3]

[1] Computer Science, East China Normal University, Shanghai, China
nellie_nw@stu.ecnu.edu.cn
[2] Electrical and Computer Engineering, University of Virginia, Charlottesville, VA, USA
[3] Computer Science, University of Virginia, Charlottesville, VA, USA

Abstract. This paper presents NeurEPDiff, a novel network to fast predict the geodesics in deformation spaces generated by a well known Euler-Poincaré differential equation (EPDiff). To achieve this, we develop a neural operator that for the first time learns the evolving trajectory of geodesic deformations parameterized in the tangent space of diffeomorphisms (a.k.a velocity fields). In contrast to previous methods that purely fit the training images, our proposed NeurEPDiff learns a nonlinear mapping function between the time-dependent velocity fields. A composition of integral operators and smooth activation functions is formulated in each layer of NeurEPDiff to effectively approximate such mappings. The fact that NeurEPDiff is able to rapidly provide the numerical solution of EPDiff (given any initial condition) results in a significantly reduced computational cost of geodesic shooting of diffeomorphisms in a high-dimensional image space. Additionally, the properties of discretiztion/resolution-invariant of NeurEPDiff make its performance generalizable to multiple image resolutions after being trained offline. We demonstrate the effectiveness of NeurEPDiff in registering two image datasets: 2D synthetic data and 3D brain resonance imaging (MRI). The registration accuracy and computational efficiency are compared with the state-of-the-art diffeomophic registration algorithms with geodesic shooting.

1 Introduction

Deformable image registration is a fundamental tool in medical image analysis, for example, computational anatomy and shape analysis [20,37], statistical analysis of groupwise images [21,31], and template-based image segmentation [4,9]. Among many applications, it is ideal that the transformation is diffeomorphic (i.e., bijective, smooth, and invertible smooth) to maintain an intact topology of objects presented in images. A bountiful literature has studied various parameterizations of diffeomorphisms, including stationary velocity fields [2,3], B-spline free-form deformations [26,27], and the large deformation diffeomorphic metric mapping (LDDMM) [7,29]. In this paper, we focus on LDDMM, which provides a rigorous mathematical definition of distance metrics on the space of diffeomorphisms. Such distance metrics play an important role in deformation-based shape analysis of images, such as geometric shape regression [13,20], longitudinal shape analysis [28], and group shape comparisons [18,22].

A. Frangi et al. (Eds.): IPMI 2023, LNCS 13939, pp. 588–600, 2023.
https://doi.org/10.1007/978-3-031-34048-2_45

Despite the advantages of LDDMM [7], the substantial time and computational cost involved in searching for the optimal diffeomorphism on high-dimensional image grids limits its applicability in real-world applications. To alleviate this problem, a geodesic shooting algorithm [29] was developed to reformulate the optimization of LDDMM (originally over a sequence of time-dependent diffeomorphisms) equivalently in the tangent space of diffeomorphisms at time point zero (a.k.a., initial velocity fields). The geodesic is then uniquely determined by integrating a given initial velocity field via the Euler-Poincaré differential equation (EPDiff) [1, 19]. This geodesic shooting algorithm demonstrated a faster convergence and avoided having to store the entire time-varying velocity fields from one iterative search to the next. A recent work FLASH has further reduced the computational cost of geodesic shooting by reparameterizing the velocity fields in a low-dimensional bandlimited space [35, 36]. The numerical solution to EPDiff was rapidly computed with much fewer parameters in a frequency domain. In spite of a significant speedup, it still takes FLASH minutes to compute when the dimension of the bandlimited space grows higher.

In this paper, we present a novel method, NeurEPDiff, that can fast predict the geodesics in deformation spaces generated by EPDiff. Inspired by the recent achievements on learning-based partial/ordinary differential equation solvers [16, 17], we develop a neural operator to learn the evolving functions/mappings of geodesics in the space of diffeomorphisms parameterized by bandlimited initial velocity fields. In contrast to current deep learning-based approaches that are tied to the resolution and discretization of training images [6, 8, 24, 30, 33], our proposed method shows consistent performances on different scales of image resolutions without the need of re-training. We also define a composition of integral operators and smooth activation functions in each layer of the NeurEPDiff network to effectively approximate the mapping functions of EPDiff over time. To summarize, our main contributions are threefold:

(i) We are the first to develop a neural operator that fast predicts a numerical solution to the EPDiff [1, 19], which is a critical step to generate geodesics of diffeomorphisms.

(ii) A composition of integral operators and smooth activation functions is defined to approximate the mapping functions of EPDiff characterized in the low-dimensional bandlimited space. This makes the training and inference of our proposed NeurEPDiff much more efficient in high-dimensional image spaces.

(iii) The performance of NeurEPDiff is invariant to image resolutions. That is to say, once it is trained, this operator can be tested on multiple image resolutions without the need of being re-trained.

To demonstrate the effectiveness of our proposed model in diffeomorphic image registration, we run experiments on both 2D synthetic data and 3D real brain MRIs. We first show that a trained NeurEPDiff is able to consistently predict the diffeomorphic transformations at multiple different image resolutions. We then compare the registration accuracy (via registration-based image segmentation) and computational efficiency of NeurEPDiff vs. the state-of-the-art LDDMM with geodesic shooting [29, 35, 36]. Experimental results show that our method outperforms the baselines in term of time consumption, while with a comparable registration accuracy.

2 Background: Geodesics of Diffeomorphisms via EPDiff

In this section, we first briefly review the basic concepts of geodesic shooting, which generates geodesics of diffeomorphisms via EPDiff [29,34]. We then show how to optimize the problem of diffeomorphic image registration in the setting of LDDMM [7] with geodesic shooting [29,34].

Geodesic Shooting via EPDiff. Let $\text{Diff}^\infty(\Omega)$ denote the space of smooth diffeomorphisms on an image domain Ω. The tangent space of diffeomorphisms is the space $V = \mathfrak{X}^\infty(T\Omega)$ of smooth vector fields on Ω. Consider a time-varying velocity field, $\{v_t\} : [0,\tau] \to V$, we can generate diffeomorphisms $\{\phi_t\}$ as a solution to the equation

$$\frac{d\phi_t}{dt} = v_t(\phi_t),\ t \in [0,\tau]. \tag{1}$$

The geodesic shooting algorithm [23] states that a geodesic path of the diffeomorphisms (in Eq. (1)) is uniquely determined by integrating a well-studied EPDiff [1,19] with an initial condition. That is, given an initial velocity, $v_0 \in V$, at $t = 0$, a geodesic path $t \mapsto \phi_t \in \text{Diff}^\infty(\Omega)$ in the space of diffeomorphisms can be computed by forward shooting the EPDiff equation

$$\frac{\partial v_t}{\partial t} = -K\left[(Dv)^T m_t + Dm_t\, v_t + m_t \operatorname{div} v_t\right], \tag{2}$$

where D denotes the Jacobian matrix. Here K is an inverse operator of $L : V \to V^*$, which is a positive-definite differential operator that maps a tangent vector $v \in V$ into the dual space $m \in V^*$. This paper employs a commonly used Laplacian operator $L = (-\alpha\Delta + e)^c$, where α is a positive weight parameter, e denotes an identity matrix, and c is a smoothness parameters. A larger value of α and c indicates more smoothness.

A recent model FLASH has later developed a Fourier variant of the EPDiff and demonstrated that the numerical solution to the original EPDiff can be efficiently computed in a low-dimensional bandlimited space with dramatically reduced computational cost [35,36]. Let $\widetilde{\text{Diff}}(\Omega)$ and $\tilde{V}$ denote the space of Fourier representations of diffeomorphisms and velocity fields respectively. The equation of EPDiff is reformulated in a bandlimited space as

$$\frac{\partial \tilde{v}_t}{\partial t} = -\tilde{K}\left[(\tilde{\mathcal{D}}\tilde{v}_t)^T \star \tilde{m}_t + \tilde{\nabla} \cdot (\tilde{m}_t \otimes \tilde{v}_t)\right], \tag{3}$$

where $\tilde{v}_t$ and $\tilde{m}_t$ are the Fourier transforms of velocity fields v_t and momentum m_t respectively. The $\tilde{\mathcal{D}}$ represents Fourier frequencies of a Jacobian matrix D with central difference approximation, $\star$ is the truncated matrix-vector field auto-correlation with zero-padding, and $\otimes$ denotes a tensor product. The Fourier coefficients of a laplacian operator L is $\tilde{L}(\xi_1,\dots,\xi_d) = \left(-2\alpha\sum_{j=1}^d (\cos(2\pi\xi_j) - 1) + 1\right)^c$, where $(\xi_1,\dots,\xi_d)$ is a d-dimensional frequency vector.

LDDMM with Geodesic Shooting. Given a source image S and a target image T defined on a d-dimensional torus domain $\Omega = \mathbb{R}^d/\mathbb{Z}^d$ ($S(x), T(x) : \Omega \to \mathbb{R}$). The space of diffeomorphisms is denoted by $\text{Diff}(\Omega)$. The problem of diffeomorphic image registration is to find the shortest path, i.e., geodesic, to generate time-varying diffeomorphisms $\{\phi_t\} : t \in [0, \tau]$, such that a deformed source image by the smooth mapping ϕ_τ, noted as $S(\phi_\tau)$, is similar to T. By parameterizing the transformation ϕ_τ with an initial velocity field $\tilde{v}_0$, we write the optimization of diffeomorphic registration in the setting of LDDMM with geodesic shooting in a low-dimensional bandlimited space as

$$E(\tilde{v}_0) = \lambda \, \mathrm{Dist}(S(\phi_\tau), T) + \frac{1}{2}\|\tilde{v}_0\|_V^2 \text{ s.t. Eq. (1) and (3)} \qquad (4)$$

Here, $\mathrm{Dist}(\cdot, \cdot)$ is a distance function that measures the dissimilarity between images and λ is a positive weighting parameter. The commonly used distance functions include the sum-of-squared intensity differences (L_2-norm) [7], normalized cross correlation (NCC) [5], and mutual information (MI) [32]. In this paper, we will use the sum-of-squared intensity differences. The $\| \cdot \|_V$ represents a Sobolev space that enforces smoothness of the velocity fields.

Despite a dramatic speed up of LDDMM with the reduced cost of Fourier EPDiff for shooting [35,36], the numerical solution operator in Eq. (3) does not scale well due to the introduced correlation and tensor convolutions.

3   Our Method: NeurEPDiff

We introduce a novel network, NeurEPDiff, that learns the solution operator to EPDiff in a low-dimensional bandlimited space. Inspired by the recent work on neural operators [15,16], we propose to develop NeurEPDiff, $\mathcal{G}_\theta$ with parameters θ, as a surrogate model to approximate the solution operator to EPDiff. Once it is trained, our NeurEPDiff can be used to fast predict the original EPDiff with a given initial condition $\tilde{v}_0$ on various resolutions of image grids.

Given a set of N observations, $\{\tilde{v}_0^n, \tilde{v}^n\}_{n=1}^N$, where $\tilde{v}^n \triangleq \{\tilde{v}_1^n, \cdots, \tilde{v}_\tau^n\}$ is a time-sequence of numerical solutions to EPDiff. The solution operator can be learned by minimizing the empirical data loss defined as

$$E(\theta) = \frac{1}{N}\sum_{n=1}^{N}\|\tilde{v}^n - \mathcal{G}_\theta(v_0^n)\|_V^2 + \beta \cdot \mathrm{Reg}(\theta), \qquad (5)$$

where β is a weighting parameter and $\mathrm{Reg}(\cdot)$ regularizes the network parameter θ.

Following the similar principles in [16], we formulate $\mathcal{G}_\theta$ as an iterative architecture $\tilde{v}_0 \mapsto \tilde{v}_1 \mapsto \cdots \mapsto \tilde{v}_\tau$, which is a sequence of functions carefully parameterized in the hidden network layers. The input $\tilde{v}_0$ is first lifted to a higher dimensional representation by a complex-valued local transformation $\tilde{P}$, which is usually parameterized by a shallow fully connected neural network. We then develop a key component, called iterative evolution layer, with operations designed in the bandlimited space to learn the nonlinear function of EPDiff.

Iterative Evolution Layer. Let $\tilde{z} \in \mathbb{C}^d$ be a d-dimensional complex-valued latent feature vector encoded from the initial velocity $\tilde{v}_0$. An iterative evolution layer (with the number of J hidden layers) to update $\tilde{z}_j \mapsto \tilde{z}_{j+1}$, where $j \in \{1, \cdots, J\}$, is defined as the composition of a nonlinear smooth activation function σ of a complex-valued linear transform $\tilde{W}_j$ with a global convolutional kernel $\tilde{\mathcal{H}}_j$:

$$\tilde{z}_{j+1} := \sigma(\tilde{W}_j \tilde{z}_j + \tilde{\mathcal{H}}_j * \tilde{z}_j). \tag{6}$$

The $*$ represents a complex convolution, which is efficiently computed by multiplying two signals in a real space. Similar to [25,30], we employ a complex-valued Gaussian error linear unit (CGeLU) that applies real-valued activation functions separately to the real and imaginary parts. With a smoothing operator $\tilde{K}$ to ensure the smoothness of the output signal, we have $\sigma := \tilde{K}(\text{GeLU}(\mathscr{R}(\cdot)) + i\text{GeLU}(\mathscr{I}(\cdot)))$, where $\mathscr{R}(\cdot)$ denotes the real part of a complex-valued vector, and $\mathscr{I}(\cdot)$ denotes an imaginary part.

We are now ready to introduce a formal definition of the NeurEPDiff $\mathcal{G}_\theta$ developed in a bandlimited space as follows.

Definition 1 (Fourier NeurEPDiff). *The neural operator $\mathcal{G}_\theta$ is defined as a composition of a complex-valued encoder $\tilde{P}$ (a decoder $\tilde{Q}$) (a.k.a. local linear transformations) to project the lower dimension function into higher dimensional space and vice versa, and a nonlinear smooth activation function σ of a local linear transform $\tilde{W}$ with a non-local convolution kernel $\tilde{\mathcal{H}}$ parameterized by network parameters, i.e.,*

$$\mathcal{G}_\theta := \tilde{Q} \circ \sigma(\tilde{W}_J, \tilde{\mathcal{H}}_J) \circ \cdots \circ \sigma(\tilde{W}_1, \tilde{\mathcal{H}}_1) \circ \tilde{P}. \tag{7}$$

The $\circ$ denotes function composition, and the parameter θ includes all parameters from $\{\tilde{P}, \tilde{Q}, \tilde{W}, \tilde{\mathcal{H}}\}$.

The network architecture of our proposed NeurEPDiff is shown in Fig. 1.

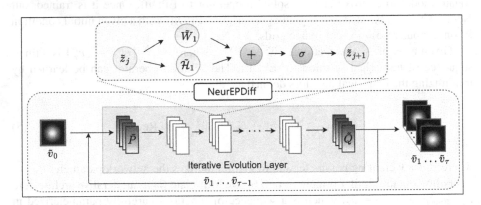

Fig. 1. An overview of our proposed NeurEPDiff. Given an initial input $\tilde{v}_0$, the iterative evolution layer updates the time sequence of $\{\tilde{v}_1, \cdots, \tilde{v}_\tau\}$.

3.1   NeurEPDiff for Diffeomorphic Image Registration

This section introduces a new diffeomorphic image registration algorithm with our developed NeurEPDiff. We will first train a neural operator $\mathcal{G}_\theta$ offline and then use it as predictive model in each iteration of the optimization of LDDMM (Eq. (4)), where the numerical solution of EPDiff is required. In this paper, we use a sum-of-squares-distance (SSD) metric for image dissimilarity. The optimization

$$E(\tilde{v}_0) = \frac{1}{2\lambda^2} \|(S(\phi_{\tau=1}) - T\|_2^2 + \frac{1}{2}\|\tilde{v}_0\|_V^2, \text{ s.t. Eq. (1) } \& \mathcal{G}_\theta(\tilde{v}_0). \qquad (8)$$

The final transformation $\phi_{\tau=1}$ is computed by Eq. (1) after predicting $\{\tilde{v}_t\}$ from a well-trained network $\mathcal{G}_\theta$.

Inference. Similar to [29,36], we employ a gradient decent algorithm to minimize the problem (8). The three main steps of our inference are described below:

(i) Forward geodesic shooting by NeurEPDiff. Given an initialized velocity field $\tilde{v}_0$, we compute the transformation ϕ_τ by first predicting $\{\tilde{v}_t\}$ and then solving Eq. (1).

(ii) Compute the gradient $\nabla_{\tilde{v}_\tau} E$ at the ending time point $t = \tau$ as

$$\nabla_{\tilde{v}_\tau} E = \tilde{K}\mathcal{F}\left(\frac{1}{\lambda^2}(S(\phi_\tau) - T) \cdot \nabla(S(\phi_\tau))\right). \qquad (9)$$

Recall that $\mathcal{F}$ denotes a Fourier transformation.

(iii) Backward integration to bring the gradient $\nabla_{\tilde{v}_\tau} E$ back to $t = 0$ by integrating the reduced adjoint Jacobi field equations defined in [35].

Figure 2 visualize the process to optimize image registration with NeurEPDiff.

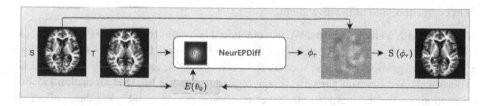

Fig. 2. An illustration of our optimization-based diffeomorphic registration with NeurEPDiff.

4 Experimental Evaluation

To evaluate the effectiveness of our proposed model, NeurEPDiff, for diffeomoprhic image registration, we compare its performance with the original LDDMM with geodesic shooting [29] and a fast variant of LDDMM (FLASH) [36] on both 2D synthetic data and 3D real brain MR images with different levels of resolution.

The training of NeurEPDiff in all our experiments is implemented on Nvidia GeForce GTX 1080Ti GPUs. We use ADAM optimizer [14] with the batch size as 20, learning rate of $1e^{-3}$, and weight decay as $1e^{-4}$, and the number of epochs as 500.

For a fair comparison with the baseline algorithms that are optimized by gradient decent on CPU, we also test our model on CPU once it is trained.

We evaluate the registration accuracy by performing registration-based segmentation and examine the resulting segmentation accuracy and runtime. To evaluate volume overlap between the propagated segmentation A and the manual segmentation B for each structure, we compute the dice similarity coefficient $\mathrm{DSC}(A, B) = 2(|A| \cap |B|)/(|A| + |B|)$, where $\cap$ denotes an intersection of two regions [10].

4.1 Data

2D Synthetic Data. We first simulate 2200 "bull-eye" synthetic data (as shown in Fig 3) with the resolution 192^2 by manipulating the width a and height b of an ellipse, formulated as $\frac{(x-96)^2}{a^2} + \frac{(y-96)^2}{b^2} = 1$. We draw the parameters a, b randomly from a Gaussian distribution $\mathcal{N}(40, 4.2^2)$ for the outer ellipse, and $a, b \sim \mathcal{N}(17, 1.6^2)$ for the inner ellipse. We then carefully down-sample the images to different resolutions of 64^3 and 128^3. The dataset is split into training (2000 volumes of resolution 64^2) and testing part (200 volumes of resolution 64^2, 128^2, and 192^2 respectively).

3D Brain MRI. We include 2200 T1-weighted 3D brain MRI scans from Open Access Series of Imaging Studies (OASIS) [12] in our experiments. We randomly select 2000 of them and down-sampled to the resolution of 64^3 for training. The remaining 200 are re-sampled to the resolution of 64^3, 128^2, and 192^2 for testing separately. The 200 testing volumes include manually delineated anatomical structures, i.e., segmentation labels, which also have been carefully down-sampled to the resolution of 64^3, 128^3, and 192^3. All of the MRI scans have undergone skull-stripping, intensity normalization, bias field correction, and affine alignment.

Ground Truth Velocity Fields. We use numerical solutions to Fourier EPDiff in Eq. (3) (i.e., computed by a Euler integrator) as the ground truth data for network training. More advanced integration methods, such as Runge-Kutta RK4, can be easily applied. In all experiments, we first run pairwise image registration using FLASH algorithm [35,36] on randomly selected images with lower resolution of 64^2 or 64^3. We then collect all numerical solutions to Fourier EPDiff as ground truth data. We set all integration steps as 10, the smoothing parameters for the operator $\tilde{K}$ as $\alpha = 3.0, c = 3.0$, the positive weighting parameter of registration as $\lambda = 0.03$, and the dimension of the bandlimited velocity field as 16.

4.2 Experiments

We first test our NeurEPDiff-based registration methods on 2D synthetic data and compare the final estimated transformation fields with the baseline algorithms. To validate whether our model is resolution-invariant, we train NeurEPDiff on images with lower resolution of 64^2, and then test its prediction accuracy on different levels of higher resolution (i.e., images with 128^2 and 192^2). We examine the optimization energy and compare it with all baseline algorithms.

We further analyze the performance of NeurEPDiff on real 3D brain MRI scans. The convergence graphs of NeurEPDiff-based registration are reported at different image resolutions. To investigate the predictive efficiency of NeurEPDiff in terms of both time-consumption and accuracy, we compare the actual runtime of solving the original EPDiff by our method and all baselines.

The registration accuracy is validated through a registration-based segmentation. Given a source image, we first run pairwise image registration with all algorithms and propagate segmentation labels of the template image to the number of 100 target images. We then examine the resulting segmentation accuracy by computing dice scores of the propagated segmentations and manually delineated labels. We report averaged dice scores of all images on four brain structures, including cortex, subcortical-gray-matter (SGM), white-matter (WM), and cerebrospinal fluid (CSF).

4.3 Results

Figure 3 visualizes examples of registration results on 2D synthetic images with the size of 192^2. The comparison is on deformed source images and transformation fields estimated by our NeurEPDiff, FLASH [36], and the original LDDMM with geodesic shooting [29]. Note that even though our method is trained on a low image resolution of 64^2, the predicted results are fairly close to the estimates from registration algorithms of both FLASH and LDDMM on a higher resolution of 192^2.

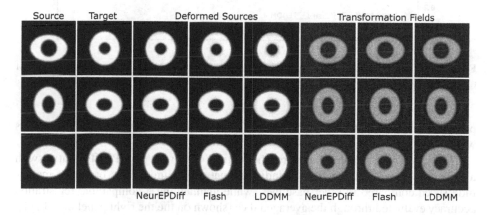

Fig. 3. Examples of registration results on 2D synthetic data. Left to right: 2D synthetic source and target images, deformed images, and transformation grids estimated from all algorithms.

The top panel of Fig. 4 displays an example of the comparison between the manually labeled segmentations and propagated segmentations on 3D brain MRI scans (at the resolution of 192^3) generated by our algorithm NeurEPDiff and the baseline models. The bottom panel of Fig. 4 reports the statistics of average dice scores over four brain structures from 100 registration pairs at different scales of image resolution. It shows that our method NeurEPDiff produces comparable dice scores. More importantly, the

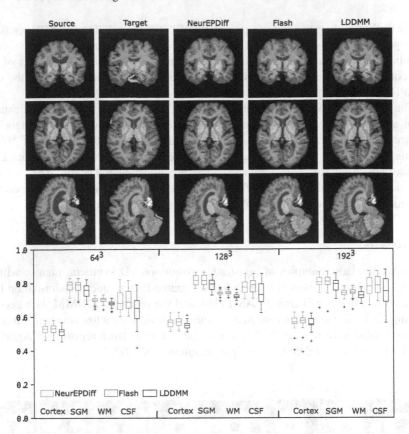

Fig. 4. Quantitative result of average dice for all methods on 3D brain MRIs at different scales of resolution: 64^3, 128^3, and 192^3.

pre-trained NeurEPDiff at a lower resolution of 64^3 also achieves comparable results when tested on higher resolutions.

The left panel of Table 1 compares the runtime of solving the EPDiff equation with all methods on a CPU machine. It indicates that our method gains orders of magnitude speed compared to other baselines, while maintaining a comparable registration accuracy evaluated through the averaged dice (shown on the the right panel of Table 1). Overall, NeurEPDiff is approximately more than 10 times faster than FLASH and 100 times faster than LDDMM with geodesic shooting.

Figure 5 demonstrates the convergence graphs of optimized total energy by our developed NeurEPDiff on different levels of resolution. It shows that our method consistently converges to a better solution than LDDMM at multiple different resolutions. However, it achieves a slightly worse but fairly close solution to FLASH [36]. Please refer to Sect. 5 for more detailed discussions.

Table 1. Left to right: a runtime on images with different scales of resolutions and comparison of average dice on all brain structures.

Methods / Time	64^3	128^3	192^3	Methods/Dice	64^3	128^3	192^3
NeurEPDiff	**0.045 s**	**0.12 s**	**0.33 s**	NeurEPDiff	0.866	0.883	0.878
Flash	0.37 s	1.24 s	21.8 s	Flash	0.867	0.884	0.879
LDDMM	3.85 s	30.76 s	103.09 s	LDDMM	0.858	0.874	0.876

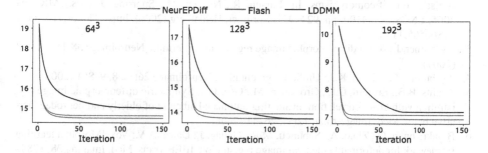

Fig. 5. A comparison of convergence graphs of all methods on 3D brain MRIs.

5 Discussion and Conclusion

We presented a novel neural operator, NeurEPDiff, to predict the geodesics of diffeomorphisms in a low-dimensional bandlimited space. Our method is the first to learn a solution operator to the geodesic evolution equation (EPDiff) with significantly improved computational efficiency. More importantly, the prediction capability of a pre-trained NeurEPDiff can be generalized to various resolutions of image grids because of its special property of resolution-invariance. To achieve this, we developed a brand-new network architecture, inspired by recent works on neural operators [15,16], to approximate the nonlinear EPDiff equations as time-series functions carefully parameterized in the network hidden layers. A composition of network operators were defined in a complex-valued space to process the bandlimited signals of the velocity fields.

We validated NeurEPDiff on both simulated 2D and real 3D images. Experimental results show that our method achieves significantly faster speed at different scales of image resolution. Interestingly, we also find out that the NeurEPDiff-based registration algorithm converges to an optimal solution with slightly higher energy compared to FLASH [36]. This may be due to a relaxed regularization on the smoothness of the velocity fields parameterized in the network, which often happens in learning-based approaches. A possible solution is to enforce a smooth convolutional kernel $\tilde{\mathcal{H}}$ [11] for each iterative evolution layer of our developed network architecture. This will be further investigated in our future work. Another potential future direction could be applying the network architectures developed in NeurEPDiff to learn another solution operator to predict transformations (Eq. (1)), which will further speed up optimization-based and learning-based diffeomorphic registration algorithms [6,29,30].

Acknowledgement. This work was supported by NSF CAREER Grant 2239977.

References

1. Arnold, V.: Sur la géométrie différentielle des groupes de lie de dimension infinie et ses applications à l'hydrodynamique des fluides parfaits. In: Annales de l'institut Fourier, vol. 16, pp. 319–361 (1966)
2. Arsigny, V., Commowick, O., Pennec, X., Ayache, N.: A log-euclidean framework for statistics on diffeomorphisms. In: Larsen, R., Nielsen, M., Sporring, J. (eds.) MICCAI 2006. LNCS, vol. 4190, pp. 924–931. Springer, Heidelberg (2006). https://doi.org/10.1007/11866565_113
3. Ashburner, J.: A fast diffeomorphic image registration algorithm. Neuroimage **38**(1), 95–113 (2007)
4. Ashburner, J., Friston, K.J.: Unified segmentation. Neuroimage **26**(3), 839–851 (2005)
5. Avants, B.B., Epstein, C.L., Grossman, M., Gee, J.C.: Symmetric diffeomorphic image registration with cross-correlation: evaluating automated labeling of elderly and neurodegenerative brain. Med. Image Anal. **12**(1), 26–41 (2008)
6. Balakrishnan, G., Zhao, A., Sabuncu, M.R., Guttag, J., Dalca, A.V.: VoxelMorph: a learning framework for deformable medical image registration. IEEE Trans. Med. Imaging **38**, 1788–1800 (2019)
7. Beg, M.F., Miller, M.I., Trouvé, A., Younes, L.: Computing large deformation metric mappings via geodesic flows of diffeomorphisms. Int. J. Comput. Vision **61**(2), 139–157 (2005)
8. Cao, T., Singh, N., Jojic, V., Niethammer, M.: Semi-coupled dictionary learning for deformation prediction. In: 2015 IEEE 12th International Symposium on Biomedical Imaging (ISBI), pp. 691–694. IEEE (2015)
9. Christensen, G.E., Rabbitt, R.D., Miller, M.I.: Deformable templates using large deformation kinematics. IEEE Trans. Image Process. **5**(10), 1435–1447 (1996)
10. Dice, L.R.: Measures of the amount of ecologic association between species. Ecology **26**(3), 297–302 (1945)
11. Feinman, R., Lake, B.M.: Learning a smooth kernel regularizer for convolutional neural networks. arXiv preprint arXiv:1903.01882 (2019)
12. Fotenos, A.F., Snyder, A., Girton, L., Morris, J., Buckner, R.: Normative estimates of cross-sectional and longitudinal brain volume decline in aging and ad. Neurology **64**(6), 1032–1039 (2005)
13. Hong, Y., Golland, P., Zhang, M.: Fast geodesic regression for population-based image analysis. In: Descoteaux, M., Maier-Hein, L., Franz, A., Jannin, P., Collins, D.L., Duchesne, S. (eds.) MICCAI 2017. LNCS, vol. 10433, pp. 317–325. Springer, Cham (2017). https://doi.org/10.1007/978-3-319-66182-7_37
14. Kingma, D.P., Ba, J.: Adam: a method for stochastic optimization. arXiv preprint arXiv:1412.6980 (2014)
15. Kovachki, N., et al.: Neural operator: learning maps between function spaces. arXiv preprint arXiv:2108.08481 (2021)
16. Li, Z., Kovachki, N., Azizzadenesheli, K., Liu, B., Bhattacharya, K., Stuart, A., Anandkumar, A.: Fourier neural operator for parametric partial differential equations. arXiv preprint arXiv:2010.08895 (2020)
17. Lu, L., Jin, P., Karniadakis, G.E.: DeepoNet: learning nonlinear operators for identifying differential equations based on the universal approximation theorem of operators. arXiv preprint arXiv:1910.03193 (2019)
18. Miller, M.I.: Computational anatomy: shape, growth, and atrophy comparison via diffeomorphisms. Neuroimage **23**, S19–S33 (2004)

19. Miller, M.I., Trouvé, A., Younes, L.: Geodesic shooting for computational anatomy. J. Math. Imaging Vis. **24**(2), 209–228 (2006)
20. Niethammer, M., Huang, Y., Vialard, F.-X.: Geodesic regression for image time-series. In: Fichtinger, G., Martel, A., Peters, T. (eds.) MICCAI 2011. LNCS, vol. 6892, pp. 655–662. Springer, Heidelberg (2011). https://doi.org/10.1007/978-3-642-23629-7_80
21. O'Donnell, L.J., Wells, W.M., Golby, A.J., Westin, C.-F.: Unbiased groupwise registration of white matter tractography. In: Ayache, N., Delingette, H., Golland, P., Mori, K. (eds.) MICCAI 2012. LNCS, vol. 7512, pp. 123–130. Springer, Heidelberg (2012). https://doi.org/10.1007/978-3-642-33454-2_16
22. Qiu, A., Younes, L., Miller, M.I., Csernansky, J.G.: Parallel transport in diffeomorphisms distinguishes the time-dependent pattern of hippocampal surface deformation due to healthy aging and the dementia of the Alzheimer's type. Neuroimage **40**(1), 68–76 (2008)
23. Risser, L., Holm, D., Rueckert, D., Vialard, F.X.: Diffeomorphic atlas estimation using karcher mean and geodesic shooting on volumetric images. In: MIUA (2011)
24. Rohé, M.-M., Datar, M., Heimann, T., Sermesant, M., Pennec, X.: SVF-Net: learning deformable image registration using shape matching. In: Descoteaux, M., Maier-Hein, L., Franz, A., Jannin, P., Collins, D.L., Duchesne, S. (eds.) MICCAI 2017. LNCS, vol. 10433, pp. 266–274. Springer, Cham (2017). https://doi.org/10.1007/978-3-319-66182-7_31
25. Scardapane, S., Van Vaerenbergh, S., Hussain, A., Uncini, A.: Complex-valued neural networks with nonparametric activation functions. IEEE Trans. Emerg. Top. Comput. Intell. **4**(2), 140–150 (2018)
26. Schnabel, J.A., et al.: A generic framework for non-rigid registration based on non-uniform multi-level free-form deformations. In: Niessen, W.J., Viergever, M.A. (eds.) MICCAI 2001. LNCS, vol. 2208, pp. 573–581. Springer, Heidelberg (2001). https://doi.org/10.1007/3-540-45468-3_69
27. Shi, W., et al.: Registration using sparse free-form deformations. In: Ayache, N., Delingette, H., Golland, P., Mori, K. (eds.) MICCAI 2012. LNCS, vol. 7511, pp. 659–666. Springer, Heidelberg (2012). https://doi.org/10.1007/978-3-642-33418-4_81
28. Singh, N., Hinkle, J., Joshi, S., Fletcher, P.T.: A hierarchical geodesic model for diffeomorphic longitudinal shape analysis. In: Gee, J.C., Joshi, S., Pohl, K.M., Wells, W.M., Zöllei, L. (eds.) IPMI 2013. LNCS, vol. 7917, pp. 560–571. Springer, Heidelberg (2013). https://doi.org/10.1007/978-3-642-38868-2_47
29. Vialard, F.X., Risser, L., Rueckert, D., Cotter, C.J.: Diffeomorphic 3D image registration via geodesic shooting using an efficient adjoint calculation. Int. J. Comput. Vision **97**(2), 229–241 (2012)
30. Wang, J., Zhang, M.: DeepFlash: an efficient network for learning-based medical image registration. In: Proceedings of the IEEE/CVF Conference on Computer Vision and Pattern Recognition (CVPR) (2020)
31. Wang, J., Zhang, M.: Geo-sic: learning deformable geometric shapes in deep image classifiers. In: The Conference on Neural Information Processing Systems (2022)
32. Wells, W., Viola, P., Atsumi, H., Nakajima, S., Kikinis, R.: Multi-modal volume registration by maximization of mutual information. Med. Image Anal. **1**, 35–51 (1996)
33. Yang, X., Kwitt, R., Styner, M., Niethammer, M.: Quicksilver: fast predictive image registration-a deep learning approach. Neuroimage **158**, 378–396 (2017)
34. Younes, L., Arrate, F., Miller, M.I.: Evolutions equations in computational anatomy. Neuroimage **45**(1), S40–S50 (2009)
35. Zhang, M., Fletcher, P.T.: Finite-dimensional lie algebras for fast diffeomorphic image registration. In: Ourselin, S., Alexander, D.C., Westin, C.-F., Cardoso, M.J. (eds.) IPMI 2015. LNCS, vol. 9123, pp. 249–260. Springer, Cham (2015). https://doi.org/10.1007/978-3-319-19992-4_19

36. Zhang, M., et al.: Frequency diffeomorphisms for efficient image registration. In: Nietham-mer, M., et al. (eds.) IPMI 2017. LNCS, vol. 10265, pp. 559–570. Springer, Cham (2017). https://doi.org/10.1007/978-3-319-59050-9_44

37. Zhang, M., Wells, W.M., Golland, P.: Low-dimensional statistics of anatomical variability via compact representation of image deformations. In: Ourselin, S., Joskowicz, L., Sabuncu, M.R., Unal, G., Wells, W. (eds.) MICCAI 2016. LNCS, vol. 9902, pp. 166–173. Springer, Cham (2016). https://doi.org/10.1007/978-3-319-46726-9_20

Non-rigid Medical Image Registration using Physics-informed Neural Networks

Zhe Min[1]([✉]), Zachary M. C. Baum[1], Shaheer U. Saeed[1], Mark Emberton[2], Dean C. Barratt[1], Zeike A. Taylor[3], and Yipeng Hu[1]

[1] Centre for Medical Image Computing and Wellcome/EPSRC Centre for Interventional and Surgical Sciences, University College London, London, UK
z.min@ucl.ac.uk
[2] Division of Surgery and Interventional Science, University College London, London, UK
[3] CISTIB Centre for Computational Imaging and Simulation Technologies in Biomedicine, Institute of Medical and Biological Engineering, University of Leeds, Leeds, UK

Abstract. Biomechanical modelling of soft tissue provides a non-data-driven method for constraining medical image registration, such that the estimated spatial transformation is considered biophysically plausible. This has not only been adopted in real-world clinical applications, such as the MR-to-ultrasound registration for prostate intervention of interest in this work, but also provides an explainable means of understanding the organ motion and spatial correspondence establishment. This work instantiates the recently-proposed physics-informed neural networks (PINNs) to a 3D linear elastic model for modelling prostate motion commonly encountered during transrectal ultrasound guided procedures. To overcome a widely-recognised challenge in generalising PINNs to different subjects, we propose to use PointNet as the nodal-permutation-invariant feature extractor, together with a registration algorithm that aligns point sets and simultaneously takes into account the PINN-imposed biomechanics. Using 77 pairs of MR and ultrasound images from real clinical prostate cancer biopsy, we first demonstrate the efficacy of the proposed registration algorithms in an "unsupervised" subject-specific manner for reducing the target registration error (TRE) compared to that without PINNs especially for patients with large deformations. The improvements stem from the intended biomechanical characteristics being regularised, e.g., the resulting deformation magnitude in rigid transition zones was effectively modulated to be smaller than that in softer peripheral zones. This is further validated to achieve low registration error values of 1.90 ± 0.52 mm and 1.94 ± 0.59 mm for all and surface nodes, respectively, based on ground-truth computed using finite element methods. We then extend and validate the PINN-constrained registration network that can generalise to new subjects. The trained network reduced the rigid-to-soft-region ratio of rigid-excluded deformation magnitude from 1.35 ± 0.15, without PINNs, to 0.89 ± 0.11 ($p < 0.001$) on unseen holdout subjects, which also witnessed decreased TREs from 6.96 ± 1.90 mm to 6.12 ± 1.95 mm ($p = 0.018$). The codes are available at https://github.com/ZheMin-1992/Registration_PINNs.

A. Frangi et al. (Eds.): IPMI 2023, LNCS 13939, pp. 601–613, 2023.
https://doi.org/10.1007/978-3-031-34048-2_46

Keywords: Medical image registration · Biomechanical constraints ·
Physics-informed neural network

1 Introduction

Multi-modal image registration enables access to clinically important information from different imaging modalities by spatially aligning them [6], in tasks such as surgical and interventional guidance [2,7]. Perhaps due to the complementary nature between cross-modality images, designing a robust objective function or an unsupervised loss function is in general highly challenging, for classical or learning-based algorithms, respectively. This work investigates an example of such cross-modality registration, for establishing spatial correspondence between preoperative MR and 3D intraoperative transrectal ultrasound (TRUS) images from the same patients. Indeed, most previously proposed approaches utilised correspondent features from both images, for either iterative optimisation algorithms [17] or neural network training [9,20]. The inevitable sparsity of these available anatomical features, such as the boundaries of prostate gland and other zonal structures, necessitates the addition of transformation smoothness constraints. Hu et al. [9] illustrated examples showing that, without imposing smoothness constraints on the registration-estimated transformation, highly distorted local deformation occurred which led to poorer target registration errors (TREs) in these areas. In addition to heuristically designed deformation regularisation, such as L^2 norm of local displacement and bending energy, displacement constraints originated from solid mechanics [17,18], have also demonstrated benefits in this application, with an arguably flexible and purposive approach through its soft tissue modelling physics.

Different from voxelised volumetric images with rectangular grids, point sets are in general unstructured and unordered [13] for efficiently yet sparsely representing geometries or shapes. PointNet was proposed to represent such point sets [13]. Originally designed for classification and segmentation tasks, PointNet was also adopted for learning-based rigid registration that either *1)* first establishes point correspondences in the feature spaces, with which then estimates the rigid transformation using closed-form solutions such as singular value decomposition [19], or *2)* directly aligns with learned feature representations to regress the rigid transformation parameters [10]. Among non-rigid registration approaches, Free Point Transformer [1,2] is an example that utilises the PointNet to extract features to predict source-point-wise displacement vectors, trained with composition of Chamfer loss [4] and/or negative log-likelihood function of Gaussian Mixture Models [2].

In [16], an adapted PointNet [13] was proposed using finite element modelling (FEM)-simulated training data to predict nodal displacement vectors for prostate meshes with unseen patients. In [5], FEM was first proposed to generate displacements for source point sets with boundary conditions established from an independent non-rigid iterative closest point (ICP) [3] procedure between prostate surfaces, before a network trained using the FEM-generated transformations [5]. Biomechanical constraints have also been investigated in motion modelling and deformable registration, for other organs, such as liver [12], brain [11] and heart [14].

This work investigates an alternative approach to encode biomechanical constraints represented by a system of partial differential equations (PDEs), which is solved simultaneously with minimising a registration loss. For registering MR and TRUS prostate images, we propose an approach that *1)* represents prostate point displacements using PointNet, previously adopted in this application [2]; *2)* develops physics-informed neural network (PINNs) for imposing elastic constraints on the estimated displacements; and *3)* formulates an end-to-end registration network training algorithm, by minimising surface distance as estimated boundary conditions in the PDEs. First, we show that the proposed PINNs effectively constrained the registration-estimated deformation with predefined elastic material properties, for registering individual point pairs. Second, with training data from as few as 75 subjects, the learned constrained registration generalised to new subjects, from which different point sets are independently sampled to represent varying sizes and geometries. We argue in this paper the significance in both results. The subject-specific algorithm incorporates elasticity or potentially other complex constraints in registration in a single network training, replacing alternative biomechnically-constrained methods requiring construction of statistical motion models [8] or finite element simulations [5,7]; whilst the second learning approach registers unseen point set pairs during efficient inference, demonstrating the generalisability over different geometries and nodal configurations - a well-recognised challenge associated with PINNs.

The contributions are summarised as follows. **1)** We developed a patient-specific registration algorithm combining PointNet and PINNs, which aligns prostate glands segmented from MR and TRUS images, subject to biomechanical constraints exerted from soft-tissue modelling PDEs (Fig. 1). **2)** We demonstrated that both the biomechanically-regularised deformation and the TRE-reducing correspondence can be generalised to unseen new patients, with the PINN-based registration network trained on a small number of training examples. **3)** We presented a set of experimental results for evaluating the theoretical and clinical efficacy in soft tissue modelling within registration algorithms, with statistical significance, using finite element (FE)-based ground-truth and independent landmark-based target registration errors (TREs), respectively.

2 Methods

Let $\mathbf{P}_\mathcal{S} \in \mathbb{R}^{N_s \times 3}$ and $\mathbf{P}_\mathcal{T} \in \mathbb{R}^{N_t \times 3}$ be a pair of source and target point sets with individual points being $\mathbf{p}_s \in \mathbb{R}^3$ and $\mathbf{p}_t \in \mathbb{R}^3$, where $N_s \in \mathbb{N}^+$ and $N_t \in \mathbb{N}^+$ are number of points, $s \in \{1, ..., N_s\}$ and $t \in \{1, ..., N_t\}$ are indexes of points. The non-rigid point set registration problem is to find point-wise displacement vectors $\mathbf{D}_\mathcal{S} \in \mathbb{R}^{N_s \times 3}$ with $\mathbf{d}_s \in \mathbb{R}^3$, such that the warped source point set $\mathsf{T}(\mathbf{P}_\mathcal{S}) = \mathbf{P}_\mathcal{S} + \mathbf{D}_\mathcal{S}$ aligns with $\mathbf{P}_\mathcal{T}$. We additionally adopt notations $\mathbf{P}_\mathcal{S}^{\text{internal}}$ and $\mathbf{P}_\mathcal{S}^{\text{surface}}$ to distinguish internal and surface points in $\mathbf{P}_\mathcal{S}$.

2.1   Physics-Informed Neural Network (PINNs) for Non-rigid Registration with Biomechanical Constraints

With the capability of universal function approximation, physics-informed neural networks (PINNs) can be utilised to model physical laws represented by nonlinear partial differential equations (PDEs) [15]. A non-rigid medical image registration problem estimating displacement vectors $\mathbf{D}_S$ is considered as the problem of seeking data-driven solutions to PDEs. The entire network $e_\theta(\mathcal{D}_k)$ where $k \in \mathbb{N}^+$ is the patient index, with trainable parameters θ, consists of two sub-networks $g_{\theta_g}(\mathcal{D}_k)$ and $h_{\theta_h}(\mathcal{D}_k)$, with completing parameter sets θ_g and θ_h, predicting displacement vectors $\mathbf{D}_S$ and stress tensors $\sigma \in \mathbb{R}^{N_s \times 6}$, respectively. Let a function $f(\mathbf{p}_s, \mathbf{d}_s, \sigma^s)$ be a PINN defining biomechanical constraints partially characterised by known material properties b_s:

$$f := f_1(\frac{\partial \sigma^s}{\partial x}, \frac{\partial \sigma^s}{\partial y}, \frac{\partial \sigma^s}{\partial z}) + f_2(\frac{\partial \mathbf{d}_s}{\partial x}, \frac{\partial \mathbf{d}_s}{\partial y}, \frac{\partial \mathbf{d}_s}{\partial z}, \sigma^s, b_s) + f_3(\frac{\partial \mathbf{d}_s}{\partial x}, \frac{\partial \mathbf{d}_s}{\partial y}, \frac{\partial \mathbf{d}_s}{\partial z}, \sigma^s),$$

(1)

where x, y and z are spatial coordinates of $\mathbf{p}_s$, the determination of b_s is detailedly described in Sect. 3, $f_1(\cdot)$, $f_2(\cdot)$ and $f_3(\cdot)$ represent norms of residuals deviating from static equilibrium, constitutive equality and null elastic energy, respectively, as defined in the remainder Sect. 2.2 and Sect. 2.3. The network parameters are optimised by minimising $\mathcal{L}^k(\theta; \mathcal{D}_k) = \mathcal{L}_R^k(\theta_g; \mathcal{D}_k) + \mathcal{L}_F^k(\theta; \mathcal{D}_k)$, where $\mathcal{L}_F^k(\theta; \mathcal{D}_k) = \sum_{s=1}^{N_s} f(\mathbf{p}_s, \mathbf{d}_s, \sigma^s)$ is the term concerning biomechanical constraints over all sampled source points, while $\mathcal{L}_R^k(\theta_g; \mathcal{D}_k)$ can be either (1) $\sum_{s=1}^{N_s} (\|\mathbf{d}_s - \mathbf{d}_s^{gt}\|_2^2)$ with $\mathbf{d}_s^{gt} \in \mathbb{R}^3$ denoting ground-truth displacement vectors of $\mathbf{p}_s$ under supervised learning (e.g., simulated data with known ground-truth deformations); or (2) $\phi(\mathsf{T}(\mathbf{P}_S), \mathbf{P}_\mathcal{T})$ being the unsupervised loss (e.g., the Chamfer loss for the purpose of aligning point sets) which measures goodness-of-prediction, resulting in a complete registration algorithm as described in Sect. 2.3 and used throughout this paper.

2.2   Governing Equations for Deforming Linear Elastic Organs Adapted for Medical Image Registration

In this section, linear elasticity is used as a specific example of prostate gland deformation between $\mathbf{P}_S$ and $\mathbf{P}_\mathcal{T}$, primarily due to contact with a moving ultrasound probe [7,8]. Adopting linear elasticity aims to demonstrate the feasibility of modelling soft tissue with the PDE-representing physics as the first step towards more complex and potentially more realistic models, such as nonlinear strain, alternative stress, time-dependent viscoelasticity and plasticity.

Strain-Displacement Equations. The strain-Displacement equation (i.e., kinematic equation) at a source point $\mathbf{p}_s$ is

$$\varepsilon^s = \frac{1}{2}(\nabla \mathbf{d}_s + \nabla \mathbf{d}_s^\mathsf{T}),$$

(2)

where ε^s is the infinitesimal second-order Cauchy strain tensor at $\mathbf{p}_s$, $\nabla \mathbf{d}_s$ is the displacement gradient w.r.t. spatial coordinates x, y, z of $\mathbf{p}_s$. Equation (2) can be rewritten explicitly as $\varepsilon_{xx}^s = \frac{\partial d_s^x}{\partial x}$, $\varepsilon_{xy}^s = \frac{1}{2}(\frac{\partial d_s^x}{\partial y} + \frac{\partial d_s^y}{\partial x})$, $\varepsilon_{yy}^s = \frac{\partial d_s^y}{\partial y}$, $\varepsilon_{yz}^s = \frac{1}{2}(\frac{\partial d_s^y}{\partial z} + \frac{\partial d_s^z}{\partial y})$, $\varepsilon_{zz}^s = \frac{\partial d_s^z}{\partial z}$, $\varepsilon_{xz}^s = \frac{1}{2}(\frac{\partial d_s^x}{\partial z} + \frac{\partial d_s^z}{\partial x})$. Equation (2) is used to compute strain tensors $\mathcal{E} \in \mathbb{R}^{N_s \times 6}$ from displacement vectors $\mathbf{D}_S$ predicted by $g_{\theta_g}(\mathcal{D}_k)$.

Static Equilibrium Equations. The spatial components of the Cauchy stress tensor σ^s at $\mathbf{p}_s$, predicted by $h_{\theta_h}(\mathcal{D}_k)$, satisfy the following equilibrium equation (i.e., equation of motion)

$$\sigma_{ji,j}^s + F_i = 0, \tag{3}$$

where $(\cdot)_{,j}^s$ is a shorthand for $\frac{\partial(\cdot)}{\partial(\mathbf{p}_s)_j}$, $F_i \in \mathbb{R}$ is the body force that is approximated to be zero at the static equilibrium, i and j denote three spatial directions. Equation (3) can be rewritten explicitly as $\frac{\partial \sigma_{xx}^s}{\partial x} + \frac{\partial \sigma_{yx}^s}{\partial y} + \frac{\partial \sigma_{zx}^s}{\partial z} = 0$, $\frac{\partial \sigma_{xy}^s}{\partial x} + \frac{\partial \sigma_{yy}^s}{\partial y} + \frac{\partial \sigma_{zy}^s}{\partial z} = 0$, $\frac{\partial \sigma_{xz}^s}{\partial x} + \frac{\partial \sigma_{yz}^s}{\partial y} + \frac{\partial \sigma_{zz}^s}{\partial z} = 0$.

Constitutive Equations. The stress and strain tensors at $\mathbf{p}_s$ are related by the constitutive equation (i.e., the generalised Hooke's law) as

$$\sigma^s = \mathsf{C} : \varepsilon^s, \tag{4}$$

where C is the fourth-order elasticity tensor. Equation (4) can be expanded as

$$\begin{bmatrix} \sigma_{xx}^s \\ \sigma_{yy}^s \\ \sigma_{zz}^s \\ \sigma_{xy}^s \\ \sigma_{xz}^s \\ \sigma_{yz}^s \end{bmatrix} = \begin{bmatrix} (\lambda+2\mu) & \lambda & \lambda & 0 & 0 & 0 \\ \lambda & (\lambda+2\mu) & \lambda & 0 & 0 & 0 \\ \lambda & \lambda & (\lambda+2\mu) & 0 & 0 & 0 \\ 0 & 0 & 0 & \mu & 0 & 0 \\ 0 & 0 & 0 & 0 & \mu & 0 \\ 0 & 0 & 0 & 0 & 0 & \mu \end{bmatrix} \begin{bmatrix} \varepsilon_{xx}^s \\ \varepsilon_{yy}^s \\ \varepsilon_{zz}^s \\ 2\varepsilon_{xy}^s \\ 2\varepsilon_{xz}^s \\ 2\varepsilon_{yz}^s \end{bmatrix}, \tag{5}$$

where $\lambda \in \mathbb{R}$ and $\mu \in \mathbb{R}$ are Lame parameters, which are computed using $\lambda = \frac{E\nu}{(1-2\nu)(1+\nu)}$ and $\mu = \frac{E}{2(1+\nu)}$ with the Young's Modulus E and Possion's ratio v.

As will be introduced in Sect. 2.3, Eq. (3) and Eq. (5) are utilised to construct PDEs that regularise $\mathbf{D}_S$ predicted by $g_{\theta_g}(\mathcal{D}_k)$, σ predicted by $h_{\theta_h}(\mathcal{D}_k)$, and $\mathcal{E}$ computed with Eq. (2).

2.3   A Non-rigid Point Set Registration Algorithm Using PINNs

Figure 1 shows the schematic of the proposed non-rigid point set registration network, with the displacement-predicting $g_{\theta_g}(\mathcal{D}_k)$ and stress-predicting $h_{\theta_h}(\mathcal{D}_k)$.

Loss Functions for Single-Pair Patient-Specific Registration. The loss function includes four terms. First, the Chamfer loss $\phi(\mathsf{T}(\mathbf{P}_S), \mathbf{P}_T)$ [4] is minimised to spatially align the two point sets, and is given by

$$\mathcal{L}_R^k(\theta_g; \mathcal{D}_k) = \frac{1}{|\widetilde{N}_t|}\left(\sum_{t \in \widetilde{N}_t} \min_{s \in \widetilde{N}_s} ||\mathsf{T}(\mathbf{p}_s) - \mathbf{p}_t||_2^2\right) + \frac{1}{|\widetilde{N}_s|}\left(\sum_{s \in \widetilde{N}_s} \min_{t \in \widetilde{N}_t} ||\mathsf{T}(\mathbf{p}_s) - \mathbf{p}_t||_2^2\right) \tag{6}$$

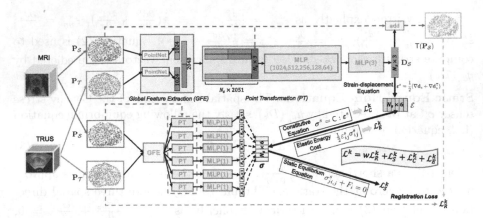

Fig. 1. The proposed non-rigid medical image registration framework using physics-informed neural networks (PINNs), whose inputs are a pair of source and target point sets $\mathbf{P}_S$ and $\mathbf{P}_T$ extracted from MRI and TRUS volumes of the same patient, respectively. PINNs consist of $g_{\theta_g}(\mathcal{D}_k)$ predicting displacement vectors $\mathbf{D}_S$ from which the point-wise strain tensors $\mathcal{E}$ are further computed with the strain-displacement equation in Eq. (2), and $h_{\theta_h}(\mathcal{D}_k)$ predicting stress tensors σ. The source point set $\mathbf{P}_S$ added by $\mathbf{D}_S$ results in the warped source point set $\mathsf{T}(\mathbf{P}_S)$, between which and $\mathbf{P}_T$ the Chamfer loss $\mathcal{L}_R^k$ is computed using Eq. (6). $\mathcal{L}_S^k$ in Eq. (7) and $\mathcal{L}_C^k$ in Eq. (8) penalise deviations from the equality of the static equilibrium equation in Eq. (3) about σ, and that of the constitutive equation in Eq. (4) about σ and $\mathcal{E}$, respectively. $\mathcal{L}_E^k$ in Eq. (9) is the elastic energy cost that shall also be minimised.

where $\tilde{N}_s \subseteq \{1, ..., N_s\}$ and $\tilde{N}_t \subseteq \{1, ..., N_t\}$ denote sets of points being either the entire organ $\mathbf{P}_S$ and $\mathbf{P}_T$ or a subset region, e.g., surface points $\mathbf{P}_S^{\text{surface}}$ and $\mathbf{P}_T^{\text{surface}}$, $|\tilde{N}_s|$ and $|\tilde{N}_t|$ are numbers of points. Second, deviation from the static equilibrium equation in Eq. (3) w.r.t. the stress σ is penalised by minimising $\mathcal{L}_S^k(\theta_h; \mathcal{D}_k)$ as

$$\mathcal{L}_S^k(\theta_h; \mathcal{D}_k) = \sum_{s=1}^{N_s} f_1\left(\frac{\partial \sigma^s}{\partial x}, \frac{\partial \sigma^s}{\partial y}, \frac{\partial \sigma^s}{\partial z}\right), \tag{7}$$

where $f_1\left(\frac{\partial \sigma^s}{\partial x}, \frac{\partial \sigma^s}{\partial y}, \frac{\partial \sigma^s}{\partial z}\right) = |\frac{\partial \sigma_{xx}^s}{\partial x} + \frac{\partial \sigma_{yx}^s}{\partial y} + \frac{\partial \sigma_{zx}^s}{\partial z}| + |\frac{\partial \sigma_{xy}^s}{\partial x} + \frac{\partial \sigma_{yy}^s}{\partial y} + \frac{\partial \sigma_{zy}^s}{\partial z}| + |\frac{\partial \sigma_{xz}^s}{\partial x} + \frac{\partial \sigma_{yz}^s}{\partial y} + \frac{\partial \sigma_{zz}^s}{\partial z}|$. Third, $\mathcal{L}_C^k(\theta; \mathcal{D}_k)$ regularises σ and strain $\mathcal{E}$ to satisfy constitutive equations in Eq. (5), and is defined as

$$\mathcal{L}_C^k(\theta; \mathcal{D}_k) = \sum_{s=1}^{N_s} f_2\left(\frac{\partial \mathbf{d}_s}{\partial x}, \frac{\partial \mathbf{d}_s}{\partial y}, \frac{\partial \mathbf{d}_s}{\partial z}, \sigma^s, b_s\right), \tag{8}$$

where $f_2\left(\frac{\partial \mathbf{d}_s}{\partial x}, \frac{\partial \mathbf{d}_s}{\partial y}, \frac{\partial \mathbf{d}_s}{\partial z}, \sigma^s, b_s\right) = (|(\lambda+2\mu)\varepsilon_{xx}^s + \lambda(\varepsilon_{yy}^s + \varepsilon_{zz}^s) - \sigma_{xx}^s| + |\lambda(\varepsilon_{xx}^s + \varepsilon_{zz}^s) + (\lambda+2\mu)\varepsilon_{yy}^s - \sigma_{yy}^s| + |\lambda\varepsilon_{xx}^s + \lambda\varepsilon_{yy}^s + (\lambda+2\mu)\varepsilon_{zz}^s - \sigma_{zz}^s| + |\sigma_{xy}^s - 2\mu\varepsilon_{xy}^s| + |\sigma_{xz}^s - 2\mu\varepsilon_{xz}^s| + |\sigma_{yz}^s - 2\mu\varepsilon_{yz}^s|)$, the strain tensor ε_{ij}^s at $\mathbf{p}_s$ is computed from network-predicted $\mathbf{d}_s$ with the automatic differentiation, according to the kinematic equation in Eq. (2). Fourth, $\mathcal{L}_E^k(\theta; \mathcal{D}_k) = \sum_{s=1}^{N_s} \frac{1}{2}\varepsilon_{ij}^s \sigma_{ij}^s$ is the elastic energy cost to be minimised

$$\mathcal{L}_E^k(\theta; \mathcal{D}_k) = \sum_s^{N_s} f_3\left(\frac{\partial \mathbf{d}_s}{\partial x}, \frac{\partial \mathbf{d}_s}{\partial y}, \frac{\partial \mathbf{d}_s}{\partial z}, \boldsymbol{\sigma}^s\right), \tag{9}$$

where $f_3\left(\frac{\partial \mathbf{d}_s}{\partial x}, \frac{\partial \mathbf{d}_s}{\partial y}, \frac{\partial \mathbf{d}_s}{\partial z}, \boldsymbol{\sigma}^s\right) = \frac{1}{2}(\varepsilon_{xx}^s \sigma_{xx}^s + \varepsilon_{yy}^s \sigma_{yy}^s + \varepsilon_{zz}^s \sigma_{zz}^s + 2\varepsilon_{xy}^s \sigma_{xy}^s + 2\varepsilon_{xz}^s \sigma_{xz}^s + 2\varepsilon_{yz}^s \sigma_{yz}^s)$.

The overall training loss $\mathcal{L}^k(\theta; \mathcal{D}_k)$ in the single-pair image registration for the given subject k is given by a $(w \in \mathbb{R}^+)$-weighted sum of these terms,

$$\mathcal{L}^k(\theta; \mathcal{D}_k) = w\mathcal{L}_R^k(\theta_g; \mathcal{D}_k) + \mathcal{L}_S^k(\theta_h, \mathcal{D}_k) + \mathcal{L}_C^k(\theta; \mathcal{D}_k) + \mathcal{L}_E^k(\theta; \mathcal{D}_k). \tag{10}$$

Optimisation for a Multi-patient Learning Algorithm. The above-described network can be adapted with minimal change in implementation, for a population-trained registration algorithm, by optimising network parameters θ with respect to an amortization loss:

$$\theta^\star = \arg \min_\theta \mathcal{L}(\theta; \mathcal{D}) = \arg \min_\theta E_k(\mathcal{L}^k(\theta; \mathcal{D}_k)), \tag{11}$$

where $\mathcal{D}$ is all the training data from multiple subjects and E_k denotes the expected value over all training examples.

2.4 Evaluation Metrics

For the experimental results described in Sect. 3, four evaluation metrics are reported. First, TREs were computed as the average distance between the geometric centroids of pairs of registered source and target landmarks, which include apex and base of the prostate, water-filled cysts, and calcifications. Further details in defining these independent landmarks followed published methods in previous studies [2,9]. Second, deformation magnitudes (DMs) were computed to measure the "pure" non-rigid part of predicted displacements of $\mathbf{P}_S$,

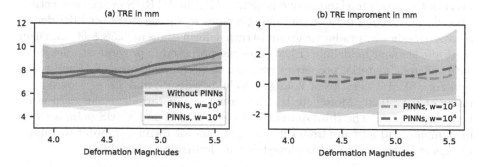

Fig. 2. (Left) TRE (mm) of MRI-TRUS fusion using patient-specific registration models with and without PINNs, w.r.t. deformation magnitudes of prostate gland. (Right) TRE improvements by incorporating PINNs compared to those without PINNs. The shaded areas are 95% confidence intervals (i.e., ± 2 standard deviations).

608   Z. Min et al.

Table 1. Quantitative results (mean ± std in mm) of patient-specific models in the first experiment. *: significantly different from results without PINNs ($p < 0.05$). °: the two with ° had no significant difference with each other but were both significantly different from the remaining result in one column ($p < 0.001$). The best results are marked in bold for CD and TRE.

Models	DM (Internal Points Rigid Region)	DM (Internal Points Soft Region)	CD	CD (Surface Points Only)	TRE
Without PINNs	4.78 ± 1.06	4.72 ± 0.95	**1.51° ± 0.23°**	**0.49° ± 0.10°**	7.52 ± 2.46
PINNs ($w = 10^4$)	4.56* ± 1.14*	4.69 ± 0.99	1.52° ± 0.26°	0.53° ± 0.19°	7.32 ± 2.60
PINNs ($w = 10^3$)	4.42* ± 1.31*	4.58 ± 1.06	1.66 ± 0.30	0.83 ± 0.42	**7.23 ± 2.60**

excluding the "largest" rigid transformation $(\mathbf{R}, \mathbf{t})$. DM was defined as residuals after solving the orthogonal Procrustes problem between $\mathbf{P}_{\mathcal{S}}$ and $\mathsf{T}(\mathbf{P}_{\mathcal{S}})$ [3]: as $\mathsf{DM} = \frac{1}{|\tilde{N}|}\sum_{s\in\tilde{N}}||\mathbf{R}\mathbf{p}_s + \mathbf{t} - \mathsf{T}(\mathbf{p}_s)||_2$, where $\tilde{N}$ can be either $\mathbf{P}_{\mathcal{S}}$ or $\mathbf{P}_{\mathcal{S}}^{\text{internal}}$, $|\tilde{N}|$ is the number of points. Third, Chamfer Distance (CD) was defined as $\mathsf{CD} = \frac{1}{2}\left(\frac{1}{|\tilde{N}_t|}\sum_{t\in\tilde{N}_t}\min_{s\in\tilde{N}_s}||\mathsf{T}(\mathbf{p}_s) - \mathbf{p}_t||_2 + \frac{1}{|\tilde{N}_s|}\sum_{s\in\tilde{N}_s}\min_{t\in\tilde{N}_t}||\mathsf{T}(\mathbf{p}_s) - \mathbf{p}_t||_2\right)$, where $\tilde{N}_t$ and $\tilde{N}_s$ are the same as those in Eq. (6). Fourth, the root-mean-square error (rmse) was defined between predicted displacement $\mathbf{D}_{\mathcal{S}}$ and ground-truth $\mathbf{D}_{\mathcal{S}}^{gt} \in \mathbb{R}^{|\tilde{N}_s|\times 3}$ as $\mathsf{rmse} = \sqrt{\frac{1}{|\tilde{N}_s|}\sum_{s\in\tilde{N}_s}||\mathbf{d}_s - \mathbf{d}_s^{gt}||_2^2}$.

3 Experiments and Results

Datasets. The first dataset contained 77 pairs of MRI and TRUS volumetric images (both were resampled to $0.8 \times 0.8 \times 0.8\,\text{mm}^3$) from prostate cancer biopsy, where the exemplar clinical application is to register pre-operative MRI images with TRUS images where prostate gland has been deformed due to surgical probe contact [8]. Each pair of point sets was extracted from the segmentations of the prostate gland in one patient's MRI and TRUS images respectively (Fig. 1). The second dataset containing 8 cases was generated over MRI-derived prostate meshes by producing ground-truth deformations in [5.58, 8.66] mm using the finite element modelling (FEM) process, proposed in previous studies [8,16], with different material properties assigned to peripheral zones (PZ) and transition zone (TZ): the ratios of Youngs' Modulus with PZ and TZ $\frac{E_{PZ}}{E_{TZ}}$ were in the range of [0.12, 0.20]. More details about zonal segmentations in this dataset can be found in [7,8]. The third dataset included 75 MRI and TRUS point-set pairs for training and 33 pairs from different patients for testing, in order to validate the generalisability of the developed population-trained model.

Implementation Details. PointNet [13] is adapted with a TNet 4-by-4 outputting 4×4 rigid transformation matrix instead of the original 3-by-3 TNet, suggested in [2]. The final global feature from a PointNet $\phi(\cdot)$ is of size 1024. In the global feature extraction (Fig. 1) module, the global features $\phi(\mathbf{P}_{\mathcal{S}})$ and $\phi(\mathbf{P}_{\mathcal{T}})$ learnt from $\mathbf{P}_{\mathcal{S}}$ and $\mathbf{P}_{\mathcal{T}}$ are concatenated. In the point transformation

Table 2. TRE statistics (mean ± std in mm) of patients whose ratios of DM between rigid and soft sub-regions were correctly modulated from > 1 without PINNs to < 1 with PINNs. (1) The 1st and 3rd rows are such cases; and (2) the 2nd and 4th rows are patients in (1) whose TRE were also improved. The column # records the number of patients per row. *: improvements of PINNs were statistically significant ($p < 0.001$).

PINNs Models	#	TRE Without PINNs	TRE With PINNs	TRE Improved With PINNs
$w = 10^4$	22	7.39 ± 2.08	**6.55 ± 1.86**	0.84 ± 2.18
$w = 10^4$ (TRE Improved)	15	8.04 ± 2.05	**6.04* ± 1.70***	2.00 ± 1.49
$w = 10^3$	20	7.95 ± 2.15	**6.68* ± 2.23***	1.27 ± 1.90
$w = 10^3$ (TRE Improved)	16	8.09 ± 2.31	**6.21* ± 2.15***	1.88 ± 1.62

module (Fig. 1), the concatenated global feature is repeated for N_s times and further concatenated with $\mathbf{P}_S$. The resulting feature map of size $N_s \times 2051$ will go through shared MLP(1024, 512, 256, 128, 64) and another shared MLP(256) without the ReLU layer. At the end, MLP(3) and 6 individual MLP(1) are used in branches $g_{\theta_g}(\mathcal{D}_k)$ predicting $\mathbf{D}_S$ and $h_{\theta_h}(\mathcal{D}_k)$ predicting $\boldsymbol{\sigma}$, respectively.

For the first and third experiments, Young's modulus E in Eq. (8) was chosen as 500 kPa and 5 kPa for points in rigid and soft compartments while Possion' ratio v was 0.49, leading to ($\lambda = 8221.48, \mu = 167.78$) and ($\lambda = 82.21, \mu = 1.68$), respectively. For the second experiment, E and v were set according to the ratio of their ground-truth values in two sub-regions. The two compartments' points were determined either by approximately taking upper $\frac{2}{3}$ and lower $\frac{1}{3}$ sub-regions in the axial view as rigid and soft compartments (as in the first and third experiments with clinical data), or taking the TZ and PZ respectively if zonal segmentations were available (as in the second experiment) [8]. All three experiments were run on an Intel(R) Xeon(R) Gold 5215 CPU with an NVIDIA Quadro GV100 32GB GPU.

Results. Table 1 and Fig. 2 include the numerical results of the first experiment. Two observations can be made from Table 1: *1)* TRE values decreased with PINNs; and more importantly *2)* DM values in the rigid sub-regions were smaller than those in the soft sub-regions with PINNs, which demonstrated biomechanical constraints are effectively preserved in the registration algorithm, i.e., $\frac{DM_{rigid}}{DM_{soft}} < 1$, significantly different from $\frac{DM_{rigid}}{DM_{soft}} > 1$ without PINNs ($p = 0.019$ for $w = 10^3$ and $p = 0.029$ for $w = 10^4$, paired t-tests at significance level α=0.05). Figure 2 shows TRE values w.r.t. varying DM thresholds. It is found from Fig. 2 that *1)* TRE values increased with larger deformation magnitudes for all methods; and *2)* PINNs reduced the TREs and demonstrated greater improvements for patients that undergo larger deformations. For example, PINNs ($w = 10^3$) significantly decreased TRE from 7.87 ± 2.03 mm without PINNs to 7.12 ± 2.21 mm ($p = 0.049$), among top 40% (31/77) patients with larger deformations.

Figure 3 shows qualitative results from two patients, with large and moderate-to-large non-rigid deformations being 6.30 mm and 5.56 mm. Take *case 1* as

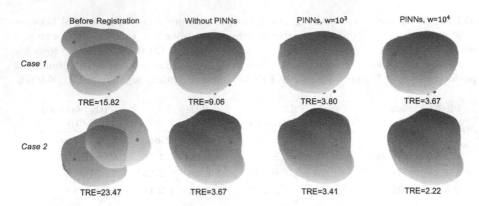

Fig. 3. Qualitative results showing meshes and TREs (mm) before and after registration, of two patient cases with large deformations. The original and warped source (i.e., MRI) meshes are depicted in blue while target (i.e., TRUS) meshes are in red. The anatomical landmarks in MRI and TRUS are denoted by blue circles and red stars. (Color figure online)

an example, the registration method with PINNs reached desired smaller DM value in the rigid compartment than that in the soft one, being 3.23 mm versus 3.82 mm for PINNs ($w = 10^4$) and 3.25 mm versus 4.04 mm for PINNs ($w = 10^3$), whereas without PINNs DM was larger (i.e., 5.26 mm) in the rigid compartment than that (i.e., 4.54 mm) in the soft one. While surface points are visually well aligned for both methods (Fig. 3) with Chamfer distances 0.83 mm, 0.45 mm and 0.48 mm for PINNs ($w = 10^4$), PINNs ($w = 10^3$) and without PINNs, PINNs greatly reduced the TRE value from that without PINNs (i.e., from 9.06 mm to 3.80 mm ($w = 10^3$) and 3.67 mm ($w = 10^4$)), which demonstrates the effectiveness of PINNs in producing more clinically meaningful deformations.

As shown in Table 2, 22 ($w = 10^4$) and 20 ($w = 10^3$) out of 77 patients achieved desired smaller DMs in the rigid sub-regions than those in the soft sub-regions with PINNs, while without PINNs for those cases DMs were larger in the rigid sub-regions than those in the soft sub-regions. The majority, 68% (15/22) and 80% (16/20) cases, obtained lower TREs than those without PINNs, for $w = 10^4$ and $w = 10^3$, respectively, where TRE improvements were statistically significant ($p < 0.001$) with mean differences being 2.00 mm and 1.88 mm, respectively. This is consistent with conclusions from previous studies, showing efficacy of imposing distinct material properties within the registration is positively correlated with more accurate registration.

Figure 4 shows results of the second experiment. The rmse values were 1.90 ± 0.52 mm and 2.11 ± 0.63 mm ($p = 0.400$) for all points, 1.94 ± 0.59 mm and 2.19 ± 0.75 mm ($p = 0.350$) for surface points, with PINNs ($w = 10^5$) and without PINNs respectively. The enhancements of the PINNs *1)* demonstrate its capability of successfully registering two point sets with lower error values; and *2)* further validate its effectiveness of producing displacement vectors that are more biomechanical compliant, considering that the ground-truth deformations are generated with FEM and thus are implicitly biomechanical encoded.

For the third experiment, compared to that without PINNs, the incorporation of PINNs ($w = 10^3$) significantly reduced the average TREs on the test subjects from 6.96 ± 1.90 mm to 6.12 ± 1.95 mm ($p = 0.018$), while Chamfer distances with and without PINNs were 2.48 ± 0.33 mm and 2.54 ± 0.37 mm ($p = 0.165$) on all points (2.96 ± 0.55 mm and 2.60 ± 0.41 mm ($p < 0.001$) on surface points), respectively. The successful imposition of biomechanical constraints on the test data was further demonstrated by 1) The ratios of DM between internal points in rigid and soft compartments $\frac{DM_{rigid}}{DM_{soft}}$ were 0.89 ± 0.11 and 1.35 ± 0.15 ($p < 0.001$) using registration methods with and without PINNs, respectively; and 2) The loss computed on the test patients using Eq. (10) was reduced from 20 to 10^{-14} after registration, which demonstrated the network's ability of inferring constraints on unseen subjects.

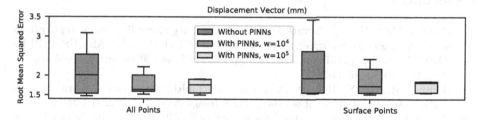

Fig. 4. Root mean squared error computed with displacement vectors $\mathbf{D}_S$ predicted by registration methods and ground-truth $\mathbf{D}_S^{gt}$ generated with finite element modelling.

4   Discussions and Conclusions

Despite the proposed model's power of regularising biomechanical constraints with predicted transformations and success of reducing registration error and generalising to unseen patients, as we showed in Sect. 3, this paper needs to be read with several limitations. First, the use of PINNs does not circumvent all limitations of biomechanical modelling shared with other approaches, such as assumptions of potentially subject-specific material properties. However, this opens up new opportunities for solutions to the material property estimation challenge, by considering an inverse data-driven discovery problem of PDEs potentially approachable with PINNs [15]. The second limitation is that our validation is focused on the MRI-TRUS prostate registration, while it is of broad interest to explore the model's effectiveness for wider clinical applications such as accurate and reliable myocardial motion tracking from cardiac cine MRI sequence [14]. The third limitation is that the linear elasticity is assumed, which is useful to demonstrate the efficacy of the methodology but both biomechanical modelling and registration performance may be further improved with more complex modelling with nonlinear materials and geometries in future studies.

To conclude, in this paper, we have presented a novel biomechanical constraining method using PINNs for non-rigid point set registration. Experimental results on FEM-produced data and clinical MRI-TRUS paired image data, using both patient-specific and multi-patient learning models, demonstrated that the proposed framework is capable of lowering registration errors with presubscribed biomechanical characteristics and generalizability, promising for clinical use and wider research in PINN-based modelling.

Acknowledgement. This work was supported by the Wellcome/EPSRC Centre for Interventional and Surgical Sciences [203145Z/16/Z] and the International Alliance for Cancer Early Detection, an alliance between Cancer Research UK [C28070/A30912; C73666/A31378], Canary Center at Stanford University, the University of Cambridge, OHSU Knight Cancer Institute, University College London and the University of Manchester.

References

1. Baum, Z.M.C., Hu, Y., Barratt, D.C.: Multimodality biomedical image registration using free point transformer networks. In: Hu, Y., et al. (eds.) ASMUS/PIPPI - 2020. LNCS, vol. 12437, pp. 116–125. Springer, Cham (2020). https://doi.org/10.1007/978-3-030-60334-2_12
2. Baum, Z.M., Hu, Y., Barratt, D.C.: Real-time multimodal image registration with partial intraoperative point-set data. Med. Image Anal. **74**, 102231 (2021)
3. Besl, P.J., McKay, N.D.: Method for registration of 3-D shapes. In: Sensor Fusion IV: Control Paradigms and Data Structures, vol. 1611, pp. 586–606. SPIE (1992)
4. Fan, H., Su, H., Guibas, L.J.: A point set generation network for 3D object reconstruction from a single image. In: Proceedings of the IEEE Conference on Computer Vision and Pattern Recognition, pp. 605–613 (2017)
5. Fu, Y., et al.: Biomechanically constrained non-rigid MR-TRUS prostate registration using deep learning based 3D point cloud matching. Med. Image Anal. **67**, 101845 (2021)
6. Haskins, G., Kruger, U., Yan, P.: Deep learning in medical image registration: a survey. Mach. Vis. Appl. **31**(1), 1–18 (2020). https://doi.org/10.1007/s00138-020-01060-x
7. Hu, Y., et al.: MR to ultrasound registration for image-guided prostate interventions. Med. Image Anal. **16**(3), 687–703 (2012)
8. Hu, Y., et al.: Modelling prostate motion for data fusion during image-guided interventions. IEEE Trans. Med. Imaging **30**(11), 1887–1900 (2011)
9. Hu, Y., et al.: Weakly-supervised convolutional neural networks for multimodal image registration. Med. Image Anal. **49**, 1–13 (2018)
10. Li, X., Pontes, J.K., Lucey, S.: PointNetLK revisited. In: Proceedings of the IEEE/CVF Conference on Computer Vision and Pattern Recognition (CVPR), pp. 12763–12772 (2021)
11. Luo, J., et al.: On the Dataset Quality Control for Image Registration Evaluation. In: Wang, L., Dou, Q., Fletcher, P.T., Speidel, S., Li, S. (eds.) Medical Image Computing and Computer Assisted Intervention - MICCAI 2022, MICCAI 2022. Lecture Notes in Computer Science, vol. 13436, pp. 36–45. Springer, Cham (2022). https://doi.org/10.1007/978-3-031-16446-0_4

12. Pfeiffer, M., et al.: Non-rigid volume to surface registration using a data-driven biomechanical model. In: Martel, A.L., et al. (eds.) MICCAI 2020. LNCS, vol. 12264, pp. 724–734. Springer, Cham (2020). https://doi.org/10.1007/978-3-030-59719-1_70

13. Qi, C.R., Su, H., Mo, K., Guibas, L.J.: PointNet: deep learning on point sets for 3D classification and segmentation. In: Proceedings of the IEEE Conference on Computer Vision and Pattern Recognition, pp. 652–660 (2017)

14. Qin, C., Wang, S., Chen, C., Bai, W., Rueckert, D.: Generative myocardial motion tracking via latent space exploration with biomechanics-informed prior. Med. Image Anal. **83**, 102682 (2023)

15. Raissi, M., Perdikaris, P., Karniadakis, G.E.: Physics-informed neural networks: a deep learning framework for solving forward and inverse problems involving nonlinear partial differential equations. J. Comput. Phys. **378**, 686–707 (2019)

16. Saeed, S.U., Taylor, Z.A., Pinnock, M.A., Emberton, M., Barratt, D.C., Hu, Y.: Prostate motion modelling using biomechanically-trained deep neural networks on unstructured nodes. In: Martel, A.L., et al. (eds.) MICCAI 2020. LNCS, vol. 12264, pp. 650–659. Springer, Cham (2020). https://doi.org/10.1007/978-3-030-59719-1_63

17. van de Ven, W.J., Hu, Y., Barentsz, J.O., Karssemeijer, N., Barratt, D., Huisman, H.J.: Biomechanical modeling constrained surface-based image registration for prostate MR guided TRUS biopsy. Med. Phys. **42**(5), 2470–2481 (2015)

18. Wang, Y.: Towards personalized statistical deformable model and hybrid point matching for robust MR-TRUS registration. IEEE Trans. Med. Imaging **35**(2), 589–604 (2016)

19. Yew, Z.J., Lee, G.H.: RPM-Net: robust point matching using learned features. In: Proceedings of the IEEE/CVF Conference on Computer Vision and Pattern Recognition, pp. 11824–11833 (2020)

20. Zeng, Q., et al.: Label-driven magnetic resonance imaging (MRI)-transrectal ultrasound (TRUS) registration using weakly supervised learning for MRI-guided prostate radiotherapy. Phys. Med. Biol. **65**(13), 135002 (2020)

POLAFFINI: Efficient Feature-Based Polyaffine Initialization for Improved Non-linear Image Registration

Antoine Legouhy[1,2]([✉]), Ross Callaghan[2], Hojjat Azadbakht[2], and Hui Zhang[1]

[1] Centre for Medical Image Computing and Department of Computer Science,
University College London, London, UK
a.legouhy@ucl.ac.uk
[2] AINOSTICS ltd., Manchester, UK

Abstract. This paper presents an efficient feature-based approach to initialize non-linear image registration. Today, nonlinear image registration is dominated by methods relying on intensity-based similarity measures. A good estimate of the initial transformation is essential, both for traditional iterative algorithms and for recent one-shot deep learning (DL)-based alternatives. The established approach to estimate this starting point is to perform affine registration, but this may be insufficient due to its parsimonious, global, and non-bending nature. We propose an improved initialization method that takes advantage of recent advances in DL-based segmentation techniques able to instantly estimate fine-grained regional delineations with state-of-the-art accuracies. Those segmentations are used to produce local, anatomically grounded, feature-based affine matchings using iteration-free closed-form expressions. Estimated local affine transformations are then fused, with the log-Euclidean polyaffine framework, into an overall dense diffeomorphic transformation. We show that, compared to its affine counterpart, the proposed initialization leads to significantly better alignment for both traditional and DL-based non-linear registration algorithms. The proposed approach is also more robust and significantly faster than commonly used affine registration algorithms such as FSL FLIRT.

Keywords: Non-linear registration · Polyaffine transformations · Feature-based registration

1 Introduction

Medical image registration is the task of finding the best transformation, over a chosen search space, mapping a moving image onto a reference one so that the anatomical structures they portray match. In the case of non-linear registration, the search space contains non-global transformations with a high number of degrees of freedom (dof) in order to capture local, subtle displacements.

Registration techniques can broadly be divided into feature-based and intensity-based methods. Feature-based methods first identify common features

A. Frangi et al. (Eds.): IPMI 2023, LNCS 13939, pp. 614–625, 2023.
https://doi.org/10.1007/978-3-031-34048-2_47

between images, then find the transformation that best spatially aligns the corresponding features. These features are often based on geometric or anatomical characteristics which confer tangibility, interpretability. However, extracting these features historically requires tedious expert annotations or computationally intensive yet inaccurate automatic algorithms. On the other hand, intensity-based methods rely on a similarity measure between voxel-wise intensities of the two images as surrogate measure of quality of alignment. These approaches now dominate the field [7], thanks to their simple formulation as an optimization problem, the solution of which is approachable by calculus techniques like gradient descent. However, intensity similarity measures induce highly non-convex cost functions that render optimization to be prone to local minima. A good starting point is therefore essential. To this end, non-linear registration is usually preceded by an affine one and the optimization often follow a coarse-to-fine pyramidal strategy where the current step is initialized by the coarser estimate at the previous step. Yet, due to the global nature and limited number of degrees of freedom (dof) of affine transformations, this initialization might be insufficient.

Recently, deep-learning architectures have shown promising results in image segmentation, to the point that convolutional neural networks like U-Net now dominate challenges for this task [8]. FastSurfer [4] is able to accurately replicate FreeSurfer's segmentation - which usually takes more than 5 h - under a minute, while SynthSeg [9] is in addition able to do so in a contrast and resolution agnostic fashion. Anatomical feature extraction, historically difficult to achieve, is now accessible at minimal cost.

We propose, using segmentation computed from deep-learning models, to quickly produce an anatomically grounded polyaffine initial transformation as starting point for non-linear registration. The polyaffine transformation has many more dof than its affine counterpart, being able to capture non-global aspects like bends, thus offering a better start for non-linear registration algorithms. Based on local affine matchings with closed-form solutions, the optimization does not require iterative processes. The fusion into a dense, overall transformation is performed through the log-Euclidean polyaffine transformation (LEPT) [1] framework which ensures a diffeomorphic result. The end-to-end computation of the polyaffine transformation including deep-learning segmentation, local matchings and fusion onto a dense transformation can be done in less time than traditional linear registration algorithms like FSL FLIRT.

We tested the proposed polyaffine initialization against its affine counterpart and showed that it improves the alignment of anatomical structures at end point for both traditional and deep-learning non-linear registration. The effect was especially pronounced in the deep-learning case, where we observed a much more stable validation overlap loss during training with our approach.

2   Method

We present here a generic method to estimate, from two sets of homologous feature points, a dense diffeomorphic transformation under the polyaffine framework [1]. To be anatomically grounded, the feature points are extracted from fine-grained segmentation maps, which can now be computed accurately at minimal cost thanks to freely available pre-trained deep-learning segmentation models like FastSurfer [4] or SynthSeg [9].

1. **Extraction of the feature points:** Considering that the reference and moving images have undergone a fine-grained segmentation process, the feature points are defined as the centroids of the n segmented regions. This leads to two paired sets of points (see Fig. 1a.):

$$\begin{cases} X = \{X_1, \ldots, X_n\} : \text{reference point set.} \\ Y = \{Y_1, \ldots, Y_n\} \ : \text{moving point set.} \end{cases}$$

2. **Estimation of the background affine transformation:** One can search for an optimal global affine transformation $\hat{A}_B$ mapping the two paired sets by formulating a linear least squares (LLS) regression problem:

$$\hat{A}_B = \begin{pmatrix} \hat{L}_B & \hat{t}_B \\ 0 & 1 \end{pmatrix} = \underset{\substack{L \in \mathrm{GL}_d(\mathbb{R}) \\ t \in \mathbb{R}^d}}{\arg\min} \sum_{i=1}^{n} \|Y_i - (LX_i + t)\|^2 \tag{1}$$

Let us consider the relative coordinates $X'_i = X_i - \bar{X}$ and $Y'_i = Y_i - \bar{Y}$. A straightforward direct solution exists:

$$\hat{L}_B = \sum_{i=1}^{n} Y'_i X'^T_i \left(\sum_{i=1}^{n} X'_i X'^T_i \right)^{-1} \quad \text{and} \quad \hat{t}_B = \bar{Y} - \hat{L}_B \bar{X} \tag{2}$$

Let us denote $\tilde{X} = \hat{L}_B X + \hat{t}_B$ the transformed reference feature points by the background transformation.

3. **Construction of a graph structure:** From two homologous points taken independently, one can only estimate local translations. We aim to estimate local affine transformations which each require at least 4 points in 3D (3 in 2D). To this end, we propose to also account for the contextual information from neighboring points by defining a graph structure on the reference set. A neighborhood, $N(X_i)$, is associated to each point X_i from the reference set which is a collection of neighbors from the same set, chosen according to some suitable criterion, e.g. spatial distance or region adjacency. Since the moving and reference sets are paired, the graph structure only has to be defined on one of them. To simplify notations, we define $N(i)$ as the set of neighbor indices of X_i such that: $p \in N(i) \equiv X_p \in N(X_i)$. In our (3D) implementation, we chose to define the graph structure using a Delaunay triangulation (see Fig. 1b) which is very easy and fast to compute. In this construction, two elements X_i and X_j are neighbors if they are connected by an edge. Since each X_i is the vertex of one or more tetrahedrons with other elements of X, there is a minimum of 4 points per neighborhood, enough to regress an affine transformation.

4. **Estimation of the local affine transformations:** For each $\tilde{X}_i$, a local optimal affine transformation $\hat{A}_i$ matching its neighborhood to the homologous neighborhood of Y_i is sought through LLS in the same vein as in Eq. 1 (see Fig. 1c):

$$\hat{A}_i = \begin{pmatrix} \hat{L}_i & \hat{t}_i \\ 0 & 1 \end{pmatrix} = \underset{\substack{L \in GL_d(\mathbb{R}) \\ t \in \mathbb{R}^d}}{\arg\min} \sum_{p \in N(i)} \|Y_p - (L\tilde{X}_p + t)\|^2 \tag{3}$$

The solution is similar to Eq. 2, except that the sum and averages are done using only the points of a neighborhood rather than using all points. At this stage we have n local affine transformations between homologous neighborhoods attached to the n reference feature points (see Fig. 1d).

5. **Creation of the weight maps:** To create an overall dense transformation, weight maps are established to spatially modulate the contribution of each local affine transformations. For any point x of the domain over which the image is defined, a set of weights associated to each X_i is defined using a smooth kernel function based on the distance between x and the center of the associated neighborhood $\tilde{X}_i$, and a parameter σ controlling the smoothness (see Fig. 1e). The kernel can typically be Gaussian of standard deviation σ. The set $\{w_i(x), i = 1, \ldots, n\}$ can be sparse if the kernel has a bounded support. In addition, a background weight w_B, uniform across the whole image domain, is chosen. It should be small enough to be negligible with respect to the weights associated to the feature points.

6. **Construction of the overall dense diffeomorphic transformation:** From the collection of local affine transformations and the weight maps, one can produce an overall dense diffeomorphic transformation through the log-Euclidean polyaffine framework [1].

   (a) A stationary velocity field (SVF) V is built by interpolating a log displacement vector for each x by averaging the logarithms of the local transformations, weighted by the associated weight maps and the background weight:

$$V(x) = \frac{\sum_{i=1}^{n} w_i(x).\log(\hat{A}_i)}{w_B + \sum_{i=1}^{n} w_i(x)} x \tag{4}$$

   (b) A diffeomorphic transformation $\exp(V)$ can be obtained by integration of a stationary ordinary differential equation (ODE) (see Fig. 1.f):

$$\exp(V) = \phi^1, \text{ where } \begin{cases} \frac{\partial \phi^t}{\partial t} = V(\phi^t) \\ \phi^0 = \text{Id} \end{cases} \tag{5}$$

   The integration can be done efficiently on regular grids using the scaling and squaring method [1] by approximating the exponential of the scaled (close to 0) field and composing recursively.

   (c) This diffeomorphic transformation composed after the background affine one forms the overall transformation: $T = \hat{A}_B \circ \exp(V)$.

A. Legouhy et al.

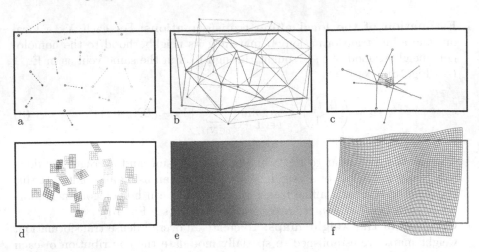

Fig. 1. Illustration of several steps of the polyaffine estimation. a) Two sets of paired points (plain disc: moving, circle: reference). b) Delaunay triangulation performed on the reference set (blue) and pattern reproduced on the moving one (red). c) Local affine regression between two homologous neighborhoods. d) Set of estimated local affine transformations. e) Color grading representing all weight maps combined. f) Overall polyaffine transformation. Black rectangles represent the frame of the reference image. (Color figure online)

Remark 1. For more robustness, one can imagine weighted or trimmed versions of the linear least squares minimization in Eq. 3 if the graph defined at step 2 is weighted, or to account for uncertainty in the feature point extraction.

Remark 2. When dealing with smaller neighborhoods, one can regress local transformations with less degrees of freedom. For a rigid transformation, i.e. L constrained to be a rotation matrix, a minimum of 2 points in each neighborhood are needed (in 3D), and a direct solution to the LLS problem can be found in [10,11]. For translations only, i.e. L an identity matrix, the solution is simply $\hat{t} = \bar{Y} - \bar{X}$. Only singleton neighborhoods are necessary, i.e. $N(X_i) = X_i$. However, this setting reduces robustness that comes with having more equations than unknowns, and the contextual information from the neighboring positions is discarded (although it still has an impact during interpolation).

Remark 3. The choice of the weight maps is crucial to shape the overall transformation. The window of the kernel modulates the amount of smoothness. For small values of σ, one can recover very local changes but there is a risk of overfitting or having sub machine precision values when the feature point set is too sparse in space (although the use of a background weight helps). As σ increases, the polyaffine result gets smoother, losing curvature, eventually converging to an affine transformation for $\sigma = \infty$. This spectrum of possible transformations is a strength of the method. For Gaussian kernels, we found empirically that taking $\sigma = \frac{2}{n} \sum_{i=1}^{n} \arg\min_{X_p \in X} \|X_i - X_p\|$ often leads to good results.

Remark 4. The role of the background transformation is to ensure stability when extrapolating in regions far away from the feature points. As this polyaffine mapping doesn't assume any pre-alignment, it cannot converge to null displacement towards the domain boundary. Instead, it should follow the overall flow in the smoothest way i.e. through a global affine. The background weight should be sufficiently small to only matter far enough from the feature points but not interfere near them.

Remark 5. In Eq. 4, we used a log-Euclidean average of the affine transformations, i.e. an Euclidean mean on their principal logarithms, to ensure an invertible overall transformation. This invertiblity wouldn't be guaranteed with a simple Euclidean average. The well-definiteness of the principal logarithm of an affine transformation matrix A only depends on its linear part L. The eigenvalues of L must not lie on the (closed) half-line of negative real numbers [20]. This, however, only concerns very large transformations that effect a rotation close to π, which is unlikely in real settings. There are only n matrix logarithms computation to perform which can be done efficiently through the inverse scaling and squaring method for matrices [2].

3   Experimental Design

To evaluate the benefit of the proposed segmentation-based polyaffine initialization compared to its affine counterpart for traditional and deep-leaning non-linear registration, we registered subjects from 3 databases onto a template and computed a structure overlap measure.

3.1   Data

We used T1-weighted images from 3 databases in order to cover various acquisition protocols and to have brains of different maturation and health conditions:

- ADNI (adni.loni.usc.edu), the Alzheimer's Disease Neuroimaging Initiative [14], is a cohort of elderly subjects divided into cognitively normal (HC), with mild cognitive impairment (MCI), and with Alzheimer's disease (AD).
- IXI dataset (brain-development.org/ixi-dataset) is composed of adult healthy subjects aged 20 years old or more.
- UK Biobank (ukbiobank.ac.uk) is a huge database of subjects from the UK, between 40 and 69 years old.

For all databases, the voxel size is around 1 mm isotropic. The reference for registration is an MNI template (ICBM 2009a [18]) with 1 mm isotropic voxel size. Subjects have been drawn randomly from those databases and distributed into training, validation and testing sets following Table 1.

3.2   Segmentation

We used FastSurfer [4] to quickly (in less than a minute) produce FreeSurfer-like segmentations of the moving and reference images into 95 anatomical regions following the Desikan-Killiany-Tourville (DKT) [5,6] protocol.

Table 1. Distribution of the subjects for the training, validation and testing sets.

Database	Training set	Validation set	Testing set
IXI	20	5	100
UK Biobank	20	5	100
ADNI (HC/MCI/AD)	60 (20/20/20)	15 (5/5/5)	150 (50/50/50)

3.3   Registration

Affine Initialization: The affine pre-registrations were performed using FSL FLIRT [15]. For 37 subjects in all sets combined (7.8%), FLIRT optimization failed, leading to severe misalignments like upside-down brains. To obtain decent outputs, we re-ran those that failed with a little "help" like resampling onto the reference image grid or constraining the search space.

Polyaffine Initialization: From the general recipe presented in Sect. 2, we opted for the following implementation details:

1. **Extraction of the feature points:** Four regions were ignored: left and right cerebral white matter (too large), white matter hypointensities and cerebrospinal fluid (not consistent between images). For each remaining region, a centroid (feature point) was computed as the average spatial coordinates of the voxels belonging to the region.
2. **Construction of the graph structure:** The Delaunay triangulation was computed only once on the template feature points using the Qhull library.
3. **Estimation of the local affine transformations:** Optimal affine transformations between homologous neighborhoods were computed using the direct solution in Eq. 2.
4. **Creation of the weight maps:** We opted for a Gaussian kernel of standard deviation $\sigma = 20$ mm roughly following the rule of thumb in Remark 3. We set the background weight to $w_B = 10^{-5}$ uniformly across the image domain.
5. **Construction of the overall dense diffeomorphic transformation:** We computed the SVF on the image grid downsampled by a factor 2 to quicken the subsequent exponentiation. The exponential of the SVF in Eq. 5 was computed through scaling and squaring using 7 integration steps and resampled onto the original grid.

All polyaffine initializations worked fine at first attempt. We have made freely available on Github[1] the implementation described here, which is based on the Python version of SimpleITK wrapper for ITK open-source software.

Traditional Non-linear Registration: Symmetric Normalization (SyN) [16], from the Advanced Normalization Tools (ANTs) suite was used. It is one of the best traditional non-linear algorithms according to the evaluation in [17]. The optimized similarity metric was the local squared correlation coefficient (LCC).

---

[1] https://github.com/CIG-UCL/polaffini.

Deep-Learning Non-linear Registration: Voxelmorph [12] style architectures with diffeomorphic implementations [13] were used. They are composed of a U-Net shaped as in [12], containing all the trainable parameters, that takes as input a moving and a reference image, and outputs a vector field. This vector field is fed to an integrator block to produce a diffeomorphism which is used to transform the moving image through a resampler block. We chose LCC as image similarity loss with weight 1. We used segmentations as auxiliary data to compute an average Dice overlap score used as segmentation loss with weight 0.3. A regularization loss, L^2 norm of the Jacobian of the vector field was also used to promote smooth estimates, with weight 1. We trained one model using images pre-registered using the proposed polyaffine initialization, and a second one using its affine counterpart. Due to limited GPU memory (8 GB), the models operated on images downsampled to $2 \times 2 \times 2$ mm grids.

3.4 Evaluation

The evaluation was performed on the subjects from the testing set that were unseen by the deep-learning models at training. For each image, the transformation estimated at initialization and the one from the non-linear registration were applied together at once so that the transformed moving image was reconstructed in the reference grid with a single interpolation. The quality of registrations was evaluated using a Dice overlap score between the segmentation labels of the reference image and the corresponding ones of the transformed moving image. For the sub-cortical results, we computed 1 dice per region and reported the average. For the cortex, in order to be comparable with [12], we regrouped the regions into one label and computed a single Dice.

Having a good overlap score between the reference and the transformed moving segmentation is meaningless if it is the result of a non topology-preserving process. Under reasonable conditions (rotations with magnitude smaller than π), polyaffine transformations under the LEPT framework [1] are diffeomorphisms. Also, the examined non-linear registration algorithms (ANTs and deep) contain a regularization term encouraging smooth transformations, yet not completely forbidding them to be improper. To evaluate the topology-preserving aspect, we also counted the number of negative values of the Jacobian determinant of each estimated final transformation (initialization transformation composed with non-linear registration output).

4 Results

Dice scores for sub-cortical areas and cortex for the various datasets are shown in Fig. 2. Statistics regarding differences between overlap scores after registration with affine against proposed polyaffine initialization are reported in Table 2 for all datasets combined.

We observe that the sub-cortical structures are generally well aligned already just with the polyaffine initialization. While non-linear registrations do not typically improve upon this initial alignment, the deep learning model with polyaffine

Table 2. Statistics about differences between overlap scores after registration with affine vs proposed polyaffine initialization. Reported p-values are for paired t-tests, Cohen's d for paired samples (with polyaffine − affine on numerator).

	affine init.	polyaffine init.	p-value	Cohen's d
sub-cortical	affine	polyaffine	$<10^{-115}$	1.864
regions	affine + ANTs	polyaffine + ANTs	$<10^{-14}$	0.447
	affine + deep	polyaffine + deep	$<10^{-108}$	1.762
cortex	affine	polyaffine	$<10^{-107}$	1.745
	affine + ANTs	polyaffine + ANTs	$<10^{-38}$	0.798
	affine + deep	polyaffine + deep	$<10^{-176}$	3.019

initialization shows clearly better overlaps. For a given approach (ANTs or deep), results are significantly better when initializing with the proposed approach, with a large effect size for the deep-learning models.

In the cortex, non-linear registration clearly improves upon the initial alignment. Using the proposed polyaffine initialization once again leads to significantly better results. While the effect appears modest for ANTs, it is consistent over almost all subjects. The difference is however striking for the deep-learning models where we report a very large effect size.

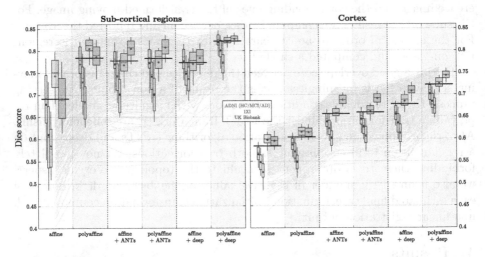

Fig. 2. Dice scores after non-linear registration initialized with affine and proposed polyaffine transformation for ADNI (red, HC/MCI/AD left to right), IXI (green) and UK Biobank (blue) subjects. Thick lines represent medians across all datasets. (Color figure online)

Part of the explanation for why the proposed polyaffine initialization leads to better results in the deep-learning case can be found by examining the evolution of the losses during training 3. While image similarity losses follow similar trajectories for both approaches, segmentation losses, which actually quantifies the alignment of anatomical structures, show a much smoother profile with the proposed polyaffine starting point. The gap between the training and validation losses is also much smaller with the proposed initialisation.

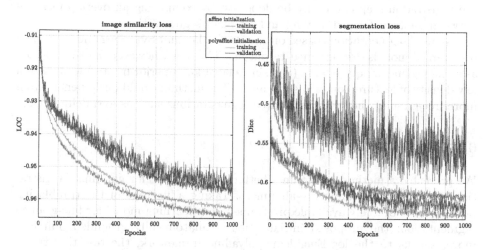

Fig. 3. Evolution of image similarity (LCC) and segmentation (average Dice on all regions) losses during training of deep-learning models with affine (blue) and proposed polyaffine (red) initialization for training (light) and validation (dark) samples. (Color figure online)

The resulting overall transformations were proper for all subjects and methods. We did not find any negative Jacobian determinant whether using an affine or the proposed polyaffine initialization and whether using ANTs or a deep-learning model for the subsequent non-linear registration.

5   Discussion

To help avoid local minima at the finest resolution, most traditional non-linear registration algorithms follow a coarse-to-fine resolution optimization and some recent deep-learning models also adopt such pyramidal strategies at training [19]. However they still rely on a prior, usually intensity-based, affine alignment to recover the largest displacement. Although it would probably allow a user to skip the first coarse levels of the pyramid, our method should not be seen as a rival to the coarse-to-fine approach but as a more robust, anatomically grounded and less constrained alternative to the affine pre-alignment.

In [3], an anatomically grounded non-linear registration scheme was proposed where optimal affine transformations between homologous delineated regions were sought by iteratively optimizing an image similarity criterion. An overall transformation is then constructed by attributing the estimated affine displacement to an eroded version of the associated regions and using a log-Euclidean interpolation for a smooth transition to enforce a diffeomorphic result. Our approach, by contrast, does not require slow iterative optimization for the matching and is intended to be taken as an alternative to affine initialization rather than a non-linear registration endpoint.

Assuming that the segmentation process was successful, quality control of the registration step can easily be done just by computing an overlap measure between the transformed moving and reference segmentations.

We used segmentations based on the DKT atlas that SynthSeg and Fastsurfer pre-trained models are designed to output. This may, however, be sub-optimal as some regions are quite large (e.g. single label for superior frontal) leading to a low density of feature points in some areas. The method would likely benefit from more fine-grained and equi-distributed segmentations such as gyri delineation.

6   Conclusion

We presented a method to obtain a better starting point for non-linear registration than the usual affine pre-alignment. Taking advantage of the trailblazing performances of freely available pre-trained deep-learning model for fine-grained segmentation, the proposed initialization is anatomically grounded. Furthermore, thanks to the log-Euclidean polyaffine framework, the resulting transformation is diffeomorphic. We showed that this polyaffine initialization leads to better structural overlap, especially in the cortex, for both traditional and deep-learning non-linear registration techniques. Our experiments revealed that deep-learning registration is more sensitive to initialization and the proposed approach provides a highly efficient and effective strategy to tackle this issue. We also verified that the overall transformation is indeed proper. Finally, the proposed polyaffine estimation was more robust and faster than using an affine registration algorithm like FLIRT.

References

1. Arsigny, V., Commowick, O., Ayache, N., et al.: A fast and log-euclidean polyaffine framework for locally linear registration. J. Math. Imaging Vis. **33**, 222–238 (2009)
2. Higham, N.J.: Functions of Matrices: Theory and Computation. Society for Industrial and Applied Mathematics (2008)
3. Commowick, O., et al.: An efficient locally affine framework for the smooth registration of anatomical structures. Med. Image Anal. **12**(4), 427–441 (2008)
4. Henschel, L., Conjeti, S., Estrada, S., Diers, L., Fischl, B., Reuter, M.: FastSurfer - a fast and accurate deep learning based neuroimaging pipeline. NeuroImage **219**, 117012 (2020)
5. Klein, A., Tourville, J.: 101 labeled brain images and a consistent human cortical labeling protocol. Front. Neurosci. **6**, 171 (2012)
6. Desikan, R.S., et al.: An automated labeling system for subdividing the human cerebral cortex on MRI scans into gyral based regions of interest. NeuroImage **31**(3), 968–980 (2006)

7. Oliveira, F.P.M., Tavares, J.M.R.S.: Medical image registration: a review. Comput. Methods Biomech. Biomed. Engin. **17**, 2 (2014)
8. Isensee, F., Jaeger, P.F., Kohl, S.A.A., et al.: nnU-Net: a self-configuring method for deep learning-based biomedical image segmentation. Nat. Methods **18**, 203–211 (2021)
9. Billot, B., et al.: SynthSeg: domain randomisation for segmentation of brain MRI scans of any contrast and resolution. arXiv:2107.09559 (2021)
10. Horn, B.K.P.: Closed-form solution of absolute orientation using unit quaternions. J. Opt. Soc. Am. A **4**(4), 629–642 (1987)
11. Horn, B.K.P., Hilden, H.M., Negahdaripour, S.: Closed-form solution of absolute orientation using orthonormal matrices. J. Opt. Soc. Am. A **5**(7), 1127–1135 (1988)
12. Balakrishnan, G., Zhao, A., Sabuncu, M.R., Guttag, J., Dalca, A.V.: VoxelMorph: a learning framework for deformable medical image registration. IEEE Trans. Med. Imaging **38**(8), 1788–1800 (2019)
13. Dalca, A.V., Balakrishnan, G., Guttag, J., Sabuncu, M.R.: Unsupervised learning of probabilistic diffeomorphic registration for images and surfaces. Med. Image Anal. **57**, 226–236 (2019)
14. Petersen, R.C., et al.: Alzheimer's disease neuroimaging initiative (ADNI): clinical characterization. Neurology **74**(3), 201–209 (2010)
15. Jenkinson, M., Bannister, P., Brady, J.M., Smith, S.M.: Improved optimisation for the robust and accurate linear registration and motion correction of brain images. Neuroimage **17**(2), 825–841 (2002)
16. Avants, B.B., Epstein, C.L., Grossman, M., Gee, J.C.: Symmetric diffeomorphic image registration with cross-correlation: evaluating automated labeling of elderly and neurodegenerative brain. Med. Image Anal. **12**(1), 26–41 (2008)
17. Klein, A., et al.: Evaluation of 14 nonlinear deformation algorithms applied to human brain MRI registration. NeuroImage **46**(3), 786–802 (2009)
18. Fonov, V., Evans, A.C., Botteron, K., Almli, C.R., McKinstry, R.C., Collins, D.L.: Unbiased average age-appropriate atlases for pediatric studies. Neuroimage **54**(1), 313–327 (2011)
19. Mok, T.C.W., Chung, A.C.S.: Large deformation diffeomorphic image registration with laplacian pyramid networks. In: Martel, A.L., et al. (eds.) MICCAI 2020. LNCS, vol. 12263, pp. 211–221. Springer, Cham (2020). https://doi.org/10.1007/978-3-030-59716-0_21
20. Cheng, S.H., Hingham, N.J., Kenny, C.S., Laub, A.J.: Approximating the logarithm of a matrix to specified accuracy. SIAM J. Matrix Anal. Appl. **22**(4), 1112–1125 (2001)

Segmentation

Separation

Better Generalization of White Matter Tract Segmentation to Arbitrary Datasets with Scaled Residual Bootstrap

Wan Liu and Chuyang Ye[(✉)]

School of Integrated Circuits and Electronics, Beijing Institute of Technology,
Beijing, China
chuyang.ye@bit.edu.cn

Abstract. *White matter* (WM) tract segmentation is a crucial step for brain connectivity studies. It is performed on *diffusion magnetic resonance imaging* (dMRI), and *deep neural networks* (DNNs) have achieved promising segmentation accuracy. Existing DNN-based methods use an annotated dataset for model training. However, the performance of the trained model on a different test dataset may not be optimal due to distribution shift, and it is desirable to design WM tract segmentation approaches that allow better generalization of the segmentation model to arbitrary test datasets. In this work, we propose a WM tract segmentation approach that improves the generalization with scaled residual bootstrap. The difference between dMRI scans in training and test datasets is most noticeably caused by the different numbers of diffusion gradients and noise levels. Since both of them lead to different *signal-to-noise ratios* (SNRs) between the training and test data, we propose to augment the training scans by adjusting the noise magnitude and develop an adapted residual bootstrap strategy for the augmentation. First, with a dictionary-based linear representation of diffusion signals, we compute the signal residuals for the training dMRI scans, which can represent samples drawn from the noise distribution. Then, we adapt the bootstrap procedure by scaling the residuals that are randomly drawn with replacement and adding the scaled residuals to the linear signal representation, where augmented dMRI scans with different SNRs are generated. Finally, the augmented and original images are jointly included in model training. Since it is difficult to know the SNR of the test data *a priori*, we choose to perform the residual scaling with multiple factors. To validate the proposed approach, two dMRI datasets were used, and the experimental results show that our method consistently improved the generalization of WM tract segmentation under various settings.

Keywords: White matter tract segmentation · residual bootstrap · generalization

1 Introduction

White matter (WM) tract segmentation on *diffusion magnetic resonance imaging* (dMRI) provides a valuable quantitative tool for various brain stud-

A. Frangi et al. (Eds.): IPMI 2023, LNCS 13939, pp. 629–640, 2023.
https://doi.org/10.1007/978-3-031-34048-2_48

ies [1,7,21,24]. Manually delineated WM tracts are generally considered the gold standard segmentation, but the annotation process can be time-consuming and requires the expertise of experienced radiologists. Therefore, automated WM tract segmentation approaches are developed, which classify fiber streamlines [4,6] obtained with tractography [2,9] or directly provide voxelwise labeling results [3,19,26]. In particular, methods based on *deep neural networks* (DNNs) have substantially improved the accuracy of WM tract segmentation [15,25,27]. For example, Zhang et al. [27] group fiber streamlines into different WM tracts with a DNN that takes the spatial coordinates of the points along a fiber streamline as input; in [25], fiber orientation maps extracted from dMRI scans are fed into a U-net [20] to directly predict the existence of WM tracts at each voxel.

The DNN-based segmentation model is generally trained on a dataset where both dMRI scans and WM tract annotations are available. However, the performance of the model on an arbitrary test dataset that is different from the training dataset may be degraded due to distribution shift, where the use of different numbers of diffusion gradients and different noise levels are two major contributing factors [18]. Since dMRI scans can be acquired with various protocols, the improvement of the generalization of WM tract segmentation models to arbitrary test data becomes an important research topic. Although domain adaptation techniques [8] may be applied to improve the generalization, they require access to the test data during model training, which is not guaranteed when arbitrary test data is considered, and thus they are out of scope for this work. To account for the different numbers of diffusion gradients between training and test datasets, in [25] additional training scans are obtained by subsampling the diffusion gradients of the training data, and this allows improved segmentation accuracy on test data. However, the segmentation accuracy may still be improved by taking the *signal-to-noise ratio* (SNR) into consideration during model training.

In this work, we seek to further improve the generalization of WM tract segmentation from the perspective of SNR.[1] We focus on volumetric WM tract segmentation that directly obtains volumes of WM tract labels without requiring the tractography step. We assume that by producing diverse SNRs for training data, the training data can better represent the test data, and the trained model can better generalize to the test data. Therefore, we propose a scaled residual bootstrap strategy that augments the training scans with adjusted noise magnitude. First, we estimate a linear dictionary-based representation of diffusion signals and compute the residuals of the representation. These residuals are considered samples drawn from the noise distribution [10]. Then, for each diffusion gradient, the residual is drawn with replacement, and we adapt the standard residual bootstrap by scaling the residual. The scaled residuals are added to the linear representation of diffusion signals to generate augmented dMRI scans with different SNRs. Finally, the augmented images are used together with the original images for model training. Since it is difficult to know the SNR of the

---

[1] Note that the use of different numbers of diffusion gradients implicitly leads to different SNRs of measures derived from dMRI as well.

test data *a priori*, we choose to perform the residual scaling with multiple factors. The proposed approach was evaluated on two brain dMRI datasets, where various experimental settings of training and test scans were considered. The results show that our method consistently improved the generalization of WM tract segmentation under these various settings.

2   Methods

2.1   Problem Formulation

Suppose we are given a set of dMRI scans from a training dataset and the set of their annotations of WM tracts. We seek to train a WM tract segmentation model with good generalization, i.e., it performs well on an arbitrary test dataset. Like existing volumetric WM tract segmentation approaches [14,25], the model input is fiber orientation maps computed from dMRI. Two major factors that cause the difference between the training and test dMRI data are the use of different numbers of diffusion gradients and different noise levels. Since increasing/decreasing the number of diffusion gradients also leads to increased/decreased SNRs in the fiber orientation maps, respectively, we assume that adjusting the SNR of the dMRI scans for model training can effectively improve the generalization of the trained model to other datasets. Although existing approaches have considered SNR manipulation in the data augmentation operations of model training [25], it is applied to the network input of fiber orientation maps. As fiber orientations are orientations with unit lengths, adding realistic noise that is consistent with imaging physics to them is nontrivial. Therefore, we seek to further explore data augmentation with SNR adjustment in model training to improve the generalization of WM tract segmentation models.

2.2   Model Training with Scaled Residual Bootstrap

To produce training data with diverse SNRs and realistic noise distributions, we propose a scaled residual bootstrap strategy for model training. For convenience, we denote the diffusion weighted signals at each voxel of a training dMRI scan by a vector y, where $y \in \mathbb{R}^{N_d}$ and N_d is the number of diffusion gradients. It has been shown that diffusion weighted signals can be linearly represented with a properly designed dictionary [16,17]:

$$y = Dx + \epsilon, \tag{1}$$

where $D \in \mathbb{R}^{N_d \times N_a}$ is the dictionary with N_a dictionary atoms, $x \in \mathbb{R}^{N_a}$ is the vector of dictionary coefficients, and $\epsilon \in \mathbb{R}^{N_d}$ represents the noise.

If the distribution of ϵ is known, different levels of realistic noise can be added to the noise-free linear representation to provide training data with different SNRs. This motivates us to adopt a residual bootstrap strategy, which provides a feasible way of approximating the noise distribution. Then, by modifying the noise distribution, we achieve the goal of augmenting the SNR levels of training data. There are two major steps in the proposed method, which are 1) residual computation and 2) data generation with scaled residuals.

Residual Computation. Like the standard residual bootstrap, we first estimate x with the pseudoinverse of $\mathbf{D}$:

$$\hat{x} = (\mathbf{D}^\mathsf{T}\mathbf{D})^{-1}\mathbf{D}^\mathsf{T}y, \tag{2}$$

where $\hat{x}$ is the estimated coefficient vector. Then, the linear representation of
the diffusion weighted signals can be estimated as

$$\hat{y} = \mathbf{D}\hat{x} = \mathbf{D}(\mathbf{D}^\mathsf{T}\mathbf{D})^{-1}\mathbf{D}^\mathsf{T}y. \tag{3}$$

The residuals $\hat{\epsilon}$ of the signal representation can be simply computed by subtracting $\hat{y}$ from y

$$\hat{\epsilon} = y - \hat{y} = (\mathbf{I} - \mathbf{D}(\mathbf{D}^\mathsf{T}\mathbf{D})^{-1}\mathbf{D}^\mathsf{T})y. \tag{4}$$

Then, to ensure that the variances of the residuals $\hat{\epsilon}$ are consistent with those of
the noise ϵ, the residuals are corrected with the following normalization [5,10]:

$$\hat{\epsilon}'_i = \frac{\hat{\epsilon}_i}{\sqrt{1 - h_{ii}}}. \tag{5}$$

Here, $\hat{\epsilon}_i$ is the i-th entry of $\hat{\epsilon}$, $\hat{\epsilon}'_i$ is the corresponding corrected residual, and h_{ii} is
the i-th diagonal entry of $\mathbf{H} = \mathbf{D}(\mathbf{D}^\mathsf{T}\mathbf{D})^{-1}\mathbf{D}^\mathsf{T}$. The set $\mathcal{E} = \{\hat{\epsilon}'_i\}_{i=1}^{N_\mathrm{d}}$ of corrected
residuals is then used in the bootstrap procedure that provides training data
with diverse SNRs, and the procedure is described next.

Data Generation with Scaled Residuals. The corrected residuals $\mathcal{E}$ can
be viewed as samples drawn from the noise distribution [5], and in the standard
residual bootstrap, they are randomly drawn with replacement and added to the
linear representation $\hat{y}$. For our purpose of better generalization, we seek to generate samples with diverse SNRs. Therefore, the standard bootstrap procedure is
modified with a scaling operation. Specifically, for the i-th diffusion gradient, we
sample from $\mathcal{E}$ with replacement, and the sampled residual is denoted by $\tilde{\epsilon}_i$. The
vector comprising the sampled residuals for all diffusion gradients is represented
as $\tilde{\epsilon} = (\tilde{\epsilon}_1, \ldots, \tilde{\epsilon}_{N_\mathrm{d}})$. Then, a bootstrap signal $\tilde{y}$ is generated as

$$\tilde{y} = \hat{y} + r\tilde{\epsilon}, \tag{6}$$

where r is the scaling factor that controls the magnitude of noise. r is selected
from a predefined candidate set $\mathcal{R}$. By repeating the scaled residual bootstrap
in Eq. (6) for each voxel, bootstrap diffusion weighted images can be generated.

Note that in dMRI acquisition, the $b0$ image without diffusion weighting is
also acquired, and when more than one $b0$ images are available, their SNR can
be adjusted as well. We denote the j-th $b0$ signal at each voxel by y_j^0, and the
number of $b0$ images is denoted by N_0. Then, the residual $\hat{\epsilon}_j^0$ for the j-th $b0$
signal is calculated by

$$\hat{\epsilon}_j^0 = y_j^0 - \bar{y}^0, \tag{7}$$

where $\bar{y}^0 = \frac{1}{N_0} \sum_{j=1}^{N_0} y_j^0$ is the mean value of all $b0$ signals. These residuals form a set $\mathcal{E}^0$. For each j, a sample is drawn from $\mathcal{E}^0$ with replacement, which is denoted by $\tilde{\epsilon}_j^0$, and the bootstrap $b0$ signal is generated as

$$\tilde{y}_j^0 = \bar{y}^0 + r\tilde{\epsilon}_j^0. \tag{8}$$

Here, r has the same value as in Eq. (6). Equation (8) is repeated for each voxel to obtain bootstrap $b0$ images.

After bootstrap $b0$ images and diffusion weighted images are generated, they are combined to obtain new dMRI scans with different SNRs. These bootstrap dMRI scans are used to train the segmentation model together with the original dMRI scans based on the WM tract annotations.

2.3 Implementation Details

Our method is agnostic to the architecture of the segmentation model. For demonstration, the state-of-the-art TractSeg architecture [25] is used as the backbone network, but other network structures [13,14] may also be applied. As in [25], we extract fiber orientation maps from dMRI scans with *constrained spherical deconvolution* (CSD) [22] (for single-shell dMRI data) or *multi-shell multi-tissue CSD* (MSMT-CSD) [11] (for multi-shell dMRI data), and use these maps as network input. At most three fiber orientations are allowed, and all WM tracts are jointly segmented [25].

We use the SHORE basis[2] [17] for the linear representation of diffusion signals, which is a common choice. To generate bootstrap training data with diverse SNRs, the set of candidate scaling factors is $\mathcal{R} = \{2, 3, 4\}$. Since it is difficult to predetermine the SNR of arbitrary test data, all values in $\mathcal{R}$ are used for bootstrap, and each value is used once for each training scan.

For model training, following [25], we use the binary cross entropy loss function, which is minimized by Adamax [12] with a batch size of 56 and 300 training epochs; the initial learning rate is set to 0.001. We select the model that has the best segmentation accuracy on a validation dataset. Traditional data augmentation implemented online in TractSeg, such as intensity perturbation and spatial transformation, is also applied online in the proposed method.

3 Results

3.1 Datasets and Experimental Settings

We used two dMRI datasets to evaluate our method. The first one is the publicly available *Human Connectome Project* (HCP) dataset [23], and the second one is an in-house dMRI dataset. A detailed description of the two datasets and their experimental settings is given below.

---

[2] The default setting given in https://dipy.org/documentation/1.4.1./reference/dipy. reconst/#dipy.reconst.shore.ShoreModel is used.

The HCP Dataset. The dMRI scans in the HCP dataset were acquired with 270 diffusion gradients ($b = 1000$, 2000, and $3000 \, \text{s/mm}^2$) and an isotropic image resolution of 1.25 mm. 18 $b0$ images were also acquired for each dMRI scan. 72 WM tracts were manually delineated for the HCP dataset[3]. We used 100 scans in our experiments, where 55 and 15 scans were used as the training set and validation set, respectively, and the remaining 30 scans were used for testing. To improve the generalization of the segmentation model to different imaging protocols, in TractSeg [25], subsampling of diffusion gradients was performed on the original training dMRI scans, where dMRI scans with 12 and 90 diffusion gradients associated with $b = 1000 \, \text{s/mm}^2$ were generated for model training together with the original dMRI scans.[4] Here, we followed [25] and performed the subsampling as well for the original and bootstrap training data for model training. For convenience, the original HCP dataset is referred to as HCP_1.25mm_270, and the subsampled datasets with 12 and 90 diffusion gradients are referred to as HCP_1.25mm_12 and HCP_1.25mm_90, respectively.

To evaluate the performance of the proposed method on test scans that were acquired with different protocols, we generated additional test sets from the 30 original test scans. First, like the training data in HCP_1.25mm_12 and HCP_1.25mm_90, the 12 and 90 diffusion gradients associated with $b = 1000 \, \text{s/mm}^2$ were selected from the 30 test scans, respectively. Second, 34 diffusion gradients associated with $b = 1000 \, \text{s/mm}^2$ were selected for the test scans, so that their imaging protocol was different from the original and subsampled training data, and the images associated with this subsampling are referred to as HCP_1.25mm_34. Only three $b0$ images were kept for HCP_1.25mm_34. Finally, another test set HCP_1.25mm_36 was generated from the test scans by selecting 18 diffusion gradients associated with $b = 1000 \, \text{s/mm}^2$ and 18 diffusion gradients associated with $b = 2000 \, \text{s/mm}^2$, which also produced dMRI scans that used a different imaging protocol than the training data. Only one $b0$ image was kept for HCP_1.25mm_36. A summary of these different datasets is listed in Table 1.

In addition, to investigate the impact of the amount of training data on the segmentation, three other experimental settings were considered, where 10, 20, or 30 training subjects were used and the other settings were not changed.

The In-House Dataset. The segmentation models trained on the HCP dataset were also applied to an in-house dataset for further evaluation. The dMRI scans in the in-house dataset were acquired with 270 diffusion gradients ($b = 1000$, 2000, and $3000 \, \text{s/mm}^2$) and one $b0$ image. The spatial resolution is 1.7 mm isotropic. These scans were acquired on a scanner that is different from that of the HCP dataset. Due to the annotation cost, only ten of the 72 annotated WM tracts of the HCP dataset were manually delineated, and the delineation was performed on 17 in-house dMRI scans. These annotations were used only to evaluate the segmentation accuracy. This in-house dataset is referred to

---

[3] The annotations can be downloaded at https://doi.org/10.5281/zenodo.1088277.

[4] All $b0$ images were kept for these two cases.

Table 1. A summary of the datasets used in the experiments

Dataset	Resolution	Diffusion gradients	Usage
HCP_1.25mm_270	1.25 mm	$90 \times b = 1000\,\text{s/mm}^2$ $90 \times b = 2000\,\text{s/mm}^2$ $90 \times b = 3000\,\text{s/mm}^2$ $18 \times b = 0\,\text{s/mm}^2$	Training & Test
HCP_1.25mm_12	1.25 mm	$12 \times b = 1000\,\text{s/mm}^2$ $18 \times b = 0\,\text{s/mm}^2$	Training & Test
HCP_1.25mm_90	1.25 mm	$90 \times b = 1000\,\text{s/mm}^2$ $18 \times b = 0\,\text{s/mm}^2$	Training & Test
HCP_1.25mm_34	1.25 mm	$34 \times b = 1000\,\text{s/mm}^2$ $3 \times b = 0\,\text{s/mm}^2$	Test
HCP_1.25mm_36	1.25 mm	$18 \times b = 1000\,\text{s/mm}^2$ $18 \times b = 2000\,\text{s/mm}^2$ $1 \times b = 0\,\text{s/mm}^2$	Test
IH_1.7mm_270	1.7 mm	$90 \times b = 1000\,\text{s/mm}^2$ $90 \times b = 2000\,\text{s/mm}^2$ $90 \times b = 3000\,\text{s/mm}^2$ $1 \times b = 0\,\text{s/mm}^2$	Test
IH_1.7mm_36	1.7 mm	$18 \times b = 1000\,\text{s/mm}^2$ $18 \times b = 2000\,\text{s/mm}^2$ $1 \times b = 0\,\text{s/mm}^2$	Test

as IH_1.7mm_270. We also synthesized another dataset IH_1.7mm_36 from IH_1.7mm_270 for evaluation, where 18 diffusion gradients of $b = 1000\,\text{s/mm}^2$ and 18 diffusion gradients of $b = 2000\,\text{s/mm}^2$ were selected from the original scans. These two datasets are also summarized in Table 1.

3.2 Evaluation of Segmentation Results on the HCP Dataset

We first present the evaluation of the segmentation results on the HCP dataset. Our method was compared with TractSeg without using bootstrap (but with the subsampling of diffusion gradients), which is referred to as the baseline method.

Examples of the segmentation results are shown in Fig. 1. For demonstration, here we show the results of representative WM tracts on HCP_1.25mm_90, HCP_1.25mm_36, and HCP_1.25mm_34 when 55 training subjects were used. For reference, the gold standard (manual delineation) is also displayed. In Fig. 1, cross-sectional views of the WM tracts are given, and regions are highlighted with zoomed views for better comparison. It can be seen that the segmented tracts of the proposed method have more similar spatial coverage to the gold standard than the baseline method.

636   W. Liu and C. Ye

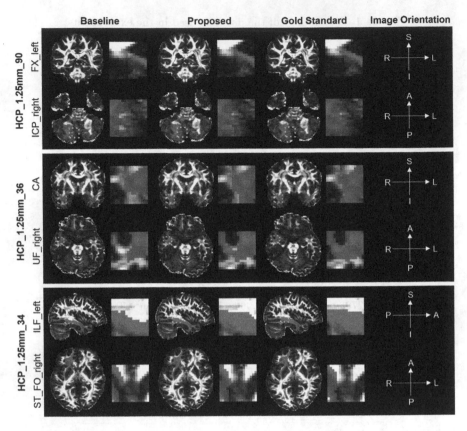

Fig. 1. Representative segmentation results (red) for the HCP dataset, together with the gold standard (manual annotation) for reference. The cross-sectional views of the segmented tracts are shown, and they are overlaid on fractional anisotropy maps. Zoomed views of the highlighted regions are also displayed for better comparison. The image orientation is shown in the rightmost column. For the meaning of the tract abbreviations, we refer readers to [25]. (Color figure online)

We then quantitatively evaluated the proposed method by computing the Dice coefficient between the segmentation results and the gold standard. The mean Dice coefficient of all 72 WM tracts for each test dataset and each number of training subjects is shown in Table 2. As some WM tracts can be more challenging to segment [14] and the improvement of the segmentation of these tracts is important, in Table 2 we also show the individual average Dice coefficients of the three most challenging WM tracts, which are the anterior commissure (CA), left fornix (FX_left), and right fornix (FX_right) [14,25]. Compared with the baseline method, the proposed method can consistently improve the Dice coefficients across the different cases, and the improvement is more prominent for the three most challenging WM tracts. In addition, the Dice coefficients of the proposed method were compared with those of the baseline method using paired Student's t-tests, and the p-values are listed in Table 2. It can be seen that the improvement of the proposed method is statistically significant in all cases.

Table 2. The mean Dice coefficient (%) of all 72 WM tracts and the individual average Dice coefficients (%) of the three most challenging tracts for the HCP dataset across different settings. The proposed method was compared with the baseline method using paired Student's t-tests, and asterisks indicate that the difference between the two methods is statistically significant (***: $p < 0.001$).

Dataset	Tract	Method	Number of training subjects							
			10		20		30		55	
HCP_1.25mm_270	All	Baseline	80.0	***	81.8	***	82.3	***	83.4	***
		Proposed	80.9		83.1		83.5		84.0	
	CA	Baseline	52.4	***	61.6	***	63.7	***	67.3	***
		Proposed	56.6		65.7		68.0		69.4	
	FX_left	Baseline	55.8	***	67.6	***	68.1	***	70.7	***
		Proposed	65.0		73.6		73.9		73.7	
	FX_right	Baseline	51.1	***	59.5	***	61.4	***	64.9	***
		Proposed	55.3		68.0		68.3		69.5	
HCP_1.25mm_90	All	Baseline	79.0	***	80.9	***	81.4	***	82.9	***
		Proposed	80.2		82.8		83.2		83.7	
	CA	Baseline	51.0	***	61.1	***	63.6	***	66.3	***
		Proposed	56.9		65.1		67.4		68.7	
	FX_left	Baseline	55.6	***	63.5	***	65.9	***	68.9	***
		Proposed	62.8		72.2		72.6		72.6	
	FX_right	Baseline	47.8	***	57.9	***	57.1	***	63.6	***
		Proposed	56.4		67.5		67.1		68.1	
HCP_1.25mm_12	All	Baseline	77.9	***	80.2	***	80.8	***	82.2	***
		Proposed	79.7		82.4		82.8		83.4	
	CA	Baseline	47.7	***	59.2	***	61.9	***	64.8	***
		Proposed	56.2		64.1		66.8		68.0	
	FX_left	Baseline	50.6	***	61.8	***	62.9	***	65.5	***
		Proposed	58.5		71.6		71.0		72.3	
	FX_right	Baseline	40.6	***	54.2	***	53.3	***	59.6	***
		Proposed	51.4		66.1		65.6		66.5	
HCP_1.25mm_36	All	Baseline	79.2	***	80.9	***	81.5	***	82.7	***
		Proposed	80.6		82.9		83.3		83.9	
	CA	Baseline	48.6	***	60.0	***	61.6	***	65.4	***
		Proposed	55.2		65.2		67.4		68.7	
	FX_left	Baseline	53.9	***	66.3	***	65.9	***	68.4	***
		Proposed	63.5		72.9		73.2		73.5	
	FX_right	Baseline	46.3	***	56.7	***	58.8	***	62.6	***
		Proposed	52.3		66.9		67.3		68.5	
HCP_1.25mm_34	All	Baseline	78.4	***	80.6	***	81.1	***	82.6	***
		Proposed	80.1		82.7		83.1		83.6	
	CA	Baseline	49.3	***	60.3	***	63.3	***	65.9	***
		Proposed	56.6		65.3		67.3		68.8	
	FX_left	Baseline	51.6	***	61.5	***	64.3	***	67.2	***
		Proposed	61.1		72.1		72.0		72.2	
	FX_right	Baseline	43.9	***	55.3	***	54.6	***	62.0	***
		Proposed	54.0		66.5		66.2		67.5	

Table 3. The mean Dice coefficient (%) of all ten annotated WM tracts and the individual average Dice coefficients (%) of two challenging tracts for the in-house dataset across different settings. The proposed method was compared with the baseline method using paired Student's t-tests, and asterisks indicate that the difference between the two methods is statistically significant (***: $p < 0.001$, **: $p < 0.01$, *: $p < 0.05$, n.s.: $p \geq 0.05$).

Dataset	Tract	Method	Number of training subjects							
			10		20		30		55	
IH_1.7mm_270	All	Baseline	58.7	n.s.	60.4	**	61.2	**	61.7	n.s.
		Proposed	59.0		62.0		61.9		61.9	
	UF_left	Baseline	48.4	***	50.7	***	53.1	**	55.1	n.s.
		Proposed	51.9		59.3		56.7		57.1	
	UF_right	Baseline	52.9	n.s.	56.5	***	57.3	***	59.1	***
		Proposed	53.4		61.3		60.1		61.4	
IH_1.7mm_36	All	Baseline	57.2	**	58.7	***	59.3	***	60.4	**
		Proposed	57.8		61.5		61.5		61.3	
	UF_left	Baseline	46.2	n.s.	46.3	***	47.3	***	51.6	**
		Proposed	47.9		58.0		55.7		55.3	
	UF_right	Baseline	48.9	***	54.2	***	53.6	***	56.3	***
		Proposed	51.2		59.1		58.5		60.5	

By comparing the results achieved with different numbers of training subjects, we observe that the overall improvement of the proposed method tends to be greater when the number is moderate (20 and 30) than when the number is small (10) or large (55). Moreover, the Dice coefficients of the proposed method obtained with 20 training subjects are comparable to or higher than the baseline performance achieved with 55 training subjects. Also, when the number of training subjects increases from 20 to 30 or 55, the Dice coefficients of the proposed method are relatively stable, whereas the Dice coefficients of the baseline method can still increase. This is possibly because the proposed method augments the training data and thus reduces the requirement for manual annotation.

3.3   Evaluation of Segmentation Results on the In-House Dataset

The proposed method was next applied to the in-house test datasets IH_1.7mm_270 and IH_1.7mm_36, and the mean Dice coefficients of all ten annotated WM tracts are summarized in Table 3. In addition, the individual average Dice coefficients of two challenging tracts, the left uncinate fasciculus (UF_left) and right uncinate fasciculus (UF_right) [14], are also shown in Table 3.[5] In each case, the proposed method achieves a higher Dice coefficient than the baseline method, and the improvement is more prominent for the two

---

[5] CA, FX_left, and FX_right were not annotated for the in-house dataset.

challenging tracts and for IH_1.7mm_36 that has a smaller number of diffusion gradients. We also performed paired Student's t-tests to compare the two methods in Table 3, and the difference between the proposed and competing methods is statistically significant in most cases.

Like the results on the HCP dataset, the improvement of the proposed method over the baseline method is greater when the number of training subjects is 20 or 30 than 10 or 55, and its performance becomes stable after the number of training subjects reaches 20. Also, the Dice coefficients of the proposed method obtained with 20 training subjects are already better than the baseline performance achieved with 55 training subjects.

4   Conclusion

We have proposed a WM tract segmentation approach that better generalizes to arbitrary test datasets. In the proposed method a scaled residual bootstrap strategy is developed, where the SNR levels of the training data are adjusted based on the residuals of a linear signal representation. This reduces the discrepancy between training and test data and thus improves the generalization of the trained segmentation model. Our method was validated on public and in-house datasets under various data settings, and the results show that it consistently improved the segmentation accuracy in the different cases.

Acknowledgements. This work is supported by the Fundamental Research Funds for the Central Universities.

References

1. Banihashemi, L., et al.: Opposing relationships of childhood threat and deprivation with stria terminalis white matter. Hum. Brain Mapp. **42**(8), 2445–2460 (2021)
2. Basser, P.J., Pajevic, S., Pierpaoli, C., Duda, J., Aldroubi, A.: In vivo fiber tractography using DT-MRI data. Magn. Reson. Med. **44**(4), 625–632 (2000)
3. Bazin, P.L., et al.: Direct segmentation of the major white matter tracts in diffusion tensor images. Neuroimage **58**(2), 458–468 (2011)
4. Cook, P.A., et al.: An automated approach to connectivity-based partitioning of brain structures. In: Duncan, J.S., Gerig, G. (eds.) MICCAI 2005. LNCS, vol. 3749, pp. 164–171. Springer, Heidelberg (2005). https://doi.org/10.1007/11566465_21
5. Davison, A.C., Hinkley, D.V.: Bootstrap Methods and Their Application. No. 1, Cambridge University Press (1997)
6. Garyfallidis, E., et al.: Recognition of white matter bundles using local and global streamline-based registration and clustering. Neuroimage **170**, 283–295 (2018)
7. Girard, G., et al.: On the cortical connectivity in the macaque brain: a comparison of diffusion tractography and histological tracing data. Neuroimage **221**, 117201 (2020)
8. Guan, H., Liu, M.: Domain adaptation for medical image analysis: a survey. IEEE Trans. Biomed. Eng. **69**(3), 1173–1185 (2021)
9. Jeurissen, B., Descoteaux, M., Mori, S., Leemans, A.: Diffusion MRI fiber tractography of the brain. NMR Biomed. **32**(4), e3785 (2019)

10. Jeurissen, B., Leemans, A., Jones, D.K., Tournier, J.D., Sijbers, J.: Probabilistic fiber tracking using the residual bootstrap with constrained spherical deconvolution. Hum. Brain Mapp. **32**(3), 461–479 (2011)
11. Jeurissen, B., Tournier, J.D., Dhollander, T., Connelly, A., Sijbers, J.: Multi-tissue constrained spherical deconvolution for improved analysis of multi-shell diffusion MRI data. Neuroimage **103**, 411–426 (2014)
12. Kingma, D.P., Ba, J.: Adam: a method for stochastic optimization. arXiv preprint arXiv:1412.6980 (2014)
13. Li, B., et al.: Neuro4Neuro: A neural network approach for neural tract segmentation using large-scale population-based diffusion imaging. Neuroimage **218**, 116993 (2020)
14. Liu, W., et al.: Volumetric segmentation of white matter tracts with label embedding. Neuroimage **250**, 118934 (2022)
15. Lu, Q., Li, Y., Ye, C.: Volumetric white matter tract segmentation with nested self-supervised learning using sequential pretext tasks. Med. Image Anal. **72**, 102094 (2021)
16. Merlet, S., Caruyer, E., Deriche, R.: Parametric dictionary learning for modeling EAP and ODF in diffusion MRI. In: Ayache, N., Delingette, H., Golland, P., Mori, K. (eds.) MICCAI 2012. LNCS, vol. 7512, pp. 10–17. Springer, Heidelberg (2012). https://doi.org/10.1007/978-3-642-33454-2_2
17. Merlet, S.L., Deriche, R.: Continuous diffusion signal, EAP and ODF estimation via compressive sensing in diffusion MRI. Med. Image Anal. **17**(5), 556–572 (2013)
18. Ning, L., et al.: Cross-scanner and cross-protocol multi-shell diffusion MRI data harmonization: algorithms and results. Neuroimage **221**, 117128 (2020)
19. Ratnarajah, N., Qiu, A.: Multi-label segmentation of white matter structures: application to neonatal brains. Neuroimage **102**, 913–922 (2014)
20. Ronneberger, O., Fischer, P., Brox, T.: U-net: convolutional networks for biomedical image segmentation. In: Navab, N., Hornegger, J., Wells, W.M., Frangi, A.F. (eds.) MICCAI 2015. LNCS, vol. 9351, pp. 234–241. Springer, Cham (2015). https://doi.org/10.1007/978-3-319-24574-4_28
21. Toescu, S.M., Hales, P.W., Kaden, E., Lacerda, L.M., Aquilina, K., Clark, C.A.: Tractographic and microstructural analysis of the dentato-rubro-thalamo-cortical tracts in children using diffusion MRI. Cereb. Cortex **31**(5), 2595–2609 (2021)
22. Tournier, J.D., Calamante, F., Connelly, A.: Robust determination of the fibre orientation distribution in diffusion MRI: non-negativity constrained super-resolved spherical deconvolution. Neuroimage **35**(4), 1459–1472 (2007)
23. Van Essen, D.C., Smith, S.M., Barch, D.M., Behrens, T.E., Yacoub, E., Ugurbil, K.: Wu-Minn HCP consortium: the WU-Minn human connectome project: an overview. Neuroimage **80**, 62–79 (2013)
24. Veraart, J., Raven, E.P., Edwards, L.J., Weiskopf, N., Jones, D.K.: The variability of MR axon radii estimates in the human white matter. Hum. Brain Mapp. **42**(7), 2201–2213 (2021)
25. Wasserthal, J., Neher, P., Maier-Hein, K.H.: TractSeg - fast and accurate white matter tract segmentation. Neuroimage **183**, 239–253 (2018)
26. Ye, C., Yang, Z., Ying, S.H., Prince, J.L.: Segmentation of the cerebellar peduncles using a random forest classifier and a multi-object geometric deformable model: Application to spinocerebellar ataxia type 6. Neuroinformatics **13**(3), 367–381 (2015)
27. Zhang, F., Karayumak, S.C., Hoffmann, N., Rathi, Y., Golby, A.J., O'Donnell, L.J.: Deep white matter analysis (DeepWMA): fast and consistent tractography segmentation. Med. Image Anal. **65**, 101761 (2020)

Bootstrapping Semi-supervised Medical Image Segmentation with Anatomical-Aware Contrastive Distillation

Chenyu You[1(✉)], Weicheng Dai[2], Yifei Min[5], Lawrence Staib[1,3,4], and James S. Duncan[1,3,4,5]

[1] Department of Electrical Engineering, Yale University, New Haven, USA
chenyu.you@yale.edu
[2] Department of Computer Science and Engineering, New York University, New York, USA
[3] Department of Biomedical Engineering, Yale University, New Haven, USA
[4] Department of Radiology and Biomedical Imaging, Yale University, New Haven, USA
[5] Department of Statistics and Data Science, Yale University, New Haven, USA

Abstract. Contrastive learning has shown great promise over annotation scarcity problems in the context of medical image segmentation. Existing approaches typically assume a balanced class distribution for both labeled and unlabeled medical images. However, medical image data in reality is commonly imbalanced (*i.e.*, multi-class label imbalance), which naturally yields blurry contours and usually incorrectly labels rare objects. Moreover, it remains unclear whether all negative samples are equally negative. In this work, we present **ACTION**, an Anatomical-aware ConTrastive dIstillatiON framework, for semi-supervised medical image segmentation. Specifically, we first develop an iterative contrastive distillation algorithm by softly labeling the negatives rather than binary supervision between positive and negative pairs. We also capture more semantically similar features from the randomly chosen negative set compared to the positives to enforce the diversity of the sampled data. Second, we raise a more important question: Can we really handle imbalanced samples to yield better performance? Hence, the *key innovation* in ACTION is to learn global semantic relationship across the entire dataset and local anatomical features among the neighbouring pixels with minimal additional memory footprint. During the training, we introduce anatomical contrast by actively sampling a sparse set of hard negative pixels, which can generate smoother segmentation boundaries and more accurate predictions. Extensive experiments across two benchmark datasets and different unlabeled settings show that ACTION significantly outperforms the current state-of-the-art semi-supervised methods.

Keywords: Contrastive Learning · Knowledge Distillation · Active Sampling · Semi-Supervised Learning · Medical Image Segmentation

© The Author(s), under exclusive license to Springer Nature Switzerland AG 2023
A. Frangi et al. (Eds.): IPMI 2023, LNCS 13939, pp. 641–653, 2023.
https://doi.org/10.1007/978-3-031-34048-2_49

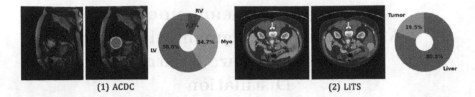

Fig. 1. Examples of two benchmarks (*i.e.*, ACDC and LiTS) showing the large variations of class distribution.

1 Introduction

Manually labeling sufficient medical data with pixel-level accuracy is time-consuming, expensive, and often requires domain-specific knowledge. To bypass the cost for labeled data, semi-supervised learning (SSL) is one of the promising, conventional ways to train models with weaker forms of supervision, given a large amount of unlabeled data. Existing SSL methods include adversarial training [12,28,32,33,37], deep co-training [21,38], mean teacher schemes [23,36], multi-task learning [4,11,16,31], and contrastive learning [3,9,29,30,34,35].

Among the aforementioned methods, contrastive learning [5,8] has recently prevailed for DNNs to rich visual representations from unlabeled data. The predominant promise of label-free learning is to capture the similar semantic relationship and anatomical structure between neighboring pixels from massive unannotated data. However, going to realistic clinical scenarios will have the following shortcomings. First, different medical images share similar anatomical structures, but prior methods follow the standard contrastive learning [5,8] in comparing positive and negative pairs by binary supervision. That naturally leads to the issues of false negatives in representation learning [10,24], which would hurt segmentation performance. Second, the underlying class distribution of medical image data is highly imbalanced, as illustrated in Fig. 1. It is well known that such imbalanced distribution will severely hurt the segmentation quality [14], which may result in blurry contours and mis-classify minority classes due to the occurrence frequencies [39]. That naturally questions whether contrastive learning can still work well in those imbalance scenarios.

In this work, we present a principled framework called **A**natomical-aware **C**on**T**rastive d**I**stillati**ON** (**ACTION**), for multi-class medical image segmentation. In contrast to prior work [3,9,35] which directly distinguish two image samples of the similar anatomical features that are in the negative pairs, the **key innovation** in ACTION is to actively learn more balanced representations by *dynamically* selecting samples that are semantically similar to the queries, and contrasting the model's *anatomical-level* features with the target model's in **imbalanced** and **unlabeled** clinical scenarios. Specifically, we introduce two strategies to improve overall segmentation quality: (1) we believe that all negative samples are not equally negative. Thus, we propose relaxed contrastive learning by using soft labeling on the negatives. In other words, we randomly sample a set of image samples as anchor points to ensure **diversity** in the set of

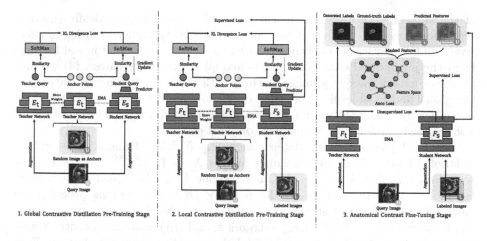

Fig. 2. Overview of the ACTION framework including three stages: (1) global contrastive distillation pre-training used in existing works, (2) our proposed local contrastive distillation pre-training, and (3) our proposed anatomical contrast fine-tuning.

sampled examples. Then the teacher model predicts the underlying probability distribution over neighboring samples by computing the anatomical similarities between the query and the anchor points in the memory bank, and the student model tries to learn from the teacher model. Such a strategy is much more regularized by mincing the same neighborhood anatomical similarity to improve the quality of the anatomical features; (2) to create strong contrastive views on anatomical features, we introduce **AnCo**, another new contrastive loss designed at the anatomical level, by sampling a set of pixel-level representation as queries, and pulling them closer to the mean feature of all representations in a class (positive keys), and pulling other representations apart from other class (negative keys). In addition to reducing the high memory footprint and computation complexity, we use active sampling to dynamically select a sparse set of queries and keys during the training. We apply ACTION on two benchmark datasets under different unlabeled settings. Our experiments show that ACTION can dramatically outperform the state-of-the-art SSL methods. We believe that our proposed ACTION can be a strong baseline for the related medical image analysis tasks in the future.

2   Method

Framework Overview. The workflow of our proposed ACTION is illustrated in Fig. 2. By default, ACTION is built on the BYOL pipeline [7] which is originally designed for image classification tasks, and for a fair comparison, we also follow the setting in [3] such as using 2D U-Net [22] as the backbone and non-linear projection heads H. The **main differences** between our proposed ACTION and [3,9] are as follows: (1) the addition of a predictor $g(\cdot)$ to the student network to avoid collapsed solutions; (2) the utilization of a slow-moving average of the student network as the teacher network for more semantically compact representations; (3) the use of the output probability rather than logits effectively and semantically constrains the distance between the anatomical features from the imbalanced data (*i.e.*, multi-class label imbalance cases); (4) we propose to contrast the query image features with other random image features at the global and local level, rather than only two augmented versions of the same image features; and (5) we design a novel unsupervised anatomical contrastive loss to provide additional supervision on hard pixels.

Let (X, Y) be a training dataset including N labeled image slices and M unlabeled image slices, with training images $X = \{x_i\}_{i=1}^{N+M}$ and the C-class segmentation labels $Y = \{y_i\}_{i=1}^{N}$. Our backbone $F(\cdot)$ (2D U-Net) consists of an encoder network $E(\cdot)$ and a decoder network $D(\cdot)$. The training procedure of ACTION includes three stages: (i) global contrastive distillation pre-training, (ii) local contrastive distillation pre-training, and (iii) anatomical contrast fine-tuning. In the first two stages, we use global contrastive distillation to train E on unlabeled data to learn global-level features, and use local contrastive distillation to train E and D on labeled and unlabeled data to learn local-level features.

Global Contrastive Distillation Pre-training. We follow a similar setting in [24]. Given an input query image $q \in \{x_i\}_{i=N+1}^{N+M}$ with the spatial size $h \times w$, we first apply two different augmentations to obtain q_t and q_s, and randomly sample a set of augmented images $\{x_j\}_{j=1}^{n}$ from a set of unlabeled image slices $\{x_i\}_{i=N+1}^{N+M}$. We believe that such relaxation enables the model to capture more rich semantic relationships and anatomical features from its neighboring images instead of only learning from the different version of the same query image. We then feed $\{x_j\}_{j=1}^{n}$ to the teacher encoder E_t, and followed by the nonlinear projection head H_t^g to generate their projection embeddings $\{H_t^g(E_t(x_j))\}_{j=1}^{n}$ as anchor points, and also feed q_t and q_s to the teacher and student (*i.e.*, E and H), creating $z_t = H_t^g(E_t(q_t))$ and $z_s = H_s^g(E_s(q_s))$. Here we utilize the probabilities after SoftMax instead of the feature embedding:

$$p_t(j) = -\log \frac{\exp\big(\mathrm{sim}\big(z_t, a_j\big)/\tau_t\big)}{\sum_{i=1}^{n} \exp\big(\mathrm{sim}\big(z_t, a_i\big)/\tau_t\big)}, \tag{1}$$

where τ_t is a temperature hyperparameter of the teacher, and $\mathrm{sim}(\cdot, \cdot)$ is the cosine similarity. Then inspired by [7], in order to avoid collapsed solutions in an unsupervised scenario, we use a shallow multi-layer perceptron (MLP) predictor $H_p^g(\cdot)$ to obtain the prediction $z_s^* = H_p^g(z_s)$. Of note, $\{a_i\}_{i=1}^{n}$, z_t, z_s, z_s^* can be generated embedding from a set of randomly chosen augmented images, teacher's projection embeddings, student's projection embeddings, and student's prediction embeddings in either Stage-i or ii. Therefore, we can calculate the similarity distance between the student's prediction and the anchor embeddings by converting them to probability distribution.

$$p_s(j) = -\log \frac{\exp\big(\mathrm{sim}\big(z_s^*, a_j\big)/\tau_s\big)}{\sum_{i=1}^{n} \exp\big(\mathrm{sim}\big(z_s^*, a_i\big)/\tau_s\big)}, \tag{2}$$

where τ_s refers to a temperature hyperparameter of the student. The unsupervised contrastive loss is computed as follows:

$$\mathcal{L}_{\text{contrast}} = \text{KL}(p_t||p_s). \tag{3}$$

Local Contrastive Distillation Pre-training. After training the teacher's and student's encoder to learn global-level image features, we attach the decoders and tune the entire models to perform pixel-level contrastive learning in a semi-supervised manner. The **distinction in the training strategy** between ours and [9] lies in Stage-ii and iii: [9] only use labeled data in training, while we use both labeled and unlabeled data in training. Considering the training procedure of Stage-ii is similar to Stage-iii, we briefly describe it here as illustrated in Fig. 2. For the labeled data, we train our model by minimizing the supervised loss (the linear combination of cross-entropy loss and dice loss) in Stage-ii and Stage-iii. As for the unlabeled input images q and $\{x_j\}_{j=1}^n$, we first apply two different augmentations to q, creating two different versions $[q_t^l, q_s^l]$, and then feed them to F_t and F_s, and their output features $[f_t, f_s]$ are fed into H_t^l and H_t^l. The student's projection embedding is subsequently fed into H_p^l to obtain the student's prediction embedding to enforce the similarity between the teacher and the student under the same loss as Eq. 3. We also include the randomly selected images to enforce such similarity because intuitively, it may be beneficial to ensure **diversity** in the set of sampled examples. It is important to note that ACTION will re-use the well-trained weight of the models F_t and F_s as initialization for Stage-iii.

Anatomical Contrast Fine-Tuning. Broadly speaking, in medical images, the same tissue types may share similar anatomical information in different patients, but different tissue types often show different class, appearance, and spatial distributions, which can be described as a complicated form of **imbalance** and **uncertainty** in real clinical data, as shown in Fig. 1. This motivates us to efficiently incorporate more useful features so the representations can be more balanced and better discriminated in such multi-class label imbalanced scenarios. Inspired by [15], we propose AnCo, a new unsupervised contrastive loss designed at the anatomical level. Specifically, we additionally attach a representation decoder head H_r to the student network, parallel to the segmentation head, to decode the multi-layer hidden features by first using multiple up-sampling layers for outputting dense features with the same spatial resolution as the query image and then mapping them into high m-dimensional query, positive key, and negative key embeddings: r_q, r_k^+, r_k^-. The AnCo loss is then defined as:

$$\mathcal{L}_{\text{anco}} = \sum_{c \in \mathcal{C}} \sum_{r_q \sim \mathcal{R}_q^c} - \log \frac{\exp(r_q \cdot r_k^{c,+}/\tau_{an})}{\exp(r_q \cdot r_k^{c,+}/\tau_{an}) + \sum_{r_k^- \sim \mathcal{R}_k^c} \exp(r_q \cdot r_k^-/\tau_{an})}, \tag{4}$$

where $\mathcal{C}$ is a set of all available classes in a mini-batch, and τ_{an} denotes a temperature hyperparameter for AnCo loss. $\mathcal{R}_q^c$ and $r_k^{c,+}$ are a set of query embeddings in class c and the positive key embedding, which is the mean representation of class c, respectively. $\mathcal{R}_k^c$ is a set of negative key embeddings which are not in class c. Suppose $\mathcal{P}$ is a set including all pixel coordinates with the same resolution with x_i, these queries and keys are then defined as:

$$\mathcal{R}_q^c = \bigcup_{[m,n]\in\mathcal{P}} \mathbb{1}(y_{[m,n]}=c)\, r_{[m,n]}, \; \mathcal{R}_k^c = \bigcup_{[m,n]\in\mathcal{P}} \mathbb{1}(y_{[m,n]}\neq c)\, r_{[m,n]}, \; r_k^{c,+} = \frac{1}{|\mathcal{R}_q^c|} \sum_{r_q\in\mathcal{R}_q^c} r_q. \tag{5}$$

In addition, we note that contrastive learning usually benefits from a large collection of positive and negative pairs, but it is usually bounded by the size of GPU memory. Therefore, we introduce two novel active hard sampling methods. To address the *uncertainty* on the most challenging pixels among all available classes (*i.e.*, close anatomical or semantic relationship), we non-uniformly sample negative keys based on relative similarity distance between the query class and each negative key class. For each mini-batch, we build a graph G to measure the pair-wise class relationship to dynamically update G.

$$G[p,q] = \left(r_k^{p,+} \cdot r_k^{q,+}\right), \quad \forall p,q \in \mathcal{C}, \text{ and } p \neq q, \tag{6}$$

where $G \in \mathbb{R}^{|\mathcal{C}|\times|\mathcal{C}|}$. Note that this process may be hard to allocate more samples. Thus, to learn a more accurate decision boundary, we first apply SoftMax function by normalizing the pair-wise relationships among all negative classes n from each query class c, yielding a distribution: $\exp(G[c,v])/\sum_{n\in\mathcal{C},n\neq c}\exp(G[c,n])$. Then we adaptively sample negative keys from each class v to help learn the corresponding query class c. To alleviate the *imbalance* issue, we sample hard queries based on a defined threshold, to better discriminate the rare classes. The easy and hard queries are computed as follows:

$$\mathcal{R}_q^{c,\,easy} = \bigcup_{r_q\in\mathcal{R}_q^c} \mathbb{1}(\hat{y}_q > \theta_s)r_q, \quad \mathcal{R}_q^{c,\,hard} = \bigcup_{r_q\in\mathcal{R}_q^c} \mathbb{1}(\hat{y}_q \leq \theta_s)r_q, \tag{7}$$

where $\hat{y}_q$ is the predicted confidence of label c corresponding to r_q after SoftMax function, and θ_s is the user-defined confidence threshold.

3   Experiments

Experimental Setup. We experiment on two benchmark datasets: ACDC 2017 dataset [1] and MICCAI 2017 Liver Tumor Segmentation Challenge (LiTS) [2].

The ACDC dataset includes 200 cardiac cine MRI scans from 100 patients with annotations including three segmentation classes (*i.e.*, left ventricle (LV), myocardium (Myo), and right ventricle (RV)). Following [16,27], we use 140, 20, and 60 scans for training, validation, and testing, respectively.

The LiTS dataset includes 131 contrast-enhanced 3D abdominal CT volumes with annotations of two segmentation classes (*i.e.*, liver and tumor). Following [13], we use the first 100 volumes for training, and the rest 31 for testing. For preprocessing, we follow the setting in [3] to normalize the intensity of each 3D scans,

Table 1. Comparison of segmentation performance (DSC[%]/ASD[voxel]) on ACDC under two unlabeled settings (3 or 7 labeled). The best results are indicated in **bold**.

Method	Average	3 Labeled RV	Myo	LV	Average	7 Labeled RV	Myo	LV
UNet-F [22]	91.5/0.996	90.5/0.606	88.8/0.941	94.4/1.44	91.5/0.996	90.5/0.606	88.8/0.941	94.4/1.44
UNet-L	51.7/13.1	36.9/30.1	54.9/4.27	63.4/5.11	79.5/2.73	65.9/0.892	82.9/2.70	89.6/4.60
EM [26]	59.8/5.64	44.2/11.1	63.2/3.23	71.9/2.57	75.7/2.73	68.0/0.892	76.5/2.70	82.7/4.60
CCT [18]	59.1/10.1	44.6/19.8	63.2/6.04	69.4/4.32	75.9/3.60	67.2/2.90	77.5/3.32	82.9/0.734
DAN [37]	56.4/15.1	47.1/21.7	58.1/11.6	63.9/11.9	76.5/3.01	75.7/2.61	73.3/3.11	80.5/3.31
URPC [17]	58.9/8.14	50.1/12.6	60.8/4.10	65.8/7.71	73.2/2.68	67.0/0.742	72.2/0.505	80.4/6.79
DCT [21]	58.5/10.8	41.2/21.4	63.9/5.01	70.5/6.05	78.1/2.64	70.7/1.75	77.7/2.90	85.8/3.26
ICT [25]	59.0/6.59	48.8/11.4	61.4/4.59	66.6/3.82	80.6/1.64	75.1/0.898	80.2/1.53	86.6/2.48
MT [23]	58.3/11.2	39.0/21.5	58.7/7.47	77.3/4.72	80.1/2.33	75.2/1.22	79.2/2.32	86.0/3.45
UAMT [36]	61.0/7.03	47.8/15.9	65.0/2.38	70.1/2.83	77.6/3.15	70.5/0.81	78.4/4.36	83.9/4.29
CPS [6]	61.0/2.92	43.8/2.95	64.5/2.84	74.8/2.95	78.8/3.41	74.0/1.95	78.1/3.11	84.5/5.18
GCL [3]	70.6/2.24	56.5/1.99	70.7/1.67	84.8/3.05	87.0/0.751	86.9/**0.584**	81.8/0.821	92.5/0.849
SCS [9]	73.6/5.37	63.5/6.23	76.6/2.42	80.7/7.45	84.2/2.01	81.4/0.850	83.0/2.03	88.2/3.12
• ACTION (ours)	**87.5/1.12**	**85.4/0.915**	**85.8/0.784**	**91.2/1.66**	**89.7/0.736**	**89.8**/0.589	**86.7/0.813**	**92.7/0.804**

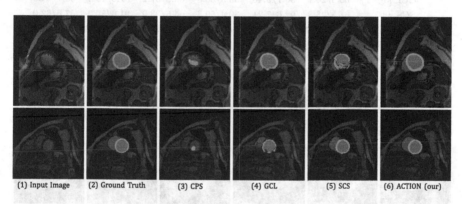

| (1) Input Image | (2) Ground Truth | (3) CPS | (4) GCL | (5) SCS | (6) ACTION (our) |

Fig. 3. Visualization of segmentation results on ACDC with 3 labeled data. As is shown, ACTION consistently produces sharper object boundaries and more accurate predictions across all methods. Different structure categories are shown in different colors. (Color figure online)

resample all 2D slices and the corresponding segmentation maps to a fixed spatial resolution (*i.e.*, 256 × 256 pixels). To quantitatively assess the performance of our proposed method, we report two popular metrics: Dice coefficient (DSC) and Average Surface Distance (ASD) for 3D segmentation results.

Implementation Details. All our models are implemented in PyTorch [19]. We train all methods with SGD optimizer (learning rate = 0.01, momentum = 0.9, weight decay = 0.0001, batch size = 6). All models are trained with two NVIDIA GeForce RTX 3090 GPUs. Stage-*i* and *ii* are trained with 100 epochs, and Stage-*iii* is with 200 epochs. We use the temperature of teacher and student as $\tau_t = 0.01$ and $\tau_s = 0.1$. The teacher is updated using the following

648   C. You et al.

Table 2. Comparison of segmentation performance (DSC [%]/ASD [voxel]) on LiTS under two unlabeled settings (5% or 10% labeled ratio). The best results are in **bold**.

| Method | Average | 5% Labeled | | Average | 10% Labeled | |
		Liver	Tumor		Liver	Tumor
UNet-F [22]	68.2/16.9	90.6/8.14	45.8/25.6	68.2/16.9	90.6/8.14	45.8/25.6
UNet-L	60.4/30.4	87.5/9.84	33.3/50.9	61.6/28.3	85.4/18.6	37.9/37.9
EM [26]	61.2/33.3	87.7/9.47	34.7/57.1	62.9/38.5	87.4/21.3	38.3/55.7
CCT [18]	60.6/48.7	85.5/27.9	35.6/69.4	63.8/31.2	90.3/7.25	37.2/55.1
DAN [37]	62.3/25.8	88.6/9.64	36.1/42.1	63.2/30.7	87.3/15.4	39.1/46.1
URPC [17]	62.4/37.8	86.7/21.6	38.0/54.0	63.0/43.1	88.1/24.3	38.9/61.9
DCT [21]	60.8/34.4	89.2/12.6	32.5/56.2	61.9/31.7	86.2/19.3	37.5/44.1
ICT [25]	60.1/39.1	86.8/12.6	33.3/65.6	62.5/32.4	88.1/16.7	36.9/48.2
MT [23]	61.9/40.0	86.7/21.6	37.2/58.4	63.3/26.2	89.7/11.6	36.9/40.8
UAMT [36]	61.0/47.0	86.9/22.1	35.2/71.8	62.3/26.0	87.4/7.55	37.3/44.4
CPS [6]	62.1/36.0	87.3/17.9	36.8/54.0	64.0/23.6	90.2/10.6	37.8/36.7
GCL [3]	63.3/20.1	90.7/9.46	35.9/30.8	65.0/37.2	91.3/10.0	38.7/64.3
SCS [9]	61.5/28.8	92.6/7.21	30.4/50.3	64.6/33.9	91.6/5.72	37.6/62.0
• ACTION (ours)	**66.8/17.7**	**93.0/6.04**	**40.5/29.4**	**67.7/20.4**	**92.8/5.08**	**42.6/35.8**

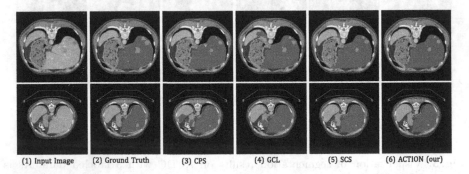

(1) Input Image   (2) Ground Truth   (3) CPS   (4) GCL   (5) SCS   (6) ACTION (our)

Fig. 4. Visualization of segmentation results on LiTS with 5% labeled ratio. As is shown, ACTION achieves consistently sharp and accurate object boundaries compared to other SSL methods. Different structure categories are shown in different colors. (Color figure online)

rule $\theta_t \leftarrow m\theta_t + (1 - m)\theta_s$, where θ refers to the model's parameters and the momentum hyperparameter m is 0.99. The memory bank size is 36. We follow the standard augmentation strategies in [7]. In Stage-i, we train E_s, E_t, H_t^g, H_s^g, and H_p^g on the unlabeled data with global-level $\mathcal{L}_{\text{contrast}}$ in Eq. 3. We follow [9] to use a MLP as heads, and the setting of the predictors is similar to [7], which has a feature dimension of 512. In Stage-ii, we train F_s, F_t, H_t^l, H_s^l, and H_p^l on the labeled and unlabeled data. We train with the supervised loss [36] on labeled data, and local-level $\mathcal{L}_{\text{contrast}}$ in Eq. 3 on unlabeled data. Given the logits output $\hat{y} \in \mathbb{R}^{C \times h \times w}$, we use the 1×1 convolutional layer to project all pixels into the latent space with the feature dimension of 512, and the output feature dimension of G is also 512. As for Stage-iii, we train F_s, F_t, H_t, H_s, and H_r on the labeled and unlabeled data. We use the supervised segmentation loss on labeled data, unsupervised cross-entropy loss (on pseudo-labels generated by a confidence threshold θ_s), and $\mathcal{L}_{\text{anco}}$ in Eq. 4 on unlabeled data. We then adaptively sample 256 query samples and 512 key samples for each mini-batch, and temperature for the student and confidence thresholds are set to $\tau_s = 0.5$ and $\theta_s = 0.97$, respectively. Of note, the projection heads, the predictor, and the representation decoder head are only utilized during the training, and will be removed during the inference.

Main Results. We compare our proposed method to previous state-of-the-art SSL methods using 2D Unet [22] as backbone, including UNet trained with full/limited supervisions (UNet-F/UNet-L), EM [26], CCT [18], DAN [37], URPC [17], DCT [21], ICT [25], MT [23], UAMT [36], CPS [6], SCS [9], and GCL [3]. Table 1 shows the evaluation results on ACDC dataset under two unlabeled settings (3 or 7 labeled cases). ACTION can substantially improve results on two unlabeled settings, greatly outperforming the previous state-of-the-art SSL methods. Specifically, our ACTION, trained on 3 labeled cases, dramatically improves the previous best averaged Dice score from 73.6% to 87.5% by a large margin, and even matches previous SSL methods using 7 labeled cases. When using 7 labeled cases, ACTION further pushes the state-of-the-art results to 89.7% in Dice. We observe that the gains are more pronounced on the two categories (*i.e.*, RV and Myo), and our ACTION achieves 89.8% and 86.7% in terms of Dice, performing competitive or even better than the supervised baseline (89.2% and 86.7%). As shown in Fig. 3, we can see the clear advantage of ACTION, where the boundaries of different regions are clearly sharper and more accurate such as RV and Myo regions. Table 2 also shows the evaluation results on LiTS dataset under two unlabeled settings (5% or 10% labeled cases). On both two labeled settings, ACTION significantly outperforms all the state-of-the-art methods by a significant margin. As shown in Fig. 4, ACTION achieves consistently sharp and accurate object boundaries compared to other SSL methods.

Ablation on Different Components. We investigate the impact of different components in ACTION. All reported results in this section are based on the ACDC dataset under the 3 labeled setting. Table 3 shows the ablation result of our model. Upon our choice of architecture, we first consider a naïve baseline (BYOL) without any random sampled images (RSI), stage-ii, and stage-iii, denoted by (1) Vanilla. Then, we consider a wide range of different settings for improved representation learning: (2) incorporating other random sampled images; (3) no stage-ii; (4) no other random sampled images and stage-ii; (5) no stage-iii; since stage-iii includes two losses, (6) no $\mathcal{L}_{\text{anco}}$, (7) no $\mathcal{L}_{\text{unsup}}$, and (8) our proposed ACTION. As shown in Table 3, it is notable that ACTION performs generally better than other evaluated baselines. We find that only applying any single component of ACTION often comes at the cost of performance degradation. The intuitions behind are as follows: (1) incorporating

Table 3. Ablation on **(a)** model component: w/o Random Sampled Images (RSI); w/o Local Contrastive Distillation (Stage-*ii*); w/o Anatomical Contrast Fine-tuning (Stage-*iii*); **(b)** loss formulation: w/o $\mathcal{L}_{anco}$; w/o $\mathcal{L}_{unsup}$;, compared to the Vanilla and our proposed ACTION. Note that $\mathcal{L}_{unsup}$ denotes cross-entropy loss (on pseudo-labels generated by a confidence threshold θ_s) together with $\mathcal{L}_{anco}$ used in Stage-*iii*.

	Method	Metrics	
		Dice [%]	ASD [voxel]
	Vanilla	60.6	6.64
	ACTION (ours)	**87.5**	**1.12**
(a)	w/o RSI	82.7	6.66
	w/o Stage-*ii*	86.4	1.69
	w/o RSI + Stage-*ii*	82.6	1.77
	w/o Stage-*iii*	76.7	2.91
(b)	w/o $\mathcal{L}_{anco}$	86.5	1.30
	w/o $\mathcal{L}_{unsup}$	83.7	2.51

Table 4. Ablation on augmentation strategies.

Method	Student Aug.	Teacher Aug.	Metrics	
			Dice[%]	ASD[voxel]
ACTION	Weak	Weak	84.6	1.78
ACTION	Strong	Weak	87.5	1.12
ACTION	Weak	Strong	85.4	2.12
ACTION	Strong	Strong	86.5	1.89

other random sampled images will enforce the diversity of the sampled data, preventing redundant anatomically and semantically similar samples; (2) using stage-*ii* leads to worse performance without considering local context; (3) using stage-*iii* enables a robust segmentation model to learn better representations with few human annotations. Using the above components confers a significant advantage at representation learning, and further illustrates the benefit of each component.

Ablation on Different Augmentations. We investigate the impact of using weak or strong augmentations for ACTION on the ACDC dataset under 3 labeled setting. We summarize the effects of different data augmentation strategies in Table 4. We apply *weak* augmentation to the teacher's input, including rotation, cropping, flipping, and *strong* augmentation to the student's input, including rotation, cropping, flipping, random contrast, and brightness changes [20]. Empirically, we find that when using weak and strong augmentation strategies on the teacher and student network, the network performance is optimal.

4   Conclusion and Limitations

In this work, we have presented ACTION, a novel anatomical-aware contrastive distillation framework with active sampling, designed specifically for medical image segmentation. Our method is motivated by two observations that all negative samples are not equally negative, and the underlying class distribution of medical images is highly unlabeled and imbalanced. Through extensive experiments across two benchmark datasets and unlabeled settings, we show that ACTION can significantly improve segmentation performance with minimal additional memory requirements, outperforming the previous state-of-the-art by a large margin. For future work, we plan to explore a more advanced contrastive learning approach for better performance when the medical data is unlabeled and imbalanced.

References

1. Bernard, O., et al.: Deep learning techniques for automatic MRI cardiac multi-structures segmentation and diagnosis: is the problem solved? IEEE Trans. Med. Imaging **37**, 2514–2525 (2018)
2. Bilic, P., et al.: The liver tumor segmentation benchmark (LiTS). arXiv preprint arXiv:1901.04056 (2019)
3. Chaitanya, K., Erdil, E., Karani, N., Konukoglu, E.: Contrastive learning of global and local features for medical image segmentation with limited annotations. In: NeurIPS (2020)
4. Chen, S., Bortsova, G., García-Uceda Juárez, A., van Tulder, G., de Bruijne, M.: Multi-task attention-based semi-supervised learning for medical image segmentation. In: Shen, D., et al. (eds.) MICCAI 2019. LNCS, vol. 11766, pp. 457–465. Springer, Cham (2019). https://doi.org/10.1007/978-3-030-32248-9_51
5. Chen, T., Kornblith, S., Norouzi, M., Hinton, G.: A simple framework for contrastive learning of visual representations. In: ICML, pp. 1597–1607. PMLR (2020)
6. Chen, X., Yuan, Y., Zeng, G., Wang, J.: Semi-supervised semantic segmentation with cross pseudo supervision. In: CVPR (2021)
7. Grill, J.B., et al.: Bootstrap your own latent-a new approach to self-supervised learning. In: NeurIPS (2020)
8. He, K., Fan, H., Wu, Y., Xie, S., Girshick, R.: Momentum contrast for unsupervised visual representation learning. In: CVPR, pp. 9729–9738 (2020)
9. Hu, X., Zeng, D., Xu, X., Shi, Y.: Semi-supervised contrastive learning for label-efficient medical image segmentation. In: de Bruijne, M., et al. (eds.) MICCAI 2021. LNCS, vol. 12902, pp. 481–490. Springer, Cham (2021). https://doi.org/10.1007/978-3-030-87196-3_45
10. Huynh, T., Kornblith, S., Walter, M.R., Maire, M., Khademi, M.: Boosting contrastive self-supervised learning with false negative cancellation. In: WACV (2022)
11. Kervadec, H., Dolz, J., Granger, É., Ben Ayed, I.: Curriculum semi-supervised segmentation. In: Shen, D., et al. (eds.) MICCAI 2019. LNCS, vol. 11765, pp. 568–576. Springer, Cham (2019). https://doi.org/10.1007/978-3-030-32245-8_63
12. Li, S., Zhang, C., He, X.: Shape-aware semi-supervised 3D semantic segmentation for medical images. In: Martel, A.L., et al. (eds.) MICCAI 2020. LNCS, vol. 12261, pp. 552–561. Springer, Cham (2020). https://doi.org/10.1007/978-3-030-59710-8_54

13. Li, X., Chen, H., Qi, X., Dou, Q., Fu, C.W., Heng, P.A.: H-DenseUNet: hybrid densely connected UNet for liver and tumor segmentation from CT volumes. IEEE Trans. Med. Imaging **37**, 2663–2674 (2018)
14. Li, Z., Kamnitsas, K., Glocker, B.: Analyzing overfitting under class imbalance in neural networks for image segmentation. IEEE Trans. Med. Imaging **40**, 1065–1077 (2020)
15. Liu, S., Zhi, S., Johns, E., Davison, A.J.: Bootstrapping semantic segmentation with regional contrast. arXiv preprint arXiv:2104.04465 (2021)
16. Luo, X., Chen, J., Song, T., Wang, G.: Semi-supervised medical image segmentation through dual-task consistency. In: AAAI (2020)
17. Luo, X., et al.: efficient semi-supervised gross target volume of nasopharyngeal carcinoma segmentation via uncertainty rectified pyramid consistency. In: de Bruijne, M. (ed.) MICCAI 2021. LNCS, vol. 12902, pp. 318–329. Springer, Cham (2021). https://doi.org/10.1007/978-3-030-87196-3_30
18. Ouali, Y., Hudelot, C., Tami, M.: Semi-supervised semantic segmentation with cross-consistency training. In: CVPR (2020)
19. Paszke, A., et al.: PyTorch: an imperative style, high-performance deep learning library. In: NeurIPS (2019)
20. Perez, F., Vasconcelos, C., Avila, S., Valle, E.: Data augmentation for skin lesion analysis. In: Stoyanov, D., et al. (eds.) CARE/CLIP/OR 2.0/ISIC -2018. LNCS, vol. 11041, pp. 303–311. Springer, Cham (2018). https://doi.org/10.1007/978-3-030-01201-4_33
21. Qiao, S., Shen, W., Zhang, Z., Wang, B., Yuille, A.: Deep co-training for semi-supervised image recognition. In: ECCV (2018)
22. Ronneberger, O., Fischer, P., Brox, T.: U-net: convolutional networks for biomedical image segmentation. In: Navab, N., Hornegger, J., Wells, W.M., Frangi, A.F. (eds.) MICCAI 2015. LNCS, vol. 9351, pp. 234–241. Springer, Cham (2015). https://doi.org/10.1007/978-3-319-24574-4_28
23. Tarvainen, A., Valpola, H.: Mean teachers are better role models: Weight-averaged consistency targets improve semi-supervised deep learning results. In: NeurIPS, pp. 1195–1204 (2017)
24. Tejankar, A., Koohpayegani, S.A., Pillai, V., Favaro, P., Pirsiavash, H.: ISD: self-supervised learning by iterative similarity distillation. In: ICCV (2021)
25. Verma, V., Kawaguchi, K., Lamb, A., Kannala, J., Bengio, Y., Lopez-Paz, D.: Interpolation consistency training for semi-supervised learning. In: IJCAI (2019)
26. Vu, T.H., Jain, H., Bucher, M., Cord, M., Pérez, P.: ADVENT: adversarial entropy minimization for domain adaptation in semantic segmentation. In: CVPR, pp. 2517–2526 (2019)
27. Wu, Y., et al.: Mutual consistency learning for semi-supervised medical image segmentation. Med. Image Anal. **81**, 102530 (2022)
28. Yang, L., et al.: NuSeT: a deep learning tool for reliably separating and analyzing crowded cells. PLoS Comput. Biol. **16**, e1008193 (2020)
29. You, C., et al.: Mine your own anatomy: revisiting medical image segmentation with extremely limited labels. arXiv preprint arXiv:2209.13476 (2022)
30. You, C., et al.: Rethinking semi-supervised medical image segmentation: a variance-reduction perspective. arXiv preprint arXiv:2302.01735 (2023)
31. You, C., et al.: Incremental learning meets transfer learning: application to multi-site prostate MRI segmentation. In: Albarqouni, S., et al. (eds.) DeCaF FAIR 2022. LNCS, vol. 13573, pp. 3–16. Springer, Cham (2022). https://doi.org/10.1007/978-3-031-18523-6_1

32. You, C., Yang, J., Chapiro, J., Duncan, J.S.: Unsupervised wasserstein distance guided domain adaptation for 3D multi-domain liver segmentation. In: Cardoso, J., et al. (eds.) IMIMIC/MIL3ID/LABELS -2020. LNCS, vol. 12446, pp. 155–163. Springer, Cham (2020). https://doi.org/10.1007/978-3-030-61166-8_17

33. You, C., et al.: Class-aware adversarial transformers for medical image segmentation. In: NeurIPS (2022)

34. You, C., Zhao, R., Staib, L.H., Duncan, J.S.: Momentum contrastive voxel-wise representation learning for semi-supervised volumetric medical image segmentation. In: Wang, L., Dou, Q., Fletcher, P.T., Speidel, S., Li, S. (eds.) MICCAI 2022. LNCS, vol. 13434, pp. 639–652. Springer, Cham (2022). https://doi.org/10.1007/978-3-031-16440-8_61

35. You, C., Zhou, Y., Zhao, R., Staib, L., Duncan, J.S.: SimCVD: simple contrastive voxel-wise representation distillation for semi-supervised medical image segmentation. IEEE Trans. Med. Imaging 41, 2228–2237 (2022)

36. Yu, L., Wang, S., Li, X., Fu, C.-W., Heng, P.-A.: Uncertainty-aware self-ensembling model for semi-supervised 3D left atrium segmentation. In: Shen, D., et al. (eds.) MICCAI 2019. LNCS, vol. 11765, pp. 605–613. Springer, Cham (2019). https://doi.org/10.1007/978-3-030-32245-8_67

37. Zhang, Y., Yang, L., Chen, J., Fredericksen, M., Hughes, D.P., Chen, D.Z.: Deep adversarial networks for biomedical image segmentation utilizing unannotated images. In: Descoteaux, M., Maier-Hein, L., Franz, A., Jannin, P., Collins, D.L., Duchesne, S. (eds.) MICCAI 2017. LNCS, vol. 10435, pp. 408–416. Springer, Cham (2017). https://doi.org/10.1007/978-3-319-66179-7_47

38. Zhou, Y., et al.: Semi-supervised 3d abdominal multi-organ segmentation via deep multi-planar co-training. In: WACV. IEEE (2019)

39. Zhu, X., Anguelov, D., Ramanan, D.: Capturing long-tail distributions of object subcategories. In: CVPR (2014)

DTU-Net: Learning Topological Similarity for Curvilinear Structure Segmentation

Manxi Lin[1], Kilian Zepf[1], Anders Nymark Christensen[1], Zahra Bashir[2],
Morten Bo Søndergaard Svendsen[3], Martin Tolsgaard[3],
and Aasa Feragen[1(✉)]

[1] Technical University of Denmark, Kongens Lyngby, Denmark
{manli,afhar}@dtu.dk
[2] Slagelse Hospital, Copenhagen, Denmark
[3] Region Hovedstaden Hospital, Copenhagen, Denmark

Abstract. Curvilinear structure segmentation is important in medical imaging, quantifying structures such as vessels, airways, neurons, or organ boundaries in 2D slices. Segmentation via pixel-wise classification often fails to capture the small and low-contrast curvilinear structures. Prior topological information is typically used to address this problem, often at an expensive computational cost, and sometimes requiring prior knowledge of the expected topology.

We present DTU-Net, a data-driven approach to topology-preserving curvilinear structure segmentation. DTU-Net consists of two sequential, lightweight U-Nets, dedicated to texture and topology, respectively. While the texture net makes a coarse prediction using image texture information, the topology net learns topological information from the coarse prediction by employing a triplet loss trained to recognize false and missed splits in the structure. We conduct experiments on a challenging multi-class ultrasound scan segmentation dataset as well as a well-known retinal imaging dataset. Results show that our model outperforms existing approaches in both pixel-wise segmentation accuracy and topological continuity, with no need for prior topological knowledge.

Keywords: Curvilinear segmentation · topology preservation · triplet loss

1 Introduction

Curvilinear structures represent thin and long objects in images, and their segmentation has extensive application in computer vision and medical imaging, e.g., segmentation of blood vessels [8], neurons [11], or organ boundaries [7].

Segmenting curvilinear structures is difficult due to their long, thin shapes: Their connectivity can be completely altered by misclassifying a small number of pixels. This is particularly challenging in images with low contrast or resolution, where the structure blends more with the background. As a result, connectivity errors due to incorrectly splitting curves (we term these "**false splits**") and incorrectly connecting them (we term these "**missed splits**"), are very common.

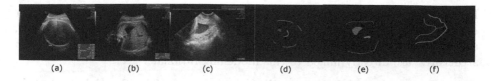

(a)   (b)   (c)   (d)   (e)   (f)

Fig. 1. Volumetric and curvilinear structures important for standard plane assessment in ultrasound of fetal head and abdomen, as well as maternal cervix are shown in (a)–(c), with manual segmentations in (d)–(f). Please zoom for details.

In this paper, we consider curvilinear segmentation problems where missed splits are just as harmful as false splits. This is in contrast with parts of the existing literature [3,13], which focuses on avoiding false splits or reproducing a prescribed topology by connecting components using prior knowledge. Our focus on missed splits is motivated by image quality assessment in fetal ultrasound [18]. Here, image quality depends on particular organs being completely visible in the image, making over-segmentation directly harmful. Examples of curvilinear structures important for fetal ultrasound quality assessment are shown in Fig. 1.

Curvilinear Structure Segmentation Is a Local Topological Problem. Curvilinear structure segmentation is often described as a topological problem [3,16]. Indeed, the topology of a subset of a topological space refers to those properties of the set that are invariant under continuous deformation. In curvilinear structure segmentation, the underlying task is to recover a curve that is visible as an elongated structure in the image. In such applications, it is often important to maintain the connectivity of the curve as accurately as possible. This is a topological problem: If $0 < a < b < 1$, then (the image of) a generic curve $c\colon [0,1] \to \Omega$ is not topologically equivalent to (the image of) its restriction to an interval with a gap (a,b) given by $c|([0,a] \cup [b,1])$.

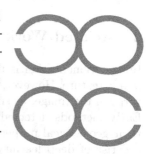

Fig. 2. A globally topologically correct segmentation with local topological errors.

However, we want to emphasize that correct connectivity is a *local* topological problem. To this end, note that topology is a global phenomenon, whereas image segmentation is inherently local. While topology, as quantified by Betti numbers, checks whether you have found the correct number of components or loops, this does not ensure a topologically correct segmentation.

To see this, consider the two curvilinear structures illustrated in Fig. 2, where the top illustrates a true segmentation and the bottom illustrates an estimated segmentation. These two segmentations are topologically equivalent, and their Betti numbers would be identical. They would also be rather similar in pixel-wise losses or measures. However, from the point of view of topology-preserving segmentation, the error made is considerable: The ground truth loop has not been detected, and the loop found in the segmentation is not supposed to be a loop.

In other words, **the segmentation needs to be topologically correct in a local sense**. As we shall iterate throughout the paper, our modelling choices, as well as how we validate the topological correctness of segmented curvilinear structures, are directly linked to this modelling rationale.

To obtain a data-driven curvilinear segmentation algorithm that is robust in the face of noisy data, we make the following contributions:

1. We build a dual-decoder neural network for curvilinear structure segmentation, generating predictions by softly fusing the two predictions produced by the two-stage prediction based on image texture and topology, respectively. The dual predictions give increased stability for low-quality images.
2. Utilizing deep contrastive learning, we leverage the latent embedding similarity to learn topological information of the images in an end-to-end scheme, which does not rely on any prior information and produces self-supervised embeddings that respect the curvilinear structure topology. This data-driven approach to curvilinear structure segmentation is particularly scalable when faced with complex topological structures.
3. We demonstrate our model on a challenging multi-class ultrasound segmentation task, and the open DRIVE retinal vessel segmentation dataset.

2   Related Work

Image segmentation is usually formulated as pixel-wise classification based on image texture. However, this is not enough to capture complete curvilinear structures in the image, as small pixel-wise errors can lead to incorrect connectivity. Early methods detected curvilinear structures by designing filters that target their geometrical properties, which are hard to detect at low contrast [4]. With the rise of deep learning, convolutional neural networks (CNNs), such as the U-net [12], have become the state-of-the-art for segmentation.

For curvilinear structure segmentation, one branch of research uses persistent topology to [3,6,7] to train networks to explicitly avoid breaking topological structure. A caveat with such approaches, especially for complex topologies, is that they require expensive topological computation. Following a similar, but less expensive route, Shit et al. [16] define a clDice loss that encourages the segmentations to maintain correct topology based on their soft skeletons.

Another branch of work [8,11] designs delicate backbones and schemes to capture the details of curvilinear features. CS-Net [11] introduces a self-attention mechanism that deals with multiple curvilinear types in a unified manner. Iter-Net [8] concatenates U-Net and multiple mini U-Nets to incorporate the structural redundancy. Nevertheless, these methods, which simply learn semantic labels per pixel, often fail to preserve the connectivity of objects. TR-GAN [2] is a generative adversarial network that improves artery/vein classification of vessels by improving segmented vessel connectivity. A topology ranking discriminator along with a contrastive loss encourages the network to generate segmentation with the expected topology ranking. The topological features learned by a pre-trained network are only involved in the computation of cost functions to preserve the image topology. These methods have in common that they are not designed to handle low-quality images. The pre-trained features used in [2,10] are less well suited for such problems, and we find experimentally that architectures such as [8] struggle with low-quality images. In this paper, we, instead, explicitly model the types of errors for which we wish the network to be robust.

A different alternative is given by post hoc approaches, which correct the broken predictions from an independent segmentation model. Sasaki et al. [13] propose an auto-encoder that detects and paints the gaps in images automatically. These models suffer from the clear disadvantage that they can only repair false splits – they are unable to handle missed splits.

All the previously mentioned models have in common that they aim to preserve topology and encode image textures in a single encoding stage. We find that in particular for images with low contrast or resolution, combining these two tasks at the same stage may make the model hard to tune. Thus, in this paper, we split the topology-preserving curvilinear segmentation into two simpler sub-tasks: First learning texture information, and next learning topological information from the texture-based coarse segmentation.

3   Method

3.1   Decoupling Topology-Preserving Curvilinear Segmentation

We first summarize our proposed **D**ual-decoder and **T**opology-aware **U-Net** (**DTU-Net**) architecture shown in Fig. 3. Our DTU-Net includes two lightweight mini U-Nets, with fewer layers than the original U-Net, referred to as the *texture net* and the *topology net*. The mini U-Net architecture is the same as that of the mini U-Nets from IterNet [8]. The texture net follows the learning scheme of most general segmentation tasks - encoding image texture information and giving pixel-wise predictions. The coarse prediction from the texture net is further fed to

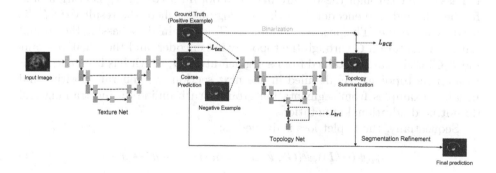

Fig. 3. Network architecture of DTU-Net. The dotted arrows refer to operations that are only performed at the training stage. Please zoom in for details.

the topology net for topology preservation. The topology net is another mini U-Net, where the encoder learns topological features using a self-supervised triplet loss, and the decoder summarizes the topology in a binary segmentation map indicating image foreground and -background. The texture-based and topology-based predictions are further softly fused to obtain the final segmentation.

Note that in practice, the texture net can be replaced with any pre-trained or randomly-initialized segmentation network. This means that our model can also be employed as an easy-to-use plugin to extend to powerful general segmentation models. Our use of lightweight mini-U-nets is motivated by the ultrasound image quality assessment task, where we aim for segmentations to serve as real-time feedback on image quality during scanning, with limited computational resources.

3.2   Topology Net

Learning Topological Embeddings with a Triplet Loss. Recent approaches to curvilinear structure segmentation [2,10] adopt a pre-trained VGG19 network to extract high-level image topological features, where contrastive losses penalize segmentations with different topological features from the ground truth. In these methods, topological information is extracted by models pre-trained on natural images, which is agnostic to the task. In addition, during training, only the feature differences are considered instead of the features themselves. We, instead, integrate contrastive learning into the latent embedding of a standard segmentation network using self-supervised learning to learn topological consistency.

Contrastive learning is often used to extract effective instance features for downstream tasks by producing embeddings that keep similar objects close, and push dissimilar objects apart. Here, we adopt a triplet loss in the bottleneck of the topology net, enabling self-supervised learning of image topology features.

The triplet loss compares an anchor point to a positive and negative example [15]. The loss function penalizes the difference between the anchor's distance to the positive and negative examples, respectively. In our formulation, the image is passed first through the texture net Θ to obtain a coarse segmentation, and further through the encoder Ψ of the topology net, where the result $\Psi(\Theta(I))$ is used as an *anchor*. The positive example $\Psi(G)$ is obtained by passing the ground truth segmentation G through the topology net encoder, and the negative example $\Psi(\hat{G})$ is similarly obtained from a segmentation $\hat{G}$, which has been corrupted to alter its topology. As detailed in the next section, we generate positive and negative examples from segmentation ground truth online during training and do not need additional annotations.

Sequentially, the triplet loss is defined as:

$$L_{tri}(\Psi(\Theta(I)), \Psi(G), \Psi(\hat{G})) = max(d^+ - d^- + \tau, 0) \qquad (1)$$

where d^+ is the averaged Euclidean distance of $\Psi(\Theta(I))$ and $\Psi(G)$, d^- is the distance of $\Psi(\Theta(I))$ and $\Psi(\hat{G})$ and τ is an empirically-set margin. By minimizing the triplet loss during training, the network forces the encoded features closer to those of the ground truth segmentation mask whose topology is correct, while pushing it further away from the embedding of the corrupted segmentation masks.

Self-supervised Training of the Triplet Loss. We generate corrupted masks from the ground truth mask by randomly breaking the image topology. Synthetic false splits in the corrupted masks are generated as follows: We assume that false splits can happen anywhere on the curves when the CNNs fail to recognize the objects. Therefore, segmentation labels are split into small patches, randomly removing λ of the patches containing foreground information, where $\lambda \in [0, 1]$ is a hyper parameter controlling the degree of corruption.

Next, we also generate corrupted masks with synthetic missed splits: We deem the missed splits tend to appear in regions with similar texture features, put simply in raw images, the pixel values. We compute the average m and standard deviation d of the foreground pixel values in the image (foreground pixels are given in the ground truth). To generate synthetic missed splits, a fixed percentage λ of the pixels with value within $[m - d, m + d]$ are selected randomly and assigned the label of the most similar foreground pixel.

In our implementation, adding missed splits and false splits is done sequentially. The introduction of missed splits brings not only incorrect connections between curves but also noise, which harms the topology as well.

During inference, the encoder only forwards $\Psi(\Theta(I))$. The triplet loss in Eq. 1 quantifies the pixel-wise feature difference. We also merge the global topological feature into the embedding by a SENet [5], which takes the average of the learned feature map and encodes them into attentive scores via a multi-layer perceptron. The scores are then multiplied by the original feature map.

Fine-to-Coarse Topology Summarization. In the DTU-Net, the texture net learns low-level information directly from raw images, which is effective in semantic segmentation tasks. Our topology net, however, abstracts the information into high-level topological features, which are less applicable for texture recognition tasks due to the loss of texture details. We expect the topology net to summarize the information from object connectivity, and contribute to the final segmentation. Consequently, we ask the topology net a binarized question: Based on the learned topology, should the structure be connected or not? For multi-class segmentation tasks such as in our ULTRASOUND dataset, these binary predictions are sequentially employed to refine the multi-class semantic segmentation from the texture net.

3.3 Segmentation Refinement and Joint Feature Learning

Our DTU-Net consists of a texture network and a topology network. We denote the texture network (a mini U-Net), the encoder, and the decoder of the topology network (another mini U-Net) as $\Theta(\cdot)$, $\Psi(\cdot)$, and $\Omega(\cdot)$ respectively. Each part of the network can be trained separately, as with the post hoc methods. We train the network in a joint way, where the gradient backpropagation from the topology net can also update the parameters in the texture net. Denoting input images I and segmentation ground truth masks G, DTU-Net is trained by a unified loss

$$L_{DTU}(I, G) = L_{tex}(\Theta(I), G) + L_{BCE}(\Omega(\Psi(\Theta(I))), \bar{G}) + L_{tri} \qquad (2)$$

where L_{tex} is the pixel-wise segmentation loss for the texture net, L_{BCE} is the binary cross entropy loss tuning the prediction from the topology net, L_{tri} is the triplet loss in Eq. 1, $\bar{G}$ and $\hat{G}$ refers to the ground truth binarized and corrupted by us. The loss function also indicates that our DTU-Net does not need additional annotations and learns image topology in a self-supervised way.

As the topology net and the texture net view the segmentation problem from different points of view, we fuse their predictions into a final prediction – this avoids sensitivity to the topology net sometimes being overconfident. Inspired by ensemble learning [14], we fuse the two predictions using weighted summation, where the hyper parameter ω indicates our confidence in the texture-based prediction. For each pixel, given p_{top} the pixel-wise probability of the topological prediction, and $p_{tex}^{(i)}$ the pixel-wise probability of the i-th category from the textural prediction, the two predictions are simply fused by:

$$p_{final}^{(i)} = \begin{cases} (1 - \omega)(1 - p_{top}) + \omega p_{tex}^{(0)}, i = 0; \\ \dfrac{(1-\omega)p_{top}+\omega \sum_{i=1}^{c} p_{tex}^{(i)}}{\sum_{i=1}^{c} p_{tex}^{(i)}} p_{tex}^{(i)}, i = 1, 2, ..., c \end{cases} \qquad (3)$$

where c is the total number of categories in the semantic segmentation task, $p_{final}^{(i)}$ is the pixel-wise final prediction, the 0-th category means the background.

4   Experiments

Baseline Models. We compare our model against a mini U-net trained with the clDice loss function, as well as with the persistent homology-based TopoLoss [7]; the data-driven IterNet [8] and the inpainting network of [13]. The input of the inpainting network is the prediction from a mini U-Net trained by a focal loss.

The clDice loss function preserves topological information by learning object skeletons, which, however, does not guarantee accurate segmentation. In the original paper [16], this is handled by mixing the clDice metric with another pixel-wise classification loss, more precisely the soft dice loss with hyper-parameter β:

$$L_{clDice} = \beta(1 - clDice) + (1 - \beta)(1 - softDice) \qquad (4)$$

In our experiments, and in particular on the more challenging ULTRA-SOUND dataset, we found the clDice loss function challenging to tune. The experimental setup from [16] did not converge with of the suggested values of $\beta = 0$ or 0.5 (see Table 3a). To improve convergence, we replaced softDice with focal loss [9]:

$$L_{clDice} = \alpha(1 - clDice) + (1 - \alpha)L_{focal}. \qquad (5)$$

Even here, however, the performance was unstable with the choice of α, as can be seen from Table 3a, and we choose to compare to the optimal observed $\alpha = 0.2$ in Table 1. For the DRIVE dataset, we choose $\alpha = 0.5$ following [16].

Table 1. Results on the ULTRASOUND dataset. Note that cIoU and cmIoU refer to IoU and mean class IoU, respectively, for curvilinear structures, whereas vIoU and vmIoU refers to volumetric structures.

Model	Loss	Frechet	Betti error	vIoU	vmIoU	cIoU	cmIoU
mini U-Net	clDice	0.7801	0.2635	97.95	29.84	31.41	20.41
mini U-Net	TopoLoss	0.9653	0.2606	98.07	26.29	25.89	18.56
IterNet	focal	**0.4329**	1.0963	98.20	17.67	5.76	3.70
inpainting	MSE	0.8189	1.1179	93.88	23.93	40.15	24.37
DTU-Net	Eq. 2	0.5818	**0.0974**	**99.51**	**44.33**	**72.30**	**52.92**

Table 2. Segmentation results on the DRIVE dataset.

Model	Loss	Frechet	Betti error	IoU
mini U-Net	clDice	3.6850	1.0890	67.95
mini U-Net	TopoLoss	3.6070	0.9232	70.52
IterNet	focal	3.5335	1.0039	70.35
inpainting	MSE	3.9001	1.3824	65.54
DTU-Net	Eq. 2	**2.9316**	**0.8597**	**73.86**

Tasks. We validate our algorithm on two different segmentation tasks.

The ULTRASOUND dataset poses a challenging fetal ultrasound multiclass segmentation task including both volumetric and curvilinear structures. These images, collected from the Danish national fetal ultrasound screening database, are obtained a typical 3rd-trimester growth scans, where four standard ultrasound planes (head, abdomen, femur, cervix) are acquired to assess preterm birth and fetal weight. Whether the images are "standard" depends on the visibility and pose of certain anatomical organs, several of which appear as curvilinear structures in the image. It is important to accurately detect whether the entire structure is present, as well as measuring its dimensions. Our motivating downstream application requires segmenting multiple different organs in real time while scanning; we have trained our models to segment both the volumetric and curvilinear structures. There are 2088 images of resolution 960×720, divided into 80% for training, 10% for validation, and 10% for testing. The dataset includes 6 curvilinear (cervix canal outline, cervix outer boundary, cervix inner boundary, outer skin boundary, outer bone boundary, thalamus) and 7 volumetric segmentation categories (bladder. femur bone, stomach bubble, umbilical vein, kidney, fossa posterior, cavum septi pellucidi). There are 468 femur, 531 abdomen, 686 head, and 403 cervix images. Images are reshaped to 224×288.

The DRIVE retinal segmentation dataset [17] induces a binary segmentation task. We used 16 images for training and 4 for testing. During training, the images are randomly cropped into 256×256 patches, while in testing patches are obtained with a 128×128 window sliding with a stride of 64.

Table 3. (a) clDice model selection on ULTRASOUND. The first two rows show the models suggested in the original paper [16]; these did not converge. The next rows show results for various parameters using the focal loss in place of the soft dice. (b) Training times on ULTRASOUND. TopoLoss and clDice were used with a mini U-Net; the inpainting included two stages.

(a)

α	β	Frechet	Betti error	vIoU	vmIoU	cIoU	cmIoU
0	0.5	-	9.8019	2.14	0.33	1.53	1.21
0	1	-	13.93	3.72	0.62	2.91	1.80
0	0	0.8418	0.4572	97.70	29.42	**32.99**	**23.14**
0.1	0	0.9500	0.3800	97.74	28.13	26.34	17.72
0.2	0	**0.7801**	**0.2635**	**97.95**	**29.84**	31.41	20.41
0.3	0	1.0798	0.4128	97.70	28.75	29.79	21.07
0.4	0	1.0727	0.3873	97.44	29.30	25.56	18.09
0.5	0	0.8969	0.3264	97.06	26.69	22.76	15.00

(b)

Model	Time/hours
TopoLoss	43.74
IterNet	20.02
clDice	6.91
inpainting	7.47
DTU-Net	17.2

Experimental Settings. The texture net is trained with a combined dice and focal loss as L_{tex} in Eq. 2. Triplet loss examples are generated incrementally by reducing λ from 50% to 10% during training, with $\omega = 0.5$ and $\tau = 0.1$.

Evaluation Metrics. To measure geometric affinity between curves, we employ the Frechet distance [1], which measures the extent to which the two curves remain close to each other throughout their course. We find this more suitable than e.g. accuracy, which is sensitive to small shifts or differences in curve thickness.

To measure the local topological correctness of the segmented structures, we apply a sliding window, in which we compare the number of components in the prediction and the ground truth annotation, respectively. This corresponds to a local Betti number computation. For each image, we compute a topological error (Betti error) which is the average difference in Betti numbers between ground truth and prediction over all patches seen, similar to what is done in [7].

For completeness, we also include the voxel-wise intersection over union (IoU), and mean class IoU for the multi-class segmentation task. For the ULTRA-SOUND dataset we report these both for the curvilinear and volumetric structure classes.

Results. Results for the ULTRASOUND dataset are found in Table 1, and for the DRIVE dataset in Table 2, with example segmentations in Fig. 4. The performance of the model used as the input for the post hoc inpainting on the ULTRASOUND dataset can be found in Table 3a with α and $\beta = 0$.

Table 4. Ablation study on different components of DTU-Net.

Model	Frechet	Betti error	vIoU	vmIoU	cIoU	cmIoU
w/o triplet loss	0.6605	0.1056	98.95	31.56	66.45	41.21
$\omega = 0$	0.5863	0.1023	98.90	43.11	70.43	52.33
$\omega = 1$	0.8601	0.1683	98.12	14.15	30.88	22.20
DTU-Net	**0.5818**	**0.0974**	**99.51**	**44.33**	**72.30**	**52.92**

Table 5. Ablation on Model size

Model	Size	Frechet	Betti error	vIoU	vmIoU	cIoU	cmIoU
full U-Net	131.8M	0.5849	0.1383	98.53	31.09	32.58	23.22
DTU-Net	65.5M	**0.5818**	**0.0974**	**99.51**	**44.33**	**72.30**	**52.92**

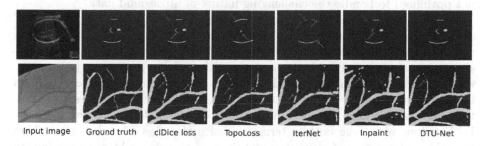

Input image   Ground truth   clDice loss   TopoLoss   IterNet   Inpaint   DTU-Net

Fig. 4. Segmentation of examples from ULTRASOUND and DRIVE. Blue arrows point at missed splits while red arrows point at false splits in the image. (Color figure online)

Running Time. Table 3b compares training times on ULTRASOUND. The models were trained on a Quadro RTX 6000 GPU with identical training settings. Note how DTU-Net has a considerably lower training time than TopoLoss, which is the closest competitor on the DRIVE dataset. While clDice and inpainting have the lowest runtimes, these do not compete in performance (Tables 1 and 2).

Ablation Study. We conducted an ablation study over different components of the DTU-Net on the ULTRASOUND dataset, see Table 4. We see that removing the triplet loss from Eq. 2 leads to increased Betti error. We also evaluate the effectiveness of feature fusion. Specifically, ω in Eq. 3 is set to be 0 and 1 respectively, meaning that the prediction of foreground and background is dominated by the topology net or the texture net. To prove that our model performance is not merely due to the increase of learnable parameters, we also include a comparison with a larger U-Net trained with the clDice loss ($\alpha = 0.2$), which has roughly two times the model size of ours (Table 5).

5   Discussion and Conclusion

We have designed a framework for curvilinear structure segmentation which is trained, in a data driven way, to avoid both false and missed splits. Our method is built around a dual-decoder network, which first makes a rough segmentation based on image texture, and next refines this optimizing for predicting correct topological structures. Here "correctness" also emphasises whether the structure is visible in the image. As the second network takes the predicted foreground probabilities as input, it learns to enhance those responses that form curvilinear structures. This is obtained using self-supervised contrastive learning, to teach the network the difference between a weak curvilinear response, and no response. On the highly challenging ULTRASOUND dataset, we see that our method outperforms all other methods. This is natural as most methods are developed using easier segmentation tasks such as e.g. vessel segmentation, and they are not optimized to handle the challenging nature of ultrasound data.

However, our method also outperforms baselines on the well-known DRIVE dataset. Here, we hypothesize that the data driven approach may have an easier time learning to handle complex topologies. It also has an advantage over persistent homology based methods, as it explicitly seeks to preserve local topology, whereas persistent homology based methods achieve locality by preserving Betti numbers within patches. This means that the locality scale becomes a parameter, which is additionally constrained by computational demands. This gives the DTU-net an advantage both in terms of choice of scale (it is inherently robust to scale) and in terms of computational time, as also shown by our experiments.

One important difference between methods is the number of parameters used. We have designed most methods around the fixed size mini U-net. Since our own method has the additional topology network, it de facto has more parameters than the mini U-net trained with clDice. To asses whether this difference in parameters was unfair, we also compared our method to a deeper U-net (with more parameters than our full architecture), trained with clDice. However, the results (see Table 5) show that we still outperform clDice on the ULTRASOUND dataset even when it is given the advantage of more parameters. The IterNet architecture also has more parameters than ours, but is not competitive.

Limitations. Our method is naturally limited by our downstream need for real-time inference, which is why we rely on mini U-Nets rather than more complex models. Allowing larger models might change the outcomes for competing methods, although this was not the case for clDice on the ULTRASOUND dataset. At the same time, our architecture's two mini U-Nets makes our method more expensive, also at inference time, than single mini U-Nets trained with topological loss functions. Finally, while the data-driven contrastive learning is clearly helpful in learning topological structure, it does not come with theoretical guarantees.

Summary. We have proposed a data-driven framework for segmenting curvilinear structures, which outperforms the baselines on DRIVE, and by far on the challenging ULTRASOUND dataset. Given that all baseline models struggle on ULTRASOUND, we see this as a cue for the community to include challenging segmentation tasks when developing curvilinear structure segmentation methods.

Acknowledgements. This work was supported by the DIREC project EXPLAIN-ME(9142-00001B), the Novo Nordisk Foundation through the Center for Basic Machine Learning Research in Life Science (NNF20OC0062606), and the Pioneer Centre for AI, DNRF grant nr P1.

References

1. Alt, H., Godau, M.: Computing the Fréchet distance between two polygonal curves. Int. J. Comput. Geom. Appl. **5**(01n02), 75–91 (1995)
2. Chen, W., et al.: TR-GAN: topology ranking GAN with triplet loss for retinal artery/vein classification. In: Martel, A.L., et al. (eds.) MICCAI 2020. LNCS, vol. 12265, pp. 616–625. Springer, Cham (2020). https://doi.org/10.1007/978-3-030-59722-1_59
3. Clough, J., Byrne, N., Oksuz, I., Zimmer, V.A., Schnabel, J.A., King, A.: A topological loss function for deep-learning based image segmentation using persistent homology. IEEE Trans. Pattern Anal. Mach. Intell. **44**, 8766–8778 (2020)
4. Frangi, A.F., Niessen, W.J., Vincken, K.L., Viergever, M.A.: Multiscale vessel enhancement filtering. In: Wells, W.M., Colchester, A., Delp, S. (eds.) MICCAI 1998. LNCS, vol. 1496, pp. 130–137. Springer, Heidelberg (1998). https://doi.org/10.1007/BFb0056195
5. Hu, J., Shen, L., Sun, G.: Squeeze-and-excitation networks. In: IEEE CVPR (2018)
6. Hu, X., Wang, Y., Li, F., Samaras, D., Chen, C.: Topology-aware segmentation using discrete Morse theory. In: ICLR (2021)
7. Hu, X., Li, F., Samaras, D., Chen, C.: Topology-preserving deep image segmentation. In: NeurIPS, vol. 32 (2019)
8. Li, L., Verma, M., Nakashima, Y., Nagahara, H., Kawasaki, R.: IterNet: retinal image segmentation utilizing structural redundancy in vessel networks. In: IEEE WACV, pp. 3656–3665 (2020)
9. Lin, T.Y., Goyal, P., Girshick, R., He, K., Dollár, P.: Focal loss for dense object detection. In: IEEE ICCV (2017)
10. Mosinska, A., Marquez-Neila, P., Koziński, M., Fua, P.: Beyond the pixel-wise loss for topology-aware delineation. In: IEEE CVPR (2018)
11. Mou, L., et al.: CS2-net: deep learning segmentation of curvilinear structures in medical imaging. Med. Image Anal. **67**, 101874 (2021)
12. Ronneberger, O., Fischer, P., Brox, T.: U-net: convolutional networks for biomedical image segmentation. In: Navab, N., Hornegger, J., Wells, W.M., Frangi, A.F. (eds.) MICCAI 2015. LNCS, vol. 9351, pp. 234–241. Springer, Cham (2015). https://doi.org/10.1007/978-3-319-24574-4_28
13. Sasaki, K., Iizuka, S., Simo-Serra, E., Ishikawa, H.: Joint gap detection and inpainting of line drawings. In: IEEE CVPR (2017)
14. Schapire, R.E.: A brief introduction to boosting. In: IJCAI, vol. 99, pp. 1401–1406 (1999)
15. Schroff, F., Kalenichenko, D., Philbin, J.: FaceNet: a unified embedding for face recognition and clustering. In: IEEE CVPR, pp. 815–823 (2015)

16. Shit, S., et al.: clDice-a novel topology-preserving loss function for tubular structure segmentation. In: IEEE CVPR (2021)
17. Staal, J., Abràmoff, M.D., Niemeijer, M., Viergever, M.A., Van Ginneken, B.: Ridge-based vessel segmentation in color images of the retina. IEEE TMI **23**(4), 501–509 (2004)
18. Wu, L., Cheng, J.Z., Li, S., Lei, B., Wang, T., Ni, D.: FUIQA: fetal ultrasound image quality assessment with deep convolutional networks. IEEE Trans. Cybern. **47**(5), 1336–1349 (2017)

HALOS: Hallucination-Free Organ Segmentation After Organ Resection Surgery

Anne-Marie Rickmann[1,2(✉)], Murong Xu[2], Tom Nuno Wolf[2],
Oksana Kovalenko[2], and Christian Wachinger[1,2]

[1] Lab for Artificial Intelligence in Medical Imaging, Ludwig Maximilians University,
Munich, Germany
[2] Department of Radiology, Technical University Munich, Munich, Germany
arickman@med.lmu.de

Abstract. The wide range of research in deep learning-based medical image segmentation pushed the boundaries in a multitude of applications. A clinically relevant problem that received less attention is the handling of scans with irregular anatomy, e.g., after organ resection. State-of-the-art segmentation models often lead to *organ hallucinations*, i.e., false-positive predictions of organs, which cannot be alleviated by oversampling or post-processing. Motivated by the increasing need to develop robust deep learning models, we propose HALOS for abdominal organ segmentation in MR images that handles cases after organ resection surgery. To this end, we combine missing organ classification and multi-organ segmentation tasks into a multi-task model, yielding a classification-assisted segmentation pipeline. The segmentation network learns to incorporate knowledge about organ existence via feature fusion modules. Extensive experiments on a small labeled test set and large-scale UK Biobank data demonstrate the effectiveness of our approach in terms of higher segmentation Dice scores and near-to-zero false positive prediction rate.

1 Introduction

Deep learning methods have become state-of-the-art for many medical image segmentation tasks, e.g. structural brain segmentation [18], tumor segmentation [12] or abdominal organ segmentation [4,5,8,17]. A challenge that remains is the generalization to unseen data, where a domain shift between training and testing data often leads to performance degradation. Research on robustness and domain adaptation [6] introduced new methods for handling domain shift, where the focus has mainly been on a shift in the intensity distribution of image data due to different imaging protocols, different scanner types or different modalities.

In contrast, a domain shift in the anatomy itself, e.g., by missing organs due to surgical organ resection has received less attention. In comparison to natural images, which can show image compositions with arbitrary objects, medical

---

A.-M. Rickmann and M. Xu—The authors contributed equally.

A. Frangi et al. (Eds.): IPMI 2023, LNCS 13939, pp. 667–678, 2023.
https://doi.org/10.1007/978-3-031-34048-2_51

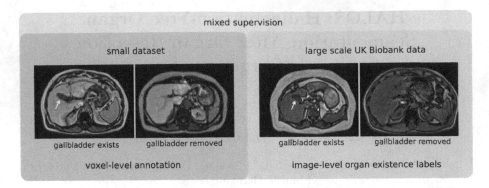

Fig. 1. Mixed supervision in HALOS using a small dataset with voxel-level annotations of multiple organs and a large-scale dataset with image-level binary labels of organ existence. The white arrow points to the gallbladder.

images of the human abdomen usually contain the same organs in the same ordering. This constraint of the human anatomy is beneficial for training networks and has, for instance, been explicitly used by incorporating shape priors [11,15,24]. However, as we move to clinical translation or to large-scale population studies, we will also encounter cases that do not follow the normal anatomy, which will yield a degradation in segmentation accuracy.

In this work, we mainly focus on gallbladder resection (cholecystectomy), as it is one of the most commonly performed abdominal surgeries. The indication for gallbladder removal is usually gallstones, which most of the time has no effect on other organs and the overall anatomy. We further evaluate our method on the cases of nephrectomy (kidney resection), where the indication can be more severe, e.g., kidney tumors, which could come with anatomical changes in other organs, like metastases. Further, kidneys are much larger than gallbladders, so that their removal can lead to post-surgical organ shift [19].

As we will demonstrate in our experiments, state-of-the-art segmentation networks often identify organs in the images, although they were removed. A phenomenon that we refer to as *organ hallucination*. We believe that organ hallucinations have so far not received more attention because publicly available segmentation datasets rarely contain cases after organ resection. This is probably due to the relatively small sample size of most segmentation datasets, as manual segmentation is time-consuming and costly. Fortunately, large-scale population imaging studies like the UK Biobank (UKB) Imaging study [10] with a targeted 100,000 subjects are becoming available that provide representative data of the population. The prevalence of cholecystectomies (gallbladder resection) in our sample of UK biobank is 3.7%, which provides enough data for studying this research question.

We introduce HALOS for the HALlucination-free Organ Segmentation after organ resection surgery. HALOS is a multi-task network that simultaneously learns classification of organ existence and segmentation of six abdominal organs (liver, spleen, kidneys, pancreas, gallbladder). HALOS is trained using mixed supervision, which accounts for the fact that we only have voxel-level annotations for a small dataset but image-level labels of organ removal on a large dataset, see Fig. 1. A key component of HALOS is a feature fusion module [22] that integrates the knowledge of organ existence into the segmentation branch. The key contributions are:

- a robust and flexible multi-task segmentation and classification model that predicts near-to-zero false positive cases on the UKB dataset
- the multi-scale feature fusion with the dynamic affine feature map transform [22] of the classification output into the segmentation branch
- a demonstration of the relevance of the missing organ problem by comparing to state-of-the-art segmentation models.

1.1 Related Work

Abdominal Multi-organ Segmentation. Nowadays, convolutional neural networks are state-of-the-art for abdominal organ segmentation in CT and MRI scans [3–5,17,21]. One method to point out is nnU-Net [8], which is an automatic pipeline to configure a U-Net to a given dataset. nnU-Net has won several medical image segmentation challenges, and has proven to be a robust and generic method. Therefore we consider nnU-Net as a baseline in our experiments.

Missing Organ Segmentation. To our best knowledge, the missing organ problem has so far only been studied for CT scans in [19], where an atlas-based approach is used. It trains a Gaussian Mixture Model on normal images and detects missing organs by analyzing fitting errors. However, this method inevitably relies on heavy simulation for parameter tuning and is therefore vulnerable to distribution shift. In a more recent method [20], the Dice loss was studied and it was argued that setting the reduction dimension over the complete batch would help to predict images with missing organs. However, the method was not tested on cases after organ resection. We compare to this approach in our experiments.

Classification-Assisted Segmentation. As image-level labels are easier to obtain than voxel-wise annotations, prior work has considered including these additional labels by extending the segmentation network with a classification branch [13,14,23]. In [14], the two branches are trained jointly using both fully-annotated and weakly-annotated data with shared layers at the beginning, for 2D brain tumor segmentation and classification of tumor existence. They showed that the additional classification significantly improved segmentation performance compared to standard supervised learning. We compare to this approach in our experiments.

Feature Fusion. Some approaches for classification-assisted segmentation use feature fusion, i.e., the interweaving of segmentation and classification branches. For example, separate segmentation and classification models are trained in [23] for Covid-19 diagnosis. Feature maps of the classification and segmentation model are merged with Squeeze-and-Excitation (SE) blocks [7]. After the feature fusion, the enhanced feature map is fed into the decoder for segmentation. An alternative for feature fusion is the combination with metadata, such as age, gender, or measurements of biomarkers. The Dynamic Affine Feature Map Transform (DAFT) [22] predicts the scales and shifts to excite or repress feature maps on a channel-level from such metadata, as seen in Fig. 2.

2   Methods

Figure 2 illustrates the dual-branch classification-assisted segmentation pipeline of HALOS that combines Multitask Learning and Feature Fusion to handle missing organs. In the following, we describe each part of our pipeline in detail.

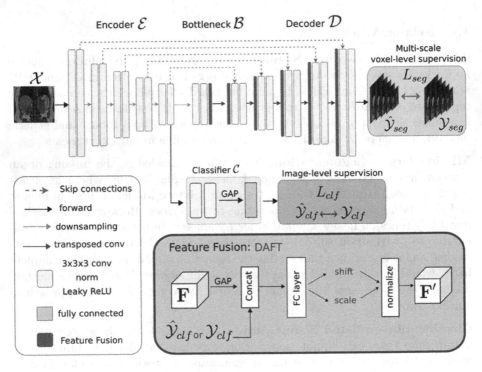

Fig. 2. Overview of the HALOS multi-task pipeline.

Segmentation Branch. In the segmentation branch, we use a U-Net architecture, based on nnU-Net [8] as the segmentation network which consists of an encoder $\mathcal{E}$, bottleneck $\mathcal{B}$, and a decoder $\mathcal{D}$. As previously mentioned, nnU-Net [8] is one of the most generic and well-performing medical image segmentation models. The nnU-Net pipeline automatically determines the best U-Net architecture and data augmentation for the given data. Therefore, we fed our segmentation dataset into the nnU-Net pipeline and took the architecture of the best-performing nnU-Net model and the data augmentation scheme as our baseline. The resulting model is a 3D U-Net with 32 starting channels and 5 downsampling levels.

The advantage of using encoder-decoder structured networks is that intermediate representations can be obtained at different scales. The U-Net is trained on input MR images $\mathcal{X}$ and voxel-level annotations $\mathcal{Y}_{seg}$ to output segmentation predictions $\hat{\mathcal{Y}}_{seg}$ under full supervision. The segmentation loss is defined as an average of Dice and Cross-Entropy loss L_{seg} with enabling of deep supervision at each feature map scale and dynamic class weights for individual images:

$$L_{seg} = L_{CE} + L_{Dice}, \qquad L_{CE} = -\frac{1}{N} \sum_i^C y_i \log(\hat{y}_i),$$

$$L_{Dice} = 1 - \frac{2 \cdot |\hat{\mathcal{Y}}_{seg} \cap \mathcal{Y}_{seg}| + \epsilon}{|\hat{\mathcal{Y}}_{seg}| + |\mathcal{Y}_{seg}| + \epsilon},$$

$$(1)$$

where we denote the class-wise ground truth y_i, class-wise predictions $\hat{y}_i$, the number of classes C and samples N, a smoothing term ϵ. Note that some implementations of the Dice loss only add ϵ to the denominator, to avoid division by zero. In our case, it is important to add ϵ to numerator and denominator, as we want to ensure a Dice loss of 0, rather than 1, for true negative predictions of gallbladders.

Classification Branch. Compared to manual voxel-level annotations, the global image-level labels are less informative but can be obtained at a substantially lower cost. Hence, we incorporate the classification task into the pipeline to study the impact of the low-dimensional prior knowledge on the final predicted segmentation. In the classification branch, classifier $\mathcal{C}$ is built on top of the encoder $\mathcal{E}$ and takes a feature map from a specific encoder block as input. The precise location of the classifier can be tuned as a hyper-parameter, but we found encoder blocks 4 and 5 promising for most models. Compared to training a standalone classification model, such a shared feature structure between $\mathcal{C}$ and $\mathcal{E}$ enables a more lightweight classification model and thus saves redundant computation. $\mathcal{C}$ consists of a convolutional block with the same structure as an encoder block, a 3D global average pooling step, and a fully connected layer for producing the final classification. The classifier is trained on MR scans with image-level surgery labels $\mathcal{Y}_{clf}$. The classification loss L_{clf} is the average cross-entropy weighted by the actual class ratio in the training set.

Feature Fusion. A key component of HALOS is the feature fusion module. The prior information about the resection of the gallbladder is fused with the feature maps of the segmentation branch at multiple locations. As shown in Fig. 2, these locations are the bottleneck and each stage of the decoder. Importantly, we can either use the ground truth image-level labels $\mathcal{Y}_{clf}$ or the classifier's prediction $\hat{\mathcal{Y}}_{clf}$ as input to the feature fusion, depending on whether the information about previous surgeries is available at test time.

We use DAFT [22] to perform feature fusion, which was originally designed to combine 3D images with low-dimensional tabular information, and can be conveniently integrated into any type of CNN. In our case, the tabular data to be concatenated is the binary classification result or ground truth label about gallbladder resection. To the best of our knowledge, DAFT has not yet been used in segmentation models or in a multi-scale fashion. We expect that information sharing at multiple scales of the decoder will emphasize the prior knowledge about the organ's presence and conduce the decoder to produce fewer false positive predictions of non-existing classes. The exact position of integrating feature fusion modules into the U-Net architecture is illustrated in Fig. 2. The classification labels are fused to the bottleneck feature map, which contains the highest-level information. Then the fused version will be forwarded to the decoder where we repeat the feature fusion blocks after each transpose convolution. We place feature fusion via DAFT before each decoder block, which avoids interaction with other normalization layers. Formally, for each item in a batch, let $\hat{\mathbf{y}} \in \mathbb{R}$, be the predicted output from the classifier, and $\mathbf{F}_{d,c} \in \mathbb{R}^{D \times H \times W}$, where D, H, W denote the depth, height, and width of the feature map, the c-th channel of the input feature map of block $d \in \{0, \ldots, 5\}$ in the decoder, as illustrated in Fig. 2. DAFT [22] learns to predict scale $\alpha_{d,c}$ and offset $\beta_{d,c}$

$$\mathbf{F}'_{d,c} = \alpha_{d,c}\mathbf{F}_{d,c} + \beta_{d,c}, \tag{2}$$

$$\alpha_{d,c} = f_c(\mathbf{F}_{d,c}, \hat{\mathbf{y}}_d), \qquad \beta_{d,c} = g_c(\mathbf{F}_{d,c}, \hat{\mathbf{y}}_d), \tag{3}$$

where f_c, g_c are arbitrary mappings from image and tabular As proposed in [22], a single fully connected neural network h_c models f_c, g_c and outputs a single α-β-pair.

During training, we randomly sample MR images with voxel-level and image-level labels to form batches and use them to update the segmentation model and classifier respectively. With the previously defined L_{seg} and L_{clf}, the final loss of HALOS is

$$L = \alpha \cdot L_{seg} + (1 - \alpha) \cdot L_{clf}, \tag{4}$$

where α indicates the weight assigned to the segmentation loss.

3   Results and Discussion

3.1   Experiment Setup

Segmentation Data. We use whole-body MRI scans with voxel-level annotations from three different sources: the German National Cohort (NAKO) [2], the Cooperative Health Research in the Region of Augsburg (KORA) [1], and UKB [10]. The samples cover a general population from Germany and the UK. All three studies acquired abdominal images with a two-point Dixon sequence, where we use the oppose-phase scans in this work. For pre-processing, we follow guidelines of other work [9,16]. The scans were manually segmented by a medical expert. The dataset contains 63 scans in total (16 NAKO, 15 KORA, 32 UKB), of which 18 are patients after gallbladder resection. We have split this data into 42(9) scans for training, 7(3) for validation and 11(6) scans for testing, the count of missing gallbladder cases is given in parentheses.

UKB Data. The UK Biobank dataset is much larger than the segmentation data, but only contains image-level annotations indicating organ presence. We use it for training the organ existence classifier in our multi-task pipeline. It can also be used for evaluating the model robustness since we can count the false positive segmentations of non-existing gallbladders. We used the information about past surgeries from the UKB database and our medical expert verified the labels for correctness. Out of 19,000 images we requested from UK Biobank, we counted 701 after gallbladder removal. We additionally randomly selected normal subjects. We split the data into two subsets, one for training and validation of models (899 scans with and 349 without gallbladder), and one which serves as an unseen test set (952 scans with and 352 without gallbladder). The ratio of no-gallbladder cases in each subset is set to be roughly 0.4.

Implementation Details and Hyperparameter Tuning. In this work, we use GPUs DGX A100 for running our experiments. The implementation is based on Python, PyTorch and MONAI. We perform hyperparameter tuning for the loss weight α, weight decay, learning rates for the segmentation model and classifier, normalization type (instance or batch normalization), batch size, and the location of the classifier using Ray Tune. We train our models using the automated mixed precision of PyTorch. Our code is publicly available at https://github.com/ai-med/HALOS.

Metrics. We evaluate our models by comparing Dice scores for all organs and false positive rate (FPR) of gallbladder segmentations. We define a sample as false positive if one or more voxels have been segmented as non-existing gallbladder. As the Dice score is not defined for non-existing organs, we define it to be 1 for true negative cases and 0 for false positive cases. Therefore, we can observe large changes in the Dice score when reducing the false positive rate.

Baselines. Apart from the nnU-Net baseline as described in Sect. 2, we further choose two alternative baselines, i.e., oversampling and post-processing. For oversampling, we oversample the cases without a gallbladder in training to achieve a balance in class frequency. Note that we are already weighting the loss functions by class frequency. The post-processing baseline is another method, where we use the prior information about gallbladder resection to remove false positives as a direct post-processing step.

Table 1. Comparison of HALOS with baseline nnU-Net, with oversampling, post-processing, Dice loss with batch reduction [20] (Dice batch red.), and multi-task model [14]. FF: feature fusion, gt: ground truth labels at test-time. We list Dice scores for all organs and false positive rate (FPR) for removed gallbladders. We provide mean and standard deviation over 5-fold cross-validation. *: best architecture for our data proposed by the nnU-Net pipeline was re-implemented.

Method	Dice Scores ↑							FPR ↓
	Mean	liver	spleen	r kidney	l kidney	pancreas	gallbl.	
nnU-Net* [8]	0.823±0.014	0.938±0.004	0.891±0.006	0.898±0.003	0.894±0.002	0.643±0.016	0.674±0.076	0.267±0.149
+ oversampling	0.832±0.008	0.940±0.006	0.894±0.005	0.901±0.005	0.891±0.005	0.655±0.011	0.712±0.052	0.233±0.091
+ post-proc. (gt)	0.847±0.005	0.938±0.004	0.891±0.006	0.898±0.003	0.894±0.002	0.643±0.016	0.819±0.009	0±0
+ batch red. [20]	0.818±0.010	0.945±0.002	0.895±0.002	0.901±0.005	0.894±0.006	0.663±0.014	0.610±0.045	0.400±0.091
multi-task [14]	0.822±0.010	0.930±0.006	0.879±0.004	0.895±0.003	0.885±0.002	0.625±0.016	0.716±0.054	0.233±0.091
HALOS w/o FF	0.825±0.010	0.941±0.002	0.892±0.009	0.898±0.004	0.892±0.005	0.657±0.013	0.668±0.073	0.3±0.139
HALOS (pred, gt)	0.853±0.002	0.939±0.003	0.899±0.005	0.899±0.003	0.893±0.004	0.649±0.021	0.840±0.015	0±0

3.2   Experiments on Cholecystectomy Cases

We train the baseline nnU-Net, oversampling and post-processing baselines, state-of-the-art methods [14,20] and HALOS using 5-fold cross-validation and report the average results over all folds on the segmentation test set in Table 1 and on the UKB test set in Table 2. The average FPR is quite high for the baseline nnU-Net on both datasets, which leads to a low gallbladder Dice score of 0.674. The segmentation performance on pancreas is also low 0.643, but the pancreas is very hard to segment, due to its shape variability. The oversampling only slightly improves performance, so we can assume that the reason for the high FPR is not only caused by class imbalance. As expected, the post-processing leads to higher gallbladder Dice scores and zero FPR, since it uses the ground truth information about cholecystectomy. A shortcoming of the post-processing is that the model's false positive prediction may appear in neighboring organs, which will result in a hole in the segmentation. The gallbladder usually lies in fossa vesicae biliaris, which is a depression on the visceral surface of the liver anteriorly, between the quadrate and the right lobes. Since the location is closely connected to the liver, we found many mistakes produced by our baseline that are either localized inside the liver or partly in the liver and partly in other tissues like visceral fat. Examples of typical organ hallucinations are shown in Fig. 3C–D, where the gallbladder is predicted in the fossa vesicae biliaris (C), inside the liver (D) and in the intestine (E). The recent work [20] proposes to set ϵ in the Dice loss to a low value, e.g. 10^{-7}, the batch size higher than 1 and to reduce the Dice loss over the batch dimension. We set the batch size to 8, which reached the limit of our GPU memory. Note that in our baseline model the batch size is set to 2, ϵ is 1 and we also reduce over the batch dimension. Interestingly, we observe an increase in FPR for both datasets. In preliminary experiments, we have removed the batch reduction in the Dice loss, but we have observed no significant difference in performance. The multi-task model proposed in [14] includes a classifier right before the segmentation output of the decoder. We use our nnU-Net model and extend it with a classifier, following the architecture

Table 2. Comparison of HALOS with baseline nnU-Net, oversampling, post-processing, Dice loss with batch reduction [20] (+ batch red.) and multi-task model [14] on the UKB dataset. FF: feature fusion, gt: ground truth labels for FF at test-time, pred: classification predictions for FF at test-time. We provide false positive (FP), false negative (FN), true positive (TP), true negative (TN), false positive rate (FPR) and F1 score for removed gallbladders, and the balanced accuracy (BAcc) of all classifiers. All values are mean and standard deviation over 5-fold cross-validation. *: best architecture for our data proposed by the nnU-Net pipeline was re-implemented.

Method	FP ↓	TN ↑	TP ↑	FN ↓	FPR ↓	F1 ↑	BAcc ↑
nnU-Net* [8]	91.2 ±30.62	260.8 ±30.62	537.2 ±16.62	62.8 ±16.62	0.259 ±0.087	0.875 ±0.009	
+ oversampling	66.6 ±6.633	285.4 ±6.633	522.6 ±13.18	77.4 ±13.18	0.189 ±0.028	0.879 ±0.011	
+ post-proc. (gt)	0 ±0	352 ±0	537.2 ±16.62	62.8 ±16.62	0 ±0	0.945 ±0.015	
+ batch red. [20]	135.2 ±57.15	216.8 ±57.15	530.2 ±24.39	69.8 ±24.39	0.384 ±0.162	0.838 ±0.017	
multi-task [14]	100.2 ±16.48	251.8 ±16.48	578.2 ±3.701	21.8 ±3.701	0.285 ±0.047	0.905 ±0.054	0.874 ±0.045
HALOS w/o FF	52.6 ±17.67	299.4 ±17.67	547.4 ±22.39	52.6 ±22.39	0.149 ±0.050	0.869 ±0.056	0.896 ±0.047
HALOS (gt)	2 ±2.550	350 ±2.550	564.8 ±14.74	35.2 ±14.74	0.006 ±0.007	0.968 ±0.010	0.933 ±0.005
HALOS (pred)	11 ±5.339	341 ±5.339	541.8 ±14.20	58.2 ±14.20	0.031 ±0.015	0.940 ±0.010	0.933 ±0.005

proposed in [14] at decoder block 5. This model leads to a slight decrease in FPR on the segmentation data, but to a higher FPR on the UKB data. To analyze the impact of multi-task learning and the feature fusion models, we train HALOS without the feature fusion, which interestingly leads to a slight increase in FPR and an decrease in gallbladder Dice score and slight decrease in FPR on UKB, compared to nnU-Net, even though the balanced accuracy of the classifier is 0.896. Therefore we argue, that multi-task training alone is not sufficient to reduce organ hallucinations. HALOS was trained using the ground truth labels $\mathcal{Y}_{clf}$ as input for feature fusion, and leads to an impressive reduction of the FPR to 0 on the segmentation data and 0.006 on the UKB data. The multi-task classifier achieves a balanced accuracy of 0.93. When we use the classifier's prediction for feature fusion at test time, we observe a slight increase in FPR over using the ground truth labels to 0.03. This shows, that our method is flexible and depending if prior information about gallbladder resection is available at test time or not, one can either fuse the ground truth labels or the classifier's predictions.

3.3 Experiments on Nephrectomy Cases

To evaluate if HALOS can be applied to other organ resection cases, we validate the effectiveness of HALOS on cases after nephrectomy. Note that we did not do any further hyper-parameter tuning in this experiment. We create a kidney segmentation dataset that contains 46(6/2) scans for training and 10(2/1) for testing, the count of missing kidney cases is given in parentheses with the format left/right. For UKB data, we split the available subjects into one training and validation set (200 scans with 17/5 missing left/right kidneys), and another hold-out test set (55 scans with 4/2 no-kidneys). Similar to gallbladder experiments, we train the baseline nnU-Net and HALOS using 5-fold cross-validation. Note

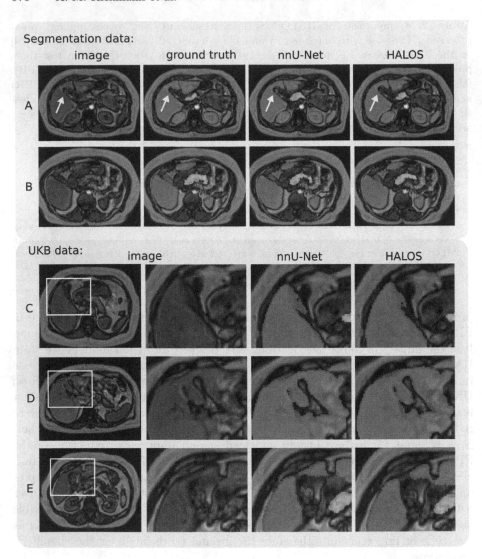

Fig. 3. Segmentation results on the segmentation data (top) and UKB (bottom). Comparison of nnU-Net and HALOS. A: scan with a resected gallbladder, nnU-Net produces a false positice. B: scan with an existing gallbladder. C: both models predict a false positive in the location where the gallbladder was resected. D: nnU-Net produces a false positive inside the liver. E: nnU-Net produces a false positive in the intestine.

that the classifier learns a multi-class classification, in contrast to the binary classification in the gallbladder experiments. We report the results of the nephrectomy experiment in the following. The baseline nnU-Net achieved an FPR of 0.2 for left kidney and 1 for right kidney on the UKB data. We observe that HALOS achieves a lower FPR of 0 for the left kidney and still high FPR of 0.7

for the right kidney, while having a significantly reduced voxel-level FP count of 16.5, compared to 129 for nnU-Net. The left kidney Dice of HALOS (0.9024) is higher than of nnU-Net (0.8413) while having no improvement on the right kidney 0.864 vs 0.867. A possible reason might be the small dataset size with severe class imbalance, of having only two cases with missing right kidneys in the training set. The balanced accuracy of the HALOS classifier is 0.93 for left kidney and 0.58 for right kidney, which also suggests that the class imbalance has more impact in this setting.

4 Conclusion

In this work, we introduced HALOS, a multi-task classification and segmentation model for hallucination-free organ segmentation. We propose to use multi-scale feature fusion, via the dynamic affine feature-map transform, to enrich the feature maps of the segmentation branch with prior information on organ existence. We have shown on cases after cholecystectom0,y and nephrectomy, that HALOS significantly reduces false positive predictions on a large scale UK Biobank test set, and increases gallbladder and left kidney Dice scores on a smaller segmentation test set, compared to nnU-Net and several additional baselines and multi-task approaches. HALOS is flexible to use ground truth organ existence labels at test-time or the prediction of the classifier, depending on the availability of such labels. In future work we would like to extend HALOS to additional cases of organ resection, e.g. hysterectomy (removal of uterus) or splenectomy (removal of spleen).

Acknowledgment. This research was partially supported by the Bavarian State Ministry of Science and the Arts and coordinated by the bidt, the BMBF (DeepMentia, 031L0200A), the DFG and the LRZ.

References

1. Bamberg, F., et al.: Subclinical disease burden as assessed by whole-body MRI in subjects with prediabetes, subjects with diabetes, and normal control subjects from the general population: the KORA-MRI study. Diabetes **66**(1), 158–169 (2017)
2. Bamberg, F., et al.: Whole-body MR imaging in the German national cohort: rationale, design, and technical background. Radiology **277**(1), 206–220 (2015)
3. Bobo, M.F., et al.: Fully convolutional neural networks improve abdominal organ segmentation. In: Medical Imaging 2018: Image Processing, vol. 10574, p. 105742V. International Society for Optics and Photonics (2018)
4. Chen, Y., et al.: Fully automated multi-organ segmentation in abdominal magnetic resonance imaging with deep neural networks. Med. Phys. **47**(10), 4971 (2020)
5. Gibson, E., et al.: Automatic multi-organ segmentation on abdominal CT with dense V-networks. IEEE Trans. Med. Imaging **37**(8), 1822–1834 (2018)
6. Guan, H., Liu, M.: Domain adaptation for medical image analysis: a survey. IEEE Trans. Biomed. Eng. **69**(3), 1173–1185 (2021)
7. Hu, J., Shen, L., Sun, G.: Squeeze-and-excitation networks. In: CVPR (2018)

8. Isensee, F., Jaeger, P.F., Kohl, S.A., Petersen, J., Maier-Hein, K.H.: nnU-Net: a self-configuring method for deep learning-based biomedical image segmentation. Nat. Methods **18**(2), 203–211 (2021)
9. Kart, T., et al.: Automated imaging-based abdominal organ segmentation and quality control in 20,000 participants of the UK Biobank and German national cohort studies. Sci. Rep. **12**(1), 1–11 (2022)
10. Littlejohns, T.J., et al.: The UK Biobank imaging enhancement of 100,000 participants: rationale, data collection, management and future directions. Nat. Commun. **11**(1), 1–12 (2020)
11. Liu, L., Wolterink, J.M., Brune, C., Veldhuis, R.N.: Anatomy-aided deep learning for medical image segmentation: a review. Phys. Med. Biol. **66**(11), 11TR01 (2021)
12. Liu, Z., et al.: Deep learning based brain tumor segmentation: a survey. Complex Intell. Syst. **9**(1), 1001–1026 (2023)
13. Mehta, S., Mercan, E., Bartlett, J., Weaver, D., Elmore, J.G., Shapiro, L.: Y-Net: joint segmentation and classification for diagnosis of breast biopsy images. In: Frangi, A.F., Schnabel, J.A., Davatzikos, C., Alberola-López, C., Fichtinger, G. (eds.) MICCAI 2018. LNCS, vol. 11071, pp. 893–901. Springer, Cham (2018). https://doi.org/10.1007/978-3-030-00934-2_99
14. Mlynarski, P., Delingette, H., Criminisi, A., Ayache, N.: Deep learning with mixed supervision for brain tumor segmentation. J. Med. Imaging **6**(3), 034002 (2019)
15. Oktay, O., et al.: Anatomically constrained neural networks (ACNNs): application to cardiac image enhancement and segmentation. IEEE Trans. Med. Imaging **37**(2), 384–395 (2017)
16. Rickmann, A.M., Senapati, J., Kovalenko, O., Peters, A., Bamberg, F., Wachinger, C.: AbdomenNet: deep neural network for abdominal organ segmentation in epidemiologic imaging studies. BMC Med. Imaging **22**(1), 1–11 (2022)
17. Roth, H.R., et al.: Hierarchical 3D fully convolutional networks for multi-organ segmentation. arXiv preprint arXiv:1704.06382 (2017)
18. Roy, A.G., Conjeti, S., Navab, N., Wachinger, C., Initiative, A.D.N.: QuickNAT: a fully convolutional network for quick and accurate segmentation of neuroanatomy. Neuroimage **186**, 713–727 (2019)
19. Suzuki, M., Linguraru, M.G., Okada, K.: Multi-organ segmentation with missing organs in abdominal CT images. In: Ayache, N., Delingette, H., Golland, P., Mori, K. (eds.) MICCAI 2012. LNCS, vol. 7512, pp. 418–425. Springer, Heidelberg (2012). https://doi.org/10.1007/978-3-642-33454-2_52
20. Tilborghs, S., Bertels, J., Robben, D., Vandermeulen, D., Maes, F.: The dice loss in the context of missing or empty labels: introducing Φ and ϵ. In: Wang, L., Dou, Q., Fletcher, P.T., Speidel, S., Li, S. (eds.) Medical Image Computing and Computer Assisted Intervention (MICCAI 2022). LNCS, vol. 13435, pp. 527–537. Springer, Cham (2022). https://doi.org/10.1007/978-3-031-16443-9_51
21. Wang, Y., Zhou, Y., Shen, W., Park, S., Fishman, E.K., Yuille, A.L.: Abdominal multi-organ segmentation with organ-attention networks and statistical fusion. Med. Image Anal. **55**, 88–102 (2019)
22. Wolf, T.N., Pölsterl, S., Wachinger, C., Initiative, A.D.N., et al.: DAFT: a universal module to interweave tabular data and 3D images in CNNs. Neuroimage **260**, 119505 (2022)
23. Wu, Y.H., et al.: JCS: an explainable COVID-19 diagnosis system by joint classification and segmentation. IEEE Trans. Image Process. **30**, 3113–3126 (2021)
24. Zhou, Y., et al.: Prior-aware neural network for partially-supervised multi-organ segmentation. In: ICCV (2019)

Human-Machine Interactive Tissue Prototype Learning for Label-Efficient Histopathology Image Segmentation

Wentao Pan[1], Jiangpeng Yan[3(✉)], Hanbo Chen[2], Jiawei Yang[4], Zhe Xu[5], Xiu Li[1(✉)], and Jianhua Yao[2(✉)]

[1] Tsinghua Shenzhen International Graduate School, Shenzhen, China
li.xiu@sz.tsinghua.edu.cn
[2] Tencent AI Lab, Shenzhen, China
yaojianhua@tencent.com
[3] Department of Automation, Tsinghua University, Beijing, China
yanjp13@tsinghua.org.cn
[4] University of California, Los Angeles, USA
[5] The Chinese University of Hong Kong, Hong Kong, China

Abstract. Deep learning have greatly advanced histopathology image segmentation but usually require abundant annotated data. However, due to the gigapixel scale of whole slide images and pathologists' heavy daily workload, obtaining pixel-level labels for supervised learning in clinical practice is often infeasible. Alternatively, weakly-supervised segmentation methods have been explored with less laborious image-level labels, but their performance is unsatisfactory due to the lack of dense supervision. Inspired by the recent success of self-supervised learning, we present a label-efficient tissue prototype dictionary building pipeline and propose to use the obtained prototypes to guide histopathology image segmentation. Particularly, taking advantage of self-supervised contrastive learning, an encoder is trained to project the unlabeled histopathology image patches into a discriminative embedding space where these patches are clustered to identify the tissue prototypes by efficient pathologists' visual examination. Then, the encoder is used to map the images into the embedding space and generate pixel-level pseudo tissue masks by querying the tissue prototype dictionary. Finally, the pseudo masks are used to train a segmentation network with dense supervision for better performance. Experiments on two public datasets demonstrate that our method can achieve comparable segmentation performance as the fully-supervised baselines with less annotation burden and outperform other weakly-supervised methods. Codes are available at https://github.com/WinterPan2017/proto2seg.

Keywords: WSI Segmentation · Label-efficient Learning · Clustering

---

W. Pan, J. Yan and H. Chen—contributed equally.

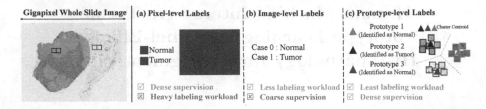

Fig. 1. Comparison of different WSI annotation types: (a) pixel-level, (b) image-level, and our (c) prototype-level labels.

1   Introduction

Visual examination of tissue sections under a microscope is a crucial step for disease assessment and prognosis. Automatic segmentation of WSIs is in high demand because it helps pathologists quantify tissue distribution. So far, supervised learning approaches have shown state-of-the-art performance on WSI segmentation with abundant annotated data [4]. Unfortunately, it takes hours to annotate pixel-wise labels for a WSI with the gigapixel scale as shown in Fig. 1(a). Thus, abundant pixel-level labels for supervised learning are often infeasible in busy daily clinical practice. Alternatively, weakly-supervised segmentation methods have been explored to supervise the model training process with less laborious image-level labels, as shown in Fig. 1(b), and estimate segmentation results. However, these weakly-supervised methods suffer from unsatisfied pixel-level predictions due to the lack of dense supervision. Therefore, we wonder **if there exist other WSI analysis pipelines that can provide pixel-level supervision without a heavy labeling workload.**

Recent years have witnessed substantial progress in the unsupervised learning of natural image analysis, especially that made by contrastive learning [7,9]. Inspired by previous works [21,22] where researchers achieved high tissue classification accuracy by performing clustering algorithms on self-supervised learned histopathology representations, in this work, we make a further step to build a bridge between contrastive learning based WSI patch pre-training and pixel-level tissue segmentation with a human-machine interactive tissue prototype learning pipeline, namely Proto2Seg. Particularly, taking advantage of self-supervised contrastive learning, we crop unlabeled WSIs into local patches and use them to train an encoder that can project these patches into a discriminative embedding space. These patches are then divided into different clusters in the embedding space via unsupervised clustering. By examining dozens of representative tissue patches in each cluster as shown in Fig. 1(c), pathologists can efficiently determine whether a cluster is a target tissue type or not. The centroids of pathologist-selected clusters are collected to build a tissue prototype dictionary, with which we can use the encoder to map the original WSIs into the embedding space and generate pseudo tissue masks by querying the nearest prototype to current local regions with flexible settings. We further adopt a refinement strategy to use the generated masks as the dense supervision for training a segmentation network from scratch. As such, pathologists only need to spend several minutes examining representative patches in every cluster for WSI segmentation model training rather than hour-counted dense annotation.

In summary, our contributions are in the following aspects: (1) We make one of the early attempts to bridge the contrastive learning-based WSI patch pre-training and dense segmentation by a low-labor-cost human-machine interactive labeling tissue prototype dictionary. (2) We propose an effective framework, namely Proto2Seg, to generate coarse tissue masks by querying the tissue prototype dictionary and designing a customized query process to further improve the coarse segmentation results. (3) By using the coarse tissue masks to supervise the training of histopathology segmentation networks, the quantitative and qualitative experimental results on two public datasets demonstrate that our approach achieves comparable segmentation performance to the fully-supervised upper bound and is superior to other weakly-supervised methods.

2   Related Works

Self-supervised WSI Analysis: Inspired by the recent success of contrastive learning [7,9] in natural image analysis, there have been some works [22,23] where researchers fine-tuned a model with contrastive learning based pre-trained weights under image-level supervision for WSI patch classification task. That is, manual labels are still required in the fine-tuning process in these works, although the model is pre-trained in a self-supervised fashion. In a recent work [21], researchers made an early attempt to distinguish different tissues with a recursive clustering algorithm, proving that integrating clustering with the contrastive learning-based histopathology representations can achieve high patch classification accuracy. In contrast to the existed works, we focus on building a bridge between the clustering-based tissue classification and the pixel-level WSI segmentation to move the topic forward for label-efficient WSI analysis.

Weakly-Supervised WSI Segmentation: Since it is difficult to obtain pixel-level labels for WSI segmentation, weakly-supervised WSI segmentation has been an active research area for years. The weakly-supervised WSI segmentation methods can be roughly categorized into CAM-based and MIL-based solutions. The CAM-based methods are built on Class Activation Map variants [5,16,24] produced by well-trained classification models. Generally, these predictions are error-prone and need to be refined by complex post-processing strategies [4]. MIL-based solutions [12,14,17] regard WSIs as a bag of local patches and can learn to predict the classification label of each local patch. The patch classification results are then merged as segmentation masks. Most MIL-based methods [12,17] are designed for binary segmentation scenario. Different from these works, our method can handle both binary and multiple tissue segmentation tasks with effective human-machine interactive steps and better results.

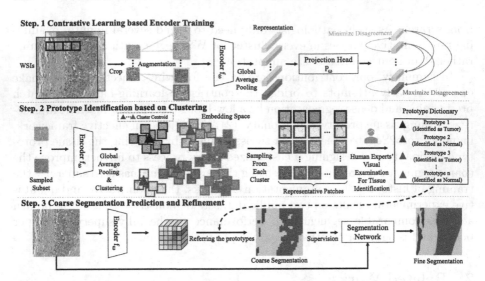

Fig. 2. Illustration of three main steps of our framework: 1) contrastive learning based encoder training (Machine), 2) prototype identification based on clustering (Machine + Human), 3) coarse segmentation prediction and refinement (Machine).

3   Methodology

Our framework, as shown in Fig. 2, includes three steps: (1) contrastive learning-based encoder training, (2) prototype identification based on clustering, and (3) coarse segmentation prediction and refinement.

3.1   Contrastive Learning Based Encoder Training

When pathologists read WSIs, they can recognize different tissues based on comparing local cells' visual appearance and surrounding micro-environments. Similarly, we need an encoder that can project local visual histopathology patterns into a discriminate space to identify different tissues. Inspired by recent studies on contrastive learning, we adopt SimCLR [7] to pre-train the encoder $f_\omega(\cdot)$ without labels in a self-supervised fashion.

The detail of this step is shown in Fig. 2 Step 1. Given a set of WSIs $\mathcal{I} = \{i_1, i_2, ..., i_m\}$, we crop them into non-overlap patches $\mathcal{P} = \{p_1, p_2, ..., p_n\}(n \gg m)$ with small resolution (empirically set as 128×128 in this study). During the training stage, given N patches in a mini-batch, we get 2N patches by applying different augmentations on each patch. Two augmented patches from the same patch are regarded as positive pairs, others are treated as negative pairs. We employ ResNet18 [11] backbone, where the final linear classification layer and global average pooling layer are removed, as the encoder $f_\omega(\cdot)$. Each patch $p_i \in \mathcal{R}^{128 \times 128 \times 3}$ is encoded by $f_\omega(\cdot)$ to get $h_i = f_\omega(p_i), h_i \in \mathcal{R}^{4 \times 4 \times 512}$, and then feed into global average pooling layer $avg(\cdot)$ to generate an embedding space with a dimension of $\mathcal{R}^{512}$. Same as [7], a projection head $g(\cdot)$ is used to map h_i to $z_i = g(avg(h_i))$, $z_i \in \mathcal{R}^{128}$ where the contrastive loss is applied. To achieve our goal, the contrastive loss is utilized to pull the positive pairs close and push the negative pairs away in the embedding space. Given a sample's embedding s and t^+, t^- as its positive and negative samples' embeddings, the formula of contrastive loss is defined as: $L_{s,t^+,t^-} = -\log \dfrac{\exp(\frac{sim(s,t^+)}{\tau})}{\exp(\frac{sim(s,t^+)}{\tau}) + \sum_{k^-} \exp(\frac{sim(s,t^-)}{\tau})}$, where $sim(\cdot)$ denotes the similarity function and τ denotes the temperature parameter. In practice, we utilize cosine similarity $cosine(x, y) = \frac{x \cdot y}{|x||y|}$ and set τ to 0.5. After training, $f_\omega(\cdot)$ is used to project histopathology patches into a discriminative embedding space of $\mathcal{R}^{512}$ for prototype identification.

It is reminded that except for SimCLR [7], other self-supervised pre-training strategies [9,10] can be adopted to obtain the encoder. Here, we use SimCLR to demonstrate the effectiveness of our framework and leave how different self-supervised strategies can affect the results as a future work.

3.2 Prototype Identification Based on Clustering

Having obtained a discriminative histopathology image encoding latent space, we then use clustering to mine potential tissue prototypes (i.e. cluster centroids), which can capture inter- and intra-class heterogeneity of different tissues and represent the entire encoding space in a dictionary. To map these prototypes with corresponding labels, only in this step do we need to integrate experts' examination. In practice, we can ask an expert to examine sampled representative patches to efficiently determine the labels of corresponding clusters without laboriously evaluating all the patches as shown in Fig. 2 Step 2. In our experiments, we simulate the visual inspection process with ground-truth labels for easy reproducibility and evaluation of the segmentation performance.

To begin with, we illustrate how to perform the clustering with the pre-trained $f_\omega(\cdot)$. To reduce computational overhead, we randomly sample a subset $\mathcal{P}_{sub}$ from the entire histopathology patch set $\mathcal{P}$. Then, $f_\omega(\cdot)$ and $avg(\cdot)$ are utilized to project patches $\mathcal{P}_{sub}$ into the discriminative embedding space to obtain the embedding set $\mathcal{E}_{sub}$. Next, we adopt unsupervised clustering on $\mathcal{E}_{sub}$ to find tissue prototypes without accessing any manual label. For computation efficiency, K-Means++ [2] is employed to generate k clusters from $\mathcal{E}_{sub}$. Naturally, the first problem comes: **how to determine a proper cluster number k?**

Intuitively, the selection of k plays a trade-off between reducing annotation workload and purifying clusters, i.e., a larger k results in smaller tissue clusters with higher inter-class similarity, but costs more labor as the pathologist needs to inspect. In order to select a good cluster number without accessing any manual labels, we adopt the elbow method [13] to determine proper k where model performance increment is no longer worth additional cost. We use an annotation-free distance-based measurement: inter-class embedding squared distance D to measure the model performance. D is given by: $D = \sum_j^k \sum_* ||centroid_j, p_{(j,*)}||_2$, where $centroid_j$ denotes the centroid of cluster j and $p_{(j,*)}$ denotes the embeddings of patches in cluster j. Generally, D decreases when k grows up. To search

684   W. Pan et al.

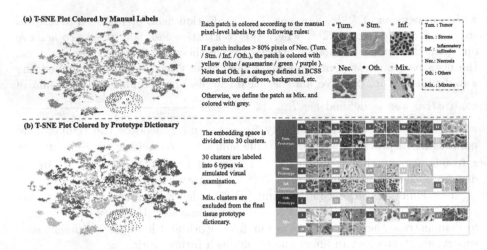

Fig. 3. Visualization of BCSS patches sampled for dictionary building in the embedding space via T-SNE and colored with: (a) manual pixel-level and (b) our clustering-based prototype-level labels. We also visualize 2 representative patches for each cluster.

a proper k, we can plot the curve of relative reduction $R_v = D_{v-1} - D_v, v \geq 2$, where v is potential values of k (e.g., $v \in [2, 3, ..., 50]$), and choose the elbow point where the reduction of D slows down as the final cluster number.

It is noted that identifying proper cluster number is an open problem, and we also evaluate other clustering methods. Our experiments show that KMeans++ is better with current setting in our following ablation study.

Based on the obtained clusters, then there comes the second problem: **how can human experts efficiently determine the tissue types for every cluster?** We propose to let human experts only evaluate t sampled representative patches to give the prototype-level labels instead of inspecting all the patches. The t representative patches are sampled by the central sampling strategy. That is, given a target cluster, we sample t patches which are closest to the cluster centroid. More sampling strategies will be discussed in the experiments. To simulate pathologists' the visual examination process, here we count the proportion of each tissue according to the pixel-level labels. If there exists one tissue type with a proportion of pixels exceeding 80%, we consider this cluster to represent this tissue and add the prototype (the cluster centroid embedding) into the prototype dictionary. Otherwise, we assume the cluster may be a mixture of different tissue types (i.e. the border area between organs), and drop it.

To help readers better understand, we in Fig. 3 illustrate the dictionary building process with the BCSS dataset [1]. We visualize the randomly sampled 20000 patches used for dictionary building with T-SNE in Fig. 3(a) which is colored according to the manual pixel-level labels. Particularly, these patches are divided into 6 types, of which 5 types (tumor, stroma, inflammatory infiltration, necrosis, and others) are defined by the dataset providers, and the mixture type defined by us. Our tissue dictionary is built on the clustering results drawn in Fig. 3(b) by K-Means++ with $k = 30$. Intuitively, we can observe that the clustering results maintain a high degree of consistency with the manual labels. By sampling 10 representative patches from each cluster center and calculating the tissue proportion, we can find that 13/5/3/1/3 clusters are identified as tumors/stroma/inflammatory infiltration/necrosis/others, correspondingly. Note that Cluster 1 is formed by blank background patches, therefore spreading in a centrosymmetric distribution. There are 5 cluster with multiple tissue mixture patterns, which are excluded from the dictionary.

After above process, we can build the prototype dictionary $\mathcal{C}$ with l prototypes after the above process. The prototype dictionary can be noted as $\mathcal{C} = \{c_1 : y_1, c_2 : y_2, ..., c_l : y_l\}$, where $\{c_1, c_2, ..., c_l\}$ are prototype embeddings, and $\{y_1, y_2, ...y_l\}$ are prototype-level labels given by the pathologist.

3.3   Coarse Segmentation Prediction and Refinement

Having obtained the pre-trained $f_\omega(\cdot)$ and prototype dictionary $\mathcal{C}$, we can generate coarse segmentation masks for a given WSI $i \in \mathcal{R}^{H \times W \times 3}$ by the process shown in Fig. 2 Step 3. We first feed WSI i into the encoder $f_\omega(\cdot)$ to have a $v \in \mathcal{R}^{\frac{H}{32} \times \frac{W}{32} \times 512}$ semantic vector map. Then, we can query $\mathcal{C}$ to determine the tissue type for each location. Intuitively, for each location, we can directly retrieve the nearest prototype to the current semantic vector from $\mathcal{C}$ and label each location with the corresponding tissue type. We termed such a query setting as Direct-Query (DQ). However, DQ treats each location independently ignoring intrinsic similarity between the embeddings. Thus, we further improve DQ with an enhanced Cluster-then-Query (CQ) setting to suppress outlier query results. Particularly, in order to explore the intrinsic similarity between the embeddings in the given feature map, we divide the $\frac{H}{32} \times \frac{W}{32}$ locations into different groups by performing K-Means++ on the $\frac{H}{32} \times \frac{W}{32}$ 512-d vector set. After that, the clustering centroids of different groups are used to query $\mathcal{C}$ for labeling each location. As for the clustering number of the target WSI, we find that it is better to decide the clustering number according to tissue types in the given WSI. According to the DQ based query results, we can roughly know that there may exist d tissue types for the target WSI. Considering the possibility of spatial dispersion of the same tissue in images, We divide the current feature map into $\gamma \times d$ clusters. $\gamma \geq 1$ is the scale factor and set to 5 empirically.

Now, we have a $\frac{H}{32} \times \frac{W}{32} \times 1$ tissue segmentation map, which can be upsampled to restore its resolution for the given WSI. The coarse map has zigzag edges but it can offer dense supervision for segmentation network training. To refine the final prediction, we further use these coarse masks as the pseudo tissue masks to train a segmentation network from scratch. By such, we finally build a bridge from the prototype learning to semantic segmentation.

Table 1. Quantitative segmentation results of different strategies. Tum./Nor./Stm./Inf./Nec./Oth. is short for Tumor/Normal/Stroma/Inflammatory infiltration/Necrosis/Others. DQ/CQ stands for the Direct-Query/Cluster-then-Query setting, correspondingly The best results are addressed in bold. Note that ABMIL can only be used for the binary prediction task.

Dataset		CAMELYON16			BCSS					
Metrics		Pix. Acc.	Tum. Dice	Nor. Dice	Pix. Acc.	Tum. Dice	Stm. Dice	Inf. Dice	Nec. Dice	Oth. Dice
Pixel-level Supervised LinkNet		0.9391	0.9078	0.9896	0.8335	0.8773	0.8162	0.7662	0.7640	0.9653
Coarse Seg.	Grad-Cam	0.8243	0.7480	0.9745	0.5799	0.4415	0.6190	0.5122	0.3251	0.7244
	Grad-Cam++	0.7462	0.6169	0.9641	0.6121	0.6327	0.6041	0.3606	0.2757	0.7638
	ABMIL	0.8455	0.8033	0.9805	N/A	N/A	N/A	N/A	N/A	N/A
	CLAM	0.8700	0.7752	0.9743	0.6441	0.7082	0.6256	0.6404	0.3571	0.9109
	Our Proto2Seg(DQ)	0.9174	0.8401	0.9810	0.7421	0.7907	0.7240	0.6590	0.6176	0.9430
	Our Proto2Seg(CQ)	0.9176	0.8491	0.9824	0.7484	0.7956	0.7281	0.6576	0.6381	0.9429
Refine. Seg.	CLAM + LinkNet	0.9189	0.8741	0.9859	0.7323	0.7797	0.7281	0.6784	0.6337	0.9054
	Our Proto2Seg(DQ) + LinkNet	**0.9266**	0.8766	0.9859	0.7491	0.8029	0.7604	**0.6800**	0.6143	**0.9479**
	Our Proto2Seg(CQ) + LinkNet	0.9231	**0.8824**	**0.9868**	**0.7750**	**0.8063**	**0.7769**	0.6799	**0.6979**	0.9477

4   Experiments

4.1   Setup

Datasets: Our experiments are conducted on two public datasets: **CAME-LYON16:** we use the train set which contains 270 WSIs focusing on sentinel lymph nodes with pixel-wise cancerous/normal region annotation [3]. **BCSS:** contains 151 Region-of-Interest cropped Images from breast cancer WSIs in TCGA with tissue-level annotations [1]. Five tissue types are labeled: tumor, stroma, inflammatory infiltration, necrosis, and 'others' (including background, blood, etc.). Similar to previous works [17, 21], we generate the $2048 \times 2048/1024 \times 1024$ image-level data from CAMELYON16/BCSS with the specimen-level pixel size of 1335 nm/250 nm to get a proper view of biological tissues as the image-level data. For BCSS, we follow the official train-test split with 2151/976 images for training/testing. For CAMELYON16, we randomly divide patients into the train-test set by 2:1 following the hold-out setting and have 1348/848 images for training/testing.

Baselines and Evaluation: We compare our framework with the pixel-level supervised upper bound established by training a LinkNet [6] with ground-truth annotations, and several weakly supervised baselines including Grad-Cam [16], Grad-Cam++ [5], ABMIL [12], and CLAM [14]. The $2048 \times 2048/1024 \times 1024$ images from Camelyon16/BCSS are cropped into 128×128 patches for MIL methods. All networks are built based on ResNet18. We report macro-average pixel accuracy and each tissue's Dice score to compare different methods.

Our Protot2Seg: We first crop the training set into the 128×128 patches without overlapping resulting in over $330k/140k$ patches for CAMELYON16/BCSS, respectively. All the patches are used to train a ResNet18-based encoder following the original SimCLR [7] setting for 200 epochs. For computation efficiency in the clustering process, 20000 patches are randomly sampled for tissue prototype clustering on both datasets. Note that, the testing set is not exposed

Table 2. Labeling Cost of different annotation types with BCSS training dataset.

Annotation Type	Quantity	Average Time	Total Time	Ratio
Pixel-level	2151 images	8 min 50 s per image	316.7 h	1583.5x
Image-level	2151 images	1 min 59 s per image	71.1 h	355.5x
Prototype-level (ours)	30 clusters	26 s per cluster	0.2 h	1x

to the encoder during the training or the dictionary building process. We set the tissue cluster number of CAMELYON16/BCSS as 15/30 through the elbow method. We employ prototype dictionary and CQ setting to generate coarse segmentation. The training of the refinement segmentation network follows the same setting as our pixel-level supervised baseline except for using coarse segmentation masks as the supervision.

4.2 Main Results

Quantitative Results: The results of segmentation performance are summarized in Table 1. Our Proto2Seg(DQ) coarse segmentation results obtained via directly querying the tissue prototype dictionary already surpass other weakly supervised baselines with significant margins. By using the Proto2Seg(CQ) coarse masks as dense supervision, the refined segmentation results achieve comparable pixel accuracy and Dice as the pixel-level supervised upper bound on CAMELYON16 for binary segmentation. For BCSS, most tissue segmentation results are further improved by Proto2Seg(DQ) + refinement training, except for the necrosis. We conjecture that the coarse masks of the necrosis provide inaccurate supervision with only 1 prototype as shown in Fig. 3(b) and mislead the refinement. Our Proto2Seg(CQ) setting, taking the embeddings' intrinsic similarity and the tissue's spatial dispersion into consideration, further improves the coarse segmentation performance. We can also observe that the CQ based coarse segmentation masks offer better supervised signals than masks generated by DQ. We also enhance the CLAM segmentation results with the refinement training step, and we can find that our solution is still superior to the CLAM + refinement. In conclusion, our solution can narrows the performance gap between pixel-level fully-supervised and weak-supervised methods with prototypes.

Note that we achieve better segmentation results with a lower annotation burden than other methods. To compare the labeling cost, we invited 3 pathologists to annotate randomly sampled 20 BCSS patches with 1024 × 1024 pixels for pixel/image-level labels and use the average time to measure the labeling cost for BCSS training set reported in Table. 2. Obviously, our method is much more labor-saving. It is also reminded that we used 1024 × 1024 cropped BCSS images but the original WSIs can be with a resolution of >10000 × 10000, i.e. it will be more time-consuming to give pixel/image-level labels for the original WSIs. But our method will not be affected by larger resolutions.

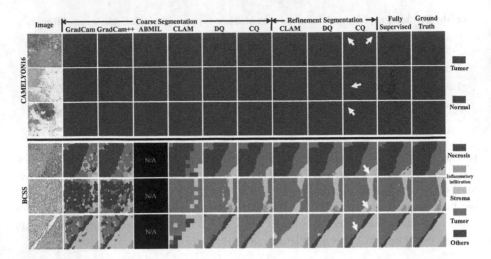

Fig. 4. Qualitative comparison of different methods. DQ/CQ stands for the Direct-Query/Cluster-then-Query setting, correspondingly. White arrows address where the coarse segmentation results are improved by the CQ setting on CAMELYON16/BCSS.

Table 3. The effects of different clustering algorithms on CAMELYON16 dataset.

Clustering Method	Kmeans	Kmeans++	SpectralClustering	DBSCAN	FINCH
Time (s)	12.82	10.67	108.9	10.85	8.08
Mean Dice (DQ)	0.9000	0.9107	0.9045	0.7802	0.8942

Qualitative Results: We visualize the qualitative results in Fig. 4. We can find that our methods can better handle binary and multi-tissue segmentation than other weakly-supervised methods. Because the DQ setting treats every location independently, there exist outlier errors in its coarse segmentation results. We use white arrows to address where the coarse segmentation results are further improved by our CQ strategy. By using the coarse segmentation as the dense supervision, it can be observed that the fine results better correct some error predictions and have smoother edges.

4.3   Ablation Study

Clustering Numbers for Prototype Identification: Figure 5(a) illustrates how the selection of cluster number k for dictionary building can affect the results. When the number of clusters $k < 10$, it's difficult for our method to distinguish different tissues, because too few clusters may not capture enough inter-/intra-class heterogeneity. When $k > 10$, the coarse segmentation masks can obtain satisfied segmentation results. As mentioned above, we follow the elbow method to set $k = 30$ without accessing any labels, while the optimal

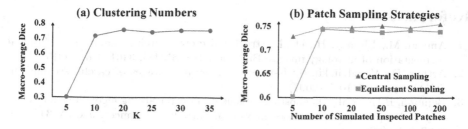

Fig. 5. Ablation studies on (a) clustering numbers and (b) representative patch sampling strategies on BCSS. The macro-average Dice of 5 tissues is reported.

Dice is achieve when $k = 20$. That is, determining proper k is an open challenge for unsupervised clustering, but our method is robust when $k > 10$.

Patch Sampling Strategies for Visual Inspection: Figure 5(b) shows how the patch sampling strategy will affect the results on BCSS. We perform experiments about central/equidistant sampling strategies with different values of t and report the macro-average Dice of 5 tissues. In equidistant sampling setting, we randomly sample t equidistant patches in each cluster form center to border for inspection. We can conclude that central sampling is constantly better. We can also observe that checking too few patches leads to performance decrease while checking more patches leads to higher accuracy but with a heavier workload. To make the trade-off, we employ central sampling with $t = 10$.

Different Clustering Algorithms: Five clustering algorithms are evaluated with our framework shown in Table 3. The clustering number of the former three algorithms are determined with the same elbow method described. DBSCAN [8] and FINCH [15] are k-free, i.e. the algorithm can automatically determine clustering number. We can observe that Kmeans++ [2] outperforms other methods.

5   Conclusion

In this paper, we make one of the early attempts to bridge the contrastive learning-based WSI patch pre-training and semantic segmentation with a human-machine interactive labeling tissue prototype dictionary. Experiments on two public datasets demonstrate that our method is comparable to the supervised upper bounds and outperform other weakly-supervised methods. The major limitations are that we simulate the visual examination of pathologists in current experiments and the coarse masks inevitably contain noises. Future work will discuss about how human factors may affect the visual injection process and incorporate techniques [18–20] to alleviate the negative impacts of label noises.

Acknowledgement. This research was partly supported by the National Key R&D Program of China (Grant No. 2020AAA0108303), Shenzhen Science and Technology Project (Grant No. JCYJ20200109143041798), Shenzhen Stable Supporting Program (Grant No. WDZC20200820200655001), and Shenzhen Key Lab of next generation interactive media innovative technology (Grant No. ZDSY S20210623092001004).

References

1. Amgad, M., Elfandy, H., et al.: Structured crowdsourcing enables convolutional segmentation of histology images. Bioinformatics **35**(18), 3461–3467 (2019)
2. Arthur, D., Vassilvitskii, S.: K-means++: the advantages of careful seeding. In: SODA, pp. 1027–1035 (2007)
3. Bejnordi, B.E., et al.: Diagnostic assessment of deep learning algorithms for detection of lymph node metastases in women with breast cancer. JAMA **318**(22), 2199–2210 (2017)
4. Chan, L., Hosseini, M.S., Rowsell, C., et al.: HistoSegNet: semantic segmentation of histological tissue type in whole slide images. In: ICCV, pp. 10662–10671 (2019)
5. Chattopadhay, A., Sarkar, A., et al.: Grad-CAM++: generalized gradient-based visual explanations for deep convolutional networks. In: WACV, pp. 839–847 (2018)
6. Chaurasia, A., Culurciello, E.: LinkNet: exploiting encoder representations for efficient semantic segmentation. In: VCIP, pp. 1–4 (2017)
7. Chen, T., Kornblith, S., Norouzi, M., Hinton, G.: A simple framework for contrastive learning of visual representations. In: ICML, pp. 1597–1607 (2020)
8. Ester, M., Kriegel, H.P., Sander, J., et al.: A density-based algorithm for discovering clusters in large spatial databases with noise. In: KDD, vol. 96, pp. 226–231 (1996)
9. Grill, J.B., Strub, F., Altché, F., et al.: Bootstrap your own latent-a new approach to self-supervised learning. In: NeurIPS, pp. 21271–21284 (2020)
10. He, K., Fan, H., Wu, Y., Xie, S., Girshick, R.: Momentum contrast for unsupervised visual representation learning. In: CVPR, pp. 9729–9738 (2020)
11. He, K., Zhang, X., Ren, S., Sun, J.: Deep residual learning for image recognition. In: CVPR, pp. 770–778 (2016)
12. Ilse, M., Tomczak, J., Welling, M.: Attention-based deep multiple instance learning. In: ICML, pp. 2127–2136 (2018)
13. Liu, F., Deng, Y.: Determine the number of unknown targets in open world based on elbow method. IEEE Trans. Fuzzy Syst. **29**(5), 986–995 (2020)
14. Lu, M.Y., et al.: Data-efficient and weakly supervised computational pathology on whole-slide images. Nat. Biomed. Eng. **5**(6), 555–570 (2021)
15. Sarfraz, S., Sharma, V., Stiefelhagen, R.: Efficient parameter-free clustering using first neighbor relations. In: CVPR, pp. 8934–8943 (2019)
16. Selvaraju, R.R., Cogswell, M., Das, A., et al.: Grad-CAM: visual explanations from deep networks via gradient-based localization. In: ICCV, pp. 618–626 (2017)
17. Xu, G., Song, Z., Sun, Z., et al.: CAMEL: a weakly supervised learning framework for histopathology image segmentation. In: CVPR, pp. 10682–10691 (2019)
18. Xu, Z., Lu, D., Luo, J., et al.: Anti-interference from noisy labels: mean-teacher-assisted confident learning for medical image segmentation. IEEE Trans. Med. Imaging **41**(11), 3062–3073 (2022)
19. Xu, Z., et al.: Noisy labels are treasure: mean-teacher-assisted confident learning for hepatic vessel segmentation. In: de Bruijne, M., et al. (eds.) MICCAI 2021. LNCS, vol. 12901, pp. 3–13. Springer, Cham (2021). https://doi.org/10.1007/978-3-030-87193-2_1
20. Xu, Z., et al.: Denoising for relaxing: unsupervised domain adaptive fundus image segmentation without source data. In: Wang, L., Dou, Q., Fletcher, P.T., Speidel, S., Li, S. (eds.) MICCAI 2022. LNCS, vol. 13435, pp. 214–224. Springer, Cham (2022). https://doi.org/10.1007/978-3-031-16443-9_21
21. Yan, J., Chen, H., Li, X., Yao, J.: Deep contrastive learning based tissue clustering for annotation-free histopathology image analysis. Comput. Med. Imaging Graph. **97**, 102053 (2022)

22. Yang, J., et al.: Towards better understanding and better generalization of low-shot classification in histology images with contrastive learning. In: ICLR (2022)
23. Yang, P., Hong, Z., Yin, X., Zhu, C., Jiang, R.: Self-supervised visual representation learning for histopathological images. In: de Bruijne, M., et al. (eds.) MICCAI 2021. LNCS, vol. 12902, pp. 47–57. Springer, Cham (2021). https://doi.org/10.1007/978-3-030-87196-3_5
24. Zhou, B., Khosla, A., Lapedriza, A., Oliva, A., Torralba, A.: Learning deep features for discriminative localization. In: CVPR, pp. 2921–2929 (2016)

Improved Segmentation of Deep Sulci in Cortical Gray Matter Using a Deep Learning Framework Incorporating Laplace's Equation

Sadhana Ravikumar[1]([✉]), Ranjit Ittyerah[1], Sydney Lim[1], Long Xie[1], Sandhitsu Das[1], Pulkit Khandelwal[1], Laura E. M. Wisse[2], Madigan L. Bedard[1], John L. Robinson[1], Terry Schuck[1], Murray Grossman[1], John Q. Trojanowski[1], Edward B. Lee[1], M. Dylan Tisdall[1], Karthik Prabhakaran[1], John A. Detre[1], David J. Irwin[1], Winifred Trotman[1], Gabor Mizsei[1], Emilio Artacho-Pérula[3], Maria Mercedes Iñiguez de Onzono Martin[3], Maria del Mar Arroyo Jiménez[3], Monica Muñoz[3], Francisco Javier Molina Romero[3], Maria del Pilar Marcos Rabal[3], Sandra Cebada-Sánchez[3], José Carlos Delgado González[3], Carlos de la Rosa-Prieto[3], Marta Córcoles Parada[3], David A. Wolk[1], Ricardo Insausti[3], and Paul A. Yushkevich[1]([✉])

[1] University of Pennsylvania, Philadelphia, USA
{ravikums,pauly2}@pennmedicine.upenn.edu
[2] Lund University, Lund, Sweden
[3] University of Castilla-La Mancha, Ciudad Real, Spain

Abstract. When developing tools for automated cortical segmentation, the ability to produce topologically correct segmentations is important in order to compute geometrically valid morphometry measures. In practice, accurate cortical segmentation is challenged by image artifacts and the highly convoluted anatomy of the cortex itself. To address this, we propose a novel deep learning-based cortical segmentation method in which prior knowledge about the geometry of the cortex is incorporated into the network during the training process. We design a loss function which uses the theory of Laplace's equation applied to the cortex to locally penalize unresolved boundaries between tightly folded sulci. Using an *ex vivo* MRI dataset of human medial temporal lobe specimens, we demonstrate that our approach outperforms baseline segmentation networks, both quantitatively and qualitatively.

Keywords: Cortical segmentation · topology correction · ex vivo MRI

1 Introduction

Segmentation of the cerebral cortex from MRI is an important first step in many neuroimaging pipelines such as quantitative morphometry analyses aimed at understanding the pathophysiology of neurological disorders. Automated

A. Frangi et al. (Eds.): IPMI 2023, LNCS 13939, pp. 692–704, 2023.
https://doi.org/10.1007/978-3-031-34048-2_53

segmentation methods applied to the cortex are challenged by various artifacts such as image noise, partial volume effects and intensity inhomogeneities which make accurate identification of the tissue boundaries difficult and result in geometrically inaccurate cortical reconstructions. The cerebral cortex or gray matter (GM) can be defined as the space between two cortical surfaces; the pial surface which separates the GM from the surrounding cerebrospinal fluid (CSF), and the white matter (WM) surface which separates the GM from WM. The cortex has a complex geometry and is often modelled as a highly folded 2D sheet, with spatially varying curvature and thickness [14]. Geometrically accurate segmentation of the cortex requires accurate reconstruction of both the WM and pial cortical surfaces, complete with all cortical folds and narrow sulci. A commonly used simplification when solving cortical surface reconstruction problems is to view the cortical surfaces as having the topology of a 3D sphere [4]. However, unless explicitly corrected for, imaging artifacts often introduce topological defects in the resulting surface reconstructions. Defects due to partial volume effects are particularly apparent in tightly folded sulci and result in opposing banks of sulci appearing fused together. This creates either 'bridged' or 'unresolved' sulci in the resulting cortical reconstruction, which cause errors in downstream quantitative brain morphometry measures such as cortical thickness. While topological defects can be corrected by manual editing, these checks can be time-consuming.

Topology-corrected reconstruction of cortical surfaces is a well studied topic in *in vivo* neuroimaging literature, and several state-of-the-art methods have been developed to address this problem [4,5,11]. The widely used Freesurfer framework employs a mesh-based approach to topology-correction which consists of two main steps [4]. First, the inner WM surface is generated by applying mesh tessellation to a volumetric WM segmentation that has been corrected for topological defects [4]. Second, this WM surface is expanded using a deformable surface model to reconstruct the outer, pial surface, while ensuring that the topology of the initial surface is preserved [4,5,11]. In recent years, elements of the FreeSurfer pipeline have been implemented as deep learning networks, resulting in significant speedups [1,7,8,12]. However, these frameworks still either require the time-consuming post-processing step of topology correction [1,7] or rely on a predefined initial mesh with the correct, spherical topology to reconstruct the cortical surface [8,12].

In this work, we are specifically interested in developing an automated cortical segmentation method that can be applied to *ex vivo* brain MRI datasets to generate geometrically valid models of the cortex. Instead of a mesh deformation-based approach, we propose a novel volumetric deep image segmentation method that learns to segment the cortex while explicitly modeling the 'sheet-like' geometry of the cortex. In *ex vivo* studies, it is common to image only a portion of the brain hemisphere, thus violating the assumption of spherical WM topology made by mesh-based approaches. Furthermore, many of the *in vivo* cortical segmentation methods contain algorithms that are optimized for data with a standard 1 mm voxel size [4], and would result in unrealistic computational times if applied to high-resolution *ex vivo* MRI datasets. As a result, existing methods are not easily applicable to *ex vivo* MRI scans. As far as we know, no prior work has focused on topology correction of *ex vivo* cortical segmentations.

Previous studies have used the concept of Laplace's equation as a tool for modelling the cortex [10,11,14]. By setting different boundary conditions at the WM/GM and GM/CSF interfaces, Laplace's equation can be solved within the GM volume to generate a laminar 'potential' field that smoothly varies in value depending on its distance between the two cortical surfaces. The gradient of the Laplacian field can be used to compute cortical thickness [10] and defines an expansion path which guarantees topology-preserving deformation between the WM and pial cortical surfaces [11,14]. Building on this idea and the success of deep convolutional neural networks (CNN) in medical image segmentation tasks, here we design a differentiable numerical solver for Laplace's equation and incorporate it within a deep segmentation framework to locally impose a Laplacian mapping between the predicted WM and pial surfaces. We train the segmentation network by comparing the predicted tissue segmentations and corresponding Laplacian field maps with the equivalent ground truth images, thus penalizing self-intersections in the predicted segmentations. Our results show that when compared to a baseline network trained without Laplacian constraints, our method is able to better reconstruct the intrinsic, layered geometry of the cortex. To our knowledge, this is the first time that an iterative numerical solver has been incorporated within a cortical segmentation network to directly compute Laplacian fields in an end-to-end setting.

2   Methods

As illustrated in Fig. 1, our proposed framework builds upon any given backbone segmentation network (Sect. 2.1) by appending a numerical solver for Laplace's equation to the output of the network. We reformulate the solver to be differentiable with respect to the input image to allow for gradient-based learning, used within standard CNN training (Sect. 2.2). In addition to the standard tissue segmentation loss, we design a loss function which compares the predicted Laplacian field to the solution of Laplace's equation applied to the ground truth cortical segmentation, which is assumed to have correct topology (Sect. 2.3).

2.1   Backbone Segmentation Network

The proposed Laplacian solver is compatible with any semantic segmentation network since it only relies on the segmentation map output by the backbone network. We conducted experiments using two backbone networks for 3D image segmentation: the state-of-art nnU-Net framework [9] based on the U-net architecture, and nnFormer [18], a variant framework based on the recently popular transformer architecture. Both frameworks use image patches, deep supervision, and a variety of data-augmentation techniques to train the network [9].

2.2   Differentiable Laplacian Solver

To compute the Laplacian field corresponding to the cortical segmentation predicted based on a given input patch, the iterative solver for Laplace's equation

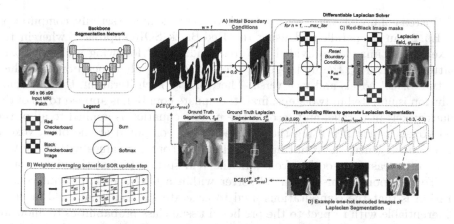

Fig. 1. Schematic illustration of the proposed framework. A differentiable numerical solver for Laplace's equation, based on the successive over relaxation (SOR) algorithm, is incorporated within end-to-end training of a segmentation network.

is appended after the final layer of the backbone network. Laplace's equation, $\Delta\varphi = 0$, is a second-order partial differential equation, where Δ is the Laplacian operator $(\partial_{xx}^2 + \partial_{yy}^2 + \partial_{zz}^2)$ and φ is a twice-differentiable, real-valued function. To solve Laplace's equation within a domain (in our case, the GM volume), specific conditions need to be set that the Laplacian field φ must satisfy at the boundaries of the domain. In our case, we set $\varphi_{x,y,z} = 0$ at the GM/WM boundary and $\varphi_{x,y,z} = 1$ at the GM/pial boundary. Voxels within the GM domain are initialized with $\varphi_{x,y,z} = 0.5$.

Given these boundary conditions, the Laplacian field can be approximated using the finite-difference method, which is solved using an iterative numerical solver. Here, we use the Successive Over Relaxation (SOR) algorithm, a variant of the Gauss-Seidel method, to solve for the Laplacian field [6]. In the Gauss-Seidel method, given initial values for all the voxels in an image, at each iteration, the new value for a particular voxel within the GM volume, $\tilde{\varphi}_{xyz}$, is computed by taking the weighted sum of the most recently updated values of its six neighboring voxels (Eq. 1). The superscript n refers to the iteration of the algorithm and the subscripts are the voxel indices.

$$\tilde{\varphi}_{xyz}^n = \frac{1}{6}(\varphi_{x-1,y,z}^n + \varphi_{x+1,y,z}^{n-1} + \varphi_{x,y-1,z}^n + \varphi_{x,y+1,z}^{n-1} + \varphi_{x,y,z-1}^n + \varphi_{x,y,z+1}^{n-1}) \quad (1)$$

The SOR algorithm accelerates this approach by taking, at each iteration n, the weighted sum of the current solution and the solution from the previous iteration (Eq. 2). The over-relaxation parameter, ω accelerates the rate of convergence of the Gauss-Seidel method when $1 < \omega < 2$ [17]. It has been shown that the optimum value, ω_{opt} is given by $\omega_{opt} = \frac{2}{1+sin(\frac{\pi}{N+1})}$ where N is the minimum dimension of the input grid [17].

$$\varphi_{x,y,z}^n = (1 - \omega_{opt})\varphi_{x,y,z}^{n-1} + \frac{\omega_{opt}}{6}(\varphi_{x-1,y,z}^n + \varphi_{x+1,y,z}^{n-1}$$
$$+ \varphi_{x,y-1,z}^n + \varphi_{x,y+1,z}^{n-1} + \varphi_{x,y,z-1}^n + \varphi_{x,y,z+1}^{n-1}) \quad (2)$$

Instead of updating the value of each voxel in the image serially, computation of Eq. 2 can be parallelized using the Red-Black SOR approach, wherein the voxels in an image are divided into 'red' and 'black' following a checkerboard pattern [3]. During the update step, the 'red' voxels only depend on the values of the 'black' voxels and vice versa. Therefore, at each iteration, the Laplacian solution is updated in two steps; first, the update equation is applied to all of the black voxels in parallel and second, the update equation is applied to all of the red voxels in parallel, using the updated values computed at the black voxels. After convergence, the resulting Laplacian field contains values within the GM volume increasing smoothly from 0 at the WM surface to 1 at the pial surface.

To incorporate this numerical solver within a CNN, an important consideration is that the computations used to generate the Laplacian field must be differentiable with respect to the predicted tissue class probabilities to allow for back-propagation of the final loss through the network. To this end, we initialized the boundary conditions for the Laplacian solver by taking a weighted sum of the GM, WM and background probability maps, with weights of 0.5, 0 and 1 respectively (Fig. 1A). Additionally, the SOR update step (Eq. 2) was reformulated as a 1×1 convolutional layer with fixed neighborhood weights specified in a $3 \times 3 \times 3$ kernel (Fig. 1B). The voxels in the image were divided into a red and black grid by generating 3D 'red' $(mod(x + y + z, 2) = 0)$, and 'black' $(mod(x + y + z, 2) = 1)$ binary checkerboard images which were applied as image masks to retain values of interest after each convolutional layer (Fig. 1C). Lastly, the maximum number of iterations for the Laplacian solver was empirically set to 60, as a trade-off between computational time and convergence of the Laplacian solution. Since the solver numerically computes the solution to Laplace's equation, it does not introduce any additional training parameters within the network.

2.3   Loss Function

To train the model, the backbone networks compare the predicted tissue segmentation, S_{pred} with the ground truth cortical segmentation, S_{gt} using a combination of Dice and cross-entropy loss, $DCE(S_{gt}, S_{pred})$ [9]. We introduce an additional loss term which compares the Laplacian field computed from the predicted tissue segmentation, φ_{pred}, with the solution of Laplace's equation applied to the ground truth segmentation, φ_{gt}. To simplify comparison between the predicted Laplacian field and the ground truth solution, we convert the Laplacian field computed by the solver to a multi-label segmentation, S^{φ}_{pred} using a series of thresholding functions. The advantage of this approach is that it enables the use of the same combined Dice and cross-entropy loss used by the backbone network on the outputs of the Laplacian solver, instead of the mean square error loss, thereby allowing us to equally weight the two loss terms: $\mathcal{L} = DCE(S_{gt}, S_{pred}) + DCE(S^{\varphi}_{gt}, S^{\varphi}_{pred})$. To threshold the Laplacian field and create a multi-label segmentation in a differentiable way, the computed Laplacian field is passed through a product of two sigmoid functions, $(1 + e^{-\beta(x - t_{lower})})^{-1} \times (1 + e^{\beta(x - t_{upper})})^{-1}$, which together create a 'band-pass' thresholding filter. β controls the steepness of the filter, and t_{lower} and t_{upper} control the domain of the filter. Each filter creates an image that has values close to 1 for voxels in the Laplacian field lying within the domain of the filter, and values close to 0 otherwise. Therefore, by varying the lower and upper threshold values, a one-hot encoded image can be created, where each channel corresponds to a different label along the laminar axis of the GM. A multi-label Laplacian segmentation is then generated by applying the argmax operation to the computed one-hot encoded image (Fig. 1D).

3   Experiments

3.1   Dataset

MRI Image Acquisition. To train and evaluate the proposed framework, we used *ex vivo* images of intact temporal lobe specimens, obtained from 27 brain donors from either the brain bank operated by the National Disease Research Interchange, or autopsies performed at the University of Pennsylvania Center for Neurodegenerative Disease Research (CNDR) and the University of Castilla-La Mancha (UCLM) Human Neuroanatomy Laboratory (HNL) in Spain. Brain specimens were obtained in accordance with the University of Pennsylvania Institutional Review Board guidelines, and the Ethical Committee of UCLM. Where possible, pre-consent during life and, in all cases, next-of-kin consent at death was given. Following 4+ weeks of fixation, the tissue specimens were scanned overnight on a 9.4 T 31 cm bore MRI scanner using a T2-weighted, multi-slice spin echo sequence (TE = 9330 ms, TR = 23 ms), with a resolution of $0.2 \times 0.2 \times 0.2$ mm^3. Following image acquisition, the images were corrected for bias field non-uniformity and normalized to a common intensity range of [0, 1000]. In our work, we are specifically interested in segmenting the medial temporal lobe (MTL), a region affected early in Alzheimer's Disease. Therefore, to facilitate semi-automated MTL segmentation, each scan was re-oriented so that the long axis of the MTL aligned with the anterior-poster direction.

Ground Truth Tissue Segmentation. To generate 3D segmentations of the MTL cortex, we adopted a semi-automatic interpolation technique [15]. The boundary of the MTL was manually traced approximately every 3 mm (i.e. 12–15 slices per dataset). Given the subset of labeled slices, the interpolation method uses contour and intensity information to compute the intermediate segmentations. This algorithm was applied iteratively, allowing the interpolated result to be reviewed and manually edited at each step to refine the segmentation. When editing, we ensured that in narrow and bridged sulci, the full extent of each sulcus was correctly labeled as background. Additionally, in a small region surrounding the MTL, the white matter and background voxels were semi-automatically labeled using a combination of intensity-based thresholding and morphological operations. The ground truth segmentations also contain a separate label for the stratum radiatum lacunosum moleculare (SRLM), which is the thin, hypo-intense layer within the hippocampus. We note that the ground truth segmentations only cover the region in the image encompassing the MTL, and not the entirety of the *ex vivo* MRI scan.

Ground Truth Laplacian Maps. The proposed framework requires the Laplacian field maps corresponding to the ground truth segmentations to train the model. To solve Laplace's equation within the ground truth GM volume, we used the iterative finite-differences approach, as implemented in [2]. This implementation employs a 26-neighbour average to compute the updated potential field and terminates the numerical solver when the Laplacian field change is below a specified threshold (sum of changes <0.001% of total volume). To initialize the solver, source and sink boundary conditions were semi-manually labeled as the WM and pial surfaces of the MTL respectively. We note that the hippocampus voxels were not included in the GM domain of Laplace's equation.

3.2   Implementation Details

We used Pytorch 1.9.1 and Nvidia Quadro RTX 500 GPUs to train the models. We implemented the differentiable Laplacian solver within the standardized training framework presented in [9] that is employed by both backbone networks. The framework includes pre-processing, automated hyper-parameter selection and fixed techniques for data augmentation. In our experiments, we made a few modifications to the default training parameters. First, since the ground truth segmentations only cover a portion of the input *ex vivo* MRI scans, we set the *ignore_label* parameter in the loss function to 0 to exclude the unlabeled background voxels from the training process. Additionally, we increased the *oversample_foreground* parameter such that only foreground patches are sampled during training. Lastly, we used an input patch size of $96 \times 96 \times 96$, to encourage the network to learn more local image features instead of larger contextual information, like the anatomical boundaries of the MTL. The networks were trained with a batch size of 2 (nnFormer) and 4 (nnU-Net), for 250 epochs in a five-fold cross validation setting. We used the results of the first fold to tune the network parameters. Consistent with the evaluation scheme used within nnU-Net, we aggregated the results across the remaining four folds for reporting test accuracy. We tested the performance of the network when using either 5 or 10 class labels (i.e. laminar layers) for generating the Laplacian segmentation. We found that increasing the number of class labels and training the network using Laplacian segmentations with a denser number of laminar layers improved the network's ability to detect obscured sulci. To convert the Laplacian fields to segmentations, we used $\beta = 10$ (softmax scaling parameter) and selected evenly spaced thresholds spanning the $[0, 1]$ range of the Laplacian field. More specifically, the following lower (t_{lower}) and upper (t_{upper}) threshold values were used: $[(-0.3, -0.2), (0, 0.1), (0.1, 0.2), (0.2, 0.3), (0.3, 0.4), (0.4, 0.5), (0.5, 0.6), (0.6, 0.7), (0.7, 0.8), (0.8, 0.95), (0.95, 1.05)]$. Additionally, we tested the effect of increasing the weight given to the Laplacian segmentation loss relative to the tissue segmentation loss and found that it had minimal effect on cortical segmentation accuracy.

3.3 Evaluation

We compared the performance of our approach with the performance of the corresponding backbone segmentation networks, trained only with the tissue segmentation loss. We measured segmentation accuracy by computing the DSC between the predicted and ground truth tissue segmentations, and Laplacian field segmentations within the MTL region of interest. We also report the symmetric Hausdorff Distance (HD) 95th percentile between the predicted and ground truth segmentation of the MTL cortex. Since the numerical solvers used to generate the ground truth Laplacian fields and embedded in the network leverage different finite-difference approximation methods, during evaluation, we re-computed the Laplacian field for both the ground truth and predicted cortical segmentations using 120 iterations of the SOR solver used by the network, and computed the corresponding Laplacian segmentation with 5 laminar layers.

In a secondary analysis we evaluated the effect of introducing the Laplacian constraint on downstream cortical thickness measures. We applied the nnU-Net models to a dataset of 36 temporal lobe specimens obtained from individuals not included in the training dataset. For each specimen, we quantified MTL thickness at 6 manually identified landmarks corresponding to the anterior and posterior locations of MTL subregions Brodmann Area (BA) 35, BA36 and the parahippocampal cortex (PHC). We chose these subregions since they typically lie along the banks of the collateral sulcus (CS), and are therefore mostly likely to be affected by topological errors in the segmentation. For each location, we extracted the GM segmentation surrounding the landmark and measured cortical thickness using the pipeline described in [16]. In brief, given the GM segmentation surrounding a landmark, a maximally inscribed sphere is computed using Voronoi skelentonization [13], and the diameter of the sphere gives the thickness at that landmark. We compared thickness measurements obtained when using automatic GM segmentations generated by the baseline nnU-Net and the proposed model (*nnU-Net+Laplacian*), and reference thickness measurements computed using semi-automatic segmentations of the GM in terms of Pearson's correlation and the average fixed-raters Intra-Class Correlation Coefficient (ICC).

4   Results and Discussion

4.1   Segmentation Accuracy

Table 1 presents the quantitative results, averaged across four cross-validation folds, evaluating the proposed framework using the corresponding backbone networks as baseline. Since sulci can be very thin structures, any improvements to mislabeled sulci would contribute minimally to the tissue segmentation DSC. Therefore, it is not surprising that we do not see statistically significant differences in the GM DSC measures. However, when looking at the Laplacian segmentation DSC, the results show that the Laplacian solver significantly improves upon the baseline performance of both nnU-Net and nnFormer in terms of better

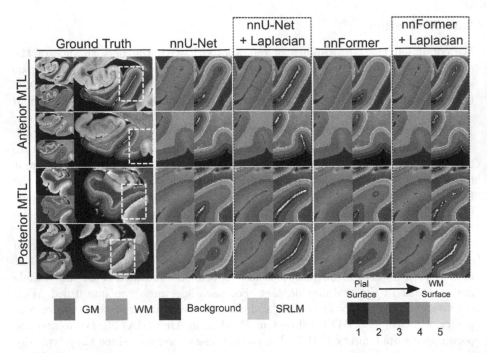

Fig. 2. Qualitative comparison of the cortical segmentations generated by the proposed method and nnU-Net. Cross-sectional views are provided through the anterior and posterior MTL, using four different specimens. The white dashed boxes are used to indicate cortical folds demonstrating improved geometric accuracy. GM: Gray Matter; WM: White Matter; SRLM: Stratum Radiatum Lacunosum Moleculare (Color figure online)

preserving the layered structure of the cortex. This is visualized in Fig. 2 which provides a qualitative comparison of the predicted segmentations generated by each method. Labels 1–3 correspond to layers closest to the pial surface, and are therefore the class labels mostly likely to reflect errors such as bridged or unresolved sulci.

We observe that in the anterior portion of the MTL, the baseline models are often able to distinguish the sulcus, even without the Laplacian constraint (Fig. 2, row 2). This is likely because in the ground truth segmentation protocol, the labeled GM extends to include both banks of the CS in the anterior MTL, but only the medial bank of the CS in the posterior MTL. As a result, in the ground truth segmentation, the sulcus is clearly labeled in the anterior MTL and therefore included in the tissue segmentation loss. Conversely, in the posterior MTL, the ground truth tissue segmentation does not explicitly enforce the presence of the sulcus. However, in this region, the Laplacian segmentation term implicitly includes information about the location of the sulcus and therefore drives the network to learn the correct pial boundary of the cortex. To further investigate the contribution of the proposed loss function in the anterior and

Table 1. Quantitative metrics comparing cortical segmentation accuracy of the proposed network and the baseline networks. We compute the Laplacian Segmentation accuracy, which reflects the networks ability to capture the layered nature of the cortex, across the whole MTL, and also separately for the anterior and posterior MTL. Metrics: Dice Score Coefficient (DSC) per label; Hausdorff Distance (HD). Standard deviations are reported in parentheses. (Statistical significance was assessed using paired t-tests; *** < 0.001, **: $p < 0.01$, *: $p < 0.05$)

	Method	DSC Laplacian Segmentation (%)					DSC Tissue Segmentation (%)				HD 95 (mm)
		Pial Surface → WM Surface					GM	WM	BG	SRLM	
Whole MTL	nnU-Net	78.3 (3.3)	78.6 (3.1)	77.4 (3.1)	77.7 (5.0)	58.7 (12.3)	94.5 (1.5)	96.0 (1.3)	95.3 (6.1)	85.8 (3.8)	0.341 (0.134)
	nnU-Net + Laplacian	80.7**** (2.6)	81.8**** (2.4)	80.4**** (3.5)	78.9* (5.8)	58.2 (13.4)	94.5 (1.7)	95.9 (1.3)	95.4 (5.9)	85.6 (3.8)	0.344 (0.205)
	nnFormer	75.5 (4.1)	75.5 (3.7)	74.6 (3.4)	75.9 (4.7)	57.0* (10.6)	93.8** (1.9)	95.1 * (2.1)	95.1 (6.1)	83.8** (4.2)	0.559 (0.885)
	nnFormer + Laplacian	78.1*** (3.4)	79.8**** (3.0)	78.4**** (3.6)	77.3*** (5.2)	55.7 (11.6)	93.5 (2.2)	94.9 (2.3)	95.1 (6.0)	83.2 (4.5)	0.589 (0.980)
Anterior MTL	nnU-Net	80.0 (3.7)	80.6 (3.4)	79.9 (3.3)	79.6 (5.0)	59.3 (12.8)	94.5 (1.6)	95.0** (1.6)	95.7 (5.1)	85.9 (4.4)	0.208 (0.025)
	nnU-Net + Laplacian	81.6** (3.0)	82.8*** (2.7)	81.8*** (3.7)	80.3 (5.9)	58.1 (13.7)	94.4 (1.9)	94.7 (1.7)	95.9 (5.0)	85.7 (4.5)	0.217 (0.049)
	nnFormer	76.8 (4.5)	76.9 (4.3)	76.5 (4.4)	78.0 (5.1)	58.4** (11.4)	93.8* (2.5)	93.9 (3.6)	95.7 (5.2)	83.7** (5.0)	0.208 (0.0205)
	nnFormer + Laplacian	79.3*** (3.4)	80.9**** (3.0)	80.0**** (3.8)	79.1** (5.2)	56.8 (12.1)	93.6 (2.7)	93.7 (3.9)	95.8 (5.1)	83.1 (5.4)	0.212 (0.030)
Posterior MTL	nnU-Net	74.8 (6.2)	74.7 (5.8)	73.2 (5.4)	74.6 (6.1)	57.2 (12.9)	94.7 (1.6)	96.9 (1.4)	94.3* (9.0)	85.3 (3.4)	0.228 (0.096)
	nnU-Net + Laplacian	78.9**** (4.7)	79.8**** (3.9)	78.1**** (4.1)	76.8*** (6.5)	58.0 (13.5)	94.6 (1.7)	97.0 (1.1)	94.1 (9.0)	85.2 (3.4)	0.223 (0.062)
	nnFormer	72.7 (8.1)	72.8 (7.0)	71.1 (6.6)	72.7 (6.9)	55.0 (11.9)	93.9*** (2.5)	96.2* (2.0)	93.8* (8.9)	84.3* (4.0)	0.348 (0.430)
	nnFormer + Laplacian	75.4*** (8.3)	77.5**** (6.9)	75.4**** (6.6)	74.3** (7.9)	54.0 (13.1)	93.2 (3.2)	96.0 (2.2)	93.6 (9.0)	83.6 (4.0)	0.436 (0.692)

posterior MTL, we computed the DSC metrics of the Laplacian segmentations separately for the anterior and posterior MTL (Table 1). We observe that even when considering the anterior MTL on its own, the proposed framework improves Laplacian segmentation accuracy compared to the baseline networks, confirming that the addition of the Laplacian term is in fact contributing towards the network better learning the layered organization of the cortex. This is further seen in the *nnU-Net+Laplacian* result in Fig. 2, row 2, where the proposed network is able to detect a buried sulcus in a cortical fold not included within the ground truth region of interest. We note that Laplacian segmentation label 5, which corresponds to the innermost cortical surface at the GM/WM boundary, forms a very thin layer and therefore has greater variation in DSC compared to the other labels.

4.2 Downstream Thickness Measurements

Figure 3 shows the results of the morphometry analysis correlating automated and manual measurements of cortical thickness in BA35, BA36 and the PHC, when using cortical segmentations generated with and without the Laplacian-based loss term. Since we found that nnU-Net achieves better segmentation performance than nnFormer, we only conducted experiments using nnU-Net in

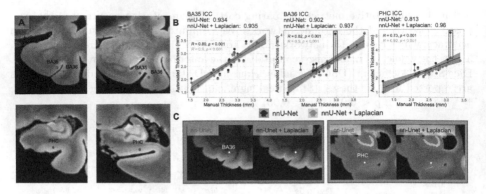

Fig. 3. A) Example scan showing the 6 landmarks where cortical thickness is measured. For each subregion, the thickness measurement is averaged across two landmark locations. B) Scatter plots showing the correlation between automated segmentation-based cortical thickness measurements and reference measurements based on semi-automatic segmentations for three MTL subregions, with (green) and without (purple) the Laplacian constraint. C) Segmentations produced by *nnUnet* and *nnU-Net+Laplacian* for BA36 (red) and PHC (blue) landmarks where thickness measures derived from the two networks differed the most. BA: Brodmann Area; PHC: parahippocampal cortex (Color figure online)

the secondary analysis. Compared to the baseline nnU-Net, we observe that the thickness measurements of BA36 and the PHC computed using the proposed network are more strongly correlated with manual measurements, in terms of both correlation coefficient and ICC. Both models achieve similar correlations in BA35. BA36 is located in the anterior MTL, whereas PHC corresponds to the posterior MTL. The strengthened correlations in both these regions further demonstrate that the proposed method is able to improve the accuracy of the predicted segmentations across the whole length of MTL.

5   Conclusions

We present a novel deep learning-based solution for cortical segmentation, applied to *ex vivo* MRI, that is able to learn the layered geometry of the cortex by locally imposing Laplacian mappings between the predicted WM and pial cortical surfaces. A limitation of this approach is the long run-time of the iterative solver during training ($\sim$9x slower/epoch relative to the backbone). However, at inference time, the input image is only passed through the backbone segmentation network which typically takes 3–5 min per scan. Another limitation is the need for the sulci to be well delineated in the training data. In the future, we will explore ways to relax this requirement, perhaps using additional geometric priors. While in this work we demonstrate the utility of our approach in the context of MTL cortical segmentation, this approach can be extended to other high-resolution neuroimaging datasets such as *ex vivo* whole hemisphere scans or other *in vivo* image segmentation tasks which involve similar sheet-like structures. Future work will focus on applying this method to *in vivo* brain MRI, thus allowing for evaluation of our approach against existing cortical surface reconstruction methods.

Acknowledgements. We gratefully acknowledge the tissue donors and their families. This work was supported by the NIH (Grants RF1 AG069474, P30 AG072979 and R01 AG056014), a UCLM travel and research grant (to R.I), and an Alzheimer's Association grant (AARF-19-615258) (to L.E.M.W).

References

1. Cruz, R.S., Lebrat, L., Bourgeat, P., Fookes, C., Fripp, J., Salvado, O.: DeepCSR: a 3D deep learning approach for cortical surface reconstruction. In: Proceedings of the IEEE/CVF Winter Conference on Applications of Computer Vision, pp. 806–815 (2021)
2. DeKraker, J., Ferko, K.M., Lau, J.C., Köhler, S., Khan, A.R.: Unfolding the hippocampus: an intrinsic coordinate system for subfield segmentations and quantitative mapping. Neuroimage **167**, 408–418 (2018)
3. Epicoco, I., Mocavero, S.: The performance model of an enhanced parallel algorithm for the SOR method. In: Murgante, B., et al. (eds.) ICCSA 2012. LNCS, vol. 7333, pp. 44–56. Springer, Heidelberg (2012). https://doi.org/10.1007/978-3-642-31125-3_4
4. Fischl, B.: Freesurfer. Neuroimage **62**(2), 774–781 (2012)
5. Han, X., Pham, D.L., Tosun, D., Rettmann, M.E., Xu, C., Prince, J.L.: CRUISE: cortical reconstruction using implicit surface evolution. Neuroimage **23**(3), 997–1012 (2004)
6. Hansen, P.B.: Numerical solution of Laplace's equation (1992)
7. Henschel, L., Conjeti, S., Estrada, S., Diers, K., Fischl, B., Reuter, M.: Fastsurfer-a fast and accurate deep learning based neuroimaging pipeline. NeuroImage **219**, 117012 (2020)
8. Hoopes, A., Iglesias, J.E., Fischl, B., Greve, D., Dalca, A.V.: TopoFit: rapid reconstruction of topologically-correct cortical surfaces. In: Medical Imaging with Deep Learning (2021)
9. Isensee, F., Jaeger, P.F., Kohl, S.A., Petersen, J., Maier-Hein, K.H.: nnU-net: a self-configuring method for deep learning-based biomedical image segmentation. Nat. Methods **18**(2), 203–211 (2021)
10. Jones, S.E., Buchbinder, B.R., Aharon, I.: Three-dimensional mapping of cortical thickness using Laplace's equation. Hum. Brain Mapp. **11**(1), 12–32 (2000)
11. Kim, J.S., et al.: Automated 3-D extraction and evaluation of the inner and outer cortical surfaces using a Laplacian map and partial volume effect classification. Neuroimage **27**(1), 210–221 (2005)
12. Ma, Q., Robinson, E.C., Kainz, B., Rueckert, D., Alansary, A.: PialNN: a fast deep learning framework for cortical pial surface reconstruction. In: International Workshop on Machine Learning in Clinical Neuroimaging, pp. 73–81 (2021)
13. Ogniewicz, R., Kübler, O.: Hierarchic Voronoi skeletons. Pattern Recognit. **28**(3), 343–359 (1995)
14. Osechinskiy, S., Kruggel, F.: Cortical surface reconstruction from high-resolution MR brain images. Int. J. Biomed. Imaging **2012** (2012)

15. Ravikumar, S., Wisse, L., Gao, Y., Gerig, G., Yushkevich, P.: Facilitating manual segmentation of 3D datasets using contour and intensity guided interpolation. In: 2019 IEEE 16th International Symposium on Biomedical Imaging (ISBI 2019), pp. 714–718 (2019)
16. Wisse, L.E., et al.: Downstream effects of polypathology on neurodegeneration of medial temporal lobe subregions. Acta Neuropathol. Commun. **9**(1), 1–11 (2021)
17. Yang, S., Matthias, K.G.: The optimal relaxation parameter for the SOR method applied to a classical model problem. Technical report, Technical Report TR2007-6, University of Maryland, Baltimore County (2007)
18. Zhou, H.Y., Guo, J., Zhang, Y., Yu, L., Wang, L., Yu, Y.: nnFormer: interleaved transformer for volumetric segmentation. arXiv preprint arXiv:2109.03201 (2021)

Med-NCA: Robust and Lightweight Segmentation with Neural Cellular Automata

John Kalkhof$^{(\boxtimes)}$ (ID), Camila González (ID), and Anirban Mukhopadhyay (ID)

Darmstadt University of Technology, Karolinenplatz 5, 64289 Darmstadt, Germany
john.kalkhof@gris.tu-darmstadt.de

Abstract. Access to the proper infrastructure is critical when performing medical image segmentation with Deep Learning. This requirement makes it difficult to run state-of-the-art segmentation models in resource-constrained scenarios like primary care facilities in rural areas and during crises. The recently emerging field of Neural Cellular Automata (NCA) has shown that locally interacting *one-cell* models can achieve competitive results in tasks such as image generation or segmentations in low-resolution inputs. However, they are constrained by high VRAM requirements and the difficulty of reaching convergence for high-resolution images. To counteract these limitations we propose Med-NCA, an end-to-end NCA training pipeline for high-resolution image segmentation. Our method follows a two-step process. Global knowledge is first communicated between cells across the downscaled image. Following that, patch-based segmentation is performed. Our proposed Med-NCA outperforms the classic UNet by 2% and 3% Dice for hippocampus and prostate segmentation, respectively, while also being **500 times smaller**. We also show that Med-NCA is by design invariant with respect to image scale, shape and translation, experiencing only slight performance degradation even with strong shifts; and is robust against MRI acquisition artefacts. Med-NCA enables high-resolution medical image segmentation even on a Raspberry Pi B+, arguably the smallest device able to run PyTorch and that can be powered by a standard power bank.

Keywords: Neural Cellular Automata · Medical Image Segmentation · Robustness

1 Introduction

State-of-the-art medical image segmentation is dominated by UNet-style architectures [18], which still perform at the top of most grand challenges in its various forms [12]. This trend of task-specific optimizations of UNet-style models is usually accompanied by diminishing returns regarding model size versus performance. The increase in model complexity raises serious concerns that machine learning cannot be leveraged in resource-constrained environments [1]. In settings

A. Frangi et al. (Eds.): IPMI 2023, LNCS 13939, pp. 705–716, 2023.
https://doi.org/10.1007/978-3-031-34048-2_54

such as primary care facilities in rural areas, only minimal computing infrastructure is available [4], so it is challenging to deploy models requiring large GPUs. UNet-style models are also particularly susceptible to the *domain shift* problem [7] and have difficulty generalising to other input resolutions. To mitigate this, shifts are often brute-forced into the training pipeline by adding augmentations like translation or simulated acquisition artefacts [5].

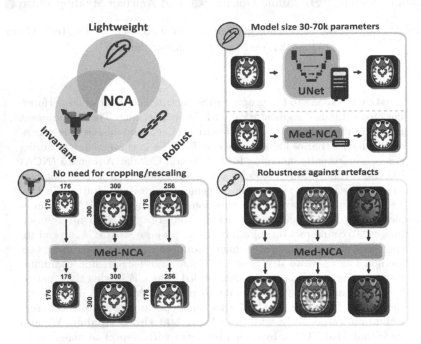

Fig. 1. NCA models are *lightweight* due to their small size and asynchronous inference, and can be run on low-powered systems. They are also, by design, *invariant* to the input scale and field of view. Further, they are *robust* against image artefacts.

Unlike most state-of-the-art methods that rely on optimising UNet-based frameworks, such as the nnUNet pipeline [12], we investigate a fundamentally different learning system that is by design **lightweight, robust** and **input-invariant**, yet achieves **reliable performance**. We introduce Neural Cellular Automata (NCA) as our base architecture, which due to the *one-cell model size*, can be *distributed across any image size* during deployment. The minimal size and asynchronous inference *requires significantly less computing power* than classical models. Additionally, due in part to the limited amount of parameters, a rule has to be learnt that *generalises well* across the problem space, which renders it *robust* by design (illustrated in Fig. 1).

Heavily inspired by the interaction between cells in living organisms, NCAs are minimal models that look at a single cell at a time and can only communicate with their direct neighbours. Global knowledge can be transmitted by deploying the model on each cell and iteratively applying the same rules. Each iteration increases the perceptive range by one cell in each direction. Recently, NCAs have made advances in tasks like robust image generation [13] and even segmentation for small-resolution natural images [20], all while learning a single local update rule that is applied incrementally to each cell.

Despite its advantages such as lightweight inference, training NCAs requires *exponentially higher video ram (VRAM)* depending on the input size during training, which quickly reaches 20 GB for a single sample with a resolution of 256×256. In medical image processing, this is prohibiting as data is typically high in resolution. The local interaction makes inference on big images difficult (greater than 100×100), as many steps are required to communicate global knowledge. In addition, *high-resolution images increase the convergence difficulty*. Due to these constraints, previous works on NCAs have focused on small-resolution computer vision benchmarks [2,9,13–15,17]. We solve these limitations with *Med-NCA*, a two-step NCA model. The model distributes global knowledge across a downscaled image in the first step. In the second step, Med-NCA combines the resulting information with high-resolution image patches to perform high-quality segmentations.

We evaluate our proposed Med-NCA on T1-weighted hippocampus and T2-weighted prostate MRI datasets. We first compare the segmentation performance of Med-NCA to classic and efficient UNet-style architectures, where Med-NCA outperforms them by at least 2% for the hippocampus and 3% for the prostate, with a *90 to 500 times smaller model size*, although there is still a 2% and 10% performance gap to the auto ML pipeline nnUNet. Secondly, we perform an in-depth analysis of three types of **input invariances**: scale, shape and translation. Med-NCA shows *consistent performance* in comparison to UNet-style models and can even outperform the nnUNet for strong shifts in shape and translation. We then investigate the influence of synthetic MRI acquisition artefacts of increasing severity on Med-NCA and UNet. Despite the vastly different one-cell local interaction setup of Med-NCA, our experiments show similar **robustness** to the UNet in terms of anisotropy and bias field and even slightly better robustness to ghosting artefacts. Lastly, we demonstrate that **deployment in low-resource environments** is possible, due to the asynchronous inference and the one-cell model size of NCAs, on a Raspberry Pi Model B+(US$ 35). There is currently a slightly stronger successor, the Raspberry Pi Zero, which costs US$ 5.

To ensure reproducibility and drive further research on NCA segmentation, *we make our complete framework available* under github.com/MECLabTUDA/Med-NCA.

2   Related Work

Recent publications have shown the applicability of NCAs to different tasks, such as robust image generation from a single cell [13] and foreground-background segmentation [20]. In this section, we review relevant related work on NCAs and medical image segmentation.

Neural Cellular Automata: NCA models are a one-cell model architecture, recently adapted to convolutional neural networks by Gilpin [6]. NCAs do not look at the whole image globally but instead only interact locally. Each cell can exclusively communicate with its direct neighbours, and all inherit the same learnt rule. By performing multiple iterations, global knowledge can be conveyed between cells. Despite their small size, NCAs have shown robustness in tasks such as image generation [13,15], where models display a high degree of resilience against perturbations. To the best of our knowledge, only one previous work explores image segmentation with NCAs [20]. The proposed method focuses on foreground-background segmentation on small images of 64 × 64. While it provides a simple up/downscaling solution for high-resolution images, performance is insufficient for medical image segmentation (see Table 1).

Medical Image Segmentation: With the improvements in graphics cards and VRAM availability [12], machine learning models are growing significantly in size. Models like the state-of-the-art nnUNet define 4GB of VRAM as their minimum requirement [12], and thus require proper infrastructure for inference. There have been several attempts to create minimal segmentation models, mainly by modifying the well-established UNet [18]. The 'Segmentation Models' python package [10] provides a collection of UNet-style models where the encoder has been replaced with smaller architectures like *EfficientNet* [22], *MobileNetV2* [19], *DenseNet* [11], *ResNet18* [8] and *VGG11* [21]. However, computational requirements are still significant (see Fig. 3).

UNets and other state-of-the-art segmentation models typically have a pyramid-like structure with multiple up- and downscaling blocks. NCAs stand in strong contrast as they are tiny models acting on a single pixel and communicating global knowledge through iteratively applying the same rule. This intrinsic change in the design allows NCAs to maintain a number of parameters several orders of magnitude lower, and consequently to be run on minimal hardware.

3 Methodology

The local architecture of NCAs, where the model deals only with a single cell and its surroundings, allows them to be *lightweight* in terms of storage space and inference time, but this does not come without limitations. Training NCAs end-to-end on images of size 256 × 256 can easily require 20GB of VRAM (for batch size 1). This is because NCAs are replicated across all input image pixels, and backpropagation is performed through all the iterative steps of the model, increasing VRAM needs. VRAM requirements are therefore dependent on model size, input size and number of iterations/steps.

An additional consideration is that increasing the difficulty of the problem (e.g. directly learning on the full-scale high-resolution image) *can result in NCAs not converging*. Med-NCA reduces VRAM needs and simplifies the segmentation problem by separating it into two steps (illustrated in Fig. 2). Its standard configuration reduces the required VRAM for training by a factor of 16 compared to a full-resolution learning setup and enough steps to increase the perceptive range of each cell to a global scale. This could be improved even further by adding one or more downsampling steps into the pipeline if the need arose.

3.1 Med-NCA

Med-NCA is our main methodological contribution defining a pipeline for training NCAs on high-resolution images. It is optimised to reduce VRAM as well as simplify training for bigger images. We illustrate the training procedure in Fig. 2. We start by describing the backbone NCAs we use in Med-NCA, and then describe the training and inference processes.

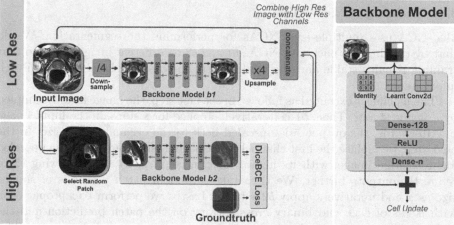

Fig. 2. The *Med-NCA* training strategy relies on cropped images in the 'High Res' step of segmentation to limit VRAM requirements. The final inference is performed on the full image and does not require patchification.

3.2 Backbone NCA

Med-NCA consists of two identical backbone NCA models that iterate over different scales of the input image. Our models are inspired by the architecture presented in *Growing Neural Cellular Automata* [13].

The backbone model is constructed of n input channels, where the first x channels are reserved for the image. The NCA can freely set the remaining $n - x$ channels. Instead of growing from a single pixel, we adapt it to the segmentation task by immediately distributing it across the whole image.

The proposed model consists of two 3×3 learnt convolutional layers and is concatenated with the current cell state, resulting in a state vector of size $3 * n$, as illustrated in Fig. 2. It is important that the learnt convolutional layers use reflect padding, as the model otherwise learns to use the image borders for 'spatial orientation', which worsens input invariance capabilities. The state vector is connected to a dense layer with hidden size h, following a ReLU and another dense layer with the output size n. The standard configuration of our model uses the following parameters: $n = 32$, $h = 128$. These parameters are set to the maximum value at which the model still converges stably, as this provides cells with more memory and allows them to learn a more advanced update rule. For a version of the model that requires 2.5 times fewer parameters, the channel size can be set to $n = 16$, resulting in a slight performance degradation, as shown in the ablation study in Table 1. Similar to other NCA approaches [13,20], we randomly activate $x = 50\%$ of the cells in each step to simulate asynchronous activation.

3.3   Training of Med-NCA

Med-NCA uses multiple-level NCAs for performing the segmentation. We do that by training two backbone NCAs $b1$ and $b2$ on different image scales so that training remains stable and the model learns to consider global and detail-rich information.

The pipeline is executed as follows: in the first step, the image is downsampled by a factor of four. Then, $b1$ is iteratively applied for s steps on the input image x. Afterwards, the output $\bar{x}$ is upscaled back to the original image size. In the next step, we replace the first channel of the output, still containing the upscaled low-resolution image, with its high-resolution counterpart, thus allowing $b2$ to refine the outputs further. We then take a random patch p that is of similar size as x and iteratively apply $b2$ s times. Lastly, we perform backpropagation with a loss of Dice and binary cross-entropy on the patch prediction and the corresponding ground truth segmentation.

This two-step process lowers VRAM requirements by a factor of 16 during training, making it possible to train Med-NCA on high-resolution images.

3.4   Inference

While training has to be specifically adapted to work well with big images, inference is very simple. Extracting a prediction is extremely lightweight and can be carried out on nearly any device that runs PyTorch. The asynchronous and iterative nature of NCAs makes this possible. Inference is only limited by the RAM size, where Med-NCA has a much smaller footprint than, e.g., a UNet. Since NCAs itself are very small, only the size of the image and number of channels are relevant. Further, *during inference patchification is not necessary* as the trained model can be applied directly to the whole image.

4   Experimental Setup

The evaluation of our Med-NCA focuses on the three main aspects *robustness*, *input invariance* and *model size*. The evaluation is performed on two segmentation tasks, namely hippocampus and prostate. We compare our proposed method with UNet-style architectures as well as the auto ML pipeline nnUNet.

Data: We use the hippocampus data released for the Medical Segmentation Decathlon (MSD) [2]. The prostate data is a mixture of two datasets, the MSD (CC-BY-SA 4.0 licence) and ISBI 2013 challenge (CC BY 3.0) [3]. The prostate images range from 320 × 320 to 384 × 384. We scale the image to a training size of 256 × 256, allowing us to perform out-of-distribution experiments for smaller and higher-resolution scales.

Evaluated Architectures: We compare the performance of our proposed Med-NCA to the well-established UNet [18] and different resource-efficient versions. In the latter case, the encoder is replaced by *EfficientNet* [22], *MobileNetV2* [19], *DenseNet* [11], *ResNet18* [8] and *VGG11* [21]. We use the version of each encoder with the least amount of parameters provided in the *Segmentation Models* repository [10]. We also compare our approach to 2D and 3D full-resolution versions of the nnUNet [12].

5   Results

The evaluation of our proposed Med-NCA model shows that it deals well with MRI acquisition artefacts and input size variations while reaching state-of-the-art performance on the hippocampus and prostate segmentation datasets.

5.1   Performance and Resource Consumption

We perform a thorough performance analysis of Med-NCA compared to other UNet-style architectures, and place it in the context to resource consumption. This relation is illustrated in Fig. 3.

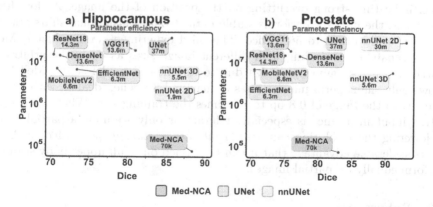

Fig. 3. We compare Med-NCA to other efficient UNet setups as well as the nnUNet in terms of performance vs. the number of parameters.

We see a general trend of UNet-style models suffering in performance from a decrease in model size. Med-NCA, on the other hand, reaches higher Dice scores than the classic UNet, outperforming it by 2% on the hippocampus data and 3% on the prostate dataset, while also requiring 500 times less trainable parameters.

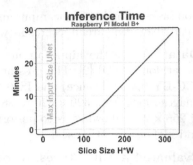

Fig. 4. Med-NCA single slice segmentation time on a Raspberry Pi B+by image size.

To show how lightweight Med-NCA is we deploy it on a Raspberry Pi B+and perform inference up to an image size of 320 × 320 (see Fig. 4). While inference is rather slow, with 30 min per slice for the maximal image size of 320 × 320, using a more recent system with an *RTX 2060 Super* and an *AMD Ryzen 5600X* inference only takes seconds for a whole MRI. In comparison, the UNet can only be executed on the Raspberry Pi up to an image size of 32 × 32.

5.2 Input Invariance

When dealing with medical images, the scale or input size may change after training. A problem with UNet-style segmentation models is that they are not good at adapting to such variability, which our experiments in Fig. 5 show. When UNet faces strong shifts in vertical shape, the Dice drops by 13%, whereas Med-NCA only loses 3%. While the auto ML pipeline nnUNet performs robustly until strong severity of shape changes, it then loses even more performance and drops to 50% Dice. We experience similar results for translational changes. While Med-NCA performs consistently across all introduced shifts, the UNet drops to 0% Dice, indicating strong overfitting on the position of the image. As for shape variations, the nnUNet performs stable until strong translational shifts appear, where it again drops to 56% Dice. The third experiment we conduct is Med-NCAs capability of dealing with different image sizes as it can be arbitrarily distributed across any image size due to its one-cell model setup. Med-NCA shows only slight performance losses of maximal 3% when dealing with images that are in the range of 0.8 up to 1.5 times the training size. When working on MRIs, input invariance is especially relevant as only scanning a partial region or lowering the resolution can *greatly increase capturing speed*. Med-NCA has the considerable advantage that it can be trained on a full upper body scan but perform equally on partial images.

5.3 Robustness

Robustness against MRI acquisition artefacts is an important trait in medical image segmentation. In our robustness analysis, illustrated in Fig. 6, we show that Med-NCA can handle anisotropy and bias field artefacts similarly well to

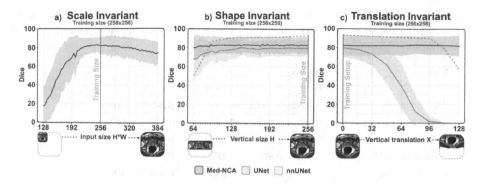

Fig. 5. Comparison of our proposed Med-NCA with UNet and the nnUNet in of out-of-distribution (a) scale, (b) shape and (c) translation scenarios.

a classic UNet. Both models experience no drop in performance with anisotropy artefacts up to a severeness factor of 4. In the case of bias field, Med-NCA and UNet drop more than 25% in performance when the severity of the artefacts becomes too strong. In cases of severe ghosting artefacts, we can see the performance of Med-NCA suffers less than a classic UNet, where Med-NCA performs 12% better in the most severe cases of ghosting.

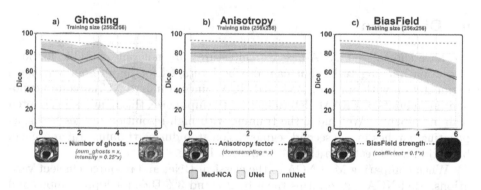

Fig. 6. Robustness analysis of our proposed Med-NCA in comparison to UNet and the nnUNet. nnUNet results are collected for fewer data points, indicated by the dashed line, as the non-flexible pipeline requires manual evaluation. Analysis is performed for synthetic MRI acquisition artefacts of increasing severity: (a) Ghosting, (b) Anisotropy and (c) Bias Field (using TorchIO [16]).

As our comparison to the nnUNet demonstrates, an auto ML pipeline can increase the robustness as nnUNet only suffers from slight performance losses across the MRI acquisition artefacts. It is plausible to include a similar setup in the training of Med-CNA and thus improve robustness.

Table 1. Comparison of different Med-NCA setups (where $c = channels$, $h = hiddensize$), as well as the Backbone-NCA, previous work on NCA segmentation and a standard 2D UNet.

Model	Hippocampus		Prostate	
	Dice ↑	# Param. ↓	Dice ↑	# Param. ↓
Med-NCA	**0.886 ± 0.042**	70016	**0.838 ± 0.083**	70016
Med-NCA c = 16	0.873 ± 0.033	**25920**	0.822 ± 0.087	**25920**
Med-NCA h = 64	0.858 ± 0.047	47530	0.808 ± 0.125	47530
Backbone-NCA 64 × 64	0.871 ± 0.870	35008	0.789 ± 0.132	35008
Seg. NCA [20]	0.805 ± 0.045	39472	0.634 ± 0.190	39472
UNet	0.858 ± 0.044	36951555	0.799 ± 0.099	36951555

5.4 Ablation Study

Lastly, we perform an ablation on our approach in Table 1. Our results show that a different trade-off of performance vs. model size can make Med-NCA 2.5x more lightweight while only sacrificing 1% of performance. Further, we can see that the previous NCA segmentation model *Seg. NCA* [20] is not suitable for the task of medical image segmentation as it performs 7.4% worse for hippocampus and even 20.2% worse for the prostate segmentation task.

6 Discussion

We have shown that NCA-based architectures are not only suitable for low-resolution imaging tasks, but can also be leveraged for high-resolution image segmentation with our proposed Med-NCA. Since standard NCAs require exponential amounts of VRAM, determined by the input size, this requires an adapted training pipeline. We enable the training with high-resolution images by incorporating patches in the second segmentation step during training. Inference can be performed directly on the full-resolution image.

When comparing Med-NCA to the classical UNet and resource-efficient variations, Med-NCA outperforms them by 2% and 3% Dice for hippocampus and prostate image segmentation, respectively. Med-NCA is also more robust to changes in image scale, shape and translation, which is directly inherited by its one-cell model size. The local interaction prevents the model from knowing where in space information is located and therefore eliminates biases based on position. Further, despite the vastly different one-cell model size with 500 times fewer parameters, Med-NCA shows similar or better robustness than a UNet to MRI acquisition artefacts. Due to the one-cell model size and asynchronous inference, NCAs are lightweight enough to be executed on any hardware that runs PyTorch, which we demonstrate by deploying Med-NCA on a Raspberry Pi B+, that can be powered by any 5 W power source like a standard power bank.

While we have solved the problem of NCAs for high-resolution inputs, one limitation of the present approach is that it only admits 2D image slices. Further optimisations could make Med-NCA work for 3D inputs by adapting the perceptive field and integrating VRAM improvements based on the three-dimensional space. 3D inputs would make Med-NCA suitable for more datasets and might further improve performance. In addition, although Med-NCA is inherently robust, we experience performance degradation when the severity of the acquisition artefacts becomes too strong, which could be improved upon in future work.

A further limitation is that there is a 10% performance difference between Med-NCA and nnUNet on the prostate task. For future work, we propose the development of a training pipeline similar to the nnUNet, which uses NCA as the core architecture and includes data augmentations and post-processing steps during training. We expect that such a framework could achieve equivalent performance while profiting from the inherited benefits of the NCA architecture like size and scale invariance and general robustness.

7   Conclusion

In this work we introduce the Med-NCA segmentation model, which solves the VRAM limitations inherited by NCAs for high-resolution images and is therefore suitable for medical images. Our approach can be run on minimal hardware (e.g. a Raspberry Pi B+), is inherently robust against scale, shape, translation and image artefacts, and reaches near state-of-the-art segmentation performance. We compare our model to UNet-style models optimised for low parameter size and show that Med-NCA not only achieves a higher Dice score, but also outperforms them in terms of input invariance, robustness and resource use. This makes Med-NCA a perfect candidate for primary care scenarios with limited infrastructure and highly variable imaging equipment.

References

1. Ajani, T.S., Imoize, A.L., Atayero, A.A.: An overview of machine learning within embedded and mobile devices-optimizations and applications. Sensors **21**(13), 4412 (2021)
2. Antonelli, M., et al.: The medical segmentation decathlon. Nat. Commun. **13**(1), 1–13 (2022)
3. Bloch, N., et al.: NCI-ISBI 2013 challenge: automated segmentation of prostate structures (2015). https://doi.org/10.7937/K9/TCIA.2015.zF0vlOPv
4. Boppart, S.A., Richards-Kortum, R.: Point-of-care and point-of-procedure optical imaging technologies for primary care and global health. Sci. Transl. Med. **6**(253), 253rv2 (2014)
5. Chlap, P., Min, H., Vandenberg, N., Dowling, J., Holloway, L., Haworth, A.: A review of medical image data augmentation techniques for deep learning applications. J. Med. Imaging Radiat. Oncol. **65**(5), 545–563 (2021)
6. Gilpin, W.: Cellular automata as convolutional neural networks. Phys. Rev. E **100**(3), 032402 (2019)
7. González, C., et al.: Distance-based detection of out-of-distribution silent failures for Covid-19 lung lesion segmentation. Med. Image Anal. **82**, 102596 (2022)

8. He, K., Zhang, X., Ren, S., Sun, J.: Deep residual learning for image recognition. In: Proceedings of the IEEE Conference on Computer Vision and Pattern Recognition, pp. 770–778 (2016)

9. Hernandez, A., Vilalta, A., Moreno-Noguer, F.: Neural cellular automata manifold. In: Proceedings of the IEEE/CVF Conference on Computer Vision and Pattern Recognition, pp. 10020–10028 (2021)

10. Iakubovskii, P.: Segmentation models PyTorch (2019). https://github.com/qubvel/segmentation_models.pytorch

11. Iandola, F., Moskewicz, M., Karayev, S., Girshick, R., Darrell, T., Keutzer, K.: DenseNet: implementing efficient convnet descriptor pyramids. arXiv preprint arXiv:1404.1869 (2014)

12. Isensee, F., Jaeger, P.F., Kohl, S.A., Petersen, J., Maier-Hein, K.H.: nnU-Net: a self-configuring method for deep learning-based biomedical image segmentation. Nat. Methods 18(2), 203–211 (2021)

13. Mordvintsev, A., Randazzo, E., Niklasson, E., Levin, M.: Growing neural cellular automata. Distill 5(2), e23 (2020)

14. Niklasson, E., Mordvintsev, A., Randazzo, E., Levin, M.: Self-organising textures. Distill 6(2), e00027–003 (2021)

15. Palm, R.B., Duque, M.G., Sudhakaran, S., Risi, S.: Variational neural cellular automata. In: International Conference on Learning Representations (2021)

16. Pérez-García, F., Sparks, R., Ourselin, S.: TorchIO: a python library for efficient loading, preprocessing, augmentation and patch-based sampling of medical images in deep learning. Comput. Methods Programs Biomed. 106236 (2021). https://doi.org/10.1016/j.cmpb.2021.106236. https://www.sciencedirect.com/science/article/pii/S0169260721003102

17. Randazzo, E., Mordvintsev, A., Niklasson, E., Levin, M., Greydanus, S.: Self-classifying MNIST digits. Distill 5(8), e00027–002 (2020)

18. Ronneberger, O., Fischer, P., Brox, T.: U-Net: convolutional networks for biomedical image segmentation. In: Navab, N., Hornegger, J., Wells, W.M., Frangi, A.F. (eds.) MICCAI 2015. LNCS, vol. 9351, pp. 234–241. Springer, Cham (2015). https://doi.org/10.1007/978-3-319-24574-4_28

19. Sandler, M., Howard, A., Zhu, M., Zhmoginov, A., Chen, L.C.: MobileNetv 2: inverted residuals and linear bottlenecks. In: Proceedings of the IEEE Conference on Computer Vision and Pattern Recognition, pp. 4510–4520 (2018)

20. Sandler, M., Zhmoginov, A., Luo, L., Mordvintsev, A., Randazzo, E., et al.: Image segmentation via cellular automata. arXiv preprint arXiv:2008.04965 (2020)

21. Simonyan, K., Zisserman, A.: Very deep convolutional networks for large-scale image recognition. arXiv preprint arXiv:1409.1556 (2014)

22. Tan, M., Le, Q.: EfficientNet: rethinking model scaling for convolutional neural networks. In: International Conference on Machine Learning, pp. 6105–6114. PMLR (2019)

Mixup-Privacy: A Simple yet Effective Approach for Privacy-Preserving Segmentation

Bach Ngoc Kim[1]([✉])[ID], Jose Dolz[1][ID], Pierre-Marc Jodoin[2][ID],
and Christian Desrosiers[1][ID]

[1] École de Technologie Supérieure, Montreal, QC H3C1K3, Canada
bachknk49@gmail.com
[2] Université de Sherbrooke, Sherbrooke, QC J1K2R1, Canada

Abstract. Privacy protection in medical data is a legitimate obstacle for centralized machine learning applications. Here, we propose a client-server image segmentation system which allows for the analysis of multi-centric medical images while preserving patient privacy. In this approach, the client protects the to-be-segmented patient image by mixing it to a reference image. As shown in our work, it is challenging to separate the image mixture to exact original content, thus making the data unworkable and unrecognizable for an unauthorized person. This proxy image is sent to a server for processing. The server then returns the mixture of segmentation maps, which the client can revert to a correct target segmentation. Our system has two components: 1) a segmentation network on the server side which processes the image mixture, and 2) a segmentation unmixing network which recovers the correct segmentation map from the segmentation mixture. Furthermore, the whole system is trained end-to-end. The proposed method is validated on the task of MRI brain segmentation using images from two different datasets. Results show that the segmentation accuracy of our method is comparable to a system trained on raw images, and outperforms other privacy-preserving methods with little computational overhead.

Keywords: Privacy · Medical imaging · Segmentation · Mixup

1 Introduction

Neural networks are the *de facto* solution to numerous medical analysis tasks, from disease recognition, to anomaly detection, segmentation, tumor resurgence prediction, and many more [4,14]. Despite their success, the widespread clinical deployment of neural nets has been hindered by legitimate privacy restrictions, which limit the amount of data the scientific community can pool together.

Researchers have explored a breadth of solutions to tap into massive amounts of data while complying with privacy restrictions. One such solution is federated

Supported by Natural Sciences and Engineering Research Council of Canada (NSERC) and Reseau de BioImagrie du Quebec (RBIQ).

learning (FL) [13,27], for which training is done across a network of computers each holding its local data. While FL has been shown to be effective, it nonetheless suffers from some limitations when it comes to medical data. First, from a cybersecurity standpoint, communicating with computers located in a highly-secured environment such as a hospital, while complying with FDA/MarkCE cybersecurity regulation, is no easy feast. Second, having computers communicate with their local PACS server is also tricky. And third, since FL is a decentralized *training* solution, it requires a decentralized set of computers to process images at test time, making it ill-suited for software as a service (SAAS) cloud services. Another solution is to train a centralized network with homomorphic data encryption [6]. While this ensures a rigorous data protection, as detailed in Sect. 2, the tremendous computational complexity of homomorphic networks prohibits their use in practice.

Recent studies have investigated centralized cloud-based solutions where data is encoded by a neural network prior being sent to the server [11]. While the encoded data is unworkable for unauthorized parties, it nonetheless can be processed by a network that was trained to deal with such encoded data. In some methods, such as Privacy-Net [11], the data sent back to the client (e.g., predicted segmentation maps) is not encoded and may contain some private information about the patient (e.g., the patient's identity or condition). To ensure that the returned data is also unworkable for non-authorized users, Kim et al. [12] proposed an encoding method based on reversible image warping, where the warping function is only known by the client.

In this paper, we propose a novel client-server cloud system that can effectively segment medical images while protecting subjects' data privacy. Our segmentation method, which relies on the hardness of blind source separation (BSS) as root problem [2,3,9,19], leverages a simple yet powerful technique based on mixup [5]. In the proposed approach, the client protects the to-be-segmented patient image by mixing it to a reference image only known to this client. This reference image can be thought as a private key needed to encode and decode the image and its segmentation map. The image mixture renders the data unworkable and unrecognizable for a non-authorized person, since recovering the original images requires to solve an intractable BSS problem. This proxy image is sent to a server for a processing task, which corresponds to semantic segmentation in this work. Instead of sending back the non-encoded segmentation map, as in [11], the server returns to the client a mixture of the target and reference segmentation maps. Finally, because the client knows the segmentation map for the reference image, as well as the mixing coefficients, it can easily recover the segmentation for the target.

Our work makes four contributions to privacy-preserving segmentation:

1. We introduce a simple yet effective method inspired by mixup, which encodes 3D patches of a target image by mixing them to reference patches with known ground-truth. Unlike FL approaches, which require a bulky training setup, or homomorphic networks which are computationally prohibitive, our method works in a normal training setup and has a low computational overhead.

2. We also propose a learning approach for recovering the target segmentation maps from mixed ones, which improves the noisy results of directly reversing the mixing function.
3. Results are further improved with a test-time augmentation strategy that mixes a target image with different references and then ensembles the segmentation predictions to achieve a higher accuracy.
4. We conduct extensive experiments on two challenging 3D brain MRI benchmarks, and show our method to largely outperform state-of-art approaches for privacy-preserving segmentation, while being simpler and faster than these approaches and yet offering a similar level of privacy.

2   Related Works

Most privacy-preserving approaches for image analysis fall in two categories: those based on homomorphic encryption and the ones using adversarial learning.

Homomorphic Encryption (HE) [1,7,18]. This type of encryption enables to compute a given function on encrypted data without having to decrypt it first or having access to the private key. Although HE offers strong guarantees on the security of the encrypted data, this approach suffers from two important limitations: 1) it has a prohibitive computational/communication overhead [22]; 2) it is limited to multiplications and additions, and non-linear activation functions have to be approximated by polynomial functions. As a result, homomorphic networks have been relatively simplistic [6], and even computing the output of a simple CNN is prohibitively slow (e.g., 30 min for a single image [18]).

Adversarial Learning (AL). This type of approach uses a neural net to encode images so that private information is discarded, yet the encoded image still holds the necessary information to perform a given image analysis task [20,26]. The encoder is trained jointly with two downstream networks taking the encoded image as input, the first one seeking to perform the target task and the other one (the discriminator) trying to recover the private information. The parameters of the encoder are updated to minimize the task-specific utility loss while maximizing the loss of the discriminator. In medical imaging tasks, where patient identity should be protected, the discriminator cannot be modeled as a standard classifier since the number of classes (e.g., patient IDs) is not fixed. To alleviate this problem, the method in [11] uses a Siamese discriminator which receives two encoded images as input and predicts if the images are from the same patient or not. While input images are encoded, the method produces non-encoded segmentation maps which may still be used to identify the patient. The authors of [12] overcome this limitation by transforming input images with a reversible non-linear warping which depends on a private key. When receiving a deformed segmentation map from the server, the client can recover the true segmentation by reversing the transformation. However, as the method in [11], this approach requires multiple scans of the same patient to train the Siamese discriminator, which may not be available in practice. Furthermore, the learned encoder is highly sensitive to the distribution of input images and fails to obfuscate identity when this distribution shifts. In contrast, our method does not require multiple scans per patient. It is also simpler to train and, because it relies on the general principle of BSS, is less sensitive to the input image distribution.

3   Methodology

We first introduce the principles of blind source separation and mixup on which our work is based, and then present the details of our Mixup-Privacy method.

3.1   Blind Source Separation

Blind source separation (BSS) is a well-known problem of signal processing which seeks to recover a set of unknown source signals from a set of mixed ones, without information about the mixing process. Formally, let $x(t) = [x_1(t), \ldots, x_n(t)]^T$ be a set of n source signals which are mixed into a set of m signals, $y(t) = [y_1(t), \ldots, y_m(t)]^T$, using matrix $A \in \mathbb{R}^{m \times n}$ as follows: $y(t) = A \cdot x(t)$. BSS can be defined as recovering $x(t)$ when given only $y(t)$. While efficient methods exist for cases where $m = n$, the problem is much harder to solve when $m < n$ as the system of equations then becomes under-determined [9]. For the extreme case of single channel separation ($n = 1$), [3] showed that traditional approaches such as Independent Component Analysis (ICA) fail when the sources have substantially overlapping spectra. Recently, the authors of [10] proposed a deep learning method for single channel separation, using the noise-annealed Langevin dynamics to sample from the posterior distribution of sources given a mixture. Although it achieves impressive results for the separation of RGB natural images, as we show in our experiments, this method does not work on low-contrast intensity images such as brain MRI. Leveraging the ill-posed nature of single source separation, we encode 3D patches of images to segment by mixing them with those of reference images.

3.2   Mixup Training

Mixup is a data augmentation technique that generates new samples via linear interpolation between random pairs of images as well as their associated one-hot encoded labels [28]. Let (x_i, y_i) and (x_j, y_j) be two examples drawn at random from the training data, and $\alpha \sim \text{Beta}(b, b)$ be a mixing coefficient sampled from the Beta distribution with hyperparameter b. Mixup generates virtual training examples $(\tilde{x}, \tilde{y})$ as follows:

$$\tilde{x} = \alpha x_i + (1 - \alpha)x_j; \quad \tilde{y} = \alpha y_i + (1 - \alpha)y_j. \tag{1}$$

While Mixup training has been shown to bring performance gains in various problems, including image classification [5] and semantic segmentation [29], it has not been explored as a way to preserve privacy in medical image segmentation.

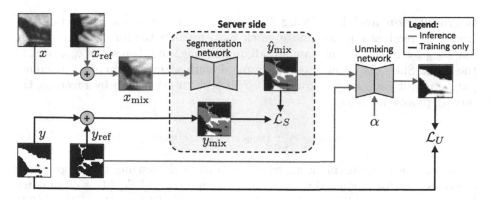

Fig. 1. Training diagram of the proposed system with the client (left and right) and the server (middle). The client mixes the input image x and segmentation map y with a reference pair $(x_{\text{ref}}, y_{\text{ref}})$. The mixed data is then fed to a segmentation network located on a server and whose output is a mixed segmentation map. The resulting segmentation map is sent back to the client, which decodes it with a unmixing network and the reference map y_{ref}.

3.3 Proposed System

As shown in Fig. 1, our method involves a client which has an image x to segment and a server which has to perform segmentation without being able to recover private information from x. During training, the client mixes an image x and its associated segmentation map y with a reference data pair x_{ref} and y_{ref}. The mixed data $(x_{\text{mix}}, y_{\text{mix}})$ is then sent to the server. Since unmixing images requires to solve an under-determined BSS problem, x cannot be recovered from x_{mix} without x_{ref}. This renders x_{mix} unusable if intercepted by an unauthorized user. During inference, the server network returns the mixed segmentation maps $\hat{y}_{\text{mix}}$ to the client, which then recovers the true segmentation maps y by reversing the mixing process. The individual steps of our method, which is trained end-to-end, are detailed below.

Data Mixing. Since 3D MR images are memory heavy, our segmentation method processes images in a patch-wise manner. Each patch $x \in \mathbb{R}^{H \times W \times D}$ is mixed with a reference patch of the same size:

$$x_{\text{mix}} = \alpha x_{\text{target}} + (1 - \alpha) x_{\text{ref}}, \tag{2}$$

where $\alpha \in [0, 1]$ is a mixing weight drawn randomly from the uniform distribution[1]. During training, the one-hot encoded segmentation ground-truths $y \in [0, 1]^{C \times H \times W \times H}$ are also mixed using the same process:

$$y_{\text{mix}} = \alpha y_{\text{target}} + (1 - \alpha) y_{\text{ref}}, \tag{3}$$

and are sent to the server with the corresponding mixed image patches x_{mix}.

---

[1] Unlike Mixup which uses the Beta distribution to have a mixing weight close to 0 or 1, we use the uniform distribution to have a broader range of values.

Segmentation and Unmixing Process. The server-side segmentation network $S(\cdot)$ receives a mixed image patch x_{mix}, predicts the mixed segmentation maps $\hat{y}_{\mathrm{mix}} = S(x_{\mathrm{mix}})$ as in standard Mixup training, and then sends $\hat{y}_{\mathrm{mix}}$ back to the client. Since the client knows the ground-truth segmentation of the reference patch, y_{ref}, it can easily recover the target segmentation map by reversing the mixing process as follows:

$$\hat{y}_{\mathrm{target}} = \frac{1}{\alpha}\big(\hat{y}_{\mathrm{mix}} - (1-\alpha)y_{\mathrm{ref}}\big). \tag{4}$$

However, since segmenting a mixed image is more challenging than segmenting the ones used for mixing, the naive unmixing approach of Eq. (4) is often noisy. To address this problem, we use a shallow network $D(\cdot)$ on the client side to perform this operation. Specifically, this unmixing network receives as input the mixed segmentation $\hat{y}_{\mathrm{mix}}$, the reference segmentation y_{ref}, and the mixing coefficient α, and predicts the target segmentation as $\hat{y}_{\mathrm{target}} = D(\hat{y}_{\mathrm{mix}}, y_{\mathrm{ref}}, \alpha)$.

3.4   Test-Time Augmentation

Test-time augmentation (TTA) is a simple but powerful technique to improve performance during inference [24]. Typical TTA approaches generate multiple augmented versions of an example x using a given set of transformations, and then combine the predictions for these augmented examples based on an ensembling strategy. In this work, we propose a novel TTA approach which augments a target patch x_{target} by mixing it with different reference patches $\{x_{\mathrm{ref}}^k\}_{k=1}^K$:

$$x_{\mathrm{mix}}^k = \alpha x_{\mathrm{target}} + (1-\alpha)x_{\mathrm{ref}}^k. \tag{5}$$

The final prediction for the target segmentation is then obtained by averaging the predictions of individual mixed patches:

$$\hat{y}_{\mathrm{target}} = \frac{1}{K}\sum_{k=1}^K D\big(\hat{y}_{\mathrm{mix}}^k, y_{\mathrm{ref}}^k, \alpha\big). \tag{6}$$

As we will show in experiments, segmentation accuracy can be significantly boosted using only a few augmentations.

4   Experimental Setup

Datasets. We evaluate our method on the privacy-preserving segmentation of brain MRI from two public benchmarks, the Parkinson's Progression Marker Initiative (PPMI) dataset [15] and the Brain Tumor Segmentation (BraTS) 2021 Challenge dataset. For the PPMI dataset, we used T1 images from 350 subjects for segmenting brain images into three tissue classes: white matter (WM), gray matter (GM) and cerebrospinal fluid (CSF). Each subject underwent one or two baseline acquisitions and one or two acquisitions 12 months later for a total of 773 images. The images were registered onto a common MNI space and resized to $144 \times 192 \times 160$ with a $1\,\text{mm}^3$ isotropic resolution. We divided the dataset into training and testing sets containing 592 and 181 images, respectively, so that images from the same subject are not included in both the training and testing sets. Since PPMI has no ground-truth annotations, as in [11,12], we employed Freesurfer to obtain a pseudo ground-truth for training. We included the PPMI dataset in our experiments because it has multiple scans per patient, which is required for some of the compared baselines [11,12].

BraTS 2021 is the largest publicly-available and fully-annotated dataset for brain tumor segmentation. It contains 1,251 multi-modal MRIs of size $240 \times 240 \times 155$. Each image was manually annotated with four labels: necrose (NCR), edema (ED), enhance tumor (ET), and background. We excluded the T1, T2 and FLAIR modalities and only use T1CE. From the 1,251 scans, 251 scans were used for testing, while the remaining constituted the training set.

Evaluation Metrics. Our study uses the 3D Dice similarity coefficient (DSC) to evaluate the segmentation performance of tested methods. For measuring the ability to recover source images, we measure the Multi-scale Structural Similarity (MS-SSIM) [25] between the original source image and the one recovered from a BSS algorithm [10]. Last, to evaluate the privacy-preserving ability of our system, we model the task of recovering a patient's identity as a retrieval problem and measure performance using the standard F1-score and mean average precision (mAP) metrics.

Implementation Details. We used patches of size $32 \times 32 \times 32$ for PPMI and $64 \times 64 \times 64$ for BraTS. Larger patches were considered for BraTS to capture the whole tumor. We adopted architectures based on U-Net [21] for both the segmentation and unmixing networks. For the more complex segmentation task, we used the U-Net++ architecture described in [30], whereas a small U-Net with four convolutional blocks was employed for the unmixing network. For the latter, batch normalization layers were replaced by adaptive instance normalization layers [8] which are conditioned on the mixing coefficient α. Both the segmentation and unmixing networks are trained using combination of multi-class cross entropy loss and 3D Dice loss [17]. End-to-end training was performed for 200,000 iterations on a NVIDIA A6000 GPU, using the Adam optimizer with a learning rate of 1×10^{-4} and a batch size of 4.

Compared Methods. We evaluate different variants of our Mixup-Privacy method for privacy-preserving segmentation. For the segmentation unmixing process, two approaches were considered: a *Naive* approach which reverses the mixing process using Eq. (4), and a *Learned* one using the unmixing network $D(\cdot)$. Both approaches were tested with and without the TTA strategy described in Sect. 3.4, giving rise to four different variants. We compared these variants against a segmentation *Baseline* using non-encoded images and two recent approaches for cloud-based privacy-preserving segmentation: *Privacy-Net* [11]

724   B. N. Kim et al.

Table 1. Main results of the proposed approach across different tasks - including segmentation, blind source separation and test-retest reliability - and two datasets (PPMI and BraTS2021).

	PPMI				BraTS2021			
	GM	WM	CSF	Avg	NCR	ED	ET	Avg
SEGMENTATION (DICE SCORE)								
Baseline	0.930	0.881	0.876	0.896	0.846	0.802	0.894	0.847
Privacy-Net [11]	0.905	0.804	0.732	0.813	—	—	—	—
Deformation-Proxy [12]	0.889	0.825	0.757	0.823	—	—	—	—
Ours (*Naive*)	0.758	0.687	0.634	0.693	0.656	0.635	0.692	0.661
Ours (*Naive + TTA*)	0.852	0.829	0.793	0.825	0.775	0.737	0.804	0.772
Ours (*Learned*)	0.893	0.833	0.795	0.840	0.805	0.763	0.842	0.803
Ours (*Learned + TTA*)	**0.925**	**0.879**	**0.863**	**0.889**	**0.841**	**0.808**	**0.872**	**0.840**
BLIND SOURCE SEPARATION (MS-SSIM)								
Separation Accuracy		0.602 ± 0.104				0.588 ± 0.127		
TEST-RETEST RELIABILITY (ICC VALUE)								
ICC	0.845	0.812	0.803	—	0.842	0.812	0.803	—
Upper bound	0.881	0.856	0.844	—	0.878	0.855	0.839	—
Lower bound	0.798	0.783	0.771	—	0.805	0.777	0.768	—

and *Deformation-Proxy* [12]. The hyperparameters of all compared methods were selected using 3-fold cross-validation on the training set.

4.1 Results

Segmentation Performance. The top section of Table 1 reports the segmentation performance of the compared models. Since Privacy-Net and Deformation-Proxy require longitudinal data to train the Siamese discriminator, we only report their results for PPMI, which has such data. Comparing the naive and learned approaches for segmentation unmixing, we see that using an unmixing network brings a large boost in accuracy. Without TTA, the learned unmixing yields an overall Dice improvement of 14.7% for PPMI and of 14.2% for BraTS2021. As shown in Fig. 2, the naive approach directly reversing the mixing process leads to a noisy segmentation which severely affects accuracy.

Results in Table 1 also demonstrate the positive impact of our TTA strategy on segmentation performance. Thus, adding this strategy to the naive unmixing approach increases the overall Dice by 13.2% for PPMI and by 11.1% for BraTS2021. Likewise, combining it with the learned unmixing approach boosts the overall Dice by 4.9% for PPMI and by 3.7% in the case of BraTS2021. Looking at the predictions for different reference patches in Fig. 2, we see a high variability, in particular for the naive unmixing approach. As can be seen in the

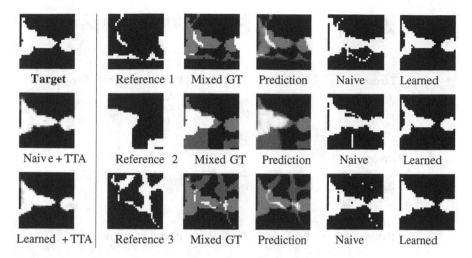

Fig. 2. Examples of segmented patches obtained by the naive and learned unmixing approaches from the same target and three different references. Naive + TTA and Learned + TTA show the mean prediction of these approaches for 30 augmentations (each one using a different reference).

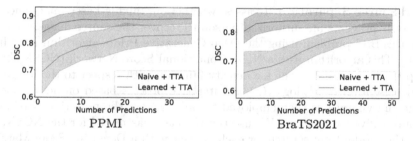

Fig. 3. Segmentation accuracy (DSC) against the number of TTA predictions.

first column of the figure (Naive + TTA and Learned + TTA), averaging multiple predictions in our TTA strategy reduces this variability and yields a final prediction very close to the ground-truth. As in other TTA-based approaches, our TTA strategy incurs additional computations since a segmentation prediction must be made for each augmented example (note that these predictions can be made in a single forward pass of the segmentation network). It is therefore important to analyze the gain in segmentation performance for different numbers of TTA augmentations. As shown in Fig. 3, increasing the number of predictions for augmented examples leads to a higher Dice, both for the naive and learned unmixing approaches. Interestingly, when using the learned unmixing (i.e., Learned + TTA), the highest accuracy is reached with only 10–15 augmentations. In summary, our TTA strategy brings considerable improvements with limited computational overhead.

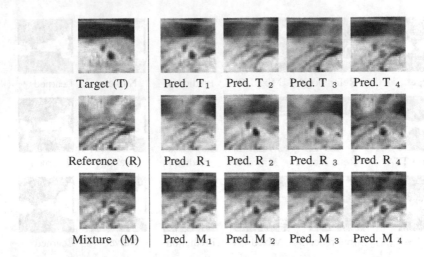

	Pred. T_1	Pred. T_2	Pred. T_3	Pred. T_4
Target (T)				
	Pred. R_1	Pred. R_2	Pred. R_3	Pred. R_4
Reference (R)				
	Pred. M_1	Pred. M_2	Pred. M_3	Pred. M_4
Mixture (M)				

Fig. 4. Examples of blind source separation (BSS) results for the mixture of given target and reference patches. Columns 2–5 correspond to results for different random initializations of the BSS algorithm.

Blind Source Separation. To assess whether our mixing-based image encoding effectively prevents an authorized person to recover the source image, we try to solve this BSS problem using the Deep Generative Priors algorithm introduced in [10]. This algorithm uses a Noise Conditional Score Network (NCSN) [23] to compute the gradient of the log density function with respect to the image at a given noise level σ, $\nabla_x \log p_\sigma(x)$. An iterative process based on noise-annealed Langevin dynamics is then employed to sample the posterior distribution of sources given a mixture. We use the U-Net++ as model for the NCSN, and train this model from scratch for each dataset with a Denoising Score Matching loss. Training is performed for 100,000 iterations on NVIDIA A6000 GPU, using the Adam optimizer with a learning rate of 5×10^{-4} and a batch size of 16.

The second section of Table 1 gives the mean ($\pm$ stdev) of MS-SSIM scores (ranging from 0 to 1) between original target images and those recovered from the BSS algorithm: 0.602 ± 0.104 for PPMI and 0.588 ± 0.127 for BraTS2021. These low values indicate that the target image cannot effectively be recovered from the mixed one. This is confirmed in Fig. 4 which shows the poor separation results of the BSS algorithm for different random initializations.

Test-Retest Reliability. One source of variability in our method (without TTA) is the choice of the reference image used for mixing. To evaluate the stability of our method with respect to this factor, we perform a test-retest reliability analysis measuring the intra-class correlation coefficient (ICC) [16] of the test DSC for two predictions using different references. A higher ICC (ranging from 0 to 1) corresponds to a greater level of consistency. The third section of Table 1 reports the ICC score obtained for each segmentation class, as well as the upper and lower bounds at 95% confidence. We see that all ICC values are above 0.75, indicating a good reliability.

Subject Re-identification. To mea- sure how well our method protects the identity of patients, we carry out a patient re-identification analysis using the PPMI dataset which has multi- ple scans for the same patient. In this analysis, we encode each image in the dataset by mixing it with a randomly chosen reference. For an encoded image x_{mix}, we predict the patient identity as the identity of the

Table 2. Subject re-identification analysis on the PPMI dataset.

Method	F1-score	mAP
No Proxy	0.988	0.998
Privacy-Net [11]	0.092	0.202
Deformation-Proxy [12]	0.122	0.147
Ours	0.284	0.352

other encoded image x'_{mix} most similar to x_{mix} based on the MS-SSIM score. Table 2 compares the F1-score and mAP performance of our method to a base- line with no image encoding (No Proxy), Privacy-Net and Deformation-Proxy. As can be seen, the re-identification of patients is quite easy when no encoding is used (mAP of 0.998), and all encoding-based methods significantly reduce the ability to recover patient identity using such retrieval approach. While our mix- ing based method does not perform as well as the more complex Privacy-Net and Deformation-Proxy approaches, it still offers a considerable protection while largely improving segmentation accuracy (see Table 1).

5   Conclusion

We introduced an efficient method for privacy-preserving segmentation of med- ical images, which encodes 3D patches of a target image by mixing them to reference patches with known ground-truth. Two approaches were investigated for recovering the target segmentation maps from the mixed output of the seg- mentation network: a naive approach reversing the mixing process directly, or using a learned unmixing model. We also proposed a novel test-time augmen- tation (TTA) strategy to improve performance, where the image to segment is mixed by different references and the predictions for these mixed augmentations are averaged to generate the final prediction.

We validated our method on the segmentation of brain MRI from the PPMI and BraTS2021 datasets. Results showed that using a learned unmixing instead of the naive approach improves DSC accuracy by more than 14% for both datasets. Our TTA strategy, which alleviates the problem of prediction vari- ability, can also boost DSC performance by 3.7%-13.2% when added on top of its single-prediction counterpart. Compared to state-of-art approaches such as Privacy-Net and Deformation-Proxy, our method combining learned unmixing and TTA achieves a significantly better segmentation, while also offering a good level of privacy.

In the future, we plan to validate our method on other segmentation tasks involving different imaging modalities. While we encoded a target image by mixing it to a reference one, other strategies could be also explored, for example, mixing more than two images. This could make the BSS more difficult, hence increasing the security of the method, at the cost of a reduced segmentation accuracy. The prediction variance of our TTA strategy could also be used as a measure of uncertainty in semi-supervised segmentation settings or to suggest annotations in an active learning system.

References

1. Avants, B.B., Tustison, N.J., Song, G., Cook, P.A., Klein, A., Gee, J.C.: A reproducible evaluation of ANTs similarity metric performance in brain image registration. Neuroimage **54**(3), 2033–2044 (2011)
2. Cardoso, J.F.: Blind signal separation: statistical principles. Proc. IEEE **86**(10), 2009–2025 (1998). https://doi.org/10.1109/5.720250
3. Davies, M., James, C.: Source separation using single channel ICA. Signal Process. **87**(8), 1819–1832 (2007)
4. Dolz, J., Desrosiers, C., Ayed, I.B.: 3D fully convolutional networks for subcortical segmentation in MRI: a large-scale study. Neuroimage **170**, 456–470 (2018)
5. Guo, H., et al.: Mixup as locally linear out-of-manifold regularization. In: Proceedings of AAAI, vol. 33, pp. 3714–3722 (2019)
6. Hardy, S., et al.: Private federated learning on vertically partitioned data via entity resolution and additively homomorphic encryption. arXiv (2017)
7. Hesamifard, E., Takabi, H., Ghasemi, M.: CryptoDL: deep neural networks over encrypted data. arXiv preprint arXiv:1711.05189 (2017)
8. Huang, X., Belongie, S.: Arbitrary style transfer in real-time with adaptive instance normalization. In: Proceedings of the IEEE International Conference on Computer Vision, pp. 1501–1510 (2017)
9. Jain, S., Rai, D.: Blind source separation and ICA techniques: a review. IJEST **4**, 1490–1503 (2012)
10. Jayaram, V., Thickstun, J.: Source separation with deep generative priors. In: International Conference on Machine Learning (ICML), pp. 4724–4735 (2020)
11. Kim, B., et al.: Privacy-Net: an adversarial approach for identity-obfuscated segmentation of medical images. IEEE Trans. Med. Imaging **40**, 1737–1749 (2021)
12. Kim, B., et al.: Privacy preserving for medical image analysis via non-linear deformation proxy. In: British Machine and Vision Conference (BMVC) (2021)
13. Konecný, J., McMahan, H.B., Ramage, D., Richtárik, P.: Federated optimization: distributed machine learning for on-device intelligence. CoRR (2016)
14. Litjens, G., et al.: A survey on deep learning in medical image analysis. Media **42**, 60–88 (2017)
15. Marek, K., et al.: The Parkinson progression marker initiative (PPMI). Prog. Neurobiol. **95**(4), 629–635 (2011)
16. Mcgraw, K., Wong, S.: Forming inferences about some intraclass correlation coefficients. Psychol. Methods **1**, 30–46 (1996)
17. Milletari, F., Navab, N., Ahmadi, S.A.: V-Net: fully convolutional neural networks for volumetric medical image segmentation. In: 2016 Fourth International Conference on 3D Vision (3DV), pp. 565–571. IEEE (2016)
18. Nandakumar, K., Ratha, N., Pankanti, S., Halevi, S.: Towards deep neural network training on encrypted data. In: Proceedings of CVPR-W (2019)

19. Nouri, A., et al.: A new approach to feature extraction in MI-based BCI systems. In: Artificial Intelligence-Based Brain-Computer Interface, pp. 75–98 (2022)

20. Raval, N., Machanavajjhala, A., Cox, L.P.: Protecting visual secrets using adversarial nets. In: Proceedings of CVPR-W, pp. 1329–1332 (2017)

21. Ronneberger, O., Fischer, P., Brox, T.: U-Net: convolutional networks for biomedical image segmentation. In: Navab, N., Hornegger, J., Wells, W.M., Frangi, A.F. (eds.) MICCAI 2015. LNCS, vol. 9351, pp. 234–241. Springer, Cham (2015). https://doi.org/10.1007/978-3-319-24574-4_28

22. Rouhani, B., Riazi, S., Koushanfar, F.: DeepSecure: scalable provably-secure deep learning. In: Proceedings of Design Automation Conference (DAC) (2018)

23. Song, Y., Ermon, S.: Generative modeling by estimating gradients of the data distribution. In: Advances in Neural Information Processing Systems, vol. 32 (2019)

24. Wang, G., Li, W., Aertsen, M., Deprest, J., Ourselin, S., Vercauteren, T.: Aleatoric uncertainty estimation with test-time augmentation for medical image segmentation with convolutional neural networks. Neurocomputing **338**, 34–45 (2019)

25. Wang, Z., Simoncelli, E.P., Bovik, A.C.: Multiscale structural similarity for image quality assessment. In: The Thirty-Seventh Asilomar Conference on Signals, Systems & Computers, vol. 2, pp. 1398–1402. IEEE (2003)

26. Xu, C., et al.: GANobfuscator: mitigating information leakage under GAN via differential privacy. IEEE TIFS **14**(9), 2358–2371 (2019)

27. Yang, Q., et al.: Federated machine learning: concept and applications. ACM Trans. Intell. Syst. Technol. (TIST) **10**(2), 12 (2019)

28. Zhang, H., Cisse, M., Dauphin, Y.N., Lopez-Paz, D.: Mixup: beyond empirical risk minimization. In: International Conference on Learning Representations (2018)

29. Zhou, Z., et al.: Generalizable medical image segmentation via random amplitude mixup and domain-specific image restoration. In: Avidan, S., Brostow, G., Cissé, M., Farinella, G.M., Hassner, T. (eds.) ECCV 2022. LNCS, pp. 420–436. Springer, Heidelberg (2022). https://doi.org/10.1007/978-3-031-19803-8_25

30. Zhou, Z., Rahman Siddiquee, M.M., Tajbakhsh, N., Liang, J.: UNet++: a nested U-Net architecture for medical image segmentation. In: Stoyanov, D., et al. (eds.) DLMIA/ML-CDS -2018. LNCS, vol. 11045, pp. 3–11. Springer, Cham (2018). https://doi.org/10.1007/978-3-030-00889-5_1

Rethinking Boundary Detection in Deep Learning Models for Medical Image Segmentation

Yi Lin, Dong Zhang, Xiao Fang, Yufan Chen, Kwang-Ting Cheng, and Hao Chen[✉]

The Hong Kong University of Science and Technology, Hong Kong, China
jhc@cse.ust.hk

Abstract. Medical image segmentation is a fundamental task in the community of medical image analysis. In this paper, a novel network architecture, referred to as Convolution, Transformer, and Operator (CTO), is proposed. CTO employs a combination of Convolutional Neural Networks (CNNs), Vision Transformer (ViT), and an explicit boundary detection operator to achieve high recognition accuracy while maintaining an optimal balance between accuracy and efficiency. The proposed CTO follows the standard encoder-decoder segmentation paradigm, where the encoder network incorporates a popular CNN backbone for capturing local semantic information, and a lightweight ViT assistant for integrating long-range dependencies. To enhance the learning capacity on boundary, a boundary-guided decoder network is proposed that uses a boundary mask obtained from a dedicated boundary detection operator as explicit supervision to guide the decoding learning process. The performance of the proposed method is evaluated on six challenging medical image segmentation datasets, demonstrating that CTO achieves state-of-the-art accuracy with a competitive model complexity.

Keywords: Medical Image Segmentation · CNNs · Vision Transformer · Boundary Detection · Network Architecture

1 Introduction

Medical Image Segmentation (MISeg) aims to locate pixel-level semantic lesion areas and/or human organs of the given image, which is one of the fundamental yet challenging tasks in the community of medical image analysis [27,33]. In the past few years, this task has been extensively studied and applied to a wide range of downstream applications, *e.g.*, robotic surgery [16], cancer diagnosis [29], and treatment design [38]. To achieve a desired MISeg result, it is critical to extract a set of rich and discriminative image feature representations.

Recently, thanks to the successful utilization of Vision Transformer (ViT) on computer vision tasks [12], ViT-based methods have greatly promoted the accuracy of medical image analysis [21]. For example, the state-of-the-art methods

---

Y. Lin and D. Zhang—Equal contribution.

for some medical image analysis tasks (*e.g.*, diagnosis [39], segmentation [8], and detection [35]) are based on the ViT framework [12]. Compared to CNNs-based methods, ViT has a stronger capacity to capture long-range dependencies, which have been shown to be beneficial for visual recognition [41]. For a canonical ViT-based MISeg model, it first partitions the input image into image patches. Then, these patches are treated as tokens for interactions via a multi-head self-attention layer, where the positional embedding is used for capturing the relative spatial information if needed. Finally, a normalization strategy and feature regulation operations are used to generate the output. The above processes are connected to form a basic transformer block, and such a block is repeated to encode semantic representations for the MISeg head network.

Despite that ViT-based methods have achieved preliminary success, they inherently suffer from two potential problems, *i.e.*, lack of translation invariance and weakness in local features [8]. To address these two problems, the advanced CNNs-ViT hybrid architectures were proposed for MISeg, *e.g.*, TransUNet [8], UNETR [21], Swin UNETR [20]. These attempts add convolutional operations in a ViT framework for local feature interactions, and strategies can improve the model convergence are also used. Particularly, the CNNs-ViT hybrid methods for MISeg are mainly based on UNet [33] and add transformer blocks in the backbone networks [21], and skip connections [23].

The explicit boundary also matters - although this information is usually overlooked in the deep learning era. Compared to the implicit learning manner (*e.g.*, CNNs, and ViT), an explicit learning model provides an immediate feature learning pattern, which has remarkable advantages of simple implementation, high efficiency, and purposeful objective. In the recent past, boundary operators are gradually valued in some pixel-level tasks and have been used to explicitly enhance the learning capacity on localization [10,13,28]. For MISeg, we believe that the boundary operator should play a more important role. Because, intuitively, a lesion region can be regarded as a kind of noise compared to normal regions. Besides, empirically, the explicit learning strategy can help the implicit feature learning model improve its representation capacity.

We propose a new network architecture, called CTO (Convolution, Transformer, and Operator), forMISeg that combines CNNs, ViT, and boundary detection operators to leverage both local semantic information and long-range dependencies in the learning process. CTO follows the canonical encoder-decoder segmentation paradigm, where the encoder network is composed of a CNNs backbone and an assistant lightweight ViT branch. To enhance boundary learning capacity, we introduce a boundary-guided decoder network that uses a self-generated boundary mask extracted by boundary detection operators as explicit supervisions to guide the decoding learning process. Our CTO architecture has higher recognition accuracy and achieves a better trade-off between accuracy and efficiency compared to the advanced MISeg architectures. We evaluate CTO on six representative yet challenging MISeg datasets, *i.e.*, two ISIC datasets [11,19], PH2 [31], CoNIC [17], LiTS17 [4], and BTCV [24]. Experimental results demonstrate that our CTO can achieve: 1) a new high accuracy on these datasets; 2) a superior performance to state-of-the-art methods; 3) and with competitive model complexity and efficiency.

2 Related Work

Medical Image Segmentation (MISeg). The existing MISeg methods can be roughly divided into the following three camps: i) CNNs-based methods; ii) ViT-based methods; and iii) CNNs-ViT hybrid methods. One of the most notable commonalities among these methods is that they are mainly based on an encoder-decoder paradigm. In the first camp, there are representative V-Net [32], U-Net [33], Attention-UNet [34]. These methods use CNNs as the backbone to extract image features, and combine some elaborate tricks (e.g., skip connection, multi-scale representation [7], feature interaction [6]) for feature enhancement. However, since convolution is inherently a local operation, methods in this camp may result in the problem of incomplete segmentation mask. In the second camp, there are Swin-UNet [5] and MissFormer [23]. Such methods use a ViT to replace CNNs as encoder/decoder to aggregate long-range feature dependencies. However, due to the limited number of medical images and the small inherent variability, such methods are difficult to optimize and have excessively high computational costs. In the third camp, there are TransUNet [8], UNETR [21] and Swin UNETR [20]. This type of method combines advantages of both CNNs and ViT, i.e., the model can capture not only local information but also long-range feature dependencies. However, an obvious disadvantage is that they are computationally intensive and have high computation complexity. In our work, we propose to use a lightweight ViT as an assistant to help the mainstream CNN capture long-range feature dependencies. Besides, a boundary-enhanced feature, which is generated by an explicit boundary detection operator, is used to guide the decoding learning process.

Operators in Image Processing. The operator is a fundamental component in traditional digital image processing, where the boundary detection operator is the most core element. The commonly used boundary detection operators can be divided into: i) the first derivative operator (e.g., Roberts, Prewitt, and Sobel), and ii) the second derivative operators (e.g., Laplacian) [25]. Recently, boundary detection operators have been revived in pixel-level computer vision tasks, such as manipulation detection [10] and camouflaged object detection [13]. In this paper, the boundary detection operator is used as an explicit mask extractor to guide an implicit feature learning model for MISeg. Our contribution is to use feature maps of the intermediate layer to synthesize a high-quality boundary prediction without requiring additional information.

3 Convolution, Transformer, and Operator (CTO)

3.1 Overview

The overall architecture of CTO is illustrated in Fig. 1. For an input image $X \in \mathbb{R}^{H \times W \times 3}$ with a spatial resolution of $H \times W$ and C channels, we aim to predict a pixel-wise labelmap Y, where each pixel has been assigned a class label. The whole model follows an encoder-decoder pattern, which also adopts

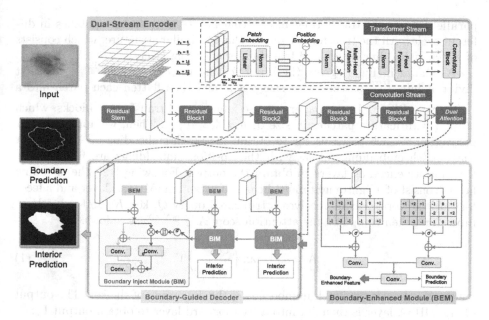

Fig. 1. Illustration of our CTO, which follows an encoder-decoder paradigm, where the encoder network consists of a mainstream CNNs and an assistant ViT. The decoder network employs a boundary detection operator to guide its learning process.

skip connections to aggregate low-level features from the encoder to the decoder. For the encoder, we design a dual-stream encoder (*ref.* Sect. 3.2), which combines a convolutional neural network (*i.e.*, Res2Net [15]) and a lightweight vision transformer to capture local feature dependencies and long-range feature dependencies between image patches, respectively. Such a combination will not bring many computational overheads. For the decoder, an operator-guided decoder (*ref.* Sect. 3.3) uses a boundary detection operator (*i.e.*, Sobel [25]) to guide the learning process via the generated boundary mask. The whole model is trained in an end-to-end manner.

3.2   Dual-Stream Encoder

The Mainstream Convolution Stream. The convolution stream is used to capture local feature dependencies. To this end, we choose the strong yet efficient Res2Net [15] as the backbone, which is composed of one convolution stem and four residual blocks, generating feature maps F_c^k with the spatial resolution of $H/4 \times W/4$, $H/8 \times W/8$, $H/16 \times W/16$, and $H/32 \times W/32$, respectively.

The Assistant Transformer Stream. The lightweight vision transformer (LightViT) is designed to capture the long-range feature dependencies between image patches in different scales. Specifically, the LightViT consists of multiple parallel lightweight transformer blocks that are fed with feature patches in different scales. All the transformer blocks share a similar structure, which consists of patch embedding layers and transformer encoding layers.

As shown in Fig. 1, given the input feature map $F_1^c \in \mathbb{R}^{\frac{H}{4} \times \frac{W}{4} \times C}$, we first divide it into $\frac{HW}{16p^2}$ patches with size $p \times p$, and then flatten each patch into a vector $\mathbf{v}_i \in \mathbb{R}^{p^2 \times C}$. In our paper, we use four parallel transformer blocks, which are fed with feature patches in size of $p = 4, 8, 16, 32$. Then, we apply a linear projection to each patch vector to obtain the patch embedding $e_i \in \mathbb{R}^C$. After that, patch embeddings along with the position embeddings are fed into the transformer encoding layers to obtain the output. Following [12], the encoding layers consist of a lightweight multi-head self-attention (MHSA) layer and a feed-forward network. MHSA receives a truncated query Q, key K, and value V as input, and then computes the attention score $A \in \mathbb{R}^{N \times N}$ as follows:

$$A = \text{softmax}\left(\frac{QK^T}{\sqrt{d_k}}\right) V, \tag{1}$$

where N is the size of patch number, d_k is the dimension of the key. The output of the MHSA layer is then fed into a feed-forward layer to obtain output F_t:

$$F_t = \text{FFN}(A), \tag{2}$$

where FFN is the feed-forward network with two linear layers with ReLU activation function. Then, F_t is reshaped into the same size as F_c^1 to obtain the output. Outputs of all the transformer blocks are concatenated along the channel dimension and fed into the convolutional layer to obtain the final output.

3.3 Boundary-Guided Decoder

The boundary-guided decoder uses a gradient operator module to extract the boundary information of foreground objects. Then, the boundary-enhanced feature F_b is integrated into multi-level encoder's features by a boundary optimization module, aiming to simultaneously characterize the intra- and inter-class consistency in the feature space, enriching the feature representative ability.

Boundary Enhanced Module (BEM). BEM takes the high-level F_c^4 and low-level features F_c^1 as inputs to extract the boundary information while filtering the trivial boundary irrelevant information. To achieve this goal, we apply Sobel operator [25] at both horizontal G_x and vertical G_y directions to obtain the gradient maps. Specifically, we utilize two 3×3 parameter-fixed convolutions and apply convolution operation with stride 1. Two convolutions are defined as:

$$K_x = \begin{bmatrix} -1 & 0 & 1 \\ -2 & 0 & 2 \\ -1 & 0 & 1 \end{bmatrix}, \quad K_y = \begin{bmatrix} -1 & -2 & -1 \\ 0 & 0 & 0 \\ 1 & 2 & 1 \end{bmatrix}. \tag{3}$$

Then, we apply the two convolutions to the input feature map to obtain the gradient maps M_x and M_y. After that, the gradient maps are normalized by a sigmoid function and then fused with the input feature map to obtain the edge-enhanced feature map F_e:

$$F_e = F_c \odot \sigma(M_{xy}), \qquad (4)$$

where $\odot$ denotes the element-wise multiplication, σ is the sigmoid function, and M_{xy} is the concatenation of M_x and M_y along the channel dimension. Then, we fuse the edge-enhanced feature maps of F_e^1 and F_e^4 with a simple stacked convolution layer in the bottleneck. Specifically, we first apply a 1×1 convolution with a bilinear upsampling operation to the feature map F_e^4 to obtain the feature map with the same size as F_e^1. Then, we separately apply 1×1 convolution operation to equate the channel size of these two features. Finally, we concatenate these two feature maps along the channel dimension and apply a two-layer convolutions to get the final feature map $\bar{F}_e$. The output is supervised by the ground truth boundary map, which in turn eliminates the edge feature inside the objects, producing the boundary-enhanced feature F_b.

Boundary Inject Module (BIM). The obtained boundary-enhanced feature from BEM can be used as a prior to improve the image representation ability of the features produced by the encoder. We propose BIM that introduces a dual path boundary fusion scheme to promote the feature representation in both foreground and background. Specifically, BIM takes two inputs: the channel-wise concatenation of the boundary-enhanced feature F_b and the corresponding feature F_c from the encoder network, and the feature from the previous decoder layer F_d^{j-1}. Then, these two inputs are fed into BIM, which contains two individual paths aiming to promote the feature representation in the foreground and background, respectively. For the foreground path, we directly concatenate the two inputs along the channel dimension, and then apply a sequential Conv-BN-ReLU (*i.e.*, convolution, batch normalization, ReLU activator) layers to obtain the foreground feature F_{fg}. For the background path, we design the background attention component to selectively focus on the background information, which is expressed as:

$$F_{bg} = \text{Convs}\left((1 - \sigma(F_d^{j-1})) \odot F_c\right), \qquad (5)$$

where Convs is a three-layer Conv-BN-ReLU layers, σ is the sigmoid function, and $\odot$ denotes the element-wise multiplication. The term $\left(1 - (\sigma F_d^{j-1})\right)$ is the background attention map, which is computed by first applying the sigmoid function to the feature map from the previous decoder layer, which will generate a foreground attention map. Then, we subtract the foreground attention map from 1 to obtain the background attention map. Finally, we concatenate the foreground feature F_{fg}, the background feature F_{bg}, and the previous decoder feature F_d^{j-1} along the channel dimension to obtain the final output F_d^j.

3.4   Overall Loss Function

Since the proposed CTO is a multi-task model (*i.e.*, interior and boundary segmentation), we define an overall loss function to jointly optimize these two tasks.

Interior Segmentation Loss. The interior segmentation loss is the weighted sum of cross-entropy loss $\mathcal{L}_{\text{CE}}$ and mean intersection-over-union (mIoU) loss $\mathcal{L}_{\text{mIoU}}$, which are defined as:

$$\mathcal{L}_{\text{CE}} = -\frac{1}{N}\sum_{i=1}^{N}\left(y_i \log(\hat{y}_i) + (1 - y_i)\log(1 - \hat{y}_i)\right), \tag{6}$$

$$\mathcal{L}_{\text{mIoU}} = 1 - \frac{\sum_{i=1}^{N}(y_i * \hat{y}_i)}{\sum_{i=1}^{N}(y_i + \hat{y}_i - y_i * \hat{y}_i)}, \tag{7}$$

where y_i and $\hat{y}_i$ are the ground truth and the predicted label for the i-th pixel, respectively, and N is the total number of pixels of the image.

Boundary Loss. Considering the class imbalance problem between the foreground and background pixels in boundary detection, we employ the Dice Loss:

$$\mathcal{L}_{\text{Dice}} = 1 - \frac{2\sum_{i=1}^{N}(y_i * \hat{y}_i)}{\sum_{i=1}^{N}(y_i + \hat{y}_i)}. \tag{8}$$

Total Loss. The total loss is composed of the major segmentation loss $\mathcal{L}_{\text{seg}}$, and the boundary loss $\mathcal{L}_{\text{bnd}}$. Note that for the boundary detection loss, we only consider the prediction from BEM, which takes encoder's feature maps from the high-level layer (*i.e.*, F_b^4) and low-level layer (*i.e.*, F_b^1) as input. As for the major image segmentation loss, we apply the deep supervision strategy to obtain the prediction from the decoder's feature at different levels. In summary, the total loss can be formulated as:

$$\mathcal{L} = \mathcal{L}_{\text{seg}} + \mathcal{L}_{\text{bnd}} = \sum_{i}^{L}(\mathcal{L}_{\text{CE}} + \mathcal{L}_{\text{mIoU}}) + \alpha \mathcal{L}_{\text{Dice}}, \tag{9}$$

where L is the number of BOMs, which is set to 3 in this work. α is the weighting factor, which is set to 3 to balance the losses.

4   Experiments

4.1   Datasets and Evaluation Metrics

Datasets. We evaluate our CTO on six public MISeg datasets, including three datasets for skin lesion segmentation, *i.e.*, ISIC [11,19] and PH2 [31], the Colon Nuclei Identification and Counting (CoNIC) challenge dataset [17], the Liver

Table 1. Comparisons with other methods on ISIC [11,19] & PH2 [31].

Methods	ISIC 2016 & PH2		Methods	ISIC 2018	
	Dice ↑	IoU ↑		Dice ↑	IoU ↑
SSLS [1]	78.38	68.16	Deeplabv3 [9]	88.4	80.6
MSCA [2]	81.57	72.33	U-Net++ [42]	87.9	80.5
FCN [30]	89.40	82.15	CE-Net [18]	89.1	81.6
Bi *et al* [3]	90.66	83.99	MedT [36]	85.9	77.8
Lee *et al* [26]	91.84	84.30	TransUNet [8]	89.4	82.2
CTO(Ours)	**91.89**	**85.18**	Ours	**91.2**	**84.5**

Tumor Segmentation (LiTS17) Challenge dataset [4], and the Beyond the Cranial Vault (BTCV) challenge dataset [37]. As in [11,19], we perform 5-fold cross-validation on ISIC 2018, and train the model on ISIC 2016 and test it on PH2 [31]. BTCV is divided into 18 cases for training and 12 cases for test [5,8]. CoNIC and LiTS17 are randomly divided into training, validation, and test sets with a radio of 7:1:2. **Evaluation Metrics.** Following [5,8,26], the commonly used Dice Coefficient (Dice), Intersection over Union (IoU), average Hausdorff Distance (HD) and Panoptic Quality (PQ) are used as the primary accuracy evaluation metrics. Besides, FLOPs and model parameters are used to evaluate the model efficiency.

4.2   Implementation Details

We optimize our model using the ADAM optimizer with an initial learning rate 1e-4. The default batch size is set to 32 with the image size of 256×256. The encoder is initialized with the pre-trained weights of Res2Net-50 [15] on ImageNet and then fine-tuned for 90 epochs on a single NVIDIA RTX 3090 GPU. All 3D volumes are inferenced in a sliding-window manner with the stride of 1, and the final segmentation results are obtained by stacking the prediction maps to reconstruct the 3D volume for evaluation. Except for a special statement, all the experimental settings follow the baseline paper [5,8,26,40].

4.3   Experimental Results

Comparisons with State-of-the-Art Methods. We compare our CTO with the state-of-the-art (SOTA) methods including U-Net [33], ResUNet [33] with ResNet-50 [22] as the backbone, VNet [32], ViT [12], TransUNet [8], and Swin-Unet [5]. On ISIC 2016 [19] & PH2 [31], we compare CTO with five related methods. The results are shown in Table 1. We can observe that CTO achieves 91.89% in Dice and 85.18% in IoU, which outperforms the SOTA methods by 0.05% and 0.88%, respectively. On ISIC 2018 [11], our CTO achieves 91.2% in

Table 2. Comparisons with other methods on CoNIC [17] and LiTS17 [4].

Methods	CoNIC			LiTS17		Model Efficiency	
	Dice ↑	IoU ↑	PQ ↑	Dice ↑	IoU ↑	Param.(M)	GFLOPs
V-Net [32]	77.46	64.94	63.59	89.20	80.71	11.84	18.54
U-Net [33]	78.42	66.39	64.44	84.66	73.63	7.78	14.59
R50-UNet [33]	77.67	65.34	63.67	91.24	84.14	33.69	20.87
Att-UNet [34]	79.48	66.06	65.25	85.88	75.40	7.88	43.35
R50-AttUNet [34]	78.21	65.86	64.02	89.98	82.13	33.25	49.25
R50-ViT [12]	75.36	62.35	58.03	83.67	72.49	110.62	26.91
UNETR [21]	71.46	57.24	52.26	81.48	69.04	87.51	26.41
Swin-UNETR [20]	70.07	55.56	51.59	84.00	72.76	6.29	4.86
CTO(Ours)	**79.77**	**66.42**	**65.58**	**91.50**	**84.59**	59.82	22.72

Table 3. Comparisons with other methods on BTCV [24].

Methods	mDice↑	HD↓	Aorta	Gallb.	Kid(L)	Kid(R)	Liver	Panc.	Spleen	Stom.
V-Net [32]	68.81	-	75.34	51.87	77.10	80.75	87.84	40.05	80.56	56.98
DARR [14]	69.77	-	74.74	53.77	72.31	73.24	94.08	54.18	89.90	45.96
U-Net [33]	76.85	39.70	89.07	**69.72**	77.77	68.60	93.43	53.98	86.67	75.58
R50-UNet [33]	74.68	36.87	84.18	62.84	79.19	71.29	93.35	48.23	84.41	73.92
Att-UNet [34]	77.77	36.02	**89.55**	68.88	77.98	71.11	93.57	58.04	87.30	75.75
R50-AttUNet [34]	75.57	36.97	55.92	63.91	79.20	72.71	93.56	49.37	87.19	74.95
R50-ViT [12]	71.29	32.87	73.73	55.13	75.80	72.20	91.51	45.99	81.99	73.95
TransUNet [8]	77.48	31.69	87.23	63.13	81.87	77.02	94.08	55.86	85.08	75.62
SwinUNet [5]	79.12	21.55	85.47	66.53	83.28	79.61	94.29	56.58	**90.66**	76.60
CTO(Ours)	**81.10**	**18.75**	87.72	66.44	**84.49**	**81.77**	**94.88**	**62.74**	90.60	**80.20**

Dice and 84.5% in IoU by 5-fold cross-validation, which outperforms the SOTA methods by 1.8% and 2.3%, respectively. On CoNIC [17], results are shown in Table 2, we can observe that our CTO achieves 79.77%, 66.42%, and 65.58% in Dice, IoU, and PQ, respectively, consistently outperforming other methods. Qualitative result comparisons are illustrated in Fig. 2. We can observe that our CTO delineates more accurate object contours than other methods regarding diverse shapes and sizes of nuclei, especially on some blurred nuclei objects.

We also conduct experiments on 3D MISeg tasks. On LiTS17 [4], as shown in Table 2, our model achieves 91.50% in Dice and 84.59% in IoU, outperforming SOTA methods by 0.26% and 0.45%, respectively. On BTCV [24], as shown in Table 3, our CTO achieves 81.10% in Dice and 18.75% in HD, which outperforms the SOTA methods. In particular, as for Dice, our CTO outperforms the second, third, and fourth best methods by 1.98%, 3.33%, and 3.62, respectively. Besides, the distinct improvements can be markedly observed for organs with blurry boundaries, e.g., the "pancreas" and the "stomach", where our model achieves significant gains over the SOTA methods, i.e., 4.70% and 3.60% in Dice, respec-

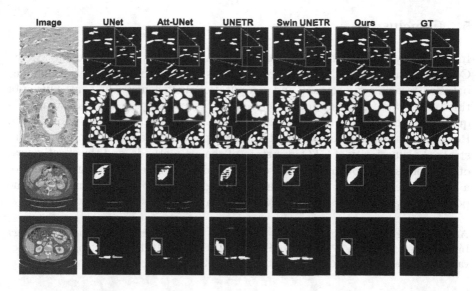

| Image | UNet | Att-UNet | UNETR | Swin UNETR | Ours | GT |

Fig. 2. Visualizations on CoNIC [17] (top two rows) and LiTS17 [4] (bottom two rows). The red boxes highlight the main difference of each method. (Color figure online)

tively. As for the model efficiency, we can observe that CTO achieves competitive performance improvements with comparable FLOPs and parameters.

Table 4. Ablation study results on ISIC 2018 [11]. * means the component achieves significant performance improvement with $p < 0.05$ via paired t-test.

CNNs	LightViT	CBM	BEM	BIM	Dice ↑	IoU ↑
✓					88.32	81.51
✓	✓				$89.31^*_{+0.99}$	$82.47^*_{+0.96}$
✓	✓	✓			$89.41^*_{+1.09}$	$82.51_{+1.00}$
✓	✓	✓	✓		$89.52_{+1.20}$	$82.81_{+1.30}$
✓	✓	✓	✓	✓	$91.21^*_{+2.89}$	$84.45^*_{+2.94}$

Ablation Study. We conduct ablation studies to explore the effectiveness of each component in CTO. In Table 4, we compare the performance of CTO variants on ISIC 2018 [11]: 1) CNNs, only the convolution stream; 2) +LightViT, the dual-stream encoder with convolution and Transformer; 3) +CBM, adding the boundary supervision with the same architecture of BEM, except the Sobel layer; 4) +BEM, the boundary-enhanced module; 5) +BIM, the boundary inject module. All the components consistently boost the performance by 0.99%, 1.09%, 1.20%, 2.89% in Dice, respectively. Especially, we observe that the boundary supervision (*i.e.*, BIM) is crucial for MISeg.

5   Conclusion

In this study, a new network architecture named CTO is proposed for MISeg. Compared to advanced MISeg architectures, CTO achieves a better balance between recognition accuracy and computational efficiency. The contribution of this paper is the utilization of intermediate feature maps to synthesize a high-quality boundary supervision mask without requiring additional information. Results from experiments conducted on six publicly available datasets demonstrate the superiority of CTO over state-of-the-art methods, and the effectiveness of each of its components. Future work includes the extension of the concept of a couple-stream encoder to various advanced backbone architectures, and the potential adaptation of CTO to a 3D manner.

Acknowledgement. This work was supported by Shenzhen Science and Technology Innovation Committee (Project No. SGDX20210823103201011) and Hong Kong Innovation and Technology Fund (Project No. ITS/028/21FP).

References

1. Ahn, E., et al.: Automated saliency-based lesion segmentation in dermoscopic images. In: International Conference of the IEEE Engineering in Medicine and Biology Society (EMBC) (2015)
2. Bi, L., Kim, J., Ahn, E., Feng, D., Fulham, M.: Automated skin lesion segmentation via image-wise supervised learning and multi-scale superpixel based cellular automata. In: IEEE International Symposium on Biomedical Imaging (ISBI) (2016)
3. Bi, L., Kim, J., Ahn, E., Kumar, A., Fulham, M., Feng, D.: Dermoscopic image segmentation via multistage fully convolutional networks. IEEE Trans. Biomed. Eng. **64**(9), 2065–2074 (2017)
4. Bilic, P., et al.: The liver tumor segmentation benchmark (LiTS). Med. Image Anal. **84**, 102680 (2023)
5. Cao, H., et al.: Swin-UNet: UNet-like pure transformer for medical image segmentation. arXiv (2021)
6. Chen, H., Dou, Q., Yu, L., Qin, J., Heng, P.A.: VoxresNet: deep voxelwise residual networks for brain segmentation from 3D MR images. Neuroimage **170**, 446–455 (2018)
7. Chen, H., Qi, X., Yu, L., Heng, P.A.: DCAN: deep contour-aware networks for accurate gland segmentation. In: Proceedings of the IEEE Conference on Computer Vision and Pattern Recognition (CVPR) (2016)
8. Chen, J., et al.: TransUnet: transformers make strong encoders for medical image segmentation. arXiv (2021)
9. Chen, L.C., Papandreou, G., Schroff, F., Adam, H.: Rethinking atrous convolution for semantic image segmentation. arXiv (2017)
10. Chen, X., Dong, C., Ji, J., Cao, J., Li, X.: Image manipulation detection by multi-view multi-scale supervision. In: Proceedings of the IEEE/CVF International Conference on Computer Vision (ICCV) (2021)
11. Codella, N., et al.: Skin lesion analysis toward melanoma detection 2018: a challenge hosted by the international skin imaging collaboration (ISIC). arXiv (2019)

12. Dosovitskiy, A., et al.: An image is worth 16×16 words: transformers for image recognition at scale. In: International Conference on Learning Representations (ICLR) (2020)

13. Fan, D.P., Ji, G.P., Sun, G., Cheng, M.M., Shen, J., Shao, L.: Camouflaged object detection. In: Proceedings of the IEEE Conference on Computer Vision and Pattern Recognition (CVPR) (2020)

14. Fu, S., et al.: Domain adaptive relational reasoning for 3D multi-organ segmentation. In: Martel, A.L., et al. (eds.) MICCAI 2020. LNCS, vol. 12261, pp. 656–666. Springer, Cham (2020). https://doi.org/10.1007/978-3-030-59710-8_64

15. Gao, S.H., Cheng, M.M., Zhao, K., Zhang, X.Y., Yang, M.H., Torr, P.: Res2Net: a new multi-scale backbone architecture. IEEE Trans. Pattern Anal. Mach. Intell. **43**(2), 652–662 (2019)

16. Gao, X., Jin, Y., Zhao, Z., Dou, Q., Heng, P.-A.: Future frame prediction for robot-assisted surgery. In: Feragen, A., Sommer, S., Schnabel, J., Nielsen, M. (eds.) IPMI 2021. LNCS, vol. 12729, pp. 533–544. Springer, Cham (2021). https://doi.org/10.1007/978-3-030-78191-0_41

17. Graham, S., et al.: CoNIC: colon nuclei identification and counting challenge 2022. arXiv (2021)

18. Gu, Z., et al.: CE-NET: context encoder network for 2D medical image segmentation. IEEE Trans. Med. Imaging **38**(10), 2281–2292 (2019)

19. Gutman, D., et al.: Skin lesion analysis toward melanoma detection: a challenge at the international symposium on biomedical imaging (ISBI) 2016, hosted by the international skin imaging collaboration (ISIC). arXiv (2016)

20. Hatamizadeh, A., Nath, V., Tang, Y., Yang, D., Roth, H.R., Xu, D.: Swin UNETR: swin transformers for semantic segmentation of brain tumors in MRI images. In: Crimi, A., Bakas, S. (eds.) BrainLes 2021. LNCS, vol. 12962, pp. 272–284. Springer, Cham (2022). https://doi.org/10.1007/978-3-031-08999-2_22

21. Hatamizadeh, A., et al.: UNETR: transformers for 3D medical image segmentation. In: Proceedings of the IEEE/CVF Winter Conference on Applications of Computer Vision (WACV) (2022)

22. He, K., Zhang, X., Ren, S., Sun, J.: Deep residual learning for image recognition. In: Proceedings of the IEEE Conference on Computer Vision and Pattern Recognition (CVPR) (2016)

23. Huang, X., Deng, Z., Li, D., Yuan, X.: MISSFormer: an effective medical image segmentation transformer. arXiv (2021)

24. Irshad, S., Gomes, D.P., Kim, S.T.: Improved abdominal multi-organ segmentation via 3D boundary-constrained deep neural networks. arXiv (2022)

25. Kanopoulos, N., Vasanthavada, N., Baker, R.L.: Design of an image edge detection filter using the Sobel operator. IEEE J. Solid-State Circ. **23**(2), 358–367 (1988)

26. Lee, H.J., Kim, J.U., Lee, S., Kim, H.G., Ro, Y.M.: Structure boundary preserving segmentation for medical image with ambiguous boundary. In: Proceedings of the IEEE Conference on Computer Vision and Pattern Recognition (CVPR) (2020)

27. Lin, Y., Liu, L., Ma, K., Zheng, Y.: Seg4Reg+: consistency learning between spine segmentation and cobb angle regression. In: de Bruijne, M., et al. (eds.) MICCAI 2021. LNCS, vol. 12905, pp. 490–499. Springer, Cham (2021). https://doi.org/10.1007/978-3-030-87240-3_47

28. Lin, Y., et al: Label propagation for annotation-efficient nuclei segmentation from pathology images. arXiv preprint arXiv:2202.08195 (2022)

29. Lin, Y., et al.: Automated pulmonary embolism detection from CTPA images using an end-to-end convolutional neural network. In: Shen, D., et al. (eds.) MICCAI

2019. LNCS, vol. 11767, pp. 280–288. Springer, Cham (2019). https://doi.org/10.1007/978-3-030-32251-9_31

30. Long, J., Shelhamer, E., Darrell, T.: Fully convolutional networks for semantic segmentation. In: Proceedings of the IEEE Conference on Computer Vision and Pattern Recognition (CVPR) (2015)

31. Mendonça, T., Ferreira, P.M., Marques, J.S., Marcal, A.R., Rozeira, J.: PH2-a dermoscopic image database for research and benchmarking. In: EMBC (2013)

32. Milletari, F., Navab, N., Ahmadi, S.A.: V-Net: fully convolutional neural networks for volumetric medical image segmentation. In: 3DV. IEEE (2016)

33. Ronneberger, O., Fischer, P., Brox, T.: U-Net: convolutional networks for biomedical image segmentation. In: Navab, N., Hornegger, J., Wells, W.M., Frangi, A.F. (eds.) MICCAI 2015. LNCS, vol. 9351, pp. 234–241. Springer, Cham (2015). https://doi.org/10.1007/978-3-319-24574-4_28

34. Schlemper, J., et al.: Attention gated networks: learning to leverage salient regions in medical images. Med. Image Anal. **53**, 197–207 (2019)

35. Shamshad, F., et al.: Transformers in medical imaging: a survey. arXiv (2022)

36. Valanarasu, J.M.J., Oza, P., Hacihaliloglu, I., Patel, V.M.: Medical transformer: gated axial-attention for medical image segmentation. arXiv (2021)

37. Vaswani, A., et al.: Attention is all you need. In: Advances in Neural Information Processing Systems (NeurIPS) (2017)

38. Wijeratne, P.A., Alexander, D.C., for the Alzheimer's Disease Neuroimaging Initiative: Learning transition times in event sequences: the temporal event-based model of disease progression. In: Feragen, A., Sommer, S., Schnabel, J., Nielsen, M. (eds.) IPMI 2021. LNCS, vol. 12729, pp. 583–595. Springer, Cham (2021). https://doi.org/10.1007/978-3-030-78191-0_45

39. Wu, J., et al.: SeATrans: learning segmentation-assisted diagnosis model via transformer. In: Wang, L., Dou, Q., Fletcher, P.T., Speidel, S., Li, S. (eds.) MICCAI 2022. LNCS, vol. 13432, pp. 677–687. Springer, Cham (2022). https://doi.org/10.1007/978-3-031-16434-7_65

40. Zhang, D., et al.: Deep learning for medical image segmentation: tricks, challenges and future directions. arXiv (2022)

41. Zhang, D., Tang, J., Cheng, K.T.: Graph reasoning transformer for image parsing. In: ACM MM (2022)

42. Zhou, Z., Rahman Siddiquee, M.M., Tajbakhsh, N., Liang, J.: UNet++: a nested U-Net architecture for medical image segmentation. In: Stoyanov, D., et al. (eds.) DLMIA/ML-CDS 2018. LNCS, vol. 11045, pp. 3–11. Springer, Cham (2018). https://doi.org/10.1007/978-3-030-00889-5_1

Token Sparsification for Faster Medical Image Segmentation

Lei Zhou[1]([✉]), Huidong Liu[1,3], Joseph Bae[2], Junjun He[4], Dimitris Samaras[1],
and Prateek Prasanna[2]

[1] Department of Computer Science, Stony Brook University, Stony Brook, NY, USA
lezzhou@cs.stonybrook.edu
[2] Department of Biomedical Informatics, Stony Brook University,
Stony Brook, NY, USA
[3] Amazon, Seattle, WA, USA
[4] Shanghai Artificial Intelligence Laboratory, Shanghai, China

Abstract. *Can we use sparse tokens for dense prediction, e.g., seg-mentation?* Although token sparsification has been applied to Vision Transformers (ViT) to accelerate classification, it is still unknown how to perform segmentation from sparse tokens. To this end, we reformu-late segmentation as a *sparse encoding → token completion → dense decoding* (SCD) pipeline. We first empirically show that naïvely applying existing approaches from classification token pruning and masked image modeling (MIM) leads to failure and inefficient training caused by inap-propriate sampling algorithms and the low quality of the restored dense features. In this paper, we propose *Soft-topK Token Pruning (STP)* and *Multi-layer Token Assembly (MTA)* to address these problems. In *sparse encoding*, STP predicts token importance scores with a lightweight sub-network and samples the topK tokens. The intractable topK gradients are approximated through a continuous perturbed score distribution. In *token completion*, MTA restores a full token sequence by assembling both sparse output tokens and pruned multi-layer intermediate ones. The last *dense decoding* stage is compatible with existing segmentation decoders, e.g., UNETR. Experiments show SCD pipelines equipped with *STP* and *MTA* are much faster than baselines without token pruning in both train-ing (up to 120% higher throughput) and inference (up to 60.6% higher throughput) while maintaining segmentation quality. Code is available here: https://github.com/cvlab-stonybrook/TokenSparse-for-MedSeg.

Keywords: Token Pruning · Multi-layer Token Assembly · Medical Image Segmentation

1 Introduction

Vision Transformers (ViT) [6] for dense prediction [20,29] have achieved impres-sive results in tasks including medical image segmentation [8]. In general, high-resolution features [26] preserving details are always desirable for precise seg-

A. Frangi et al. (Eds.): IPMI 2023, LNCS 13939, pp. 743–754, 2023.
https://doi.org/10.1007/978-3-031-34048-2_57

mentation. However, because of the quadratic computation complexity in self-attention [25], doubling the resolution per dimension in a 3D volume can lead to an 8× longer sequence and hence 64× more computation. This growing computing burden can quickly surpass limited computation budgets. Considering ViT's flexibility and great potential in masked image modeling [9,14], we explore acceleration algorithms based on the standard ViT. Recently, token sparsification [15,16,21] has been proposed to accelerate inference in ViT for classification by dropping less important tokens. However, *to the best of our knowledge, there are no ViT token sparsification approaches for segmentation.* This leads us to ask the question: *Can we use sparse tokens for dense prediction, e.g., segmentation?*

To answer the question, we reformulate segmentation as a *sparse encoding* → *token completion* → *dense decoding* (SCD) pipeline. Unlike a standard *dense encoding* → *dense decoding* (DD) pipeline, *sparse encoding* and *token completion* are required in SCD. *Sparse encoding* requires learning a sparse token representation for speed and *token completion* is needed to restore the full set of tokens for dense prediction. We first examine a naïve realization of *sparse encoding* and *token completion* by applying existing approaches. Specifically, we adapt sampling methods in classification, e.g., EViT [15] and DynamicViT [21], to *sparse encoding*, and masked image modeling (MIM) [2,9] to *token completion*. However, we observe significantly inferior results in this SCD pipeline (See Table 1). Next, we provide more insight into the problems of existing methods.

Problems in *Sparse Encoding*. There are two steps in this step, i.e., token score estimation and token sampling. We show that EViT's token score estimation is inappropriate for segmentation and DynamicViT's token sampling leads to training inefficiency: *i)* **EViT** [15] uses the attention weights between spatial tokens and the [CLS] token to estimate scores. While this is sound for classification since [CLS] is used for prediction, it is sub-optimal for segmentation because [CLS] is deprecated in the segmentation decoder. *ii)* **DynamicViT** [21] estimates token scores with a sub-network. DynamicViT frames token sampling as a series of independent binary decisions to keep or drop tokens. This does not guarantee a fixed number of sampled tokens for each training input. To fit in batch training, DynamicViT keeps all tokens in memory and masks self-attention entries, leading to training inefficiency.

Problems in *Token Completion*. Previous sparse token classification models [15,21] do not require *token completion*. Thus, we borrow the design from MIM. MIM reconstructs full tokens from a partial token sequence by padding it to full length with learnable mask tokens and then hallucinating the masked regions from their context. While MIM is useful for pre-training, it cannot accurately restore detailed information, resulting in inferior segmentation results.

We propose *Soft-topK Token Pruning (STP)* and *Multi-layer Token Assembly (MTA)* to implement *sparse encoding* and *token completion*. *i)* In *sparse encoding*, STP predicts token importance scores with a sub-network, avoiding the limitation of [CLS] in segmentation. STP then samples topK-scored tokens instead of making binary decisions per token separately, accelerating training by retaining only the sampled tokens in memory and computing. Motivated by subset sampling [5, 12, 28], the intractable gradients of the topK operation are approximated through a perturbed continuous score distribution. *ii)* In *token completion*, the *MTA* restores a full token sequence by assembling both sparse output tokens and pruned intermediate tokens from multiple layers. Compared to MIM that fills the pruned positions with identical mask tokens, *MTA* produces more informative, position-specific representations. For *dense decoding*, the SCD pipeline is compatible with existing segmentation decoders, such as UNETR.

We evaluate our method on two relatively sparse 3D medical image segmentation datasets, the CT Abdomen Multi-organ Segmentation (BTCV [11], $N = 30$) dataset and the MRI Brain Tumor Segmentation (MSD BraTS [1], $N = 484$) dataset. On both tasks, STP+MTA+UNETR matches the UNETR baseline while providing significant computing savings with large token pruning ratios. On BraTS, STP+MTA+UNETR accelerates segmentation inference-/training throughput by 60.6%/120% and achieves the same segmentation accuracy. On BTCV, STP+MTA+UNETR increases inference/training throughput by 24.1%/97.36% while maintaining performance. In summary, our contributions are:

- To the best of our knowledge, we are the first to use token pruning/dropping for ViT-based medical image segmentation.
- Based on subset sampling, our proposed *Soft-topK Token Pruning (STP)* module can be flexibly incorporated into a standard ViT to prune tokens with greater efficiency while maintaining accuracy.
- We propose *Multi-layer Token Assembly (MTA)* to recover a full set of tokens, i.e., a dense representation, from a sparse set. *MTA* preserves high-detail information for accurate segmentation.
- We show that STP+MTA+UNETR maintains performance compared with UNETR with much less computation on two 3D medical image datasets.

2   Methodology

Generally, a segmentation model consists of an encoder and a decoder. Our goal is to accelerate the ViT segmentation encoder. To this end, we reformulate segmentation as a *sparse encoding* → *token completion* → *dense decoding* (SCD) pipeline. *Sparse encoding* learns a sparse token representation for acceleration; *token completion* restores the full tokens for dense prediction; *dense decoding* predicts the segmentation mask from dense features. We first recap Vision Transformers and then illustrate the three components in the SCD pipeline.

Preliminary: Vision Transformers. Vision Transformers treat an image/volume as a sequence of tokens. In the case of 3D medical images, a 3D volume $\mathbf{x} \in \mathbb{R}^{H \times W \times D \times C_{in}}$ is first reshaped to a sequence of flattened patches $\mathbf{x}_p \in \mathbb{R}^{N \times (P^3 \times C_{in})}$ where $H \times W \times D$ is the spatial size, C_{in} is the input channel, $P \times P \times P$ is the patch size, and $N = HWD/P^3$ is the sequence length, i.e., the number of patches. All the patches are then projected linearly to a C-dimensional token space, with position embeddings added to the projected patches. These

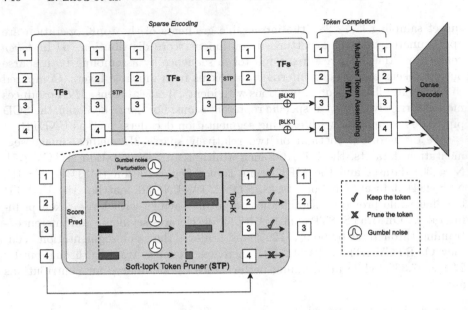

Fig. 1. Sparse Token Segmentation Pipeline. We reformulate segmentation as a *sparse encoding* → *token completion* → *dense decoding* pipeline. In *sparse encoding*, we design a *Soft-topK Token Pruning (STP)* module. In the forward pass, *STP* performs topK sampling on perturbed scores. In the backward pass, *STP* approximates the intractable gradient with a continuous Gumbel Softmax estimation. In *token completion*, we propose *Multi-layer Token Assembly (MTA)* to assemble both the output sparse tokens and the pruned intermediate ones to restore the complete tokens. In *dense decoding*, we avoid the intermediate sparse tokens by taking all inputs from the output of *MTA*. In this simplified figure, we visualize token pruning as dropping the last token. However, in practice pruned tokens are selected according to predicted scores.

patch tokens, together with a learnable prepended [CLS] token, are denoted as $\mathbf{z}_0 \in \mathbb{R}^{(1+N) \times C}$. $\mathbf{z}_0$ are further processed by L Transformer blocks sequentially. Each block consists of a multi-head self-attention (MSA) module and an MLP. We denote the tokens output from the ith Transformer block as $\mathbf{z}_i \in \mathbb{R}^{(1+N) \times C}$. For the segmentation task, before feeding the output $\mathbf{z}_L$ of the encoder to the decoder, we drop the [CLS] token and project the non-[CLS] token sequence $\mathbf{z}_L^{[1:N]} \in \mathbb{R}^{N \times C}$ back to the original 3D feature map $\mathbf{x}_L \in \mathbb{R}^{H/P \times W/P \times D/P \times C}$.

2.1 Sparse Encoding: Soft-topK Token Pruning (STP)

We build our sparse encoder on a ViT without modifying the self-attention module. Instead, we propose a learnable plug-and-play *Soft-topK Token Pruning (STP)* module. Compared to EViT & DynamicViT, our *STP*, as shown in the lower half of Fig. 1, estimates token scores more effectively and can be trained efficiently. *STP* can be inserted between two Transformer blocks TF_i and TF_{i+1}. Receiving as input the token sequence $\mathbf{z}_i \in \mathbb{R}^{N_i \times C}$ from TF_i, *STP* prunes tokens with a ratio r and passes the remaining tokens $\mathbf{z}'_i \in \mathbb{R}^{\lfloor (1-r)N_i \rfloor \times C}$ to TF_{i+1}. In particular, STP consists of token-wise score estimation and token sampling. To be concise, we change the notation of number of tokens from N_i to n.

Token Score Estimation. To decide which tokens to keep or prune, we introduce a lightweight sub-network $s_\theta : \mathbb{R}^{n \times C} \to \mathbb{R}^n$ to predict the token importance scores $\mathbf{s}$, where θ are the network parameters. The architecture of s_θ is designed to aggregate both the local and global features, similarly to [21]. The global feature is simply obtained by average pooling over all the tokens.

$$\mathbf{s} = s_\theta(\mathbf{z}) = \texttt{Sigmoid}\Big(\texttt{MLP}_2\big([\mathbf{z}, \texttt{AvgPool}(\texttt{MLP}_1(\mathbf{z}))]\big)\Big) \tag{1}$$

Straight-Through Gumbel Soft TopK Sampling. Given a token pruning ratio r, STP needs to select $K = \lfloor (1-r)n \rceil$ tokens out of n to keep. After predicting the scores $\mathbf{s}$, we re-interpret each score value $\mathbf{s}_i$ as the probability of the i-th token ranking in the topK. We formulate this process as sampling a binary policy mask $\mathbf{M} \in \{0, 1\}^n$ from the predicted probabilities where $\mathbf{M}$ is subject to $\texttt{sum}(\mathbf{M}) = K$. $\mathbf{M}_i = 1$ indicates keeping the i-th token while $\mathbf{M}_i = 0$ indicates pruning. However, such discrete sampling is non-differentiable. To overcome the problem, we relax the sampling of discrete topK masks to a continuous approximation, the Gumbel-Softmax distribution:

$$\underbrace{\mathbf{M}_i = \mathbb{1}_{\text{topK}}(\log(s_i) + g_i)}_{\text{forward}} \xleftarrow{\text{approx}} \underbrace{\tilde{\mathbf{M}}_i = \frac{\exp((\log(s_i) + g_i)/\tau)}{\sum_{j=1}^n \exp((\log(s_j) + g_j)/\tau)}}_{\text{backward}} \tag{2}$$

where $\mathbb{1}_{\text{topK}}$ is an indicator function of whether the input perturbed score is among the topK of all n perturbed scores, $\{g\}_n$ are i.i.d samples from the Gumbel$(0, 1)$ distribution[1]. While training, we forward STP to sample the topK tokens based on the discrete $\mathbf{M}$ but backward with the gradient approximated by the continuous $\tilde{\mathbf{M}}$. We call this Straight-through (ST) Gumbel Soft TopK Sampling. During inference, we perform normal topK selection based on predicted scores without Gumbel noise perturbation for deterministic inference.

2.2   Token Completion: Multi-layer Token Assembly (MTA)

The output of the STP-ViT encoder is sparse. Thus, before passing the output to the decoder, we need to first restore the complete tokens. A straightforward solution can be obtained from Masked Image Modeling (MIM) [2,9]. MIM reconstructs an image from random partial image patches. It first pads the sparse token set with learnable [MASK] tokens up to its full length. Then the padded tokens are forwarded through Transformer blocks to reconstruct the masked regions. However, MIM is mostly utilized for pre-training which focuses more on semantic hallucination rather than accurate detail restoration. Thus,

---

[1] Gumbel$(0, 1)$ samples are drawn by sampling $-\log(-\log u)$ where $u \sim \text{Uniform}(0, 1)$.

it is sub-optimal for segmentation tasks that require assigning labels to pixels accurately.

We propose *Multi-layer Token Assembly (MTA)* to restore dense features by assembling both the outputted sparse tokens and the pruned intermediate tokens from multiple layers. Suppose we insert three STPs, $\{STP_1, STP_2, STP_3\}$, after different Transformer blocks in a ViT. We denote the token sets pruned by the three STPs as $\{\bar{z}_1, \bar{z}_2, \bar{z}_3\}$. We concatenate these pruned tokens with the final output z_L and rearrange them to their original spatial order. Then, we add three learnable block tokens $\{[\text{BLK}_1], [\text{BLK}_2], [\text{BLK}_3]\}$ to the corresponding pruned tokens to indicate which block each token is pruned from. Finally, we introduce sin-cos position embeddings $\mathbf{E}_{pos}$ to all the tokens and forward them through Transformer blocks. The completion process can be summarized as follows:

$$z_{\text{compl}} = \texttt{TF}(\texttt{rearrange}([\bar{z}_1 + [\text{BLK}_1], \bar{z}_2 + [\text{BLK}_2], \bar{z}_3 + [\text{BLK}_3], z_L]) + \mathbf{E}_{pos}) \quad (3)$$

2.3 Dense Decoding and Optimization

As our goal is to design an acceleration method that is agnostic to decoder designs, designing a new segmentation decoder is beyond the scope of this paper. Thus, we couple the SCD pipeline with existing segmentation decoders. However, certain segmentation decoders, e.g., UNETR, require inputs from multiple layer outputs from the encoder, which causes problems because intermediate features are still sparse. Motivated by recent research on the non-hierarchical feature pyramid [13], we use the output z_{compl} of the completion network to replace all the intermediate features required by the segmentation head, as shown in Fig. 1.

Unlike DynamicViT, we do not introduce additional loss functions for token pruning. We optimize all segmentation models by segmentation loss. We adopt a combination of cross entropy and Dice loss. Both loss weights are set to 1.

3 Experiments

3.1 Dataset Description

We evaluate on two benchmark 3D medical segmentation datasets with sparse targets. The tasks are CT multi-organ and MRI Brain tumor segmentation.

CT Multi-organ Segmentation (BTCV). The BTCV [11] (Multi Atlas Labeling Beyond The Cranial Vault) dataset consists of 30 subjects with abdominal CT scans where 13 organs were annotated under the supervision of board-certified radiologists. Each CT volume has 85–198 slices of 512×512 pixels, with a voxel spatial resolution of $(0.54 \times 0.98 \times [2.5\text{--}5.0] \text{ mm}^3)$. For comparison convenience, we follow [3,4] to split the 30 cases into 18 for training and 12 for validation. Hyper-parameters are selected via 3-fold cross validation in the training set. We report the average DSC (Dice Similarity Coefficient) and 95% Hausdorff Distance (HD95) on 8 abdominal organs (aorta, gallbladder, spleen, left kidney, right kidney, liver, pancreas, spleen, stomach) to align with [4].

Table 1. Performance of existing approaches on BTCV. We first examine the performance of the naïve combination of existing approaches. For a large pruning ratio $r = 0.9$ on BTCV, MIM fails to perform segmentation effectively. Even with our proposed MTA instead of MIM, EViT and DynamicViT still perform worse than our STP. We report the **mean** and **std** on three random runs unless otherwise stated. Please see Sec. 3 for more analysis.

DSC(%) on BTCV (pruning ratio $r = 0.9$)		sparse encoding		
		DynamicViT [21]	EViT [15]	STP (ours)
token completion	MIM [2,9]	24.35 (single run)	18.64 (single run)	44.71 (single run)
	MTA (ours)	80.24 ± 0.34	78.62 ± 0.10	82.18 ± 0.12

MRI Brain Tumor Segmentation (BraTS). The Medical Segmentation Decathlon (MSD) [1] BraTS dataset has 484 multi-modal (FLAIR, T1w, T1-Gd and T2w) MRI scans. The ground-truth segmentation labels include peritumoral edema, GD-enhancing tumor and the necrotic/non-enhancing tumor core. The performance is measured on three recombined regions, i.e., tumor core, whole tumor and enhancing tumor. We randomly split the dataset into training (80%), validation (15%), and test (5%) sets. We report average DSC and HD95.

3.2 Implementation Details

Our method is implemented in PyTorch [19] and MONAI [18] on a single NVIDIA A100. Our encoder is based on a ViT-Base model. Three *STP* modules are inserted after the 3rd, 6th, and 9th Transformer blocks in ViT-B. We follow UNETR [8] on data processing. For BTCV, we clip the raw values between -958 and 326, and re-scale the range between -1 and 1. For BraTS, we perform an instance-wise normalization over the non-zero region per channel. For training, we set the batch size to 2 and the initial learning rate to 1.3e−4. We use AdamW as the optimizer and adopt layer-wise learning rate decay (ratio $= 0.75$) to improve training. For inference, we use a sliding window with an overlap of 50%.

3.3 Results

Naïve Combination of EViT/DynamicViT+MIM. We first test the straightforward approach of applying EViT/DynamicViT to *sparse encoding* and MIM to *token completion*. We use UNETR as the segmentation decoder. In Table 1, EViT/DynamicViT + MIM fails to perform dense prediction for a very high pruning ratio $r = 0.9$ on BTCV. This justifies our efforts in this paper to accelerate sparse token segmentation models while maintaining performance.

Our Approach: STP+MTA. We evaluate the efficiency of our *Soft-topK Token Pruning (STP)* and *Multi-layer Token Assembly (MTA)* on the BTCV

Table 2. STP+MTA+UNETR vs. UNETR performance comparison. Based on the same ViT scale and patch size, our proposed STP+MTA+UNETR can maintain performance while significantly reducing computation by a large margin. We report the mean and std of three random runs on BTCV. Please refer to Sect. 3.3 for more details on the experimental setting and analysis.

Method	MSD BraTS		Encoder Throughput(img/s)	Throughput (img/s)	MACs(G)
	DSC↑	HD95↓			
UNETR	75.44	8.89	7.10	4.85	824.38
STP+MTA+UNETR	**75.79**	**8.31**	20.04	7.79 (+60.6%)	428.28

Method	BTCV		Encoder Throughput(img/s)	Throughput (img/s)	MACs(G)
	DSC↑	HD95↓			
UNETR	80.78 ± 0.34	**15.90** ± 1.01	30.30	16.18	273.45
STP+MTA+UNETR	**82.18** ± 0.12	19.85 ± 1.12	57.31	20.08 (+24.1%)	146.63

and BraTS datasets based on UNETR. We measure the efficiency by profiling the throughput(image/s) and MAC number (Multiply-accumulate operations) for each model variant. The throughput is measured on a NVIDIA A100 GPU with batch size 1. MACs are computed by measuring the forward complexity of a single image. We present the results in Table 2. On BraTS, with an input size of $(128 \times 128 \times 128)$, our STP+MTA+UNETR $(r = 0.75)$ maintains performance while significantly increasing inference throughput by 60.8%. On BTCV, with an input size of $(96 \times 96 \times 96)$, STP+MTA+UNETR $(r = 0.9)$ can maintain performance while the corresponding inference throughput increases by 24.1%. Our method also increases training efficiency. The training throughput on BTCV increases from 2.65 imgs/s to 5.23 imgs/s by 97.36%. The training throughput on BraTS increases from 0.75 imgs/s to 1.65 imgs/s by 120%.

Sparse Encoding: STP vs. EViT/DynamicViT. EViT [15] and DynamicViT [21] were initially designed for classification. Thus, we need to adapt EViT/DynamicViT for comparison. To constrain the pruning ratio in DynamicViT, we add the ratio loss function $\mathcal{L}_{ratio}$ with a weight of $\lambda_{ratio} = 2$ following [21]. In EViT, we take the [CLS] attention weights from the Transformer block as the token scores and use topK for sampling. As shown in Table 4a, our STP-ViT performs the best. The inferiority of DynamicViT could be caused by i) mismatch between the training (variable number of pruned tokens) and testing phases (fixed number of pruned tokens) and ii) more hyper-parameters (e.g., λ_{ratio}). The performance drop in EViT indicates that the [CLS] attention scores are not suitable for representing the true token importance in segmentation.

Token Completion: MTA vs. MIM. We implement a baseline inspired by MIM [2,9]. As Table 4b shows, MIM-style completion fails (44.71%) with a high pruning ratio $r = 0.9$. Our results suggest that pruned token reuse in MTA plays an important role in a highly sparse token segmentation framework.

Token Pruning Ratio in STP. We ablate the pruning ratio in Table 3. STP is robust to a wide range of pruning ratios $[0.25, 0.9]$. Thus, our STP+MTA+UNETR can adopt a high pruning ratio to reduce computation by a large margin.

Table 3. Ablation on the Pruning Ratio r**.** STP shows robustness to a wide range of pruning ratios (0.25 → 0.9) in terms of DSC. Different datasets have different optimal pruning ratios. Refer to Sect. 3.3 for more details. We report the **mean** and **std** of three random runs on BTCV unless otherwise stated.

Pruning Ratio r	BTCV			BraTS		Encoder Throughput	Throughput	MACs(G)
	DSC↑	HD95↓		DSC↑	HD95↓			
baseline	80.78 ± 0.34	15.90 ± 1.01		75.44	8.89	7.10	4.85	824.38
0.25	81.56 ± 0.16	19.65 ± 3.25		75.50	7.98	11.77	6.12	631.75
0.50	81.81 ± 0.59	**15.78** ± 1.01		75.02	**7.40**	17.34	7.35	497.97
0.75	81.95 ± 0.18	16.37 ± 5.41		**75.79**	8.31	20.04	7.79	428.28
0.9	**82.18** ± 0.12	19.85 ± 1.12		75.32	8.04	21.63	8.04	404.14

Table 4. Ablation studies on BTCV. In (a), we compare STP with DynamicViT and EViT. STP achieves better performance. In (b), we compare our proposed MTA with MIM where MIM performs much worse than MTA. In (c), we demonstrate that Gumbel perturbation is beneficial. In (d), we ablate different τ values. $\tau = 0.1$ and $\tau = 1$ perform similarly while $\tau = 0.01$ performs worse. We report the **mean** and **std** of three random runs unless otherwise stated.

Encoder	DSC
DynamicViT	80.24 ± 0.34
EViT	78.62 ± 0.10
STP-ViT (Ours)	**82.18** ± 0.12

(a) Comparison with DynamicViT&EViT

Token Completion	DSC
MIM	44.71 (single run)
MTA (ours)	**82.18** ± 0.12

(b) Token Completion Methods

Perturbation	DSC
No (ST TopK)	81.67 ± 0.21
Yes (ours)	**82.18** ± 0.12

(c) Gumbel Perturbation

τ	DSC
0.01	81.36 ± 0.15
0.1	82.06 ± 0.22
1 (ours)	**82.18** ± 0.12

(d) Temperature τ in STP

Although our method achieves higher DSC on BTCV than UNETR, the HD95 is worse. We speculate that HD95 is more sensitive to the boundary segmentation results and that token pruning may lead to sub-optimal boundary prediction.

Temperature τ **in STP.** We ablate temperature τ in Eq. 2 in Table 4d. According to [10], a small temperature leads to a large variance of gradients and vice versa. We tried three different τ values $\{0.01, 0.1, 1\}$. Experiments show $\tau = 0.1$ and $\tau = 1$ perform similarly while $\tau = 0.01$ performs worse.

Noise Perturbation in STP. In *Soft-topK Token Pruning (STP)*, we design a straight-through (ST) Gumbel soft topK algorithm for sampling. STP forward process can be split into three steps, i.e., score prediction, Gumbel perturbation, and topK sampling. In Table 4c, we ablate the Gumbel perturbation on BTCV by evaluating a straight-through (ST) topK variant. Note that we do not add Gumbel noise during inference, to ensure that the model performs deterministically for inference. For the ST topK variant, we also remove the Gumbel noise

Table 5. Comparison with other methods on BTCV.

Framework	DSC↑/HD95↓	Aorta	Gallbladder	Kidney(L)	Kidney(R)	Liver	Pancreas	Spleen	Stomach
V-Net [17]	68.81/–	75.34	51.87	77.10	80.75	87.84	40.05	80.56	56.98
DARR [7]	69.77/–	74.74	53.77	72.31	73.24	94.08	54.18	89.90	45.96
U-Net(R50) [22]	74.68/36.87	84.18	62.84	79.19	71.29	93.35	48.23	84.41	73.92
AttnUNet(R50) [23]	75.57/36.97	55.92	63.91	79.20	72.71	93.56	49.37	87.19	74.95
TransUNet [4]	77.48/31.69	87.23	63.13	81.87	77.02	94.08	55.86	85.08	75.62
UNETR (PatchSize = 16)	78.83/25.59	85.46	70.88	83.03	82.02	95.83	50.99	88.26	72.74
UNETR (PatchSize = 8)	80.78/**15.90**	88.59	70.97	83.38	83.76	95.52	59.76	88.53	74.30
STP+MTA+UNETR (PatchSize = 8)	**82.18**/19.85	89.23	73.60	85.66	83.65	95.59	62.17	88.84	77.37

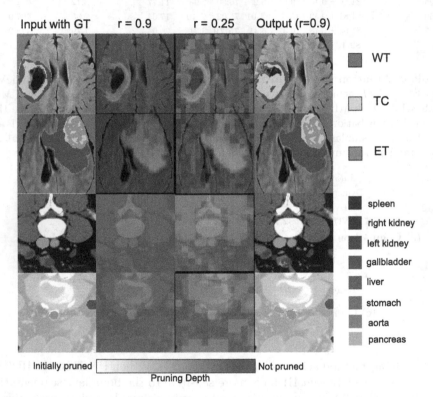

Fig. 2. Ground truth and model outputs on BraTS (first two rows) and BTCV (last two rows). We visualize the depth at which tokens are pruned under high (r = 0.9) and low (r = 0.25) pruning ratios (red shading in columns 2 and 3). Tokens that are immediately dropped are not shaded, whereas darker red shading indicates the pruning of tokens in later layers. (Color figure online)

perturbation from the training phase. With a pruning ratio $r = 0.9$, results show that the Gumbel perturbation is beneficial. It is worth noting that the ST topK variant without perturbation also achieves a competitive result.

Pruning Policy Visualization. We visualize the pruning policy for both brain tumors and abdominal organs in Fig. 2 under two extreme pruning ratios, the highest one at $r = 0.9$ and the lowest at $r = 0.25$. We use shades of red to denote the depth at which tokens are pruned. Patches (tokens in ViT) with no red overlap are pruned by the very first *STP*, whereas patches with the deepest red color are kept in ViT until the last. In Fig. 2, with $r = 0.9$, most tokens are dropped at a very early stage. Some tokens around the brain tumor, especially at tumor boundaries, are never pruned. When the ratio decreases to $r = 0.25$, more patches are kept and still cluster around the target tumor region.

Class-Wise Comparison with Others on BTCV. We show class-wise results of UNETR, STP+MTA+UNETR, and other methods in Table 5.

STP+MTA+UNETR shows improvement over a series of methods on BTCV. Note that current SOTA methods [24,27,30] rely on either stronger priors (window attention) or SSL pre-training. However, our goal is accelerating standard ViT-based segmentation instead of purely pursuing increased performance.

4 Conclusion and Future Work

We introduced a ViT-based sparse token segmentation framework for medical images. First, we proposed a *Soft-topK Token Pruning* (STP) module to prune tokens in ViT. STP can speed up ViTs in both training and inference phases. To produce a full set of tokens for dense prediction, we proposed *Multi-layer Token Assembly (MTA)* that recovers a complete set of tokens by assembling both output and intermediate tokens from multiple layers. In our 3D medical image experiments STP+MTA+UNETR speeds up the UNETR baseline significantly while maintaining segmentation performance. Accelerating the decoder, which also plays a big role in the inference speed, is left for future work.

Acknowledgement. The reported research was partly supported by NIH award # 1R21CA258493-01A1, NSF awards IIS-2212046 and IIS-2123920, and Stony Brook OVPR seed grants. The content is solely the responsibility of the authors and does not necessarily represent the official views of the National Institutes of Health.

References

1. Antonelli, M., et al.: The medical segmentation decathlon. arXiv preprint arXiv:2106.05735 (2021)
2. Bao, H., Dong, L., Wei, F.: BEiT: BERT pre-training of image transformers. arXiv preprint arXiv:2106.08254 (2021)
3. Chen, J.N.: Transunet. https://github.com/Beckschen/TransUNet
4. Chen, J., et al.: TransUNet: transformers make strong encoders for medical image segmentation. arXiv preprint arXiv:2102.04306 (2021)
5. Cordonnier, J.B., Mahendran, A., Dosovitskiy, A., Weissenborn, D., Uszkoreit, J., Unterthiner, T.: Differentiable patch selection for image recognition. In: CVPR, pp. 2351–2360 (2021)
6. Dosovitskiy, A., et al.: An image is worth 16×16 words: transformers for image recognition at scale. arXiv preprint arXiv:2010.11929 (2020)
7. Fu, S., et al.: Domain adaptive relational reasoning for 3D multi-organ segmentation. In: Martel, A.L., et al. (eds.) MICCAI 2020. LNCS, vol. 12261, pp. 656–666. Springer, Cham (2020). https://doi.org/10.1007/978-3-030-59710-8_64

8. Hatamizadeh, A., et al.: UNETR. In: WACV (2022)
9. He, K., Chen, X., Xie, S., Li, Y., Dollár, P., Girshick, R.: Masked autoencoders are scalable vision learners. arXiv preprint arXiv:2111.06377 (2021)
10. Jang, E., Gu, S., Poole, B.: Categorical reparameterization with gumbel-softmax. arXiv preprint arXiv:1611.01144 (2016)
11. Landman, B., Xu, Z., Igelsias, J., Styner, M., Langerak, T., Klein, A.: MICCAI multi-atlas labeling beyond the cranial vault-workshop and challenge. In: Proceedings of the MICCAI Multi-Atlas Labeling Beyond Cranial Vault-Workshop Challenge (2015)
12. Li, J., Cotterell, R., Sachan, M.: Differentiable subset pruning of transformer heads. Trans. Assoc. Comput. Linguist. **9**, 1442–1459 (2021)
13. Li, Y., Mao, H., Girshick, R., He, K.: Exploring plain vision transformer backbones for object detection. arXiv preprint arXiv:2203.16527 (2022)
14. Li, Y., Xie, S., Chen, X., Dollar, P., He, K., Girshick, R.: Benchmarking detection transfer learning with vision transformers. arXiv preprint arXiv:2111.11429 (2021)
15. Liang, Y., Chongjian, G., Tong, Z., Song, Y., Wang, J., Xie, P.: Evit: expediting vision transformers via token reorganizations. In: ICLR (2021)
16. Meng, L., et al.: AdaViT: adaptive ViTs for efficient image recognition. arXiv preprint arXiv:2111.15668 (2021)
17. Milletari, F., Navab, N., Ahmadi, S.A.: V-net: fully convolutional neural networks for volumetric medical image segmentation. In: 3DV, pp. 565–571. IEEE (2016)
18. MONAI Consortium: MONAI: Medical Open Network for AI (2020). https://doi. org/10.5281/zenodo.4323058, https://github.com/Project-MONAI/MONAI
19. Paszke, A., et al.: PyTorch: an imperative style, high-performance deep learning library. In: NeurIPS, vol. 32 (2019)
20. Ranftl, R., Bochkovskiy, A., Koltun, V.: Vision transformers for dense prediction. In: ICCV (2021)
21. Rao, Y., Zhao, W., Liu, B., Lu, J., Zhou, J., Hsieh, C.J.: DynamicViT: efficient vision transformers with dynamic token sparsification. In: NeurIPS, vol. 34 (2021)
22. Ronneberger, O., Fischer, P., Brox, T.: U-net: convolutional networks for biomedical image segmentation. In: Navab, N., Hornegger, J., Wells, W.M., Frangi, A.F. (eds.) MICCAI 2015. LNCS, vol. 9351, pp. 234–241. Springer, Cham (2015). https://doi.org/10.1007/978-3-319-24574-4_28
23. Schlemper, J., et al.: Attention gated networks: learning to leverage salient regions in medical images. Med. Image Anal. **53**, 197–207 (2019)
24. Tang, Y., et al.: Self-supervised pre-training of swin transformers for 3D medical image analysis. In: CVPR (2022)
25. Vaswani, A., et al.: Attention is all you need. In: NeurIPS, vol. 30 (2017)
26. Wang, J., Sun, K., Cheng, T., Jiang, B., Deng, C., Zhao, Y., Liu, D., Mu, Y., Tan, M., Wang, X., et al.: Deep high-resolution representation learning for visual recognition. IEEE Trans. PAMI **43**(10), 3349–3364 (2020)
27. Wu, Y., et al.: D-former: a U-shaped dilated transformer for 3D medical image segmentation. arXiv preprint arXiv:2201.00462 (2022)
28. Xie, S.M., Ermon, S.: Reparameterizable subset sampling via continuous relaxations. arXiv preprint arXiv:1901.10517 (2019)
29. Zheng, S., et al.: Rethinking semantic segmentation from a sequence-to-sequence perspective with transformers. In: CVPR (2021)
30. Zhou, H.Y., Guo, J., Zhang, Y., Yu, L., Wang, L., Yu, Y.: nnFormer: interleaved transformer for volumetric segmentation. arXiv preprint arXiv:2109.03201 (2021)

blob loss: Instance Imbalance Aware Loss Functions for Semantic Segmentation

Florian Kofler[1,2,3,7(✉)], Suprosanna Shit[1,2], Ivan Ezhov[1,2], Lucas Fidon[4],
Izabela Horvath[1,5], Rami Al-Maskari[1,5], Hongwei Bran Li[1,13],
Harsharan Bhatia[5,6], Timo Loehr[1,3], Marie Piraud[7], Ali Erturk[5,6,8,9],
Jan Kirschke[3], Jan C. Peeken[10,11,12], Tom Vercauteren[4], Claus Zimmer[3],
Benedikt Wiestler[3], and Bjoern Menze[1,13]

[1] Department of Informatics, Technical University Munich, Munich, Germany
florian.kofler@tum.de
[2] TranslaTUM - Central Institute for Translational Cancer Research, Technical
University of Munich, Munich, Germany
[3] Department of Diagnostic and Interventional Neuroradiology, School of Medicine,
Klinikum rechts der Isar, Technical University of Munich, Munich, Germany
[4] School of Biomedical Engineering and Imaging Sciences, King's College London,
London, UK
[5] Insitute for Tissue Engineering and Regenerative Medicine, Helmholtz Institute
Munich (iTERM), Oberschleißheim, Germany
[6] Institute for Stroke and Dementia research (ISD), University Hospital, LMU
Munich, Munich, Germany
[7] Helmholtz AI, Helmholtz Munich, Neuherberg, Germany
[8] Graduate School of Neuroscience (GSN), Munich, Germany
[9] Munich Cluster for Systems Neurology (Synergy), Munich, Germany
[10] Department of Radiation Oncology, Klinikum rechts der Isar, Technical University
of Munich, Munich, Germany
[11] Institute of Radiation Medicine (IRM), Department of Radiation Sciences (DRS),
Helmholtz Zentrum, Munich, Germany
[12] Deutsches Konsortium für Translationale Krebsforschung (DKTK), Partner Site
Munich, Munich, Germany
[13] Department of Quantitative Biomedicine, University of Zurich, Zürich, Switzerland

Abstract. Deep convolutional neural networks (CNN) have proven to
be remarkably effective in semantic segmentation tasks. Most popular
loss functions were introduced targeting improved volumetric scores, such
as the Dice coefficient (DSC). By design, DSC can tackle class imbalance,
however, it does not recognize instance imbalance within a class. As a
result, a large foreground instance can dominate minor instances and still
produce a satisfactory DSC. Nevertheless, detecting tiny instances is crucial for many applications, such as disease monitoring. For example, it
is imperative to locate and surveil small-scale lesions in the follow-up of

B. Wiestler and B. Menze—Equal contribution.

Supplementary Information The online version contains supplementary material
available at https://doi.org/10.1007/978-3-031-34048-2_58.

multiple sclerosis patients. We propose a novel family of loss functions, *blob loss*, primarily aimed at maximizing instance-level detection metrics, such as *F1* score and *sensitivity*. *Blob loss* is designed for semantic segmentation problems where detecting multiple instances matters. We extensively evaluate a DSC-based *blob loss* in five complex 3D semantic segmentation tasks featuring pronounced instance heterogeneity in terms of texture and morphology. Compared to soft Dice loss, we achieve *5%* improvement for MS lesions, *3%* improvement for liver tumor, and an average *2%* improvement for microscopy segmentation tasks considering *F1* score.

Keywords: semantic segmentation loss function · instance imbalance awareness · multiple sclerosis · lightsheet microscopy

1 Introduction

In recent years convolutional neural networks (CNN) have gained increasing popularity for complex machine learning tasks, such as *semantic segmentation*. In *semantic segmentation*, one segments object from different classes without differentiating multiple instances within a single class. In contrast, *instance segmentation* explicitly takes multiple instances into account, which involves simultaneous localization and segmentation. While U-net variants [23] still represent the state-of-the-art to address semantic segmentation, *Mask-RCNN* and its variants dominate *instance segmentation* [11]. The scarcity of training data often hinders the application of back-bone-dependent Mask RCNNs, while U-Nets have proven to be less data-hungry [5].

However, many semantic segmentation tasks feature relevant instance imbalance, where large instances dominate over smaller ones within a class, as illustrated in Fig. 1. Instances can vary not only with regard to size but also texture and other morphological features. U-nets trained with existing loss functions, such as Soft Dice [6,18,19,24,28], cannot address this. Instance imbalance is particularly pronounced and significant in medical applications: For example, even a single new multiple sclerosis (MS) lesion can impact the therapy decision. Despite many ways to compensate for class-imbalance [2,9,22,28], there is a notable void in addressing instance imbalance in semantic segmentation settings. Additionally, established metrics have been shown to correlate insufficiently with expert assessment [16].

Contribution: We propose *blob loss*, a novel framework to equip semantic segmentation models with instance imbalance awareness. This is achieved by dedicating a specific loss term to each instance without the necessity of instance-wise prediction. *Blob loss* represents a method to convert any loss function into a novel instance imbalance aware loss function for semantic segmentation problems designed to optimize detection metrics. We evaluate its performance on five complex three-dimensional (3D) semantic segmentation tasks, for which the discovery of miniature structures matters. We demonstrate that extending soft

Dice loss to a *blob loss* improves detection performance in these multi-instance semantic segmentation tasks significantly. Furthermore, we also achieve volumetric improvements in some cases.

Related Work: Sirinukunwattana et al. [27] suggested an instance-based Dice metric for evaluating segmentation performance. Salehi et al. [24] were among the first to propose a loss function, called *Tversky loss*, for semantic segmentation of multiple sclerosis lesions in magnetic resonance imaging (MR), trying to improve detection metrics. Similarly, Zhu et al. [32] introduced Focal Loss, initially designed for object detection tasks [17], into medical semantic segmentation tasks.

There have been few recent attempts aiming for a solution to instance imbalance. Zhang et al. [30] propose an auxiliary lesion-level sphere prediction task. However, they do not explicitly consider each instance separately. Shirokikh et al. [25] propose an instance-weighted loss function where a global weight map is inversely proportional to the size of the instances. However, unlike size, not all types of imbalance, such as morphology or texture, can be quantified easily, limiting the method's applicability.

2   Methods

First, we introduce the problem of instance imbalance in semantic segmentation tasks. Then we present our proposed *blob loss* functions.

Problem Statement: Large foreground areas dominate the calculation of established volumetric metrics (or losses); see Fig. 1. This is because the volumetric measures only accumulate true or false predictions on a voxel level but not at the instance level. Therefore, training models with volumetry-based loss functions, such as soft Dice loss (*dice*), often leads to unsatisfactory instance detection performance. To achieve a better instance detection performance, it is necessary to take instance imbalance into account. Instance imbalance can be of many categories, such as morphology and texture. Importantly, instance imbalance often cannot be easily specified and quantified for use in CNN training, for example, as instance weights in the loss function. Thus, using conventional methods, it is difficult to incorporate instance imbalance in CNN training. Our objective is to design loss functions to compensate for the instance imbalance while being agnostic to the instance imbalance type. Therefore, we aim to dissect the image domain in an instance-wise fashion:

blob loss Formulation: Consider a generic volumetric loss function $\mathcal{L}$ and image domain Ω and foreground domain P. Formally our objective is to find an instance-specific subdomain $\Omega_n \subseteq \Omega$ corresponding to the nth instance such that $\mathcal{L}$ acting on Ω_n is aware of instance imbalance. The criteria to obtain these subsets $\{\Omega_n\}_{n=1}^N$ are such that $\Omega_i \cap \Omega_j \cap P = \phi; \forall (i,j), \ s.t. \ 1 \leq i, j \leq N, i \neq j$ and $\cup_{n=1}^N \Omega_n = \Omega$. In simple terms, the subsets $\{\Omega_n\}_{n=1}^N$ need to be mutually exclusive regarding foreground and collectively exhaustive with regard to the whole image domain.

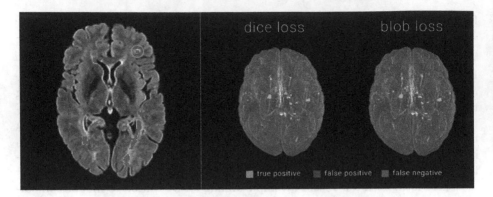

Fig. 1. *Problem statement (left):* The Dice coefficient (DSC) for the segmentation with vs. without a lesion, encircled in green, is: *0.9806.* Therefore, the segmentations are hardly distinguishable in terms of DSC. However, from a clinical perspective, the difference is important as the detection of a single lesion can affect treatment decisions. *Comparison of segmentation performance (right):* Maximum intensity projections of the FLAIR images overlayed with segmentations for *dice* and *blob dice.* Lesions are colored according to their detection status: Green for *true positive*; Blue for *false positive*; Red for *false negative.* For this particular patient, applying the transformation to a *blob loss* improves *F1* from *0.74* to *1.0* and the volumetric Dice coefficient from *0.56* to *0.70* and the latter is caused by an increase in *volumetric precision* from *0.48* to *0.75*, while the *volumetric sensitivity* remains constant at *0.66.*

To formalize *blob loss*, we address instance imbalance within a binary semantic segmentation framework. At the same time, we remain agnostic towards particular instance attributes and do not incorporate these in the loss function. To this extent, we propose to leverage the existing reference annotations and formally propose a novel family of instance-aware loss functions.

Consider a segmentation problem with N instances; for different input images, N can vary from few to many. Specifically, we propose to compute the instance-specific domain Ω_n by excluding all but the nth foreground from the whole image domain Ω, see Eq. (1):

$$\Omega_n = \Omega \setminus \cup_{j=1,\, j\neq n}^{N} P_j \qquad (1)$$

where P_j is the foreground domain for j^{th} instances of P. This masking process is illustrated by Fig. 2. It is worth noting that the background voxels are included in every Ω_n.

We propose to convert any loss function $\mathcal{L}$ for binary semantic segmentation into an instance-aware loss function $\mathcal{L}_{blob}$ defined as:

$$\mathcal{L}_{blob}\left((p_i)_{i\in\Omega}, (g_i)_{i\in\Omega}\right) = \frac{1}{N} \sum_{n=1}^{N} \mathcal{L}\left((p_i)_{i\in\Omega_n}, (g_i)_{i\in\Omega_n}\right) \qquad (2)$$

where $\{g_i\}_{i\in\Omega}$ is the ground-truth segmentation, $\{p_i\}_{i\in\Omega}$ is the predicted segmentation, N is the number of instances in the foreground.

Fig. 2. Masking process described in Eq. (2). *Left:* the global ground truth label (GT), with the nth instance highlighted in green. *Middle:* The loss mask Ω_n for the nth instance (MASK) for multiplication with the network outputs. *Right:* the label used for the computation of the local *blob loss* for the nth instance. This process is repeated for every instance.

As our goal is to assign equal importance to all instances irrespective of their size, shape, texture, and other topological attributes, we average over all instances.

To compute the total loss for a volume, we combine the instance-wise Loss component from Eq. (2) with a global component to obtain the final Loss:

$$\mathcal{L}_{total} = \alpha \mathcal{L}_{global} + \beta \mathcal{L}_{blob} \tag{3}$$

where α and β denote the weights for the global and instance constraint $\mathcal{L}_{blob}$. We (anonymously) provide a sample Pytorch implementation of a *dice*-based *blob loss* on GitHub. In order to accelerate our training, we precompute the instances, here defined as connected components using *cc3d* [26], version *3.2.1*.

Model Training: For all our experiments, we use a basic 3D U-Net implemented via MONAI inspired by [8] and further depicted in supplementary materials. Furthermore, we use a dropout ratio of *0.1* and employ *mish* as activation function [20]. Otherwise, we stick to the default parameters of the U-Net implementation.

Loss Functions for Comparison: As baselines we use the MONAI implementations of soft Dice loss (*dice*) and Tversky loss (*tversky*) [24]. For *tversky*, we always use the standard parameters of $\alpha = 0.3$ and $\beta = 0.7$ suggested by the authors in the original publication [24]. For comparison we create *blob dice*, by transforming the standard *dice* into a *blob loss* using our conversion method Eq. (2). The final loss is obtained by employing *dice* in the $\mathcal{L}_{global}$ and $\mathcal{L}_{blob}$ terms of the proposed total loss Eq. (3). In analog fashion, we derive *blob tversky*. Furthermore, we compare against *inverse weighting (iw)*, the globally weighted loss function of Shirokikh et al. [25]. For this, we use the official GitHub implementation to compute the weight maps and loss and deploy these in our training pipelines.

Training Procedure: Our CNNs are trained on multiple cuboid-shaped crops per batch element, with higher resolution in the axial dimension, enabling the learning of contextual image features. The crops are randomly sampled around a center voxel that consists of *foreground* with a *95%* probability. We consider one epoch as one full iteration of forward and backward passes through all batches of the training set. For all training, *Ranger21* [29] serves as our optimizer. For each experiment, we keep the initial learning rate (lr) constant between training runs. Depending on the segmentation task, we deploy varying suitable image normalization strategies. For comparability, we keep all training parameters except for the loss functions constant on a segmentation task basis and stick to this standard training procedure.

Training-Test Split and Model Selection: Given the high heterogeneity of our bio-medical datasets and the limited availability of high-quality ground truth annotations due to the very costly labeling procedures requiring domain experts, we do not set aside data for validation and therefore do not conduct model selection. Instead, inspired by [13], we split our data *80:20* into training and test set and evaluate on the last checkpoint of the model training. As an exception, the MS dataset comes with predefined training, validation, and test set splits; therefore, we additionally evaluate the *best* model checkpoint, meaning the model with the lowest loss on the validation set. As we are more interested in *blob loss'* generalization capabilities than exact quantification of improvements on particular datasets, we prioritize a broad validation on multiple datasets over cross-validation.

Technical Details: Our experiments were conducted using NVIDIA RTX8000, RTX6000, RTX3090, and A6000 GPUs using CUDA version *11.4* in conjunction with Pytorch version *1.9.1* and MONAI version *0.7.0*.

2.1   Evaluation Metrics and Interpretation

Metrics: We obtain global, volumetric performance measures from *pymia* [14]. In addition to DSC, we also evaluate *volumetric sensitivity (S), volumetric precision (P)*, and the *Surface Dice similarity coefficient (SDSC)*. To compute instance-wise detection metrics, namely instance F1 (*F1*), *instance sensitivity (IS)* and *instance precision (IP)*, we employ a proven evaluation pipeline from Pan et al. [21].

Interpretation: By design, human annotators tend to overlook tiny structures. For comparison, human annotators initially missed *29%* of micrometastases when labeling the DeepMACT light-sheet microscopy dataset [21]. Therefore, the likelihood of a structure being correctly labeled in the ground truth is much higher for foreground than for background structures. Additionally, human annotators have a tendency to label a structure's center but do not perfectly trace its contours. Both phenomena are illustrated in Fig. 3. These effects are particularly pronounced for microscopy datasets, which often feature thousands of blobs. These factors are important to keep in mind when interpreting the results. Consequently, volumetric - and instance sensitivity are much more informative than volumetric and instance precision.

3 Experiments

To validate *blob loss*, we train segmentation models on a selection of datasets from different 3D imaging modalities, namely brain MR, thorax CT, and light-sheet microscopy. We select datasets featuring a variety of fragmented semantic segmentation problems. For simplicity, we use the default values $\alpha = 2.0$ and $\beta = 1.0$ across all experiments.

Multiple Sclerosis (MRI): The Multiple Sclerosis (MS) dataset, comprising 521 single timepoint MRI examinations of patients with MS, was collected for internal validation of MS lesion segmentation algorithms. The patients come from a representative, institutional cohort covering all stages (in terms of time from disease onset) and forms (relapsing-remitting, progressive) of MS. A 3D T1w and a 3D FLAIR sequence were acquired on a 3 T *Philips Achieva* scanner. All 3D volumes feature $193 \times 193 \times 229$ voxels in 1mm isotropic resolution. The dataset divides into a fixed training set of 200, a validation set of 21, and a test set of 200 cases. The annotations feature a total of *4791* blobs, with *25.69* $\pm$ *23.01* blobs per sample. Expert neuroradiologists annotated the MS lesions manually and ensured pristine ground truth quality with consensus voting.

For all training runs of *500* epochs, we set the initial learning rate to *1e−2* and the batch size to *4*. The networks are trained on a single GPU using *2* random crops with a patch size of *192 × 192 × 32* voxels per batch element after applying a *min/max* normalization. As the MS dataset comes with a predefined validation set of 21 images, we also save the checkpoint with the lowest loss on the validation set and compare it to the respective last checkpoint of the training. In addition to the standard *dice*, we also compare against *tversky*. Furthermore, we conduct an ablation study to find out how the performance metrics are affected by choosing different values for α and β.

Liver Tumors - LiTS (CT): To develop an understanding of *blob loss* performance on other imaging modalities, we train a model for segmenting liver tumors on CT images of the *LiTS* challenge [4]. The dataset consists of varying high-resolution CT images of the abdomen. The challenge's original task was segmenting liver and liver tumor tissue. As we are primarily interested in segmenting small fragmented structures, we limit our experiments to the liver area and segment only liver tumor tissue (in contrast to tumors, the liver represents a huge solid structure, and we are interested in blobs). We split the publicly available training set into 104 images for training and 27 for testing. The annotations were created by expert radiologists and feature a total of *908* blobs, with *12.39 ± 14.92* blobs per sample.

For all training runs of *500* epochs, we set the initial learning rate to *1e−2* and the batch size to *2*. The networks are trained on two GPUs in parallel using *2* random crops with a patch size of *192 × 192 × 64* voxels per batch element. We apply normalization based on windowing on the Hounsfield (HU) scale. Therefore, we define a normalization window suitable for liver tumor segmentation around center *30 HU* with a width of *150 HU*, and *20%* added tolerance.

DISCO-MS (Light-Sheet Microscopy). To develop an understanding for *blob loss* performance on other imaging modalities, we train a model for segmenting Amyloid plaques in light-sheet microscopy images of the *DISCO-MS* dataset [3].

The volumes of *300 × 300 × 300* voxels resolution contain cleared tissue of mouse brain. We split the publicly available dataset into 41 volumes for training and six for testing. The annotations feature a total of *988* blobs, with *28.32 ± 24.44* blobs per sample. Even though the label quality is very high, the results should still be interpreted with care following the guidelines in Sect. 2.1.

For all training runs of *800* epochs, we set the initial learning rate to *1e−3* and the batch size to *6*. As our initial model trained with *dice* does not produce satisfactory results, we furthermore try learning rates of *1e−2*, *3e−4* and *1e−4*, following the heuristics suggested by [1] without success. The networks are trained on two GPUs in parallel using *2* random crops with a patch size of *192 × 192 × 64* per batch element. The images are globally normalized, using a minimum and maximum threshold defined by the *0.5* and *99.5* percentile.

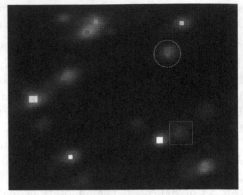

Fig. 3. Zoomed in 2D view on a volume of the SHANEL [31] dataset. The overlayed labels are colored according to a 3D connected component analysis. The expert biologists did not label each foreground object in every slice, e.g., the magenta-colored square. Furthermore, the contours of the structures are imperfectly segmented, for instance, the red label within the bright green circle. These effects can partially be attributed to the ambiguity of the light-sheet microscopy signal [15]. However, they are also observed in the human annotations of the MS and LiTS dataset. (Color figure online)

SHANEL (Light-Sheet Microscopy). For further validation, we evaluate neuron segmentation in light-sheet microscopy images of the *SHANEL* dataset [31]. The volumes of *200 × 200 × 200* voxels resolution contain cleared human brain tissue from the primary visual cortex, the primary sensory cortex, the primary motor cortex, and the hippocampus. We split this publicly available dataset into nine volumes for training and three for testing. The annotations feature a total of *20684* blobs, with *992.14 ± 689.39* blobs per sample. As the data is more sparsely annotated than DISCO-MS, F1 and especially DSC should be interpreted with great care, as described in Sect. 2.1.

For all training runs of *1000* epochs, we set the initial learning rate to *1e−3* and the batch size to *3*. The networks are trained on two GPUs in parallel using *6* random crops with a patch size of *128 × 128 × 32* per batch element, with min/max normalization.

DeepMACT (Light-Sheet Microscopy). For further validation, we evaluate the segmentation of micrometastasis in light-sheet microscopy images of the *DeepMact* dataset [21]. The volumes of *350 × 350 × 350* resolution contain cleared tissue featuring different body parts of a mouse. We split the publicly available dataset into *115* images for training and *19* for testing. The annotations feature a total of *484* blobs, with *6.99 ± 8.14* blobs per sample. As the data is sparsely annotated, *F1* and especially *DSC* should be interpreted with great care, as described in Sect. 2.1.

For all training runs of *500* epochs, we set the initial learning rate to *1e−2* and the batch size to *4*. The networks are trained on a single GPU using *2* random crops with a patch size of *192 × 192 × 48*. The images are globally normalized based using a minimum and maximum threshold defined by the *0.0* and *99.5* percentile.

4 Results

Table 1 summarizes the results of our experiments. Across all datasets, we find that extending *dice* to a *blob loss* helps to improve detection metrics. Furthermore, in some cases, we also observe improvements in volumetric performance measures. While model selection seems not beneficial on this dataset, employing *blob loss* produces more robust results, as both the *dice* and *tversky* models suffer performance drops for the *best* checkpoints. Notably, even though *tversky* was explicitly proposed for MS lesion segmentation, it is clearly outperformed by *dice*, as well as *blob dice* and *blob tversky*. Further, even with the mitigation strategies suggested by the authors, *inverse weighting* produced over-segmentations.

Table 2 summarizes the results of the ablation study on α and β parameters of *blob loss*. We find that assigning higher importance to the global parameter by choosing $\alpha = 2$ and $\beta = 1$ seems to produce the best results. Overall, we find that *blob loss* seems quite robust regarding the choice of hyperparameters as long as the global term remains included by choosing a α greater than *0*.

5 Discussion

Contribution: *blob loss* can be employed to provide existing loss functions with instance imbalance awareness. We demonstrate that the application of *blob loss* improves detection- and in some cases, even volumetric segmentation performance across a diverse set of complex 3D bio-medical segmentation tasks. We evaluate *blob loss'* performance in the segmentation of multiple sclerosis (MS) lesions in MR, liver tumors in CT, and segmentation of different biological structures in 3D light-sheet microscopy datasets. Depending on the dataset, it achieves these improvements either due to better detection of foreground objects, better suppression of background objects, or both. We provide an implementation of blob loss leveraging on a precomputed connected component analysis for fast processing times.

Limitations: Certainly, the biggest disadvantage of *blob loss* is the dependency on instance segmentation labels; however, in many cases, these can be simply

Table 1. Experimental results for five datasets. For all training runs with *blob loss* we use $\alpha = 1$ and $\beta = 2$. Note that the results for LiTS are based on a different, more challenging test set and are therefore not comparable with the public leaderboard of the LiTS challenge. For DISCO-MS, the *dice* model completely over-segments and produces dissatisfactory results. Therefore, we try two additional training runs with reduced learning rates following the heuristics suggested by [1], resulting in similar over-segmentation. The same problem is observed for *inverse weighting (iw)*. Shirokikh et al. [25] themselves note the stability problems of the method and suggest lowering the learning rate to $1 - e3$.

dataset	loss	lr	**DSC**	SDSC	F1	IS	IP
MS	blob dice	1e−2	**0.680**	**0.848**	**0.810**	0.822	**0.828**
	dice	1e−2	0.660	0.820	0.758	**0.854**	0.711
	iw [25]	1e−2	0.153	0.167	0.278	0.801	0.188
	iw [25]	1e−3	0.243	0.273	0.282	0.819	0.189
	blob tversky	1e−2	**0.690**	**0.852**	**0.804**	0.829	**0.804**
	tversky	1e−2	0.601	0.697	0.566	**0.854**	0.459
LiTS	blob dice	1e−2	**0.663**	0.542	**0.657**	**0.861**	**0.631**
	dice	1e−2	0.660	**0.546**	0.623	0.801	0.599
SHANEL	blob dice	1e−3	**0.543**	**0.808**	**0.792**	**0.874**	**0.724**
	dice	1e−3	0.539	0.794	0.783	0.854	0.723
DISCO−MS	blob dice	1e−3	**0.546**	**0.678**	**0.589**	0.760	**0.481**
	dice	1e−3	0.095	0.083	0.012	0.870	0.006
	dice	3e−4	0.016	0.036	0.379	**0.896**	0.240
	dice	1e−4	0.007	0.011	0.228	0.825	0.132
DeepMACT	blob dice	1e−2	**0.357**	**0.393**	**0.391**	**0.871**	**0.276**
	dice	1e−2	0.353	0.372	0.367	0.801	0.254

Table 2. Ablation analysis on the *blob loss'* hyperparameters α and β for the MS lesions dataset. We observe that *blob loss* seems to be quite robust with regard to hyperparameter choice, as long as the global term remains present, compare Eq. (3). The default parameters $\alpha = 2$ and $\beta = 1$ provide the best results.

loss	α	β	DSC	S	P	SDSC	F1	IS	IP
blob dice	3	1	0.674	0.629	0.765	0.833	0.790	0.796	0.815
blob dice	2	1	**0.680**	0.626	0.782	**0.848**	**0.810**	0.822	**0.828**
blob dice	1	1	0.658	0.580	0.802	0.839	0.804	0.840	0.801
blob dice	1	2	0.630	0.552	0.803	0.819	0.792	0.832	0.786
dice	1	0	0.660	**0.704**	0.656	0.820	0.758	**0.854**	0.711
blob	0	1	0.522	0.409	**0.837**	0.728	0.744	0.805	0.727

obtained by a connected component analysis, as demonstrated in our experiments. Another disadvantage of *blob loss* compared to other loss functions are the more extensive computational requirements. By definition, the user is required to run computations with large patch sizes that feature multiple instances. This results in an increased demand for GPU memory, especially when working with 3D data (as in our experiments). However, larger patch sizes have proven helpful for bio-medical segmentation problems, in general, [12]. Furthermore, according to our formulation, *blob loss* possesses an interesting mathematical property, it penalizes false positives proportionally to the number of instances in the volume. Additionally, even though *blob loss* can easily be reduced to a single hyperparameter, and it proved quite robust in our experiments, it might be sensitive to hyperparameter tuning. Moreover, by design *blob loss* can only improve performance for multi-instance segmentation problems.

Interpretation: One can only speculate why *blob loss* improves performance metrics. CNNs learn features that are very sensitive to texture [10]. Unlike conventional loss functions, *blob loss* adds attention to every single instance in the volume. Thus the network is forced to learn the instance imbalanced features such as, but not limited to morphology and texture, which would not be well represented by optimizing via *dice* and alike. Such instance imbalance was observed in the medical field, as it has been shown that MS lesions change their imaging phenotype over time, with recent lesions looking significantly different from older ones [7]. These aspects might explain the gains in instance sensitivity. Furthermore, adding the multiple instance terms leads to heavy penalization on background, which might explain why we often observe an improvement in precision, see supplementary materials.

Outlook: Future research will have to reveal to which extent transformation to *blob loss* can be beneficial for other segmentation tasks and loss functions. A first and third place in recent public segmentation challenges using a compound-based variant *blob loss* indicate that *blob loss* might possess broad applicability towards other instance imbalanced semantic segmentation problems.

References

1. Bengio, Y.: Practical recommendations for gradient-based training of deep architectures. In: Montavon, G., Orr, G.B., Müller, K.-R. (eds.) Neural Networks: Tricks of the Trade. LNCS, vol. 7700, pp. 437–478. Springer, Heidelberg (2012). https://doi.org/10.1007/978-3-642-35289-8_26
2. Berman, M., et al.: The lovász-softmax loss: a tractable surrogate for the optimization of the intersection-over-union measure in neural networks. In: Proceedings of the IEEE Conference on Computer Vision and Pattern Recognition, pp. 4413–4421 (2018)
3. Bhatia, et al.: Proteomics of spatially identified tissues in whole organs. arXiv (2021)
4. Bilic, P., et al.: The liver tumor segmentation benchmark (LiTS) (2019)
5. Caicedo, J.C., et al.: Nucleus segmentation across imaging experiments: the 2018 data science bowl. Nat. Methods **16**(12), 1247–1253 (2019)

6. Eelbode, T., et al.: Optimization for medical image segmentation: theory and practice when evaluating with dice score or Jaccard index. IEEE Trans. Med. Imaging **39**(11), 3679–3690 (2020)
7. Elliott, C., et al.: Slowly expanding/evolving lesions as a magnetic resonance imaging marker of chronic active multiple sclerosis lesions. Mult. Scler. J. **25**(14), 1915–1925 (2019)
8. Falk, T., et al.: U-net: deep learning for cell counting, detection, and morphometry. Nat. Methods **16**(1), 67–70 (2019)
9. Fidon, L., et al.: Generalised wasserstein dice score for imbalanced multi-class segmentation using holistic convolutional networks. In: Crimi, A., Bakas, S., Kuijf, H., Menze, B., Reyes, M. (eds.) BrainLes 2017. LNCS, vol. 10670, pp. 64–76. Springer, Cham (2018). https://doi.org/10.1007/978-3-319-75238-9_6
10. Geirhos, R., et al.: ImageNet-trained CNNs are biased towards texture; increasing shape bias improves accuracy and robustness. arXiv preprint arXiv:1811.12231 (2018)
11. He, K., et al.: Mask R-CNN. In: Proceedings of the IEEE International Conference on Computer Vision, pp. 2961–2969 (2017)
12. Isensee, F., Jaeger, P.F., Kohl, S.A., Petersen, J., Maier-Hein, K.H.: nnU-net: a self-configuring method for deep learning-based biomedical image segmentation. Nat. Methods **18**(2), 203–211 (2021)
13. Isensee, F., et al.: nnU-net: breaking the spell on successful medical image segmentation. arXiv preprint arXiv:1904.08128, vol. 1, pp. 1–8 (2019)
14. Jungo, A., et al.: pymia: a python package for data handling and evaluation in deep learning-based medical image analysis. Comput. Methods Programs Biomed. **198**, 105796 (2021)
15. Kofler, F., et al.: Approaching peak ground truth. arXiv preprint arXiv:2301.00243 (2022)
16. Kofler, F., et al.: Are we using appropriate segmentation metrics? Identifying correlates of human expert perception for CNN training beyond rolling the dice coefficient (2021)
17. Lin, T.Y., et al.: Focal loss for dense object detection. In: Proceedings of the IEEE International Conference on Computer Vision, pp. 2980–2988 (2017)
18. Ma, J., et al.: Loss odyssey in medical image segmentation. Med. Image Anal. **71**, 102035 (2021)
19. Milletari, F., Navab, N., Ahmadi, S.A.: V-net: fully convolutional neural networks for volumetric medical image segmentation. In: 2016 Fourth International Conference on 3D Vision (3DV), pp. 565–571, IEEE (2016)
20. Misra, D.: Mish: a self regularized non-monotonic neural activation function. arXiv preprint arXiv:1908.08681 (2019)
21. Pan, C., et al.: Deep learning reveals cancer metastasis and therapeutic antibody targeting in the entire body. Cell **179**(7), 1661–1676 (2019)
22. Rahman, M.A., Wang, Y.: Optimizing intersection-over-union in deep neural networks for image segmentation. In: Bebis, G., et al. (eds.) ISVC 2016. LNCS, vol. 10072, pp. 234–244. Springer, Cham (2016). https://doi.org/10.1007/978-3-319-50835-1_22
23. Ronneberger, O., Fischer, P., Brox, T.: U-net: convolutional networks for biomedical image segmentation. In: Navab, N., Hornegger, J., Wells, W.M., Frangi, A.F. (eds.) MICCAI 2015. LNCS, vol. 9351, pp. 234–241. Springer, Cham (2015). https://doi.org/10.1007/978-3-319-24574-4_28

24. Salehi, S.S.M., Erdogmus, D., Gholipour, A.: Tversky loss function for image segmentation using 3D fully convolutional deep networks. In: Wang, Q., Shi, Y., Suk, H.-I., Suzuki, K. (eds.) MLMI 2017. LNCS, vol. 10541, pp. 379–387. Springer, Cham (2017). https://doi.org/10.1007/978-3-319-67389-9_44

25. Shirokikh, B., et al.: Universal loss reweighting to balance lesion size inequality in 3D medical image segmentation. In: Martel, A.L., et al. (eds.) MICCAI 2020. LNCS, vol. 12264, pp. 523–532. Springer, Cham (2020). https://doi.org/10.1007/978-3-030-59719-1_51

26. Silversmith, W.: seung-lab/connected-components-3d: Zenodo release v1. Zenodo (2021). https://doi.org/10.5281/zenodo.5535251

27. Sirinukunwattana, K., Snead, D.R., Rajpoot, N.M.: A stochastic polygons model for glandular structures in colon histology images. IEEE Trans. Med. Imaging **34**(11), 2366–2378 (2015)

28. Sudre, C.H., Li, W., Vercauteren, T., Ourselin, S., Jorge Cardoso, M.: Generalised dice overlap as a deep learning loss function for highly unbalanced segmentations. In: Cardoso, M.J., et al. (eds.) DLMIA/ML-CDS -2017. LNCS, vol. 10553, pp. 240–248. Springer, Cham (2017). https://doi.org/10.1007/978-3-319-67558-9_28

29. Wright, L., Demeure, N.: Ranger21: a synergistic deep learning optimizer. arXiv preprint arXiv:2106.13731 (2021)

30. Zhang, H., et al.: All-net: Anatomical information lesion-wise loss function integrated into neural network for multiple sclerosis lesion segmentation. NeuroImage: Clin. **32**, 102854 (2021)

31. Zhao, S., et al.: Cellular and molecular probing of intact human organs. Cell **180**(4), 796–812 (2020)

32. Zhu, W., et al.: AnatomyNet: deep learning for fast and fully automated whole-volume segmentation of head and neck anatomy. Med. Phys. **46**(2), 576–589 (2019)

Self Supervised Learning

Noise2Contrast: Multi-contrast Fusion Enables Self-supervised Tomographic Image Denoising

Fabian Wagner[1]([⊠]), Mareike Thies[1], Laura Pfaff[1,2], Noah Maul[1,2],
Sabrina Pechmann[3], Mingxuan Gu[1], Jonas Utz[4], Oliver Aust[5],
Daniela Weidner[5], Georgiana Neag[5], Stefan Uderhardt[5], Jang-Hwan Choi[6],
and Andreas Maier[1]

[1] Pattern Recognition Lab, FAU Erlangen-Nürnberg, Erlangen, Germany
fabian.wagner@fau.de
[2] Siemens Healthcare GmbH, Erlangen, Germany
[3] Fraunhofer Institute for Ceramic Technologies and Systems IKTS,
Hermsdorf, Germany
[4] Department AIBE, FAU Erlangen-Nürnberg, Erlangen, Germany
[5] Department of Rheumatology and Immunology, FAU Erlangen-Nürnberg,
Erlangen, Germany
[6] Division of Mechanical and Biomedical Engineering,
Ewha Womans University, Seoul, Korea

Abstract. Self-supervised image denoising techniques emerged as convenient methods that allow training denoising models without requiring ground-truth noise-free data. Existing methods usually optimize loss metrics that are calculated from multiple noisy realizations of similar images, e.g., from neighboring tomographic slices. However, those approaches fail to utilize the multiple contrasts that are routinely acquired in medical imaging modalities like MRI or dual-energy CT. In this work, we propose the new self-supervised training scheme Noise2Contrast that combines information from multiple measured image contrasts to train a denoising model. We stack denoising with domain-transfer operators to utilize the independent noise realizations of different image contrasts to derive a self-supervised loss. The trained denoising operator achieves convincing quantitative and qualitative results, outperforming state-of-the-art self-supervised methods by 4.7–11.0%/4.8–7.3% (PSNR/SSIM) on brain MRI data and by 43.6–50.5%/57.1–77.1% (PSNR/SSIM) on dual-energy CT X-ray microscopy data with respect to the noisy baseline. Our experiments on different real measured data sets indicate that Noise2Contrast training generalizes to other multi-contrast imaging modalities.

Keywords: Self-Supervised Denoising · Known Operator Learning · Contrast Fusion

1 Introduction

Measured data is inherently affected by uncertainty determined by the measurement process and its related physics. In image data, that uncertainty appears

© The Author(s), under exclusive license to Springer Nature Switzerland AG 2023
A. Frangi et al. (Eds.): IPMI 2023, LNCS 13939, pp. 771–782, 2023.
https://doi.org/10.1007/978-3-031-34048-2_59

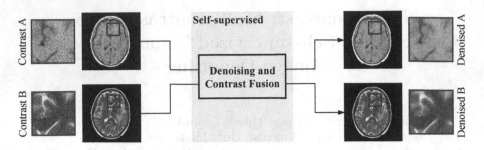

Fig. 1. Noise2Contrast: Fusion of image contrasts A and B enables self-supervised denoising, e.g., using T1 and T2 weighted MRI scans.

as image noise, disturbing an underlying ground truth image signal. Whereas imaging parameters like acquisition time, detector sensitivity, or illumination can be chosen to keep noise levels low, realistic imaging settings usually require a trade-off between acquisition parameters and image quality. In fact, some measurements, e.g., in clinical workflows, can only be carried out by accepting severe amounts of noise due to radiation exposure, acquisition time, or patient motion. Therefore, image processing algorithms were developed to reduce noise levels and extract the underlying noise-free signal. Conventional algorithms robustly denoise image data but require expert knowledge to adapt the algorithm to domain-specific conditions [14]. Unlike conventional filters, learning-based models can learn task-specific features purely from a training data distribution without domain-specific knowledge. However, deep neural networks inherently lack interpretability and were shown to be prone to prediction artifacts on out-of-domain samples [15]. Different hybrid approaches tried to combine data-driven optimization with conventional image filters to create reliable denoising operators with close to state-of-the-art performance [16].

Recently, multiple self-supervised denoising methods were proposed, circumventing the need for ground truth noise-free data during training [1,6–8]. Noise2Noise [8] and Noise2Void [7] allow self-supervised image denoising using two noisy representations of the same image or pixel-wise masking to calculate loss metrics that do not require a ground truth. Different other works applied these concepts to medical imaging modalities, e.g., by using neighboring volumetric slices [5,19] or time frames [18] as training targets following the Noise2Noise scheme.

Although self-supervised training on individual medical scans showed promising results, most existing approaches are not capable of using all available data. Many used medical imaging modalities like Magnetic Resonance Imaging (MRI) or dual-energy Computed Tomography (DECT) routinely acquire multiple image contrasts of the same scanned object that remain so far unused in self-supervised denoising approaches. In this work, we present the novel denoising method Noise2Contrast which is capable of using multiple image contrasts to train a denoising model in a fully self-supervised manner. An overview of Noise2Contrast is illustrated in Fig. 1. Our method is able to employ the independent noise realizations in different image contrasts of medical imaging modalities to train a robust denoising operator. We confirm our theoretical considerations with extensive experiments on real medical data. Our contributions are three-fold.

- We present the self-supervised denoising method Noise2Contrast combining image information from different acquired image contrasts.
- We demonstrate how to train a robust denoising operator using our proposed scheme by simultaneously learning denoising and domain transformation.
- Extensive experiments quantitatively and qualitatively confirm the applicability of our method on different real medical data sets.

2   Methods

2.1   Self-supervised Image Denoising

Each image acquisition j introduces noise n through the image formation and detection processes to the ground truth object y

$$x_i^{(j)} = y_i + n_j. \tag{1}$$

Image denoising then aims to find an operator f_w that maps noise-affected images $x_i^{(j)}$ to a denoised prediction $\hat{y}_i$ close to the noise-free ground truth y_i by minimizing

$$\operatorname*{argmin}_w \sum_i \mathcal{L}\left(f_w\left(x_i^{(1)}\right), y_i\right) \tag{2}$$

based on a loss metric $\mathcal{L}$ and parameters w. Supervised learning methods typically use a training set of N paired samples $\left(x_i^{(1)}, y_i\right)$ with $i \in \{1, \ldots, N\}$ to train a neural network data driven to predict denoised images from the learned training data distribution. As paired ground truth images are often difficult to obtain in real applications, self-supervised training methods aim to find an optimal denoising operator while having solely access to noisy images. Lehtinen et al. [8] demonstrated that learning the mapping of the noisy measurement to a second image with the same content but a different noise realization $x_i^{(2)}$, e.g., a second photo taken, is similar to solving the supervised problem in Eq. 2

$$\operatorname*{argmin}_w \sum_i \mathcal{L}\left(f_w\left(x_i^{(1)}\right), x_i^{(2)}\right). \tag{3}$$

Although many works adopt this so-called Noise2Noise training scheme, the method requires at least two images with equivalent content and contrast per sample during training which might not be available in reality. Other works, e.g., Noise2Void [7], propose masking individual pixels of noisy images to create pseudo-paired training samples $x_i^{(1\star)}$. Subsequently, a denoising model can be trained by learning to predict the correct intensity values at the masked positions. However, Noise2Void demands pixel-wise statistically independent noise which is often not satisfied in particular on real detector data and for medical imaging modalities [16].

2.2 Denoising Using Known Operators

Including prior knowledge in neural network architectures has been shown beneficial in terms of model performance, generalizability, and prediction robustness
[9,15]. We adapt the known operator learning concept in our proposed method
by separating denoising and domain-transfer tasks through the network architecture as described in Sect. 2.3. As the denoising operator, we use trainable
bilateral filter layers [16] that can be trained via gradient-based optimization
like any other neural network layer. The filter forward operation smooths image
content in homogeneous regions (spatial kernel) while preserving edges through
a range kernel

$$\hat{Y}_k = \frac{1}{\alpha_k} \sum_{n \in \mathcal{N}} G_{\sigma_s}(\|k - n\|) G_{\sigma_r}(X_k - X_n) X_n \tag{4}$$

with

$$\alpha_k = \sum_{n \in \mathcal{N}} G_{\sigma_s}(\|k - n\|) G_{\sigma_r}(X_k - X_n) \tag{5}$$

and G_{σ_s} and G_{σ_r} denoting Gaussian spatial and range kernel of width σ_s and σ_r
respectively. $\|\ldots\|$ indicates the spatial distance between pixels of index k and
n and $\mathcal{N}$ is the filter window. The differentiable implementation of Wagner et
al. [16] allows optimizing all filter parameters σ_s and σ_r data driven using deep
learning frameworks. The algorithmic filter design from Eq. 4 proves that the
bilateral filter can solely act as a denoising operator as it is not able to extract
complex features or modify the images besides local pixel intensity averaging.

In addition to data-driven optimization of a known denoising algorithm, we
demonstrate how to employ a neural network as an independent denoising operator. By training denoising and domain-transfer networks subsequently, different
image processing tasks can be entirely separated into independent network parts
to enable self-supervised denoising of multi-contrast data. The proposed training
schemes are presented in the following section.

2.3 Multi-contrast Fusion Through Domain Transfer

Measuring two images with the same content to perform Noise2Noise denoising
is often infeasible in medical imaging due to radiation and time constraints. However, modalities like MRI or DECT routinely acquire multiple image contrasts
that show the same anatomical structures but highlight different biological features. A second noisy image contrast indicated by † (e.g., T1 and T2 weighting
in MRI imaging)

$$x_i^{\dagger(j)} = y_i^{\dagger(j)} + n \tag{6}$$

could be used as a noise-affected target image in the setting of Eq. 3. In such
a setting, a network would learn to predict a denoised image with the target

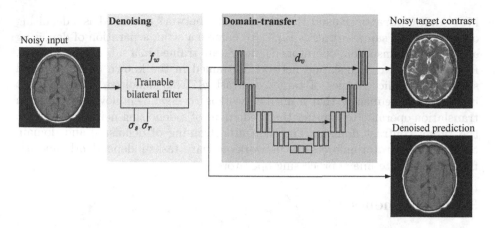

Fig. 2. Illustration of the proposed domain-transfer-based self-supervised denoising approach Noise2Contrast on the example of MRI T1 and T2-weighted contrasts. A noisy input of contrast one is processed by subsequent denoising (blue) and domain-transfer (green) operators. This allows deriving a self-supervised loss metric $\mathcal{L}$ using the noisy target with contrast two. The denoised input image is obtained by removing the domain-transfer operator. (Color figure online)

contrast. However, it is not possible to extract a solely denoised image $\hat{y}_i$ from the network prediction with preserved contrast. To avoid mixing both tasks, we propose separating the trained model into known operators to allow using the network parts individually during inference as we are only interested in the denoised prediction but want to preserve the original image contrast. An illustration of the presented training scheme is illustrated in Fig. 2. We present two solutions how to separate the denoising and domain translation tasks to enable self-supervised denoising.

Known Operator-Based. First, a known denoising operator is used in combination with a domain translation neural network d_v and trained self supervised. We use a trainable bilateral filter layer (Sect. 2.2) as the filter operation can not perform complex domain translations or intensity shifts by design and thus can be considered as a known denoising operator. Therefore, denoising and domain translation are inherently separated through the pipeline's architecture when training the chained operators d_v and f_w. The following training task results

$$\operatorname*{argmin}_{w,v} \sum_i \mathcal{L}\left(d_v\left(f_w\left(x_i^{(1)}\right)\right), x_i^{\dagger(1)}\right) \tag{7}$$

containing the domain translation operator d_v represented by a U-Net [11] with trainable parameters v. A self-supervised mean squared error loss is calculated between the denoised and domain-translated input image and the target contrast image $x_i^{\dagger(1)}$ with independent noise.

Network Operator-Based. Second, a neural network is trained as a denoising operator in the same setting as Eq. 7. To enforce a strict separation of denoising and domain translation, operators d_v and f_w are trained in a subsequent fashion. First, the domain translation network is trained in the known operator-based setting to predict images of target contrast $y_i^{\dagger}$ from denoised input contrast images $\hat{y}_i$. Subsequently, that trained network is frozen and employed as a domain translation operator to transfer the predictions of a denoising neural network to the target contrast domain. The sequential training of denoising and domain translation operator enforces the networks to learn tasks independently and use them as separate image processing operators.

3   Experiments

3.1   Data

We perform multiple experiments to investigate how noise can be effectively reduced in multi-contrast medical data without requiring noise-free ground truth data. First, we evaluate our method on three different MRI contrasts that are routinely used to identify tissue-specific properties and abnormalities: T1, T2, and Fluid Attenuated Inversion Recovery (FLAIR)-weighting. We use the public Brain-Tumor-Progression data set [12] consisting of clinical MRI head scans of 20 brain tumor patients and split it into twelve training, two validation, and six test patients. Each scan contains T1, T2, and FLAIR-weighted reconstructions that are used as input and target data to evaluate the proposed self-supervised denoising method. We simulate Gaussian noise as present in the real and imaginary part of complex-valued reconstructed MR images or in the phase-corrected magnitude images [10] and choose the noise standard deviation as 5 % of the maximum scan intensity.

In a second experiment, we compare denoising methods on a mouse tibia bone sample scanned in a dual-energy Zeiss Xradia 620 Versa X-ray Microscope (XRM). Tomographic XRM imaging is instructive for investigating bone-remodeling and bone-related diseases on the micrometer scale due to its high bone-to-soft tissue contrast. Here, dual-energy acquisitions allow quantitative measurements of bone density and sample composition [4]. However, dual-energy XRM measurements contain severe noise levels due to finite scan times and dose concerns in potential in vivo measurements [17]. We denoise a 1.5 h dual-energy scan (50 kV and 70 kV) and compare the predictions with a 14 h high-SNR acquisition that is regarded as ground truth. XRM scans are reconstructed using the pipeline of Thies et al. [13]. The two settings LE (low-energy) → HE (high-energy) and HE → LE are investigated.

3.2   Networks

Three stacked trainable bilateral filter layers [16] are employed as known operator-based denoising model f_w. The domain translation network d_v is represented by a standard U-Net [11] with 16 input features and around 1.1 Mio trainable parameters v. We use the Adam optimizer with learning rate $5 \cdot 10^{-5}$ in all our experiments. Models are trained until convergence of the self-supervised training loss computed on the validation scans in each epoch (MRI data) or on the training scan (XRM data).

3.3 Denoising Experiments

Different contrast combinations are investigated for the MRI data set to evaluate the generalizability of our proposed self-supervised denoising approach. We chose the settings T1 → T2, T2 → T1, and T2 → FLAIR for our experiments with the respective input x_i and target $x_i^{\dagger}$ contrast domains input → target and trained a model for each setting individually.

We compare our methods to the state-of-the-art blind-spot training scheme Noise2Void (N2V) [7]. In addition, we compare to a different reference method where a target image $x_i^{(2)}$ is chosen as the neighboring slice of the input image $x_i^{(1)}$. Multiple related works apply this or similar principles to create pseudo-pairs of noisy images [2,5,19]. We denote the reference approach as Noise2Neighbor (N2N) in the following as it comes close to the initial Noise2Noise idea where two noisy images of the same contrast are available. Our known operator and network operator-based methods are denoted as Noise2Contrast (BFs) and Noise2Contrast (U-Net) respectively.

4   Results

Table 1. Quantitative denoising results on the Brain-Tumor-Progression [12] MRI test data set. (mean ± std) is calculated over the patients. The best-performing method is highlighted in bold.

Setting	Method	PSNR (mean ± std)	SSIM (mean ± std)
T1 → T2	Noisy baseline	26.02 ± 0.01	0.384 ± 0.059
	Noise2Contrast (BFs)	**36.76 ± 1.40**	**0.869 ± 0.021**
	Noise2Contrast (U-Net)	30.43 ± 0.20	0.385 ± 0.119
	Noise2Void (BFs) [7]	35.81 ± 1.56	0.847 ± 0.023
	Noise2Neighbor (BFs) [2, 19]	32.49 ± 4.36	0.867 ± 0.047
T2 → T1	Noisy baseline	26.02 ± 0.02	0.444 ± 0.071
	Noise2Contrast (BFs)	**34.69 ± 1.91**	**0.865 ± 0.023**
	Noise2Contrast (U-Net)	33.19 ± 1.03	0.698 ± 0.070
	Noise2Void (BFs) [7]	34.30 ± 1.69	0.841 ± 0.024
	Noise2Neighbor (BFs) [2, 19]	29.91 ± 3.80	0.828 ± 0.067
T2 → FLAIR	Noisy baseline	26.02 ± 0.02	0.444 ± 0.071
	Noise2Contrast (BFs)	**35.21 ± 1.70**	**0.871 ± 0.022**
	Noise2Contrast (U-Net)	21.74 ± 0.05	0.221 ± 0.135
	Noise2Void (BFs) [7]	34.30 ± 1.69	0.842 ± 0.023
	Noise2Neighbor (BFs) [2, 19]	29.90 ± 3.80	0.828 ± 0.067

Table 2. Quantitative denoising results on the dual-energy XRM bone scan. (mean ± std) is calculated over the z-slices. The best-performing method is highlighted in bold.

Setting	Method	PSNR (mean ± std)	SSIM (mean ± std)
LE → HE	Noisy baseline	22.17 ± 0.29	0.158 ± 0.008
	Noise2Contrast (BFs)	**29.86 ± 0.19**	**0.622 ± 0.015**
	Noise2Void (BFs) [7]	27.28 ± 0.22	0.420 ± 0.015
HE → LE	Noisy baseline	23.15 ± 0.29	0.178 ± 0.010
	Noise2Contrast (BFs)	**30.63 ± 0.22**	**0.610 ± 0.015**
	Noise2Void (BFs) [7]	28.36 ± 0.24	0.453 ± 0.015

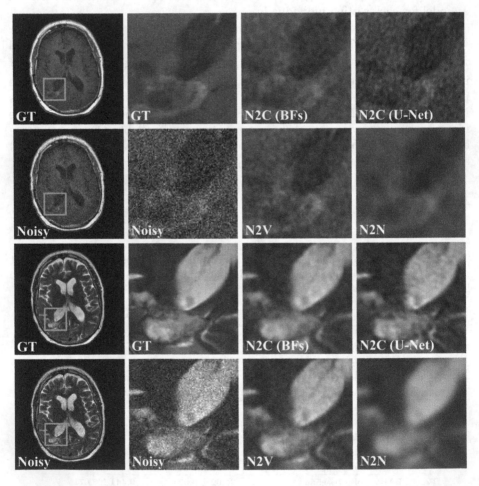

Fig. 3. Qualitative denoising results on the Brain-Tumor-Progression [12] MRI test data set of T1 → T2 (top) and T2 → T1 predictions. The images are displayed in equal windows.

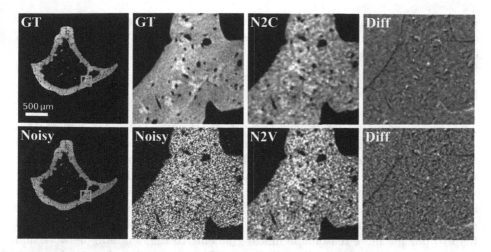

Fig. 4. Qualitative denoising results on the dual-energy XRM bone scan in the LE →
HE setting. The images are displayed in equal windows. Diff denotes the difference
images between the respective method and the high-dose ground truth.

We compute the quantitative image quality metrics peak signal-to-noise
ratio (PSNR) and structural similarity index measure (SSIM) for all model
predictions. Results on the Brain-Tumor-Progression data are presented in
Table 1. Our proposed multi-contrast training scheme using known operators Noise2Contrast (BFs) quantitatively outperforms all comparison methods.
Noise2Contrast (BFs) improves the results of Noise2Void by 4.7–11.0% PSNR
and by 4.8–7.3% SSIM with respect to the noisy baseline. Exemplary predictions visualized in Fig. 3 confirm the quantitative findings and show that our
training scheme converges to a solution that preserves features while removing the image noise. Predictions of our additional experiment using a U-Net
for denoising (Noise2Contrast (U-Net)) exhibit lower noise removal compared to
the known operator-based method. State-of-the-art Noise2Void training achieves
similar visual results compared to our method, however, predictions contain a
slightly higher noise level. Noise2Neighbor fails to predict reasonable images and
blurs high-frequency features. The lower half of the magnified regions in Fig. 3
contains a brain lesion that allows comparing perceptual noise levels on a clinical
pathology.

Results on the dual-energy XRM data are presented in Table 2 and Fig. 4.
Here, Noise2Contrast (BFs) improves the results of Noise2Void by 43.6–50.5%
PSNR and by 57.1–77.1% SSIM with respect to the noisy baseline. On par
with the quantitative metrics, the visual predictions of Noise2Void contain considerably more noise than our presented Noise2Contrast (BFs) training. This is
particularly visible in the provided difference images that are calculated between
the model predictions and the 14 h high-SNR XRM acquisitions. Note that the
low-dose network input and the high-dose ground-truth scans are independently

acquired scans. Despite the high mechanical precision of the used XRM, subsequent scanning results in small micrometer-scale shifts that are visible as thin edges in the difference images.

5   Discussion

The training configuration using a U-Net-based denoising model in combination with a domain-transfer model Noise2Contrast (U-Net) achieved promising visual results. However, the quantitative performance left room for improvement compared to the best-performing methods. We recognized that the trained denoising U-Net predicted visually appealing results but did not always fully preserve all input contrast intensities which led to poor quantitative metrics. We believe that a better-designed and more thoroughly trained domain-transfer model would help to provide more reasonable image gradients to the denoising network and improve the overall denoising performance. Pre-trained domain transfer models that are trained on ground truth data [3] can be employed here to improve the domain transfer operation. Alternatively, a regularizing loss term calculated between denoised input contrast and noisy input image can be investigated to enforce preserved intensities.

In the case of abnormalities being highlighted by the former image contrast but not being visible in the latter one or vice versa, the question arises if the Noise2Contrast training scheme can handle such samples and provide meaningful gradients to the denoising operator. We believe that as long as the domain transfer operator can learn a reasonable contrast transformation in particular a known and stable denoising operator like trainable bilateral filters that focus on local image properties can learn a reasonable denoising. In fact, the presented experiments on the Brain-Tumor-Progression MRI data contain such samples as tumors greatly vary in their visibility for different MRI contrasts. However, the generalizability of this finding must be proven on more clinical data and evaluated individually on the given imaging modality and set of pathologies.

We performed additional experiments directly mapping the input contrast to the target contrast image with a single model following the standard Noise2Noise approach. Although such models learned to simultaneously denoise and map to the target domain, their clinical application remains very limited as the model predictions inherently alter the image contrast which is usually not desired. In this Noise2Noise setting, image quality metrics calculated between model prediction and input contrast ground truth yielded poor scores as expected due to the modified prediction contrast. Additionally, we investigated a setting with a known denoising operator like the trainable bilateral filter used to predict the denoised input contrast by mapping on the target contrast without using a domain translation network. This yielded poor results likewise as the known denoising operator is not capable of learning the contrast mapping such that it only predicted blurred images to minimize the training loss.

Only a few fully self-supervised denoising techniques exist that can remove noise while preserving high-frequency image features. Blind-spot methods like

Noise2Void can achieve impressive results on certain data sets but are limited to pixel-wise independent noise statistics by design. Therefore, compelling results can be achieved on imaging modalities with simple noise characteristics and simulated data like the Brain-Tumor-Progression MRI scans in the first part of our study. Real measured data and computed tomography scans generally contain correlated noise caused by the detection process and the image reconstruction algorithm. Our experiments on real measured dual-energy XRM data confirm this limitation of Noise2Void. In contrast, our proposed known operator-based training scheme Noise2Contrast achieves considerably better quantitative and qualitative results as it relaxes prerequisites for specific noise properties like pixel-wise independent signals in the measured and reconstructed image data as demonstrated by the XRM experiments. Therefore, we conclude that Noise2Contrast is better suited to train models on modalities with correlated noise patterns like dual-energy CT compared to state-of-the-art Noise2Void training.

6 Conclusion

In this work, we presented the Noise2Contrast training scheme that allows self-supervised image denoising using multi-contrast data. Noise2Contrast combines information from independently measured image contrasts through an operator-based pipeline to train a denoising model. Our experiments on routine clinical MRI contrasts and on a pre-clinical dual-energy tomographic X-ray Microscope bone scan demonstrate superior performance of Noise2Contrast compared to the few other existing self-supervised denoising techniques. We believe that the universal Noise2Contrast training scheme can be applied on data from many more multi-contrast imaging modalities like photon-counting-CT, confocal microscopy, or hyperspectral imaging.

Acknowledgements. This work was supported by the European Research Council (ERC Grant No. 810316) and a GPU donation through the NVIDIA Hardware Grant Program.

F.W. conceived and conducted the experiments. M.T., L.P., N.M., M.G., J.U., and J.-H.C. provided valuable technical feedback during development. S.P., O.A., D.W., G.N., and S.U. prepared and scanned the bone samples. A.M. supervised the project. All authors reviewed the manuscript. L.P. and N.M. are employees of Siemens Healthcare GmbH.

References

1. Batson, J., Royer, L.: Noise2Self: blind denoising by self-supervision. In: Proceedings of the ICML, pp. 524–533. PMLR (2019)
2. Choi, K., Lim, J.S., Kim, S.: Self-supervised inter-and intra-slice correlation learning for low-dose CT image restoration without ground truth. Expert Syst. Appl. **209**, 118072 (2022)

3. Denck, J., Guehring, J., Maier, A., Rothgang, E.: Enhanced magnetic resonance image synthesis with contrast-aware generative adversarial networks. J. Imaging **7**(8), 133 (2021)
4. Genant, H.K., Boyd, D.: Quantitative bone mineral analysis using dual energy computed tomography. Investig. Radiol. **12**(6), 545–551 (1977)
5. Jeon, S.Y., Kim, W., Choi, J.H.: MM-net: multi-frame and multi-mask-based unsupervised deep denoising for low-dose computed tomography. IEEE TRPMS 1–12 (2022)
6. Kim, K., Kwon, T., Ye, J.C.: Noise distribution adaptive self-supervised image denoising using tweedie distribution and score matching. In: Proceedings of the CVPR, pp. 2008–2016 (2022)
7. Krull, A., Buchholz, T.O., Jug, F.: Noise2Void-learning denoising from single noisy images. In: Proceedings of the CVPR, pp. 2129–2137 (2019)
8. Lehtinen, J., et al.: Noise2Noise: learning image restoration without clean data. In: Proceedings of the PMLR, vol. 80, pp. 2965–2974. PMLR (2018)
9. Maier, A.K., et al.: Learning with known operators reduces maximum error bounds. Nat. Mach. Intell. **1**(8), 373–380 (2019)
10. Prah, D.E., Paulson, E.S., Nencka, A.S., Schmainda, K.M.: A simple method for rectified noise floor suppression: phase-corrected real data reconstruction with application to diffusion-weighted imaging. Magn. Reson. Med. **64**(2), 418–429 (2010)
11. Ronneberger, O., Fischer, P., Brox, T.: U-net: convolutional networks for biomedical image segmentation. In: Navab, N., Hornegger, J., Wells, W.M., Frangi, A.F. (eds.) MICCAI 2015. LNCS, vol. 9351, pp. 234–241. Springer, Cham (2015). https://doi.org/10.1007/978-3-319-24574-4_28
12. Schmainda, K.M., Prah, M.A.: Data from brain-tumor-progression. Technical report Version 1, The Cancer Imaging Archive (2018). https://doi.org/10.7937/K9/TCIA.2018.15quzvnb
13. Thies, M., et al.: Calibration by differentiation-self-supervised calibration for X-ray microscopy using a differentiable cone-beam reconstruction operator. J. Microsc. **287**(2), 81–92 (2022)
14. Tomasi, C., Manduchi, R.: Bilateral filtering for gray and color images. In: Proceedings of the ICCV, pp. 839–846. IEEE (1998)
15. Wagner, F., et al.: Trainable joint bilateral filters for enhanced prediction stability in low-dose CT. Sci. Rep. **12**(1), 1–9 (2022)
16. Wagner, F., et al.: Ultralow-parameter denoising: trainable bilateral filter layers in computed tomography. Med. Phys. **49**(8), 5107–5120 (2022)
17. Wagner, F., et al.: Monte Carlo dose simulation for in-vivo X-ray nanoscopy. In: Maier-Hein, K., Deserno, T.M., Handels, H., Maier, A., Palm, C., Tolxdorff, T. (eds.) Bildverarbeitung für die Medizin 2022. Informatik aktuell, pp. 107–112. Springer, Wiesbaden (2022). https://doi.org/10.1007/978-3-658-36932-3_22
18. Wu, D., Ren, H., Li, Q.: Self-supervised dynamic CT perfusion image denoising with deep neural networks. IEEE Trans. Radiat. Plasma Med. Sci. **5**(3), 350–361 (2020)
19. Zhang, Z., Liang, X., Zhao, W., Xing, L.: Noise2Context: context-assisted learning 3D thin-layer for low-dose CT. Med. Phys. **48**(10), 5794–5803 (2021)

Precise Location Matching Improves Dense Contrastive Learning in Digital Pathology

Jingwei Zhang[1(✉)], Saarthak Kapse[1], Ke Ma[2], Prateek Prasanna[1],
Maria Vakalopoulou[3], Joel Saltz[1], and Dimitris Samaras[1]

[1] Stony Brook University, Stony Brook, USA
{jingwezhang,samaras}@cs.stonybrook.edu,
{saarthak.kapse,prateek.prasanna}@stonybrook.edu,
Joel.Saltz@stonybrookmedicine.edu
[2] Snap Inc., New York, USA
kemma@cs.stonybrook.edu
[3] CentraleSupélec, University of Paris-Saclay, Paris, France
maria.vakalopoulou@centralesupelec.fr

Abstract. Dense prediction tasks such as segmentation and detection of pathological entities hold crucial clinical value in computational pathology workflows.
However, obtaining dense annotations on large cohorts is usually tedious and
expensive. Contrastive learning (CL) is thus often employed to leverage large
volumes of unlabeled data to pre-train the backbone network. To boost CL for
dense prediction, some studies have proposed variations of dense matching objectives in pre-training. However, our analysis shows that employing existing dense
matching strategies on histopathology images enforces invariance among incorrect pairs of dense features and, thus, is *imprecise*. To address this, we propose
a *precise location-based matching mechanism* that utilizes the overlapping information between geometric transformations to precisely match regions in two augmentations. Extensive experiments on two pretraining datasets (TCGA-BRCA,
NCT-CRC-HE) and three downstream datasets (GlaS, CRAG, BCSS) highlight
the superiority of our method in semantic and instance segmentation tasks. Our
method outperforms previous dense matching methods by up to 7.2% in average
precision for detection and 5.6% in average precision for instance segmentation
tasks. Additionally, by using our matching mechanism in the three popular contrastive learning frameworks, MoCo-v2, VICRegL, and ConCL, the average precision in detection is improved by 0.7% to 5.2%, and the average precision in
segmentation is improved by 0.7% to 4.0%, demonstrating generalizability. Our
code is available at https://github.com/cvlab-stonybrook/PLM_SSL.

Keywords: Dense contrastive learning · Self-supervised learning ·
Segmentation · Detection · Computational Pathology

1 Introduction

In computational pathology, dense prediction tasks such as segmentation and detection
are essential in analyzing digitized histology scans [24,25]. However, unlike classifica-

---

J. Zhang and S. Kapse—These authors contributed equally to this paper.

A. Frangi et al. (Eds.): IPMI 2023, LNCS 13939, pp. 783–794, 2023.
https://doi.org/10.1007/978-3-031-34048-2_60

tion, obtaining labels from pathologists for dense prediction tasks is very tedious and expensive.

Contrastive learning (CL) is being increasingly adopted as a self-supervised learning (SSL) strategy in computational pathology [4,5,17] to reduce the need for annotations. In standard CL, two augmented views are obtained from the input image, and the key idea is to pull the representations of these views closer while pushing apart representations from any other image. Popular CL methods such as SimCLR [6], MoCo [7,8,12], BYOL [11], and VICReg [2], generalize well to multiple computer vision and medical imaging tasks. In CL, traditionally, a spatial pooling operation is applied to the output feature map of the backbone network to encode each view into a global representation. These global CL approaches [6,7,11] work well when the downstream tasks involve classification problems; however, for dense prediction tasks, such representations are not optimal since they require detailed local descriptors of the images. Towards this direction, DenseCL [26] and VICRegL [3] propose incorporating local details in pre-training through leveraging dense matching objectives between the feature maps across both views. In particular, representations from local patches are extracted, and their correspondences across the different views are investigated through the dense matching operation. DenseCL utilizes feature space cosine similarity matching (denoted by M_{ft}) between local representations across the views to find the correspondence pairs. Whereas VICRegL employs spatial location of local patches of the feature maps (denoted by M_{loc}) to find the closest spatial distance between corresponding pairs across the views.

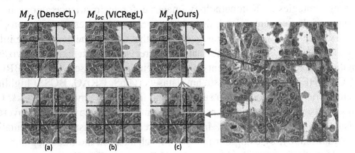

Fig. 1. Local feature matching using three methods for two random augmentations of the same image. Green arrows indicate the matching operation, and orange boxes indicate the same regions in the two views. (a) **Feature similarity-based matching** (M_{ft}), used by DenseCL [26]. The network erroneously pairs patches with multiple nuclei to a local patch containing mainly stroma and non-tissue. (b) **Location-based matching** (M_{loc}), proposed by VICRegL [3]. It matches a patch consisting of multiple nuclei to only one patch in another view containing fewer nuclei due to zooming augmentation. (c) **Precise location-based matching** (M_{pl}). We match a patch to multiple patches in the other view by incorporating exact overlapping weights between the orange boxes across the views. (Color figure online)

Need for Precise Matching in Pathology: The pitfalls of both M_{ft} and M_{loc} can be observed in Fig. 1(a) & (b) respectively. The feature-based matching, M_{ft}, erroneously

matches the local patch consisting of multiple nuclei to a patch in another view predominantly consisting of stroma and non-tissue regions. This is because their similarity
is defined on the features, which is significantly affected by the model. This could critically hamper the representation learning in histopathology as the invariance between
these two patches could force the model to focus on stroma-based descriptors while
ignoring other crucial information about cells and their morphology. Location-based
matching M_{loc} avoids this error by storing the location information after the geometric
transformations to find the pairs. However it can be observed in Fig. 1(b) that the local
patch consisting of multiple nuclei is matched with a patch in the other view containing
fewer nuclei. This is because they allow matching to strictly one patch. This invariance
could potentially enforce the network to ignore crucial information regarding cell density. Due to the *zooming* and *cropping* augmentations, a local patch in a given view may
overlap with multiple local patches in another. Formulation of M_{loc} thus has the undesired constraint that a local patch in a view can only match to one corresponding local
patch in the other, which is not precise and sub-optimal. Since histopathology images
consist of numerous fine-grained individual entities/objects, there is a need for more
precise dense matching across the views to overcome the limitations encountered and
provide better representations for dense prediction tasks.

To this end, we propose a precise location-based matching strategy, denoted by M_{pl},
which matches a local patch in a view to multiple corresponding overlapping patches in
another, as shown in Fig. 1(c). By relaxing the previous constraint, M_{pl} enables *precise
matching* between the views. We demonstrate the efficacy of our precise matching strategy in dense prediction tasks involving detection and segmentation on multiple datasets
across colon and breast cancer. Experiment results show that our precise location-based
matching outperforms previous local matching strategies, improving average precision
by up to 7.2% for detection and 5.6% for instance segmentation. We further demonstrate the generalizability of our approach by adopting M_{pl} in three popular contrastive
learning frameworks: MoCo-v2 [7], VICRegL [3], and ConCL [28]. M_{pl} shows a relative improvement in average precision by 5.2%, 1.5%, 0.7% for detection, and by 4.0%,
2.9%, 0.7% for instance segmentation with the three aforementioned CL frameworks,
respectively.

2   Method

Our method consists of a global contrastive learning part similar to MoCo, which learns
a global feature representation of an input image, and a local dense contrastive learning
part that learns the local feature representations of small local patches in an input image.
These two parts share the same backbone, while the projection heads are different. For
the rest of the paper, we use x to represent the input image and x^q and x^k to represent
the query and key images, respectively. When x^q and x^k are two randomly augmented
views of the same image, we optimize the network to pull their feature representations
closer. When x^q and x^k come from two different images, we optimize the network to
push their feature representations apart.

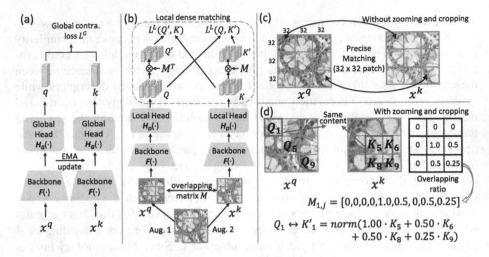

Fig. 2. Overview of the proposed method. (a) Overall structure of our global contrastive framework, the same as MoCo-v2 [7]. For each view of the image, we generate a single global feature representation that represents the entire image. (b) Overall structure of our local dense contrastive framework. For each view of the image, we generate several local feature representations. Each local feature represents a local patch in the image. (c) Without zooming and cropping, patches from two augmentations (for e.g., color jitter) precisely match with each other. (d) If the augmentation contains zooming and cropping, Q_1 matches the weighted sum of K_5, K_6, K_8 and K_9. The weights are calculated as the overlapping ratio of K_i and the red boxed region in x^k corresponding to Q_1. $M_{1,j}$ is the first row of overlapping matrix M.

2.1 Global Contrastive Learning

As shown in Fig. 2(a), in the left branch (query branch), image x^q is passed to a backbone $F(\cdot)$ and a global head $H_G(\cdot)$ to produce a global representation q as shown in Eq. 1. In the right branch (key branch), the same operation is performed on input image x^k to produce a global representation k, as shown in Eq. 1.

$$q = H_G(F(x^q)); \quad k = H_G(F(x^k)) \tag{1}$$

We then calculate the global contrastive loss $\mathcal{L}^G$ between the two global feature q and k as follows:

$$\mathcal{L}^G = \mathcal{L}_{con}(q, k) \tag{2}$$

where the $\mathcal{L}_{con}$ represents the vanilla MoCo-v2 loss [7].

2.2 Local Dense Contrastive Learning

Apart from the global head, we also have a local dense head similar to the DenseCL [26]. As shown in Fig. 2(b), in the left query branch, backbone $F(\cdot)$ and a dense head $H_D(\cdot)$ map the input image x^q to a set of local query features:

$$Q = H_D(F(x^q)) = \{Q_i\}, i = 1, \ldots, n \tag{3}$$

where n is the number of local features. Similarly, the right key branch produces a set of local key features:

$$K = H_D(F(x^k)) = \{K_i\}, i = 1, \ldots, n \tag{4}$$

Each feature Q_i or K_i in the two sets corresponds to a local patch in the original image x; we use P_i^q and P_i^k to represent such regions. As shown in Fig. 2(c), assuming x^q is a 96×96 patch and $H_D(F(x^q))$ outputs a 3×3 feature map $\{Q_i\}$, $i = 1, \ldots, 9$, each Q_i corresponds to a 32×32 P_i^q patch in x^q.

We calculate a local dense contrastive loss $\mathcal{L}^L$ between the two groups of local features Q and K as:

$$\mathcal{L}^L = \mathcal{L}^{pre}(Q, K) + \mathcal{L}^{pre}(K, Q) \tag{5}$$

where the loss is calculated by matching both Q to K and K to Q and $\mathcal{L}^{pre}(\cdot)$ is the precise location-based feature matching loss we will introduce in the next subsection.

2.3 Precise Location-Based Feature Matching

The key problem in the local dense branch is that random zooming and cropping augmentations lead to spatially mismatched features. For example, as shown in Fig. 2(c), if the augmentation operation does not contain any zooming or cropping, Q_i should precisely match to K_i, since they represent the same 32×32 patch context ($P_i^q = P_i^k, \forall i$).

However, zooming and cropping are among the key augmentations in contrastive schemes [6, 12]. To solve this problem, in this study we propose a method to address this limitation. As demonstrated in Fig. 2(d), when augmentation operations include zooming and cropping, Q_1 and K_1 are spatially mismatched, since the represented regions are different. Instead, Q_1 should match entire K_5 and part of K_6, K_8 and K_9 in the example presented in Fig. 2(d). Observing this, we use a weighted sum of $K_{5,6,8,9}$ to match Q_1, where the weights are calculated from the extent of the overlapping areas between Q_1 and $K_{5,6,8,9}$. To achieve this, we define a $n \times n$ overlapping matrix M between two augmentations, as shown in Fig. 2(b). The elements of M are defined as:

$$M_{i,j} = A(P_i^q \cap P_j^k) \tag{6}$$

where P_i^q represents the ith patch in the query augmentation and $A(x)$ is an area function that calculates the area of x. $M_{i,j}$ can be easily calculated using the bounding boxes of P^q and P^k generated during data augmentation. $M_{i,j}$ represents the overlapped area in the original image x between i^{th} patch in the left query augmentation and j^{th} patch in the right key augmentation. $M_{i,j}$ can be easily calculated from the position and size of the patches P_i^q and P_j^k.

To match the local features Q_i to features K, we need to find out the overlapping area between Q_i and all possible K. This overlapping area is the ith row of the overlapping matrix M, thus the multiplication $M_{i,*} \cdot m_K$ represents the weighted sum of all K overlapped with Q_i. Considering all i, it is (in matrix format):

$$K' = M \cdot m_K / ||M \cdot m_K||, \quad m_K = [K_1, K_2, \ldots, K_n]^\top \tag{7}$$

where $|| \cdot ||$ represents the row-wise L^2 norms. For simplicity, K' is also viewed as a set of its rows $\{K'_{i,*} | i = 1, 2, \ldots, n\}$. The same process is repeated for matching K_i to all possible Q.

$$Q' = M^\top \cdot m_Q / ||M^\top \cdot m_Q||, \quad m_Q = [Q_1, Q_2, \ldots, Q_n]^\top \tag{8}$$

We then define the weights of each pair matching. Matches are not equally important since they have different overlaps. For example, as shown in Fig. 2(d), Q_1 is covered by $K_{5,6,8,9}$. However, Q_5 only overlaps with K_9 by a small area and Q_9 does not have any overlapping with the K. We thus define the weight of matching as follows:

$$w_i^q = \sum_j M_{i,j} / A(P_i^q); \quad w_i^k = \sum_j (M^T)_{i,j} / A(P_i^k) \tag{9}$$

where $A(P_i^q)$ is the area of the ith patch in x^q in the original image x, and $A(P_i^q)$ is the area of the ith patch in x^k in the original image x, and $w_i^q, w_i^k \in [0, 1]$. The final local loss between Q and K can then be formalized as:

$$\mathcal{L}^L = \mathcal{L}^{pre}(Q, K) + \mathcal{L}^{pre}(K, Q) \tag{10}$$

$$= \frac{1}{\sum_i (w_i^q + w_i^k)} \sum_i (w_i^q \cdot \mathcal{L}_{con}(Q_i, K_i') + w_i^k \cdot \mathcal{L}_{con}(K_i, Q_i')) \tag{11}$$

where the $\mathcal{L}_{con}$ represents the contrastive loss function which can be any contrastive loss applicable to the problem.

2.4 Optimization

The joint loss $\mathcal{L}$ is defined as the sum of global and local losses as follows:

$$\mathcal{L} = (1 - \lambda)\mathcal{L}^G + \lambda\mathcal{L}^L \tag{12}$$

where $\lambda \in [0, 1]$ is a weight hyper-parameter. The parameters in the left query branch θ_q are optimized end-to-end using the gradients calculated by loss $\mathcal{L}$. The parameters in the right key branch θ_k are optimized using exponential moving average (EMA) as follows:

$$\theta_k = m\theta_k + (1 - m)\theta_q \tag{13}$$

where $m \in [0, 1)$ is a momentum coefficient. We use the MoCo-v2 contrastive learning loss, InfoNCE [16], for $\mathcal{L}_{con}$ in our experiments, given by:

$$\mathcal{L}_{con}^{q,k^+,k^-} = -\log \frac{\exp(q \cdot k^+ / \tau)}{\exp(q \cdot k^+ / \tau) + \sum_{k^-} \exp(q \cdot k^- / \tau)} \tag{14}$$

where q is a query representation, k^+ represents the positive (similar) key samples, and k^- represents the negative (dissimilar) key samples. τ is a temperature hyper-parameter. A query and a key form a positive pair if they are augmented from the same image, and otherwise form a negative pair. Our method does not have any requirement on the choice of contrastive loss for $\mathcal{L}_{con}$, making our framework generalizable to other contrastive learning frameworks.

3 Experiments and Discussion

3.1 Datasets

In our experiments, we use 5 datasets. Two of them, NCT-CRC-HE-100K [15] and TCGA-BRCA-100K [19], are used as pretraining datasets. The other three, GlaS [23], CRAG [10], and BCSS [1] are used to evaluate downstream tasks.

NCT-CRC-HE-100K. The NCT-CRC-HE-100K dataset [15] has 100,000 patches of the size 224×224 cropped from 86 H&E stained colorectal adenocarcinoma cancer and normal tissue slides. All the patches are extracted at $20\times$ magnification. The patches are annotated into nine classes. However, these patch-level labels are not utilized as we use this dataset for self-supervised pre-training. This dataset is used for pretraining followed by the downstream segmentation on the GlaS and CRAG dataset.

TCGA-BRCA-100K. The TCGA-BRCA dataset [19] has 1133 slides from patients diagnosed with either Invasive Ductal (IDC) or Invasive Lobular Breast Carcinoma (ILC). We create a dataset by randomly sampling 100,000 tissue patches at $20\times$ magnification and denote this dataset as TCGA-BRCA-100K. Since this dataset is used for pre-training followed by the downstream segmentation on the BCSS [1] dataset, we ensure the slides for the pre-training and downstream tasks do not have any patient-level overlap.

GlaS. The Gland Segmentation in Colon Histology Images (GLaS) dataset [23] has 165 images of size 775×522 cropped from 16 H&E histological sections of stage T3 or T4 colorectal adenocarcinoma. The digitization of slides is done at $20\times$ magnification. Each image contains object-instance-level annotations of both the benign and malignant glands.

CRAG. The Colorectal adenocarcinoma gland (CRAG) dataset [10] has 213 images of the size mostly around 1512×1516 collected from 38 H&E whole slide images (WSIs). The images are sampled at $20\times$ magnification. The annotations include the instance-level segmentation masks of the adenocarcinoma and benign glands in colon cancer.

BCSS. The Breast Cancer Semantic Segmentation (BCSS) dataset [1] has over 20,000 semantic segmentation annotations of tissue regions sampled from 151 H $\times$ E stained breast cancer images at $40\times$ magnification from TCGA-BRCA [19]. The annotations include the segmentation masks of 21 classes, such as Tumor, Stroma, Inflammatory, Necrosis, etc.

Table 1. Quantitative results of object detection, instance segmentation, and semantic segmentation. For GlaS and CRAG, model is pre-trained for 200 epochs on the NCT dataset. For BCSS, model is pre-trained for 200 epochs on randomly sampled patches from TCGA-BRCA. VICRegL^m corresponds to a dense matching extension of VICRegL [3] in MoCo-v2 framework.

Dataset	GlaS		CRAG		BCSS	
Metric	AP_{det}	AP_{seg}	AP_{det}	AP_{seg}	Jaccard	Dice
MoCo-v2	52.3	55.3	50.0	50.3	0.6529	0.7771
w/M_{ft} (DenseCL [26])	53.9	56.5	52.3	52.2	0.6547	0.7778
w/M_{ft} & M_{loc} (VICRegL^m [3])	51.3	56.0	53.5	51.1	0.6554	0.7783
Ours (M_{pl})	**55.0**	**57.5**	**54.5**	**54.0**	**0.6559**	**0.7787**

3.2 Implementation Details

In all the experiments, we use ResNet18 [14] as our backbone network and generate $n = 7 \times 7 = 49$ local features for Q and K. We compare pre-training ResNet18 with the multiple baseline methods and our precise location-based SSL method for 200 epochs with a batch size 256. We use the SGD optimizer with a learning rate of 0.03, weight decay of 0.0001, momentum of 0.9 and apply a cosine annealing learning rate decay policy. For downstream instance segmentation tasks on the GlaS and CRAG datasets, we use MaskRCNN [13] with Resnet18 [14] backbone. We train the network on the CRAG dataset for 15000 iterations and GlaS for 5000 iterations. We use a batch size of 16 and a base learning rate of 0.02. The other hyperparameters are the default ones in Detectron2 [27]. For downstream semantic segmentation on the BCSS dataset, we train a ResNet18 based UNet [22] using the AdamW [20] optimizer with a batch size of 32, a learning rate of $5e^{-4}$, and a cosine annealing learning rate decay. We use the PyTorch library [21], adopting the OpenSelfSup [9] code base. We train our models on NVIDIA Tesla A100 and Nvidia Quadro RTX 8000 GPUs.

3.3 Results

We evaluate the performance of our method on the three downstream datasets involving colorectal and breast cancers. After pre-training, the model is used as the backbone for the downstream segmentation tasks. Since the tasks in the GlaS and CRAG datasets involve instance segmentation, we use the COCO-style [18] metrics to evaluate the model: mean average precision for detection and segmentation, denoted by AP_{det} and AP_{seg}, respectively. For the BCSS dataset, we use the Jaccard index, and the Dice score to evaluate the quality of predictions.

Segmentation Performance Evaluation. *Pre-training:* We use vanilla MoCo-v2 as the base SSL framework. To evaluate the dense contrastive learning baselines, we use the feature-based matching DenseCL and location-based matching VICRegL in MoCo-v2. DenseCL corresponds to MoCo-v2 w/ M_{ft}, whereas MoCo-v2 w/ M_{ft} & M_{loc} corresponds to adoption of VICRegL [3] in the MoCo-v2 framework (denoted as

Table 2. Experiments on the generalizability of our proposed method on the GlaS dataset: 1) MoCo-v2 (global contrastive learning), 2) VICRegL (global + local dense contrastive learning), 3) ConCL (global + clustering-based contrastive learning).

CL method	MoCo-v2 [7]		VICRegL [3]		ConCL [28]	
Metric	AP_{det}	AP_{seg}	AP_{det}	AP_{seg}	AP_{det}	AP_{seg}
vanilla method	52.3	55.3	48.3	52.3	56.8	58.7
vanilla method w/M_{pl}	**55.0**	**57.5**	**49.0**	**53.8**	57.2	59.1

VICRegL^m). Our precise location-based matching M_{pl} in MoCo-v2 is denoted in the rest of the paper as MoCo-v2 w/ M_{pl}. In Table 1, we observe that pre-training MoCo-v2 with dense matching techniques such as M_{ft} or M_{loc} results in better performance. Our proposed dense matching method M_{pl}, unlike M_{loc}, uses better zooming and cropping augmentations and thus is a more reliable as a complete dense matching strategy. In Table 1, we empirically verify this claim by showing a consistent improvement across multiple datasets and downstream tasks. Compared to vanilla MoCo-v2, our method achieves a relative improvement in average precision of 5.2% and 9% in detection, 4.0% and 7.3% in instance segmentation on GlaS and CRAG, respectively. Compared to DenseCL, we see a consistent relative improvement of 2.0% and 4.2% in detection, 1.8% and 3.5% in instance segmentation; compared to VICRegL^m, we see a relative improvement of 7.2% and 1.9% in detection, 2.7% and 5.7% in instance segmentation on GlaS and CRAG respectively. For semantic segmentation, compared to vanilla MoCo-v2, relative improvement in Jaccard index is up to 0.45%.

Evaluation of Generalizability. We demonstrate the generalizability of our precise location-based matching by incorporating M_{pl} into popular SSL frameworks, including vanilla MoCo-v2 [7], VICRegL [3], and ConCL [28]. All experiments are performed on the GlaS dataset in this study. In Table 2, we observe that our matching method consistently boosts the performance of all the SSL frameworks. This shows that our precise location-based matching can be easily adopted by a diverse set of SSL frameworks to boost the representation learning abilities for the dense prediction tasks such as detection and segmentation.

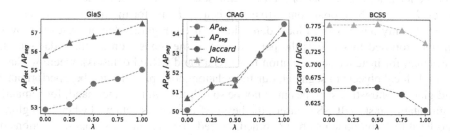

Fig. 3. Illustration of the effect of loss weight λ on model performance The optimal λ for the GlaS and the CRAG dataset is 1.0, and the optimal λ for the BCSS dataset is 0.5.

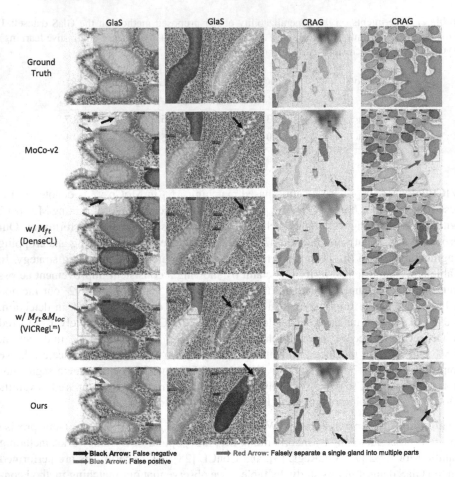

Fig. 4. Qualitative comparison on the GlaS and the CRAG datasets. Our method has fewer false negative and false positive segmentations, outperforming other methods. (Color figure online)

Ablation Study on the Hyperparameter λ. To study the optimal weight of the global and the local losses, we perform an ablation study on $\lambda \in \{0.0, 0.25, 0.5, 0.75, 1.0\}$. When $\lambda = 0.0$, the global contrastive loss is used alone for the training, whereas for $\lambda = 1.0$, only our proposed local dense loss is used. Figure 3 demonstrates that overall our proposed loss boosts the performance of the models on all three datasets. In particular, for instance segmentation tasks (GlaS and CRAG datasets), where we have multiple local objects to segment, our formulation alone provides the best performance and global contrastive schemes may not be so helpful. On the other hand, for semantic segmentation tasks (BCSS dataset), the best performance is achieved when both global and local loss components are combined. Indeed, for such tasks, global interactions of different regions are important to capture different structures.

Qualitative Comparison. To qualitatively compare the performance of our method against others, we visualize the segmentation masks and detection boxes in Fig. 4.

Different detection errors (false positives, false negatives and falsely separating a single gland into multiple parts) are indicated by arrows in different colors. Overall, our method has fewer false positive and false negative errors, outperforming previous methods and providing more robust segmentations.

4 Conclusion

In this paper, we introduced a precise location-based matching for SSL frameworks that matches a local patch in a view to multiple corresponding overlapping patches in the other view. We applied our proposed matching on two pre-training datasets and evaluated on three downstream tasks. Our method consistently outperforms state-of-the-art local matching strategies, showing substantial improvement in average precision in both detection and instance segmentation. Moreover, by using our matching mechanism, the average precision in detection and segmentation was improved in the three popular contrastive learning frameworks, demonstrating the method's generalizability. Our proposed approach shows the promising potential of local matching in self supervised learning. In future work, we will perform extensive cross-validation on the current datasets and further explore better matching mechanisms and their application to a diverse set of computational pathology tasks.

Acknowledgements. This work was partially supported by the ANR Hagnodice ANR-21-CE45-0007, the NSF IIS-2212046, the NSF IIS-2123920, the NCI UH3CA225021, the NIH 1R21CA258493-01A1, the Stony Brook Provost Venture Fund (ProFund) and generous donor support from Bob Beals and Betsy Barton.

References

1. Amgad, M., et al.: Structured crowdsourcing enables convolutional segmentation of histology images. Bioinformatics **35**(18), 3461–3467 (2019)
2. Bardes, A., Ponce, J., LeCun, Y.: VICReg: variance-invariance-covariance regularization for self-supervised learning. In: International Conference on Learning Representations (2022)
3. Bardes, A., Ponce, J., LeCun, Y.: VICRegL: self-supervised learning of local visual features. In: NeurIPS (2022)
4. Boyd, J., Liashuha, M., Deutsch, E., Paragios, N., Christodoulidis, S., Vakalopoulou, M.: Self-supervised representation learning using visual field expansion on digital pathology. In: Proceedings of the IEEE/CVF ICCV, pp. 639–647 (2021)
5. Chen, R.J., et al.: Scaling vision transformers to gigapixel images via hierarchical self-supervised learning. In: Proceedings of the IEEE CVPR, pp. 16144–16155 (2022)
6. Chen, T., Kornblith, S., Norouzi, M., Hinton, G.: A simple framework for contrastive learning of visual representations. In: ICML, pp. 1597–1607. PMLR (2020)
7. Chen, X., Fan, H., Girshick, R., He, K.: Improved baselines with momentum contrastive learning. arXiv preprint arXiv:2003.04297 (2020)
8. Chen, X., Xie, S., He, K.: An empirical study of training self-supervised vision transformers. In: Proceedings of the IEEE/CVF ICCV, pp. 9640–9649 (2021)
9. Contributors, M.: MMSelfSup: OpenMMLab self-supervised learning toolbox and benchmark (2021). https://github.com/open-mmlab/mmselfsup

10. Graham, S., et al.: MILD-net: minimal information loss dilated network for gland instance segmentation in colon histology images. Med. Image Anal. **52**, 199–211 (2019)
11. Grill, J.B., et al.: Bootstrap your own latent-a new approach to self-supervised learning. Adv. Neural. Inf. Process. Syst. **33**, 21271–21284 (2020)
12. He, K., Fan, H., Wu, Y., Xie, S., Girshick, R.: Momentum contrast for unsupervised visual representation learning. In: Proceedings of the IEEE/CVF CVPR, pp. 9729–9738 (2020)
13. He, K., Gkioxari, G., Dollár, P., Girshick, R.: Mask R-CNN. In: ICCV, pp. 2961–2969 (2017)
14. He, K., Zhang, X., Ren, S., Sun, J.: Deep residual learning for image recognition. In: Proceedings of the IEEE CVPR, pp. 770–778 (2016)
15. Kather, J., Halama, N., Marx, A.: 100,000 histological images of human colorectal cancer and healthy tissue. 1214456 (2018) https://doi.org/10.5281/zenodo
16. Lai, C.I.: Contrastive predictive coding based feature for automatic speaker verification. arXiv preprint arXiv:1904.01575 (2019)
17. Li, B., Li, Y., Eliceiri, K.W.: Dual-stream multiple instance learning network for whole slide image classification with self-supervised contrastive learning. In: CVPR (2021)
18. Lin, T.-Y., et al.: Microsoft COCO: common objects in context. In: Fleet, D., Pajdla, T., Schiele, B., Tuytelaars, T. (eds.) ECCV 2014. LNCS, vol. 8693, pp. 740–755. Springer, Cham (2014). https://doi.org/10.1007/978-3-319-10602-1_48
19. Lingle, W., et al.: Radiology data from the cancer genome atlas breast invasive carcinoma [TCGA-BRCA] collection. Cancer Imaging Arch. **10**, K9 (2016)
20. Loshchilov, I., Hutter, F.: Decoupled weight decay regularization. In: ICLR (2018)
21. Paszke, A., et al.: PyTorch: an imperative style, high-performance deep learning library. Adv. Neural Inf. Process. Syst. **32** (2019). https://proceedings.neurips.cc/paper/2019/hash/bdbca288fee7f92f2bfa9f7012727740-Abstract.html
22. Ronneberger, O., Fischer, P., Brox, T.: U-net: convolutional networks for biomedical image segmentation. In: Navab, N., Hornegger, J., Wells, W.M., Frangi, A.F. (eds.) MICCAI 2015. LNCS, vol. 9351, pp. 234–241. Springer, Cham (2015). https://doi.org/10.1007/978-3-319-24574-4_28
23. Sirinukunwattana, K., et al.: Gland segmentation in colon histology images: the GlaS challenge contest. Med. Image Anal. **35**, 489–502 (2017)
24. Tellez, D., Litjens, G., van der Laak, J., Ciompi, F.: Neural image compression for gigapixel histopathology image analysis. IEEE TPAMI **43**(2), 567–578 (2019)
25. Wang, S., Yang, D.M., Rong, R., Zhan, X., Xiao, G.: Pathology image analysis using segmentation deep learning algorithms. Am. J. Pathol. **189**(9), 1686–1698 (2019)
26. Wang, X., Zhang, R., Shen, C., Kong, T., Li, L.: Dense contrastive learning for self-supervised visual pre-training. In: Proceedings of the IEEE CVPR, pp. 3024–3033 (2021)
27. Wu, Y., Kirillov, A., Massa, F., Lo, W.Y., Girshick, R.: Detectron2 (2019)
28. Yang, J., Chen, H., Liang, Y., Huang, J., He, L., Yao, J.: ConCL: concept contrastive learning for dense prediction pre-training in pathology images. In: Avidan, S., Brostow, G., Cissé, M., Farinella, G.M., Hassner, T. (eds.) ECCV 2022. LNCS, vol. 13681, pp. 523–539. Springer, Cham (2022). https://doi.org/10.1007/978-3-031-19803-8_31

Surface Analysis and Segmentation

A Surface-Normal Based Neural Framework for Colonoscopy Reconstruction

Shuxian Wang[✉], Yubo Zhang, Sarah K. McGill, Julian G. Rosenman,
Jan-Michael Frahm, Soumyadip Sengupta, and Stephen M. Pizer

University of North Carolina at Chapel Hill, Chapel Hill, USA
{shuxian,zhangyb,jmf,ronisen,pizer}@cs.unc.edu, mcgills@email.unc.edu,
rosenmju@med.unc.edu

Abstract. Reconstructing a 3D surface from colonoscopy video is challenging due to illumination and reflectivity variation in the video frame that can cause defective shape predictions. Aiming to overcome this challenge, we utilize the characteristics of surface normal vectors and develop a two-step neural framework that significantly improves the colonoscopy reconstruction quality. The normal-based depth initialization network trained with self-supervised normal consistency loss provides depth map initialization to the normal-depth refinement module, which utilizes the relationship between illumination and surface normals to refine the frame-wise normal and depth predictions recursively. Our framework's depth accuracy performance on phantom colonoscopy data demonstrates the value of exploiting the surface normals in colonoscopy reconstruction, especially on en face views. Due to its low depth error, the prediction result from our framework will require limited post-processing to be clinically applicable for real-time colonoscopy reconstruction.

Keywords: Colonoscopy · 3D reconstruction · surface normal

1 Introduction

Reconstructing the 3D model of colon surfaces concurrently during colonoscopy improves polyp (lesion) detection rate by lowering the percentage of the colon surface that is missed during examination [7]. Often surface regions are missed due to oblique camera orientations or occlusion by the colon folds. By reconstructing the surveyed region, the unsurveyed part can be reported to the physician as holes in the 3D surface (as in Fig. 1). This approach makes it possible to guide the physician back and examine the missing region without delay.

S. Wang and Y. Zhang—These authors contributed equally to this work.

Supplementary Information The online version contains supplementary material available at https://doi.org/10.1007/978-3-031-34048-2_61.

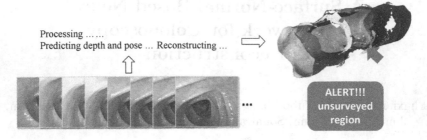

Fig. 1. Reconstructing the 3D mesh from a colonoscopy video in real-time according to the predicted depth and camera pose, allowing holes in the mesh to alert the physician to unsurveyed regions on the colon surface.

To reconstruct colon surfaces from the colonoscopy video, a dense depth map and camera position need to be predicted from each frame. Previous work [12,14] trained deep neural networks to predict the needed information in real time. With the proper help from post-processing [13,15], these methods often are able to reconstruct frames with abundant photometric and geometric features such as in "down-the-barrel" (axial) views where the optical axis is aligned with the organ axis. However, they often fail to reconstruct from frames where the optical axis is perpendicular to the surface ("en face" views). We address the problem of reconstruction from these en face views. In our target colonoscopy application, the geometry of scenes in these two viewpoints are significantly different, manifesting as a difference in depth ranges. In particular, the en face views have near planar geometry, resulting in limited geometric structures informing the photometric cues. As a result, dense depth estimation is challenging using photometric cues alone. However, the characteristics of the endoscopic environment (with a co-located light source and camera located in close proximity to the highly reflective mucus layer coating the colon) mean that illumination is a strong cue for understanding depth and the surface geometry. We capitalize upon this signal to improve reconstruction in en face views. We also aim to yield the reconstruction from frame-wise predictions with minimal post-integration to achieve near real-time execution, which requires strong geometric awareness of the network.

In this work we build a neural framework that fully exploits the surface normal information for colonoscopy reconstruction. Our approach is two-fold, 1) **normal-based depth initialization** (Sect. 3.1) followed by 2) **normal-depth refinement** (Sect. 3.2). Trained with a large amount of clinical data, the normal-based depth initialization network alone can already provide good-quality reconstructions of "down-the-barrel" video segments. To improve the performance on en face views, we introduced the normal-depth refinement module to refine the depth prediction. We find that the incorporation of surface normal-aware losses improves both frame-wise depth estimation and 3D surface reconstruction from the C3DV [3] and clinical datasets, as indicated by both measurements and visualization.

2   Background

Here we describe prior work on 3D reconstruction from endoscopic video, particularly focusing on colonoscopic applications. They usually start with a neural module to provide frame-wise depth and camera pose estimation, followed by an integration step that combines features across a video sequence to generate a 3D surface. With no ground truth from clinical data to supervise the frame-wise estimation network training, some methods transferred the prior learned from synthetic data to real data [4,16,17] while others utilized the self-consistent nature of video frames to conduct unsupervised training [12]. In order to incorporate optimization-based methods to calibrate the results from learning-based methods, Ma et al. [14,15] introduced the system with a SLAM component [5] and a post-averaging step to correct potential camera pose errors; Bae et al. [1] and Liu et al. [13] integrated Structure-from-Motion [18] with the network, trading off time efficiency for better dense depth quality.

When using widely-applied photometric and simple depth consistency objectives [2,25] in training, networks frequently fail to predict high quality and temporally consistent results due to the low geometric texture of endoscopic surfaces and time-varying lighting [23]. The corresponding reconstructions produced by these methods have misaligned or unrealistic shapes as a result. Meanwhile, recent work in computer vision has shown surface normals to be useful for enforcing additional geometric constraints in refining depth predictions [8,20,22] while the relationship between surface normals and scene illumination has been exploited in photometric stereo [9–11,19]. The success of utilizing surface normals in complex scene reconstruction inspires us to explore this property in the endoscopic environment.

3   Methods

Surface normal maps describe the orientation of the 3D surface and reflect local shape knowledge. We incorporate this information in two ways: first, to enhance unsupervised consistency losses in our normal-based depth initialization (Fig. 2a) and second, to allow us to use illumination information in our normal-depth refinement (Fig. 2b). We use this framework as initialization for a SLAM-based pipeline that fuses the frame-wise output into a 3D mesh following Ma et al. [15].

3.1   Normal-Based Depth Initialization

In order to fully utilize the large amount of unlabeled clinical data, our initialization network is trained with self-supervision signals based on the scene's consistency of frames from the same video. We particularly exploit the surface normal consistency in training to deal with the challenges of complicated colon topology in addition to applying the commonly used photometric consistency losses [2,6,25], which are less reliable due to lighting complexity in our application. Trained with the scheme described below, this network produces good depth and camera pose initialization for later reconstruction. We refer to this model as "NormDepth" or "ND" in Sect. 4.

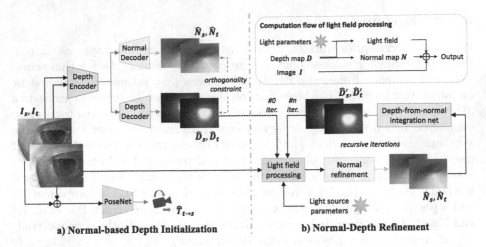

Fig. 2. Our two-fold framework of colonoscopy reconstruction. a) **Normal-based depth initialization** network is trained with self-supervised surface normal consistency loss to produce depth map and camera pose initialization. b) **Normal-depth refinement** framework utilizes the relation between illumination and surface geometry to refine depth and normal predictions.

Background - Projection. The self-supervised training losses discussed in this section are built upon the pinhole camera model and the projection relation between a source view s and a target view t [25]. Given the camera intrinsic K, a pixel p_t in a target view can be projected into the source view according to the predicted depth map $\hat{D}_t$ and the relative camera transformation $\hat{T}_{t\to s}$. This process yields the pixel's homogeneous coordinates $\hat{p}_s$ and its projected depth $\hat{d}_s^t$ in the source view, as in Eq. 1:

$$\hat{p}_s, \hat{d}_s^t \sim K\hat{T}_{t\to s}\hat{D}_t(p_t)K^{-1}p_t \tag{1}$$

Normal Consistency Objective. As the derivative of vertices' 3D positions, surface normals can be sensitive to the error and noise on the predicted surface. Therefore, when the surface normal information is appropriately connected with the network's primary predictions, i.e., the depth and camera pose, utilizing surface normal consistency during training can further correct the predictions and improve the shape consistency.

Let $\hat{N}_t$ be the object's surface normals in the target coordinate system. In the source view's coordinate system, the direction of those vectors depends on the relative camera rotation $\hat{R}_{t\to s}$ (the rotation component of $\hat{T}_{t\to s}$) and should agree with the source view's own normal prediction $\hat{N}_s$; using this correspondence we form the normal consistency objective as

$$L_{norm} = ||\hat{N}_s \langle \hat{p}_s \rangle - \hat{R}_{t \to s} \hat{N}_t ||_1 \tag{2}$$

Here, we use the numerical difference between the two vectors (L1 loss) for error. In practice, we find that using angular difference has similar performance.

Surface Normal Prediction. We found that when training with colonoscopy data, computing normals directly from depths as in some previous work [20,21] is less stable and tends to result in unrealistic shapes. Instead, we built the network to output the initial surface normal information individually, and trained it in consensus with depth prediction using L_{orth}:

$$\hat{V}(p) = \hat{D}(p_a)K^{-1}p_a - \hat{D}(p_b)K^{-1}p_b \tag{3}$$

$$L_{orth} = \sum_p \hat{N}(p) \cdot \hat{V}(p) \tag{4}$$

where $\hat{V}(p)$ is the approximate surface vector around p, which is computed from the depths of p_a and p_b, p's nearby pixels. In practice, we apply two pairs of $p_{a/b}$ position combinations, i.e., p's top-left/bottom-right and top-right/bottom-left neighboring pixels. This orthogonality constraint bridges the surface normal and depth outputs so that the geometric consistency constraint on the normal will in turn regularize the depth prediction.

Training Overview. We adapt our depth initialization network from Godard et al. [6] with an additional decoder to produce per-pixel normal vectors besides depths, and apply their implementation of photometric consistency loss L_{photo} and depth smoothness loss L_{sm}. Besides the surface normal consistency, we also enforce the prediction's geometric consistency by minimizing the difference between the predicted depths of the same scene in different frames, as in [2]:

$$L_{depth} = \frac{\left| \hat{D}_s \langle \hat{p}_s \rangle - \hat{D}_s^t \right|}{\hat{D}_s \langle \hat{p}_s \rangle + \hat{D}_s^t} \tag{5}$$

With the per-pixel mask M to mask out the stationary [6], invalid projected or specular pixels, the final training loss to supervise this initialization network is the weighted sum of the above elements, where λ_{1-4} are the hyper-parameters:

$$L^{init} = (L_{photo} + \lambda_1 L_{norm} + \lambda_2 L_{depth}) \odot M + \lambda_3 L_{orth} + \lambda_4 L_{sm} \tag{6}$$

3.2 Normal-Depth Refinement

In the endoscopic environment, there is a strong correlation between the scene illumination from the point light source and the scene geometry characterized by the surface normals. Our normal-depth refinement framework uses a combination of the color image, scene illumination as represented by the light field, and an initial surface normal map as input. We use both supervised and self-supervised consistency losses to simultaneously enforce improved normal map refinement and consistent performance across varying scene illumination scenarios.

Light Field Computation. We use the light field to approximate the amount of light each point on the viewed surface receives from the light source. As in Lichy et al. [10] we parameterize our light source by its position relative to the camera, light direction, and angular attenuation. In the endoscopic environment, the light source and camera are effectively co-located so we take the light source position and light direction to be fixed at the origin O and parallel to the optical axis $\vec{z}$, respectively. Thus for attenuation μ and depth map $\hat{D}$, we define the point-wise light field $\hat{F}$ and the point-wise attenuation $\hat{A}$ as

$$\hat{F} = \frac{O - \hat{D}}{||O - \hat{D}||}, \quad \hat{A} = \frac{(-\sum \hat{F} \cdot \vec{z})^\mu}{||O - \hat{D}||^2} \tag{7}$$

For our model input, we concatenate the RGB image, $\hat{F}$, $\hat{A}$, and normal map $\hat{N}$ (computed from the gradient of the depth map) along the channel dimension.

Training Overview. In order to use illumination in colonoscopy reconstruction, we adapt our depth-normal refinement model from Lichy et al. [10] with additional consistency losses and modified initialization. We use repeated iterations for refinement; in order to reduce introduced noise, we use a multi-scale network as in many works in neural photometric stereo [9–11]. After each recursive iteration, we upsample the depth map to compute normal refinement at a higher resolution. We denote n iterations with "$n \times$ NR".

We compute the following losses for each scale, rescaling the ground truth where necessary to match the model output. For the supervised loss L_{gt} for iteration i, we minimize L1 loss on the normal refinement module output $\hat{N}_i$ and the matching ground truth normal map N as well as the L1 loss on the depth-from-normal model output $\hat{D}_i$ and the matching ground truth depth map D. We define a scaling factor $\alpha_i = \frac{\text{median}(D)}{\text{median}(\hat{D}_i)}$.

$$L_{gt} = \sum_i ||N - \hat{N}_i||_1 + ||D - \alpha_i \hat{D}_i||_1 \tag{8}$$

For the depth-from-normal integration module, we compute a normal map $\hat{N}_i'$ from its depth output and minimize L1 loss between it and the input normal map $\hat{N}_i$; this has the effect of imposing the orthogonality constraint between the depth and surface normal maps.

$$L_{dfn} = \sum_i ||\hat{N}_i' - \hat{N}_i||_1 \tag{9}$$

We use a multi-phase training regime for stability. In an iteration, we first train the normal refinement module and substitute an analytical depth-from-normal integration method. For the second phase, we freeze the normal refinement module and train only the depth-from-normal integration module. For the third and final phase of training, we use the normal refinement module and neural integration, optimizing a weighted sum of all losses with hyperparameters λ_1 and λ_2. Thus we define the losses for each phase respectively as follows:

$$L_{refine}^{(1)} = L_{gt} + \lambda_1 L_{norm} \tag{10}$$

$$L_{refine}^{(2)} = L_{dfn} \tag{11}$$

$$L_{refine}^{(3)} = L_{gt} + \lambda_1 L_{norm} + \lambda_2 L_{dfn} \tag{12}$$

4 Experiments

In our experiments, we demonstrate that incorporating surface normal information improves both frame-wise depth estimation and 3D surface reconstruction. We describe the frame-wise depth map improvement over baseline and the effect of various ablations in Sect. 4.1. To evaluate the effect of the frame-wise depth estimation on surface reconstruction, we compare the reconstructions obtained from initializing the SLAM pipeline [15] with the outputs from various methods of frame-wise depth estimation against initialization with ground truth depth maps. We provide a comparison of Chamfer distance [24] on aligned mesh reconstructions in Table 1 and a qualitative comparison in Sect. 4.2. We also provide a qualitative comparison of the surfaces reconstructed from clinical video in Sect. 4.3.

Dataset. To train the normal-based depth initialization network, as well as the self-supervised baseline Monodepth2 [6], we collected videos from 85 clinical procedures and randomly sampled 185k frames as the training set and another 5k for validation. We used the Colonoscopy 3D Video Dataset (C3DV) [3] for the normal-depth refinement module, which provides ground truth depth maps and camera poses from a colonoscopy of a silicone colon phantom. We divide this dataset into 5 randomly-drawn cross-validation partitions with 20 training and 3 testing sequences such that the test sequences do not overlap. The results reported in Sect. 4.1 are the methods' average performance across all folds.

4.1 Frame-Wise Depth Evaluation

We compared our method's depth prediction with several ablations and the baseline against the ground truth in C3DV. Following the practice in Godard

Table 1. Error averaged over 5-fold cross validation test sets of C3DV, ± standard deviation. Best performance in bold. "NormDepth" and "ND" stand for normal-based depth initialization and "$n \times$ NR" stands for normal-depth refinement for n iterations. "NormDepth $-L_{norm}$" denotes NormDepth trained without L_{norm}. "flat init" denotes refinement initialized with planar depth rather than NormDepth output.

Method	Depth Error ↓				Chamfer Distance ↓
	Abs Rel	Sq Rel	RMSE	log RMSE	
Monodepth2 [6]	0.189	2.878	11.779	0.232	0.057 ± 0.039
NormDepth $-L_{norm}$	0.137	1.328	7.411	**0.168**	0.044 ± 0.018
NormDepth	0.141	1.373	7.447	0.173	0.046 ± 0.019
ND init + 1×NR	**0.136**	**1.271**	**7.376**	0.170	**0.038 ± 0.017**
ND init + 4×NR	0.141	1.353	7.479	0.173	0.044 ± 0.018
flat init + 4×NR	0.166	1.927	8.955	0.201	0.047 ± 0.022

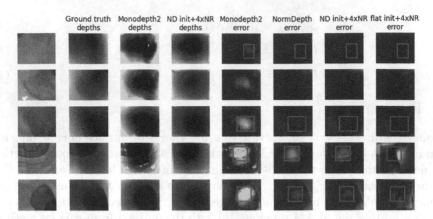

Fig. 3. Example depth predictions and RMSE from C3DV. For the depth maps, darker colors denote more distant depths. For the RMSE, brighter colors denote higher error. Some areas of improvement are highlighted in boxes. (Color figure online)

et al. [6], we rescaled our depth output to match the median of the ground truth and reported 4 pixel-wise aggregated error metrics in Table 1.

Comparing depth prediction errors (Fig. 3), both models using our two-stage method significantly outperform the photometric-based baseline Monodepth2, demonstrating the merit of emphasising geometric features (specifically surface normals) in colonoscopic depth estimation. Meanwhile, although each individual stage of our two-stage method (NormDepth and flat init+NR) already produces relatively good performance, our combined system performs even better and generates the best quantitative result on this dataset (from ND init $+ 1\times$NR). Notice that although based on the results from ablation models, the normal consistency loss L_{norm} and multi-iteration of normal refinement quantitatively do not boost performance here due to the nature of C3DV dataset, they are critical for generating better 3D reconstruction shapes (Sects. 4.2 and 4.3).

4.2   C3DV Reconstruction

In this section, we demonstrate the improvement in reconstructions of the C3DV data using our normal-aware methods. In particular, we examine the effects of initializing our SLAM pipeline with the various depth and pose estimation methods. Although C3DV provides a digital model of the phantom, here we compare against the reconstruction produced by using the ground truth depths and poses as initialization to our SLAM pipeline (and refer to this as the ground truth below). In this way, we can control for the impact of the SLAM pipeline in our reconstruction comparison. In Table 1, we measure the Chamfer distance from the ground truth to the reconstructed mesh after ordinary Procrustes alignment and optimizing the scaling factor for Chamfer distance from the ground truth to the reconstruction.

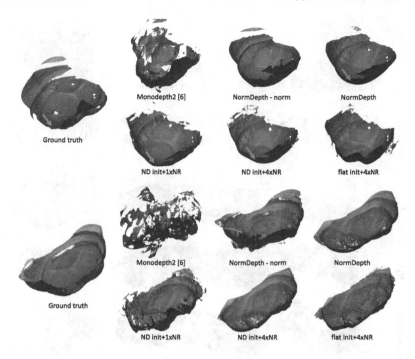

Fig. 4. Example reconstructed sequences from C3DV using various methods of initialization for SLAM pipeline. The more planar shapes observed in the ND init+$n\times$NR compared to NormDepth variations are closer to the ground truth reconstruction while the noisy reconstructions using Monodepth2 and flat init+4$\times$NR are farther from the ground truth. Select areas of improvement highlighted with arrows.

Overall, we find that the performance improvements observed in the framewise depth estimation are reflected in the reconstructions as well. Similarly, the weaknesses observed in the frame-wise inference also transfer to the reconstructions. In particular, we note that where significant noise is present in the frame-wise depth estimation for ND init+1xNR and reduced in ND init+4xNR, the corresponding reconstructions reflect the difference in noise as well.

In Fig. 4, we visualize the reconstructions corresponding to example video sequences. In these sequences, we observe that our normal-aware methods significantly outperform the baseline Monodepth2 in qualitative similarity to the ground truth. In addition, we notice that the high curvature of the surface observed in the NormDepth and NormDepth-L_{norm} reconstructions is reduced after refinement, bringing the overall reconstructed result closer to the ground truth.

4.3   Clinical Reconstructions

We tested our trained depth estimation models on clinical colonoscopy sequences and generated 3D reconstruction with the SLAM pipeline. Figure 5 shows the

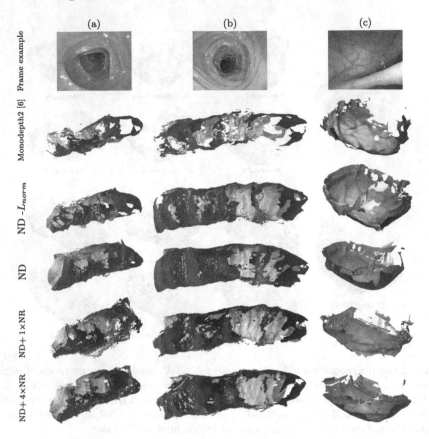

Fig. 5. 3D reconstruction results on clinical colonoscopy data. Our combined system can handle both "down-the-barrel" and en face views, outperforming the photometric baseline Monodepth2.

reconstructed meshes of two "down-the-barrel" segments (Fig. 5a and b) and an en face segment (Fig. 5c).

The reconstruction quality from the two stages of our method ("ND" and "ND+$n\times$NR") significantly outperforms the photometric baseline Monodepth2. For "down-the-barrel" sequences where features are relatively rich, we expect a generalized cylinder shape with limited sparsity. For sequence (a), we expect two large blind spots due to occlusion by ridges and a slightly curved center-line. For sequence (b), we also expect two large blind spots due to the camera position but fairly dense surface coverage elsewhere. For these sequences, our predictions' shapes are more cylindrical and have surface coverage that more accurately reflect the quantity of surface surveyed compared to the reconstruction produced using Monodepth2. The results also indicate that when trained without the normal consistency loss ($-L_{norm}$), NormDepth tends to predict more artifacts such as the skirt-shape outlier in sequence (a). This demonstrates the benefit of surface normal information in network training for improved consistency between frames. Meanwhile, using multi-scale iterations of normal-depth refinement can reduce the noise and sparsity of reconstructed meshes compared to a single iteration.

For the en-face sequence (c), we expect a nearly planar surface. Similar to the observations made in reconstructing sequences from C3DV, the high surface curvature produced from the initialization network is reduced after refinement, resulting in a more realistic reconstruction.

5   Conclusion

In this work we introduced the use of surface normal information to improve frame-wise depth and camera pose estimation in colonoscopy video and found that this in turn improves our ability to reconstruct 3D surfaces from videos with low geometric texture. We used a combination of supervised and unsupervised losses to train our multi-stage framework and found significant performance improvements over methods that do not consider surface geometry. We have also shown that the incorporation of normal-aware losses allows us to reconstruct clinical videos of low-texture en face views.

Limitations and Future Work. In this work, we have treated "down-the-barrel" and en face views separately. In practice, colonoscopy videos transition between these two view types, so constructing a framework that can also transition between view types would have significant clinical application; we leave this investigation to future work.

Acknowledgements. We thank Zhen Li and his team at Olympus, Inc. for support and collaboration and Taylor Bobrow for early access to the C3DV dataset.

References

1. Bae, G., Budvytis, I., Yeung, C.-K., Cipolla, R.: Deep multi-view stereo for dense 3D reconstruction from monocular endoscopic video. In: Martel, A.L., et al. (eds.) MICCAI 2020. LNCS, vol. 12263, pp. 774–783. Springer, Cham (2020). https://doi.org/10.1007/978-3-030-59716-0_74
2. Bian, J., et al.: Unsupervised scale-consistent depth and ego-motion learning from monocular video. Adv. Neural. Inf. Process. Syst. **32**, 35–45 (2019)
3. Bobrow, T.L., Golhar, M., Vijayan, R., Akshintala, V.S., Garcia, J.R., Durr, N.J.: Colonoscopy 3D video dataset with paired depth from 2D–3D registration. arXiv preprint arXiv:2206.08903 (2022)
4. Cheng, K., Ma, Y., Sun, B., Li, Y., Chen, X.: Depth estimation for colonoscopy images with self-supervised learning from videos. In: de Bruijne, M., et al. (eds.) MICCAI 2021. LNCS, vol. 12906, pp. 119–128. Springer, Cham (2021). https://doi.org/10.1007/978-3-030-87231-1_12
5. Engel, J., Koltun, V., Cremers, D.: Direct sparse odometry. IEEE Trans. Pattern Anal. Mach. Intell. **40**(3), 611–625 (2017)

6. Godard, C., Mac Aodha, O., Firman, M., Brostow, G.J.: Digging into self-supervised monocular depth estimation. In: Proceedings of the IEEE/CVF International Conference on Computer Vision, pp. 3828–3838 (2019)
7. Hong, W., Wang, J., Qiu, F., Kaufman, A., Anderson, J.: Colonoscopy simulation. In: Medical Imaging 2007: Physiology, Function, and Structure from Medical Images, vol. 6511, p. 65110R. International Society for Optics and Photonics (2007)
8. Li, B., Huang, Y., Liu, Z., Zou, D., Yu, W.: StructDepth: leveraging the structural regularities for self-supervised indoor depth estimation. In: Proceedings of the IEEE/CVF International Conference on Computer Vision, pp. 12663–12673 (2021)
9. Li, Z., Xu, Z., Ramamoorthi, R., Sunkavalli, K., Chandraker, M.: Learning to reconstruct shape and spatially-varying reflectance from a single image. In: SIGGRAPH Asia 2018 Technical Papers, p. 269. ACM (2018)
10. Lichy, D., Sengupta, S., Jacobs, D.W.: Fast light-weight near-field photometric stereo. In: Proceedings of the IEEE/CVF Conference on Computer Vision and Pattern Recognition (2022)
11. Lichy, D., Wu, J., Sengupta, S., Jacobs, D.W.: Shape and material capture at home. In: Proceedings of the IEEE/CVF Conference on Computer Vision and Pattern Recognition (2021)
12. Liu, X., et al.: Dense depth estimation in monocular endoscopy with self-supervised learning methods. IEEE Trans. Med. Imaging **39**(5), 1438–1447 (2019)
13. Liu, X., et al.: Reconstructing sinus anatomy from endoscopic video – towards a radiation-free approach for quantitative longitudinal assessment. In: Martel, A.L., et al. (eds.) MICCAI 2020. LNCS, vol. 12263, pp. 3–13. Springer, Cham (2020). https://doi.org/10.1007/978-3-030-59716-0_1
14. Ma, R., Wang, R., Pizer, S., Rosenman, J., McGill, S.K., Frahm, J.-M.: Real-time 3D reconstruction of colonoscopic surfaces for determining missing regions. In: Shen, D., et al. (eds.) MICCAI 2019. LNCS, vol. 11768, pp. 573–582. Springer, Cham (2019). https://doi.org/10.1007/978-3-030-32254-0_64
15. Ma, R., Wang, R., Zhang, Y., Pizer, S., McGill, S.K., Rosenman, J., Frahm, J.M.: Rnnslam: Reconstructing the 3d colon to visualize missing regions during a colonoscopy. Med. Image Anal. **72**, 102100 (2021)
16. Mahmood, F., Chen, R., Durr, N.J.: Unsupervised reverse domain adaptation for synthetic medical images via adversarial training. IEEE Trans. Med. Imaging **37**(12), 2572–2581 (2018)
17. Mathew, S., Nadeem, S., Kumari, S., Kaufman, A.: Augmenting colonoscopy using extended and directional CycleGAN for lossy image translation. In: Proceedings of the IEEE/CVF Conference on Computer Vision and Pattern Recognition, pp. 4696–4705 (2020)
18. Schonberger, J.L., Frahm, J.M.: Structure-from-motion revisited. In: Proceedings of the IEEE/CVF Conference on Computer Vision and Pattern Recognition, pp. 4104–4113 (2016)
19. Xie, W., Nie, Y., Song, Z., Wang, C.C.L.: Mesh-based computation for solving photometric stereo with near point lighting. IEEE Comput. Graphics Appl. **39**(3), 73–85 (2019). https://doi.org/10.1109/MCG.2019.2909360
20. Yang, Z., Wang, P., Wang, Y., Xu, W., Nevatia, R.: LEGO: learning edge with geometry all at once by watching videos. In: Proceedings of the IEEE/CVF Conference on Computer Vision and Pattern Recognition, pp. 225–234 (2018)
21. Yang, Z., Wang, P., Xu, W., Zhao, L., Nevatia, R.: Unsupervised learning of geometry from videos with edge-aware depth-normal consistency. In: Thirty-Second AAAI conference on artificial intelligence (2018)

22. Yu, Z., Peng, S., Niemeyer, M., Sattler, T., Geiger, A.: MonoSDF: exploring monocular geometric cues for neural implicit surface reconstruction. In: Advances in Neural Information Processing Systems (2022)
23. Zhang, Y., Wang, S., Ma, R., McGill, S.K., Rosenman, J.G., Pizer, S.M.: Lighting enhancement aids reconstruction of colonoscopic surfaces. In: Feragen, A., Sommer, S., Schnabel, J., Nielsen, M. (eds.) IPMI 2021. LNCS, vol. 12729, pp. 559–570. Springer, Cham (2021). https://doi.org/10.1007/978-3-030-78191-0_43
24. Zhou, Q.Y., Park, J., Koltun, V.: Open3D: a modern library for 3D data processing. arXiv:1801.09847 (2018)
25. Zhou, T., Brown, M., Snavely, N., Lowe, D.G.: Unsupervised learning of depth and ego-motion from video. In: Proceedings of the IEEE/CVF Conference on Computer Vision and Pattern Recognition, pp. 1851–1858 (2017)

Hierarchical Geodesic Polynomial Model for Multilevel Analysis of Longitudinal Shape

Ye Han[1], Jared Vicory[1], Guido Gerig[2], Patricia Sabin[3],
Hannah Dewey[3], Silvani Amin[3], Ana Sulentic[3], Christian Hertz[3],
Matthew Jolley[3], Beatriz Paniagua[1], and James Fishbaugh[1](✉)

[1] Kitware, Inc., Clifton Park, NY 12065, USA
james.fishbaugh@kitware.com
[2] NYU Tandon School of Engineering, Brooklyn, NY 11201, USA
[3] Children's Hospital of Philadelphia, Philadelphia, PA 19104, USA

Abstract. Longitudinal analysis is a core aspect of many medical applications for understanding the relationship between an anatomical subject's function and its trajectory of shape change over time. Whereas mixed-effects (or hierarchical) modeling is the statistical method of choice for analysis of longitudinal data, we here propose its extension as hierarchical geodesic polynomial model (HGPM) for multilevel analyses of longitudinal shape data. 3D shapes are transformed to a non-Euclidean shape space for regression analysis using geodesics on a high dimensional Riemannian manifold. At the subject-wise level, each individual trajectory of shape change is represented by a univariate geodesic polynomial model on timestamps. At the population level, multivariate polynomial expansion is applied to uni/multivariate geodesic polynomial models for both anchor points and tangent vectors. As such, the trajectory of an individual subject's shape changes over time can be modeled accurately with a reduced number of parameters, and population-level effects from multiple covariates on trajectories can be well captured. The implemented HGPM is validated on synthetic examples of points on a unit 3D sphere. Further tests on clinical 4D right ventricular data show that HGPM is capable of capturing observable effects on shapes attributed to changes in covariates, which are consistent with qualitative clinical evaluations. HGPM demonstrates its effectiveness in modeling shape changes at both subject-wise and population levels, which is promising for future studies of the relationship between shape changes over time and the level of dysfunction severity on anatomical objects associated with disease.

Keywords: geodesic regression · statistical shape analysis · hierarchical modeling · longitudinal data

1 Introduction

Studying change over time is a core aspect of many medical applications. Trajectories of change are followed in studies of childhood development, aging, and

© The Author(s), under exclusive license to Springer Nature Switzerland AG 2023
A. Frangi et al. (Eds.): IPMI 2023, LNCS 13939, pp. 810–821, 2023.
https://doi.org/10.1007/978-3-031-34048-2_62

disease development. This can involve sampling a cross-sectional population to estimate a possible time course. However, such cross-sectional studies may show large variability across the population and may not properly reflect the nature of longitudinal changes of individuals. On the other hand, longitudinal studies involve following subjects over time, allowing to capture subject-wise trajectories as well as at the population level. Longitudinal study design comes with data challenges such as staggered time points, missing time points, and subjects with different number of observations. Dedicated modeling schemes are needed to correctly account for the correlated measurement within subjects. These models are known as mixed-effects or hierarchical models and have shown great promise for modeling derived measure in medical imaging studies [1,17].

Several models have been explored for longitudinal analysis of higher dimension data [3,4,11,13,16,18]. Two main directions have been followed. First, dedicated methods with a specific data representation in mind such as diffeomorphisms on images or shapes. Image or shape change is represented as continuous diffeomorphic deformations at the subject and population level. The second methodological direction are intrinsic Riemannian models which may be adapted to new data representations from a variety of manifold representations. These models require the definition of a few key manifold specific operations, to be discussed in Sect. 2, in order to be applied to new data types. In this work, we favor this second approach due to the potential for extension to new data representations and thus a variety of clinical problems.

In this paper we propose to extend the hierarchical multi-geodesic model in [11] by applying polynomial expansion at both subject-wise and population levels, with the aim of modeling shape trajectories associated to changes in related covariates, such as sex, cognitive scores, or disease severity. Our development of polynomial regression allows for more flexibility in data-matching than the traditional geodesic model, while still enabling the choice of geodesic as a polynomial of degree 1. The polynomial expansions at different model levels are inherently compatible with the hierarchical modeling framework where subjects may have a different number of observations. Due to the non-Euclidean nature of the shape space, a fast and efficient model estimation algorithm similar to the computation of the Fréchet mean is implemented. We validate our method on synthetic data as well as clinical data right ventricle shape change over the cardiac cycle as it relates to covariates such as dysfunction severity. This promises to meet a currently unmet need of clinical researchers to correlate geometry with function.

2   Methods

2.1   Shape Space and Geodesics

We define shape space as the pre-shape space of Kendall space [12] with rotations removed. As such, the final shapes are obtained by removing their translation, rotation, and similarity components through partial procrustes alignment. The shape space is formed as a hyper-sphere and can be treated as a high dimensional Riemannian manifold M. Performing geodesic regressions in the shape space allows for efficient computation with proven Kendall-space equivalence as indicated in [8,15]. A geodesic on M is a zero-acceleration curve with the minimizing property that there is no curve shorter than a geodesic between any two points within a small neighborhood. Three geodesic-related operations are extensively used in this work: exponential map, log map, and parallel transport. An exponential map $Exp(p, v) = q$ maps a shape $p \in M$ to another shape $q \in M$ in the direction and magnitude of a tangent vector v. A log map $Log(p, q) = v$ is the inverse of the exponential map in which two shapes p and q are given and the unique tangent vector that maps p to q is obtained. The Riemannian distance between the two shapes is then defined as the L2-norm of their log map $d(p, q) = ||Log(p, q)||$. The parallel transport operation $\psi_{p \to q}(u)$ transports a tangent vector $u \in T_p M$ from p to q while maintaining angle and scale preservation properties. For rigorous and complete definitions, please see [2].

2.2   Hierarchical Geodesic Model for Manifold-Valued Data

Geodesic regression is very similar to linear regression in Euclidean space, with analogies of the anchor point to the intercept and tangent vector to the slope. In a similar way, multilevel models could be constructed in a "geodesic" way by following the framework of hierarchical linear models [19]. In this study, we further extend geodesic regressions in [11] to higher order polynomial versions at both the subject-specific trajectory level and the population level for better adaptability in the modeling of longitudinal data with covariate induced variability. Unlike the Riemannian polynomial described in [9,10] where the polynomial is defined in a differential manner with covariant derivatives, our polynomial expansions are applied to the tangent vectors under the geodesic regression model in an algebraic form, making them straightforward and consistent to fit into different levels of the hierarchical model. Thus, geodesic polynomial regression in this study refers to expanding the composition of tangent vectors in their hypertangent space with polynomials of different orders.

Subject-Wise Level Model. We first perform geodesic polynomial regression at the subject-wise level (level 1). The nth order polynomial model on subject-specific trajectory Y_k is formulated as

$$Y_k = Exp(\hat{a_k}, \sum_{p=1}^{n} \hat{b_{kp}} t^p) \tag{1}$$

where $\hat{a}_k$ is the anchor point of subject-specific trajectory k, $\hat{b}_{kp}$ is the tangent vector of the pth polynomial term and t is the independent time variable. Given the input observations $\boldsymbol{y_k}$, $\hat{a}_k$ and $\hat{b}_{kp}$'s are estimated by least squares geodesic regression

$$(\hat{a}_k, \hat{\boldsymbol{b}}_{\boldsymbol{k}}) = \underset{a_k, \boldsymbol{b_k}}{\operatorname{argmin}} \sum_{i=1}^{N_{obs}} d^2(y_{ki}, Exp(a_k, \sum_{p=1}^{n} b_{kp} t_{ki}^p)) \tag{2}$$

where $\hat{b}_k(b_k)$ is the combined representation of all $\hat{b}_{kp}(b_{kp})$, y_{ki} and t_{ki} are the ith observation and corresponding time variable in y_k. Note that due to the number of free parameters in the regressing polynomial, $N_{obs} \geq n+1$ is required to avoid singularity in the solution.

Population Level Model. At population level (level 2), let the subject trajectories be associated with a set of m covariates $\eta = \{\eta_1, \eta_2, \ldots, \eta_m\}$, the final form of the hierarchical geodesic polynomial model can be written as

$$Y = Exp(Exp(f(\eta), \sum_{p=1}^{n} g_p(\eta)t^p), \epsilon) \tag{3}$$

where $f(\eta)$ and $g_p(\eta)$ are the models for the anchor point and the basis tangent vector of polynomial order p respectively. Technically speaking, the two models do not necessarily share the same set of covariates to allow for more flexible regression. However, in most cases, the same set of covariates would be used for both models if there is no compelling indication that a specific covariate is solely associated with one of the models. We promote both the anchor point model and the tangent vector model with quadratic expansion on the covariates. The aim is to associate each model with higher order terms of the covariates as well as cross terms to accurately model their combined effects on subject trajectories.

Anchor Point Model. The anchor point model with quadratic expansion on m covariates is written as

$$f(\eta) = Exp(\hat{\beta}_0, \sum_{i=1}^{m}(\hat{\beta}_i\eta_i + \hat{\beta}_{ii}\eta_i^2) + \sum_{i=1}^{m-1}\sum_{j=i+1}^{m} \hat{\beta}_{ij}\eta_i\eta_j) \tag{4}$$

where $\hat{\beta}_0 \in M$ is a base anchor point and $\hat{\beta}_i, \hat{\beta}_{ii}, \hat{\beta}_{ij} \in T_{\hat{\beta}_0}M$ are basis vectors of the anchor point polynomial. These coefficients can be estimated from the results of subject-specific trajectory regression as

$$\hat{\beta} = \underset{\beta}{\operatorname{argmin}} \sum_{k=1}^{N_s} d^2(f(\eta_k), \hat{a}_k) \tag{5}$$

where η_k and $\hat{a}_k$ are the covariates and regressed anchor point of subject k, N_s is the total number of input subject trajectories and $\hat{\beta}$ is the combined representation of $\hat{\beta}_0, \hat{\beta}_i, \hat{\beta}_{ii}, \hat{\beta}_{ij}$ in the anchor point model.

Tangent Vector Model. Recall that the $\hat{b}_k$ is the combined representation of the n tangent vector bases of different orders from the level 1 regression. Thus, there are n corresponding tangent vector models at the population level $G(\eta) = \{g_1(\eta), g_2(\eta), \ldots, g_n(\eta)\}$. Each tangent vector model with quadratic expansion on m associated covariates can be formulated as

$$g(\eta) = \psi_{\hat{\beta}_0 \to f(\eta)}(g_{\hat{\beta}_0}(\eta)) \tag{6}$$

$$g_{\hat{\beta}_0}(\boldsymbol{\eta}) = \hat{\gamma}_0 + \sum_{i=1}^{m}(\hat{\gamma}_i\eta_i + \hat{\gamma}_{ii}\eta_i^2) + \sum_{i=1}^{m-1}\sum_{j=i+1}^{m}\hat{\gamma}_{ij}\eta_i\eta_j \qquad (7)$$

where $g \in T_{f(\boldsymbol{\eta})}M$ and $g_{\hat{\beta}_0} \in T_{\hat{\beta}_0}M$ are the tangent vectors at $f(\boldsymbol{\eta})$ and $\hat{\beta}_0$ respectively, $\hat{\gamma}_0, \hat{\gamma}_i, \hat{\gamma}_{ii}, \hat{\gamma}_{ij} \in T_{\hat{\beta}_0}M$ are bases of the tangent vector polynomial at $\hat{\beta}_0$. Note that the subscript referring to the order of the polynomial model p is omitted here for readability purposes. From the above formulation, the final tangent vector basis is obtained by calculating a tangent vector at $\hat{\beta}_0$ from the polynomial model and then transporting it to the corresponding anchor point $f(\boldsymbol{\eta})$ by a parallel transport function $\psi_{\hat{\beta}_0 \to f(\boldsymbol{\eta})}$ defined on M. This is due to the consideration that subject-specific tangent vectors must be comparable with each other to perform regression in a consistent manner. Therefore, we need to transport all of them to the same tangent vector space $T_{\hat{\beta}_0}M$ for the regression calculation, as well as to transport them to their corresponding anchor point $f(\boldsymbol{\eta})$ in the forward calculation. Due to the existence of the parallel transport functions and the fact that regressed subject-specific anchor points $\hat{a}_k$ do not necessarily lie on $f(\boldsymbol{\eta})$, the actual tangent vector $\tilde{b}_k$ being used for regression calculation is obtained as

$$\tilde{b}_k = \psi_{f(\boldsymbol{\eta_k}) \to \hat{\beta}_0}(\psi_{\hat{a}_k \to f(\boldsymbol{\eta_k})}(\hat{b}_k)) \qquad (8)$$

so that all $\tilde{b}_k \in T_{\hat{\beta}_0}M$. Stop-over transport avoids arbitrary rotation from direct transport $\psi_{\hat{a}_k \to \hat{\beta}_0}$, explained in [11]. Regression on the basis tangent vector of a specific polynomial order is then formulated as

$$\hat{\boldsymbol{\gamma}} = \underset{\gamma}{\mathrm{argmin}} \sum_{k=1}^{N_s} ||g_{\hat{\beta}_0}(\boldsymbol{\eta_k}) - \tilde{b}_k||^2 \qquad (9)$$

where $\hat{\boldsymbol{\gamma}}$ is the combined representation of $\hat{\gamma}_0, \hat{\gamma}_i, \hat{\gamma}_{ii}$, and $\hat{\gamma}_{ij}$.

Iterative Optimization Scheme. Since our shape space is constructed as a Riemannian manifold and thus not Euclidean, there exists no closed-form solution for such geodesic polynomial regressions on subject-wise trajectory and anchor point model. Similar to the calculation of Fréchet mean, we employ an iterative solution scheme for obtaining the optimal parameters in Eq. (2) and Eq. (5). The Algorithm 1 illustrates how the parameters are updated over iterations. Note that the function $LeastSquaresPolynomialFitting$ depends on ω_0, which means that all points in $\boldsymbol{y}$ are transformed to the hyper-tangent space at ω_0 using a log map for calculating new parameters ω_{new}.

R^2 and Hypothesis Testing. At subject-wise level, the R^2 is calculated using the Fréchet variance [14], intrinsically defined by

$$var(y_i) = \min_{y \in M} \frac{1}{N} \sum_{i=1}^{N} d(\bar{y}, y_i)^2 \qquad (10)$$

Algorithm 1: Iterative solution scheme

---

input : x: timestamps in Eq. (2) or covariates in Eq. (5)

  y: observation points in Eq. (2) or regressed anchor points in Eq. (5)

output: $\omega = \{\omega_0, \omega_t\}$: anchor point (intercept) and tangent vectors (slope)

  from model fitting, (a_k, b_k) in Eq. (2) or β in Eq. (5)

---

initialization: $i \leftarrow 0$; $E \leftarrow \infty$; $\omega_0 \leftarrow y_0$

while $i < MaxIterations$ **do**

  for $j \leftarrow 1$ to N_{entry} **do**

  | $\omega_{new,j} \leftarrow LeastSquaresGeodesicPolynomialFitting(x, y, \omega_0)$

  end

  $E_{new} \leftarrow ComputeRegressionError(x, y, \omega_{new})$;

  if $E_{new} \geq E$ **then**

  | return ω;

  else

  | $E = E_{new}$;

  | $\omega = \omega_{new}$

  end

  $i = i + 1$

end

$$var(\epsilon_i) = \min_{y \in M} \frac{1}{N} \sum_{i=1}^{N} d(\hat{y}_i, y_i)^2 \qquad (11)$$

$$R^2 = 1 - \frac{var(\epsilon_i)}{var(y_i)} \qquad (12)$$

where $\bar{y}$ is the Fréchet mean, $\hat{y}_i$ is the regressed point i, and ϵ_i is the error between $\hat{y}_i$ and observation point y_i. To test statistical significance of fitting a geodesic polynomial model with respect to time, hypothesis test is conducted against the null hypothesis H_0: *t is irrelevant to change in shape* using the permutation approach described in [6,7].

3   Results and Discussion

3.1   Test on Low Dimensional Synthetic Data

In order to validate as well as to visualize our hierarchical geodesic polynomial model, we first test our implementation on points on the unit sphere. Points on the sphere represent shapes with only one 3D point and the sphere is the corresponding shape space. On the left of Fig. 1 shows the result from fitting four input points (red) with geodesic model (green) and 3rd order geodesic polynomial model (blue) respectively. The four input points have integer times steps ranging from 0 to 3. The 3rd order geodesic polynomial is able to fit the input points almost perfectly, whereas as the linear geodesic model can only fit the inputs in a least square sense, which is similar to the case in Euclidean space.

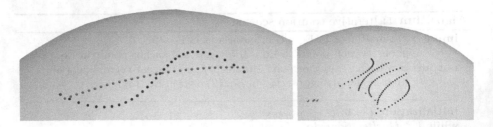

Fig. 1. Left) Results from fitting input points (red) with geodesic model (green) and 3rd order polynomial model (blue). **Right**) Fitting three input trajectories (red) with hierarchical geodesic polynomial model. Green points represent changes in anchor point location with respect to the covariate values ranging from 0 to 2, and the blue points show changes in regressed subject-specific trajectory with respect to the covariate values. (Color figure online)

The right of Fig. 1 shows regression results from fitting three input trajectories with a hierarchical geodesic polynomial model. Each input trajectory contains three points with integer timestamps ranging from 0 to 2. The three input trajectories represent trajectories with integer covariate values ranging from 0 to 2 respectively. A quadratic model is used at both the subject-specific and the population level. The green points on the left of Fig. 1 also show the changes in anchor points from fitting a quadratic model, with covariate values ranging from 0 to 2 with an interval of 0.25. The blue points demonstrate the changes in regressed trajectories for covariates ranging from 0 to 2 with an interval of 0.5. Given the three input trajectories, our hierarchical geodesic polynomial model is capable of fitting the inputs perfectly with a quadratic model at both subject-wise level and population level.

3.2   Analysis of 4D Pediatric Right Ventricular Data

The shape of the right ventricle (RV) is known to influence the function of the tricuspid valve, but precise shape-based characterization of the RV in Hypoplastic left heart syndrome (HLHS) has not been described. The 4D trajectories of 94 pediatric RVs are acquired from 3D echocardiogram-based speckle tracking, and then transferred into the TOMTEC imaging system for the computation of the 3D models of the RV chamber. Each acquisition contains approximately one cardiac cycle, with 10 to 30 captured frames. Due to some incomplete cardiac cycle acquisitions as well as the fact that some obtained cardiac cycles do not perfectly repeat themselves from the end to the start point, the data is divided into two subsets, the systolic and diastolic phases. Since lengths of individual cardiac cycles can be different, we standardize trajectories for each systolic and diastolic phase with equally-spaced 50 frames in each trajectory [5]. As such, we finally obtain 58 systolic and 36 diastolic trajectories for hierarchical geodesic regression analysis. To evaluate the impact our polynomial model, we compare

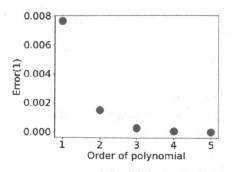 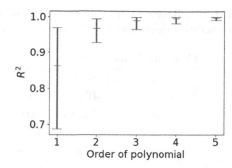

Fig. 2. Left) Error in fitting a single subject trajectory with respect to the order of the polynomial model. **Right**) Means and ranges of R^2 from fitting systole population with models of different orders.

to the geodesic model [11], as to our knowledge it is the only longitudinal shape model that incorporates multiple covariates.

Subject-Wise Model. We first fit a polynomial model to subject-specific trajectories with the left of Fig. 2 showing the resulting error from fitting models to a representative single subject trajectory. As the order of polynomial model increases, the regression error decreases significantly from 7.65×10^{-3} (1st order) to 3.82×10^{-5} (5th order) with a 99.5% reduction (unitless because shapes are normalized to unit size). Figure 3 shows that the trajectory from a higher order polynomial regression exhibits more nonlinearity than the geodesic model, as expected. Though not obvious, it is observed that the maximum distance point shifts from the left to the right side of the shape in the polynomial model, whereas the most distant point remains the same point in the geodesic model throughout time. The right of Fig. 2 shows the ranges and the mean values of the R^2 across the systole population, which indicates that trajectories of shape change over time better matches observed data with models of higher order.

Anchor Point Model. As more samples are desirable for regression analysis, we first test our anchor point model on all end systolic and end diastolic shapes in both systolic and diastolic trajectories. Two covariates from the demographics are chosen for level 2 models: tricuspid regurgitation severity (TRS) and right ventricle function (RVF) take on values shown in Table 1.

We fit both geodesic linear models and quadratic polynomial models to the end systolic and end diastolic shapes, with the RVF and/or TRS as covariates. Table 2 shows regression errors. It can be seen that (i) quadratic models lead to smaller errors than geodesic models, and (ii) model fitting with respect to TRS leads to smaller errors than the ones with respect to RVF, which indicates the shape changes are better aligned with changes in TRS, and (iii) regressions using both covariates outperform those using single covariate, in terms of model fitting, which is expected as more independent covariates are taken into account.

Table 1. Relationship between qualitative clinical assessments in the demographics to numerical values used in polynomial model fitting.

Covariate Value	TRS	RVF
0	Trivial	Normal
1	Mild	Low normal
1.5	Mild to moderate	Low normal to mildly diminished
2	Moderate	Mildly diminished
2.5	Moderate to severe	Mildly to moderately diminished
3	Severe	Moderately diminished
3.5		Moderately to severely diminished
4		Severely diminished

From visual observations of the shape changes, similar to the results from fitting polynomial model to subject specific trajectories, the higher order polynomial model fitting including covariates yield more nonlinear changes in the end systolic and end diastolic shapes. While RV shape changes with respect to RVF is smoother, shape changes associated with TRS shows more local variability over time, leading to an obvious compressed edge between the RV top region and the septal wall at the most severe level of tricuspid regurgitation.

Fig. 3. A geodesic and a 5th order polynomial model on the systolic trajectory of a representative subject. Color indicates distance to the initial shape.

Table 2. Error from fitting multivariate polynomial model to anchor points.

Phase	Covariates	Linear model errors (1)	Quadratic model errors (1)
End diastole	RVF	5.521	5.487
	TRS	5.515	5.442
	RVF & TRS	5.454	5.208
End systole	RVF	5.178	5.138
	TRS	5.147	5.062
	RVF & TRS	5.102	4.840

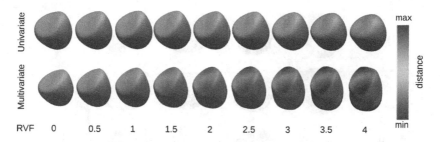

Fig. 4. Univariate and multivariate geodesic polynomial regression results for end diastolic shapes at different RVF severities.

It is also observed that changes in shape with respect to the same covariates from multivariate regression are more prominent. Figure 4 and 5 show that the multivariate regressions yield more observable changes in shapes as the covariate-specific changes of the shape are co-captured by different covariates separately.

Figure 6 shows the full spectrum of how the shape of the RV changes with respect to different values of the covariates at end diastole. Due to the sparsity and large variability in the input data set, extrapolating RV shape to extreme values of both covariates leads to a non-feasible real world shape as no such combination appeared in the input data set.

3.3 Future Work

There are a few aspects that can be further extended to our current work. First, the current parallel transport of tangent vectors is computed along the geodesic between the start and end points. Meanwhile at the population level, it is also possible that the directions of the tangent vectors are dependent on the anchor points' trajectories, in which parallel transport should be computed along a certain path on the Riemannian manifold (c.g. the regressed anchor point's trajectory in the single covariate case). In the case with multiple covariates, the choice of the path requires further study. Second, if scale is a key factor to consider

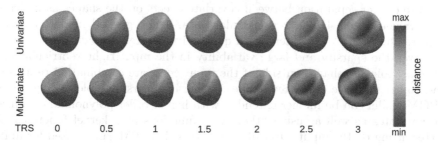

Fig. 5. Univariate and multivariate geodesic polynomial regression results for end diastolic shapes at different levels of TRS.

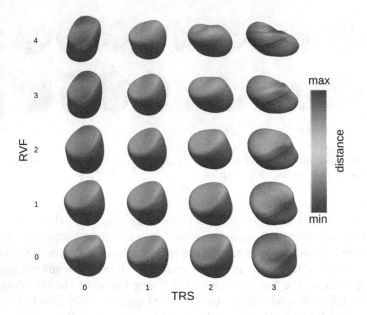

Fig. 6. Full spectrum of end diastolic shapes at various levels of RVF and TRS.

in the shape model, it is also feasible to append a scale factor to the existing model, which is regarded as an additional entry in shape space, and the solution process would be almost identical to the other entries in the anchor point or tangent vector. As we collect more data, we will also investigate modeling growth or pathology models of HLHS over a larger time period (i.e. years) instead of the cardiac cycle.

4   Conclusions

In comparison with previous geodesic models, polynomial regression leads to more accurate and flexible data-matching results at both subject-wise and population levels. Population-level regression with respect to multiple covariates leads to clearer separation between covariate effects on the shapes as indicated from validation on 4D right ventricular data. The regressed model is able to yield results that are consistent with qualitative clinical evaluations.

Given the sparsity and large variability in the input right ventricular data set, extrapolating shapes outside of the input covariate combination range may lead to irregular reconstructed shapes, which is understandable. The proposed HPGM model can be further extended with higher order polynomial expansion on covariates as well as using other basis functions (e.g. kernel functions) for better fitting on the input. Overall, the proposed HGPM can be used for multi-level analysis of longitudinal shape data, leading to interpretable results relating functions (covariates) with shape trajectories, thus being promising for a variety of relevant clinical research in the future.

References

1. Bernal-Rusiel, J.L., et al.: Statistical analysis of longitudinal neuroimage data with linear mixed effects models. Neuroimage **66**, 249–260 (2013)
2. do Carmo, M.P.: Differential Geometry of Curves and Sur4. Prentice Hall (1976)
3. Durrleman, S., Pennec, X., Trouvé, A., Braga, J., Gerig, G., Ayache, N.: Toward a comprehensive framework for the spatiotemporal statistical analysis of longitudinal shape data. Int. J. Comput. Vision **103**(1), 22–59 (2013)
4. Durrleman, S., Pennec, X., Trouvé, A., Gerig, G., Ayache, N.: Spatiotemporal atlas estimation for developmental delay detection in longitudinal datasets. In: Yang, G.-Z., Hawkes, D., Rueckert, D., Noble, A., Taylor, C. (eds.) MICCAI 2009. LNCS, vol. 5761, pp. 297–304. Springer, Heidelberg (2009). https://doi.org/10.1007/978-3-642-04268-3_37
5. Fishbaugh, J., Gerig, G.: Acceleration controlled diffeomorphisms for nonparametric image regression. In: ISBI, pp. 1488–1491 (2019)
6. Fletcher, P.T.: Geodesic regression on riemannian manifolds. In: MICCAI MFCA, pp. 75–86 (2011). https://hal.inria.fr/inria-00623920
7. Fletcher, P.T.: Geodesic regression and the theory of least squares on Riemannian manifolds. IJCV **105**(2), 171–185 (2013)
8. Guigui, N., Maignant, E., Trouvé, A., Pennec, X.: Parallel transport on kendall shape spaces. In: GSI, pp. 103–110 (2021)
9. Hinkle, J., Muralidharan, P., Fletcher, P.T., Joshi, S.: Polynomial regression on Riemannian manifolds. In: ECCV, pp. 1–14 (2012)
10. Hinkle, J., Muralidharan, P., Fletcher, P.T., Joshi, S.: Intrinsic polynomials for regression on Riemannian manifolds. J. Math. Imaging Vision (2014)
11. Hong, S., Fishbaugh, J., Wolff, J.J., Styner, M.A., Gerig, G.: Hierarchical multi-geodesic model for longitudinal analysis of temporal trajectories of anatomical shape and covariates. In: Shen, D., et al. (eds.) MICCAI 2019. LNCS, vol. 11767, pp. 57–65. Springer, Cham (2019). https://doi.org/10.1007/978-3-030-32251-9_7
12. Klingenberg, C.P.: Walking on Kendall's shape space: understanding shape spaces and their coordinate systems. Evol. Biol. **47**, 1–19 (2020)
13. Lorenzi, M., Pennec, X., Frisoni, G.B., Ayache, N., Initiative, A.D.N., et al.: Disentangling normal aging from Alzheimer's disease in structural magnetic resonance images. Neurobiol. Aging **36**, S42–S52 (2015)
14. Lou, A., Katsman, I., Jiang, Q., Belongie, S., Lim, S.N., De Sa, C.: Differentiating through the fréchet mean. In: ICML (2020)
15. Nava-Yazdani, E., Hege, H.C., Sullivan, T.J., von Tycowicz, C.: Geodesic analysis in Kendall's shape space with epidemiological applications. J. Math. Imaging Vision **62**(4), 549–559 (2020)
16. Nava-Yazdani, E., Hege, H.C., von Tycowicz, C.: A hierarchical geodesic model for longitudinal analysis on manifolds. J. Math. Imaging Vis. **64**(4), 395–407 (2022)
17. Sadeghi, N., Prastawa, M., Fletcher, P.T., Wolff, J., Gilmore, J.H., Gerig, G.: Regional characterization of longitudinal DT-MRI to study white matter maturation of the early developing brain. Neuroimage **68**, 236–247 (2013)
18. Singh, N., Hinkle, J., Joshi, S., Fletcher, P.T.: A hierarchical geodesic model for diffeomorphic longitudinal shape analysis. In: Gee, J.C., Joshi, S., Pohl, K.M., Wells, W.M., Zöllei, L. (eds.) IPMI 2013. LNCS, vol. 7917, pp. 560–571. Springer, Heidelberg (2013). https://doi.org/10.1007/978-3-642-38868-2_47
19. Woltman, H., Feldstain, A., MacKay, J.C., Rocchi, M.: An introduction to hierarchical linear modeling. Tutor. Quant. Methods Psychol. **8**(1), 52–69 (2012)

Model-Informed Deep Learning for Surface Segmentation in Medical Imaging

Xiaodong Wu[1,2]($\boxtimes$), Leixin Zhou[1], Fahim Zaman[1], Bensheng Qiu[3],
and John M. Buatti[2]

[1] Department of Electrical and Computer Engineering, The University of Iowa,
Iowa City, IA 52242, USA
{leixin-zhou,fahim-zaman}@uiowa.edu
[2] Department of Radiation Oncology, University of Iowa, Iowa City, IA 52242, USA
{xiaodong-wu,john-buatti}@uiowa.edu
[3] School of Information Science and Technology, The University of Science and
Technology of China, Hefei 230027, China
bqiu@ustc.edu.cn

Abstract. Automated surface segmentation is an important tool for utilizing medical image data in modern precision medicine for routine clinical practice and research. Deep-learning based methods have been developed for various medical image segmentation tasks. The inherent classification nature of those methods yet limits their capability of modeling global spatial dependency, which poses great challenges in incorporating geometric priors for segmentation, such as surface shape and surface smoothness, significantly compromising the accuracy and robustness of segmentation performance. To solve this problem, we propose integrating the graph-based optimal surface segmentation model into a new form of Convolutional Neural Networks (CNNs) that unifies the strengths of both deep learning and the graph segmentation model. To this end, we propose to parameterize the graph-based surface segmentation model and formulate the optimal surface segmentation as a quadratic programming problem, which admits an efficient inference for globally optimal solutions. The resulting network fully unifies graph segmentation modeling with CNNs, making it possible to train the whole deep network end-to-end with the usual back-propagation algorithm. Our experiments on two medical image segmentation applications demonstrated high performance of the proposed method with respect to segmentation accuracy, demands for annotated training data, and robustness to adversarial noise.

1 Introduction

Highly-automated and consistently accurate quantitative analysis of volumetric medical image data is a pre-requisite to utilize medical image data in modern precision medicine. Surface segmentation, which aims to accurately define the boundary surfaces of tissues captured by image data, is becoming increasingly necessary in quantitative image analysis. Many surface segmentation methods have been developed, including parametric deformable models, geometric deformable models, and atlas-guided approaches.

As one of the prominent surface segmentation approaches, the graph-based optimal surface segmentation method (Graph-OSSeg) [1] has demonstrated efficacy in the medical imaging field [2]. It is capable of simultaneously detecting multiple interacting surfaces with global optimality with respect to the energy function designed for the target surfaces with geometric constraints, which define the surface smoothness and interrelations. It also enables sub-pixel accurate surface segmentation [3]. The method solves the surface segmentation problem by transforming it to compute a minimum s-t cut in a derived arc-weighted directed graph, which can be solved optimally with a low-order polynomial time complexity. The major limitation of Graph-OSSeg is associated with the need for handcrafted features to define the parameters of the underlying graph model.

Armed with superior data representation learning capability, deep learning (DL) methods are emerging as powerful alternatives to current segmentation algorithms for many medical image segmentation tasks [4]. The state-of-the-art DL segmentation methods in medical imaging include fully convolutional networks (FCNs) [5] and U-net based frameworks [6,7], which model the segmentation problem as a pixel-wise or voxel-wise classification problem. Those convolutional neural network (CNN) methods have some critical limitations that restrict their use in the medical setting: (i) *Training data demand:* current schemes often need extensive training data, which is an almost insurmountable obstacle due to the risk to patients and high cost. (ii) *Difficulty in exploiting prior information* (shape, boundary smoothness and interaction): the methods are classification-based in nature, and the output probability maps are relatively unstructured. (iii) *Vulnerability to adversarial perturbations:* recent research has demonstrated that, compared to the segmentation CNNs alone, the integration of a graphical model such as conditional random fields (CRFs) into CNNs enhances the robustness of the method to adversarial perturbations [8].

To address those limitations, many model-based attempts have been proposed. One natural way is to use CNNs to learn the probability maps and then apply the traditional model-based methods such as graph cuts and deformable models to incorporate the prior information for segmentation [9,10]. In this scheme, feature learning by CNNs is, in fact, disconnected from the segmentation model; the learned features thus may not be truly appropriate for the model. Recent works introduce the energy function of a segmentation model into the loss function to guide CNNs for more model-specific feature learning, and improved segmentation performance has been demonstrated [11,12]. The model is not yet explicitly enforced while inferring the segmentation solutions with the trained network. In Zheng et al.'s work [13], the CRFs model is implemented as a recurrence neural network (RNN) and is integrated with an FCN for feature learning in a single neural network to achieve end-to-end learning. Arnab et al. [14] and Vemulapalli et al. [15] have demonstrated that the CRF-RNN framework outperforms other DL methods for semantic segmentation in computer vision. However, the CRFs inference is computationally intractable, thus no optimal solutions can be guaranteed – the solutions can be far from the optimal one at any scale, which may confuse the network during training and may contribute to its known high training complexity. In fact, the CRF-RNN method has not been widely used in medical image segmentation.

In this study, we propose unifying the powerful feature learning capability of DL with the successful graph-based optimal surface segmentation (Graph-OSSeg) model in a single deep neural network for end-to-end learning to achieve globally optimal segmentation. In this model-informed deep-learning segmentation method for optimal surface segmentation (MiDL-OSSeg), the known model is integrated into the DL network, which provides an advanced "attention" mechanism to the network. The network does not need to learn the prior information encoded in the model, reducing the demand of labeled data, which is critically important for medical imaging where scarcity of labeled data is common. Our major contributions are, as follows. (i) We model the graph-based optimal surface segmentation as a quadratic programming, blending learning and inference in a deep structured model while achieving global optimality of the segmentation solutions. (ii) The parameters of the graph-based optimal surface segmentation model are parameterized and learned by leveraging deep learning with a U-net as the backbone. (iii) Our experiments have demonstrated the high performance of our proposed method with high segmentation accuracy, less labeled data demand, and high robustness to adversarial perturbations.

2   Method

In this section, we present our MiDL-OSSeg method, merging the strength of both DL and Graph-OSSeg. We first formally define the optimal surface segmentation problem, which is formulated as a quadratic programming problem by parameterizing the Graph-OSSeg model. The proposed MiDL-OSSeg network is then depicted in detail, followed by its training strategy.

2.1   Quadratic Programming Formulation of Surface Segmentation

To present our method in a comprehensible manner, we consider a task of single *terrain-like* surface segmentation while incorporating the shape priors. Note that this simple principle used for this illustration is directly applicable to more complex surface segmentation (see Sect. 3.1 for prostate segmentation).

Let $\mathcal{I}(X, Y, Z)$ of size $X \times Y \times Z$ be a given 3-D volumetric image. For each (x, y) pair (i.e., $(x, y) \in X \times Y$), the voxel subset $\{\mathcal{I}(x, y, z) | 0 \leq z < Z\}$ forms a column parallel to the z-axis, denoted by $p(x, y)$. Each column has a set of neighboring columns for a certain neighboring setting $\mathcal{N}$, e.g., the four-neighbor relationship. Our goal is to seek a terrain-like surface $\mathcal{S}$, which intersects each column $p(x, y)$ at exactly one voxel. Thus, the terrain-like surface $\mathcal{S}$ can be defined as a function $\mathcal{S}(x, y)$, mapping $p(x, y)$ pairs to their z-values z_p.

In the Graph-OSSeg model [1], each voxel $\mathcal{I}(x, y, z)$ is associated with an on-surface cost $c(x, y, z)$ for the sought surface $\mathcal{S}$, which is inversely related to the likelihood that the desired surface $\mathcal{S}$ contains the voxel, and is computed based on handcrafted image features. The on-surface cost function $c(x, y, z)$ for each column $p(x,y)$ (i.e., $z = 0, 1, \ldots, Z-1$) can be an arbitrary function in the Graph-OSSeg model (Fig. 1a). However, an ideal cost function $c(x,y,z)$ should express a certain type of convexity: as we aim to formulate surface segmentation as a minimization problem, $c(x,y,z)$ should be low at the surface location for the column $p(x,y)$; while the distance increases from the surface location along the column, the cost should increase proportionally. We propose to make use of a Gaussian distribution $\mathcal{G}(\mu_p, \sigma_p)$ to model the likelihood of the column voxels on the target surface $\mathcal{S}$, and to define the on-surface cost function $c(x,y,z)$ for each column $p(x,y)$ as $c(x,y,z) = \frac{(z-\mu_p)^2}{2\sigma_p^2}$ $(0 \le z \le Z-1)$ (Fig. 1b). Thus, the on-surface cost functions for all columns are parameterized with (μ, σ). In the Graph-OSSeg model, it is at least nontrivial to determine (μ, σ) based on the handcrafted features. In this work, we propose to leverage DL for the on-surface cost parameterization with Gaussians.

It is critically important to incorporate shape priors in the segmentation model. In the Graph-OSSeg model [2], the shape changes of surface $\mathcal{S}$ are defined as the surface position changes between pairs of neighboring columns. Specifically, for any pair of neighboring columns p and q, the shape change of $\mathcal{S}$ between the column pair (p,q) is $d_{p,q} = (z_p - z_q)$ (note that surface $\mathcal{S}$ cuts the columns p and q at z_p and z_q, respectively). Then, $\boldsymbol{d} = (d_{p,q})_{(p,q)\in\mathcal{N}}$ forms a parameterization of the shape prior, which will be dynamically learned with DL for the input image during the inference. We use a quadratic function $((z_p - z_q) - d_{p,q})^2$ to penalize the deviation of the shape change to the prior model $\boldsymbol{d}$.

Thus, the MiDL-OSSeg problem is to find a terrain-like surface $\mathcal{S}$, such that $\mathcal{S}$ intersects each columns $p(x,y)$ at exactly one location z_p $(0 \le z_p \le Z-1)$ while minimizing the energy function $\mathbb{E}(\boldsymbol{z})$, with

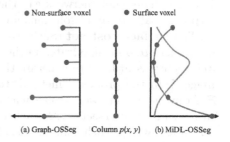

(a) Graph-OSSeg   Column $p(x,y)$ (b) MiDL-OSSeg

Fig. 1. On-surface cost parameterization with Gaussians. (a) The on-surface cost function in the Graph-OSSeg model defined on voxels for each column based on handcrafted features. The green line segments indicate the magnitudes for the corresponding voxels. (b) The on-surface cost function in the MiDL-OSSeg model (green curve) is computed based on the Gaussian-parameterized likelihood function (grey curve) over the column voxels. (Color figure online)

$$\mathbb{E}(\boldsymbol{z}) = \sum_{p\in\mathcal{C}} \frac{(z_p - \mu_p)^2}{2\sigma_p^2} + w \sum_{(p,q)\in\mathcal{N}} ((z_p - z_q) - d_{p,q})^2, \qquad (1)$$

where $\mathcal{C}$ is the set of all columns, $\mathcal{N}$ is the set of neighboring column pairs, and w is the coefficient. In the problem formulation (1), the surface location vector $\boldsymbol{z}$ is relaxed as continuous variables, that is, $0 \le z_p \le Z-1$ for each $p \in \mathcal{C}$. Hence, instead of keeping the target surface passing the center of a voxel, we allow the target surface $\mathcal{S}$ intersecting each column at any location, which may alleviates the partial volume effect.

2.2   The MiDL-OSSeg Model

The proposed MiDL-OSSeg model consists of two integrative components – a data representation learning network (DRLnet) and an optimal surface inference network (OSInet) (Fig. 2). The DRLnet is a DL network aiming to learn data representations

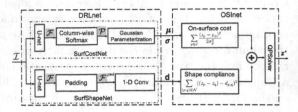

Fig. 2. Inference pipeline of the proposed method.

in the form of those in the MiDL-OSSeg model, that is, the on-surface cost parameterization $(\boldsymbol{\mu}, \boldsymbol{\sigma})$ and the shape prior parameterization $\boldsymbol{d}$. The OSInet strikes to solve the optimal surface interference by optimizing the energy function $\mathbb{E}(\boldsymbol{z})$. The whole network can then be trained in an end-to-end fashion and output globally optimal solutions for surface segmentation.

The surface cost net (SurfCostNet) for learning the on-surface cost parameterization $(\boldsymbol{\mu}, \boldsymbol{\sigma})$ is illustrated in the upper left panel of Fig. 2. A common U-net architecture is utilized to generate the discrete probability map $\mathcal{P}$ for the input image $\mathcal{I}$. In the proposed method, the softmax layer, taking the feature maps $\mathcal{F}$ from the U-net, works on each *column*, instead of for each voxel. As the target surface intersects with each column exactly once, the probabilities are normalized within each column $q(x, y)$ to obtain the probability vector $\mathcal{P}_q$. Each element $\mathcal{P}_q[z]$ indicates the probability of voxel $\mathcal{I}(x, y, z)$ being on the target surface $\mathcal{S}$, and the total sum of the probabilities of all voxels on the column q equals to 1. Then, $\mathcal{P} = \{\mathcal{P}_q | q \in \mathcal{C}\}$ forms the probability map of the input image. As we intend to parameterize the on-surface costs, the probability vector $\mathcal{P}_q$ for each column q is expected to be in a Gaussian distribution. To regularize the probability map $\mathcal{P}$ output from the U-net with a Gaussian, which mimics the Bayesian learning for each column and shares merits with knowledge distillation and distillation defense.

The *Gaussian parameterization* block is then applied to compute a Gaussian $\mathcal{G}(\mu_q, \sigma_q)$ to best fit to the discrete probability vector $\mathcal{P}_q$ for each column $q(x, y) \in \mathcal{C}$. $\mathcal{P}_q$ can be viewed as a discrete sample of the continuous Gaussian probability density function $\mathcal{G}(\mu_q, \sigma_q)$. We can estimate μ_q and σ_q from the probability vector $\mathcal{P}_q$ by minimizing a weighted mean square error, which admits an analytic solution for backpropogation [16].

The surface shape net (SurfShapeNet) for learning the parameterized shape model $\boldsymbol{d}$ is illustrated in the lower left panel of Fig. 2. It consists of a common U-net for the extraction of representative features $\mathcal{F}$, a padding layer to enable sufficient context information, and one 1-D convolution layer to generate the shape model $\boldsymbol{d}$.

To compute the surface position change $d_{p,q}$ between two adjacent columns p and q in the shape model $\boldsymbol{d}$, we consider a 4-neighborhood setting for the purpose of comprehensible illustration, in which each column $p(x, y)$ has four adjacent columns: $p(x - 1, y)$ and $p(x + 1, y)$ in the x-dimension, and $p(x, y - 1)$ and

$p(x, y + 1)$ along the y-dimension. This simple illustrative principle is directly applicable to an arbitrary neighborhood setting.

Consider two adjacent columns $p(x, y)$ and $p(x+1, y)$ along the x-dimension, denoted by p and $p + 1$, respectively. In general, we use $p + j$ to denote the column $q(x + j, y)$. For a robust inference of the surface position change $d_{p,p+1}$ between columns p and $p + 1$, we consider $N_c > 0$ consecutive neighboring columns of p and $p + 1$. The set $\mathcal{F}_p^{pad}$ of feature maps output from U-net for those columns with possible padding are used to infer $d_{p,p+1}$. Then, a 1-D convolution layer with a kernel size 1 and a stride of 1 is applied to the padded feature map $\mathcal{F}_p^{pad}$ to generate the surface position change $d_{p,p+1}$ between any two adjacent columns $p(x, y)$ and $p(x + 1, y)$ along the x-dimension.

Similarly, the surface position change between any two adjacent columns $p(x, y)$ and $p(x, y + 1)$ in the y-dimension can be computed. Thus, the parameterized shape model d can be dynamically generated for the input image $\mathcal{I}$.

The optimal surface inference network (OSInet) aims to solve the optimization problem in Eq. (1) with a globally optimal solution. To minimize the energy function $\mathbb{E}(z)$, we convert it to a standard quadratic form. For the purpose of comprehensible illustration, we consider a 4-neighborhood setting $\mathcal{N}$ for the adjacency of columns. Then, the grid $X \times Y$ defines the domain of all the columns, that is, every pair (x, y) corresponds an image column. The sought surface positions on $X \times Y$ thus form a matrix $z \in \mathbb{R}^{X \times Y}$. To convert $\mathbb{E}(z)$ to a quadratic form, we flatten the matrix z to a vector $z' \in \mathbb{R}^{XY}$, as follows. Each element $z(x, y)$ ($x = 0, 1, \ldots, X - 1$ and $y = 0, 1, \ldots, Y - 1$) corresponds to $z'(x * Y + y)$. It is equivalent to do a column-major order traversal of z. We explicitly maintain the adjacency relationship of each $z(x, y)$ in the flattened vector z' with $\mathcal{N}'$ for the corresponding elements. That is, for any two adjacent columns $p(x, y)$ and $q(x, y)$ with $(p, q) \in \mathcal{N}$, let $k = x * Y + y$ and $\bar{k} = x' * Y + y'$, then $(k, \bar{k}) \in \mathcal{N}'$. The matrix μ and σ are flattened into the vectors μ' and σ', respectively, in the same way as z. We then have the following form of the objective function $\mathbb{E}(z)$.

$$\mathbb{E}(z) = \mathbb{E}(z') = \frac{1}{2} z'^T H z' + c^T z' + \text{CONST.}, \tag{2}$$

where H is a Hessian matrix of $\mathbb{E}(z')$. It can be proved that H is positive definite by using the Gershgorin circle theorem [17]. The energy function $\mathbb{E}(z')$ is thus convex. Let the gradient $\nabla = H z' + c$ to be zero, we have the global optimal solution $z'^* = -H^{-1} c$. We thus do not need to make use of a recurrent neural network (RNN) to implement OSInet for a globally optimal solution.

2.3 Training Strategy

Pre-training the Surface Cost Net. To pre-train SurfaceCostNet to obtain the probability map $\mathcal{P}$ for on-surface cost parameterization (Fig. 2), we make use of the ground truth $\mathcal{S}_{gt}$ of the surface segmentation. For each column q, if the voxel is on $\mathcal{S}_{gt}$, then the probability of the voxel is 1; otherwise, it is 0. Thus, the probabilities of all the voxels on q form a delta function, which is Gaussianized by setting the standard deviation σ to be 0.1 times of the column length to obtain a Gaussian distribution $\hat{\mathcal{P}}_q$ as the ground truth for each column q. Let $\mathcal{P}_q$ be the output probability vector from the column-wise softmax layer for column q. The loss for the pre-training, $Loss_{pre}$, is formulated as the Kullback-Leibler divergence of $\hat{\mathcal{P}}_q$ and $\mathcal{P}_q$.

Training the Surface Shape Net. The reference surface $\mathcal{S}_{gt}$ is first used to generate the ground truth $\hat{d}$ to train the surface shape net (SurfShapeNet). For each pair of adjacent columns $(p, q) \in \mathcal{N}$, compute the surface position change $\hat{d}_{p,q}$ between columns p and q from $\mathcal{S}_{gt}$. Let d be the output of Surf-ShapeNet. The mean square error of the surface position changes between d and $\hat{d}$, $Loss_{shape}$, is then utilized for the loss function. Note that the surface position changes could be highly erratic, especially when the ground truth surface positions are defined in the discrete voxel space. This hinders SurfShapeNet from learning useful representation and usually the trained SurfShapeNet just generates a constant prediction that is not much useful. We propose smoothing the ground truth $\hat{d}$ by using the sliding window average method for the training of SurfShapeNet. The predicted shape model d by the network trained with the smoothed ground truth $\hat{d}$ is much more accurate.

Fine Tuning. The L_1-loss on surface positioning errors is used for the fine tuning of the whole network. The fine tuning proceeds alternatively between the training of SurfaceCostNet and OSInet. The training data is used for the Sur-faceCostNet training, while the validation data is utilized to train OSInet. As OSInet only has one parameter (w) that needs to be trained, the chance of over-fitting is low. Note that SurfaceCostNet is not trained on the validation data, the learned parameter w should be more representative in the wild. Otherwise, if we use the training data for the fine tuning of both SurfaceCostNet and OSInet, the learned w tends to be small due to the pre-training process of SurfaceCostNet, which may marginalize the shape term in the energy function $\mathbb{E}(z)$. As the shape priors are relatively stable, we freeze the pre-trained SurfaceShapeNet to obtain the shape model during the network fine tuning.

3   Performance Assessment

The performance of the proposed MiDL-OSSeg method was evaluated to determine: segmentation accuracy, annotated data demands for model training, and robustness to adversarial perturbations. The experiments were carried out on medical images from spectral domain optical coherence tomography (SD-OCT) and magnetic resonance imaging (MRI). Assessments of terrain-like and closed surface segmentation were performed.

3.1   Application Experiments

Automated Retinal Layer Segmentation in SD-OCT Images. To demonstrate the utility of our MiDL-OSSeg method in segmenting terrain-like surfaces, automated retinal layer segmentation in SD-OCT images was performed.

Data. 382 SD-OCT scans (114 normal eyes and 268 eyes with intermediate age-related macular degeneration (AMD)) and their respective manual tracings by an expert were obtained from the publicly available repository of datasets [18]. Each OCT volume consists of $400 \times 60 \times 512$ voxels with a size of$6.54 \times 67 \times 3.23$ μm^3. The dataset was randomly divided into 3 sets: 1) training set - 266 volumes (79 normal, 187 AMD), 2) validation set - 57 volumes (17 normal, 40 AMD), and 3) testing set - 59 volumes (18 normal, 41 AMD). The surfaces considered are the Inner aspect of Retinal Pigment Epithelium drusen complex (IRPE) and the Outer aspect of Bruch Membrane (OBM) (Fig. 3). The proposed MiDL-OSSeg model was trained and tested on the 2D B-scans of the OCT volumes.

Prostate Segmentation in MR Images. The proposed MiDL-OSSeg method was evaluated on automated prostate segmentation in 3D MR images to demonstrated its applicability of segmenting irregular surfaces in 3D.

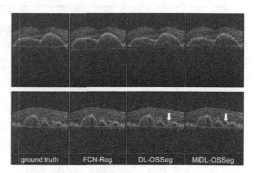

Data. The dataset is provided by the NCI-ISBI 2013 Challenge - Automated Segmentation of Prostate Structures [19]. This dataset has two labels: peripheral zone (PZ) and central gland (CG). We treat both of them as prostate for single surface segmentation. The ground truth sur-

Fig. 3. Illustrations of SD-OCT segmentation results. Red: IRPE; Green: OBM. (Color figure online)

face of the prostate boundary in each image was generated from the PZ and CG labels. The challenge data set consists of the training set (60 cases), the leader board set (10 cases) and the test set (10 cases). 70 cases in total were used as the test set was not available. Ten-fold cross validation was applied on that dataset. For each fold, the training, validation and test sets consist of 58, 5 and 7 cases, respectively. The shape-aware patch generation method [20] was adopted to divide each MRI scan into 6 volumetric patches. Each patch contains a portion of prostate boundary, which is a terrain-like surface in 3D. Our MiDL-OSSeg model was trained and validated on the volumetric patches.

3.2 Segmentation Accuracy

OCT Retinal Layer Segmentation. Unsigned mean surface positioning error (UMSP) was utilized for accuracy assessment of retina OCT segmentation. We compared the proposed MiDL-OSSeg method to the Graph-OSSeg method [2] as well as Shah *et al.*'s FCN-based regression model (denoted by FCN-Reg) [21]. To ensure a fair comparison, we reimplemented Shah *et al.*'s method to make sure that the training, validation and test data splitting was the same for the two compared methods. For the purpose of an ablation study, we showed the segmentation results of our method without incorporating the shape priors, that is, the means of Gaussians μ output from the Gaussian Parameterization block in Fig. 2 are treated as the predicted surface positions. The method is marked as DL-OSSeg in Table 1. Our MiDL-OSSeg method significantly outperformed all other methods for each surface with the p-value less than 0.05. Specifically, MiDL-OSSeg incorporating the shape priors which was implemented with OSInet yielded significant improvement compared to DL-OSSeg. Sample segmentation results are illustrated in Fig. 3.

Table 1. UMSP errors and standard deviations in μm evaluated on the SD-OCT dataset. Depth resolution is 3.23 μm. Numbers in bold are the best in that row.

Surfaces	Normal				AMD			
	Graph-OSSeg	FCN-Reg	DL-OSSeg	MiDL-OSSeg	Graph-OSSeg	FCN-Reg	DL-OSSeg	MiDL-OSSeg
IRPE	4.55±0.36	3.70±0.69	2.16±0.67	**1.89±0.68**	9.30±1.74	6.45±2.11	3.09±1.52	**2.96±1.91**
OBM	5.59±1.20	3.58±0.38	3.28±0.71	**2.55±0.40**	10.14±5.30	6.43±2.82	5.74±2.51	**4.29±1.71**

Prostate MRI Segmentation. The proposed MiDL-OSSeg method for prostate segmentation was compared to the Graph-OSSeg method [2] and other two CNN-based approaches, U-net [6] and PSNet [22]. PSNet is the state-of-the-art method on the dataset. The Dice similarity coefficient (DSC), Hausdroff distance, and the average surface distance (ASD) between predicted prostate boundary surface and manual delineation for each method are shown in Table 2. With respect to all three metrics, the proposed MiDL-OSSeg significantly outperformed all the compared methods, especially for the surface-based ASD and HD metrics. Figure 4 shows an example segmentation results by MiDL-OSSeg for a 3D prostate MR image in the transverse, sagittal and coronal views.

Table 2. The DSC, ASD and HD with standard deviations evaluated on the prostate dataset. Numbers in bold are the best in that column among all the methods.

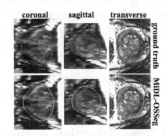

Fig. 4. Example segmentation of a prostate MR image.

Methods	DSC	ASD (mm)	HD (mm)
U-net	0.84±0.05	3.3±1.0	10.1±3.2
PSNet	0.85±0.04	3.0±0.9	9.3±3.5
Graph-OSSeg	0.80±0.04	2.7±0.6	13.9±1.8
MiDL-OSSeg	**0.89±0.03**	**1.36±0.34**	**7.28±3.20**

3.3   Annotated Data Demands for Training

We evaluated the segmentation performance changes of the proposed method with respect to the different training data sizes. The validation and test datasets were fixed and the training dataset for model training was randomly sampled with different rates. Each trained segmentation model was applied to the same test dataset for performance evaluation. For the OCT retina layer segmentation, 50%, 30%, and 10% of the training set were randomly generated for model training. The proposed MiDL-OSSeg method was compared to Shah *et al.*'s FCN-Reg model [21] while trained on different sampled training sets. The UMSP errors evaluated on the SD-OCT dataset are shown in Fig. 5 for the two compared methods. The proposed MiDL-OSSeg model trained on each of the reduced training sets (50%, 30%, and 10%) significantly outperformed the FCN-Reg model trained on the same training set for each target surface of normal and AMD subjects. Of note: while the MiDL-OSSeg model was trained on 10% of the whole training set it achieved an even better accuracy, compared to the FCN-Reg trained on the entire training data set.

3.4   Robustness to Adversarial Perturbations

Robustness of the proposed MiDL-OSSeg model was evaluated against adversarial samples [23], which are legitimate samples with human-imperceptible perturbations that attempt to fool a trained model to make incorrect predictions with high confidence. To push the model to its limit for performance degeneration, we adopted the

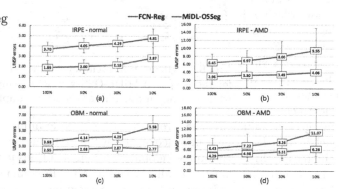

Fig. 5. Segmentation accuracy evaluated on the SD-OCT dataset for the proposed MiDL-OSSeg model, compared to FCN-Reg [21], while trained on 100%, 50%, 30%, and 10% of the training set.

white-box attack methods [24], in which the full knowledge of the network architecture and the model parameters is used to generate adversarial noises. In our experiments, the fast gradient sign method (FGSM) [24] was utilized.

Our robustness experiments were conducted on the retinal OCT dataset for retinal layer segmentation. For each attack level $\epsilon = 0.02, 0.04, 0.06, 0.08, 0.10$, an adversarial sample $\mathcal{I}_{adv}$ was generated for each OCT image $\mathcal{I}$ in the test set, all of which form an adversarial sample set for the corresponding attack level ϵ. The MiDL-OSSeg model trained with the original training and validation sets (without using adversarial samples) was then tested on the adversarial sample set of each ϵ for segmentation accuracy. For comparison, Shah *et al.*'s FCN-Reg method [21] was also evaluated for its segmentation performance on the adversarial sample sets. The segmentation accuracy measured with UMSP errors and standard deviations for both IRPE and OBM surfaces of normal and

AMD subjects are summarized in Fig. 6 for each adversarial attack level ϵ. The proposed MiDL-OSSeg method showed higher robustness to adversarial noise than FCN-Reg, as the UMSP errors increased much slower with respect to the increased attack levels than those of FCN-Reg consistently in all four cases. We attribute this MiDL-OSSeg robustness to the incorporation of the graph-based segmentation model.

4   Conclusion

In this paper, we developed a model-informed deep learning segmentation method for optimal surface segmentation, which unifies DL with the Graph-OSSeg model in a single deep neural network for end-to-end learn-

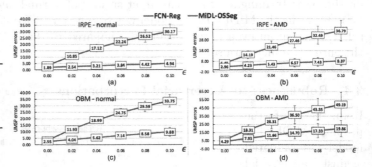

Fig. 6. Segmentation accuracy evaluated on the SD-OCT dataset for the proposed MiDL-OSSeg model, compared to FCN-Reg [21], while testing on different adversarial sample sets.

ing, greatly enhancing the strengths of both while minimizing the drawbacks of each. To the best of our knowledge, this is the first study for surface segmentation which can achieve guaranteed globally optimal solutions using deep learning. The proposed method has been validated on two medical image segmentation tasks, demonstrating its efficacy with respect to segmentation accuracy, demands for annotated training data, and robustness to adversarial noise.

References

1. Li, K., Wu, X., Chen, D.Z., Sonka, M.: Optimal surface segmentation in volumetric images-a graph-theoretic approach. IEEE Trans. Pattern Anal. Mach. Intell. **28**(1), 119–134 (2006)
2. Song, Q., Bai, J., Garvin, M.K., Sonka, M., Buatti, J.M., Wu, X.: Optimal multiple surface segmentation with shape and context priors. IEEE Trans. Med. Imaging **32**(2), 376–386 (2013)
3. Shah, A., Abramoff, M.D., Wu, X.: Optimal surface segmentation with convex priors in irregularly sampled space. Med. Image Anal. **54**, 63–75 (2019)
4. Litjens, G., et al.: A survey on deep learning in medical image analysis. Med. Image Anal. **42**, 60–88 (2017)
5. Long, J., Shelhamer, E., Darrell, T.: Fully convolutional networks for semantic segmentation. In: CVPR 2015, pp. 3431–3440 (2015)

6. Ronneberger, O., Fischer, P., Brox, T.: U-net: convolutional networks for biomedical image segmentation. In: Navab, N., Hornegger, J., Wells, W.M., Frangi, A.F. (eds.) MICCAI 2015. LNCS, vol. 9351, pp. 234–241. Springer, Cham (2015). https://doi.org/10.1007/978-3-319-24574-4_28

7. Isensee, F., Jaeger, P.F., Kohl, S.A.A., Petersen, J., Maier-Hein, K.H.: nnU-Net: a self-configuring method for deep learning-based biomedical image segmentation. Nat. Methods 18(2), 203–211 (2021)

8. Arnab, A., Miksik, O., Torr, P.H.: On the robustness of semantic segmentation models to adversarial attacks. In: Proceedings of the IEEE Conference on Computer Vision and Pattern Recognition, pp. 888–897 (2018)

9. Lu, F., Wu, F., Hu, P., Peng, Z., Kong, D.: Automatic 3d liver location and segmentation via convolutional neural network and graph cut. Int. J. Comput. Assist. Radiol. Surg. 12(2), 171–182 (2017)

10. Liu, F., Zhou, Z., Jang, H., Samsonov, A., Zhao, G., Kijowski, R.: Deep convolutional neural network and 3d deformable approach for tissue segmentation in musculoskeletal magnetic resonance imaging. Magn. Reson. Med. 79(4), 2379–2391 (2018)

11. Milletari, F., Rothberg, A., Jia, J., Sofka, M.: Integrating statistical prior knowledge into convolutional neural networks. In: Descoteaux, M., Maier-Hein, L., Franz, A., Jannin, P., Collins, D.L., Duchesne, S. (eds.) MICCAI 2017. LNCS, vol. 10433, pp. 161–168. Springer, Cham (2017). https://doi.org/10.1007/978-3-319-66182-7_19

12. Ravishankar, H., Venkataramani, R., Thiruvenkadam, S., Sudhakar, P., Vaidya, V.: Learning and incorporating shape models for semantic segmentation. In: Descoteaux, M., Maier-Hein, L., Franz, A., Jannin, P., Collins, D.L., Duchesne, S. (eds.) MICCAI 2017. LNCS, vol. 10433, pp. 203–211. Springer, Cham (2017). https://doi.org/10.1007/978-3-319-66182-7_24

13. Zheng, S., et al.: Conditional random fields as recurrent neural networks. In: Proceedings of the IEEE International Conference on Computer Vision, pp. 1529–1537 (2015)

14. Arnab, A., et al.: Conditional random fields meet deep neural networks for semantic segmentation: combining probabilistic graphical models with deep learning for structured prediction. IEEE Signal Process. Mag. 35(1), 37–52 (2018)

15. Vemulapalli, R., Tuzel, O., Liu, M.-Y., Chellapa, R.: Gaussian conditional random field network for semantic segmentation. In: Proceedings of the IEEE Conference on Computer Vision and Pattern Recognition, pp. 3224–3233 (2016)

16. Guo, H.: A simple algorithm for fitting a gaussian function [DSP tips and tricks]. IEEE Signal Process. Mag. 28(5), 134–137 (2011)

17. Horn, R.A., Johnson, C.R.: Matrix Analysis, 2nd edn. Cambridge University Press, Cambridge (2012)

18. Farsiu, S., et al.: Quantitative classification of eyes with and without intermediate age-related macular degeneration using optical coherence tomography. Ophthalmology 121(1), 162–172 (2014)

19. Bloch, N., Madabhushi, A., Huisman, H., Freymann, J., Kirby, J., Grauer, M.: NCI-ISBI 2013 challenge: automated segmentation of prostate structures. Cancer Imaging Arch. 370 (2015)

20. Zhou, L., Zhong, Z., Shah, A., Qiu, B., Buatti, J., Wu, X.: Deep neural networks for surface segmentation meet conditional random fields (2019). https://arxiv.org/abs/1906.04714

21. Shah, A., Zhou, L., Abrámoff, M.D., Wu, X.: Multiple surface segmentation using convolution neural nets: application to retinal layer segmentation in oct images. Biomed. Opt. Express **9**(9), 4509–4526 (2018)
22. Tian, Z., Liu, L., Zhang, Z., Fei, B.: PSNet: prostate segmentation on MRI based on a convolutional neural network. J. Med. Imaging **5**(2), 021208 (2018)
23. Szegedy, C., et al.: Intriguing properties of neural networks. In: International Conference on Learning Representations (2014)
24. Goodfellow, I., Shlens, J., Szegedy, C.: Explaining and harnessing adversarial examples. In: International Conference on Learning Representations (2015)

Author Index

A. Frangi et al. (Eds.): IPMI 2023, LNCS 13939, pp. 835–839, 2023.
https://doi.org/10.1007/978-3-031-34048-2

Printed in the United States
by Baker & Taylor Publisher Services

Printed in the United States
by Baker & Taylor Publisher Services